AIRCRAFT GAS TURBINE ENGINES OF THE WORLD

AND

DICTIONARY OF THE GAS TURBINE

By
Charles E. Otis, M.Ed.
and Peter A. Vosbury, M. Ed.

International Standard Book Number 0-89100-390-8
For sale by: IAP, Inc., A Hawks Industries Company
Mail To: P.O. Box 10000, Casper, WY 82602-1000
Ship To: 7383 6WN Road, Casper, WY 82604-1835
(800) 443-9250 ❖ (307) 266-3838 ❖ FAX: (307) 472-5106

HBC0493 Printed in the USA

© 1991 by Otis / Vosbury
All Rights Reserved

Except as permitted under the United States Copyright Act of 1976, no part of this publication may be reproduced or distributed in any form or by any means, or stored in a database or retrieval system, without the prior written permission of the publisher.

Printed in the United States of America

Library of Congress Cataloging-in-Publication Data

Otis, Charles E. (Charles Edward), 1932-
    Aircraft gas turbine engines of the world and dictionary of the gas turbine / by Charles E. Otis and Peter A. Vosbury.
       p. cm.
    "Order number EA-390."
    Includes index.
    ISBN 0-89100-390-8 (pbk.) : $22.95
    1. Aircraft gas-turbines—Catalogs.   I. Aircraft gas-turbines-
-Dictionaries.   I. Vosbury, Peter A., 1949-   .   II. Title.
TL709.O78 1991
629.134'353'0294—dc20
                                                                                     91-28478
                                                                                     CIP

TABLE OF CONTENTS

	PAGE
TABLE OF CONTENTS	III
PREFACE	IV
AUTHOR'S PROFILE	IV
BOOK OVERVIEW	V
ACKNOWLEDGMENTS	VI
SECTION I - GAS TURBINE ENGINE DATA SHEETS	
Section I Table of Contents	1
International Engines, listed by Country of Manufacture	13
Multi-National Engines, listed by Manufacturer	142
USA Manufactured Engines, listed by Manufacturer	159
Auxiliary Engines (Power Units)	244
SECTION II - AIRCRAFT DIRECTORY (Gas Turbine Powered)	251
SECTION III - DICTIONARY OF THE GAS TURBINE ENGINE	279
APPENDIX-1 - LISTING OF EARLY TURBINE ENGINES	437
APPENDIX-2 - LISTING OF TURBINE ENGINE INSPECTION TERMS	440
APPENDIX-3 - LISTING OF BEARING INSPECTION TERMS	444
APPENDIX-4 - CONVERSION TABLES	446
APPENDIX-5 - AIRCRAFT, ENGINE SYMBOLS	448
INDEX	450

PREFACE

This book is the result of a few thousand hours of research and digging, to try and find information on all the Gas Turbine Engines and gas turbine powered aircraft produced in the world. Since the days when Paul H. Wilkinson produced the book "Aircraft Engines of the World", there hasn't been a book which provides a comprehensive listing of gas turbine engines and their specifications. The authors, sorely missing the Wilkinson books and recognizing the need, were motivated to produce this text.

Along the way, it became apparent that more and more of the terminology used in the gas turbine engine field also needed to be identified and defined. There are thousands of "terms" which are very specific to the Gas Turbine Engine. Many of the words have similar meanings, and are often used interchangeably. In most cases, however, there are discrete and very important differences. The dictionary found in this book should make the terminology of the turbine engine field easier to understand and easier to relate to. It will define many words that are rarely seen outside of manufacturer's technical manuals or trade journals, and offer some insight as to how these words apply to the aviation professional and other interested readers.

The information found in the turbine engine "Data Sheets", in the "Aircraft Section", and in the "Dictionary", is always in a state of change. Seldom does a week pass without a new word appearing in a magazine article, a new engine application being developed, or a new aircraft appearing on the drawing boards. These changes are the things that make aviation exciting but, perhaps, a little confusing. It is the hope of the authors that this book will add much to the excitement of aviation, and hopefully, through new and updated editions, relieve some of the confusion.

A reader card is provided in the book for your use. If you would like to see any additions to the three sections of this book, or have comments about the information which is provided, please fill out the card and forward it to the publisher.

AUTHOR'S PROFILE

Charles E. Otis is a Professor of Aviation Maintenance at Embry-Riddle Aeronautical University, Daytona Beach, Florida.

Peter A. Vosbury is a Professor of Aviation Maintenance at Embry-Riddle Aeronautical University, Daytona Beach, Florida.

TEXT OVERVIEW

The book is arranged in three sections, as follows:

Section I contains:

Data Sheets for the current Gas Turbine Engines in worldwide use, including:

... A side view line drawing of each Gas Turbine Engine.
... A side view photograph of each Gas Turbine Engine.
... A table of specifications for each Gas Turbine Engine.

Section II contains:

... A listing of Gas Turbine Powered Aircraft, to include U.S. and international airplanes and rotorcraft. This section includes both civilian and military aircraft.

Section III contains:

... A Dictionary of terms relating to the gas turbine engine, both new and old, with cross references to similar terms. The definitions are designed specifically around the Gas Turbine Engine, its systems, its basic aerodynamic and thermodynamic functions, and its interface with the aircraft.
... Abbreviations are included in alphabetical order in the listing, with a cross reference to the basic term.

Appendices 1 thru 5 Contain:

... A Listing of Early Turbine Engines.
... A Listing of Engine Inspection and Distress Terms.
... A Listing of Bearing Inspection and Distress Terms.
.. Conversion Tables.
... Symbols and Micron to Sieve Number Table.

Note: It is the intention of the authors and publisher to update and publish revised editions at regular intervals to provide the reader with the latest information concerning the Gas Turbine Engine.

ACKNOWLEDGMENTS

Without the help of many companies and many people, this book would not have been possible. The authors would like to thank the following companies for providing the appropriate "photographs" and "technical data" of their Gas Turbine Engines.

ALFA ROMEO AVIO SPA, AERO ENGINE DIVISION
ALLISON TURBINE DIVISION OF GENERAL MOTORS
ATLAS AIRCRAFT CORP.
BET-SHEMESH ENGINES LTD. (BSEL)
CENTRAL INSTITUTE OF AVIATION MOTORS TSIAM
CENTRAL NATIONAL AERONAUTICE
CHINA AERO-TECHNOLOGY IMPORT CORP. (CATIC)
CHINA NATIONAL ENGINE CORP. (CAREC)
FIAT AVIAZIONE SPA.
GARRETT AND ALLISON (ALLIED SIGNAL AEROSPACE/CFE)
GARRETT TURBINE ENGINE COMPANY
GENERAL ELECTRIC AIRCRAFT ENGINES, Large Engine Group
GENERAL ELECTRIC AIRCRAFT ENGINES, Small Engine Group
HAWKER SIDDELEY CANADA INC. (ORENDA DIVISION)
HINDUSTAN AERONAUTICS LTD.
INDUSTRIE AERONAUTICHE MECCANICHE (RINALDO PIAGGIO SPA)
ISHIKAWAJIMA-HARIMA HEAVY INDUSTRIES (IHI), JET ENGINE DIVISION
ISRAEL AIRCRAFT INDUSTRIES (IAI)
KAWASAKI AKASHI HEAVY INDUSTRIES (KHI), JET ENGINE DIVISION
LIGHT HELICOPTER TURBINE ENGINE COMPANY-LHTEC
MICROTURBO CORP. OF NORTH AMERICA
MITSUBISHI NAGASAKI HEAVY INDUSTRIES (MHI), JET ENGINE DIVISION
MOTORLET KONCERN
MOTOREN-UND TURBINEN-UNION (MTU)
NATIONAL AEROSPACE LABORATORY
NOEL PENNY TURBINES LTD.
ORAO AIR FORCE DEPOT
POSKIE ZAKLADY LOTNICZE (PZL)
PRATT & WHITNEY AIRCRAFT OF CANADA INC.
ROLLS-ROYCE BRISTOL, AERO DIVISION
ROLLS-ROYCE DERBY, AERO DIVISION
ROLLS-ROYCE LEAVENSDEN, AERO DIVISION
SNECMA
SUNDSTRAND CORP.
TELEDYNE CAE CORP.
TELEDYNE CONTINENTAL MOTORS, AIRCRAFT PRODUCTS
TEXTRON-LYCOMING CORP.
TURBINE TECHNOLOGIES LTD
TURBOMECA CORP.
TUSAS ENGINE INDUSTRIES (TEI)
UNITED TECHNOLOGIES (PRATT & WHITNEY) COMMERCIAL PRODUCTS DIVISION
UNITED TECHNOLOGIES, (PRATT & WHITNEY) GOVERNMENT PRODUCTS DIVISION
VOLVO FLYGMOTOR AB
WILLIAMS INTERNATIONAL CORP.
ZAVODY NAVYROBU LOZISK KONCERN

SECTION I

GAS TURBINE ENGINE

DATA SHEETS

CONTENTS OF SECTION I

INTERNATIONAL TURBINE ENGINE COMPANIES

CANADA:
PAGE

HAWKER SIDDELEY CANADA INC. (ORENDA DIVISION)
3160 DERRY ROAD EAST, MISSISSAUGA, ONTARIO, CANADA, L4T 1A9
Tele: (416) 677 3250

ENGINE DESIGNATIONS:
J79-11A	13
J85-CAN-15	14
J85-CAN-40	15
ORENDA	16

PRATT & WHITNEY
AIRCRAFT OF CANADA INC.[†]
1000 MARIE-VICTORIN, LONGUEUIL, QUEBEC, CANADA, J4G1A1
Tele: (514) 677-9411

ENGINE DESIGNATIONS:
JT15D-5	17
PT6A-27	18
PT6A-65AG	19
PT6B-36A	20
PT6T-6 (T-400)	21
PW118	22
PW126	23
PW205	24
PW305	25

CZECHOSLOVAKIA:

MOTORLET KONCERN
NEKAZ ANKA II, PRAHA, 1, CZECHOSLOVAKIA, 11221
Tele: Prague 521119

ENGINE DESIGNATIONS:
WALTER M601	26
WALTER M602	27
WALTER M701c	28

ZAVODY NAVYROBU LOZISK KONCERN
01701 POVAZSKA BYSTRICA, CZECHOSLOVAKIA.

ENGINE DESIGNATIONS:
PROGRESS DV-2	29

FRANCE:

MICROTURBO CORP. OF NORTH AMERICA
55 ORVILLE DRIVE, HOHEMIA, N.Y., 11716, USA. Tele: (516) 567-3780
CHEMIN duPONT deRUPE, BP2089.31019, TOLOUSE, FRANCE. Tele: (61) 70 1127

ENGINE DESIGNATIONS:
TRI 60-1	30
TRS 18-1	31

SNECMA (Society Nationale d'Etude et de Construction de Moteurs d'Aviation)
Address: 2 BOULEVARD VICTOR, 75724 PARIS, CADEX 15, FRANCE
Tele: 40-60-8080

ENGINE DESIGNATIONS:
ATAR 9K50	32
CF6-50 (SNECMA-GENERAL ELECTRIC IN MULTI-NATIONAL LISTING)	155
LARZAC (SNECMA-TURBOMECA IN MULTI-NATIONAL LISTING)	156
M53-P2	33
M88-2	34
OLYMPUS-593 (SNECMA-ROLLS ROYCE IN MULTI-NATIONAL LISTING)	150
TYNE-20 (SNECMA-ROLLS ROYCE MTU IN MULTI-NATIONAL LISTING)	151

TURBOMECA CORP.
BORDES, 64320, BIZANDS, FRANCE   Tele: (59) 32 8437
2709 FORUM DRIVE, GRAND PRARIE, TX 75051 USA. Tele: (214) 641-6645

ENGINE DESIGNATIONS:
ARRIEL 1B	35
ARTOUSTE 3B	36
ASTAFAN IV	37
ASTAZOU 3A	38
ASTAZOU 16G	39
AUBISQUE-1A	40
LARZAC (SNECMA-TURBOMECA IN MULTI-NATIONAL LISTING)	156
MAKILA 1A	41
ARRIUS (FORMERLY TM-319)	42
TM 333-1A	43
TURMO 4C	44

GERMANY:

MOTOREN-UND TURBINEN-UNION (MTU)
DACHAUER STRABE 665, P.O. BOX 500640
D-8000 MUNCHEN 50, GERMANY
Tele: (089) 1489-0

ENGINE DESIGNATIONS:
250-MTU-C20B	45
J79-MTU-17A	46
MTR-390 (MTU, RR, TM IN MULTI-NATIONAL LISTING)	148
T64-MTU-7	47
TYNE-MTU-MK.21/22	48

GREAT BRITAIN:

NOEL PENNY TURBINES LTD.
SISKIN DRIVE, TOLL BAR END, CONVENTRY, ENGLAND, CV3 4FE
Tele: 0230-301528

ENGINE DESIGNATIONS:
NPT-401B	49

ROLLS-ROYCE BRISTOL, AERO DIVISION
BOX 3, BRISTOL, ENGLAND, BS12 7QE
Tele: 0272-791234

ENGINE DESIGNATIONS:
OLYMPUS-593 (SNECMA-ROLLS ROYCE IN MULTI-NATIONAL LISTING)	150
ORPHEUS 805	50
PEGASUS	51
VIPER 601	52

ROLLS-ROYCE DERBY, AERO DIVISION
BOX 31, MOORE LANE, DERBY, ENGLAND, DE2 8BJ
Tele: 0332 242424

PAGE

ENGINE DESIGNATIONS:
ADOUR 811 (RR, TM IN MULTI-NATIONAL LISTING)	152
ADOUR 871 (RR, TM IN MULTI-NATIONAL LISTING)	153
CONWAY 43	53
DART MK. 540	54
GAZELLE MK. 165	55
RB.183-555-15P	56
RB.199 (RR, TURBO-UNION IN MULTI-NATIONAL LISTING)	157
RB.211-524H	57
RB.211-535C	58
SPEY MK.512-14DW	59
SPEY RB.168-66 (RR, ALLISON IN MULTI-NATIONAL LISTING)	149
TAY 620-15	60
TRENT 600	61
TRENT 700	62
TRENT 800	63
TYNE-12 MK.515	64
TYNE-20 (RR,SNECMA,MTU IN MULTI-NATIONAL LISTING)	151

ROLLS-ROYCE LEAVENSDEN, AERO DIVISION
WATFORD, HERTS, ENGLAND, WD2 7BZ
Tele: 0923 674000

ENGINE DESIGNATIONS:
GEM-60-3	65
GEM/RR1004	66
GNOME H.1400-1	67
NIMBUS MK103	68
RTM-322 (RR,TM,MTU IN MULTI-NATIONAL LISTING)	154

INDIA:

HINDUSTAN AERONAUTICS LTD.
Post Office Bag 9301, Bangalore, India, 560 093
Tele: 565201

ENGINE DESIGNATIONS:
ADOUR MK. 811	69
ASTAZOU IIIB	70

ISRAEL:

BET-SHEMESH ENGINES LTD (BSEL)
MOBILE POST, HA'ELA, BETH SHEMESH, ISRAEL, 99000
Tele: 972-2911661-6

ENGINE DESIGNATIONS:
MARBORE-6	71

ISRAEL AIRCRAFT INDUSTRIES (IAI)
BENGURION INT'L AIRPORT, ISRAEL, 70100
Tele: 03 971 3111

ENGINE DESIGNATIONS:
J79-IAI-J1E	72

PAGE

ITALY:

ALFA ROMEO AVIO SPA, AERO ENGINE DIVISION
80038 POMIGILIANO D'ARCO, NAPLES, ITALY
Tele: 081843 0111

ENGINE DESIGNATIONS:
AR.318	73
J85-GE-13-1A	74
T58-GE-10	75

FIAT AVIAZIONE SPA.
VIA NIZZA 312, 10127 TURIN, ITALY
Tele: (011) 69311

ENGINE DESIGNATIONS:
J79-GE-19	76
SPEY MK. 807	77
T64-P4D	78

**INDUSTRIE AERONAUTICHE MECCANICHE
(RINALDO PIAGGIO SPA)**
VIA BRIGATO BISARGNO 14, 16129, GENOA, ITALY
Tele: 010-60041

ENGINE DESIGNATIONS:
GEM/RR 1004	79
T53-L-13 (A,B)	80
T55-L-712	81
VIPER 632-43	82

JAPAN:

ISHIKAWAJIMA-HARIMA HEAVY INDUSTRIES (IHI), JET ENGINE DIVISION
SHIN OHTEMACHI BLDG. 2-1, OHTEMACHI 2-CHOME,
CHIYODA-KU TOKYO, JAPAN
Tele: 03-244-5331

ENGINE DESIGNATIONS:
CT58-IHI-140	83
CT700-700	84
F100-IHI-100	85
F3-IHI-30	86
T56-IHI-14	87
T64-IHI-10J	88
TF40-IHI-801A	89

KAWASAKI AKASHI HEAVY INDUSTRIES (KHI), JET ENGINE DIVISION
1-1 KAWASAKI-CHO, AKASHI 673, JAPAN
Tele: KOBE 078-921-1515

ENGINE DESIGNATIONS:
T53-K-703	90
T55-K-712	91

MITSUBISHI NAGASAKI HEAVY INDUSTRIES (MHI), JET ENGINE DIVISION
1-1 AKUNOURA-MACHI, NAGASAKI 850, JAPAN
Tele: (0568) 79-2111

ENGINE DESIGNATIONS:
CT63-M-5A	92

NATIONAL AEROSPACE LABORATORY
1880 JINDDAIJI-MACHI, CHOFU CITY, TOKYO, JAPAN, 0422
Tele: 47-5911

ENGINE DESIGNATIONS:
FJR-710-600S	93

PAGE

PEOPLES REPUBLIC OF CHINA:

CHINA NATIONAL ENGINE CORP. (CAREC)
P.O. BOX 1671, BEIJING, P.R.CHINA
Tele: 442-444

ENGINE DESIGNATIONS:
CHENGDU WP-7A	94
CHENGDU WZ-6	94
HARBIN WJ-5	94
HARBIN WZ-8	94
LIYANG WP-7B	94
LIYANG WP-13A	94
SHENYANG WP-6/6A	94
SHENYANG WS-6	94
XIAN WP-8	94
XIAN WS-9 (SPEY 202)	94
ZHUZHOW WJ-6	94

POLAND:

POSKIE ZAKLADY LOTNICZE (PZL)
MIODO WA5, 0025, WARSAW, POLAND
Tele: WARSAW 261441

ENGINE DESIGNATIONS:
ISOTOV GTD-350P	95
K-15	96
PZL-10W	97
SO-1/2	98
SO-3W	99
TWD-10B	100

ROMANIA:

CENTRAL NATIONAL AERONAUTICE
BOULEVARD DALIA 13, CASUTA POSTALA 22-149, 2-70185
BUCHAREST, RUMANIA
Tele: 12-08-78

ENGINE DESIGNATIONS:
SPEY 512-14DW	101
VIPER 632	102
TURMO IVC	103

SOUTH AFRICA:

ATLAS AIRCRAFT CORP.
P.O. BOX 11, ATLAS ROAD, TRANSVALL, SOUTH AFRICA
Tele: 011-927-9111

ENGINE DESIGNATIONS:
VIPER 540	104

SWEDEN:

VOLVO FLYGMOTOR AB
SR-461 81 TROLLHAETTAN, SWEDEN, R46181
Tele: 0520-94000

ENGINE DESIGNATIONS:
RM8B	105
RM12	106

  PAGE

TURKEY:

TUSAS ENGINE INDUSTRIES (TEI)
MUTTALIP MEVKIS MRK, PK 610, ESHISEHIR, TURKEY
Tele: 904-324-2140

ENGINE DESIGNATIONS:
F110-GE-100 107

USSR Presently known as, CIS (COMMONWEALTH OF INDEPENDENT STATES):[†††]

CENTRAL INSTITUTE OF AVIATION MOTORS TSIAM
2 AVIAMOTORNAYA ST. 111250
MOSCOW, RUSSIA, CIS

GLUSHENKOV/OMSK ENGINE DESIGN BUREAU

ENGINE DESIGNATIONS:
GTD-3BM 108
TVD-10B 109
TVD-20 140
TVD-1500 110

KLIMOV ENGINE DESIGN BUREAU (Formerly Isotov) OF:
ST. PETERSBURG SCIENTIFIC & PRODUCTION ASSOCIATION, RUSSIA, CIS

ENGINE DESIGNATIONS:
GTD-350 111
RD-33 112
TV2-117A 113
TV3-117V 114
TV7-117V 115
VK-1 140

KOLIESOV ENGINE DESIGN BUREAU OF RYBINSK MOTORS (RKBM)
RYBINSK, RUSSIA, CIS.

ENGINE DESIGNATIONS:
RD-36 140
VD-7 140
VD-57 140

KUZNETSOV-TRUD ENGINE DESIGN BUREAU OF:
SAMARA STATE SCIENTIFIC PRODUCTION ENTERPRISE, SAMARA, RUSSIA, CIS

ENGINE DESIGNATIONS:
NK-8-4 117
NK-12VM 118
NK-86 140
NK-144 119
NK-144 (Upgrade) 120

LYULKA-SATURN ENGINE DESIGN BUREAU
MOSCOW, RUSSIA, CIS

ENGINE DESIGNATIONS:
AL-7F-1-100 121
AL-21F-3 122
AL-31F 140

PERM SCIENTIFIC INDUSTRIAL ASSOCIATION "PNPP AVIADVIGATEL"
(formerly Soloviev), PERM, RUSSIA, CIS

ENGINE DESIGNATIONS:
SOLOVIEV D-15	140
SOLOVIEV D-20	124
SOLOVIEV D-25V	125
SOLOVIEV D-30	126
SOLOVIEV D-30F6	140
SOLOVIEV D-30-KP	127
PS-90A	128

PROGRESS ENGINE DESIGN BUREAU (formerly Ivchenko and Lotarev Design Bureaus)
OF: ZAPOROZHYE ENGINE CENTER, ZAPROZHYE, UKRAINE, CIS

ENGINE DESIGNATIONS:
AI-20M	129
AI-24VT	130
AI-25A	131
D-27	140
D-18T	132
D-136	133
D-436T	134

SOYUZ DESIGN BUREAU OF:
MOSCOW SCIENTIFIC PRODUCTION CORP., MOSCOW, RUSSIA, CIS
(FORMERLY TUMANSKY AND MIKULIN DESIGN BUREAUS)

KOPTCHVENKO ENGINE DESIGNS
ENGINE DESIGNATIONS:
TV-O-100	116

MIKULIN ENGINE DESIGNS

ENGINE DESIGNATIONS:
RD-3M	123

SOYUZ-ISOTOV ENGINE DESIGNS

ENGINE DESIGNATIONS:
RD-41	140

TUMANSKY ENGINE DESIGNS

ENGINE DESIGNATIONS:
M701C500	140
R-11-300	135
R-13F2-300	136
R-15BD-300	140
R-25	140
R-27-22A	140
R-29B	137
R-31	138
R-31 (Upgrade)	139
R-35F-300	140
RD-45F	140
R-79 & 195	140

YUGOSLAVIA:

ORAO AIR FORCE DEPOT
9 NEMANJINA ST. 11005 BELGRADE 9, YUGOSLAVIA Tele: 11-621-522

ENGINE DESIGNATIONS:
VIPER 632	141

MULTI-NATIONAL TURBINE ENGINE COMPANIES

BMW, ROLLS ROYCE GmbH
OBERURSEL WORK, HOHENMARKES STRRASSE 60-70, OBERUUESL, GERMANY Tele: 06171-5000

	PAGE
ENGINE DESIGNATIONS:	
T53-L-13	142
T117	143

CFM INTERNATIONAL (G.E. AND SNECMA)
BOX 15514, INTERSTATE HWY 75, CINCINNATI, OHIO, 45215, USA
Tele: (513) 552-3311

ENGINE DESIGNATIONS:
CFM56-5C2 144

EUROJET TURBO GMBH (R-R, FIAT, MTU, ITP SPAIN)
ARABELLA STRSSE 13, D-8000, MUNICH, 81 F.R. GERMANY
Tele: (089) 921 00050

ENGINE DESIGNATIONS:
EJ 200 145

INTERNATIONAL AERO ENGINES LTD.
(P&W, R-R, Fiat, MTU, Japanese Aero)
287 Main Street, East Hartford, CT, 06108, USA
Tele: (203) 280-1800

ENGINE DESIGNATIONS:
V2500-A1 146

INTERNATIONAL TURBINE ENGINE COMPANY
(ALLIED SIGNAL AND AEROSPACE INDUSTRY DEVELOPMENT CENTER, TAIWAN)
P.O. BOX 5217, Phoenix, AR. 85010, USA.
Tele: (602) 231-1000

ENGINE DESIGNATIONS:
TFE 1042-70 147

MTU, ROLLS-ROYCE, TURBOMECA Ltd.
65 Buckingham Gate, London SW1 E6AT, England
Tele: 01-222-9020

ENGINE DESIGNATIONS:
MTR 390 (Formerly MTM 385) 148

ROLLS-ROYCE, ALLISON TURBINE COMPANY
65 Buckingham Gate, London SW1 E6AT, England
Tele: 01-222-9020

ENGINE DESIGNATIONS:
SPEY RB. 168-66 149

ROLLS-ROYCE, SNECMA Ltd.
65 Buckingham Gate, London SW1 E6AT, England
Tele: 01-222-9020

ENGINE DESIGNATIONS:
OLYMPUS 593 150

ROLLS-ROYCE, SNECMA, MTU Ltd.
65 Buckingham Gate, London SW1 E6AT, England
Tele: 01-222-9020

ENGINE DESIGNATIONS:
TYNE-20 151

PAGE

ROLLS-ROYCE, TURBOMECA LTD.
4-5 Grosvenor Place, London SW1X 7HH, England
Tele: 01-235-3641

ENGINE DESIGNATIONS:
ADOUR MK.811/815	152
ADOUR MK.871	153
RTM 322-01	154

SNECMA, GENERAL ELECTRIC USA
CINCINNATI, OH, 45215, USA
Tele: (513) 243-2000

CF6-50	155

TURBOMECA, SNECMA, MTU Ltd.
1 Rue Beaujon, BP 37 75008 Paris, France
Tele: (33/1) 45-61-4895

ENGINE DESIGNATIONS:
LARZAC 04-C20	156

TURBO-UNION (FIAT, MTU, R-R) Ltd.
P. O. Box 3, Filton, Bristol BS12 7QE, England
Tele: 0272-79-12134

ENGINE DESIGNATIONS:
RB.199-34R (MK.103)	157

WILLIAMS INTERNATIONAL CORP.
2280 W. MAPLE ROAD, WALLED LAKE, MI, 48088
Tele: (313) 624-5200

ENGINE DESIGNATIONS:
FJ44	158

UNITED STATES TURBINE ENGINE COMPANIES

ALLISON TURBINE DIVISION OF GENERAL MOTORS
P.O.BOX 894, INDIANAPOLIS, IN, 46206-0894 - USA
Tele: (317) 230-2000

ENGINE DESIGNATIONS:
250-B17E	159
250-C20R	160
250-C34	161
501-D22A & G	162
GMA-2100	163
GMA 3007	164
T56-A-425	165
T63-A-720	166
T406-AD-400	167
T703-A-700	168
TF41-A-400	169

GARRETT AND ALLISON (ALLIED SIGNAL AEROSPACE/CFE)
P.O. BOX 5217, PHOENIX, AR, 85010 - USA
Tele: (602) 231-3285

ENGINE DESIGNATIONS:
CFE 738	170

GARRETT TURBINE ENGINE COMPANY
(DIVISION OF ALLIED SIGNAL CORP)
111 SOUTH 34 th Street, PHOENIX, AR, 85034 - USA
Tele: (602) 231-1000

ENGINE DESIGNATIONS:	PAGE
ATF3-6A	171
F-109	172
F124-GA-101	173
T76-G10	174
TFE-731-5	175
TPE 331-14	176
TPF-351-20	177

GENERAL ELECTRIC AIRCRAFT ENGINES, Large Engine Group
CINCINNATI, OH, 45215 - USA
Tele: (513) 243-2000

ENGINE DESIGNATIONS:	
CF6-45A2	178
CF6-50E2F	179
CF6-80C2	180
CJ805-3	181
CJ805-23	182
GE-36-B22A (UDF)	183
J79-GE-119	184
TF39-GE-1C	185

GENERAL ELECTRIC AIRCRAFT ENGINES, Small Engine Group
LYNN, MASS, 01901 - USA
Tele: (617) 594-0100

ENGINE DESIGNATIONS:	
CF34-1A	186
CF700-2D-2	187
CJ610-8	188
CT7-6	189
CT7-9	190
CT58-140	191
CT64-800-4	192
F101-GE-102	193
F110-GE-400	194
F118-GE-100	195
F404-GE-400	196
J85-GE-17A	197
J85-GE-21	198
T58-GE-16	199
T64-GE-419	200
T700-GE-701A	201

LIGHT HELICOPTER TURBINE ENGINE COMPANY-LHTEC (Allison and Garrett)
(LH Tech,) 111 South 34th Street, Mail Drop 76 503-1T
Phoenix, AR, 85034 - USA
Tele: (602) 231-1000

ENGINE DESIGNATIONS:	
CTS-800	202
T800-LHT-800	203

TELEDYNE CAE CORP.
P.O. BOX 6971, TOLEDO, OH, 43612 - USA
Tele: (419) 470-3826

ENGINE DESIGNATIONS:	
J69-T-25	204
J402-CA-702 (373-8B)	205

PAGE

TELEDYNE CONTINENTAL MOTORS AIRCRAFT PRODUCTS
P.O. BOX 90, MOBILE, ALABAMA - USA
Tele: (205) 438-3411

ENGINE DESIGNATIONS:
TP-500	206

TEXTRON-LYCOMING CORP.
550 S. MAIN STREET, STRATFORD, CT, 06497 - USA
Tele: (203) 385-2000

ENGINE DESIGNATIONS:
AL5512	207
ALF502L	208
LF-507	209
LTP101-700A-1	210
LTS101-750C-1	211
T53-L-701A	212
T53-L-703	213
T5508D	214
T55-L-712	215
T55-L-714	216

TURBINE TECHNOLOGIES LTD
430 Phillips Street, Chetek, WI, 54728 - USA
Tele: (715) 924-2436

ENGINE DESIGNATIONS:
SR-30	217

UNITED TECHNOLOGIES (PRATT & WHITNEY) [††]
COMMERCIAL PRODUCTS DIVISION
400 MAIN STREET, HARTFORD, CT, 06108 - USA
Tele: (203) 565-4321

ENGINE DESIGNATIONS:
J60-P-6	218
JT3C-7	219
JT3D-7	220
JT4A-11	221
JT8D-17AR	222
JT8D-219	223
JT9D-7R4G2	224
JT12A-8	225
JTFD12A-5A	226
PW 2037	227
PW 4152	228
PW 4460	229

UNITED TECHNOLOGIES, (PRATT & WHITNEY)
GOVERNMENT PRODUCTS DIVISION
P.O. BOX 109600, WEST PALM BEACH, FL, 33410-9600 - USA
Tele: (407) 796-2000

ENGINE DESIGNATIONS:
F100-PW-220	230
F117-PW-100	231
J52-P-408	232
J57-P-59W	233
J57-P-420	234
J58	235
J75-P-19W	236
PW 1115	237
PW 1120	238
PW 1212	239
PW 1216	240

	PAGE
PW 3005	241
TP30-P-100	242
TF33-PW-102	243

GAS TURBINE AUXILIARY POWER UNITS (USA)

GARRETT TURBINE ENGINE COMPANY	244
UNITED TECHNOLOGIES USA (HAMILTON STANDARD)	247
UNITED TECHNOLOGIES USA (PRATT & WHITNEY)	248
SUNDSTRAND CORP	248
WILLIAMS INTERNATIONAL CORP.	249

† **NOTE**

...The later model Turbine Engines manufactured by Pratt & Whitney of Canada (P&WC) have designations which indicate engine type and power class as follows:
The first digit "1" indicates Turboprop, "2" indicates Turboshaft, "3" indicates Turbofan. The last two digits indicate power class. Example: In the PW125, (1) indicates Turboprop Engine, (25) indicates 2,500 shp class.

†† **NOTE**

...The later model Turbine Engines manufactured by United Technologies (Pratt and Whitney of Hartford, Connecticut, USA), have designations which indicate engine type, customer, and power class.

...The first digit, when "odd", identifies a Military Engine. The first digit, when "even", identifies a Civil Engine.

...The second digit identifies the aircraft manufacturer as follows: 0 - Indicates Boeing, 1 - Indicates Airbus, 2 - Is not currently used, 3 - Is no longer used, 4 - Indicates McDonnell-Douglas.

...The last two digits indicate the power class. Example: In the PW2037, (2) indicates Civil Application, (0) indicates Boeing, (37) indicates 37,000 lbt. power class.

††† **NOTE**

...The authors have used USSR as the country of origin for this edition because it was felt that more people are familiar with this term of reference at this writing. The abbreviation "CIS" will be used when referring to the Commonwealth of Independent States in future editions.

...Russian Engines generally carry the name of the Design Bureau (often the name of the original Chief Design Engineer). This listing is in alphabetical order by the name of the current Design Bureau. However, the name of the original designer will appear on many engines.

...Compiling data concerning aircraft engines produced in the USSR (now CIS) and the Peoples Republic of China (PRC) can be a difficult task because in these countries, aircraft engine programs are not set up in a manner that lends itself to dissemination of information to the general public. Therefore, some of the data herein is incomplete.

GAS TURBINE ENGINES FOR AIRCRAFT

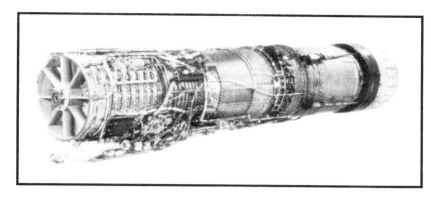
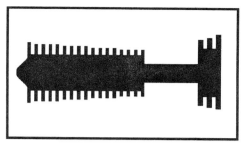

INTERNATIONAL TURBINE ENGINE COMPANIES

COUNTRY: CANADA

MANUFACTURER: HAWKER SIDDELEY CANADA INC. (ORENDA DIVISION)
MISSISSAUGA, ONTARIO

DESIGNATION (military): J79-11A.

BACKGROUND: Engine manufactured under license, General Electric Corp. (USA).
Date First J79-11A Produced: 1960.
Date If Out-of-Production: 1966. Number of J79-11A Engines Produced to Date: 480.

ENGINE TYPE: Turbojet, Single-Shaft with Afterburning.

COMPRESSOR TYPE: 17-Stage Single-Spool, Axial Flow Compressor.

COMPRESSOR DATA: Compressor Pressure Ratio: 12.0 : 1, Total Mass Airflow: 169 lb/sec (76,6 kg/sec).

TURBINE TYPE: Three-Stage Axial Flow Turbine.

COMBUSTOR TYPE: Through-Flow, Can-Annular with 10 liners.

AFTERBURNER: Four-Stage.

POWER (take-off) RATING: 10,900 lbt (48,5 kN) [dry], 15,800 lbt (66,7 kN) [A/B]. Rating Approved to: 59°F (15°C).

TAKE-OFF SPECIFIC FUEL CONSUMPTION (SFC): 0.84 lb/hr/lbt (23,79 mg/Ns) [dry], 1.97 lb/hr/lbt (55,8 mg/Ns) [A/B].

BASIC DRY WEIGHT: 3,560 lbs (1,615 kg).

POWER/WEIGHT RATIO: 3.06 : 1 [dry], 4.44 : 1 [A/B] - lbt/lb.

APPLICATION:
CF-104, Fighter.

OTHER ENGINES IN SERIES:
J79-OEL-7 also in CF-104.
Locate other 79 models under General Electric Aircraft Engines, USA.

GAS TURBINE ENGINES FOR AIRCRAFT

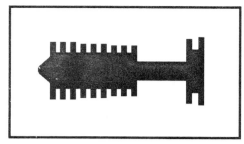

INTERNATIONAL TURBINE ENGINE COMPANIES

COUNTRY: CANADA

MANUFACTURER: HAWKER SIDDELEY CANADA INC. (ORENDA DIVISION)
MISSISSAUGA, ONTARIO

DESIGNATION (military): J85-CAN-15.

BACKGROUND: Engine manufactured under license, General Electric Corp. (USA).
Date First J85-CAN-15 Produced: 1966.
Date If Out-of-Production: 1974. Number of J85-CAN-15 Engines Produced to Date: 610.

ENGINE TYPE: Turbojet, Single-Shaft with Afterburning.

COMPRESSOR TYPE: 9-Stage Single-Spool, Axial Flow Compressor.

COMPRESSOR DATA: Compressor Pressure Ratio: 6.9 : 1, Total Mass Airflow: 44.0 lb/sec (19,96 kg/sec).

TURBINE TYPE: 2-Stage Axial Flow Turbine.

COMBUSTOR TYPE: Annular, Through-Flow.

AFTERBURNER: 1-Stage.

POWER (take-off) RATING: 2,925 lbt [dry]. 4,300 lbt [A/B], Rating Approved to: 59°F (15°C).

TAKE-OFF SPECIFIC FUEL CONSUMPTION (SFC): 0.92 lb/hr/lbt (26,06 mg/Ns) [dry], 2.18 lb/hr/lbt (61,75 mg/Ns) [A/B].

BASIC DRY WEIGHT: 627 lbs (284 kg).

POWER/WEIGHT RATIO: 4.67: 1 [dry], 6.86 : 1 [A/B] - lbt/lb.

APPLICATION:
Canadair-Northrop CF-5A/D "Freedom Fighter and Netherlands NF-5A/B; Fighter/Trainer Aircraft.

OTHER ENGINES IN SERIES:
J85-CAN-40 (Non-Afterburning model) in Canadair CL-41 "Tutor" Trainer Aircraft.
Locate other J85 models under General Electric Aircraft Engines, USA.

GAS TURBINE ENGINES FOR AIRCRAFT

 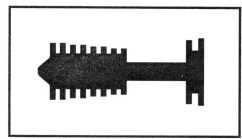

INTERNATIONAL TURBINE ENGINE COMPANIES

COUNTRY: CANADA

MANUFACTURER: HAWKER SIDDELEY CANADA INC. (ORENDA DIVISION)
MISSISSAUGA, ONTARIO

DESIGNATION (military): J85-CAN-40.

BACKGROUND: Manufactured under license, General Electric Corp. (USA).
Date First J85-CAN-40 Produced: 1962.
Date If Out-of-Production: 1965. Number of J85-CAN-40 Engines Produced to Date: 230.

ENGINE TYPE: Turbojet (non-afterburning). Single-Shaft.

COMPRESSOR TYPE: 8-Stage Single-Spool, Axial Flow Compressor.

COMPRESSOR DATA: Compressor Pressure Ratio: 6.9 : 1, Total Mass Airflow: 44.0 lb/sec (19,96 kg/sec).

TURBINE TYPE: 2-Stage Axial Flow Turbine.

COMBUSTOR TYPE: Annular, Through Flow.

POWER (take-off) RATING: 2,850 lbt (12,68 kN thrust). Rating Approved to: 59°F (15°C).

TAKE-OFF SPECIFIC FUEL CONSUMPTION (SFC): 0.99 lb/hr/lbt (28,04 mg/Ns).

BASIC DRY WEIGHT: 399 lbs (182 kg).

POWER/WEIGHT RATIO: 7.14 : 1 lbt/lb.

APPLICATION:
Canadair CL-41 "Tutor" (also called CT-114 as an RCAF Trainer Aircraft).

OTHER ENGINES IN SERIES:
Locate Afterburning model under: J85-CAN-15 listing.
Locate other J85 models under General Electric Aircraft Engines, USA.

GAS TURBINE ENGINES FOR AIRCRAFT

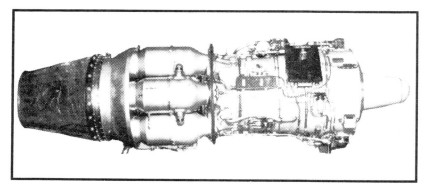 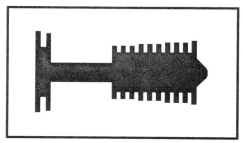

INTERNATIONAL TURBINE ENGINE COMPANIES

COUNTRY: CANADA

MANUFACTURER: HAWKER SIDDELEY CANADA INC. (ORENDA DIVISION)
MISSISSAUGA, ONTARIO

DESIGNATION (military): ORENDA.

BACKGROUND: Originally Manufactured by the Orenda Engine Company.
Date First Orenda Produced: 1952.
Date If Out-of-Production: 1958. Number of Orenda Engines Produced to Date: 3,800.

ENGINE TYPE: Turbojet (non-afterburning). Single-Shaft.

COMPRESSOR TYPE: 10-Stage Single-Spool, Axial Flow Compressor.

COMPRESSOR DATA: Compressor Pressure Ratio: 10 : 1, Total Mass Airflow: 106 lb/sec (48 kg/sec).

TURBINE TYPE: Two-Stage Axial Flow Turbine.

COMBUSTOR TYPE: Can-Annular, Through Flow, with 6 liners.

POWER (take-off) RATING: 7,500 lbt (33,36 kN thrust). Rating Approved to: 59°F (15°C).

TAKE-OFF SPECIFIC FUEL CONSUMPTION (SFC): 0.85 lb/hr/lbt (24,1 mg/Ns).

BASIC DRY WEIGHT: 2,560 lbs (1,160 kg).

POWER/WEIGHT RATIO: 2.92 : 1 lbt/lb.

APPLICATION:
CF-100, F-86A Fighter Aircraft.

OTHER ENGINES IN SERIES:
ORENDA -1,-2,-3,-8,-9,-10,-11.

GAS TURBINE ENGINES FOR AIRCRAFT

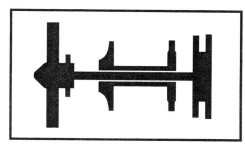

INTERNATIONAL TURBINE ENGINE COMPANIES

COUNTRY: CANADA

MANUFACTURER: PRATT & WHITNEY OF CANADA, INC.
LONGUEUIL, QUEBEC

DESIGNATION (civil/military): JT15D-5.

BACKGROUND: Originally Developed for commercial aircraft.
Date First JT15D Produced: 1975. Date first JT15D-5 Produced: 1983.
Date If Out-of-Production: N/A. Number of JT15D Engines Produced to Date: 3,850.

ENGINE TYPE: Turbofan. Dual-Shaft, Medium-Bypass Front Fan.

COMPRESSOR TYPE: Combination Axial-Centrifugal Flow Compressor including: 1-Stage Fan followed by I Axial Stage in the Low Pressure Compressor. 1-Stage Centrifugal Flow Impeller in the High Pressure Compressor.

COMPRESSOR DATA: Compressor Pressure Ratio: 13 : 1, Fan Pressure Ratio: 1.6 : 1, Fan Bypass Ratio: 2.1 : 1, Total Mass Airflow: 85 lb/sec (38,6 kg/sec).

TURBINE TYPE: 3-Stage Axial Flow Turbine including: 1-Stage High Pressure Turbine, 2-Stage Low Pressure Turbine.

COMBUSTOR TYPE: Annular, Through-Flow.

POWER (take-off) RATING: 2,900 lbt (12,9 kN thrust). Rating Approved to: 80°F (27°C).
Thermodynamic Thrust (Std. Day) 3,200 lbt.

TAKE-OFF SPECIFIC FUEL CONSUMPTION (SFC): 0.56 lb/hr/lbt (15,86 mg/Ns).

BASIC DRY WEIGHT: 642 lbs (291 kg).

POWER/WEIGHT RATIO: 4.52 : 1 lbt/lb.

APPLICATION:
Cessna Citation II and V.
Beech Corp. Beechjet-400 (formerly Mitsubishi Diamond) Business Aircraft.
Navy T-47A "Citation".
USAF T-1A Trainer and Transport C-550 "BeechJet-400".

OTHER ENGINES IN SERIES:
JT15D-1, -1A/B in Citation (formerly Fanjet-500), Citation-I (formerly Citation-500) and Siai-Marchetti S-211, Italy.
JT15D-4, -4B/C/D in Citation-II and Aerospatiale SN-601, France.

GAS TURBINE ENGINES FOR AIRCRAFT

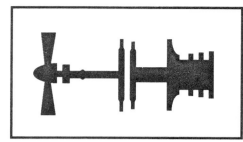

INTERNATIONAL TURBINE ENGINE COMPANIES

COUNTRY: CANADA

MANUFACTURER: PRATT & WHITNEY OF CANADA, INC.
LONGUEUIL, QUEBEC

DESIGNATION (civil): PT6A-27.

BACKGROUND: Military version designated T-74-CP-701. The PT6 Series ranges in power output from 500 eshp (373 ekw) to 1,875 eshp (1,398 ekw). It is the most widely used Small Gas Turbine Engine in world service.
Date First PT6 Produced: 1963. Date First PT6A-27 Produced: 1967.
Date If Out-of-Production: N/A. Number of PT6A Series Engines Produced to Date: 29,000 total; including 22,500 turboprops and 6,500 turboshafts.

ENGINE TYPE: Turboprop (Rear Drive) Dual-Shaft (Compressor Drive & Power Output Drive).

COMPRESSOR TYPE: Combination, Axial-Centrifugal Flow Compressor including: 3-Stages of Axial Flow Compression, 1-Stage of Centrifugal Flow Compression.

COMPRESSOR DATA: Compressor Pressure Ratio: 6.5 : 1, Total Mass Airflow: 6.5 lb/sec (2,95 kg/sec).

TURBINE TYPE: 2-Stage Axial Flow Turbine including: 1-Stage Gas Producer (compressor drive) Turbine, 1-Stage Power (free) Turbine.

COMBUSTOR TYPE: Annular, Reverse Flow.

POWER (take-off) RATING: 715 eshp (533 ekw). Includes 80 shp (59,7 kw) from 200 lbt (0,89 kN thrust). Rating Approved to: 71°F (22°C).

TAKE-OFF SPECIFIC FUEL CONSUMPTION (SFC): 0.59 lb/hr/eshp (0,36 kg/hr/ekw).

BASIC DRY WEIGHT: 328 lbs (149 kg).

POWER/WEIGHT RATIO: 2.18 : 1 eshp/lb (3,58 :1 ekw/kg).

APPLICATION:
Beech 99A, DeHavilland Twin Otter, Embraer EMB-110, Equitor P-550, Frakes Mallard Conversion, McKinnon Turbo-Goose and Turbo-Beaver Conversions and Pilatus PC-6.

OTHER ENGINES IN SERIES: PT6A Series: -10, -11, -11AG, -112, -15AG, -21, -25, -25A, -25C, -27, -28, -34, -34B, -34AG, 110, -112, 114, -135, -135A, -36, -41, -41AG, -42, -45, -45R, -50, -60, -65B, -65R, -66, -67. PT6B Series: -35F, -36. PT6T Series: -3B, -6. ST6L-73. In many models of fixed-wing aircraft including: Avtec, IAI Arava, Ayres, Basler, Beech, Cessna, Cinair, Commuter, DeHavilland, Dornier, Embraer, Gates, Harbin, IMP, LearFan, Maul, Mooney, New-Cal, Norman, Omac, Omni, Pezetel, Piaggio, Pilatus, Pox, iper, Shorts, TBM (Aerospatiale-Mooney), Valmet. Also Rotor-Wing aircraft including: Agusta-Bell, Bell, Sikorsky.
T-74 US Army applications include: RC-12, UV-18, UV-20, RU-21, UV-21, UV-24 UV-27.
T-74 US Navy Applications include: T-34, T-44, UC-12.

GAS TURBINE ENGINES FOR AIRCRAFT

 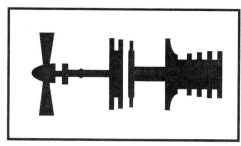

INTERNATIONAL TURBINE ENGINE COMPANIES

COUNTRY: CANADA

MANUFACTURER: PRATT & WHITNEY OF CANADA, INC.
LONGUEUIL, QUEBEC

DESIGNATION (civil)**:** PT6A-65AG.

BACKGROUND: The AG Series Denotes use for Agricultural Aircaft.
Date First PT6A-65 Produced: 1982.
Date If Out-of-Production: N/A. Number of -65 Series Engines Produced to Date: 1,000.

ENGINE TYPE: Turboprop. Dual-Shaft (Compressor Drive & Power Output Drive).

COMPRESSOR TYPE: Combination, Axial-Centrifugal Flow Compressor including: 4-Stages of Axial Flow Compression, 1-Stage of Centrifugal Flow Compression.

COMPRESSOR DATA: Compressor Pressure Ratio: 9.7 : 1, Total Mass Airflow: 10.3 lb/sec (4,67 kg/sec).

TURBINE TYPE: 3-Stage Axial Flow Turbine including: 1-Stage Gas Producer (compressor drive) Turbine, 2-Stage Power (free) Turbine.

COMBUSTOR TYPE: Annular, Reverse Flow.

POWER (take-off) RATING: 1,459 eshp (1,088 ekw). Includes 82 shp (61 kw) from 205 lbt (0,91 kN thrust). Rating Approved to: 71°F (22°C).

TAKE-OFF SPECIFIC FUEL CONSUMPTION (SFC): 0.521 lb/hr/eshp (0.32 kg/hr/ekw).

BASIC DRY WEIGHT: 486 lbs (220 kg).

POWER/WEIGHT RATIO: 3.0 : 1 shp/lb (4,95 :1 kw/kg).

APPLICATION:
Air Tractor AT-400.
Ayres S2R "Turbo Thrush".
Frakes "Turbo-Cat".
Schweizer G-164B "Ag-Cat".

OTHER ENGINES IN SERIES:
Ag Models: PT6A-11 in Weatherly 620TP, -15 and -34 in Air Tractor AT-402/502, Ayres S2R, USA and Turbo-Kruk, Poland.
Non-Ag model PT6A-45 in Aerospatiale N-262, France; Turbo-Dromander, Poland; Air Tractor-503, USA.
Non-Ag model PT6A-65 in AMI Corp. DC-3, Beech 1900, Frakes Mohawk, USA; Shorts 360, Shorts Super-Sherpa, Great Britain.

GAS TURBINE ENGINES FOR AIRCRAFT

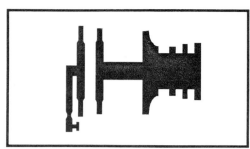

INTERNATIONAL TURBINE ENGINE COMPANIES

COUNTRY: CANADA

MANUFACTURER: PRATT & WHITNEY OF CANADA, INC.
LONGUEUIL, QUEBEC

DESIGNATION (civil): PT6B-36A.

BACKGROUND: The PT6B Series Denotes A Single Turboshaft Engine as opposed to a "T" Series Twin-Pac™ in which two engines connect to a common Gearbox.
Date First PT6B Produced: 1984.
Date If Out-of-Production: N/A. Number of PT6B Series Engines Produced to Date: 3,000.

ENGINE TYPE: Turboshaft. Dual-Shaft (Compressor Drive & Power Output Drive).

COMPRESSOR TYPE: Combination, Axial-Centrifugal Flow Compressor including: 3-Stages of Axial Flow Compression, 1-Stage of Centrifugal Flow Compression.

COMPRESSOR DATA: Compressor Pressure Ratio: 7.1 : 1, Total Mass Airflow: 6.2 lb/sec (2,81 kg/sec).

TURBINE TYPE: 2-Stage Axial Flow Turbine including: 1-Stage Gas Producer (compressor drive) Turbine, 1-Stage Power (free) Turbine.

COMBUSTOR TYPE: Annular, Reverse Flow.

POWER (take-off) RATING: 981 shp (732 kw). Rating Approved to: 99°F (37°C).

TAKE-OFF SPECIFIC FUEL CONSUMPTION (SFC): 0.58 lb/hr/shp (0.35 kg/hr/kw).

BASIC DRY WEIGHT: 378 lbs (171 kg).

POWER/WEIGHT RATIO: 2.6 : 1 shp/lb (4,28 :1 kw/kg).

APPLICATION:
Sikorsky S-76B Twin-Engine Helicopter.

OTHER ENGINES IN SERIES:
PT6B-3 in Rogerson-Hiller 1099.
PT6B-3B in Agusta AB-212 and AB-412, Italy.
PT6B-35F proposed for LearFan Turboprop Business Jet.
PT6B-67 IN TW-68, Tilt-Wing Aircraft.

GAS TURBINE ENGINES FOR AIRCRAFT

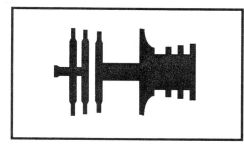

INTERNATIONAL TURBINE ENGINE COMPANIES

COUNTRY: CANADA

MANUFACTURER: PRATT & WHITNEY OF CANADA, INC.
LONGUEUIL, QUEBEC

DESIGNATION (civil)**:** PT6T-6 (Twin-Pac™) Civil version of Military T400-CP-400. CP denotes manufactured in Canada.

BACKGROUND: Twin-Pac™ is a PW trade mark. "T" in the designation reflects a Twinned-Engine (two engines powering a combined P&WC Gearbox with automatic power sharing for Rotorcraft). The PT6 Series ranges in power output from 500 eshp (373 ekw). to 1,875 eshp (1,398 ekw). The PT6 is the most widely used Small Gas Turbine Engine in world service.
Date First PT6T Produced: 1974.
Date If Out-of-Production: N/A. Number of PT6T-Series Engines Produced to Date: 3,000.

ENGINE TYPE: Turboshaft. Dual-Shaft (Compressor Drive & Power Output Drive).

COMPRESSOR TYPE: Combination, Axial-Centrifugal Flow Compressor including: 3-Stages of Axial Flow Compression, 1-Stage of Centrifugal Flow Compression.

COMPRESSOR DATA: Compressor Pressure Ratio: 7.3 : 1, Total Mass Airflow: 6.8 lb/sec (3,08 kg/sec).

TURBINE TYPE: 3-Stage Axial Flow Turbine including: 1-Stage Gas Producer (compressor drive) Turbine, 2-Stage Power (free) Turbines coupled to a combined gearbox.

COMBUSTOR TYPE: Annular, Reverse Flow.

POWER (take-off) RATING: 1,875 shp (1,398 kw), Twin-Pac™ total SHP. Rating Approved to: 70°F (21°C).

TAKE-OFF SPECIFIC FUEL CONSUMPTION (SFC): 0.59 lb/hr/shp (0,36 kg/hr/kw).

BASIC DRY WEIGHT: 657 lbs (298 kg).

POWER/WEIGHT RATIO: 2.85: 1 shp/lb (4,69 : 1 kw/kg).

APPLICATION:
Agusta 212 and 412, Sikorsky S-58T Re-Engine of CH-34 "Choctaw" Piston-Engine Helicopter.

OTHER ENGINES IN SERIES:
PT6T-3 and -3B in Sikorsky S-58T, Agusta 212, Bell 412.
PT6T-6 in Lockheed Model-86 Helicopter.
Military version known as T400-WV-402 in Bell AH-1J/ 1T, "Sea Cobra" and "Super Cobra".
Note: WV denotes manufactured by P&WC in West Virginia, USA.

GAS TURBINE ENGINES FOR AIRCRAFT

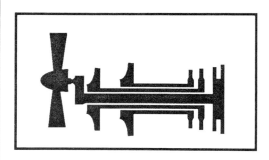

INTERNATIONAL TURBINE ENGINE COMPANIES

COUNTRY: CANADA

MANUFACTURER: PRATT & WHITNEY OF CANADA, INC.
LONGUEUIL, QUEBEC

DESIGNATION (civil): PW 118A

BACKGROUND: No military counterpart. Follow on Model of PW 115.
Date First PW 100 Series Engine Produced: 1983.
Date If Out-of-Production: N/A. Number of PW 100 Series Engines Produced to Date: 1,900.

ENGINE TYPE: Turboprop. Three-Shaft (dual compressor drive & single power output drive).

COMPRESSOR TYPE: Dual Impeller, Centrifugal Flow Compressor including: 1-Stage Low Pressure Compressor, 1-Stage High Pressure Compressor.

COMPRESSOR DATA: Compressor Pressure Ratio: 14.1 : 1, Total Mass Airflow: 14.7 lb/sec (6,7 kg/sec).

TURBINE TYPE: 4-Stage Axial Flow Turbine includes: Gas Producer Section consisting of a 1-Stage High Pressure Turbine and a 1-Stage Low Pressure Turbine and a Power Output Section consisting of a 2-Stage Power (free) Turbine.

COMBUSTOR TYPE: Annular, Reverse Flow.

POWER (take-off) RATING: 1,892 eshp (1,411 ekw). Includes 92 shp (69 kw) from 230 lbt (1,02 kN thrust). Rating Approved to: 108°F (42°C).

TAKE-OFF SPECIFIC FUEL CONSUMPTION (SFC): 0.504 lb/hr/eshp (30,6 kg/hr/kw).

BASIC DRY WEIGHT: 861 lbs (391 kg).

POWER/WEIGHT RATIO: 2.2 : 1 eshp/lb (3,60 :1 ekw/kg).

APPLICATION:
Embraer EMB-120 Commuter Aircraft, Brazil. (PW-115 in early EMB-120).

OTHER ENGINES IN SERIES:
PW 115 in early EMB-120, Commuter Aircraft.
PW 119 in Dornier 328, Commuter Aircraft.
PW 120,121 in DeHavilland Dash 8, in Aerospatiale-Aleina ATR 42, Regional Airliners.
PW 123 DeHavilland Dash-8 and PW 123AF in Canadair CL-215T Transport.
PW 124, 126, 127 in BAe ATP, and Aerospatiale-Aleina ATR 72 Regional Airliners.
PW 125 in Fokker 50 Regional Airliner.

GAS TURBINE ENGINES FOR AIRCRAFT

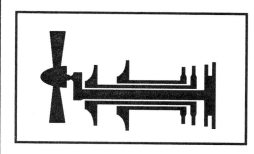

INTERNATIONAL TURBINE ENGINE COMPANIES

COUNTRY: CANADA

MANUFACTURER: PRATT & WHITNEY OF CANADA, INC.
LONGUEUIL, QUEBEC

DESIGNATION (civil)**:** PW126A.

BACKGROUND: No military counterpart.
Date First PW126 Produced: 1987.
Date If Out-of-Production: N/A. Number of PW100 Series Engines Produced to Date: 1,600.

ENGINE TYPE: Turboprop. Three-Shaft (dual compressor drive & single power output drive).

COMPRESSOR TYPE: Dual Impeller, Centrifugal Flow Compressor including: 1-Stage Centrifugal Flow Compressor in the Low Pressure Compressor, 1-Stage Centrifugal Flow Compressor in the High Pressure Compressor.

COMPRESSOR DATA: Compressor Pressure Ratio: 14.4 : 1, Total Mass Airflow: 17.7 lb/sec (8,03 kg/sec).

TURBINE TYPE: 4-Stage Axial Flow Turbine includes: Gas Producer Section consisting of a 1-Stage High Pressure Turbine, 1-Stage Low Pressure Turbine, and a Power Output Section consisting of a 2-Stage Power (free) Turbine.

COMBUSTOR TYPE: Annular, Reverse Flow.

POWER (take-off) RATING: 2,779 eshp (2,072 ekw). Includes 134.4 shp (98,7 kw) from 331 lbt (147 kN thrust). Rating Approved to: 90°F (32°C).

TAKE-OFF SPECIFIC FUEL CONSUMPTION (SFC): 0.462 lb/hr/eshp (0,28 kg/hr/kw)

BASIC DRY WEIGHT: 1,060 lbs (481 kg).

POWER/WEIGHT RATIO: 2.63 : 1 eshp/lb (4,31 :1 ekw/kg).

APPLICATION:
British Aerospace ATP Regional Airliner.

OTHER ENGINES IN SERIES:
PW 120,120A in DeHavilland Dash 8, Aerospatiale ATR 42.
PW 124 & 126 in BAe ATP.
PW 125 in Fokker 50 and Aerospatiale-Aleina ATR 72 Regional Airliner.

GAS TURBINE ENGINES FOR AIRCRAFT

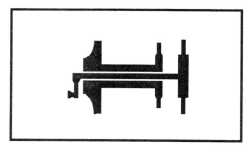

INTERNATIONAL TURBINE ENGINE COMPANIES

COUNTRY: CANADA

MANUFACTURER: PRATT & WHITNEY OF CANADA, INC.
LONGUEUIL, QUEBEC

DESIGNATION (civil): PW 205B.

BACKGROUND: Date First PW205 Produced: In Development.
Date If Out-of-Production: N/A. Number of PW 205 Engines Produced to Date: N/A.

ENGINE TYPE: Turboshaft. Dual-Shaft (Compressor Drive & Power Output Drive).

COMPRESSOR TYPE: Single Impeller, Centrifugal Flow Compressor.

COMPRESSOR DATA: Compressor Pressure Ratio: 8.0 : 1, Total Mass Airflow: Unknown at this time.

TURBINE TYPE: 2-Stage Axial Flow Turbine including: 1-Stage Gas Producer (compressor drive) Turbine, 1-Stage Power (free) Turbine.

COMBUSTOR TYPE: Annular, Reverse Flow.

POWER (take-off) RATING: 590 shp (440 kw). Rating Approved to: 59°F (15°C).

TAKE-OFF SPECIFIC FUEL CONSUMPTION (SFC): 0.556 lb/hr/shp (0,34 kg/hr/kw).

BASIC DRY WEIGHT: 218 lbs (98.9 kg).

POWER/WEIGHT RATIO: 2.07: 1 shp/lb (4,45 :1 kw/kg).

APPLICATION:
MBB BO-105LS and BO-108 Helicopters (Germany).

OTHER ENGINES IN SERIES:
PW 206A in McDonnell-Douglas MD-900 "MDX" Helicopter.
PW 209T Twin-Pac™ in Bell 400 Helicopter under development.

GAS TURBINE ENGINES FOR AIRCRAFT

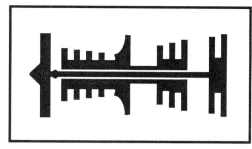

INTERNATIONAL TURBINE ENGINE COMPANIES

COUNTRY: CANADA

MANUFACTURER: PRATT & WHITNEY OF CANADA, INC.
LONGUEUIL, QUEBEC

DESIGNATION (civil)**:** PW 305.

BACKGROUND: Date First PW305 Produced: 1990.
Date If Out-of-Production: N/A. Number of PW 305 Engines Produced to Date: In Development.
Engine produced in collaboration with MTU, Germany.

ENGINE TYPE: Turbofan, Dual-Shaft, High Bypass Ratio.

COMPRESSOR TYPE: 6-Stage, Dual-Spool Compressor, including one Fan stage in the Low Pressure Rotor and a Combination High Pressure Rotor including a 4-Stage Axial and 1- Stage Centrifugal Flow Compressor.

COMPRESSOR DATA: Compressor Pressure Ratio: 19.4 : 1, Fan Pressure Ratio: 1.624 : 1
Fan Bypass Ratio 4.5 : 1. Total Mass Airflow: 170 lb/sec (77,1 kg/sec).

TURBINE TYPE: 5-Stage Axial Flow Turbine including: 3-Stage High Pressure Turbine,
2-Stage Low Pressure Turbine.

COMBUSTOR TYPE: Annular, Through Flow.

POWER (take-off) RATING: 5,225 lbt (23,24 kN) Rating Approved to: 72°F (22°C).

TAKE-OFF SPECIFIC FUEL CONSUMPTION (SFC): 0.407 lb/hr/lbt (11,53 mg/Ns).

BASIC DRY WEIGHT: 964 lbs (437 kg).

POWER/WEIGHT RATIO: 5.4 : 1 lbt/lb.

APPLICATION:
BAe-1000, follow on model of BAe-125 Commuter/Business Aircraft.
Learjet-60 Business Jet.
Lockheed T-33 "Shooting Star" Export Model.
Vopar PW305-F20 (Dassault Falcon Conversion) Commuter Aircraft.

OTHER ENGINES IN SERIES:
PW-304, down-rated version of PW-305 in development.

GAS TURBINE ENGINES FOR AIRCRAFT

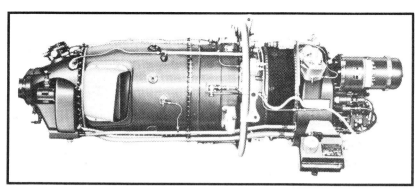 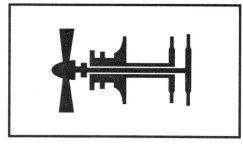

INTERNATIONAL TURBINE ENGINE COMPANIES

COUNTRY: CZECHOSLOVAKIA

MANUFACTURER: MOTORLET CORP.
PRAGUE.

DESIGNATION (civil & military): Walter M601.

BACKGROUND: Date First M601 Produced: 1975.
Date If Out-of-Production: N/A. Number of M601 Engines Produced to Date: 4,200.

ENGINE TYPE: Turboprop. Dual-Shaft (Compressor Drive & Power Output Drive).

COMPRESSOR TYPE: Combination Compressor; including: 2-Stage Axial Flow Compressor, 1-Stage Centrifugal Flow Compressor.

COMPRESSOR DATA: Compressor Pressure Ratio: 7 : 1, Total Mass Airflow: 7.9 lb/sec (3,60 kg/sec).

TURBINE TYPE: 2-Stage Axial Flow Turbine including: 1-Stage Gas Producer (compressor drive) Turbine, 1-Stage Power (free) Turbine.

COMBUSTOR TYPE: Annular, Through Flow.

POWER (take-off) RATING: 780 eshp (582 ekw). Includes 33 shp (24,6 kw) from 84 lbt (0.37 kN thrust). Rating Approved to: 104°F (40°C).

TAKE-OFF SPECIFIC FUEL CONSUMPTION (SFC): 0.64 lb/hr/eshp (0,395 kg/hr/ekw).

BASIC DRY WEIGHT: 425 lbs (193 kg).

POWER/WEIGHT RATIO: 1.84 : 1 eshp/lb (3,02 :1 ekw/kg).

APPLICATION:
LET Corp. L-410 Airliner-Transport and Zlin Corp. Z-37T Agro-Turbo Aircraft, Poland.

OTHER ENGINES IN SERIES:
M-601B, M-601D proposed for future aircraft.

GAS TURBINE ENGINES FOR AIRCRAFT

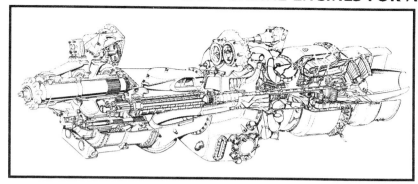 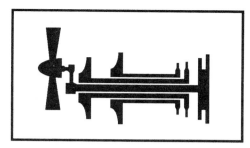

INTERNATIONAL TURBINE ENGINE COMPANIES

COUNTRY: CZECHOSLOVAKIA

MANUFACTURER: MOTORLET CORP.
PRAGUE.

DESIGNATION (civil & military): Walter M602.

BACKGROUND: Date First M602 Produced: In Development.
Date If Out-of-Production: N/A. Number of M602 Engines Produced to Date: N/A.

ENGINE TYPE: Turboprop. Three-Shaft (Compressor Drive & Power Output Drive).

COMPRESSOR TYPE: Dual Impeller Compressor; including: 1-Stage Low Pressure Centrifugal Flow, and a 1-Stage High Pressure Centrifugal Flow Compressor.

COMPRESSOR DATA: Compressor Pressure Ratio: 8 : 1, Total Mass Airflow: 16.1 lb/sec (7,3 kg/sec).

TURBINE TYPE: 4-Stage Axial Flow Turbine including: 1-Stage Low Pressure Compressor Drive Turbine, 1-Stage High Pressure Compressor Drive Turbine, 2-Stage Power (free) Turbine.

COMBUSTOR TYPE: Annular, Reverse Flow.

POWER (take-off) RATING: 1,823 eshp (1360 ekw). Rating Approved to: 104°F (40°C).

TAKE-OFF SPECIFIC FUEL CONSUMPTION (SFC): 0.534 lb/hr/eshp (0,324 kg/hr/ekw).

BASIC DRY WEIGHT: 1100 lbs (500 kg).

POWER/WEIGHT RATIO: 1.66 : 1 eshp/lb (2,72 :1 ekw/kg).

APPLICATION:
LET Corp. L-410 Airliner-Transport Aircraft, Poland.

OTHER ENGINES IN SERIES:
None.

GAS TURBINE ENGINES FOR AIRCRAFT

INTERNATIONAL TURBINE ENGINE COMPANIES

COUNTRY: CZECHOSLOVAKIA

MANUFACTURER: MOTORLET CORP.
PRAGUE.

DESIGNATION (civil/military): Walter M701c.

BACKGROUND: Date First M701c Produced: 1963.
Date If Out-of-Production: N/A. Number of M701 Engines Produced to Date: 7,000.

ENGINE TYPE: Turbojet, (non-afterburning) Single Shaft.

COMPRESSOR TYPE: One-Stage Centrifugal Flow.

COMPRESSOR DATA: Compressor Pressure Ratio: 4.34 : 1, Total Mass Airflow: 36.82 lb/sec (16,7 kg/sec).

TURBINE TYPE: 1-Stage Axial Flow.

COMBUSTOR TYPE: Multiple-Can with liners.

POWER (take-off) RATING: 1,960 lbt (8,72 Kn). Rating Approved to: 104°F (40°C).

TAKE-OFF SPECIFIC FUEL CONSUMPTION (SFC): 0.64 lb/hr/lbt (18,13 mg/Ns).

BASIC DRY WEIGHT: 425 lbs (193 kg).

POWER/WEIGHT RATIO: 4.6 : 1 lbt/lb.

APPLICATION:
Aero L-29 "Maya", Basic Trainer Aircraft.
Aero L-29A "Akrobat", Acrobatic Aircraft.

OTHER ENGINES IN SERIES:
None.

GAS TURBINE ENGINES FOR AIRCRAFT

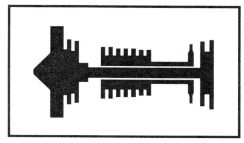

INTERNATIONAL TURBINE ENGINE COMPANIES

COUNTRY: CZECHOSLOVAKIA

MANUFACTURER: ZAVODY NA VYROBU LOZISK KONCERN
POVAZSKA BYSTRICA.

DESIGNATION (military): DV-2.

BACKGROUND: Licensed production from Progress Engine Bureau, Zaporozhye Motorworks, USSR.
Date First DV-2 Produced: 1990.
Date If Out-of-Production: N/A. Number of DV-2 Engines Produced to Date: 31.

ENGINE TYPE: Turbofan. Dual-Shaft, Low Bypass, Front Fan (non-afterburning).

COMPRESSOR TYPE: Dual Spool, 10-Stage, Axial Flow Compressor with a single stage fan and two booster stages in the Low Pressure Compressor and seven stages in the High Pressure Compressor.

COMPRESSOR DATA: Compressor Pressure Ratio: 13.5 : 1, Total Mass Airflow: 109 lb/sec (49,5 kg/sec). Fan Bypass Ratio: 1.5 : 1. Fan Pressure Ratio: 1 : 1.

TURBINE TYPE: Three-Stage Axial Flow with a two-stage Low Pressure Turbine and a single stage High Pressure Turbine.

COMBUSTOR TYPE: Annular, Through Flow.

POWER (take-off) RATING: 4,852 lbt (21,58 Kn).   Rating Approved to: 75°F (24°C).

TAKE-OFF SPECIFIC FUEL CONSUMPTION (SFC): 0.59 lb/hr/lbt (16,7 mg/Ns).

BASIC DRY WEIGHT: 992 lbs (450 kg).

POWER/WEIGHT RATIO: 4.89 : 1 lbt/lb.

APPLICATION:
Czec Aero L-59, formerly Aero L-39MS, Basic Trainer Aircraft, single-engine.
Ilyushin IL-10B, twin-engine Business Jet (in development).

OTHER ENGINES IN SERIES:
PV-22, 7,900 lbt. derivative in development for future IL-10B aircraft.

GAS TURBINE ENGINES FOR AIRCRAFT

 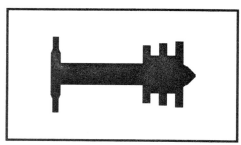

INTERNATIONAL TURBINE ENGINE COMPANIES

COUNTRY: FRANCE

MANUFACTURER: MICROTURBO CORPORATION.
Toulouse, France.

DESIGNATION (military): TRI 60-1.

BACKGROUND: Date First TRI-60 Produced: 1974.
Date If Out-of-Production: N/A. Number of TRI 60 Engines Produced to Date: 1,300.

ENGINE TYPE: Turbojet, (non-afterburning) Single-Shaft.

COMPRESSOR TYPE: Axial Flow Compressor, 3-Stage.

COMPRESSOR DATA: Compressor Pressure Ratio: 3.7 :1, Mass Airflow: 13 lb/sec (5,9 kg/sec).

TURBINE TYPE: 1-Stage Axial Flow Turbine.

COMBUSTOR TYPE: Can Type with one Combustor.

POWER (take-off) RATING: 787 lbt (3,50 kN thrust). Rating Approved to: 59°F (15°C).

TAKE-OFF SPECIFIC FUEL CONSUMPTION (SFC): 1.3 lb/hr/lbt (36,83 mg/Ns).

BASIC DRY WEIGHT: 99 lbs (45 kg).

POWER/WEIGHT RATIO: 7.95 :1 lbt/lb.

APPLICATION:
Beech and Northrop Target Drones.
Future Aircraft Applications.

OTHER ENGINES IN SERIES:
TRI 60-2, 3, 5, 20, 30 (Drone and Missile Applications).

GAS TURBINE ENGINES FOR AIRCRAFT

INTERNATIONAL TURBINE ENGINE COMPANIES

COUNTRY: FRANCE

MANUFACTURER: MICROTURBO CORPORATION.
Toulouse, France.

DESIGNATION (civil & military)**:** TRS 18-1.

BACKGROUND: Date First TRS Produced: 1976.
Date If Out-of-Production: N/A. Number of TRS 18 Engines Produced to Date: 335.

ENGINE TYPE: Turbojet, Single-Shaft.

COMPRESSOR TYPE: Centrifugal Compressor, I-Stage.

COMPRESSOR DATA: Compressor Pressure Ratio: 4.7 :1, Mass Airflow: 5.4 lb/sec (2,5 kg/sec).

TURBINE TYPE: 1-Stage Axial Flow Turbine.

COMBUSTOR TYPE: Annular, Reverse Flow.

POWER (take-off) RATING: 326 lbt (1,45 kN), Rating Approved to: 59°F (15°C).

TAKE-OFF SPECIFIC FUEL CONSUMPTION (SFC): 1.2 lb/hr/lbt (33,99 mg/Ns).

BASIC DRY WEIGHT: 84.9 lbs (38,5 kg).

POWER/WEIGHT RATIO: 3.84 :1 lbt/lb.

APPLICATION:
Microjet 200B, France.
Agusta-Caproni C-22J "Vantura" Acrobatic Aircraft, Italy.
Caproni C-22J , "TwinJet Trainer", Italy.
Caproni-Calif A-21SJ "Powered Glider", Italy.

OTHER ENGINES IN SERIES:
TRS-046 in Bede BD-5J (USA).
TRS-046-1 in ENSAE Research Aircraft (France).
Caproni-Calif A21SJ (Italy).
Prometheus Glider (Switzerland).
NASA AD-1 Slew Wing Research Aircraft, (USA).

GAS TURBINE ENGINES FOR AIRCRAFT

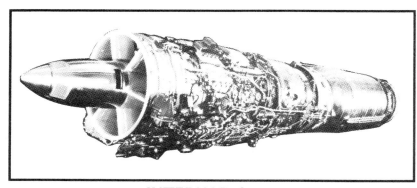
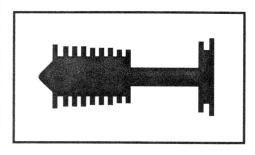

INTERNATIONAL TURBINE ENGINE COMPANIES

COUNTRY: FRANCE

MANUFACTURER: SNECMA, PARIS, FRANCE.
(Society Nationale d'Etude et de Construction de Moteurs d'Aviation)

DESIGNATION (military): Atar 9K-50.

BACKGROUND: The original Atar series engine first flew in 1,952. The 9K50 Turbojet is the latest version and the most powerful of the Atar family.
Date First Atar Produced: 1972.
Date If Out-of-Production: N/A. Number of Atar Engines Produced to Date: 1,060.

ENGINE TYPE: Turbojet, Single-Shaft with Afterburning.

COMPRESSOR TYPE: 9-Stage Single-Spool, Axial Flow Compressor.

COMPRESSOR DATA: Compressor Pressure Ratio: 6.15 : 1, Total Mass Airflow: 159 lb/sec (72 kg/sec).

TURBINE TYPE: 2-Stage Axial Flow Turbine.

COMBUSTOR TYPE: Annular, Through Flow.

AFTERBURNER: 2-Stage.

POWER (take-off) RATING: 11,022 lbt (49 kN) [dry], 15,870 lbt (71kN) [A/B], Rating Approved to: 59°F (15°C).

TAKE-OFF SPECIFIC FUEL CONSUMPTION (SFC): 0.96 lb/hr/lbt (27,19 mg/Ns) [dry], 1.96 lb/hr/lbt (55,52 mg/Ns) [A/B]

BASIC DRY WEIGHT: 3,490 lbs (1,583 kg).

POWER/WEIGHT RATIO: 3.16 : 1. [dry], 4.55 : 1 [A/B] - lbt/lb.

APPLICATION:
Dassault-Breguet Mirage F1 Fighter, G.4 & G.8 Attack/Fighter Aircraft.
Dassault-Breguet Mirage-50 Attack/Fighter Aircraft.

OTHER ENGINES IN SERIES:
Atar 8C in Etendard IV. Early Atar Series also flew in "Mystere" Fighter and "Mirage" III and IV.
Atar 8K-50 (non-afterburning version) powers the French Navy Super Etendard.
Atar 9B in "Mirage III C", Atar 9C in "Mirage III E" and M5.
Atar 9C in IAI "Nesher", Israel and Atlas "Cheetah", So. Africa.
Atar 9K in "Mirage IV".
Atar 101D in "Mystere II", Atar 101E in "Vautour" Early Aircraft, Atar 101G in "Super Mystere".

GAS TURBINE ENGINES FOR AIRCRAFT

 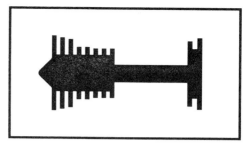

INTERNATIONAL TURBINE ENGINE COMPANIES

COUNTRY: FRANCE

MANUFACTURER: SNECMA, PARIS, FRANCE.
(Society Nationale d'Etude et de Construction de Moteurs d'Aviation)

DESIGNATION (military): M53-P2.

BACKGROUND: No Civil Counterpart. Built to replace Atar Series Engines.
Date First M53-P2 Produced: 1985.
Date If Out-of-Production: N/A. Number of M53 Engines Produced to Date: 293.

ENGINE TYPE: Turbofan. Single-Shaft, Low-Bypass Front Fan with Afterburning.

COMPRESSOR TYPE: 8-Stage, Single-Spool, Axial Flow Compressor including: 3-Stage Front Fan and 5-Stage Compressor.

COMPRESSOR DATA: Compressor Pressure Ratio: 9.8:1, Fan Bypass Ratio: 0.36:1, Total Mass Airflow: 207 lb/sec (94 kg/sec).

TURBINE TYPE: 2-Stage Axial Flow Turbine.

COMBUSTOR TYPE: Annular, Through-Flow.

AFTERBURNER: 6-Stage, 1-Stage in Fan Bypass Exhaust, 5-Stages in the Core Exhaust.

POWER (take-off) RATING: 14,500 lbt (64,5 kN thrust) [dry], 21,360 (95 kN thrust) [A/B]. Rating Approved to: 59°F (15°C).

TAKE-OFF SPECIFIC FUEL CONSUMPTION (SFC): 0.92 lb/hr/lbt (26,06 mg/Ns) [dry], 2.07 lb/hr/lbt (58,64 mg/Ns) [A/B].

BASIC DRY WEIGHT: 3,300 lbs (1,497 kg).

POWER/WEIGHT RATIO: 4.39:1 [dry], 6.47:1 [A/B].

APPLICATION:
Dassault-Brequet Mirage 2000 Fighter Aircraft.
Dassault-Breguet Super-Mirage 4000 Fighter Aircraft.

OTHER ENGINES IN SERIES:
M53-5 in Mirage 2000, with 19,840 lbt in A/B (88.23 kN thrust).

GAS TURBINE ENGINES FOR AIRCRAFT

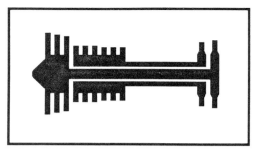

INTERNATIONAL TURBINE ENGINE COMPANIES

COUNTRY: FRANCE

MANUFACTURER: SNECMA, PARIS FRANCE.
(Society Nationale d'Etude et de Construction de Moteurs d'Aviation)

DESIGNATION (military): M88-2.

BACKGROUND: Date First M88-2 Produced: In Development (Expected 1995).
Date If Out-of-Production: N/A. Number of M88 Engines Produced to Date: 11 developmental units.

ENGINE TYPE: Turbofan. Dual-Shaft, Low-Bypass Front Fan with Afterburning.

COMPRESSOR TYPE: 9-Stage Dual-Spool, Axial Flow Compressor including: 3-Stage Front Fan making up the Low Pressure Compressor, 6-Stage High Pressure Compressor.

COMPRESSOR DATA: Compressor Pressure Ratio: 24:1, Fan Pressure Ratio: 3.8:1, Fan Bypass Ratio: 0.2:1, Total Mass Airflow: 140 lb/sec (65 kg/sec).

TURBINE TYPE: 2-Stage Axial Flow Turbine including: 1-Stage High Pressure Turbine, 1-Stage Low Pressure Turbine.

COMBUSTOR TYPE: Annular, Through-Flow.

AFTERBURNER: Combined flow (fan and core).

POWER (take-off) RATING: 11,200 lbt (50 kN thrust) [dry], 16,800 (75 kN thrust) [A/B]. Rating Approved to: 59°F (15°C).

TAKE-OFF SPECIFIC FUEL CONSUMPTION (SFC): 0.78 lb/hr/lbt (22,09 mg/Ns) [dry], 1.77 lb/hr/lbt (50 mg/Ns) [A/B].

BASIC DRY WEIGHT: 1,984 lbs (900 kg).

POWER/WEIGHT RATIO: 5.65:1 [dry], 8.5:1 [A/B].

APPLICATION:
Dassault-Breguet "Rafael-D", Fighter Aircraft.

OTHER ENGINES IN SERIES:
M88-2S, non-afterburning version of 14,000 lbt in development at CFM Int'l Corp.
CFM-88, civil version, 12,000 to 14,000 lbt thrust range in development.

GAS TURBINE ENGINES FOR AIRCRAFT

 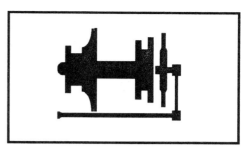

INTERNATIONAL TURBINE ENGINE COMPANIES

COUNTRY: FRANCE

MANUFACTURER: TURBOMECA (Society Turbomeca)
Bordes, Bizanos, France. (USA Division Located, Grand Prairie, TX)

DESIGNATION (civil & military): Arriel 1B.

BACKGROUND: Date First Arriel Produced: 1976.
Date If Out-of-Production: N/A. Number of Arriel Engines Produced to Date: 2500.

ENGINE TYPE: Turboshaft. Dual-Shaft (Compressor Drive & Power Output Drive).

COMPRESSOR TYPE: Combination Compressor: 1-Stage Axial, 1-Stage Centrifugal.

COMPRESSOR DATA: Compressor Pressure Ratio: 8.0:1. Mass Airflow: 5.3 lb/sec (2,4 kg/sec).

TURBINE TYPE: 3-Stage Axial Flow Turbine including: 2-Stage Gas Producer (compressor drive) Turbine, 1-Stage Power (free) Turbine.

COMBUSTOR TYPE: Can-Type with one Combustor.

POWER (take-off) RATING: 640 shp (477 KW). Rating Approved to: 59°F (15°C).

TAKE-OFF SPECIFIC FUEL CONSUMPTION (SFC): 0.58 lb/hr/shp (0,35 kg/hr/kw).

BASIC DRY WEIGHT: 240 lbs (109 kg).

POWER/WEIGHT RATIO: 2.67:1 shp/lb (4,38:1 kw/kg).

APPLICATION:
Aerospatiale AS.350B "Ecureuil" Helicopter.
Aerospatiale SA.365 "Dauphin-II" Helicopter.

OTHER ENGINES IN SERIES:
Arriel-1C in Aerospatiale SA.365N.
Arriel-1C1 in Aerospatiale SA.365N1 and AS 565.
Arriel-1C1 in HH-65 USCG Dolphin, USA.
Arriel-1C2 in Aerospatiale SA.365N2.
Arriel-1D in Aerospatiale AS.350B1.
Arriel-1D1 in Aerospatiale AS.350B2.
Arriel-1E in MBB BK.117, Germany.
Arriel-1S in Sikorsky S-76A Plus, USA.
Arriel-1S1 in Sikorsky S-76C, USA.
Arriel-1k in Agusta A-109K, Italy.

GAS TURBINE ENGINES FOR AIRCRAFT

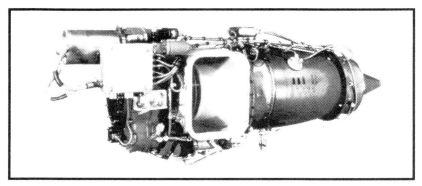 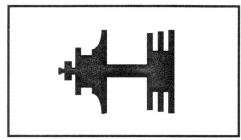

INTERNATIONAL TURBINE ENGINE COMPANIES

COUNTRY: FRANCE

MANUFACTURER: TURBOMECA (Society Turbomeca)
Bordes, Bizanos, France. (USA Div. Grand Prairie, TX)

DESIGNATION (civil & military): Artouste 3B.

BACKGROUND: Date First Artouste 3B Produced: 1955.
Date If Out-of-Production: N/A. Number of Artouste Engines Produced to Date: 3,929.

ENGINE TYPE: Turboshaft. Single-Shaft Design.

COMPRESSOR TYPE: Combination Compressor: 1-Stage Axial, 1-Stage Centrifugal.

COMPRESSOR DATA: Compressor Pressure Ratio: 5.35 :1. Mass Airflow: 9.5 lb/sec (4,3 kg/sec).

TURBINE TYPE: 3-Stage Axial Flow Turbine.

COMBUSTOR TYPE: Can Type with one Combustor.

POWER (take-off) RATING: 562 shp (419 KW). Rating Approved to: 130°F (55°C).

TAKE-OFF SPECIFIC FUEL CONSUMPTION (SFC): 0.66 lb/hr/shp (0,40 kg/hr/kw).

BASIC DRY WEIGHT: 401 lbs (182 kg).

POWER/WEIGHT RATIO: 1.4 :1 shp/lb (2,3 :1 kw/kg).

APPLICATION:
Aerospatiale SA.315 "Lama" Helicopter.
Aerospatiale SA.316 "Alouette" Helicopter.
Hindustan SA.316 "Chetak", Hindustan Aircraft, India.

OTHER ENGINES IN SERIES:
Artouste-II and IIID in Aerospatiale SA.318 "Alouette-II".

GAS TURBINE ENGINES FOR AIRCRAFT

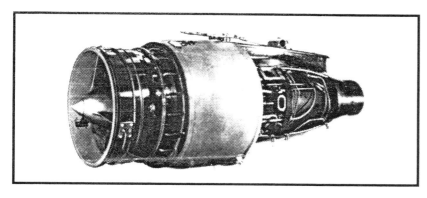 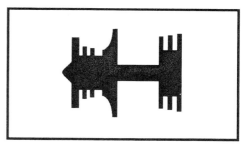

INTERNATIONAL TURBINE ENGINE COMPANIES

COUNTRY: FRANCE

MANUFACTURER: TURBOMECA (Society Turbomeca)
Bordes, Bizanos, France. (USA Div. Grand Prairie, TX)

DESIGNATION (military): AstaFan IV

BACKGROUND: One of a very few variable pitch, constant speed turbofan engines ever produced.
Date First AstaFan IV Produced: 1980.
Date If Out-of-Production: 1985. Number of Engines Produced to Date: 4.

ENGINE TYPE: Turbofan. Single-Shaft with Variable Pitch Fan (non-afterburning).

COMPRESSOR TYPE: Combination Compressor: 1-Stage Variable Pitch Fan followed by two booster stages and a 1-stage centrifugal compressor.

COMPRESSOR DATA: Compressor Pressure Ratio: 10.3 :1. Mass Airflow: 140 lb/sec (65,5 kg/sec). Fan Compression Ratio: 1.25 : l. Fan Bypass Ratio: 4 :1.

TURBINE TYPE: 3-Stage Axial Flow Turbine.

COMBUSTOR TYPE: Annular, Through Flow.

POWER (take-off) RATING: 2,530 lbt (11,25 kN thrust). Rating Approved to: 59°F (15°C).

TAKE-OFF SPECIFIC FUEL CONSUMPTION (SFC): 0.31 lb/hr/lbt (8,78 mg/Ns).

BASIC DRY WEIGHT: 485 lbs (220 kg).

POWER/WEIGHT RATIO: 5.22 : 1 lbt/lb.

APPLICATION:
Aerospatiale Fouga-90 Prototype Trainer Aircraft.

OTHER ENGINES IN SERIES:
None.

GAS TURBINE ENGINES FOR AIRCRAFT

 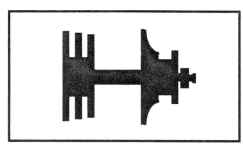

INTERNATIONAL TURBINE ENGINE COMPANIES

COUNTRY: FRANCE

MANUFACTURER: TURBOMECA (Society Turbomeca)
Bordes, Bizanos, France. (USA Div. Grand Prairie, TX)

DESIGNATION (civil/military): Astazou 3A.

BACKGROUND: Date First Astazou 3A Produced: 1972.
Date If Out-of-Production: N/A . Number of Engines Produced to Date: 2,900.

ENGINE TYPE: Turboshaft. Single-Shaft Design.

COMPRESSOR TYPE: Combination Compressor: 1-Stage Axial, 1-Stage Centrifugal.

COMPRESSOR DATA: Compressor Pressure Ratio: 5.7 :1. Mass Airflow: 5.5 lb/sec (2,5 kg/sec).

TURBINE TYPE: 3-Stage Axial Flow Turbine.

COMBUSTOR TYPE: Can Type with one Combustor.

POWER (take-off) RATING: 590 shp (440 kw). Rating Approved to: 122°F (50°C).

TAKE-OFF SPECIFIC FUEL CONSUMPTION (SFC): 0.64 lb/hr/shp (0,39 kg/hr/kw).

BASIC DRY WEIGHT: 324 lbs (147 kg).

POWER/WEIGHT RATIO: 1.82 : 1 shp/lb (2,99 : 1 kw/kg).

APPLICATION:
Aerospatiale SA.319" Alouette-III" Helicopter.
Aerospatiale SA.341 "Gazelle" Helicopter.

OTHER ENGINES IN SERIES:
Astazou-IIA in SA.318 "Alouette-II" Helicopter.
Astazou-14H in SA.342 "Gazelle" Helicopter.
Astazou-14M in Egyptian and Yugoslavian SA.342 "Gazelle" Helicopter.
Astazou-18 in SA.360 "Dauphin" Helicopter.
Locate Turboprop Models under Astazou-16G.

GAS TURBINE ENGINES FOR AIRCRAFT

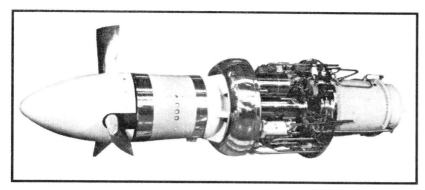 

INTERNATIONAL TURBINE ENGINE COMPANIES

COUNTRY: FRANCE

MANUFACTURER: TURBOMECA (Society Turbomeca)
Bordes, Bizanos, France. (USA Div. Grand Prairie, TX)

DESIGNATION (civil & military): Astazou 16G.

BACKGROUND: Date First Astazou 16G Produced: 1969.
Date If Out-of-Production: 1987. Number of Astazou Engines Produced to Date: 250.

ENGINE TYPE: Turboprop. Single-Shaft Design.

COMPRESSOR TYPE: Combination Compressor: 2-Stage Axial, 1-Stage Centrifugal.

COMPRESSOR DATA: Compressor Pressure Ratio: 8.0 :1. Mass Airflow: 7.34 lb/sec (3,33 kg/sec).

TURBINE TYPE: 3-Stage Axial Flow Turbine.

COMBUSTOR TYPE: Can Type with one Combustor.

POWER (take-off) RATING: 1,021 eshp (761 ekw)., including 63.4 shp (47,28 kw) from 158.5 lbt (0,705 kN thrust). Rating Approved to: 59°F (15°C).

TAKE-OFF SPECIFIC FUEL CONSUMPTION (SFC): 0.53 lb/hr/eshp (0,32 kg/hr/ekw).

BASIC DRY WEIGHT: 503 lbs (228 kg).

POWER/WEIGHT RATIO: 2.03 :1 eshp/lb (3,34 :1 ekw/kg).

APPLICATION:
FAMA IA-58 "Pacara", Transport, Argentina.

OTHER ENGINES IN SERIES:
Astazou-16D/F in early BAe Jetstream I and II Commuter Aircraft.
Locate Turboshaft Models under Astazou-3A.

GAS TURBINE ENGINES FOR AIRCRAFT

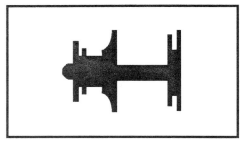

INTERNATIONAL TURBINE ENGINE COMPANIES

COUNTRY: FRANCE

MANUFACTURER: TURBOMECA (Society Turbomeca)
Bordes, Bizanos, France. (USA Div. Grand Prairie, TX)

DESIGNATION (military): Aubisque 1A.

BACKGROUND: Date First Aubisque Produced: 1964.
Date If Out-of-Production: 1969. Number of Aubisque Engines Produced to Date: 374.

ENGINE TYPE: Turbofan. Single-Shaft, Low Bypass, Front Fan (non-afterburning).

COMPRESSOR TYPE: Combination Compressor: 2-Stage Axial including a 1-Stage Fan and one Booster stage, plus a 1-Stage Centrifugal Impeller all on one shaft.

COMPRESSOR DATA: Compressor Pressure Ratio: 6.9 :1. Mass Airflow: 49 lb/sec (22,2 kg/sec). Fan Pressure Ratio: 1.5 : 1. Fan Bypass Ratio: 2 : 1.

TURBINE TYPE: 2-Stage Axial Flow Turbine.

COMBUSTOR TYPE: Annular, Through Flow.

POWER (take-off) RATING: 1,635 Lbt (7,27 kN), Rating Approved to: 122°F (50°C).

TAKE-OFF SPECIFIC FUEL CONSUMPTION (SFC): 0.6 lb/hr/lbt (17 mg/Ns).

BASIC DRY WEIGHT: 644 lbs (292 kg).

POWER/WEIGHT RATIO: 2.5 :1 lb/lbt.

APPLICATION:
Saab SK-60A Trainer Aircraft, Sweden.
Saab SK-60B Attack Aircraft, Sweden.
Saab SK-60C Attack Aircraft, Sweden.
Saab SK-105 Trainer Aircraft, Sweden.

OTHER ENGINES IN SERIES:
Aubisque-6 in Hispano HA-230, Business Jet; Spain.

GAS TURBINE ENGINES FOR AIRCRAFT

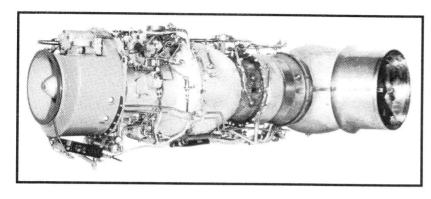 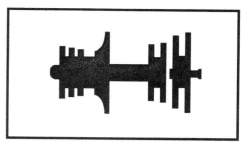

INTERNATIONAL TURBINE ENGINE COMPANIES

COUNTRY: FRANCE

MANUFACTURER: TURBOMECA (Society Turbomeca)
Bordes, Bizanos, France. (USA Div. Grand Prairie, TX)

DESIGNATION (civil & military): Makila 1A.

BACKGROUND: Date First Makila 1A Produced: 1980.
Date If Out-of-Production: N/A. Number of Makila Engines Produced to Date: 927.

ENGINE TYPE: Turboshaft. Dual-Shaft (Compressor Drive & Power Output Drive).

COMPRESSOR TYPE: Combination Compressor: 3-Stage Axial, 1-Stage Centrifugal.

COMPRESSOR DATA: Compressor Pressure Ratio: 10.8:1. Mass Airflow: 12.35 lb/sec (5,6 kg/sec).

TURBINE TYPE: 4-Stage Axial Flow Turbine including: 2-Stage Gas Producer (compressor drive) Turbine, 2-Stage Power (free) Turbine.

COMBUSTOR TYPE: Can-Type with one Combustor.

POWER (take-off) RATING: 1,662 shp (1,240 kw), Rating Approved to: 59°F (15°C).

TAKE-OFF SPECIFIC FUEL CONSUMPTION (SFC): 0.48 lb/hr/shp (0,29 kg/hr/kw).

BASIC DRY WEIGHT: 525 lbs (238 kg).

POWER/WEIGHT RATIO: 3.17:1 shp/lb (5,21:1 kw/kg).

APPLICATION:
Aerospatiale SA.332 C/L "Super Puma" Civil Helicopter.
Aerospatiale AS.332 B/M "Supa Puma" military Helicopter.

OTHER ENGINES IN SERIES:
Makila 1A1 in Aerospatiale AS.332 L1 civil Helicopter and AS.532UC "Cougar" military Helicopter.
Makila 1A2 in Aerospatiale AS.332 L2 civil Helicopter and AS.532 "Cougar" Military Helicopter.

GAS TURBINE ENGINES FOR AIRCRAFT

 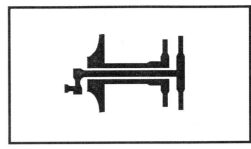

INTERNATIONAL TURBINE ENGINE COMPANIES

COUNTRY: FRANCE

MANUFACTURER: TURBOMECA (Society Turbomeca)
Bordes, Bizanos, France. (USA Div. Grand Prairie, TX)

DESIGNATION (civil & military): ARRIUS-1A (formerly TM-319).

BACKGROUND: Date First TM 319 Engine Produced: 1989.
Date If Out-of-Production: N/A. Number of TM 319 Engines Produced to Date: 60.

ENGINE TYPE: Turboshaft. Dual-Shaft (Compressor Drive & Power Output Drive).

COMPRESSOR TYPE: 1-Stage Centrifugal.

COMPRESSOR DATA: Compressor Pressure Ratio: 8.8 :1. Mass Airflow: 3.5 lb/sec (1,6 kg/sec).

TURBINE TYPE: 2-Stage Axial Flow Turbine including: 1-Stage Gas Producer (compressor drive) Turbine, 1-Stage Power (free) Turbine.

COMBUSTOR TYPE: Annular, Reverse Flow.

POWER (take-off) RATING: 479 shp (357 kW). Rating Approved to: 59°F (15°C).

TAKE-OFF SPECIFIC FUEL CONSUMPTION (SFC): 0.555 lb/hr/shp (0,337 kg/hr/kw).

BASIC DRY WEIGHT: 265 lbs (120 kg).

POWER/WEIGHT RATIO: 1.81 : 1 shp/lb (2,98 : 1 kw/kg).

APPLICATION:
Aerospatiale AS.355N "Ecureuil" Helicopter.

OTHER ENGINES IN SERIES:
ARRIUS-1M (formerly TM 319) in Aerospatiale AS.555-UN military Helicopter.
ARRIUS-TP (formerly TP-319) Turboprop Model in Aerospatiale-Socata Division "Omega" Trainer.
ARRIUS-2C in McDonnell "Explorer" (alternate engine to PW-206).

GAS TURBINE ENGINES FOR AIRCRAFT

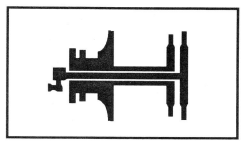

INTERNATIONAL TURBINE ENGINE COMPANIES

COUNTRY: FRANCE

MANUFACTURER: TURBOMECA (Society Turbomeca)
Bordes, Bizanos, France. (USA Div. Grand Prairie, TX)

DESIGNATION (civil & military): TM 333-1A.

BACKGROUND: Date First TM333 Produced: 1986.
Date If Out-of-Production: N/A Dual-Shaft (Compressor Drive & Power Output Drive). Dual-Shaft (Compressor Drive & Power Output Drive).   Number of TM 333 Engines Produced to Date: 20.

ENGINE TYPE: Turboshaft. Dual-Shaft Design.

COMPRESSOR TYPE: Combination Compressor: 2-Stage Axial, 1-Stage Centrifugal.

COMPRESSOR DATA: Compressor Pressure Ratio: 11 : 1. Mass Airflow: 6.5 lb/sec (2,95 kg/sec).

TURBINE TYPE: 2-Stage Axial Flow Turbine including: 1-Stage Gas Producer (compressor drive) Turbine, 1-Stage Power (free) Turbine.

COMBUSTOR TYPE: Annular, Reverse Flow.

POWER (take-off) RATING: 846 shp (631 kw). Rating Approved to: 59°F (15°C).

TAKE-OFF SPECIFIC FUEL CONSUMPTION (SFC): 0.52 lb/hr/shp (0,32 kg/hr/kw).

BASIC DRY WEIGHT: 297 lbs (135 kg).

POWER/ WEIGHT RATIO: 2.85 : 1 shp/lb (4,67 :1 kw/kg).

APPLICATION:
Uprated Aerospatiale SA.365M "Dauphin II" Helicopter.

OTHER ENGINES IN SERIES:
TM 333-2B in Hindustan, ALH military Helicopter; India.

GAS TURBINE ENGINES FOR AIRCRAFT

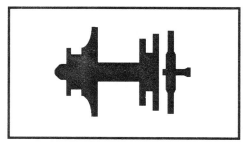

INTERNATIONAL TURBINE ENGINE COMPANIES

COUNTRY: FRANCE

MANUFACTURER: TURBOMECA (Society Turbomeca)
Bordes, Bizanos, France. (USA Division, Grand Prairie, TX)

DESIGNATION (civil & military): Turmo 4C.

BACKGROUND: Date First Turmo 4C Produced: 1970.
Date If Out-of-Production: N/A. Number of Turmo Engines Produced to Date: 2,250.

ENGINE TYPE: Turboshaft. Dual-Shaft (Compressor Drive & Power Output Drive).

COMPRESSOR TYPE: Combination Compressor: 1-Stage Axial, 1-Stage Centrifugal.

COMPRESSOR DATA: Compressor Pressure Ratio: 5.9 :1. Mass Airflow: 13 lb/sec (5,9 kg/sec).

TURBINE TYPE: 3-Stage Axial Flow Turbine including: 2-Stage Gas Producer (compressor drive) Turbine, 1-Stage Power (free) Turbine.

COMBUSTOR TYPE: Can Type with one Combustor.

POWER (take-off) RATING: 1,495 shp (1,114 kw), Rating Approved to: 59°F (15°C).

TAKE-OFF SPECIFIC FUEL CONSUMPTION (SFC): 0.64 lb/hr/shp (0,39 kg/hr/kw).

BASIC DRY WEIGHT: 515 lbs (234 kg).

POWER/WEIGHT RATIO: 2.90 : 1 shp/lb (4,76 :1 kw/kg).

APPLICATION:
Aerospatiale SA.330 "Puma" Helicopter.
Aerospatiale SA.330 J/G civil Helicopter.
Aerospatiale SA.330L military Helicopter.

OTHER ENGINES IN SERIES:
Turmo-IIIC in Aerospatiale SA.321 "Super Frelon" Helicopter.

GAS TURBINE ENGINES FOR AIRCRAFT

 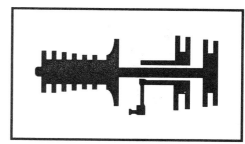

INTERNATIONAL TURBINE ENGINE COMPANIES

COUNTRY: GERMANY

MANUFACTURER: MOTOREN-UND TURBINEN-UNION GMBH (MTU). MUNICH, WEST GERMANY

DESIGNATION (civil & military): 250-MTU-C20B.

BACKGROUND: Manufactured under Allison 250 License, General Motors, USA.
Date First 250-MTU Engine Produced: 1979.
Date If Out-of-Production: 1983. Number of Engines Built under License to Date: 714.

ENGINE TYPE: Turboshaft. Dual-Shaft (Compressor Drive & Power Output Drive).

COMPRESSOR TYPE: Combination Compressor: 6-Stage Axial, 1-Stage Centrifugal.

COMPRESSOR DATA: Compressor Pressure Ratio: 7.2:1, Total Mass Airflow: 3.6 lb/sec (1,63 kg/sec).

TURBINE TYPE: 4-Stage Axial Flow Turbine, including a 2-Stage Gas Producer (compressor drive) Turbine, 2-Stage Power (free) Turbine.

COMBUSTOR TYPE: Can-Type, one combustor.

POWER (take-off) RATING: 420 shp (313 kw). Rating Approved to: 59°F (15°C).

TAKE-OFF SPECIFIC FUEL CONSUMPTION (SFC): 0.65 lb/hr/shp (0,395 kg/hr/kw).

BASIC DRY WEIGHT: 158 lbs (72 kg).

POWER/WEIGHT RATIO: 2.66:1 shp/lb (4,35:1 kw/kg).

APPLICATION:
MBB (Messerschmitt-Boelkow-Blohm), BO.105 Helicopter; Germany.

OTHER ENGINES IN SERIES:
Locate other Models Allison 250 under Allison, General Motors (USA).

GAS TURBINE ENGINES FOR AIRCRAFT

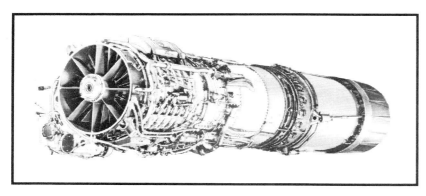
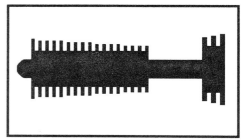

INTERNATIONAL TURBINE ENGINE COMPANIES

COUNTRY: GERMANY

MANUFACTURER: MOTOREN-UND TURBINEN-UNION GMBH (MTU). MUNICH, WEST GERMANY

DESIGNATION (military): J79-MTU-17A.

BACKGROUND: Manufactured under J79 License, General Electric Co., (USA).
Date First J79-MTU-17A Produced: 1972.
Date If Out-of-Production: 1976. Number of J79-MTU-17A Engines Produced to Date: 458.

ENGINE TYPE: Turbojet, Single-Shaft with Afterburning.

COMPRESSOR TYPE: Axial Flow: 17-Stage.

COMPRESSOR DATA: Compressor Pressure Ratio: 13.5 : 1, Total Mass Airflow: 170 lb/sec (77 kg/sec).

TURBINE TYPE: 3-Stage Axial Flow Turbine.

COMBUSTOR TYPE: Can-Annular, Through Flow, with 10 liners.

AFTERBURNER: Four-Stage.

POWER (take-off) RATING: 11,870 lbt (52,8 kN thrust) [Dry], 17,900 lbt (79,62 kN thrust) [A/B]. Rating Approved to: 59°F (15°C).

TAKE-OFF SPECIFIC FUEL CONSUMPTION (SFC): 0.84 lb/hr/lbt (23,79 mg/Ns) [dry], 1.97 lb/hr/lbt (55,8 mg/Ns) [A/B].

BASIC DRY WEIGHT: 3,860 lbs (1,741 kg).

POWER/WEIGHT RATIO: 3.08 :1 [dry], 4.64: 1 [A/B].

APPLICATION:
German Built, McDonnell-Douglas F-4F "Phantom-II" Fighter Aircraft.

OTHER ENGINES IN SERIES:
J79-MTU-J1K in German, F-104G "StarFighter".
Locate other J79 models under General Electric Aircraft Engines, USA.

GAS TURBINE ENGINES FOR AIRCRAFT

 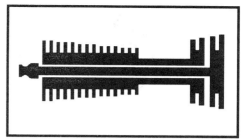

INTERNATIONAL TURBINE ENGINE COMPANIES

COUNTRY: GERMANY

MANUFACTURER: MOTOREN-UND TURBINEN-UNION GMBH (MTU). MUNICH, WEST GERMANY

DESIGNATION (military)**:** T64-MTU-7.

BACKGROUND: Manufactured under T64 License, General Electric Aircraft Engines, USA.
Date First T64-MTU-7 Produced: 1972.
Date If Out-of-Production: 1975. Number of Engines Built under License to Date: 267.

ENGINE TYPE: Turboshaft. Dual-Shaft (single compressor shaft plus power output shaft).

COMPRESSOR TYPE: 14-Stage Single-Spool, Axial Flow Compressor.

COMPRESSOR DATA: Compressor Pressure Ratio: 13.0 : 1, Total Mass Airflow: 27 lb/sec (12 kg/sec).

TURBINE TYPE: 4-Stage Axial Flow Turbine including: 2-Stage Gas Producer (compressor drive) Turbine, 2-Stage Power (free) Turbine.

COMBUSTOR TYPE: Annular, Through Flow.

POWER (take-off) RATING: 3,925 shp, (2,927 kw). Rating Approved to: 59°F (15°C).

TAKE-OFF SPECIFIC FUEL CONSUMPTION (SFC): 0.48 lb/hr/shp (0,29 kg/hr/kw).

BASIC DRY WEIGHT: 712 lbs (323 kg).

POWER/WEIGHT RATIO: 5.51 : 1 shp/lb (9.0 :1 kw/kg).

APPLICATION:
German Built, Sikorsky CH-53G Helicopter.

OTHER ENGINES IN SERIES:
Locate other T-64 models under General Electric Aircraft Engines, USA.

GAS TURBINE ENGINES FOR AIRCRAFT

 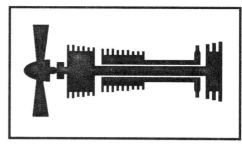

INTERNATIONAL TURBINE ENGINE COMPANIES

COUNTRY: GERMANY

MANUFACTURER: MOTOREN-UND TURBINEN-UNION GMBH.
MUNICH, WEST GERMANY

DESIGNATION (civil & military): Tyne-MTU MK. 21/22.

BACKGROUND: Manufactured under Rolls-Royce License, Great Britain.
Date First Tyne-MTU MK. 21/22 Produced: 1965.
Date If Out-of-Production: N/A .
Number of Tyne-MTU Mk. 21/22 Engines Produced to Date: 470.

ENGINE TYPE: Turboprop. Dual-Shaft (Compressor Drive & Power Output Drive).

COMPRESSOR TYPE: 15-Stage Dual-Spool, Axial Flow Compressor including: 6-Stage Low Pressure Compressor, 9-Stage High Pressure Compressor.

COMPRESSOR DATA: Compressor Pressure Ratio: 13.5 : 1, Total Mass Airflow: 46.5 lb/sec. (21,1 kg/sec).

TURBINE TYPE: 4-Stage Axial Flow Turbine including: 1-Stage High Pressure Turbine, 3-Stage Low Pressure Turbine.

COMBUSTOR TYPE: Can-Annular, Through-Flow, with 10 liners.

POWER (take-off) RATING: 5,665 eshp (4,224 ekw). Includes shp 435 (324 kw) from 1,124 lbt (5,0 kN thrust). Rating Approved to: 59°F (15°C).

TAKE-OFF SPECIFIC FUEL CONSUMPTION (SFC): 0.47 lb/hr/eshp (0,29 kg/hr/ekw).

BASIC DRY WEIGHT: 2,194 lbs (995 kg).

POWER/WEIGHT RATIO: 2.58 : 1 eshp/lb (4,24 : 1 ekw/kg).

APPLICATION:
Transporter-Allianz (Franco-German), Transall C-160, Transport Aircraft.
Dassault-Breguet "Atlantic", Maritime Aircraft; France.

OTHER ENGINES IN SERIES:
Locate Tyne-20 under Multi-National Listing.
Locate other Tyne models also under Rolls-Royce Listing, Great Britain.

GAS TURBINE ENGINES FOR AIRCRAFT

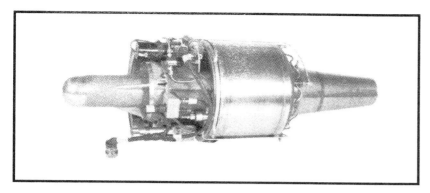

INTERNATIONAL TURBINE ENGINE COMPANIES

COUNTRY: GREAT BRITAIN

MANUFACTURER: NOEL PENNY TURBINES LTD.
Coventry, England

DESIGNATION (military): NPT-401B.

BACKGROUND: Date First NPT-401B Produced: 1989.
Date If Out-of-Production: N/A. Number of NPT-401B Engines Produced to Date: 4.

ENGINE TYPE: Turbojet. Single-Shaft (non-afterburning).

COMPRESSOR TYPE: Single Stage, Centrifugal Flow Compressor.

COMPRESSOR DATA: Compressor Pressure Ratio: 4.5 : 1, Total Mass Airflow: 5.77 lb/sec (2,62 kg/sec).

TURBINE TYPE: 1-Stage Axial Flow.

COMBUSTOR TYPE: Annular, Reverse-Flow.

POWER (take-off) RATING: 400 lbt (1,78 kN thrust). Rating Approved to: 59°F (15°C).

TAKE-OFF SPECIFIC FUEL CONSUMPTION (SFC): 1.18 lb/hr/lbt (33,4 mg/Ns).

BASIC DRY WEIGHT: 100 lbs (45,4 kg).

POWER/WEIGHT RATIO: 4.0 : 1 lbt/lb.

APPLICATION:
Reusable Aircraft Target Drones.
Future Aircraft Applications.

OTHER ENGINES IN SERIES:
NPT 301 Series (300 lbt) proposed for small Business Jet Aircraft.

GAS TURBINE ENGINES FOR AIRCRAFT

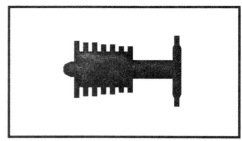

INTERNATIONAL TURBINE ENGINE COMPANIES

COUNTRY: GREAT BRITAIN

MANUFACTURER: ROLLS-ROYCE LTD.
Bristol, England

DESIGNATION (military): Orpheus 805.

BACKGROUND: Date First Orpheus Produced: 1959
Date If Out-of-Production: 1986. Number of Orpheus Engines Produced to Date: 2,400.

ENGINE TYPE: Turbojet. Single-Shaft (non-afterburning).

COMPRESSOR TYPE: 7-Stage Single-Spool, Axial Flow Compressor.

COMPRESSOR DATA: Compressor Pressure Ratio: 4.4 : 1, Total Mass Airflow: 84 lb/sec (38,1 kg/sec).

TURBINE TYPE: 1-Stage Axial Flow.

COMBUSTOR TYPE: Annular, Through-Flow.

POWER (take-off) RATING: 4,000 lbt (17,79 kN thrust). Rating Approved to: 59°F (15°C).

TAKE-OFF SPECIFIC FUEL CONSUMPTION (SFC): 1.04 lb/hr/lbt (29,46 mg/Ns).

BASIC DRY WEIGHT: 860 lbs (390 kg).

POWER/WEIGHT RATIO: 4.65 : 1 lbt/lb.

APPLICATION:
Fuji F-2/T-1, Trainer Aircraft; Japan.

OTHER ENGINES IN SERIES:
Orpheus-101 in Hawker Siddeley, Gnat Fighter; Great Britain.
Orpheus-500 in Fairchild C-119, "Flying Boxcar" as Auxiliary Engine; USA.
Orpheus-701 in H.S Gnat, Finland and Hindustan Ajeet and Kiran Fighters, India.
Orpheus-703 in Hindustan HF-24 "Mahut" Transport, India.
Orpheus-803 in Fiat G.91 Attack/Trainer Aircraft, Italy.

GAS TURBINE ENGINES FOR AIRCRAFT

 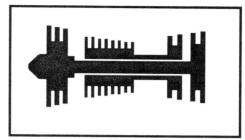

INTERNATIONAL TURBINE ENGINE COMPANIES

COUNTRY: GREAT BRITAIN

MANUFACTURER: ROLLS-ROYCE LTD.
Bristol, England

DESIGNATION (military): Pegasus F402-RR-406/Mk.105.

BACKGROUND: One of a very few successful Vectored-Thrust Engines produced. Originally developed by the Bristol Company. Presently used in military V/STOL Fighter Aircraft. Two Variable Angle Fan Exhaust Nozzles, and two Variable Angle Hot Exhaust Nozzles, all of which move in unison.
Date First Pegasus F402 Produced: 1960.
Date If Out-of-Production: N/A. Number of Pegasus Engines Produced to Date: 1,000.

ENGINE TYPE: Turbofan. Dual-Shaft, Lift-Fan. Non-Afterburning.

COMPRESSOR TYPE: 11-Stage Dual-Spool, Axial Flow Compressor including: 3-Stage Front Fan acting also as a Low Pressure Compressor 8-Stage High Pressure Compressor.

COMPRESSOR DATA: Compressor Pressure Ratio: 14.0 : 1, Fan Pressure Ratio: 2.15 : 1, Fan Bypass Ratio: 1.5 : 1, Total Mass Airflow: 450 lb/sec (204 kg/sec).

TURBINE TYPE: 4-Stage Axial Flow Turbine including: 2-Stage High Pressure Turbine, 2-Stage Low Pressure Turbine.

COMBUSTOR TYPE: Annular, Through-Flow.

POWER (take-off) RATING: 22,000 lbt (97,86 kN thrust). Rating Approved to: 59°F (15°C).

TAKE-OFF SPECIFIC FUEL CONSUMPTION (SFC): 0.67 lb/hr/lbt (18,98 mg/Ns).

BASIC DRY WEIGHT: 3,240 lbs (1,470 kg).

POWER/WEIGHT RATIO: 6.79 : 1 lbt/lb.

APPLICATION:
Hawker-Siddeley "Harrier" GR Mk.5, Attack/Reconnaissance Aircraft.
McDonnell-Douglas AV-8B "Matador" Attack/Reconnaissance Aircraft.
McDonnell-Douglas TAV-8B Trainer Aircraft.

OTHER ENGINES IN SERIES:
Pegasus-11 in the Hawker-Siddeley Sea Harrier.
Pegasus F402-RR-408 (24,000 LBT model) in AV-8B.

GAS TURBINE ENGINES FOR AIRCRAFT

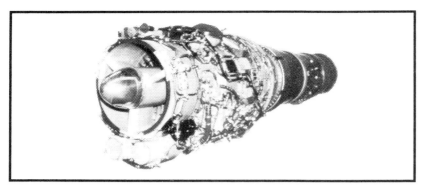 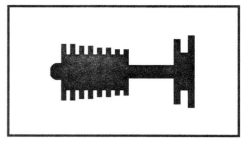

INTERNATIONAL TURBINE ENGINE COMPANIES

COUNTRY: GREAT BRITAIN

MANUFACTURER: ROLLS-ROYCE LTD.
Bristol, England

DESIGNATION (civil & military)**:** Viper Mk.601.

BACKGROUND: Date First Viper Engine Produced: 1951.
Date If Out-of-Production: N/A. Number of Viper Engines Produced to Date: 5,000.

ENGINE TYPE: Turbojet. Single-Shaft (non-afterburning).

COMPRESSOR TYPE: 8-Stage Single-Spool, Axial Flow Compressor.

COMPRESSOR DATA: Compressor Pressure Ratio: 5.6:1, Total Mass Airflow: 58.4 lb/sec (26,5 kg/sec).

TURBINE TYPE: 2-Stage Axial Flow.

COMBUSTOR TYPE: Through-Flow, Annular.

POWER (take-off) RATING: 3,750 lbt (16,68 kN thrust). Rating Approved to: 59°F (15°C).

TAKE-OFF SPECIFIC FUEL CONSUMPTION (SFC): 0.92 lb/hr/lbt (26 mg/Ns).

BASIC DRY WEIGHT: 831 lbs (377 kg).

POWER/WEIGHT RATIO: 4.51 : 1 lbt/lb.

APPLICATION: Hawker Siddeley, HS 125-600 Commuter Aircraft.

OTHER ENGINES IN SERIES:
Viper MK.202 in BAC-145 "Jet Provost".
Viper MK.500 in H.S. T-1 "Dominie".
Viper MK.521 in H.S. DH-125-1A.
Viper Mk.522 in H.S.125.
Viper MK.526 in Piggio-Douglas PD-808, Italy.
Viper Mk.535/540 in BAC-167 "StrikeMaster" and Aermacchi MB.326G, Italy.
Viper Mk.540 in Atlas MB.326, So. Africa and Embraer AT-26, EMB.326; Brazil.
Viper Mk.632 in Aermacchi MB.326 and MB.339, Italy.
Viper MK.632 in Soko Industries Gastreb, Jastreb and Yurom, Yugoslavia.
Viper Mk.680 in Aermacchi MB-339K, Italy and Hindustan "Kiran", India.
Locate other Viper Models under Rumania and Yugoslavia listings.

GAS TURBINE ENGINES FOR AIRCRAFT

 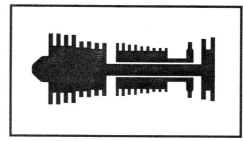

INTERNATIONAL TURBINE ENGINE COMPANIES

COUNTRY: GREAT BRITAIN

MANUFACTURER: ROLLS-ROYCE LTD.
Derby, England

DESIGNATION (civil & military)**:** Conway 43

BACKGROUND: Developed for Commercial and Military Aircraft.
Date First Conway Engine Produced: 1958.
Date If Out-of-Production: 1969. Number of Conway Engines Produced to Date: 904.

ENGINE TYPE: Turbofan. Dual-Shaft, Low-Bypass Front Fan (non-afterburning).

COMPRESSOR TYPE: 17-Stage Dual-Spool, Axial Flow Compressor including: 4-Stage Front Fan plus four booster stages in the Low Pressure Compressor, 9-Stage High Pressure Compressor.

COMPRESSOR DATA: Compressor Pressure Ratio: 15.8 :1, Fan Pressure Ratio: 2.76 : 1, Fan Bypass Ratio: 0.5 :1, Total Mass Airflow: 375 lb/sec (170 kg/sec).

TURBINE TYPE: 3-Stage Axial Flow Turbine including: 1-Stage High Pressure Turbine, 2-Stage Low Pressure Turbine.

COMBUSTOR TYPE: Can-Annular, Through Flow, with ten liners.

POWER (take-off) RATING: 21,800 lbt (96,97 kN). Rating Approved to: 59°F (15°C).

TAKE-OFF SPECIFIC FUEL CONSUMPTION (SFC): 0.59 lb/hr/lbt (16,7 mg/Ns).

BASIC DRY WEIGHT: 5,226 lbs (2,371 kg).

POWER/WEIGHT RATIO: 4.17 : 1 lbt/lb.

APPLICATION:
BAC Super VC.10 Airliner.
RAF VC-10 Transport.

OTHER ENGINES IN SERIES:
Conway-12 in McDonnell/Douglas DC-8-40 and Boeing B-707-420 Airliners.
Conway-17 in Hanley-Page, Victor-B.2 Bomber and Victor-K Tanker.
Conway-43 in Military VC.10 Transport and VC-10K Tanker.

GAS TURBINE ENGINES FOR AIRCRAFT

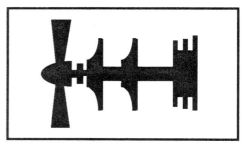

INTERNATIONAL TURBINE ENGINE COMPANIES

COUNTRY: GREAT BRITAIN

MANUFACTURER: ROLLS-ROYCE LTD.
Derby, England

DESIGNATION (civil & military): Dart Mk.540 Series.

BACKGROUND: The Dart Series of Turboprop Engines has had the longest production run of any Turboprop ever built. The first engine test occurred in 1946.
First Test Flight Date of Original Dart Turboprop: 1948, rated at 1,540 shp.
Date First Dart Engine Produced: 1953.
Date If Out-of-Production: 1986. Number of Dart Engines Produced to Date: 7,094.

ENGINE TYPE: Turboprop. Single-Shaft Design.

COMPRESSOR TYPE: Centrifugal Compressor: Dual-Stage, Single Entry.

COMPRESSOR DATA: Compressor Pressure Ratio: 6.35:1. Mass Airflow: 27 lb/sec (12,25 kg/sec).

TURBINE TYPE: Three-Stage Axial Flow Turbine.

COMBUSTOR TYPE: Multiple-Can Type with Seven Combustors.

POWER (take-off) RATING: 3,025 eshp (2,256 ekw), including 285 shp (213 kw) from 715 lbt (3,18 kN thrust). Rating Approved to: 59°F (15°C).

TAKE-OFF SPECIFIC FUEL CONSUMPTION (SFC): 0.69 lb/hr/eshp (0,42 kg/hr/ekw).

BASIC DRY WEIGHT: 1,388 lbs (630 kg).

POWER/WEIGHT RATIO: 2.18:1 eshp/lb (3,58:1 ekw/kg).

APPLICATION:
Fokker F-27 Commuter/Airliner, Netherlands.

OTHER ENGINES IN SERIES:
Dart MK.520 in Vickers-Armstrong Viscount, Airliner.
Dart Mk.526 in BAe Herald, Transport.
Dart MK.532 in BAe 748 "Andover", Fokker F-27, US Navy C-4; Transports.
Dart MK.536 in Gulfstream-1 Transport.
Dart MK.534 in Hawker-Siddeley Argosy-220 Transport.
Dart MK.542 in Convair 600 and 640 Airliner.
Dart Mk.542 in Nihon YS-11, Transport, Japan.

GAS TURBINE ENGINES FOR AIRCRAFT

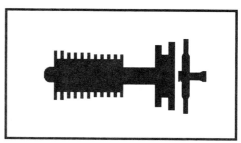

INTERNATIONAL TURBINE ENGINE COMPANIES

COUNTRY: GREAT BRITAIN

MANUFACTURER: ROLLS-ROYCE LTD.
Derby, England

DESIGNATION (military)**:** Gazelle NGa.22 Mk. l65.

BACKGROUND: Date First Gazelle Engine Produced: 1955.
Date If Out-of-Production: 1970. Number of Gazelle Engines Produced to Date: 218.

ENGINE TYPE: Turboshaft. Dual-Shaft (Compressor Drive & Power Output Drive).

COMPRESSOR TYPE: 11-Stage Single-Spool, Axial Flow Compressor.

COMPRESSOR DATA: Compressor Pressure Ratio: 5.9 : 1, Mass Airflow: 17.0 lb/sec (7,71 kg/sec).

TURBINE TYPE: 3-Stage Axial Flow Turbine including: 2-Stage Gas Producer (compressor drive) Turbine, 1-Stage Power (free) Turbine.

COMBUSTOR TYPE: Multiple-Can, with six Combustors.

POWER (take-off) RATING: 1,600 shp (1,193 kw). Rating Approved to: 59°F (15°C).

TAKE-OFF SPECIFIC FUEL CONSUMPTION (SFC): 0.69 lb/hr/shp (0,42 kg/hr/kw).

BASIC DRY WEIGHT: 884 lbs (401 kg).

POWER/WEIGHT RATIO: 1.81 : 1 shp/lb (2,98 :1 kw/kg).

APPLICATION:
Early Westland Wessex Helicopters.

OTHER ENGINES IN SERIES:
Gazelle Mk.161 in Westland Wessex Mark-1.
Gazelle Mk.162 in Westland Wessex Mark-31.
Later Wessex Helicopters fitted with Rolls-Royce Gnome Turboshaft engines.

GAS TURBINE ENGINES FOR AIRCRAFT

 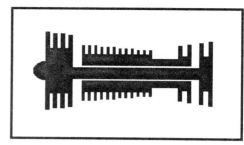

INTERNATIONAL TURBINE ENGINE COMPANIES

COUNTRY: GREAT BRITAIN

MANUFACTURER: ROLLS-ROYCE LTD.
Derby, England

DESIGNATION (civil): RB.183-555-15P.

BACKGROUND: Referred to as Junior Spey, developed from the Spey Series.
Date First Spey Engine Produced: 1968.
Date If Out-of-Production: 1986. Number of RB-183 Engines Produced to Date: 613.

ENGINE TYPE: Turbofan. Dual-Shaft, High-Bypass Front Fan.

COMPRESSOR TYPE: 16-Stage Dual-Spool, Axial Flow Compressor including: 4-Stage Front Fan acting also as a Low Pressure Compressor, 12-Stage High Pressure Compressor.

COMPRESSOR DATA: Compressor Pressure Ratio: 15.4 : 1, Fan Pressure Ratio: 1.42 : 1, Fan Bypass Ratio: 1 : 1, Total Mass Airflow: 199 lb/sec (90,3 kg/sec).

TURBINE TYPE: 4-Stage Axial Flow Turbine including: 2-Stage High Pressure Turbine, 2-Stage Low Pressure Turbine.

COMBUSTOR TYPE: Can-Annular, Through-Flow With 10 Liners.

POWER (take-off) RATING: 9,900 lbt.(44 kN). Rating Approved to: 85.5°F (29.7°C).

TAKE-OFF SPECIFIC FUEL CONSUMPTION (SFC): 0.56 lb/hr/lbt (15,86 mg/Ns).
Cruise SFC: 0.80 lb/hr/lbt.

BASIC DRY WEIGHT: 2,257 lbs (1,024 kg).

POWER/WEIGHT RATIO: 4.4 : 1 lbt/lb.

APPLICATION:
Fokker F-28 Mk.4000 Commuter/Airliner, Netherlands.

OTHER ENGINES IN SERIES:
Locate other Spey models under Rolls-Royce Listing, Great Britain.

GAS TURBINE ENGINES FOR AIRCRAFT

 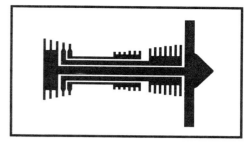

INTERNATIONAL TURBINE ENGINE COMPANIES

COUNTRY: GREAT BRITAIN

MANUFACTURER: ROLLS-ROYCE LTD.
Derby, England

DESIGNATION (civil): RB.211-524H.

BACKGROUND: No military counterpart. Originally developed for Lockheed TriStar (L-1011).
Date First RB.211 Engine Produced: 1972.
Date If Out-of-Production: N/A. Number of RB-211 Engines Produced to Date: 1864.

ENGINE TYPE: Turbofan. Three-Shaft, High-Bypass Front Fan.

COMPRESSOR TYPE: 14-Stage Triple-Spool, Axial Flow Compressor including: 1-Stage Front Fan acting also as a Low Pressure Compressor, 7-Stage Intermediate Pressure Compressor, 6-Stage High Pressure Compressor.

COMPRESSOR DATA: Compressor Pressure Ratio: 34.7 : 1, Fan Pressure Ratio: 1.85 : 1, Fan Bypass Ratio: 4 : 1, Total Mass Airflow: 1,658 lb/sec (752 kg/sec).

TURBINE TYPE: 5-Stage Axial Flow Turbine including: 1-Stage High Pressure Turbine, 1-Stage Intermediate Pressure Turbine, 3-Stage Low Pressure Turbine.

COMBUSTOR TYPE: Annular, Through-Flow.

POWER (take-off) RATING: 63,000 lbt (280,2 kN thrust). Rating Approved to: 77°F (25°C).

TAKE-OFF SPECIFIC FUEL CONSUMPTION (SFC): 0. 33 lb/hr/lbt (9,35 mg/Ns),
Cruise SFC: 0.802 lb/hr/lbt.

BASIC DRY WEIGHT: 9,874 lbs (4,479 kg).

POWER/WEIGHT RATIO: 6.38: 1 lbt/lb.

APPLICATION:
Boeing B-747, B-767 Wide-Bodied Airliners.
Lockheed L-1011 "Tristar" Wide Bodied Airliner.

OTHER ENGINES IN SERIES:
RB-211-22B in Lockheed L-1011.
RB-211-524B2/C2/D4/D4C/D4D/G in B-747.
RB-211-524E4 in B-747 and B-767.
RB-211-535C/E in B-757.
Follow On Engines designated as Trent-600, Trent-700, Trent-800, proposed for: Airbus A-330, France, and MD-11, B-767, USA. Trent engines were originally known as RB.211-524L.SA.

GAS TURBINE ENGINES FOR AIRCRAFT

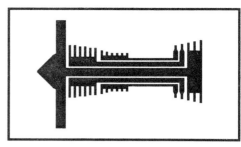

INTERNATIONAL TURBINE ENGINE COMPANIES

COUNTRY: GREAT BRITAIN

MANUFACTURER: ROLLS-ROYCE LTD.
Derby, England

DESIGNATION (civil): RB.211-535C.

BACKGROUND: No military counterpart. Developed for Commercial Aircraft.
Date First RB.211 Engine Produced: 1972.
Date If Out-of-Production: N/A. Number of RB-211 Engines Produced to Date: 1864.

ENGINE TYPE: Turbofan. Three-Shaft, High-Bypass Front Fan.

COMPRESSOR TYPE: 13-Stage Triple-Spool, Axial Flow Compressor including: 1-Stage Front Fan acting also as a Low Pressure Compressor, 6-Stage Intermediate Pressure Compressor, 6-Stage High Pressure Compressor.

COMPRESSOR DATA: Compressor Pressure Ratio: 21.0 : 1, Fan Pressure Ratio: 1.68 : 1, Fan Bypass Ratio: 4.48 : 1, Total Mass Airflow: 1,133 lb/sec (514 kg/sec).

TURBINE TYPE: 5-Stage Axial Flow Turbine including: 1-Stage High Pressure Turbine, 1-Stage Intermediate Pressure Turbine, 3-Stage Low Pressure Turbine.

COMBUSTOR TYPE: Annular, Through-Flow.

POWER (take-off) RATING: 37,400 lbt (166,4 kN thrust). Rating Approved to: 84°F (29°C).

TAKE-OFF SPECIFIC FUEL CONSUMPTION (SFC): 0.35 lb/hr/lbt (9,91 mg/Ns).

BASIC DRY WEIGHT: 7,294 lbs (3,309 kg).

POWER/WEIGHT RATIO: 5.13: 1 lbt/lb.

APPLICATION:
Boeing B-757 Airliner.

OTHER ENGINES IN SERIES:
RB-211-535E4 in Boeing B-757.

GAS TURBINE ENGINES FOR AIRCRAFT

 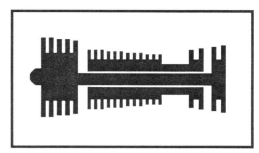

INTERNATIONAL TURBINE ENGINE COMPANIES

COUNTRY: GREAT BRITAIN

MANUFACTURER: ROLLS-ROYCE LTD.
Derby, England

DESIGNATION (civil & military): Spey Mk.512-14DW.

BACKGROUND: Known also as: RB-168 (RAF) and F113-RR-100 (USAF).
Date First Spey Engine Produced: 1966.
Date If Out-of-Production: 1980 in U.K. Number of Spey Engines Produced to Date: 4050.
Production continues in Rumania under R-R License.

ENGINE TYPE: Turbofan. Dual-Shaft, High-Bypass Front Fan (non-afterburning).

COMPRESSOR TYPE: 17-Stage Dual-Spool, Axial Flow Compressor including: 5-Stage Front Fan acting also as a Low Pressure Compressor, 12-Stage High Pressure Compressor.

COMPRESSOR DATA: Compressor Pressure Ratio: 21.0 : 1, Fan Pressure Ratio: 1.5 : 1, Fan Bypass Ratio: 0.71 : 1, Total Mass Airflow: 208 lb/sec (94,4 kg/sec).

TURBINE TYPE: 4-Stage Axial Flow Turbine including: 2-Stage High Pressure Turbine, 2-Stage Low Pressure Turbine.

COMBUSTOR TYPE: Can-Annular, Through-Flow With 10 Liners.

POWER (take-off) RATING: 12,550 lbt (55,8 kN thrust). Rating Approved to: 77°F (25°C).

TAKE-OFF SPECIFIC FUEL CONSUMPTION (SFC): 0.612 lb/hr/lbt (17,34 mg/Ns).
Cruise SFC: 0.82 lb/hr/lbt.

BASIC DRY WEIGHT: 2,609 lbs (1,183 kg).

POWER/WEIGHT RATIO: 4.8 : 1 lbt/lb.

APPLICATION: BAC One-Eleven-500 (BAe), Rombac One-Eleven, Romania; Airliners.

OTHER ENGINES IN SERIES:
Spey Mk.101 in BAe Buccaneer, Attack/Recon Aircraft.
Spey Mk.202 in H-7 Attack Aircraft, P.R.China.
Spey MK.250/251 in BAe Nimrod, Maritime Aircraft.
Spey MK.505-/512W in BAe-111-200 Airliner.
Spey Mk.511-5/MK.512 in Trident-2, Fokker F-28; Airliners.
Spey Mk. 511-8 in Gulfstream G2 and G3, Commuter Aircraft.
Spey MK.807 in Italian-Brazilian AM-X, Fighter/Trainer.
Spey F113-RR-100, in U.S. Air Force version, Gulfstream C-20.
Spey RB 168 RAF version in F-4M "Phantom" Fighter.
BR-700 series, follow-on turbofan to Spey series (Rolls-Royce & BMW Germany).

GAS TURBINE ENGINES FOR AIRCRAFT

 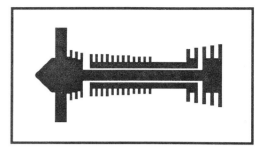

INTERNATIONAL TURBINE ENGINE COMPANIES

COUNTRY: GREAT BRITAIN

MANUFACTURER: ROLLS-ROYCE LTD.
Derby, England

DESIGNATION (civil): Tay 620-15.

BACKGROUND: No military counterpart. Developed for Commercial Aircraft.
Date First Tay Produced: 1986.
Date If Out-of-Production: N/A . Number of Tay Engines Produced to Date : 413.

ENGINE TYPE: Turbofan. Dual-Shaft, High-Bypass Front Fan.

COMPRESSOR TYPE: 16-Stage Dual-Spool, Axial Flow Compressor including: 1-Stage Front Fan followed by 3-low pressure stages and a 12-Stage High Pressure Compressor.

COMPRESSOR DATA: Compressor Pressure Ratio: 16.0 : 1, Fan Pressure Ratio: 1.5 : 1, Fan Bypass Ratio: 3 : 1, Total Mass Airflow: 414 lb/sec (188 kg/sec).

TURBINE TYPE: 5-Stage Axial Flow Turbine including: 2-Stage High Pressure Turbine, 3-Stage Low Pressure Turbine.

COMBUSTOR TYPE: Can-Annular, Through Flow, with 10 liners.

POWER (take-off) RATING: 13,850 lbt (61,1 kN thrust). Rating Approved to: 86°F (30°C).

TAKE-OFF SPECIFIC FUEL CONSUMPTION (SFC): 0.44 lb/hr/lbt (12,46 mg/Ns).
Cruise SFC: 0.69 lb/hr/lbt.

BASIC DRY WEIGHT: 3,185 lbs (1,445 kg).

POWER/WEIGHT RATIO: 4.35 : 1 lbt/lb.

APPLICATION:
Fokker-70 and Fokker-100 Commuter/Airliners, Netherlands.

OTHER ENGINES IN SERIES:
Tay 611 in Gulfstream IV.
Tay 650 in Fokker-100 also BAe 1-11 and Boeing, B-727 Re-Engine Programs.
Tay 670 (18,000 lbt), McDonnell/D DC-9 & MD-95, Boeing B-727, B-737 Re-Engine Programs.

GAS TURBINE ENGINES FOR AIRCRAFT

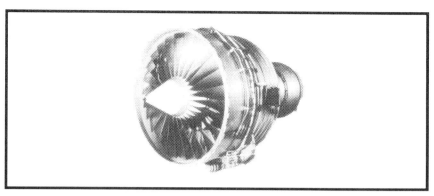
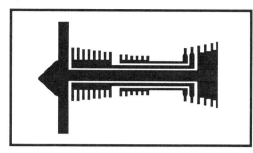

INTERNATIONAL TURBINE ENGINE COMPANIES

COUNTRY: GREAT BRITAIN

MANUFACTURER: ROLLS-ROYCE LTD.
Derby, England

DESIGNATION (civil): Trent 665 & Trent 668.

BACKGROUND: Follow-on turbofan of the RB211 series.
Date first Trent engine produced: In Development.

ENGINE TYPE: Turbofan. Three Shaft, High-Bypass Front Fan.

COMPRESSOR TYPE: 15-Stage triple spool, Axial Flow: 1-Stage Front Fan acting as a Low Pressure Compressor; 8-Stage Intermediate Compressor; 6-Stage High Pressure Compressor.

COMPRESSOR DATA: Compressor Pressure Ratio: 33.05:1 (665), 35.04:1 (668), Fan Pressure Ratio: 1.786:1 (665), 1.792:1 (668), Fan Bypass Ratio: 4.79:1 (665), 4.66:1 (668), Total Mass Airflow: 1,826 (665), 1,862 (668) lb/sec take-off.

TURBINE TYPE: 6-Stage Axial Flow Turbine including: 1-Stage High Pressure Turbine, 1-Stage Intermediate Pressure Turbine, 4-Stage Low Pressure Turbine.

COMBUSTOR TYPE: Annular, Through-Flow.

POWER (TAKE-OFF) RATING: 64,150 lbt. (665), 67,950 lbt. (668). Rating approved to: 86°F (30°C) for model 665 and 89.5°F (32°C) for model 668.

TAKE-OFF SPECIFIC FUEL CONSUMPTION: 0.314 (665), 0.323 (668) lb/hr/lbt.

BASIC DRY WEIGHT: 13,738 lbs. (model 665 and 668).

POWER/WEIGHT RATIO: 4.67 : 1 (665) and 4.95 : 1 (668) lbt./lb.

APPLICATION:
Proposed for McDonnell-Douglas MD-11 and MD-12 Airliners.

OTHER ENGINES IN SERIES:
Trent 700, 800 Series engines currently in development to 85,000 LBT.

GAS TURBINE ENGINES FOR AIRCRAFT

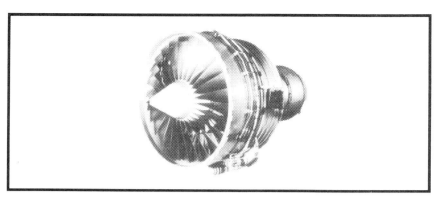 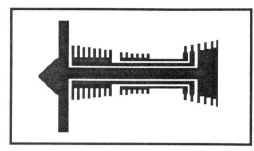

INTERNATIONAL TURBINE ENGINE COMPANIES

COUNTRY: GREAT BRITAIN

MANUFACTURER: ROLLS-ROYCE LTD.
Derby, England

DESIGNATION (civil)**:** Trent 768 and Trent 772.

BACKGROUND: Follow-on turbofan of the RB211 series.
Date first Trent engine produced: In Development.

ENGINE TYPE: Turbofan. Three Shaft, High-Bypass Front Fan.

COMPRESSOR TYPE: 15-Stage triple spool, Axial Flow: 1-Stage Front Fan acting as a Low Pressure Compressor; 8-Stage Intermediate Compressor; 6-Stage High Pressure Compressor.

COMPRESSOR DATA: Compressor Pressure Ratio: 35.55:1 (768), 37.42:1 (772), Fan Pressure Ratio: 1.853:1 (768), 1.853:1 (772), Fan Bypass Ratio: 4.75:1 (768), 4.66:1 (772), Total Mass Airflow: 1,970 (768), 2,011 (772) lb/sec take-off.

TURBINE TYPE: 6-Stage (768), 7-Stage (772) Axial Flow Turbine including: 1-Stage High Pressure Turbine, 1-Stage Intermediate Pressure Turbine, 4-Stage (768), 5-Stage (772) Low Pressure Turbine.

COMBUSTOR TYPE: Annular, Through-Flow.

POWER (TAKE-OFF) RATING: 67,500 lbt. (768), 71,100 lbt. (772). Rating approved to: 86°F (30°C) for models 768 and 772.

TAKE-OFF SPECIFIC FUEL CONSUMPTION: 0.319 (768), 0.327 (772) lb/hr/lbt.

BASIC DRY WEIGHT: 13,669 lbs. (model 768 and 772).

POWER/WEIGHT RATIO: 4.94 : 1 (768) and 5.2 : 1 (772) lbt./lb.

APPLICATION:
Airbus A330 Airliner.

OTHER ENGINES IN SERIES:
Trent 764 in McDonnel/D, MD-12.
Trent 600, 800 Series engines currently in development to 85,000 LBT.

GAS TURBINE ENGINES FOR AIRCRAFT

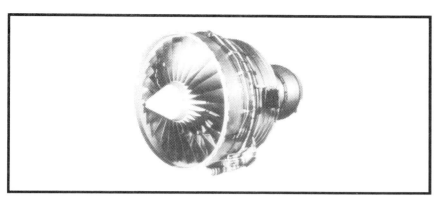 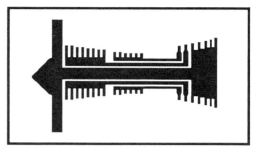

INTERNATIONAL TURBINE ENGINE COMPANIES

COUNTRY: GREAT BRITAIN

MANUFACTURER: ROLLS-ROYCE LTD.
Derby, England

DESIGNATION (civil): Trent 871 and Trent 884.

BACKGROUND: Follow-on turbofan of the RB211 series.
Date first Trent engine produced: In Development.

ENGINE TYPE: Turbofan. Three Shaft, High-Bypass Front Fan.

COMPRESSOR TYPE: 15-Stage triple spool, Axial Flow: 1-Stage Front Fan acting as a Low Pressure Compressor; 8-Stage Intermediate Compressor; 6-Stage High Pressure Compressor.

COMPRESSOR DATA: Compressor Pressure Ratio: 34.88:1 (871), 39.15:1 (884), Fan Pressure Ratio: 1.782:1 (871), 1.782:1 (884), Fan Bypass Ratio: 6.27:1 (871), 6.01:1 (884), Total Mass Airflow: 2,503 (871), 2,639 (884) lb/sec take-off.

TURBINE TYPE: 7-Stage Axial Flow Turbine including: 1-Stage High Pressure Turbine, 1-Stage Intermediate Pressure Turbine, 5-Stage Low Pressure Turbine.

COMBUSTOR TYPE: Annular, Through-Flow.

POWER (TAKE-OFF) RATING: 74,900 lbt. (871), 84,700 lbt. (884). Rating approved to: 86°F (30°C) for models 871 and 884.

TAKE-OFF SPECIFIC FUEL CONSUMPTION: 0.308 (871), 0.323 (884) lb/hr/lbt.

BASIC DRY WEIGHT: 16,000 lbs. (model 871 and 884).

POWER/WEIGHT RATIO: 4.68 : 1 (871) and 5.29 : 1 (884) lbt./lb.

APPLICATION:
Boeing B-777A, Airlner (Trent 871).
Boeing B-777B, Airlner (Trent 884).

OTHER ENGINES IN SERIES:
Trent 600, 700 Series engines also currently in development.

GAS TURBINE ENGINES FOR AIRCRAFT

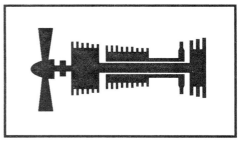

INTERNATIONAL TURBINE ENGINE COMPANIES

COUNTRY: GREAT BRITAIN

MANUFACTURER: ROLLS-ROYCE LTD.
Derby, England

DESIGNATION (civil)**:** Tyne-12, Mk.515.

BACKGROUND: No military counterpart. Developed for Commercial Aircraft.
Date First Tyne RTY12 Mk.515 Produced: 1959.
Date If Out-of-Production: 1965. Number of Tyne Engines Produced to Date: 225.

ENGINE TYPE: Turboprop. Dual-Shaft (Compressor Drive & Power Output Drive).

COMPRESSOR TYPE: 15-Stage Dual-Spool, Axial Flow Compressor including: 6-Stage Low Pressure Compressor, 9-Stage High Pressure Compressor.

COMPRESSOR DATA: Compressor Pressure Ratio: 13.5 : 1, Total Mass Airflow: 46.5 lb/sec (21,1 kg/sec).

TURBINE TYPE: 4-Stage Axial Flow Turbine including: 1-Stage High Pressure Turbine, 3-Stage Low Pressure Turbine.

COMBUSTOR TYPE: Can-Annular, Through-Flow, with 10 liners.

POWER (take-off) RATING: 5,505 eshp (4,105 ekw). Rating Approved to: 59°F (15°C).

TAKE-OFF SPECIFIC FUEL CONSUMPTION (SFC): 0.45 lb/hr/eshp (0,27 kg/hr/ekw).

BASIC DRY WEIGHT: 2,219 lbs (1,007 kg).

POWER/WEIGHT RATIO: 2.48 : 1 eshp/lb (4,08 :1 ekw/kg).

APPLICATION:
Shorts "Belfast", Canadair CL-44, "Yukon", Transports.

OTHER ENGINES IN SERIES:
Tyne-12 MK.101 in Canadair CL-44.
Tyne-20 MK.21/22 in Dassault-Breguet "Atlantic" and Franco-German, C-160 Transall.
Tyne-20 MK.801 in Alenia G.222T (Formerly Aeritalia).

GAS TURBINE ENGINES FOR AIRCRAFT

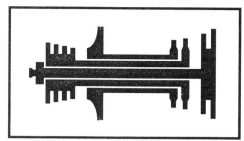

INTERNATIONAL TURBINE ENGINE COMPANIES

COUNTRY: GREAT BRITAIN

MANUFACTURER: ROLLS-ROYCE LTD.
Leavesden, England

DESIGNATION (civil & military): GEM-60-3.

BACKGROUND: Date First GEM Engine Produced: 1969.
Date If Out-of-Production: N/A. Number of Gem Engines Produced to Date: 1,100.

ENGINE TYPE: Turboshaft. Three-Shaft (Dual Compressor Drive & Single Power Output Drive).

COMPRESSOR TYPE: Combination Axial-Centrifugal Compressor. Dual-Compressor of 5-Stages Total Compression, including: 4-stage Low Pressure Axial Compressor followed by a 1-Stage independently rotating Centrifugal High Pressure Compressor.

COMPRESSOR DATA: Compressor Pressure Ratio: 14.4 : 1, Total Mass Airflow: 9 lb/sec (4,1 kg/sec).

TURBINE TYPE: 4-Stage Axial Flow includes: Gas Producer Section consisting of a 1-stage High Pressure Turbine and a 1-Stage Low Pressure Turbine, and Power Output Section consisting of a 2-Stage Power (free) Turbine.

COMBUSTOR TYPE: Annular, Reverse-Flow.

POWER (take-off) RATING: 1,203 shp (897 kw). Rating Approved to: 86°F (30°C).

TAKE-OFF SPECIFIC FUEL CONSUMPTION (SFC): 0.49 lb/hr/shp (0,30 kg/hr/kw).

BASIC DRY WEIGHT: 346 lbs (157 kg).

POWER/WEIGHT RATIO: 3.48 : 1 shp/lb (5,7 :1 kw/kg).

APPLICATION:
Westland W30-100, W30-160, Super Lynx Helicopters.

OTHER ENGINES IN SERIES:
GEM-2 in Westland "Lynx".
GEM-41 in Westland 30-140 "Super-Lynx".
GEM-2 Mk.1004 in Agusta A.129 "Mangusta", Italy.

GAS TURBINE ENGINES FOR AIRCRAFT

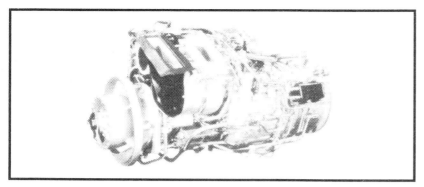 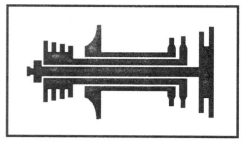

INTERNATIONAL TURBINE ENGINE COMPANIES

COUNTRY: GREAT BRITAIN

MANUFACTURER: ROLLS-ROYCE LTD.
Leavesden, England.

DESIGNATION (civil & military): Gem/RR 1004.

BACKGROUND: This engine is a direct drive version of the Gem Turboshaft.
Date First RR 1004 Engine Produced: 1985.
Date If Out-of-Production: N/A. Number of RR 1004 Engines Produced to Date: Unknown.

ENGINE TYPE: Turboshaft. Three-Shaft (Dual Compressor Drive & Single Power Output Drive).

COMPRESSOR TYPE: Combination Axial-Centrifugal Compressor. Dual-Compressor of 5-Stages Total Compression, including: 4-stage Low Pressure Axial Compressor followed by a 1-Stage independently rotating Centrifugal High Pressure Compressor.

COMPRESSOR DATA: Compressor Pressure Ratio: 10.8 : 1, Total Mass Airflow: 7.0 lb/sec (3,1 kg/sec).

TURBINE TYPE: 4-Stage Axial Flow includes: Gas Producer Section consisting of a 1-stage High Pressure Turbine and a 1-Stage Low Pressure Turbine, and Power Output Section consisting of a 2-Stage Power (free) Turbine.

COMBUSTOR TYPE: Annular, Reverse-Flow.

POWER (take-off) RATING: 1,018 shp (759 kw). Rating Approved to: 59°F (15°C).

TAKE-OFF SPECIFIC FUEL CONSUMPTION (SFC): 0.53 lb/hr/shp (0,32 kg/hr/kw).

BASIC DRY WEIGHT: 313 lbs (142 kg).

POWER/WEIGHT RATIO: 3.25 : 1 shp/lb (5,34 : 1 kw/kg).

APPLICATION:
Agusta A.129 "Mangusta" Helicopter.

OTHER ENGINES IN SERIES:
Refer also to Gem 60-3 under Rolls-Royce, Great Britain.

GAS TURBINE ENGINES FOR AIRCRAFT

 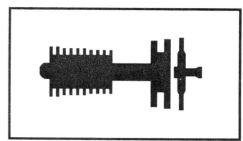

INTERNATIONAL TURBINE ENGINE COMPANIES

COUNTRY: GREAT BRITAIN

MANUFACTURER: ROLLS-ROYCE LTD.
Leavesden, England

DESIGNATION (civil & military): Gnome H.1400-1.

BACKGROUND: Manufactured under General Electric (USA) T-58 license.
Date First Gnome Engine Produced: 1959.
Date If Out-of-Production: N/A. Number of Gnome Engines Produced to Date: 2,300.

ENGINE TYPE: Turboshaft. Dual-Shaft (Compressor Drive & Power Output Drive).

COMPRESSOR TYPE: Axial Flow, 10-Stage Compressor.

COMPRESSOR DATA: Compressor Pressure Ratio: 8.5 : 1, Total Mass Airflow: 14 lb/sec (6,35 kg/sec).

TURBINE TYPE: 3-Stage Axial Flow including: 2-stage, Gas Producer (compressor drive) Turbine, 1-Stage Power (free) Turbine.

COMBUSTOR TYPE: Annular, Through-Flow.

POWER (take-off) RATING: 1,660 shp (1,238 kw). Rating Approved to: 59°F (15°C).

TAKE-OFF SPECIFIC FUEL CONSUMPTION (SFC): 0.61 lb/hr/shp (0,37 kg/hr/kw).

BASIC DRY WEIGHT: 326 lbs (148 kg).

POWER/WEIGHT RATIO: 5.09 : 1 shp/lb (8,36 :1 kw/kg).

APPLICATION:
Westland Commando and Sea King Helicopter.
Kawasaki Vertol-107 Helicopter, Japan.

OTHER ENGINES IN SERIES:
H.1000 in Westland "Whirlwind", Agusta Bell 204B "Iroquois".
H.1200 Series in Wessex 1,2,5,52,53,60, Boeing Vertol 107-2.
H.1400 in Westland Sea King , Sikorsky SH-3D, Agusta 101G, Aerospatiale HC MK.2.

GAS TURBINE ENGINES FOR AIRCRAFT

 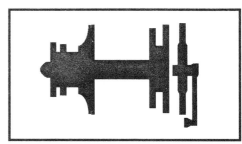

INTERNATIONAL TURBINE ENGINE COMPANIES

COUNTRY: GREAT BRITAIN

MANUFACTURER: ROLLS-ROYCE LTD.
Leavesden, England

DESIGNATION (civil & military): Nimbus Mk.103/503.

BACKGROUND: Date First Nimbus Engine Produced: 1958.
Date If Out-of-Production: 1973. Number of Nimbus Engines Produced to Date: 472.

ENGINE TYPE: Turboshaft. Dual-Shaft (Compressor Drive & Power Output Drive).

COMPRESSOR TYPE: Combination Axial-Centrifugal Flow Compressor including: 2-Axial Stages, 1-Centrifugal Stage.

COMPRESSOR DATA: Compressor Pressure Ratio: 6.0 : 1, Total Mass Airflow: 10.5 lb/sec (4,76 kg/sec).

TURBINE TYPE: 3-Stage Axial Flow including: 2-stage, Gas Producer (compressor drive) Turbine, 1-Stage Power (free) Turbine.

COMBUSTOR TYPE: Annular, Through-Flow.

POWER (take-off) RATING: 710 shp (529 kw). Rating Approved to: 113°F (45°C).

TAKE-OFF SPECIFIC FUEL CONSUMPTION (SFC): 0.89 lb/hr/shp (0,54 kg/hr/kw).

BASIC DRY WEIGHT: 607 lbs (275 kg).

POWER/WEIGHT RATIO: 1.17 :1 shp/lb (1,92 :1 kw/kg).

APPLICATION:
Westland Wasp Helicopter.

OTHER ENGINES IN SERIES:
Nimbus 105/501 in Westland Scout.

GAS TURBINE ENGINES FOR AIRCRAFT

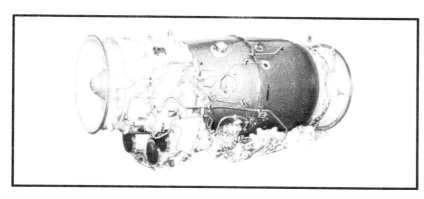
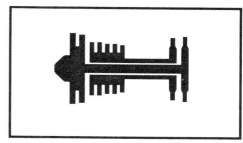

INTERNATIONAL TURBINE ENGINE COMPANIES

COUNTRY: INDIA

MANUFACTURER: HINDUSTAN AERONAUTICS LTD. (H.A.L.)
Bangalore, India

DESIGNATION (military): Adour Mk.811.

BACKGROUND: No Civil Counterpart.
Date First Adour Mk.811 Engine Produced: 1972.
Date If Out-of-Production: N/A. Number of Adour Mk.811 Engines Produced to Date: Unknown.

ENGINE TYPE: Turbofan. Dual-Shaft, Low-Bypass Front Fan with Afterburning.

COMPRESSOR TYPE: 7-Stage Dual-Spool, Axial Flow Compressor including: 2-Stage Front Fan acting as a Low Pressure Compressor, 5-Stage High Pressure Compressor.

COMPRESSOR DATA: Compressor Pressure Ratio: 11.3: 1, Fan Pressure Ratio: 1.5 : 1, Fan Bypass Ratio: 0.8 : 1, Total Mass Airflow: 94.8 lb/sec (43,1 kg/sec).

TURBINE TYPE: 2-Stage Axial Flow Turbine including: 1-Stage High Pressure Turbine, 1-Stage Low Pressure Turbine.

COMBUSTOR TYPE: Annular, Through-Flow.

AFTERBURNER: 1-Stage.

POWER (take-off) RATING: 5,700 lbt (25,35 kN thrust) dry, 8,400 lbt (37,36 kN thrust) A/B. Rating Approved to: 59°F (15°C).

TAKE-OFF SPECIFIC FUEL CONSUMPTION (SFC): 0.78 lb/hr/lb (22,1 mg/Ns) dry; 1.98 lb/hr/lbt (56 mg/Ns) A/B.

BASIC DRY WEIGHT: 1,720 lbs (780 kg).

POWER/WEIGHT RATIO: 3.3 : 1 lbt/lb (dry), 4.88 : 1 lbt/lb (A/B).

APPLICATION:
Jaguar, Attack Aircraft.

OTHER ENGINES IN SERIES:
Locate other Adour Models in Multi-National Listing under Rolls-Royce.

GAS TURBINE ENGINES FOR AIRCRAFT

INTERNATIONAL TURBINE ENGINE COMPANIES

COUNTRY: INDIA

MANUFACTURER: HINDUSTAN AERONAUTICS LTD. (H.A.L.)
Bangalore, India

DESIGNATION (civil & military): Astazou IIIB

BACKGROUND: Built under Turbomeca License, France.
Date First Astazou IIIB Produced: 1969.
Date If Out-of-Production: N/A. Number of Astazou Engines Produced to Date: Unknown.

ENGINE TYPE: Turboshaft. Single-Shaft Design.

COMPRESSOR TYPE: Combination Compressor: 1-Stage Axial, 1-Stage Centrifugal.

COMPRESSOR DATA: Compressor Pressure Ratio: 5.2 : 1. Mass Airflow: 10 lb/sec (4,55 kg/sec).

TURBINE TYPE: 3-Stage Axial Flow Turbine.

COMBUSTOR TYPE: Can Type with one Combustor.

POWER (take-off) RATING: 550 shp (410 kw). Rating Approved to: 59°F (15°C).

TAKE-OFF SPECIFIC FUEL CONSUMPTION (SFC): 0.76 lb/hr/shp (0,46 kg/hr/kw).

BASIC DRY WEIGHT: 402 lbs (182 kg).

POWER/WEIGHT RATIO: 1.37 :1 eshp/lb (2,25 :1 ekw/kg).

APPLICATION:
Hindustan SA.315 "Cheetah" Helicopter.
Hindustan SA.16 "Chetak" Helicopter.

OTHER ENGINES IN SERIES:
Locate other Astazou models under Turbomeca, France.

GAS TURBINE ENGINES FOR AIRCRAFT

INTERNATIONAL TURBINE ENGINE COMPANIES

COUNTRY: ISRAEL

MANUFACTURER: BET-SHEMESH ENGINES LTD. (BSEL)
ISRAEL

DESIGNATION (military)**:** Marbore-6.

BACKGROUND: Manufactured under license from Turbomeca Corp., France. Date First Marbore Engine Produced: 1955. Date If Out-of-Production: N/A. Number of Astazou Engines Produced to Date: Unknown.

ENGINE TYPE: Turbojet. Single-Shaft (non-afterburning).

COMPRESSOR TYPE: Centrifugal Flow: 1-Stage, Single Entry.

COMPRESSOR DATA: Compressor Pressure Ratio: 3.8:1, Total Mass Airflow: 21 lb/sec (9,6 kg/sec).

TURBINE TYPE: 1-Stage Axial Flow Turbine.

COMBUSTOR TYPE: Annular, Through Flow.

POWER (take-off) RATING: 1,060 lbt (4,71 kN thrust). Rating Approved to: 59°F (15°C).

TAKE-OFF SPECIFIC FUEL CONSUMPTION (SFC): 1.11 lb/hr/lbt (31,44 mg/Ns).

BASIC DRY WEIGHT: 370 lbs (168 kg).

POWER/WEIGHT RATIO: 2.86 : 1 lbt/lb.

APPLICATION:
Aerospatiale CM-170 "Magister" Attack Aircraft.
CASA HA-220 "Super Saeta", Spain Attack Aircraft.

OTHER ENGINES IN SERIES:
Marbore-4 in Aerospatiale CM-170 Trainer Aircraft.
Marbore-4 in Cessna T-37 "Tweet" Trainer Aircraft.

GAS TURBINE ENGINES FOR AIRCRAFT

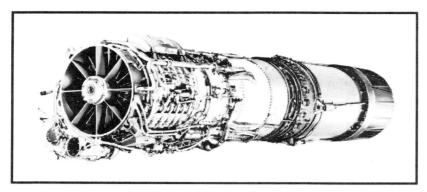 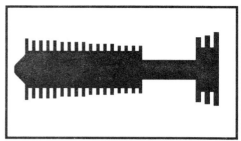

INTERNATIONAL TURBINE ENGINE COMPANIES

COUNTRY: ISRAEL

MANUFACTURER: ISRAEL AIRCRAFT INDUSTRIES LTD.
Bengurion Int'l. Airport

DESIGNATION (military)**:** J79-IAT-J1E.

BACKGROUND: Manufactured under General Electric (USA) J79 license.
Date First J79 Engine Produced in Israel : 1972.
Date If Out-of-Production: 1987. Number of J79 Engines Built under License to Date: 300.

ENGINE TYPE: Turbojet. Single-Shaft with Afterburning.

COMPRESSOR TYPE: Axial Flow: 17-Stage.

COMPRESSOR DATA: Compressor Pressure Ratio: 12.5 : 1, Total Mass Airflow: 170 lb/sec (77 kg/sec).

TURBINE TYPE: 3-Stage Axial Flow Turbine.

COMBUSTOR TYPE: Can-Annular, Through Flow, with 10 liners.

AFTERBURNER: Four-Stage.

POWER (take-off) RATING: 11,110 lbt (49,4 kN thrust) dry, 18,750 lbt (83,4 kN thrust) A/B. Rating Approved to: 59°F (15°C).

TAKE-OFF SPECIFIC FUEL CONSUMPTION (SFC): 0.84 lb/hr/lbt (23,79 mg/Ns) [dry]1.96 lb/hr/lbt (55,52 mg/Ns) [A/B].

BASIC DRY WEIGHT: 3,746 lbs (1,699 kg).

POWER/WEIGHT RATIO: 2.96 : 1 [dry], 5,0 : 1 [A/B] - lbt/lb.

APPLICATION:
Kfir Series Fighter Aircraft.

OTHER ENGINES IN SERIES:
Locate other J-79 models under General Electric Aircraft Engines, USA.

GAS TURBINE ENGINES FOR AIRCRAFT

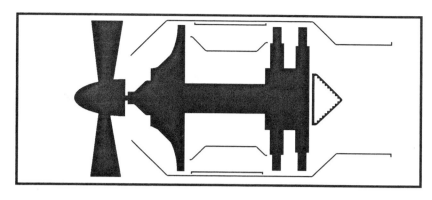

INTERNATIONAL TURBINE ENGINE COMPANIES

COUNTRY: ITALY

MANUFACTURER: ALFA ROMEO AVIO S.p.A.
Naples, Italy

DESIGNATION (civil): AR.318

BACKGROUND: Date First AR.318 Engine Produced: In Development.
Date If Out-of-Production: N/A . Number of AR.318 Engines Produced to Date: N/A.

ENGINE TYPE: Turboprop. Single-Shaft.

COMPRESSOR TYPE: Centrifugal Flow, 1-Stage Compressor.

COMPRESSOR DATA: Compressor Pressure Ratio: 5.4 : 1, Total Mass Airflow: 13.7 lb/sec (6,2 kg/sec).

TURBINE TYPE: 2-Stage Axial Flow.

COMBUSTOR TYPE: Annular, Reverse-Flow.

POWER (take-off) RATING: 608 eshp (453 ekw).

TAKE-OFF SPECIFIC FUEL CONSUMPTION (SFC): 0.58 lb/hr/eshp (0,35 kg/hr/ekw).

BASIC DRY WEIGHT: 310 lbs (140.6 kg).

POWER/WEIGHT RATIO: 1.96 : 1 eshp/lb (3,24 :1 ekw/kg).

APPLICATION:
Proposed for ßBusiness Aircraft.

OTHER ENGINES IN SERIES:
None.

GAS TURBINE ENGINES FOR AIRCRAFT

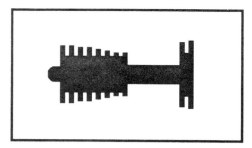

INTERNATIONAL TURBINE ENGINE COMPANIES

COUNTRY: ITALY

MANUFACTURER: ALFA ROMEO AVIO S.p.A.
Naples, Italy

DESIGNATION (military): J85-GE-13A.

BACKGROUND: Manufactured under License, General Electric Aircraft Engines (USA).
Date First J85-GE-13A Engine Produced: 1970.
Date If Out-of-Production: 1975. Number of Engines Built under License to Date: 110.

ENGINE TYPE: Turbojet. Single-Shaft (afterburning).

COMPRESSOR TYPE: Axial Flow: 8-Stage.

COMPRESSOR DATA: Compressor Pressure Ratio: 6.8: 1, Total Mass Airflow: 44.0 lb/sec (19,96 kg/sec).

TURBINE TYPE: 2-Stage Axial Flow Turbine.

COMBUSTOR TYPE: Annular, Through Flow.

AFTERBURNER: Single-Stage.

POWER (take-off) RATING: 2,720 lbt (12,1 kN thrust) [dry], 4,080 lbt (18,15 kN thrust) [A/B]. Rating Approved to: 59°F (15°C).

TAKE-OFF SPECIFIC FUEL CONSUMPTION (SFC): 1.03 lb/hr/lbt (29,18 mg/Ns) [dry], 2.22 lb/hr/lbt (62,89 mg/Ns) [A/B].

BASIC DRY WEIGHT: 597 lbs (271 kg).

POWER/WEIGHT RATIO: 4.56:1 [dry], 6.83 : 1 [A/B] - lbt/lb.

APPLICATION:
Fiat G.91Y Fighter Reconnaissance Aircraft.
Fiat G.91Y-T Trainer Aircraft.

OTHER ENGINES IN SERIES:
Locate other J-85 models under General Electric Aircraft Engines, USA.

GAS TURBINE ENGINES FOR AIRCRAFT

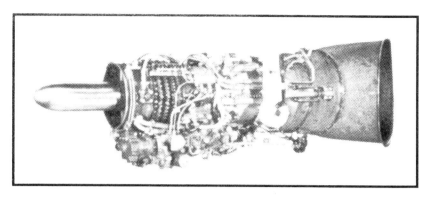 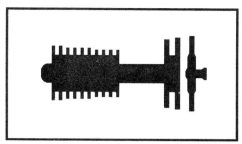

INTERNATIONAL TURBINE ENGINE COMPANIES

COUNTRY: ITALY

MANUFACTURER: ALFA ROMEO AVIO S.P.A.
Naples, Italy

DESIGNATION (military): T58-GE-10.

BACKGROUND: Manufactured under License, General Electric Aircraft Engines (USA).
Date First T58-GE-10 Engine Produced: 1967.
Date If Out-of-Production: 1985. Number of Engines Built under License to Date: 170.

ENGINE TYPE: Turboshaft. Dual-Shaft (Compressor Drive & Power Output Drive).

COMPRESSOR TYPE: Axial Flow, 10-Stage Compressor.

COMPRESSOR DATA: Compressor Pressure Ratio: 8.4 : 1, Total Mass Airflow: 13.7 lb/sec (6,2 kg/sec).

TURBINE TYPE: 3-Stage Axial Flow including: 2-stage, Gas Producer (compressor drive) Turbine, 1-Stage Power (free) Turbine.

COMBUSTOR TYPE: Annular, Through-Flow.

POWER (take-off) RATING: 1,400 eshp (1,044 ekw), includes 57 shp (43 kw) from 142 lbt (63,2 kN thrust). Rating Approved to: 59°F (15°C).

TAKE-OFF SPECIFIC FUEL CONSUMPTION (SFC): 0.61 lb/hr/eshp (0,37 kg/hr/ekw).

BASIC DRY WEIGHT: 350 lbs (159 kg).

POWER/WEIGHT RATIO: 4.0 : 1 eshp/lb (6,57 :1 ekw/kg).

APPLICATION:
Agusta-Sikorsky SH-3D Helicopter.

OTHER ENGINES IN SERIES:
T58-GE-3 in Agusta-Bell 204B "Huey" Helicopter.
T58-GE-5 in Agusta-Sikorsky HH-3F "Pelican" Helicopter.
Locate other T58 models under General Electric Aircraft Engines, USA.

GAS TURBINE ENGINES FOR AIRCRAFT

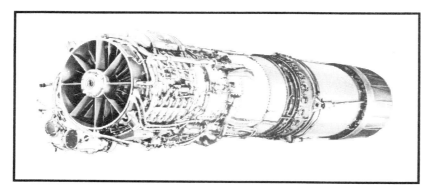 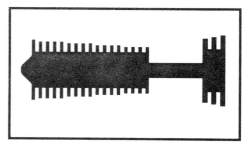

INTERNATIONAL TURBINE ENGINE COMPANIES

COUNTRY: ITALY

MANUFACTURER: Fiat Aviazione S.P.A.
Turin, Italy

DESIGNATION (military): J79-GE-19.

BACKGROUND: Manufactured under License General Electric Aircraft Engines (USA).
Date First J79-GE-19 Engine Produced: 1954.
Date If Out-of-Production: N/A. Number of Engines Built under License to Date: Unknown.

ENGINE TYPE: Turbojet. Single-Shaft with Afterburning.

COMPRESSOR TYPE: Axial Flow: 17-Stage.

COMPRESSOR DATA: Compressor Pressure Ratio: 13.5 : 1, Total Mass Airflow: 170 lb/sec (77 kg/sec).

TURBINE TYPE: 3-Stage Axial Flow Turbine.

COMBUSTOR TYPE: Can-Annular, Through Flow, with 10 liners.

AFTERBURNER: Four-Stage.

POWER (take-off) RATING: 11,870 lbt (52,8 kN thrust) dry, 17,900 lbt (79,6 kN thrust) A/B. Rating Approved to: 59°F (15°C).

TAKE-OFF SPECIFIC FUEL CONSUMPTION (SFC): 0.84 lb/hr/lbt (23,79 mg/Ns) [dry], 1.96 lb/hr/lbt (55,52 mg/Ns) [A/B].

BASIC DRY WEIGHT: 3,845 lbs (1,744 kg).

POWER/WEIGHT RATIO: 3.1 :1 [dry], 4.66 : 1 [A/B] - lbt/lb.

APPLICATION:
Fiat-Lockheed F-104S StarFighter.

OTHER ENGINES IN SERIES:
J-79-11 in Fiat-Lockheed F-104G Fighter Aircraft.
Locate other J-79 models under General Electric Aircraft Engines, USA.

GAS TURBINE ENGINES FOR AIRCRAFT

 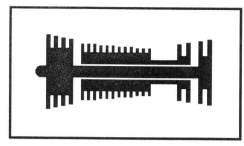

INTERNATIONAL TURBINE ENGINE COMPANIES

COUNTRY: ITALY

MANUFACTURER: Fiat Aviazione S.P.A.
Turin, Italy

DESIGNATION (civil): Spey MK. 807.

BACKGROUND: Built Under License, Rolls Royce (G.B.).
Date First Spey Engine Produced: In Development.
Date If Out-of-Production: N/A. Number of Spey 807 Engines Produced to Date: N/A.

ENGINE TYPE: Turbofan, Dual-Shaft, High-Bypass Front Fan.

COMPRESSOR TYPE: 16-Stage Dual-Spool, Axial Flow Compressor including: 4-Stage Front Fan acting also as a Low Pressure Compressor, 12-Stage High Pressure Compressor.

COMPRESSOR DATA: Compressor Pressure Ratio: 16.3.0 : 1, Fan Pressure Ratio: 1.5 : 1, Fan Bypass Ratio: 0.64 : 1, Total Mass Airflow: 202 lb/sec (91,6 kg/sec).

TURBINE TYPE: 4-Stage Axial Flow Turbine including: 2-Stage High Pressure Turbine, 2-Stage Low Pressure Turbine.

COMBUSTOR TYPE: Can-Annular, Through-Flow With 10 Liners.

POWER (take-off) RATING: 11,030 lbt. (49,06 kN thrust). Rating Approved to: 59°F (15°C).

TAKE-OFF SPECIFIC FUEL CONSUMPTION (SFC): 0.638 lb/hr/lbt (18,1 mg/Ns).

BASIC DRY WEIGHT: 2,456 lbs (1,114 kg).

POWER/WEIGHT RATIO: 4.49 : 1 lbt/lb.

APPLICATION:
Alenia-Aermacchi-Embraer "AM-X" Fighter Aircraft, Italy and Brazil.
Alenia-Aermacchi-Embraer "AM-XT" Trainer Aircraft, Italy and Brazil.

OTHER ENGINES IN SERIES:
Locate other Spey models under Rolls-Royce, Great Britain listing.

GAS TURBINE ENGINES FOR AIRCRAFT

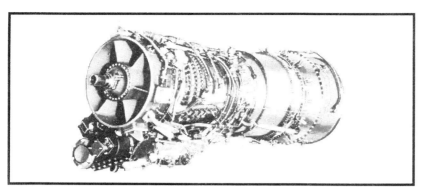 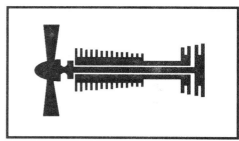

INTERNATIONAL TURBINE ENGINE COMPANIES

COUNTRY: ITALY

MANUFACTURER: FIAT AVIAZIONE S.P.A.
Turin, Italy

DESIGNATION (civil/military): T64-P4D.

BACKGROUND: Manufactured under License, General Electric Aircraft Engines (USA). Civil designation CT-64-820. General Electric Aircraft Engines also manufactures this engine in both Turboprop and Turboshaft Engine models.
Date First T64-P4D Engine Produced: 1975.
Date If Out-of-Production: N/A. Number of T64-P4D Engines Produced to Date: Unknown.

ENGINE TYPE: Turboprop. Dual-Shaft (Compressor Drive & Power Output Drive).

COMPRESSOR TYPE: Axial Flow, 14-Stage Compressor.

COMPRESSOR DATA: Compressor Pressure Ratio: 14.0 : 1, Total Mass Airflow: 27.0 lb/sec (12,0 kg/sec).

TURBINE TYPE: 4-Stage Axial Flow including: 2-stage, Gas Producer (compressor drive) Turbine, 2-Stage Power (free) Turbine.

COMBUSTOR TYPE: Annular, Through-Flow.

POWER (take-off) RATING: 3,400 eshp (2,535 ekw), includes 84 shp (75 kw) from 210 lbt (0,93 kN thrust). Rating Approved to: 59°F (15°C).

TAKE-OFF SPECIFIC FUEL CONSUMPTION (SFC): 0.48 (0,29 kg/hr/ekw).

BASIC DRY WEIGHT: 1,198 lbs (543 kg).

POWER/WEIGHT RATIO: 2.8 : 1 eshp/lb (4,67:1 ekw/kg).

APPLICATION:
Alenia (formerly Aeritalia) G.222, STOL Military Turboprop Aircraft.

OTHER ENGINES IN SERIES:
Locate other T-64 models under General Electric Aircraft Engines (USA) listing.

GAS TURBINE ENGINES FOR AIRCRAFT

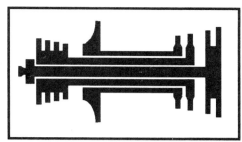

INTERNATIONAL TURBINE ENGINE COMPANIES

COUNTRY: ITALY

MANUFACTURER: INDUSTRIE AERONAUTICHE MECCANICHE RINALDO PIAGGIO S.p.A. GENOA, ITALY

DESIGNATION (civil & military): Gem/RR 1004.

BACKGROUND: Derived from Rolls Royce, Gem Turboshaft.
Date First RR 1004 Engine Produced: 1985.
Date If Out-of-Production: N/A. Number of RR 1004 Engines Built under License to Date: 45.

ENGINE TYPE: Turboshaft. Three-Shaft (Dual Compressor Drive & Single Power Output Drive).

COMPRESSOR TYPE: Combination Axial-Centrifugal Compressor. Dual-Compressor of 5-Stages Total Compression, including: 4-stage Low Pressure Axial Compressor followed by a 1-Stage independently rotating Centrifugal High Pressure Compressor.

COMPRESSOR DATA: Compressor Pressure Ratio: 10.8 : 1, Total Mass Airflow: 7.0 lb/sec (3,1 kg/sec).

TURBINE TYPE: 4-Stage Axial Flow includes: Gas Producer Section consisting of a 1-stage High Pressure Turbine and a 1-Stage Low Pressure Turbine, and Power Output Section consisting of a 2-Stage Power (free) Turbine.

COMBUSTOR TYPE: Annular, Reverse-Flow.

POWER (take-off) RATING: 881 shp (657 kw). Rating Approved to: 59°F (15°C).

TAKE-OFF SPECIFIC FUEL CONSUMPTION (SFC): 0.53 lb/hr/shp (0,318 kg/hr/kw).

BASIC DRY WEIGHT: 361 lbs (164 kg).

POWER/WEIGHT RATIO: 2.44 : 1 shp/lb (4,01 :1 kw/kg).

APPLICATION:
Agusta A-129 "Mangusta" Helicopter.

OTHER ENGINES IN SERIES:
Locate other Gem models under Rolls-Royce, Great Britain listing.

GAS TURBINE ENGINES FOR AIRCRAFT

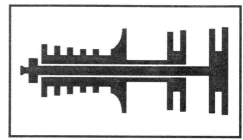

INTERNATIONAL TURBINE ENGINE COMPANIES

COUNTRY: ITALY

MANUFACTURER: INDUSTRIE AERONAUTICHE MECCANICHE RINALDO PIAGGIO S.p.A. GENOA, ITALY

DESIGNATION (civil/military): T53-L-13 (A,B).

BACKGROUND: Manufactured under License, Textron-Lycoming Company (USA). Date First T53L-13 Engine Produced: 1968. Date If Out-of-Production: 1975. Number of T53-L-13 Engines Built under License to Date: 160.

ENGINE TYPE: Turboshaft. Dual-Shaft (Compressor Drive & Power Output Drive).

COMPRESSOR TYPE: Combination Compressor: 5-Stage Axial, 1-Stage Centrifugal.

COMPRESSOR DATA: Compressor Pressure Ratio: 7:1. Mass Airflow: 12.8 lb/sec (5,8 kg/sec).

TURBINE TYPE: 4-Stage Axial Flow Turbine including: 2-Stage Gas Producer (compressor drive) Turbine, 2-Stage Power (free) Turbine.

COMBUSTOR TYPE: Annular, Reverse Flow.

POWER (take-off) RATING: 1,400 shp (1,044 kw). Rating Approved to: 59°F (15°C).

TAKE-OFF SPECIFIC FUEL CONSUMPTION (SFC): 0.58 lb/hr/shp (0,35 kg/hr/kw).

BASIC DRY WEIGHT: 540 lbs (245 kg).

POWER/WEIGHT RATIO: 2.59:1 shp/lb (4,26:1 kw/kg).

APPLICATION:
Agusta-Bell 205 "Iroquois" Helicopter. (Bell, USA)

OTHER ENGINES IN SERIES:
Locate other T53 models under Textron-Lycoming (USA) listing.

GAS TURBINE ENGINES FOR AIRCRAFT

 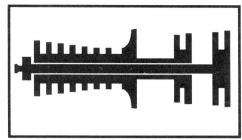

INTERNATIONAL TURBINE ENGINE COMPANIES

COUNTRY: ITALY

MANUFACTURER: INDUSTRIE AERONAUTICHE MECCANICHE RINALDO PIAGGIO S.p.A. GENOA, ITALY

DESIGNATION (civil/military): T55-L-712.

BACKGROUND: Manufactured under License, Textron-Lycoming Company (USA).
Date First T55-L-712 Engine Produced: 1984.
Date If Out-of-Production: N/A. Number of T55L-712 Engines Built under License to Date: 82.

ENGINE TYPE: Turboshaft. Dual-Shaft (Compressor Drive & Power Output Drive).

COMPRESSOR TYPE: Combination Compressor: 7-Stage Axial, 1-Stage Centrifugal.

COMPRESSOR DATA: Compressor Pressure Ratio: 6.2:1. Mass Airflow: 27.8 lb/sec (12,6 kg/sec).

TURBINE TYPE: 4-Stage Axial Flow including: 2-Stage Gas Producer (compressor drive) Turbine, 2-Stage Power (free) Turbine.

COMBUSTOR TYPE: Annular, Reverse Flow.

POWER (take-off) RATING: 3,750 shp (2,796 kw). Rating Approved to: 59°F (15°C).

TAKE-OFF SPECIFIC FUEL CONSUMPTION (SFC): 0.53 lb/hr/shp (0,322 kg/hr/kw).

BASIC DRY WEIGHT: 710 lbs (322 kg).

POWER/WEIGHT RATIO: 5.28 :1 shp/lb (8,68 :1 kw/kg).

APPLICATION:
Agusta-Vertol CH-47D "Chinook" Helicopter. (Boeing License, USA)

OTHER ENGINES IN SERIES:
Locate other T55 models under Textron-Lycoming (USA) listing.

GAS TURBINE ENGINES FOR AIRCRAFT

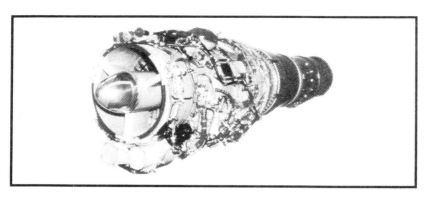 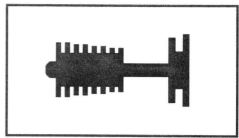

INTERNATIONAL TURBINE ENGINE COMPANIES

COUNTRY: ITALY

MANUFACTURER: INDUSTRIE AERONAUTICHE MECCANICHE RINALDO PIAGGIO S.p.A. GENOA, Italy

DESIGNATION (military): Viper 632-43.

BACKGROUND: Manufactured under License, Rolls-Royce, Bristol (G.B.).
Date First Viper 632-43 Engine Produced: 1975.
Date If Out-of-Production: N/A. Number of Viper 632 Engines Built under License to Date: 150.

ENGINE TYPE: Turbojet. Single-Shaft (non-afterburning).

COMPRESSOR TYPE: Axial Flow: 8-Stage.

COMPRESSOR DATA: Compressor Pressure Ratio: 5.95 : 1, Total Mass Airflow: 59 lb/sec (26,5 kg/sec).

TURBINE TYPE: 2-Stage Axial Flow Turbine.

COMBUSTOR TYPE: Annular, Through Flow.

POWER (take-off) RATING: 3,970 lbt (17,61 kN thrust). Rating Approved to: 59°F (15°C).

TAKE-OFF SPECIFIC FUEL CONSUMPTION (SFC): 0.96 lb/hr/lbt (27,19 mg/Ns).

BASIC DRY WEIGHT: 810 lbs (367 kg).

POWER/WEIGHT RATIO: 4.9 :1 lbt/lb.

APPLICATION:
Aermacchi MB.326K Attack Aircraft.
Aermacchi MB.329A Trainer Aircraft
Aermacchi MB 339K "Veltro-II" Attack Aircraft.

OTHER ENGINES IN SERIES:
Viper-11 in Aermacchi MB.326C.
Viper-500 in Piaggio-Douglas PD-808 Transport.
Viper-540 in Aermacchi MB.326G,K.
Locate other Viper models under Rolls-Royce, Great Britain listing.

GAS TURBINE ENGINES FOR AIRCRAFT

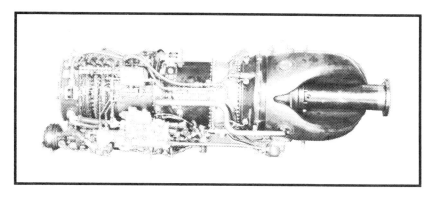 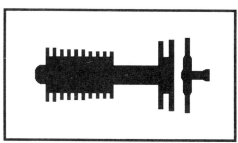

INTERNATIONAL TURBINE ENGINE COMPANIES

COUNTRY: JAPAN

MANUFACTURER: ISHIKAWAJIMA-HARIMA HEAVY INDUSTRIES (IHI)
TOKYO, JAPAN

DESIGNATION (civil)**:** CT58-IHI-140-1.

BACKGROUND: Commercial version of T58-GE-10.
Date First CT58-IHI-140-1 Engine Produced: 1960.
Date If Out-of-Production: N/A. Number of CT58-IHI Engines Built under License to Date: Unknown.

ENGINE TYPE: Turboshaft. Dual-Shaft (Compressor Drive & Power Output Drive).

COMPRESSOR TYPE: 10-Stage Single-Spool, Axial Flow Compressor.

COMPRESSOR DATA: Compressor Pressure Ratio: 8.4 : 1, Total Mass Airflow: 13.7 lb/sec (6,2 kg/sec).

TURBINE TYPE: 3-Stage Axial Flow Turbine including: 2-Stage Gas Producer (compressor drive) Turbine, 1-Stage Power (free) Turbine.

COMBUSTOR TYPE: Annular, Through Flow.

POWER (take-off) RATING: 1,521 shp (1,134 kw). Rating Approved to: 59°F (15°C).

TAKE-OFF SPECIFIC FUEL CONSUMPTION (SFC): 0.61 lb/hr/shp (0,37 kg/hr/kw).

BASIC DRY WEIGHT: 340 lbs (154 kg).

POWER/WEIGHT RATIO: 4.47 : 1 shp/lb (7,36 :1 kw/kg).

APPLICATION:
Kawasaki V-107A, KV-107 IIA Helicopters. (Boeing License, USA)
Mitsubishi S-61 Helicopter.

OTHER ENGINES IN SERIES:
T58-IHI-10-M1 in Kawasaki KV-107 IIA, HSS-2A, Shin Meiwa PS-1, US-1.
T58-IHI-10-M2 in Mitsubishi S-61A.
Locate other T58 models under General Electric Aircraft Engines, USA.

GAS TURBINE ENGINES FOR AIRCRAFT

 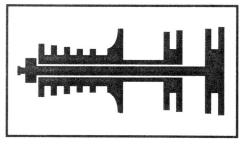

INTERNATIONAL TURBINE ENGINE COMPANIES

COUNTRY: JAPAN

MANUFACTURER: ISHIKAWAJIMA-HARIMA HEAVY INDUSTRIES (IHI)
TOKYO, JAPAN

DESIGNATION (military)**:** CT700-700.

BACKGROUND: Licensed Production from G.E. Company, USA.
Date First CT700 Engine Produced: 1990 in Japan.
Date If Out-of-Production: N/A. Number of T700-700 Engines Built under License to Date: Unknown.

ENGINE TYPE: Turboshaft. Dual-Shaft (Compressor Drive & Power Output Drive).

COMPRESSOR TYPE: 6-Stage Dual Compressor including a 5-Stage, Axial Flow Compressor, and a 1-Stage Centrifugal Flow Compressor.

COMPRESSOR DATA: Compressor Pressure Ratio: 17 : 1, Total Mass Airflow: 12 lb/sec (5,4 kg/sec).

TURBINE TYPE: 4-Stage Axial Flow Turbine including: 2-Stage Gas Producer (compressor drive) Turbine, 2-Stage Power (free) Turbine.

COMBUSTOR TYPE: Annular, Through Flow.

POWER (take-off) RATING: 1,800 shp (1,343 kw). Rating Approved to: 59°F (15°C).

TAKE-OFF SPECIFIC FUEL CONSUMPTION (SFC): 0.46 lb/hr/shp (0,28 kg/hr/kw).

BASIC DRY WEIGHT: 458 lbs (208 kg).

POWER/WEIGHT RATIO: 3.93 : 1 shp/lb (6,46 :1 kw/kg).

APPLICATION:
Mitsubishi ISH-60 and OH-60 Helicopters (Sikorsky S-70 License, USA).

OTHER ENGINES IN SERIES:
Locate other CT700 and T700 models under General Electric Aircraft Engines, USA.

GAS TURBINE ENGINES FOR AIRCRAFT

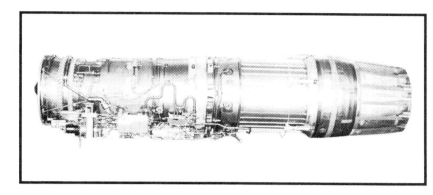
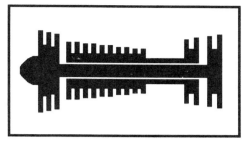

INTERNATIONAL TURBINE ENGINE COMPANIES

COUNTRY: JAPAN

MANUFACTURER: ISHIKAWAJIMA-HARIMA HEAVY INDUSTRIES (IHI)
TOKYO, JAPAN

DESIGNATION (military): F100-IHI-100.

BACKGROUND: No Civil Counterpart.
Date First F100-IHI-100 Engine Produced: 1978.
Date If Out-of-Production: N/A. Number of F100-IHI-100 Engines Built to Date: Unknown.

ENGINE TYPE: Turbofan. Dual-Shaft, Low-Bypass Front Fan with Afterburning.

COMPRESSOR TYPE: 13-Stage Dual-Spool, Axial Flow Compressor including: 3-Stage Front Fan with no other Axial Stages in the Low Pressure Compressor, 10-Stage High Pressure Compressor.

COMPRESSOR DATA: Compressor Pressure Ratio: 24:1, Fan Pressure Ratio: 0.63:1, Fan Bypass Ratio: 0.60:1, Total Mass Airflow: 227 lb/sec (103 kg/sec).

TURBINE TYPE: 4-Stage Axial Flow Turbine including: 2-Stage High Pressure Turbine, 2-Stage Low Pressure Turbine.

COMBUSTOR TYPE: Annular, Through-Flow.

AFTERBURNER: 7-Zone, 2-Zones in Fan Bypass Exhaust, 5-Zones in the Core Exhaust.

POWER (take-off) RATING: 14,375 lbt (63,9 kN thrust) [dry], 23,800 (105,9 kN thrust) [A/B].
Rating Approved to: 59°F (15°C).

TAKE-OFF SPECIFIC FUEL CONSUMPTION (SFC): 0.735 lb/hr/lbt (20,8 mg/Ns) [dry], 2.2 lb/hr/lbt (62,3 mg/Ns) [A/B].

BASIC DRY WEIGHT: 3,075 lbs (1,395 kg).

POWER/WEIGHT RATIO: 4.67:1 [dry], 7.74:1 [A/B] - lbt/lb.

APPLICATION:
F-15 and F-15J Fighter Aircraft, Japan. (McDonnell-Douglas License, USA)

OTHER ENGINES IN SERIES:
Locate other F100 Models under United Technologies (USA) listing.

GAS TURBINE ENGINES FOR AIRCRAFT

 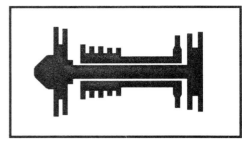

INTERNATIONAL TURBINE ENGINE COMPANIES

COUNTRY: JAPAN

MANUFACTURER: ISHIKAWAJIMA-HARIMA HEAVY INDUSTRIES (IHI)

DESIGNATION (military)**:** F3-IHI-30.

BACKGROUND: Date First F3-IHI-30 Engine Produced: 1987.
Date If Out-of-Production: N/A. Number of F3-IHI-30 engines built to date: Unknown.

ENGINE TYPE: Turbofan, Dual Shaft non-afterburning.

COMPRESSOR TYPE: 7-Stage, Dual-Spool, Axial Flow Compressor, including a 2-Stage Front Fan in the Low Pressure Compressor and a 5-Stage High Pressure Compressor.

COMPRESSOR DATA: Compressor Pressure Ratio: 11.1: 1, Fan Pressure Ratio: 2.6 : 1, Fan Bypass Ratio: 0.9 : 1, Total Mass Airflow: 75 lb/sec (34 kg/sec).

TURBINE TYPE: Three-Stage Axial Flow Turbine, including a 1-Stage High Pressure Turbine and a 2-Stage Low Pressure Turbine.

COMBUSTOR TYPE: Annular, Through-Flow.

POWER (take-off) RATING: 3,755 lbt (16,7 kN). Rating Approved to: 59°F (15°C).

TAKE-OFF SPECIFIC FUEL CONSUMPTION (SFC): 0.68 lb/hr/lbt (19,26 mg/Ns).

BASIC DRY WEIGHT: 750 lbs (340 kg).

POWER/WEIGHT RATIO: 5.0 : 1 lb/lbt.

APPLICATION:
Kawasaki T-4, Twin-Engine Trainer.

OTHER ENGINES IN SERIES:
None.

GAS TURBINE ENGINES FOR AIRCRAFT

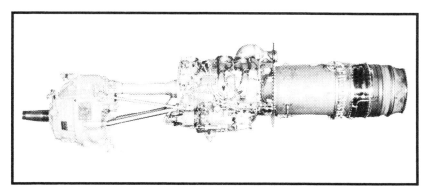
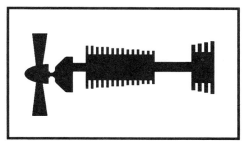

INTERNATIONAL TURBINE ENGINE COMPANIES

COUNTRY: JAPAN

MANUFACTURER: ISHIKAWAJIMA-HARIMA HEAVY INDUSTRIES (IHI)
TOKYO, JAPAN

DESIGNATION (military): T56-IHI-14.

BACKGROUND: Manufactured under General Motors (USA) T-56 license.
Date First T56-IHI-14 Engine Produced: 1978.
Date If Out-of-Production: N/A. Number of T56-IHI-14 engines built to date: Unknown.

ENGINE TYPE: Turboprop. Single-Shaft Design.

COMPRESSOR TYPE: 14-Stage Single-Spool Axial Flow Compressor.

COMPRESSOR DATA: Compressor Pressure Ratio: 9.5: 1, Total Mass Airflow: 32.4 lb/sec (14,7 kg/sec).

TURBINE TYPE: 4-Stage Axial Flow Turbine.

COMBUSTOR TYPE: Can-Annular, Through Flow, with 6 liners.

POWER (take-off) RATING: 4,910 eshp (3,661 ekw), includes 320 shp (239 kw) from 797 lbt (3,54 kN thrust). Rating Approved to: 59°F (15°C).

TAKE-OFF SPECIFIC FUEL CONSUMPTION (SFC): 0.50 lb/hr/eshp (0,30 kg/hr/ekw).

BASIC DRY WEIGHT: 1,885 lbs (855 kg).

POWER/WEIGHT RATIO: 2.6 :1 eshp/lb (4,28 :1 ekw/kg).

APPLICATION:
Kawasaki P-3C "Orion" Maritime Aircraft, Japan. (Lockheed License, USA)

OTHER ENGINES IN SERIES:
Locate other T56 models under Allison Turbine Engines (USA) listing.

GAS TURBINE ENGINES FOR AIRCRAFT

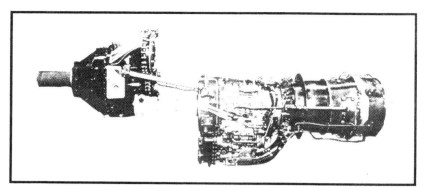 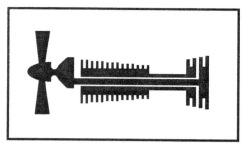

INTERNATIONAL TURBINE ENGINE COMPANIES

COUNTRY: JAPAN

MANUFACTURER: ISHIKAWAJIMA-HARIMA HEAVY INDUSTRIES (IHI)
TOKYO, JAPAN

DESIGNATION (military): T64-IHI-10J.

BACKGROUND: Date First T64-IHI Engine Produced: 1965.
Date If Out-of-Production: 1982. Number of T64-IHI-10J engines built to date: Unknown.

ENGINE TYPE: Turboprop. Dual-Shaft (Compressor Drive & Power Output Drive).

COMPRESSOR TYPE: 14-Stage Single-Spool, Axial Flow Compressor.

COMPRESSOR DATA: Compressor Pressure Ratio: 12.5 : 1, Total Mass Airflow: 24.9 lb/sec (11,3 kg/sec).

TURBINE TYPE: 4-Stage Axial Flow Turbine including: 2-Stage Gas Producer (compressor drive) Turbine, 2-Stage Power (free) Turbine.

COMBUSTOR TYPE: Annular, Through Flow.

POWER (take-off) RATING: 3,490 eshp (2,602 ekw), includes 80 shp (60 kw) from 200 lbt (0,89 kN). Rating Approved to: 59°F (15°C).

TAKE-OFF SPECIFIC FUEL CONSUMPTION (SFC): 0.48 lb/hr/eshp (0,29 kg/hr/ekw).

BASIC DRY WEIGHT: 1,240 lbs (563 kg).

POWER/WEIGHT RATIO: 2.8 :1 eshp/lb (4,62 :1 ekw/kg).

APPLICATION:
Shin Meiwa Industries US-1A, SAR Aircraft and PS-1 "ASW" Aircraft.
Kawasaki P-2J "Neptune" Patrol Aircraft.

OTHER ENGINES IN SERIES:
Locate other T64 Models under General Electric Aircraft Engines, USA.

GAS TURBINE ENGINES FOR AIRCRAFT

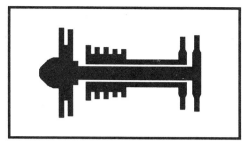

INTERNATIONAL TURBINE ENGINE COMPANIES

COUNTRY: JAPAN

MANUFACTURER: ISHIKAWAJIMA-HARIMA INDUSTRIES (IHI)
TOKYO, JAPAN

DESIGNATION (military)**:** TF40-IHI-801A.

BACKGROUND: No Civil Counterpart. T-40 Adour, License Rolls-Royce.
Date First TH40-IHI-801A Engine Produced: 1977.
Date If Out-of-Production: 1987. Number of TF40-IHI-801A engines built to date: Unknown.

ENGINE TYPE: Turbofan. Dual-Shaft, Low-Bypass Front Fan (non-afterburning).

COMPRESSOR TYPE: 7-Stage Dual-Spool, Axial Flow Compressor including: 2-Stage Front Fan with no other Axial Stages in the Low Pressure Compressor, 5-Stage High Pressure Compressor.

COMPRESSOR DATA: Compressor Pressure Ratio: 9.6 : 1, Fan Pressure Ratio: 1.9 : 1, Fan Bypass Ratio: 0.8 : 1, Total Mass Airflow: 120 lb/sec (54,5 kg/sec).

TURBINE TYPE: 2-Stage Axial Flow Turbine including: 1-Stage High Pressure Turbine, 1-Stage Low Pressure Turbine.

COMBUSTOR TYPE: Annular, Through-Flow.

POWER (take-off) RATING: 7,300 lbt. (32,47 kN thrust). Rating Approved to: 59°F (15°C).

TAKE-OFF SPECIFIC FUEL CONSUMPTION (SFC): 0.74 lb/hr/lbt (20,96 mg/Ns).

BASIC DRY WEIGHT: 1,633 lbs (741 kg).

POWER/WEIGHT RATIO: 4.47 : 1 lbt/lb.

APPLICATION:
Mitsubishi T-2 Trainer and F-1 Attack Aircraft.

OTHER ENGINES IN SERIES:
Locate other TF40 models under designation "Adour" in Rolls-Royce, Turbomeca; Multi-National listing.

GAS TURBINE ENGINES FOR AIRCRAFT

 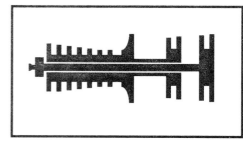

INTERNATIONAL TURBINE ENGINE COMPANIES

COUNTRY: JAPAN

MANUFACTURER: KAWASAKI HEAVY INDUSTRIES LTD (KHI)
KOBE, JAPAN

DESIGNATION (military)**:** T53-K-703.

BACKGROUND: Manufactured under License, Textron-Lycoming (USA).
Date First T53-K-703 Engine Produced: 1985.
Date If Out-of-Production: N/A. Number of T53-K Engines Produced to Date: 68.

ENGINE TYPE: Turboshaft. Dual-Shaft (Compressor Drive & Power Output Drive).

COMPRESSOR TYPE: Combination Compressor: 7-Stage Axial, 1-Stage Centrifugal.

COMPRESSOR DATA: Compressor Pressure Ratio: 8:1. Mass Airflow: 12.8 lb/sec (5,8 kg/sec).

TURBINE TYPE: 4-Stage Axial Flow Turbine including: 2-Stage Gas Producer (compressor drive) Turbine, 2-Stage Power (free) Turbine.

COMBUSTOR TYPE: Annular, Reverse Flow.

POWER (take-off) RATING: 1,485 shp (1,107 kw). Rating Approved to: 59°F (15°C).

TAKE-OFF SPECIFIC FUEL CONSUMPTION (SFC): 0.60 lb/hr/shp (0,365 kG/hr/kW).

BASIC DRY WEIGHT: 545 lbs (247 kg).

POWER/WEIGHT RATIO: 2.72:1 shp/lb (4,48:1 kw/kg).

APPLICATION:
Fuji AH-1S "Huey-Cobra" Helicopter. (Bell AH-1 License, USA)

OTHER ENGINES IN SERIES:
T53-K-13B in Fuji-Bell UH-1H "Iroquois" and Fuji-Bel 204B Civil Helicopter.
Locate other T-53 Models under Textron-Lycoming (USA) listing.

GAS TURBINE ENGINES FOR AIRCRAFT

 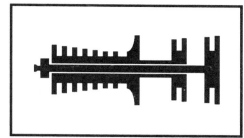

INTERNATIONAL TURBINE ENGINE COMPANIES

COUNTRY: JAPAN

MANUFACTURER: KAWASAKI HEAVY INDUSTRIES LTD (KHI)
KOBE, JAPAN

DESIGNATION (military): T55-K-712.

BACKGROUND: Manufactured under License, Textron-Lycoming (USA).
Date First T55-K Engine Produced: 1987.
Date If Out-of-Production: N/A. Number of T55-K-712 Engines Produced to Date: 60.

ENGINE TYPE: Turboshaft. Dual-Shaft (Compressor Drive & Power Output Drive).

COMPRESSOR TYPE: Combination Axial-Centrifugal Compressor: 7-Stage Axial, 1-Stage Centrifugal.

COMPRESSOR DATA: Compressor Pressure Ratio: 8.2 : 1. Mass Airflow: 29.3 lb/sec (13,3 kg/sec).

TURBINE TYPE: 4-Stage Axial Flow including: 2-Stage Gas Generator (compressor drive) Turbine, 2-Stage Power (free) Turbine.

COMBUSTOR TYPE: Annular, Reverse Flow.

POWER (take-off) RATING: 4,378 shp (3,265 kw). Rating Approved to: 59°F (15°C).

TAKE-OFF SPECIFIC FUEL CONSUMPTION (SFC): 0.52 lb/hr/shp (0,32 kg/hr/kw).

BASIC DRY WEIGHT: 750 lbs (340 kg).

POWER/WEIGHT RATIO: 5.84 : 1 shp/lb (9,60 :1 kw/kg).

APPLICATION:
Kawasaki CH-47C "Chinook" Helicopter. (Bell CH-47 License, USA)

OTHER ENGINES IN SERIES:
Locate other T-53 Models under Textron-Lycoming (USA) listing.

GAS TURBINE ENGINES FOR AIRCRAFT

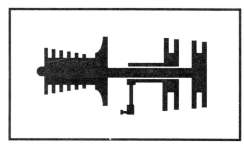

INTERNATIONAL TURBINE ENGINE COMPANIES

COUNTRY: JAPAN

MANUFACTURER: MITSUBISHI HEAVY INDUSTRIES LTD (MHI)
TOKYO, JAPAN

DESIGNATION (military)**:** CT63-M-5A.

BACKGROUND: Manufactured under T63 License. Military version of Allison-250 (USA).
Date First CT63-M-5A Engine Produced: 1967.
Date If Out-of-Production: N/A. Number of CT63-M-5A Engines Produced to Date: Unknown.

ENGINE TYPE: Turboshaft. Dual-Shaft (Compressor Drive & Power Output Drive).

COMPRESSOR TYPE: Combination Compressor: 6-Stage Axial, 1-Stage Centrifugal.

COMPRESSOR DATA: Compressor Pressure Ratio: 6.2 :1. Mass Airflow: 3.6 lb/sec (1,63 kg/sec).

TURBINE TYPE: 4-Stage Axial Flow Turbine including: 2-Stage Gas Producer (compressor drive) Turbine, 2-Stage Power (free) Turbine.

COMBUSTOR TYPE: Can-Type, Through-Flow with a single liner.

POWER (take-off) RATING: 317 shp (236 kw). Rating Approved to: 59°F (15°C).

TAKE-OFF SPECIFIC FUEL CONSUMPTION (SFC): 0.65 lb/hr/shp, (0,395 kg/hr/kw).

BASIC DRY WEIGHT: 139 lbs (63 kg).

POWER/WEIGHT RATIO: 2.28 :1 shp/lb (3,75 :1 kw/kg).

APPLICATION:
Kawasaki HK-500 Helicopter. (MDHC MD-500 License, USA)

OTHER ENGINES IN SERIES:
Locate other CT-63 and T-63 Models under Allison, General Motors (USA).

GAS TURBINE ENGINES FOR AIRCRAFT

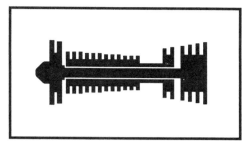

INTERNATIONAL TURBINE ENGINE COMPANIES

COUNTRY: JAPAN

MANUFACTURER: NATIONAL AEROSPACE LABORATORY (NAL)
TOKYO, JAPAN

DESIGNATION (military)**:** FJR-710-600S.

BACKGROUND: No Civil Counterpart.
Date First FJR-710-600S Engine Produced: 1976.
Date If Out-of-Production: N/A. Number of FJR-710-600S Engines Produced to Date: Unknown.

ENGINE TYPE: Turbofan. Dual-Shaft, High-Bypass Front Fan.

COMPRESSOR TYPE: 14-Stage Dual-Spool, Axial Flow Compressor including: 2-Stage Front Fan with no additional Axial Stages in the Low Pressure Compressor, 12-Stage High Pressure Compressor.

COMPRESSOR DATA: Compressor Pressure Ratio: 22.0 : 1, Fan Pressure Ratio: 1 : 1, Fan Bypass Ratio: 6.5 : 1, Total Mass Airflow: 320 lbs/sec (145 kg/sec).

TURBINE TYPE: 6-Stage Axial Flow Turbine including: 2-Stage High Pressure Turbine, 4-Stage Low Pressure Turbine.

COMBUSTOR TYPE: Annular, Through-Flow.

POWER (take-off) RATING: 14,330 lbt. (63,74 kN thrust). Rating Approved to: 59°F (15°C).

TAKE-OFF SPECIFIC FUEL CONSUMPTION (SFC): 0.34 lb/hr/lbt (9,6 mg/Ns).

BASIC DRY WEIGHT: 2,160 lbs (980 kg).

POWER/WEIGHT RATIO: 6.63 :1 lbt/lb.

APPLICATION:
Proposed Kawasaki and National Aerospace STOL Aircraft.

OTHER ENGINES IN SERIES:
FJR-710-10/20 Experimental Engine.

GAS TURBINE ENGINES FOR AIRCRAFT

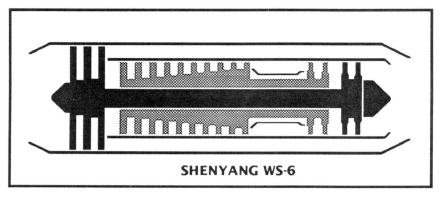

SHENYANG WS-6

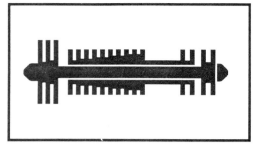

INTERNATIONAL TURBINE ENGINE COMPANIES

COUNTRY: PEOPLE'S REPUBLIC OF CHINA

GOVERNMENT CORP: CHINA NATIONAL AERO ENGINE CORP. (CAREC)

MANUFACTURER	DESIGNATION	TYPE	POWER	APPLICATION
CHENGDU	WP-7	TURBOJET	9,700 LBT	CHENGDU F-7, FIGHTER
CHENGDU	WZ-6	TURBOSHAFT	1,550 SHP	CHENGDU Z-8 (FRELON, SA-321), HELICOPTER‡
HARBIN	WJ-5A	TURBOPROP	3,100 SHP	HARBIN PS-5, SH-5; XAC Y-7, Y-14; TRANSPORTS
HARBIN	WZ-8 (ARIEL)‡	TURBOSHAFT	700 SHP	CHENGDU Z-6, Z-9 (DAUPHIN SA-365); HELICOPTERS‡
LIYANG (WOPEN)	WP-7B	TURBOJET	13,450 LBT	CHENGDU FT-7, J-7 (MIG-21); FIGHTERS
LIYANG (WOPEN)	WP-13A	TURBOJET	14,800 LBT	SAC F8-11, J-8, JH-7; FIGHTERS
SHENYANG (SAC)	WP-6	TURBOJET	7,165 LBT	SAC F-6, J-6; FIGHTERS
SHENYANG (SAC)	WP-6A	TURBOJET	8,260 LBT	NAF A-5, Q-5; ATTACK-FIGHTERS
SHENYANG (SAC)	WP-6Z	TURBOJET	8,900 LBT	NAF J-12; ATTACK-FIGHTER
SHENYANG (SAC)	WS-6/6A	TURBOJET	27,000 LBT	XAC B-7, ATTACK-FIGHTER
XIAN (XAC)	RB-199*	TURBOFAN	17,000 LBT	F-7M "AIRGUARD", FIGHTER
XIAN (XAC)	WP-8	TURBOJET	21,000 LBT	XAC B-6, H-6, (TUMANSKY TU-16); BOMBERS
XIAN (XAC)	WS-9 (SPEY 202)*	TURBOFAN	20,500 LBT	XAC (HONGZHA) B-7, H-7 (SU-16); ATTACK-FIGHTER
ZHUZHOU/SAC	WJ-6	TURBOPROP	4,250 ESHP	SHAANXI Y-8, (ANTONOV AN-12) TRANSPORT

* BRITISH LICENSE   ‡ FRENCH LICENSE

GAS TURBINE ENGINES FOR AIRCRAFT

 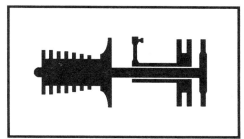

INTERNATIONAL TURBINE ENGINE COMPANIES

COUNTRY: POLAND

MANUFACTURER: PANSTWOWE ZAKLADY LOTNICZE (PZL) RZESZOW

DESIGNATION (military & civil): Isotov GTD-350P.

BACKGROUND: Date First GTD-350P Engine Produced: 1966.
Date If Out-of-Production N/A. Number of GTD-350P Engines Produced to Date: 18,900.

ENGINE TYPE: Turboshaft. Dual-Shaft (Compressor Drive & Power Output Drive).

COMPRESSOR TYPE: Combination Compressor: 7-Stage Axial, 1-Stage Centrifugal.

COMPRESSOR DATA: Compressor Pressure Ratio: 6.05 :1. Mass Airflow: 4.83 lb/sec (2,19) kg/sec).

TURBINE TYPE: 3-Stage Axial Flow Turbine including: 1-Stage Gas Producer (compressor drive) Turbine, 2-Stage Power (free) Turbine.

COMBUSTOR TYPE: Annular, Reverse Flow.

POWER (take-off) RATING: 400 shp (298 kw). Rating Approved to: 59°F (15°C).

TAKE-OFF SPECIFIC FUEL CONSUMPTION (SFC): 0.80 lb/hr/shp (0,486 kg/hr/kW).

BASIC DRY WEIGHT: 307 lbs (139,3 kg).

POWER/WEIGHT RATIO: 1.3 : 1 shp/lb (2,14 : 1 kw/kg).

APPLICATION:
Mil Mi-2 "Hoplite", Twin-Engine Multi-Role Utility Helicopter.

OTHER ENGINES IN SERIES:
Locate Other Mil Mi series Helicopters under USSR.

GAS TURBINE ENGINES FOR AIRCRAFT

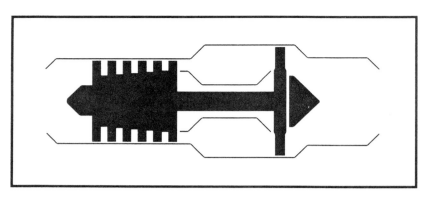 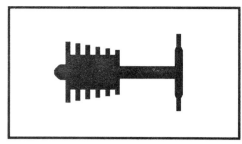

INTERNATIONAL TURBINE ENGINE COMPANIES

COUNTRY: POLAND

MANUFACTURER: PANSTWOWE ZAKLADY LOTNICZE (PZL) RZESZOW

DESIGNATION (military)**:** K-15.

BACKGROUND: Date First K-15 Engine Produced: 1986. Date If Out-of-Production N/A. Number of K-15 Engines Produced to Date: Unknown.

ENGINE TYPE: Turbojet. Single-Shaft (non-afterburning).

COMPRESSOR TYPE: Axial Flow, 6-Stage.

COMPRESSOR DATA: Compressor Pressure Ratio: 5.3 : 1, Mass Airflow: 50.7 lb/sec (23 kg/sec).

TURBINE TYPE: 1-Stage Axial Flow Turbine.

COMBUSTOR TYPE: Annular, Through Flow.

POWER (take-off) RATING: 3310 lbt (14,7 kN thrust), Rating Approved to: 59°F (15°C).

TAKE-OFF SPECIFIC FUEL CONSUMPTION (SFC): 1.02 lb/hr/lbt (28,9 mg/Ns).

BASIC DRY WEIGHT: 722 lbs (327 kg).

POWER/WEIGHT RATIO: 4.58 : 1 lbt/lb.

APPLICATION:
WSK PZL "Iryda" and "Iskra" Trainers.

OTHER ENGINES IN SERIES:
None.

GAS TURBINE ENGINES FOR AIRCRAFT

 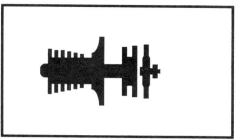

INTERNATIONAL TURBINE ENGINE COMPANIES

COUNTRY: POLAND

MANUFACTURER: PANSTWOWE ZAKLADY LOTNICZE (PZL)
RZESZOW, POLAND

DESIGNATION (civil): PZL-10W.

BACKGROUND: Date First PZL-10W Engine Produced: 1991 (twinned version of Glushenkov TWD-10B (USSR).

ENGINE TYPE: Turboshaft. Dual-Shaft.

COMPRESSOR TYPE: Combination Compressor: 6-Stage Axial Flow, 1-Stage Centrifugal Flow.

COMPRESSOR DATA: Compressor Pressure Ratio: 7.4 : 1, Mass Airflow: 10.14 lb/sec (4,6 kg/sec).

TURBINE TYPE: 3-Stage Axial Flow Turbine including: 2-Stage Gas Producer (compressor drive) Turbine, 1-Stage Power (free) Turbine.

COMBUSTOR TYPE: Annular, Through-Flow.

POWER (take-off) RATING: 890 eshp (664 ekw). Rating Approved to: 59°F (15°C).

TAKE-OFF SPECIFIC FUEL CONSUMPTION (SFC): 0.56 lb/hr/shp (0,34 Kg/hr/kw).

BASIC DRY WEIGHT: 287 lbs (130 kg).

POWER/WEIGHT RATIO: 3 : 1 eshp/lb (5,0 : 1 kw /kg).

APPLICATION:
WSK-PZL W-3 Helicopter.

OTHER ENGINES IN SERIES:
TVD-10 single turboshaft (USSR).
TVD-10B turboprop (USSR)

GAS TURBINE ENGINES FOR AIRCRAFT

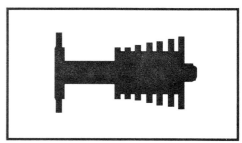

INTERNATIONAL TURBINE ENGINE COMPANIES

COUNTRY: POLAND

MANUFACTURER: PANSTWOWE ZAKLADY LOTNICZE (PZL) RZESZOW

DESIGNATION (military): SO-1/2.

BACKGROUND: Date First SO-1/2 Engine Produced: 1966. Date If Out-of-Production 1978. Number of SO-1/2 Engines Produced to Date: 19,181.

ENGINE TYPE: Turbojet. Single-Shaft (non-afterburning).

COMPRESSOR TYPE: Axial Flow, 7-Stage.

COMPRESSOR DATA: Compressor Pressure Ratio: 4.69 : 1, Total Mass Airflow: 39.24 lb/sec (17,8 kg/sec).

TURBINE TYPE: 1-Stage Axial Flow Turbine.

COMBUSTOR TYPE: Annular, Through Flow.

POWER (take-off) RATING: 2,204 lbt (9,81 kN thrust), Rating Approved to: 59°F (15°C).

TAKE-OFF SPECIFIC FUEL CONSUMPTION (SFC): 1.04 lb/hr/lbt (29,6 mg/Ns).

BASIC DRY WEIGHT: 743 lbs (337 kg).

POWER/WEIGHT RATIO: 2.96 : 1 lbt/lb.

APPLICATION:
WSK-PZL TS-11 Trainer Aircraft.
"Iskra" (Spark) Trainer Aircraft.

OTHER ENGINES IN SERIES:
None.

GAS TURBINE ENGINES FOR AIRCRAFT

 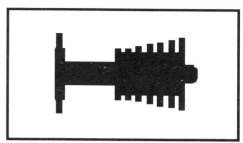

INTERNATIONAL TURBINE ENGINE COMPANIES

COUNTRY: POLAND

MANUFACTURER: PANSTWOWE ZAKLADY LOTNICZE (PZL)
RZESZOW, POLAND

DESIGNATION (military): SO-3W (PZL-5).

BACKGROUND: Upgraded SO-1.
Date First SO-3 Engine Produced: 1970
Date If Out-of-Production: N/A. Number of SO-3W Engines Produced to Date: Unknown.

ENGINE TYPE: Turbojet. Single-Shaft (non-afterburning).

COMPRESSOR TYPE: Axial Flow, 7-Stage.

COMPRESSOR DATA: Compressor Pressure Ratio: 4.69 : 1, Total Mass Airflow: 39.24 lb/sec (17,8 kg/sec).

TURBINE TYPE: 1-Stage Axial Flow Turbine.

COMBUSTOR TYPE: Annular, Through Flow.

POWER (take-off) RATING: 2,225 lbt (9,9 kN thrust),   Rating Approved to: 59°F (15°C).

TAKE-OFF SPECIFIC FUEL CONSUMPTION (SFC): 1.06 lb/hr/lbt (30,2 mg/Ns).

BASIC DRY WEIGHT: 765 lbs (347 kg).

POWER/WEIGHT RATIO: 2.9 :1 lbt/lb.

APPLICATION:
WSK-PZL TS-11, Trainer Aircraft.
"Iskra" (Spark), Trainer Aircraft.

OTHER ENGINES IN SERIES:
SO-3W22 in PZL I-22, "Iryd" Trainer Aircraft.

GAS TURBINE ENGINES FOR AIRCRAFT

 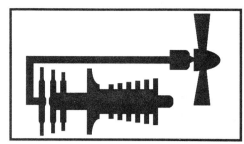

INTERNATIONAL TURBINE ENGINE COMPANIES

COUNTRY: POLAND

MANUFACTURER: PANSTWOWE ZAKLADY LOTNICZE (PZL)
RZESZOW, POLAND

DESIGNATION (civil & military): TWD-10B.

BACKGROUND: This engine was originally designed for the Kamov Ka-25 Helicopter; its military turboshaft designation was GTD-3.
Date If Out-of-Production: N/A. Number of TWD-10B Engines Produced to Date: Unknown.

ENGINE TYPE: Turboprop. Dual-Shaft.

COMPRESSOR TYPE: Combination Compressor: 6-Stage Axial Flow, 1-Stage Centrifugal Flow.

COMPRESSOR DATA: Compressor Pressure Ratio: 7.4 : 1, Mass Airflow: 10.14 lb/sec (4,6 kg/sec).

TURBINE TYPE: 2-Stage Axial Flow Turbine including: 1-Stage Gas Producer (compressor drive) Turbine, 1-Stage Power (free) Turbine.

COMBUSTOR TYPE: Annular, Through-Flow.

POWER (take-off) RATING: 990 eshp (738 ekw). Rating Approved to: 59°F (15°C).

TAKE-OFF SPECIFIC FUEL CONSUMPTION (SFC): 0.57 lb/hr/eshp (0,347 kg/hr/ekw).

BASIC DRY WEIGHT: 507 lbs (230 kg).

POWER/WEIGHT RATIO: 1.95 : 1 eshp/lb (3,2 : 1 ekw /kg).

APPLICATION:
Be-30 "Cuff" Light Transport.
Twin Engine Antonov An-28 "Cash" STOL Transport.
AN-3 Agricultural Biplane.
PZL-Mielec AN-28 (Poland).

OTHER ENGINES IN SERIES:
TVD-10B Turboprop (USSR).

GAS TURBINE ENGINES FOR AIRCRAFT

 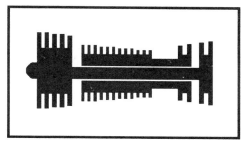

INTERNATIONAL TURBINE ENGINE COMPANIES

COUNTRY: RUMANIA

MANUFACTURER: CENTRAL NATIONAL AERONAUTICE
BUCHAREST, RUMANIA.

DESIGNATION (civil & military)**:** Spey Mk.512-14DW.

BACKGROUND: Built under Rolls-Royce License, Great Britain.

ENGINE TYPE: Turbofan. Dual-Shaft, High-Bypass Front Fan.

COMPRESSOR TYPE: 17-Stage Dual-Spool, Axial Flow Compressor including: 5-Stage Front Fan acting also as a Low Pressure Compressor, 12-Stage High Pressure Compressor.

COMPRESSOR DATA: Compressor Pressure Ratio: 21.0 : 1, Fan Pressure Ratio: 1.5 : 1, Fan Bypass Ratio: 0.71 : 1, Total Mass Airflow: 208 lb/sec (94,4 kg/sec).

TURBINE TYPE: 4-Stage Axial Flow Turbine including: 2-Stage High Pressure Turbine, 2-Stage Low Pressure Turbine.

COMBUSTOR TYPE: Can-Annular, Through-Flow With 10 Liners.

POWER (take-off) RATING: 12,550 lbt (55,8 kN thrust). Rating Approved to: 77°F (25°C).

TAKE-OFF SPECIFIC FUEL CONSUMPTION (SFC): 0.612 lb/hr/lbt (17,34 mg/Ns).
Cruise SFC: 0.82 lb/hr/lbt.

BASIC DRY WEIGHT: 2,609 lbs (1,183 kg).

POWER/WEIGHT RATIO: 4.8 : 1 lbt/lb.

APPLICATION:
Rombac One-Eleven Airliner.

OTHER ENGINES IN SERIES:
Locate other Spey models under Rolls-Royce, Great Britain listing.

GAS TURBINE ENGINES FOR AIRCRAFT

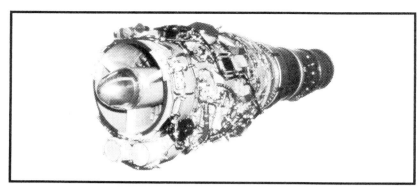 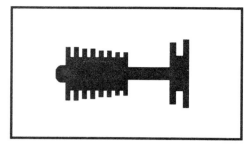

INTERNATIONAL TURBINE ENGINE COMPANIES

COUNTRY: RUMANIA

MANUFACTURER: CENTRAL NATIONAL AERONAUTICE
BUCHAREST, RUMANIA.

DESIGNATION (military): Viper 632.

BACKGROUND: Built under Rolls-Royce License, Great Britain.

ENGINE TYPE: Turbojet. Single-Shaft (non-afterburning).

COMPRESSOR TYPE: 8-Stage Single-Spool, Axial Flow Compressor.

COMPRESSOR DATA: Compressor Pressure Ratio: 5.6:1, Total Mass Airflow: 58.4 lb/sec (26,5 kg/sec).

TURBINE TYPE: 2-Stage Axial Flow.

COMBUSTOR TYPE: Through-Flow, Annular.

POWER (take-off) RATING: 3,750 lbt (16,68 kN thrust). Rating Approved to: 59°F (15°C).

TAKE-OFF SPECIFIC FUEL CONSUMPTION (SFC): 0.92 lb/hr/lbt (26 mg/Ns).

BASIC DRY WEIGHT: 831 lbs (377 kg).

POWER/WEIGHT RATIO: 4.51 : 1 lbt/lb.

APPLICATION:
Cinair IAR.93 "Yurom" Trainer Aircraft.
Cinair IAR.93B "Yurom" Afterburning Model Viper 632.

OTHER ENGINES IN SERIES:
Locate other Viper models under Rolls-Royce, Great Britain listing.

GAS TURBINE ENGINES FOR AIRCRAFT

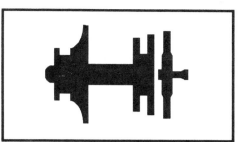

INTERNATIONAL TURBINE ENGINE COMPANIES

COUNTRY: RUMANIA

MANUFACTURER: CENTRAL NATIONAL AERONAUTICE
BUCHAREST, RUMANIA.

DESIGNATION (civil & military)**:** Turmo IVC.

BACKGROUND: Built under Turbomeca License, France.

ENGINE TYPE: Turboshaft. Dual-Shaft (Compressor Drive & Power Output Drive).

COMPRESSOR TYPE: Combination Compressor: 1-Stage Axial, 1-Stage Centrifugal.

COMPRESSOR DATA: Compressor Pressure Ratio: 5.9 :1. Mass Airflow: 13 lb/sec (5,9 kg/sec).

TURBINE TYPE: 3-Stage Axial Flow Turbine including: 2-Stage Gas Producer (compressor drive) Turbine, 1-Stage Power (free) Turbine.

COMBUSTOR TYPE: Can Type with one Combustor.

POWER (take-off) RATING: 1,495 shp (1,114 kw), Rating Approved to: 59°F (15°C).

TAKE-OFF SPECIFIC FUEL CONSUMPTION (SFC): 0.64 lb/hr/shp (0,39 kg/hr/kw).

BASIC DRY WEIGHT: 515 lbs (234 kg).

POWER/WEIGHT RATIO: 2.90 :1 shp/lb (4,76 :1 kw/kg).

APPLICATION:
Aerospatiale SA.321 "Super Frelon" Helicopter. (Rumanian Air Force)

OTHER ENGINES IN SERIES:
Locate other Turmo models under Turbomeca, France listing.

GAS TURBINE ENGINES FOR AIRCRAFT

 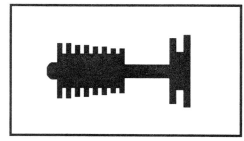

INTERNATIONAL TURBINE ENGINE COMPANIES

COUNTRY: SOUTH AFRICA

MANUFACTURER: ATLAS AIRCRAFT CORP.
TRANSVALL, SOUTH AFRICA

DESIGNATION (military): Viper 540.

BACKGROUND: Built under Rolls-Royce License, Great Britain.

ENGINE TYPE: Turbojet. Single-Shaft (non-afterburning).

COMPRESSOR TYPE: 8-Stage Single-Spool, Axial Flow Compressor.

COMPRESSOR DATA: Compressor Pressure Ratio: 5.6:1, Total Mass Airflow: 58.4 lb/sec (26,5 kg/sec).

TURBINE TYPE: 2-Stage Axial Flow.

COMBUSTOR TYPE: Through-Flow, Annular.

POWER (take-off) RATING: 3,750 lbt (16,68 kN thrust). Rating Approved to: 59°F (15°C).

TAKE-OFF SPECIFIC FUEL CONSUMPTION (SFC): 0.92 lb/hr/lbt (26 mg/Ns).

BASIC DRY WEIGHT: 831 lbs (377 kg).

POWER/WEIGHT RATIO: 4.51 : 1 lbt/lb.

APPLICATION:
Aermacchi MB.326 Trainer Aircraft. (South African Air Force)

OTHER ENGINES IN SERIES:
Locate other Viper models under Rolls-Royce, Great Britain listing.

GAS TURBINE ENGINES FOR AIRCRAFT

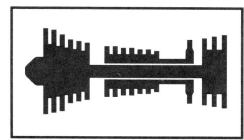

INTERNATIONAL TURBINE ENGINE COMPANIES

COUNTRY: SWEDEN

MANUFACTURER: VOLVO FLYGMOTOR AB
TROLLHAETTAN, SWEDEN

DESIGNATION (military): RM8B.

BACKGROUND: Manufactured under JT8D License, Pratt & Whitney Co., USA.
Date First RM8B Engine Produced: 1977.
Date If Out-of-Production 1988. Number of RM8 Engines Produced to Date: Unknown.

ENGINE TYPE: Turbofan. Dual-Shaft, Low-Bypass Front Fan with Afterburning.

COMPRESSOR TYPE: 13-Stage Dual-Spool, Axial Flow Compressor including: 3-Stage Front Fan with 3-additional Axial Stages in the Low Pressure Compressor, 7-Stage High Pressure Compressor.

COMPRESSOR DATA: Compressor Pressure Ratio: 16.5 : 1, Fan Pressure Ratio: 2 : 1, Fan Bypass Ratio: 1 : 1, Total Mass Airflow: 322 lb/sec (146 kg/sec).

TURBINE TYPE: 4-Stage Axial Flow Turbine including: 1-Stage High Pressure Turbine, 3-Stage Low Pressure Turbine.

COMBUSTOR TYPE: Can-Annular, Through-Flow with 9 liners.

AFTERBURNER: 3-Zones; 1-Zone in Fan Bypass Exhaust, 2-Zones in the Core Exhaust.

POWER (take-off) RATING: 16,200 lbt (72,1 kN thrust) [dry], 28,110 (125 kN thrust) [A/B].
Rating Approved to: 59°F (15°C).

TAKE-OFF SPECIFIC FUEL CONSUMPTION (SFC): 0.64 lb/hr/lbt (18,1 mg/Ns) [dry], 2.52 lb/hr/lbt (71,38 mg/Ns) [A/B].

BASIC DRY WEIGHT: 4,894 lbs (2,220 kg).

POWER/WEIGHT RATIO: 3.3 : 1 [dry], 5.7 : 1 [A/B] - lbt/lb.

APPLICATION:
Saab JA37 "Viggen" Attack/Fighter Aircraft.

OTHER ENGINES IN SERIES:
RM6 in JA35 "Draken" Attack/Fighter Aircraft.
RM8A in early JA37 Attack/Fighter and SK37 Reconnaissance Aircraft.
RM8A Military version of Pratt & Whitney, JT8D-22 Commercial Turbofan.
RM8A has a Swedish designed afterburner. In service in 1971.

GAS TURBINE ENGINES FOR AIRCRAFT

 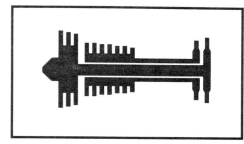

INTERNATIONAL TURBINE ENGINE COMPANIES

COUNTRY: SWEDEN

MANUFACTURER: VOLVO FLYGMOTOR AB
TROLLHAETTAN, SWEDEN

DESIGNATION (military): RM12.

BACKGROUND: Manufactured under F-404 License, General Electric Co., USA.
Date First RM12 Engine Produced: In Development.
Date If Out-of-Production N/A. Number of RM12 Engines Produced to Date: N/A.

ENGINE TYPE: Turbofan. Dual-Shaft, Low-Bypass Front Fan with Afterburning.

COMPRESSOR TYPE: 10-Stage Dual-Spool, Axial Flow Compressor including: 3-Stage Front in the Low Pressure Compressor, 7-Stage High Pressure Compressor.

COMPRESSOR DATA: Compressor Pressure Ratio: 27 : 1, Fan Pressure Ratio: 4.2 : 1, Fan Bypass Ratio: 0.28 : 1, Total Mass Airflow: 150 lb/sec (68 kg/sec).

TURBINE TYPE: 2-Stage Axial Flow Turbine including: 1-Stage High Pressure Turbine, 1-Stage Low Pressure Turbine.

COMBUSTOR TYPE: Annular, Through-Flow.

AFTERBURNER: 1-Zone fully modulated.

POWER (take-off) RATING: 12,140 lbt (54 kN thrust) [dry], 18,100 (80,5 kN thrust) [A/B]. Rating Approved to: 59°F (15°C).

TAKE-OFF SPECIFIC FUEL CONSUMPTION (SFC): 0.84 lb/hr/lbt (23,9 mg/Ns) [dry], 1.78 lb/hr/lbt (50,6 mg/Ns) [A/B].

BASIC DRY WEIGHT: 2,315 lbs (1,050 kg).

POWER/WEIGHT RATIO: 5.24 : 1 [dry], 7.8 : 1 [A/B] - lbt/lb.

APPLICATION:
Saab JAS 39 "Gripen" Attack/Fighter Aircraft.

OTHER ENGINES IN SERIES:
RM12 Upgrade to 18,000 lbt proposed for IAI "Nammer" Fighter Aircraft, Israel.

GAS TURBINE ENGINES FOR AIRCRAFT

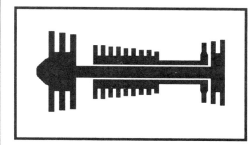

INTERNATIONAL TURBINE ENGINE COMPANIES

COUNTRY: TURKEY

MANUFACTURER: TUSAS ENGINE INDUSTRIES (TEI)
ESHISEHIR, TURKEY.

DESIGNATION (military): F110-GE-100.

BACKGROUND: A derivative of the F101 Turbofan Engine.
Date First F110-GE-100 Engine Produced: 1985.
Date If Out-of-Production: N/A. Number of F110 Engines Produced to Date: 1062.

ENGINE TYPE: Turbofan. Dual-Shaft, Low-Bypass Front Fan with Afterburning.

COMPRESSOR TYPE: 12-Stage Dual-Spool, Axial Flow Compressor including: 3-Stage Front Fan acting as the Low Pressure Compressor, 9-Stage High Pressure Compressor.

COMPRESSOR DATA: Compressor Pressure Ratio: 31.2 : 1, Fan Pressure Ratio: 3.3 : 1, Fan Bypass Ratio: 0.76 : 1, Total Mass Airflow: 270 lb/sec (122 kg/sec).

TURBINE TYPE: 3-Stage Axial Flow Turbine including: 1-Stage High Pressure Turbine, 2-Stage Low Pressure Turbine.

COMBUSTOR TYPE: Annular, Through-Flow.

AFTERBURNER: Fully modulating 3-stage augmentor including: 1-stage local (core), 1-stage core, and 1-stage fan airstream.

POWER (take-off) RATING: 16,760 lbt (74,5 kN thrust) [dry], 28,000 lbt (124,4 kN thrust) [A/B]. Rating Approved to: 59°F (15°C).

TAKE-OFF SPECIFIC FUEL CONSUMPTION (SFC): 0.675 lb/hr/lbt (19,1 mg/Ns), 2,063 lb/hr/lbt (A/B).

BASIC DRY WEIGHT: 3923 lbs (1779.5 kg).

POWER/WEIGHT RATIO: 4.3 : 1 [dry], 7.1 :1 lbt/lb (A/B).

APPLICATION:
General Dynamics F-16 C/D/N "Fighting Falcon". (Turkish Air Force)

OTHER ENGINES IN SERIES:
F100-GE-129 in F-16C/D.
F-100-GE-400 in F-14 A/D. See General Electric listing under USA manufacturers.
Locate other F100 Models under General Electric Aircraft Engines (USA) listing.

GAS TURBINE ENGINES FOR AIRCRAFT

 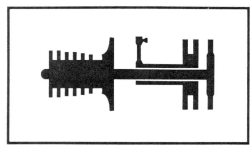

INTERNATIONAL TURBINE ENGINE COMPANIES

COUNTRY: USSR (COMMONWEALTH OF INDEPENDENT STATES)

MANUFACTURER: GLUSHENKOV ENGINE DESIGN BUREAU

DESIGNATION: GLUSHENKOV GTD-3BM.

BACKGROUND: This engine uses the core of the TVD-10 Turboprop.

ENGINE TYPE: Turboshaft. Dual-Shaft.

COMPRESSOR TYPE: Combination: Axial-Centrifugal Flow, including a 6-stage axial and a 1-stage centrifugal.

COMPRESSOR DATA: Compressor Pressure Ratio: 7.4 : 1, Mass Airflow: 10.34 lb/sec (4,6 kg/sec).

TURBINE TYPE: 3-Stage Axial Flow Turbine. 1-Stage Gas Generator (compressor drive) Turbine, 2-Stage Power Turbine.

COMBUSTOR TYPE: Annular, Reverse Flow.

POWER (take-off) RATING: 990 shp (738 kw). Rating Approved to: 59°F (15°C).

TAKE-OFF SPECIFIC FUEL CONSUMPTION (SFC): 0.53 lb/hr/shp (0,32 kg/hr/kw).

BASIC DRY WEIGHT: 503 lbs (228 kg).

POWER/WEIGHT RATIO: 1.97 : 1 shp/lb, (3,24 : 1 kw/kg).

APPLICATION:
Kamov Ka-25 "Hormone", Twin Engine Helicopter.

OTHER ENGINES IN SERIES:
GTD-3F in 900 SHP Early model Kamov Ka-25.
GTD-350 Polish Built, 397 shp (296 kw), in the Twin Engine, WSK-PZL Swidnik Mi-2 Helicopter.

GAS TURBINE ENGINES FOR AIRCRAFT

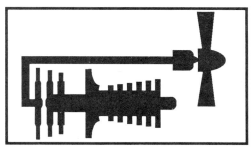

INTERNATIONAL TURBINE ENGINE COMPANIES

COUNTRY: USSR (COMMONWEALTH OF INDEPENDENT STATES)

MANUFACTURER: GLUSHENKOV ENGINE DESIGN BUREAU

DESIGNATION: GLUSHENKOV TVD-10B.

BACKGROUND: This engine was originally designed for the Kamov Ka-25 Helicopter, its military turboshaft designation was GTD-3.

ENGINE TYPE: Turboprop. Dual-Shaft.

COMPRESSOR TYPE: Combination Compressor: 6-Stage Axial Flow, 1-Stage Centrifugal Flow.

COMPRESSOR DATA: Compressor Pressure Ratio: 7.4 : 1, Mass Airflow: 10.14 lb/sec (4,6 kg/sec).

TURBINE TYPE: 3-Stage Axial Flow Turbine including: 2-Stage Gas Producer (compressor drive) Turbine, 1-Stage Power (free) Turbine.

COMBUSTOR TYPE: Annular, Through-Flow.

POWER (take-off) RATING: 990 eshp (738 ekw). Rating Approved to: 59°F (15°C).

TAKE-OFF SPECIFIC FUEL CONSUMPTION (SFC): 0.57 lb/hr/eshp (0,347 kg/hr/ekw).

BASIC DRY WEIGHT: 507 lbs (230 kg).

POWER/WEIGHT RATIO: 1.95 : 1 eshp/lb (3,2 : 1 ekw /kg).

APPLICATION:
Antonov An-28 "Cash" Airliner/Transport , developed from An-14 "Clod".
PZL-Mielec An-3 Agricultural Biplane, Poland.
Antonov An-28, Poland.

OTHER ENGINES IN SERIES:
TWD-10B Turboprop (Poland).

GAS TURBINE ENGINES FOR AIRCRAFT

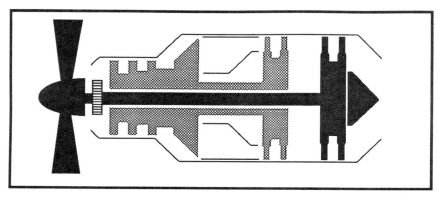 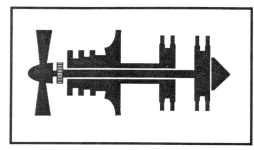

INTERNATIONAL TURBINE ENGINE COMPANIES

COUNTRY: USSR (COMMONWEALTH OF INDEPENDENT STATES)

MANUFACTURER: KOKIESOV ENGINE DESIGN BUREAU OF RYBINSK MOTORS. RYBINSK, RUSSIA

DESIGNATION: GLUSHENKOV TVD-1500.

ENGINE TYPE: Turboprop. Dual-Shaft.

COMPRESSOR TYPE: Combination Compressor: 3-Stage Axial Flow, 1 Stage Centrifugal Flow.

COMPRESSOR DATA: Compressor Pressure Ratio: 14.4 : 1, Mass Airflow: 10.1 lb/sec (4,58 kg/sec).

TURBINE TYPE: 4-Stage Axial Flow Turbine including: 2-Stage Gas Producer (compressor drive) Turbine, 2-Stage Power (free) Turbine.

COMBUSTOR TYPE: Annular, Reverse Flow.

POWER (take-off) RATING: 1,725 eshp (1,286 ekw). Rating Approved to: 59°F (15°C).

TAKE-OFF SPECIFIC FUEL CONSUMPTION (SFC): 0.443 lb/hr/eshp (0,269 kg/hr/ekw).

BASIC DRY WEIGHT: 542 lbs (246 kg).

POWER/WEIGHT RATIO: 3.18 :1 eshp/lb (5,2 : 1 ekw/kg).

APPLICATION:
Antonov An-38 Twin Turboprop.

OTHER ENGINES IN SERIES:
TVD-1500V Turboshaft in Ka-67 Helicopter
TVD-1500 Turbofan Proposed for Business Jets.

GAS TURBINE ENGINES FOR AIRCRAFT

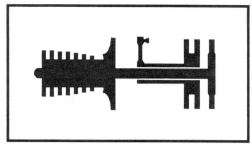

INTERNATIONAL TURBINE ENGINE COMPANIES

COUNTRY: USSR (COMMONWEALTH OF INDEPENDENT STATES)

MANUFACTURER: KLIMOV ENGINE DESIGN BUREAU (Formerly Isotov) of:
ST. PETERSBURG SCIENTIFIC & PRODUCTION ASSOCIATION

DESIGNATION: Isotov GTD-350.

ENGINE TYPE: Turboshaft. Dual-Shaft (Compressor Drive & Power Output Drive).

COMPRESSOR TYPE: Combination Compressor: 7-Stage Axial, 1-Stage Centrifugal.

COMPRESSOR DATA: Compressor Pressure Ratio: 6.0 :1. Mass Airflow: 4.8 lb/sec (2,19 kg/sec).

TURBINE TYPE: 3-Stage Axial Flow Turbine including: 1-Stage Gas Producer (compressor drive) Turbine, 2-Stage Power (free) Turbine.

COMBUSTOR TYPE: Annular, Reverse Flow.

POWER (take-off) RATING: 396 shp (295 kw). Rating Approved to: 59°F (15°C).

TAKE-OFF SPECIFIC FUEL CONSUMPTION (SFC): 0.80 lb/hr/shp (0,486 kg/hr/kw).

BASIC DRY WEIGHT: 297 lbs (135 kg).

POWER/WEIGHT RATIO: 1.34 :1 shp/lb (2,16 :1 kw/kg).

APPLICATION:
Mil Mi-2 "Hoplite", Twin-Engine Multi-Role Utility Helicopter.

OTHER ENGINES IN SERIES:
None.

GAS TURBINE ENGINES FOR AIRCRAFT

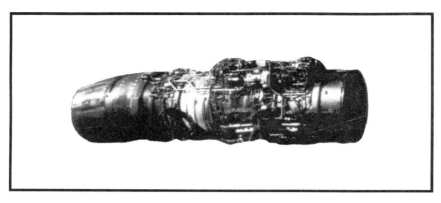
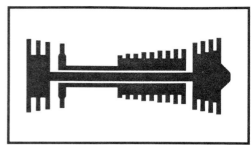

INTERNATIONAL TURBINE ENGINE COMPANIES

COUNTRY: USSR (COMMONWEALTH OF INDEPENDENT STATES)

MANUFACTURER: KLIMOV ENGINE DESIGN BUREAU (Formerly Isotov) of:
ST. PETERSBURG SCIENTIFIC & PRODUCTION ASSOCIATION

DESIGNATION: RD-33.

ENGINE TYPE: Turbofan. Dual-Shaft, with Afterburning.

COMPRESSOR TYPE: Axial Flow, 13-Stage Dual-Spool, including a 4-Stage Fan in the Low Pressure Compressor and 9-Stage High Pressure Compressor.

COMPRESSOR DATA: Compressor Pressure Ratio: 20 : 1, Mass Airflow: 170 lb/sec (77,1 kg/sec), Fan Pressure Ratio: 3.15 : 1, Bypass Ratio: 0.9 : 1.

TURBINE TYPE: 4-Stage Axial Flow Turbine including: 1-stage High Pressure Turbine, 3-stage Low Pressure Turbine.

COMBUSTOR TYPE: Annular, Through Flow.

AFTERBURNER: 4-Stage.

POWER (take-off) RATING: 11,000 lbt (48,9 kN thrust) [dry], 18,300 lbt (81,4 kN thrust) [A/B]. Rating Approved to: 59°F (15°C).

TAKE-OFF SPECIFIC FUEL CONSUMPTION (SFC): 0.77 lb/hr/lbt (21,8 mg/Ns) [dry], 2.05 lb/hr/lbt (58,07 mg/Ns) [A/B].

BASIC DRY WEIGHT: 2,300 lbs (1,043 kg).

POWER/WEIGHT RATIO: 4.78:1 [dry], 7.96 : 1 [A/B] - lbt/lb.

APPLICATION:
Mikoyan MIG-29 "Fulcrum", Twin Engine Fighter.

OTHER ENGINES IN SERIES:
None.

GAS TURBINE ENGINES FOR AIRCRAFT

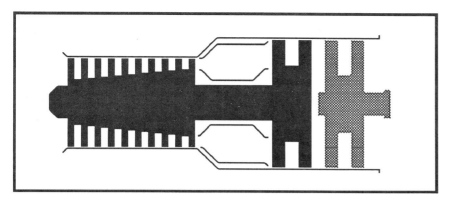 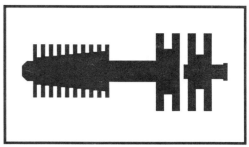

INTERNATIONAL TURBINE ENGINE COMPANIES

COUNTRY: USSR (COMMONWEALTH OF INDEPENDENT STATES)

MANUFACTURER: KLIMOV ENGINE DESIGN BUREAU (Formerly Isotov) of:
ST. PETERSBURG SCIENTIFIC & PRODUCTION ASSOCIATION

DESIGNATION: ISOTOV TV2-117A.

BACKGROUND: Date First TV2 Engine Produced: 1968.

ENGINE TYPE: Turboshaft. Dual-Shaft.

COMPRESSOR TYPE: Ten-Stage, Single Spool, Axial Flow.

COMPRESSOR DATA: Compressor Pressure Ratio: 6.6 : 1, Mass Airflow: 17.8 lb/sec (8,1 kg/sec).

TURBINE TYPE: 4-Stage Axial Flow Turbine including: 2-stage Gas Generator (compressor drive) Turbine, 2-stage Power (free) Turbine.

COMBUSTOR TYPE: Annular, Through-Flow.

POWER (take-off) RATING: 1,700 shp (1,268 kw). Rating Approved to: 59°F (15°C).

TAKE-OFF SPECIFIC FUEL CONSUMPTION (SFC): 0.6 lb/hr/shp (0,36 kg/hr/kw).

BASIC DRY WEIGHT: 730 lb (330 kg).

POWER/WEIGHT RATIO: 2.32 : 1 shp/lb (3,84 : 1 kw/kg).

APPLICATION:
Mil Mi-8T "Haze" Anti-Submarine Helicopter.
Mil Mi-14 and Mi-24 "Hind" Helicopters. (with twin-engines attached to one common gearbox).
Polish Built Mi-8, "Hip", (WSK-PZL, Swidnik) Utility Helicopter.

OTHER ENGINES IN SERIES:
TV2-117TG in military Mi-38 helicopter.

GAS TURBINE ENGINES FOR AIRCRAFT

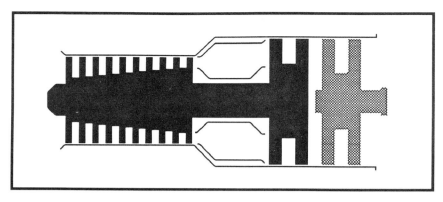
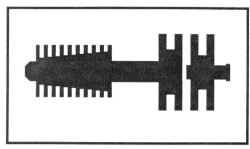

INTERNATIONAL TURBINE ENGINE COMPANIES

COUNTRY: USSR (COMMONWEALTH OF INDEPENDENT STATES)

MANUFACTURER: KLIMOV ENGINE DESIGN BUREAU (Formerly Isotov) of:
ST. PETERSBURG SCIENTIFIC & PRODUCTION ASSOCIATION

DESIGNATION: ISOTOV TV3-117V.

BACKGROUND: Upgrade Version of TV3 Turboshaft Engine.
Date First TV3 Engine Produced: 1978.

ENGINE TYPE: Turboshaft. Dual-Shaft.

COMPRESSOR TYPE: 10-Stage Single Spool Axial Flow Compressor.

COMPRESSOR DATA: Compressor Pressure Ratio: 7.5 : 1, Mass Airflow: 17.8 lb/sec (8,1 kg/sec).

TURBINE TYPE: 4-Stage Axial Flow Turbine including: 2-Stage Gas Generator (compressor drive) Turbine, 2-Stage Power (free) Turbine.

COMBUSTOR TYPE: Annular, Through-Flow.

POWER (take-off) RATING: 2,225 shp (1,660 kw). Rating Approved to: 59°F (15°C).

TAKE-OFF SPECIFIC FUEL CONSUMPTION (SFC): 0.925 lb/hr/shp (0,56 kg/hr/kw).

BASIC DRY WEIGHT: 1,080 lbs (490 kg).

POWER/WEIGHT RATIO: 2.06:1 shp/lb (3,39 : 1 kw/kg).

APPLICATION:
Military Kamov Ka-27 "Helix-A" (Twin Engine Helicopter).
Civil Ka-32 "Helix-B" (Twin Engine Helicopter).

OTHER ENGINES IN SERIES:
TV3-113 in Mil MI-18 Helicopter.
TV3-117 in the Twin-Engine Mil Mi-14 (V-14), Mil Mi-24 and Mi-28 Helicopters.
TV3-117MT in the Twin Engine Mil Mi-17 Helicopter.

GAS TURBINE ENGINES FOR AIRCRAFT

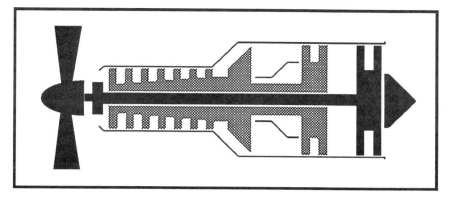 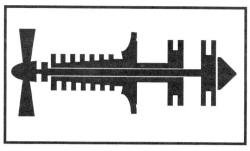

INTERNATIONAL TURBINE ENGINE COMPANIES

COUNTRY: USSR (COMMONWEALTH OF INDEPENDENT STATES)

MANUFACTURER: KLIMOV ENGINE DESIGN BUREAU (Formerly Isotov) of:
ST. PETERSBURG SCIENTIFIC & PRODUCTION ASSOC.

DESIGNATION: (military): TV7-117V.

ENGINE TYPE: Turboprop. Dual-Shaft (Compressor Drive & Power Output Drive).

COMPRESSOR TYPE: Combination Compressor: 7-Stage Axial, 1-Stage Centrifugal.

COMPRESSOR DATA: Compressor Pressure Ratio: 16 :1. Mass Airflow: 14 lb/sec (6,35) kg/sec).

TURBINE TYPE: 4-Stage Axial Flow Turbine including: 2-Stage Gas Producer (compressor drive) Turbine, 2-Stage Power (free) Turbine.

COMBUSTOR TYPE: Annular, Reverse Flow.

POWER (take-off) RATING: 2,368 shp (1,766 kw). Rating Approved to: 86°F (30°C).

TAKE-OFF SPECIFIC FUEL CONSUMPTION (SFC): 0.408 lb/hr/shp (0,248 kg/hr/kw).

BASIC DRY WEIGHT: 922 lbs (418 kg).

POWER/WEIGHT RATIO: 2.57 :1 shp/lb (4,2 : 1 kw/kg).

APPLICATION:
Ilyushin IL-114 Transport/Airliner.

OTHER ENGINES IN SERIES:
TV-117C Turboprop (2,500 SHP) in IL-114 Airliner.
TV-117TG Turboshaft version in MI-38 Helicopter.

GAS TURBINE ENGINES FOR AIRCRAFT

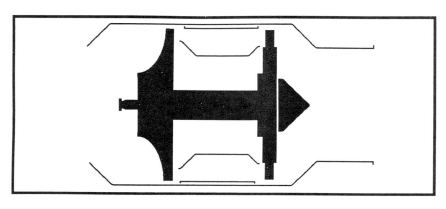
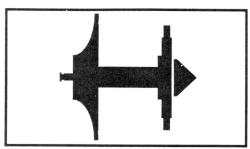

INTERNATIONAL TURBINE ENGINE COMPANIES

COUNTRY: USSR (COMMONWEALTH OF INDEPENDENT STATES)

MANUFACTURER: KOPTCHYENKO ENGINE DESIGN BUREAU
(presently part of Moscow Scientific Production Corp.)

DESIGNATION: ISOTOV TV-0-100.

ENGINE TYPE: Turboshaft. Single-Shaft.

COMPRESSOR TYPE: 1-Stage Centrifugal Flow Compressor.

COMPRESSOR DATA: Compressor Pressure Ratio: 10.2 : 1, Mass Airflow: 5.9 lb/sec (2,68 kg/sec).

TURBINE TYPE: 1-Stage Axial Flow Turbine.

COMBUSTOR TYPE: Annular, Reverse-Flow.

POWER (take-off) RATING: 720 shp (537 kw). Rating Approved to: 59°F (15°C).

TAKE-OFF SPECIFIC FUEL CONSUMPTION (SFC): 0.646 lb/hr/shp (0,39 kg/hr/kw).

BASIC DRY WEIGHT: 344 lbs (156 kg).

POWER/WEIGHT RATIO: 2.09 :1 shp/lb (3,44 : 1 kw/kg).

APPLICATION:
Military Kamov Ka-26 and KA-126 Helicopters.

OTHER ENGINES IN SERIES:
None.

GAS TURBINE ENGINES FOR AIRCRAFT

 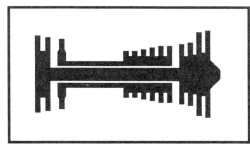

INTERNATIONAL TURBINE ENGINE COMPANIES

COUNTRY: USSR (COMMONWEALTH OF INDEPENDENT STATES)

MANUFACTURER: KUZNETSOV ENGINE DESIGN BUREAU OF:
SAMARA STATE SCIENTIFIC PRODUCTION ENTERPRISE

DESIGNATION: KUZNETSOV NK8-4.

BACKGROUND: Date First NK-8-4 Engine Produced: 1972.

ENGINE TYPE: Turbofan. Dual-Shaft, Low-Bypass Front Fan (non-afterburning).

COMPRESSOR TYPE: 10-Stage Dual-Spool, Axial Flow Compressor including: Two-Stage Front Fan with two additional Axial Stages in the Low Pressure Compressor, Six-Stage High Pressure Compressor.

COMPRESSOR DATA: Compressor Pressure Ratio: 23.2 :1, Fan Pressure Ratio: 2.15 : 1, Fan Bypass Ratio: 1.02 : 1, Total Mass Airflow: 260 lb/sec (118 kg/sec).

TURBINE TYPE: 3-Stage Axial Flow Turbine including: 1-Stage High Pressure Turbine, 2-Stage Low Pressure Turbine.

COMBUSTOR TYPE: Annular, Through-Flow.

POWER (take-off) RATING: 22,273 lbt (99 kN thrust). Rating Approved to: 59°F (15°C).

TAKE-OFF SPECIFIC FUEL CONSUMPTION (SFC): 0.78 lb/hr/lbt (22,1 mg/Ns).

BASIC DRY WEIGHT: 4,629 lbs (2100 kg).

POWER/WEIGHT RATIO: 4.81: 1 lbt/lb.

APPLICATION:
Early Ilyushin Il-62 "Classic" Military Transport and Airliner.
Tupolev Tu-154 "Careless" Tri-Jet Airliner.

OTHER ENGINES IN SERIES:
NK8-2 in Tupolev TU-154 "Careless" Transport and Airliner.
NK8-6 in Ilyushin IL-86 "Camber" Airliner.

GAS TURBINE ENGINES FOR AIRCRAFT

 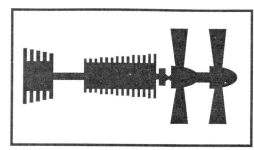

INTERNATIONAL TURBINE ENGINE COMPANIES

COUNTRY: USSR (COMMONWEALTH OF INDEPENDENT STATES)

MANUFACTURER: KUZNETSOV ENGINE DESIGN BUREAU OF:
SAMARA STATE SCIENTIFIC PRODUCTION ENTERPRISE

DESIGNATION: Kuznetsov NK-12MV.

ENGINE TYPE: Turboprop. Single-Shaft. Counter-Rotating Propellers.

COMPRESSOR TYPE: 14-Stage, Single-Spool, Axial Flow Compressor.

COMPRESSOR DATA: Compressor Pressure Ratio: 13 : 1, Total Mass Airflow: 143 lb/sec. (65 kg/sec).

TURBINE TYPE: 5-Stage Axial Flow Turbine.

COMBUSTOR TYPE: Can-Annular, Through-Flow, with 12 liners.

POWER (take-off) RATING: 14,795 eshp (11,033 ekw). Rating Approved to: 59°F (15°C).

TAKE-OFF SPECIFIC FUEL CONSUMPTION (SFC): 0.78 lb/hr/eshp (0,47 kg/hr/ekw).

BASIC DRY WEIGHT: 5,181 lbs (2,350 kg).

POWER/WEIGHT RATIO: 2.86 : 1 eshp/lb (4,69 :1 ekw/kg).

APPLICATION:
Tupolev Tu-95 "Bear" Bomber.
Tupolev Tu-114 "Cleat" Airliner.
Tu-126 "Moss" Airliner and AWACS Aircraft.
Tu-142 "Bear-F" (Four-Engine Reconnaissance and Early Warning Aircraft).
Super Seaplane Transport (wing in ground effect).

OTHER ENGINES IN SERIES:
Kuznetsov NK-12MA in Antonov An-22 "Cock" Military Transport and Airliner.
NK-12MV in Tu-114 "Cleat", Four-Engine Airliner.
NK-93 Ducted PropFan engine in development (45,000 LBT).
NK-104 Derated NK-93 at 24,500 lbt in development.
NK-112 Derated NK-93 at 18,700 lbt in development.
NK-114 Derated NK-93 at 30,380 lbt in development.

GAS TURBINE ENGINES FOR AIRCRAFT

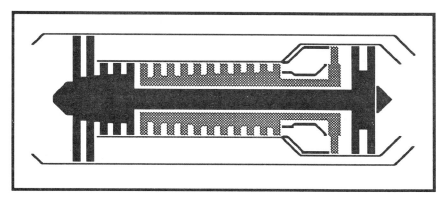 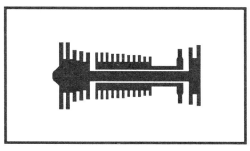

INTERNATIONAL TURBINE ENGINE COMPANIES

COUNTRY: USSR (COMMONWEALTH OF INDEPENDENT STATES)

MANUFACTURER: KUZNETSOV ENGINE DESIGN BUREAU OF:
SAMARA STATE SCIENTIFIC PRODUCTION ENTERPRISE

DESIGNATION: Kuznetsov NK-144.

BACKGROUND: Derived from the NK-8 Turbojet. The only Commercial Supersonic Gas Turbine Engine ever built other than the British/French equivalent engine in the Concorde SST.

ENGINE TYPE: Turbofan. Dual-Shaft, Low Bypass, with Afterburning.

COMPRESSOR TYPE: 16-Stage Dual-Spool, Axial Flow Compressor including: 2-Stage Fan followed by three booster stages in the Low Pressure Compressor, 11-Stage High Pressure Compressor.

COMPRESSOR DATA: Compressor Pressure Ratio: 15 : 1, Total Mass Airflow: 375 lb/sec (170 kg/sec), Fan Bypass Ratio: 1 : 1, Fan Pressure Ratio: 2:1.

TURBINE TYPE: 3-Stage Axial Flow Turbine, including: 1-Stage High Pressure Turbine, 2-Stage Low Pressure Turbine.

COMBUSTOR TYPE: Annular, Through Flow.

AFTERBURNER: 1-Stage.

POWER (take-off) RATING: 20,800 lbt (92,5 kN thrust) [dry], 28,775 lbt (128 kN thrust) [A/B]. Rating Approved to: 59°F (15°C).

TAKE-OFF SPECIFIC FUEL CONSUMPTION (SFC): 0.70 lb/hr/lbt (19,8 mg/Ns) [dry], 1.2 lb/hr/lbt (34 mg/Ns) [A/B].

BASIC DRY WEIGHT: 6,280 lbs (2,849 kg).

POWER/WEIGHT RATIO: 3.3 :1[dry], 4.58 : 1 [A/B] - lbt/lb.

APPLICATION:
Tupolev Tu-26 "Backfire" Bomber.
Tu-144D "Charger" SST Airliner. Afterburning required during cruise (now retired).

OTHER ENGINES IN SERIES:
None.

GAS TURBINE ENGINES FOR AIRCRAFT

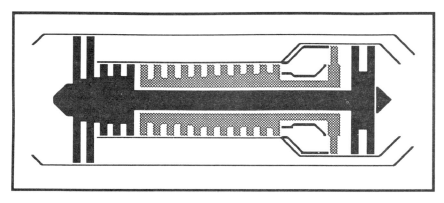 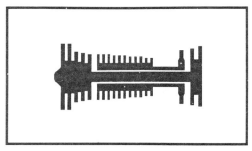

INTERNATIONAL TURBINE ENGINE COMPANIES

COUNTRY: USSR (COMMONWEALTH OF INDEPENDENT STATES)

MANUFACTURER: KUZNETSOV ENGINE DESIGN BUREAU OF:
SAMARA STATE SCIENTIFIC PRODUCTION ENTERPRISE

DESIGNATION: Kuznetsov NK-144 (upgraded).

ENGINE TYPE: Turbofan. Dual-Shaft, Low Bypass Front Fan with Afterburning.

COMPRESSOR TYPE: 16-Stage Dual-Spool, Axial Flow Compressor including: 2-Stage Front Fan with three additional Axial Stages in the Low Pressure Compressor, 11-Stage High Pressure Compressor.

COMPRESSOR DATA: Compressor Pressure Ratio: 15:1, Fan Pressure Ratio: 2:1, Fan Bypass Ratio: 1:1, Total Mass Airflow: 350 lb/sec (170 kg/sec).

TURBINE TYPE: 3-Stage Axial Flow Turbine including: 1-Stage High Pressure Turbine, 2-Stage Low Pressure Turbine.

COMBUSTOR TYPE: Annular, Through-Flow.

POWER (take-off) RATING: 20,500 lbt (91,1 kN thrust) [dry], 44,090 lbt (196 kN thrust) [A/B]. Rating Approved to: 59°F (15°C).

TAKE-OFF SPECIFIC FUEL CONSUMPTION (SFC): 0.71 lb/hr/lbt (20 mg/Ns) [dry], 1.2 lb/hr/lbt (34,5 mg/Ns) [A/B].

BASIC DRY WEIGHT: 7,600 lbs (3,447 kg).

POWER/WEIGHT RATIO: 2.69 : 1 [dry], 5.8 : 1 [A/B] - lbt/lb.

APPLICATION:
Tupolev Tu-26, "Backfire" (formerly Tu-22M), Bomber.
Tupolev Tu-160 "Blackjack" Bomber.

OTHER ENGINES IN SERIES:
NK-321, 55,000 lbt. in TU-160, Bomber.

GAS TURBINE ENGINES FOR AIRCRAFT

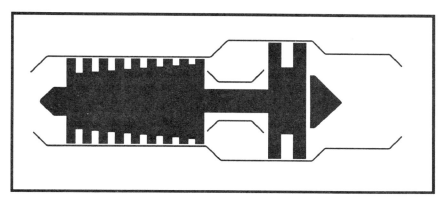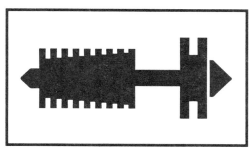

INTERNATIONAL TURBINE ENGINE COMPANIES

COUNTRY: USSR (COMMONWEALTH OF INDEPENDENT STATES)

MANUFACTURER: LYULKA-SATURN ENGINE DESIGN BUREAU

DESIGNATION: Lyulka AL-7F-1-100.

BACKGROUND: Date First AL-7F Engine Produced: 1954.

ENGINE TYPE: Turbojet. Single-Shaft, with Afterburning.

COMPRESSOR TYPE: Axial Flow, 9-Stage, Single Spool.

COMPRESSOR DATA: Compressor Pressure Ratio: 8 : 1, Mass Airflow: 283 lb/sec (128 kg/sec).

TURBINE TYPE: 2-Stage Axial Flow Turbine.

COMBUSTOR TYPE: Annular, Through Flow.

AFTERBURNER: 1-Stage.

POWER (take-off) RATING: 15,430 lbt (68,7 kg thrust) [dry], 22,046 lbt (98,1 kn thrust) [A/B]. Rating Approved to: 59°F (15°C).

TAKE-OFF SPECIFIC FUEL CONSUMPTION (SFC): 0.82 lb/hr/lbt (23,5 mg/Ns) [dry], 2.2 lb/hr/lbt (63 mg/Ns) [A/B].

BASIC DRY WEIGHT: 6,100 lbs (2767 kg).

POWER/WEIGHT RATIO: 2.53 : 1 [dry], 3.61 :1 [A/B] - lbt/lb.

APPLICATION:
Sukhoi Su-7 "Fritter", Single Engine Fighter.

OTHER ENGINES IN SERIES: AL-7F Upgrade in:
Sukhoi Su-7U "Moujik" Trainer Aircraft.
Sukhoi SU-9U "Maiden" Trainer Aircraft.
Sukhoi SU-9 "Fishpot" Fighter Aircraft.
Sukhoi SU-11 "Fishpot-B-C" Fighter Aircraft.
Sukhoi SU-15 "Flagon", Fighter Aircraft.

GAS TURBINE ENGINES FOR AIRCRAFT

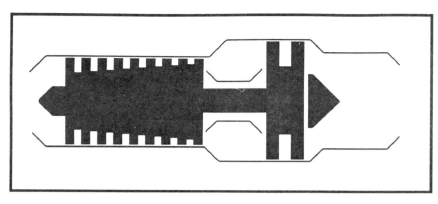 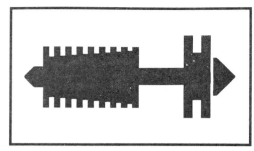

INTERNATIONAL TURBINE ENGINE COMPANIES

COUNTRY: USSR (COMMONWEALTH OF INDEPENDENT STATES)

MANUFACTURER: LYULKA-SATURN ENGINE DESIGN BUREAU

DESIGNATION: Lyulka AL-21F-3.

BACKGROUND: Based on the AL-7 Turbojet.
Date First AL-21F Engine Produced: 1965.

ENGINE TYPE: Turbojet. Single-Shaft, with Afterburning.

COMPRESSOR TYPE: Axial Flow, 9-Stage, Single-Spool.

COMPRESSOR DATA: Compressor Pressure Ratio: 8:1, Mass Airflow: 302 lb/sec (140 kg/sec).

TURBINE TYPE: Two-Stage Axial Flow Turbine.

COMBUSTOR TYPE: Annular, Through Flow.

AFTERBURNER: 1-Stage

POWER (take-off) RATING: 18,000 lbt (80,1 kN thrust) [dry], 24,700 lbt (110 kN thrust) [A/B]. Rating Approved to: 59°F (15°C).

TAKE-OFF SPECIFIC FUEL CONSUMPTION (SFC): 0.85 lb/hr/lbt (24 mg/Ns) [dry], 2.1 lb/hr/lbt, (59 mg/Ns) [A/B].

BASIC DRY WEIGHT: 6285 lbs (2850 kg).

POWER/WEIGHT RATIO: 2.86:1 [dry], 3.92:1 [A/B] - lbt/lb.

APPLICATION:
Sukhoi Su-15 "Flagon" Twin-Engine Fighter Aircraft.
Sukhoi Su-19 "Fencer" SU-24 Twin-Engine Fighter Aircraft.
Sukhoi SU-17, Su-20, Su-22 Single Engine "Fitter" Fighters.
Yakovlev Yak-36 Vtol Fighter Aircraft.

OTHER ENGINES IN SERIES:
Lyulka AL-21, Non-afterburning model with Vectored Thrust Exhaust Nozzle; in Yakovlev Yak-38.
AL-34, Turboprop in development.

GAS TURBINE ENGINES FOR AIRCRAFT

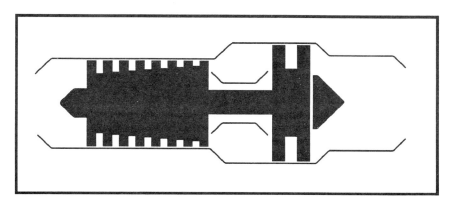 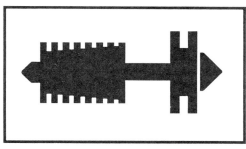

INTERNATIONAL TURBINE ENGINE COMPANIES

COUNTRY: USSR (COMMONWEALTH OF INDEPENDENT STATES)

MANUFACTURER: MIKULIN ENGINE DESIGN BUREAU
(presently part of Moscow Scientific Production Corp.)

DESIGNATION: Milulkin RD-3M (AM-3M).

BACKGROUND: Date First RD-3M Engine Produced: 1952.

ENGINE TYPE: Turbojet. Single-Shaft (non-afterburning).

COMPRESSOR TYPE: 8-Stage Single-Spool, Axial Flow Compressor.

COMPRESSOR DATA: Compressor Pressure Ratio: 6.4 : 1. Mass Airflow: 320 lb/sec (145 kg/sec).

TURBINE TYPE: 2-Stage Axial Flow Turbine.

COMBUSTOR TYPE: Annular, Through-Flow.

POWER (take-off) RATING: 20,500 lbt (91,2 kN thrust), Rating Approved to: 59°F (15°C).

TAKE-OFF SPECIFIC FUEL CONSUMPTION (SFC): 0.85 lb/hr/lbt (24 mg/Ns).

BASIC DRY WEIGHT: 5,280 lbs (2,395 kg).

POWER/WEIGHT RATIO: 3.97 : 1 lbt/lb.

APPLICATION:
Myasishchev M-4 "Bison-A" (Four Engine Bomber, Reconnaissance, ECM and Maritime).
Tupolev Tu-16 "Badger" Twin Engine Bomber.
Tupolev TU-104 "Camel" Twin-Engine Bomber.

OTHER ENGINES IN SERIES:
Mikulin AM-3D in Myasishchev M-4 "Bison".

GAS TURBINE ENGINES FOR AIRCRAFT

 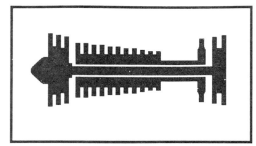

INTERNATIONAL TURBINE ENGINE COMPANIES

COUNTRY: USSR (COMMONWEALTH OF INDEPENDENT STATES)

MANUFACTURER: PERM SCIENTIFIC INDUSTRIAL ASSOCIATION (formerly Soloviev)

DESIGNATION: Soloviev D-20P.

BACKGROUND: Date First D-20P Engine Produced: 1957. PERM Bureau formerly known as the Soloviev Design Bureau.

ENGINE TYPE: Turbofan. Dual-Shaft. Low By-pass Front Fan (non-afterburning).

COMPRESSOR TYPE: 14-Stage Dual Spool Axial Flow, with a 3-Stage Front Fan acting as the Low Pressure Compressor, and an 11-Stage High Pressure Compressor.

COMPRESSOR DATA: Compressor Pressure Ratio: 13.6 :1, Mass Airflow: 249 lb/sec (113 kg/sec). Fan Pressure Ratio: 2.6 : 1 , Fan Bypass Ratio: 1.0 : 1.

TURBINE TYPE: 3-Stage Axial Flow Turbine with a 1-Stage High Pressure Compressor and a 2-Stage Low Pressure Compressor.

COMBUSTOR TYPE: Can-Annular, Through-Flow, with 12 Liners.

POWER (take-off) RATING: 11,900 lbt (52,9 kg thrust). Rating Approved to: 59°F (15°C).

TAKE-OFF SPECIFIC FUEL CONSUMPTION (SFC): 0.88 lb/hr/lbt (24,9 mg/Ns).

BASIC DRY WEIGHT: 3,190 lbs (1,450 kg).

POWER/WEIGHT RATIO: 3.73 :1 lbt/lb.

APPLICATION:
Myasishchev M-4 "Bison-B-C" (Heavy Bomber, Reconnaissance and ECM).
Tupelov TU-104 Airliner.
Tupolev TU-124 Airliner.

OTHER ENGINES IN SERIES:
None.

GAS TURBINE ENGINES FOR AIRCRAFT

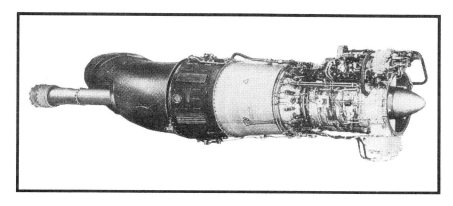 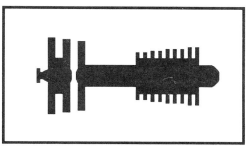

INTERNATIONAL TURBINE ENGINE COMPANIES

COUNTRY: USSR (COMMONWEALTH OF INDEPENDENT STATES)

MANUFACTURER: PERM SCIENTIFIC INDUSTRIAL ASSOCIATION (formerly Soloviev)

DESIGNATION: Soloviev D-25V.

BACKGROUND: Date First D-25 Engine Produced: 1957. Date Out of Production: 1978. The D-20P Turbofan engine's gas generator is utilized as the core of the D-25. Perm Bureau formerly known as the Soloviev Design Bureau.

ENGINE TYPE: Turboshaft. Dual-Shaft.

COMPRESSOR TYPE: 9-Stage Axial-Flow.

COMPRESSOR DATA: Compressor Pressure Ratio: 5.6 : 1, Mass Airflow: 58 lb/sec (26,5 kg/sec).

TURBINE TYPE: Three-Stage Axial Flow Turbine includes: Single-stage Gas Generator (compressor drive) Turbine, Two-stage Power (free) Turbine.

COMBUSTOR TYPE: Can-Annular, Through-Flow with 12 combustor liners.

POWER (take-off) RATING: 5,500 shp (4,100 kw). Rating Approved to: 59°F (15°C).

TAKE-OFF SPECIFIC FUEL CONSUMPTION (SFC): 0.64 lb/hr/shp (0,39 kg/hr/kw).

BASIC DRY WEIGHT: 2,620 lbs (1,188 kg).

POWER/WEIGHT RATIO: 2.09 :1 shp/lb (3,5 : 1 kw/kg).

APPLICATION:
Mil MI-6 "Hook" Heavy Transport Helicopter.
Mil MI-10 "Harke" Heavy-Lift Crane-Type Helicopter.
Both Models are Twin Engines Attached to a Common Gearbox.

OTHER ENGINES IN SERIES:
D-25VF (6,500 shp Upgrade Model) in the Mil MI-12 "Homer", Heavy-Lift Helicopter.

GAS TURBINE ENGINES FOR AIRCRAFT

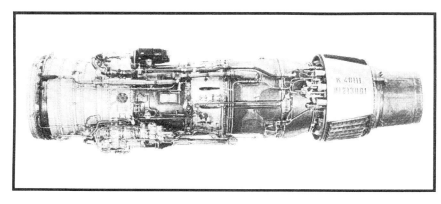 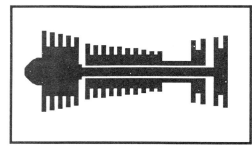

INTERNATIONAL TURBINE ENGINE COMPANIES

COUNTRY: USSR (COMMONWEALTH OF INDEPENDENT STATES)

MANUFACTURER: PERM SCIENTIFIC INDUSTRIAL ASSOCIATION (formerly Soloviev)

DESIGNATION: Soloviev D-30.

BACKGROUND: Date First D-30 Engine Produced: 1968. This was the first turbofan ever to reach a 1 : 1 Bypass Ratio. PERM Bureau formerly known as Soloviev Design Bureau.

ENGINE TYPE: Turbofan. Dual-Shaft, Low-Bypass Front Fan (non-afterburning).

COMPRESSOR TYPE: 15-Stage Dual-Spool, Axial Flow Compressor including: 5-Stage Front Fan in the Low Pressure Compressor, 10-Stage High Pressure Compressor.

COMPRESSOR DATA: Compressor Pressure Ratio: 17.4 : 1, Fan Pressure Ratio: 2.6 : 1, Fan Bypass Ratio: 1 : 1, Total Mass Airflow: 265 lb/sec (125 kg/sec).

TURBINE TYPE: 4-Stage Axial Flow Turbine including: 2-Stage High Pressure Turbine, 2-Stage Low Pressure Turbine.

COMBUSTOR TYPE: Can-Annular, Through-Flow with 12 combustor liners.

POWER (take-off) RATING: 15,000 lbt (66,7 kN thrust). Rating Approved to: 59°F (15°C).

TAKE-OFF SPECIFIC FUEL CONSUMPTION (SFC): 0.78 lb/hr/lbt (19,8 mg/Ns).

BASIC DRY WEIGHT: 3,417 lbs (1,550 kg).

POWER/WEIGHT RATIO: 4.39 : 1 lbt/lb.

APPLICATION:
Tumansky Tu-134A "Crusty" Twin-Engine Airliner.
Ilyushin IL-76 "Candid" Four-Engine Airliner/Transport/Tanker.

OTHER ENGINES IN SERIES:
D-30-KP in IL-76 "Candid" Airliner and Military Transport.
D-30-KU in Ilyushin IL-62M "Classic" Airliner and Tupelov TU-154M "Careless" Airliner.

GAS TURBINE ENGINES FOR AIRCRAFT

 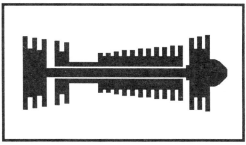

INTERNATIONAL TURBINE ENGINE COMPANIES

COUNTRY: USSR (COMMONWEALTH OF INDEPENDENT STATES)

MANUFACTURER: PERM SCIENTIFIC INDUSTRIAL ASSOCIATION (formerly Soloviev)

DESIGNATION: Soloviev D-30-KP.

BACKGROUND: Date First D-30 Engine Produced: 1971.
PERM Bureau formerly known as the Soloviev Design Bureau.

ENGINE TYPE: Turbofan. Dual-Shaft, Medium-Bypass Front Fan (non-afterburning).

COMPRESSOR TYPE: 14-Stage Dual-Spool, Axial Flow Compressor including a 3-Stage Front Fan in the form of a Low Pressure Compressor and an 11-Stage High Pressure Compressor.

COMPRESSOR DATA: Compressor Pressure Ratio: 21.0 : 1, Fan Pressure Ratio: 2.4 : 1, Fan Bypass Ratio: 2.41 : 1, Total Mass Airflow: 624 lb/sec (283 kg/sec).

TURBINE TYPE: 5-Stage Axial Flow Turbine including: 2-Stage High Pressure Turbine, 3-Stage Low Pressure Turbine.

COMBUSTOR TYPE: Can-Annular, Through-Flow with 12 combustion liners.

POWER (take-off) RATING: 26,455 lbt (117,7 kN thrust), Rating Approved to: 59°F (15°C).

TAKE-OFF SPECIFIC FUEL CONSUMPTION (SFC): 0.50 lb/hr/lbt (14,16 mg/Ns).

BASIC DRY WEIGHT: 5,070 lbs (2,300 kg).

POWER/WEIGHT RATIO: 5.22 : 1 lbt/lb.

APPLICATION:
Ilyushin IL-76 " Candid" Transport with mission similar to Lockheed, USAF C-141 Transport.
Ilyushin IL-76 "Mainstay" Reconnaissance Aircraft.
Ilyushin IL-78 "Midas" Tanker.

OTHER ENGINES IN SERIES:
D-30-KU in the IL-62M "Classic" and TU-154 "Careless", Airliners.
D-30F6 Afterburning Model in MIG-31 Fighter.

GAS TURBINE ENGINES FOR AIRCRAFT

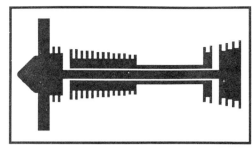

INTERNATIONAL TURBINE ENGINE COMPANIES

COUNTRY: USSR (COMMONWEALTH OF INDEPENDENT STATES)

MANUFACTURER: PERM SCIENTIFIC INDUSTRIAL ASSOCIATION (formerly Soloviev)

DESIGNATION: PS-90A.

BACKGROUND: Date First D-90 Engine Produced: 1989 as the D-90A.
PERM Bureau formerly known as the Soloviev Design Bureau.

ENGINE TYPE: Turbofan. Dual-Shaft, High-Bypass Front Fan (non-afterburning).

COMPRESSOR TYPE: 16-Stage Dual-Spool, Axial Flow Compressor including: 1-Stage Front Fan and 2 booster stages in the Low Pressure Compressor, 13-Stage High Pressure Compressor.

COMPRESSOR DATA: Compressor Pressure Ratio: 25 : 1, Fan Pressure Ratio: 1.6 : 1, Fan Bypass Ratio: 4.6 : 1, Total Mass Airflow: 1,100 lb/sec (500 kg/sec).

TURBINE TYPE: 6-Stage Axial Flow Turbine, including a 2-Stage High Pressure Turbine and a 4-Stage Low Pressure Turbine.

COMBUSTOR TYPE: 12 Flame Tubes and an Annular, Through-Flow Collector.

POWER (take-off) RATING: 35,275 lbt (156,9 kN thrust). Rating Approved to: 86°F (30°C).

TAKE-OFF SPECIFIC FUEL CONSUMPTION (SFC): 0.58 lb/hr/lbt (16,43 mg/Ns).

BASIC DRY WEIGHT: 6,173 lbs (2,800 kg).

POWER/WEIGHT RATIO: 5.7 : 1 lbt/lb.

APPLICATION:
Ilyushin IL-96-300 Four-Engine Wide-Body Airliner and Military Transport.
Tupelov TU-204 Twin-Engine Wide-Body Airliner and Military Transport.

OTHER ENGINES IN SERIES:
None.

GAS TURBINE ENGINES FOR AIRCRAFT

 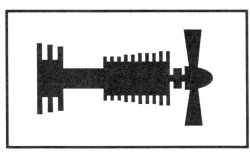

INTERNATIONAL TURBINE ENGINE COMPANIES

COUNTRY: USSR (COMMONWEALTH OF INDEPENDENT STATES)

MANUFACTURER: PROGRESS ENGINE DESIGN BUREAU (formerly Ivchenko & Lotarev)

DESIGNATION: IVCHENKO AI-20M.

BACKGROUND: This engine was originally developed as the NK-4 at the Kuznetsov. Progress Bureau was formerly known as the Ivchenko Design Bureau.

ENGINE TYPE: Turboprop. Single-Shaft Design.

COMPRESSOR TYPE: Ten-Stage, Single-Spool, Axial Flow Compressor.

COMPRESSOR DATA: Compressor Pressure Ratio: 9.4 : 1, Total Mass Airflow: 45.6 lb/sec. (20,7 kg/sec).

TURBINE TYPE: Three-Stage Axial Flow Turbine.

COMBUSTOR TYPE: Can-Annular, Through-Flow with 10 liners.

POWER (take-off) RATING: 4,250 eshp (3,169 ekw), Includes 212 shp (158 kw) from 529 lbt (2,35 kN thrust). Rating Approved to: 59°F (15°C).

TAKE-OFF SPECIFIC FUEL CONSUMPTION (SFC): 0. 62 lb/hr/eshp (0,38 kg/hr/ekw).

BASIC DRY WEIGHT: 2,292 lbs (1,040 kg).

POWER/WEIGHT RATIO: 1.85 : 1 eshp/lb (3,05 :1 ekw/kg).

APPLICATION:
Ilyushin Il-18, Il-20, Il-22 "Coot" Four-Engine Airliners.
Ilyushin Il-38 "May" Four-Engine Maritime/Patrol Aircraft.

OTHER ENGINES IN SERIES:
AI-20D in the Twin Engine Beriev M-12 "Mail" Maritime Aircraft.
AI-20DM Upgraded Model in the Twin-Engine Antonov An-32 "Cline" Airliner/Transport.
AI-20K in the Antonov An-12BP "CUB" Four Engine Transport.
AI-20 Engine also manufactured in Poland for An-28 "Cash" Airliner/Transport.

GAS TURBINE ENGINES FOR AIRCRAFT

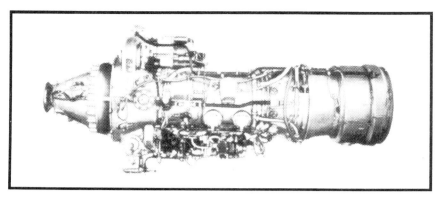 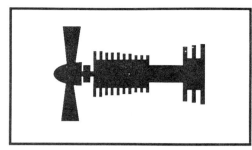

INTERNATIONAL TURBINE ENGINE COMPANIES

COUNTRY: USSR (COMMONWEALTH OF INDEPENDENT STATES)

MANUFACTURER: PROGRESS ENGINE DESIGN BUREAU (formerly Ivchenko & Lotarev)

DESIGNATION: IVCHENKO AI-24VT.

BACKGROUND: Progress Bureau was formerly known as the Ivchenko Design Bureau.

ENGINE TYPE: Turboprop. Single-Shaft Design.

COMPRESSOR TYPE: 10-Stage, Single-Spool, Axial Flow Compressor.

COMPRESSOR DATA: Compressor Pressure Ratio: 7.85:1, Total Mass Airflow: 31.7 lb/sec. (14,4 kg/sec).

TURBINE TYPE: Three-Stage Axial Flow Turbine.

COMBUSTOR TYPE: Annular, Through-Flow.

POWER (take-off) RATING: 2,820 eshp. (2,100 ekw), Includes 211 shp (157 kw) from 578 lbt (2,57 kN thrust). Rating Approved to: 59°F (15°C).

TAKE-OFF SPECIFIC FUEL CONSUMPTION (SFC): 0.51 lb/hr/eshp (0,31 kg/hr/ekw).

BASIC DRY WEIGHT: 1,320 lbs (600 kg).

POWER/WEIGHT RATIO: 2.14:1 eshp/lb (3,5:1 ekw/kg).

APPLICATION:
Antonov An-24 "Coke" Civil/Military Transport.
An-26 "Curl" Civil/Military Transport.
An-30 "Clank" Civil/Military Transport.

OTHER ENGINES IN SERIES:
AI-24T In Antonov An-24 and An-30 Twin-Engine Civil/Military Transports.

GAS TURBINE ENGINES FOR AIRCRAFT

 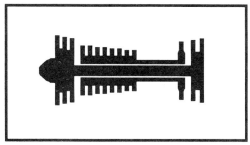

INTERNATIONAL TURBINE ENGINE COMPANIES

COUNTRY: USSR (COMMONWEALTH OF INDEPENDENT STATES)

MANUFACTURER: PROGRESS ENGINE DESIGN BUREAU (formerly Ivchenko & Lotarev)

DESIGNATION: IVCHENKO AI-25A.

BACKGROUND: Progress Bureau was formerly known as the Ivchenko Design Bureau.

ENGINE TYPE: Turbofan. Dual-Shaft, Low-Bypass Front Fan.

COMPRESSOR TYPE: 11-Stage Dual-Spool, Axial Flow Compressor including: 3-Stage Front Fan as the Low Pressure Compressor, 8-Stage High Pressure Compressor.

COMPRESSOR DATA: Compressor Pressure Ratio: 13.6 : 1, Fan Pressure Ratio: 1.7 : 1, Fan Bypass Ratio: 2 :1, Total Mass Airflow: 130 lb/sec (59 kg/sec).

TURBINE TYPE: Three-Stage Axial Flow Turbine including: 1-Stage High Pressure Turbine, 2-Stage Low Pressure Turbine.

COMBUSTOR TYPE: Annular, Through-Flow.

POWER (take-off) RATING: 3,307 lbt (14,71 kN thrust). Rating Approved to: 59°F (15°C).

TAKE-OFF SPECIFIC FUEL CONSUMPTION (SFC): 0.56 lb/hr/lbt (15,86 mg/Ns).

BASIC DRY WEIGHT: 639 lbs (290 kg).

POWER/WEIGHT RATIO: 5.18 : 1 lbt/lb.

APPLICATION:
Yakovlev Yak-40 Airliner.
Yakovlev Yak-40B "Codling" (Military Transport).

OTHER ENGINES IN SERIES:
Ivchenko AL-25TL in Aero L-39 "Albatros", Single-Engine Trainer (replacement for Aero L-29 "Maya" Trainer), and L-39Z Attack Aircraft, Czechoslovakia.
Also in WSK-PZL Mielec M-15 "Belphegor", single engine agricultural aircraft, Poland.

GAS TURBINE ENGINES FOR AIRCRAFT

 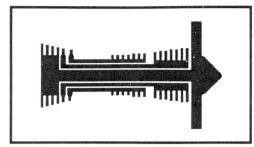

INTERNATIONAL TURBINE ENGINE COMPANIES

COUNTRY: USSR (COMMONWEALTH OF INDEPENDENT STATES)

MANUFACTURER: PROGRESS ENGINE DESIGN BUREAU (formerly Ivchenko & Lotarev)

DESIGNATION: LOTAREV D-18T.

BACKGROUND: Date First D-18 Engine Produced: 1982.
Progress Bureau was formerly known as the Lotarev Design Bureau.

ENGINE TYPE: Turbofan. Three-Shaft, High-Bypass Front Fan.

COMPRESSOR TYPE: 15-Stage Triple-Spool, Axial Flow Compressor including: 1-Stage Front Fan acting as a Low Pressure Compressor, 7-Stage Intermediate Pressure Compressor, 7-Stage High Pressure Compressor.

COMPRESSOR DATA: Compressor Pressure Ratio: 27.5 : 1, Fan Pressure Ratio: 1 : 1, Fan Bypass Ratio: 5.7 : 1, Total Mass Airflow: 1,687 lb/sec (765 kg/sec).

TURBINE TYPE: 6-Stage Axial Flow Turbine including: 1-Stage High Pressure Turbine, 1-Stage Intermediate Pressure Turbine , 4-Stage Low Pressure Turbine.

COMBUSTOR TYPE: Annular, Through-Flow.

POWER (take-off) RATING: 51,600 lbt (229,5 kN thrust), Rating Approved to: 59°F (15°C).

TAKE-OFF SPECIFIC FUEL CONSUMPTION (SFC): 0.36 lb/hr/lbt (10,2 mg/Ns).

BASIC DRY WEIGHT: 9,039 lbs (4,100 kg)

POWER/WEIGHT RATIO: 5.7 : 1 lbt/lb.

APPLICATION:
Antonov An-124 "Condor" Four-Engine Civil/Military Transport.
Antonov AN-225 "Mriya" Six-Engine Heavy-Lift Military Transport similar to the Lockheed C-5.

OTHER ENGINES IN SERIES:
None.

GAS TURBINE ENGINES FOR AIRCRAFT

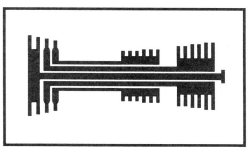

INTERNATIONAL TURBINE ENGINE COMPANIES

COUNTRY: USSR (COMMONWEALTH OF INDEPENDENT STATES)

MANUFACTURER: PROGRESS ENGINE DESIGN BUREAU (formerly Ivchenko & Lotarev)

DESIGNATION: LOTAREV D-136.

BACKGROUND: Date First D-136 Engine Produced: 1982.
Progress Bureau was formerly known as the Lotarev Design Bureau.

ENGINE TYPE: Turboshaft. Three-Shaft.

COMPRESSOR TYPE: 12-Stage Axial Flow Dual-Spool with a 6-Stage Low Pressure Compressor and a 6-Stage High Pressure Compressor.

COMPRESSOR DATA: Compressor Pressure Ratio: 18.3 : 1, Mass Airflow: 79.4 lb/sec (36 kg/sec).

TURBINE TYPE: 4-Stage Axial Flow Turbine including: 1-stage Low Pressure compressor drive turbine, 1-stage High Pressure compressor drive turbine, 2 stage free power turbine.

COMBUSTOR TYPE: Annular, Through Flow.

POWER (take-off) RATING: 11,400 shp (8,500 kw). Rating Approved to: 59°F (15°C).

TAKE-OFF SPECIFIC FUEL CONSUMPTION (SFC): 0.44 lb/hr/shp (0,268 kg/hr/kw).

BASIC DRY WEIGHT: 2,315 lbs (1,050 kg).

POWER/WEIGHT RATIO: 4.9 : 1 shp/lb (8,1 : 1 kw/kg).

APPLICATION:
MIL MI-26 "Halo" Twin Engine, Heavy Lift Helicopter.

OTHER ENGINES IN SERIES:
D-236 Experimental Propfan.

GAS TURBINE ENGINES FOR AIRCRAFT

 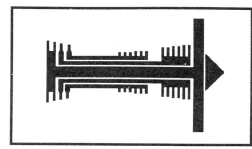

INTERNATIONAL TURBINE ENGINE COMPANIES

COUNTRY: USSR (COMMONWEALTH OF INDEPENDENT STATES)

MANUFACTURER: PROGRESS ENGINE DESIGN BUREAU (formerly Ivchenko & Lotarev)

DESIGNATION: LOTAREV D-436.

BACKGROUND: Derivative of Earlier D-36 Turbofan Engine.
Progress Bureau was formerly known as the Lotarev Design Bureau.

ENGINE TYPE: Turbofan. Three-Shaft, High-Bypass Front Fan.

COMPRESSOR TYPE: 13-Stage Triple-Spool, Axial Flow Compressor including: 1-Stage Front Fan acting as a Low Pressure Compressor, 6-Stage Intermediate Pressure Compressor, 6-Stage High Pressure Compressor.

COMPRESSOR DATA: Compressor Pressure Ratio: 19:1, Fan Pressure Ratio: 1:1, Fan Bypass Ratio: 5.34:1, Total Mass Airflow: 260 lb/sec (120 kg/sec).

TURBINE TYPE: 4-Stage Axial Flow Turbine including: 1-Stage High Pressure Turbine, 1-Stage Intermediate Pressure Turbine, 2-Stage Low Pressure Turbine.

COMBUSTOR TYPE: Annular, Through-Flow.

POWER (take-off) RATING: 16,575 lbt (73,73 kN thrust). Rating Approved to: 59°F (15°C).

TAKE-OFF SPECIFIC FUEL CONSUMPTION (SFC): 0.36 lb/hr/lbt (10,2 mg/Ns).

BASIC DRY WEIGHT: 2,380 lbs (1,080 kg).

POWER/WEIGHT RATIO: 6.03 : 1 lbt/lb.

APPLICATION:
Antonov An-72 "Coaler" Military Transport.
AN-74 "Coaler-B" Military Transport.

OTHER ENGINES IN SERIES:
Lotarev D-36 (14,500 lbt) in early AN-72, AN-74 Military Transports & Yak-42 Airliner.
Lotarev D-436K in AN-72/74 Civil/Military Transports.
Lotarev D-436M in Yak-42M Airliner.
Lotarev D-436T in TU-334 Civil/Military Transport.
Lotarev D-436T1/2 16,000/18000 lbt. in development.
Lotarev D-136 Turboshaft derivative in MIL Mi-26 Helicopter.
Lotarev D-236 experimental PropFan (11,000 eshp).

GAS TURBINE ENGINES FOR AIRCRAFT

 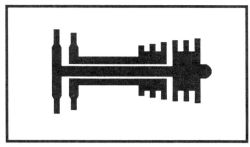

INTERNATIONAL TURBINE ENGINE COMPANIES

COUNTRY: USSR (COMMONWEALTH OF INDEPENDENT STATES)

MANUFACTURER: SOYUZ ENGINE DESIGN BUREAU (formerly Tumansky)

DESIGNATION: Tumansky R-11-300.

BACKGROUND: Date First R-11 Engine Produced: 1956.
Date If Out-of-Production: 1966. Number of R-11 Engines Produced to Date: 20,000.

ENGINE TYPE: Turbojet. Dual-Shaft, with Afterburning.

COMPRESSOR TYPE: 6-Stage Axial Flow including a 3-Stage Low Pressure Compressor, 3-Stage High Pressure Compressor.

COMPRESSOR DATA: Compressor Pressure Ratio: 8 : 1, Mass Airflow: 160 lb/sec (72 kg/sec).

TURBINE TYPE: 2-Stage Axial Flow Turbine including a: 1-Stage High Pressure Turbine, 1-Stage Low Pressure Turbine.

COMBUSTOR TYPE: Can-Annular, Through Flow with 10 liners.

AFTERBURNER: Single Stage.

POWER (take-off) RATING: 9,700 lbt (43,15 kN thrust) [dry], 13,120 lbt (58,36 kN thrust). [A/B]. Rating Approved to: 59°F (15°C).

TAKE-OFF SPECIFIC FUEL CONSUMPTION (SFC): 1.02 lb/hr/lbt (28,61 mg/Ns) [dry], 1.99 lb/hr/lbt (56,37 mg/Ns) [A/B].

BASIC DRY WEIGHT: 2770 lbs (1,256 kg).

POWER/WEIGHT RATIO: 3.5 : 1[dry], 4.7 : 1 [A/B] - lbt/lb.

APPLICATION:
Mikoyan MIG-21PFMA "Fishbed" Fighter Twin-Engine Fighter.
Yakovlev Yak-28P "Firebar" Twin-Engine Fighter.
Yakovlev Yak-28R "Brewer" Reconnaissance Aircraft.
Yakovlev Yak-28U "Maestro" Two-Place Trainer Aircraft.

OTHER ENGINES IN SERIES:
Tumansky R-11 in Early MIG-21 "Fishbed" Fighter Aircraft.

GAS TURBINE ENGINES FOR AIRCRAFT

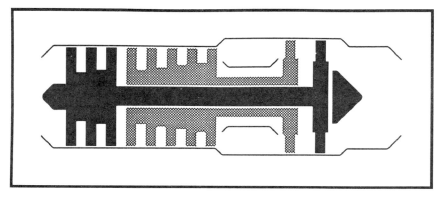 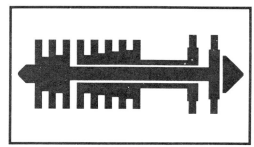

INTERNATIONAL TURBINE ENGINE COMPANIES

COUNTRY: USSR (COMMONWEALTH OF INDEPENDENT STATES)

MANUFACTURER: SOYUZ ENGINE DESIGN BUREAU (formerly Tumansky)

DESIGNATION: Tumansky R-13F2-300.

BACKGROUND: Date First R-13 Engine Produced: 1956.

ENGINE TYPE: Turbojet. Dual-Shaft, with Afterburning.

COMPRESSOR TYPE: 8-Stage Dual-Spool Axial Flow including a 3-Stage Low Pressure Compressor, 5-Stage High Pressure Compressor.

COMPRESSOR DATA: Compressor Pressure Ratio: 10 : 1, Mass Airflow: 156 lb/sec (71 kg/sec).

TURBINE TYPE: 2-Stage Axial Flow Turbine including a 1-Stage High Pressure Turbine and a 1-Stage Low Pressure Turbine.

COMBUSTOR TYPE: Annular, Through Flow.

AFTERBURNER: 1-Stage.

POWER (take-off) RATING: 9,340 lbt (41,55 kN thrust) [dry], 15,875 lbt (70,6 kN thrust) [A/B]. Rating Approved to: 59°F (15°C).

TAKE-OFF SPECIFIC FUEL CONSUMPTION (SFC): 0.96 lb/hr/lbt (27,19 mg/Ns) [dry], 2.25 lb/hr/lbt (63,73 mg/Ns) [A/B].

BASIC DRY WEIGHT: 2,670 lbs (1,211 kg).

POWER/WEIGHT RATIO: 3.5 :1[dry], 5.9 :1 [A/B] - lbt/lb.

APPLICATION:
Sukhoi Su-15/21 "Flagon" Twin Engine Fighter.
Sukhoi SU-25 "Frogfoot" Twin-Engine Attack Aircraft, non-afterburning.

OTHER ENGINES IN SERIES:
Tumansky R-13 in MiG-21 "Mongol" Trainer.
Tumansky R-13-300 in the early MIG-21, 14,550 lbt with Afterburning.
Tumansky RU-19-300 in Antonov An-26 "Curl", An-30 "Clank".
Tumansky RU-19-300 in YAK 30/32 Turboprop Aircraft as an auxiliary engine.

GAS TURBINE ENGINES FOR AIRCRAFT

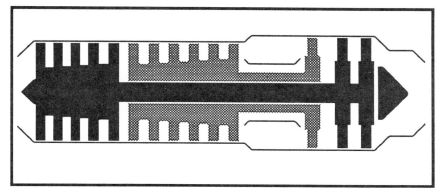
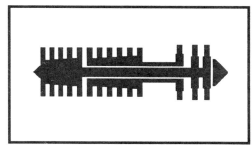

INTERNATIONAL TURBINE ENGINE COMPANIES

COUNTRY: USSR (COMMONWEALTH OF INDEPENDENT STATES)

MANUFACTURER: SOYUZ ENGINE DESIGN BUREAU (formerly Tumansky)

DESIGNATION: Tumansky R-29B.

BACKGROUND: Date First R-29 Engine Produced: 1970.

ENGINE TYPE: Turbojet. Dual-Shaft, with Afterburning.

COMPRESSOR TYPE: 11-Stage Dual-Spool, Axial Flow Compressor, including a 5-Stage Low Pressure Compressor and a 6-Stage High Pressure Compressor.

COMPRESSOR DATA: Compressor Pressure Ratio: 12.6 : 1, Mass Airflow: 235 lb/sec (97,1 kg/sec).

TURBINE TYPE: 3-Stage Axial Flow Turbine, including a 1-Stage high Pressure Turbine and a 2-Stage Low Pressure Turbine.

COMBUSTOR TYPE: Annular, Through Flow.

AFTERBURNER: Variable-Stage.

POWER (take-off) RATING: 17,635 lbt (78,5 kN thrust) [dry], 21,825 lbt (97,1 kN thrust) [A/B]. Rating Approved to: 59°F (15°C).

TAKE-OFF SPECIFIC FUEL CONSUMPTION (SFC): 0.71 lb/hr/lbt (20 mg/Ns) [dry], 1.2 lb/hr/lbt (34 mg/Ns) [A/B].

BASIC DRY WEIGHT: 3,880 lbs (1,760 kg).

POWER/WEIGHT RATIO: 4.5 : 1 [dry], 5.6 :1 [A/B] - lbt/lb.

APPLICATION:
MIG-23MF "Flogger" and MIG-27 "Flogger D/J" Single-Engine Fighters.
Sukhoi SU-19 and SU-22 "Fitter-F" Twin-Engine Attack Aircraft.

OTHER ENGINES IN SERIES:
R-29-300 in MIG series.

GAS TURBINE ENGINES FOR AIRCRAFT

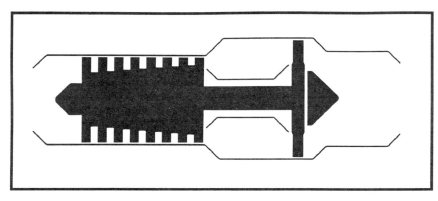 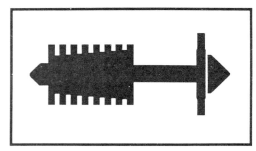

INTERNATIONAL TURBINE ENGINE COMPANIES

COUNTRY: USSR (COMMONWEALTH OF INDEPENDENT STATES)

MANUFACTURER: SOYUZ ENGINE DESIGN BUREAU (formerly Tumansky)

DESIGNATION: Tumansky R-31.

BACKGROUND: Date First R-31 Engine Produced: 1963.

ENGINE TYPE: Turbojet. Single-Shaft, with Afterburning.

COMPRESSOR TYPE: Axial Flow, 8-Stage Single-Spool.

COMPRESSOR DATA: Compressor Pressure Ratio: 7:1, Mass Airflow: 295 lb/sec (134 kg/sec).

TURBINE TYPE: 1-Stage Axial Flow Turbine.

COMBUSTOR TYPE: Annular, Through Flow.

AFTERBURNER: 1-Stage.

POWER (take-off) RATING: 20,500 lbt (90 kN thrust) [dry], 27,000 lbt (120 kN thrust) [A/B]. Rating Approved to: 59°F (15°C).

TAKE-OFF SPECIFIC FUEL CONSUMPTION (SFC): 0.84 lb/hr/lbt (24 mg/Ns) [dry], 2.2 lb/hr/lbt (63,4 mg/Ns) [A/B].

BASIC DRY WEIGHT: 6,050 lbs (2,722 kg).

POWER/WEIGHT RATIO: 3.42 : 1 [dry], 4.5 : 1 [A/B] - lbt/lb.

APPLICATION:
Mikoyan MIG-25 "Foxbat".
Early MIG-31 "Foxhound".
Early SU-27 "Flanker" (Twin-Engine Fighters).

OTHER ENGINES IN SERIES:
Tumansky R-31(R-266) in the MIG-25 "Foxbat".

GAS TURBINE ENGINES FOR AIRCRAFT

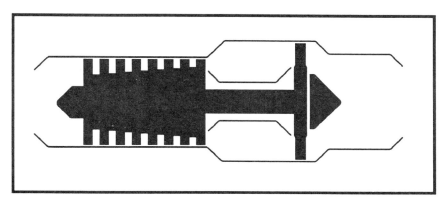
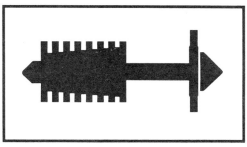

INTERNATIONAL TURBINE ENGINE COMPANIES

COUNTRY: USSR (COMMONWEALTH OF INDEPENDENT STATES)

MANUFACTURER: SOYUZ ENGINE DESIGN BUREAU (formerly Tumansky)

DESIGNATION: Tumansky R-31 (upgraded).

BACKGROUND: Upgraded RD-7.

ENGINE TYPE: Turbojet. Single-Shaft, with Afterburning and Water Injection.

COMPRESSOR TYPE: Axial Flow, 8-Stage, Single-Spool.

COMPRESSOR DATA: Compressor Pressure Ratio: 8 : 1, Mass Airflow: 295 lb/sec (134 kg/sec).

TURBINE TYPE: 1-Stage Axial Flow Turbine.

COMBUSTOR TYPE: Annular, Through Flow.

AFTERBURNER: 1-Stage.

POWER (take-off) RATING: 20,230 lbt (90 kN thrust) [dry], 30,000 lbt (138 kN thrust) [A/B]. Rating Approved to: 59°F (15°C).

TAKE-OFF SPECIFIC FUEL CONSUMPTION (SFC): 0.81 lb/hr/lbt (23 mg/Ns) [dry], 2.1 lb/hr/lbt (59 mg/Ns) [A/B].

BASIC DRY WEIGHT: 6,105 lbs (2,770 kg).

POWER/WEIGHT RATIO: 3.3 : 1 [dry], 4.9 :1 [A/B] - lbt/lb.

APPLICATION:
Sukhoi Su-27 "Flanker" Twin-Engine Fighter.

OTHER ENGINES IN SERIES:
None.

GAS TURBINE ENGINES FOR AIRCRAFT

INTERNATIONAL TURBINE ENGINE COMPANIES

COMMONWEALTH OF INDEPENDENT STATES (CIS) FORMERLY USSR!

ADDITIONAL GAS TURBINE ENGINES - For which complete engine data - unavailable.

MANUFACTURER	DESIG.	TYPE	POWER	AIRCRAFT
GLUSHENKOV	TVD-20	TURBOPROP	1,450 SHP	AN-3, AGRICULTURAL AIRCRAFT
KLIMOV (ISOTOV)	VK-1	TURBOJET	5,950 LBT	IL-28, BOMBER
KOLIESOV (RYBINSK)	RD-38	LIFT-JET	7,165 LBT	YAK-38, YAK-41
KOLIESOV (RYBINSK)	RD-41	LIFT-JET	UNKNOWN	IN DEVELOP. (YAK-141)
KOLIESOV (RYBINSK)	RD-60	BOOST-JET	6,065 LBT	A-40, AMPHIBIAN
KOLIESOV (RYBINSK)	VD-7	TURBOJET	31,000 LBT	M-4, M-50, TU-22, BOMBERS
KOLIESOV (RYBINSK)	VD-57	TURBOJET	44,000 LBT	TU-160, BOMBER
KUZNETSOV (SAMARA)	D-227	TURBOFAN	27,650 LBT	IN DEVELOP. (TU-334, AN-180)
KUZNETSOV (SAMARA)	NK-86	TURBOFAN	28,660 LBT	IL-86, Airliner
KUZNETSOV (SAMARA)	NK-87	TURBOFAN	28,660 LBT	IN DEVELOP. (SUPER-SEAPLANE)
KUZNETSOV (SAMARA)	NK-88	TURBOFAN	23,150 LBT	IN DEVELOP. (TU-155)
KUZNETSOV (SAMARA)	NK-92	PROPFAN	46,000 LBT	IN. DEVELOP. (MILITARY AIRCRAFT)
KUZNETSOV (SAMARA)	NK-93	PROPFAN	46,000 LBT	IN DEVELOP. (IL-96, TU-204)
KUZNETSOV (SAMARA)	NK-104	TURBOFAN	24,250 LBT	IN DEVELOPMENT
KUZNETSOV (SAMARA)	NK-112	TURBOFAN	18,740 LBT	IN DEVELOPMENT
KUZNETSOV (SAMARA)	NK-321	TURBOFAN	55,100 LBT	TU-160, BOMBER
LYULKA-SATURN	AL-31F	TURBOFAN	27,500 LBT	SU-27, FIGHTER
PERM (SOLOVIEV)	D-15	TURBOJET	28,650 LBT	M-4 BISON, BOMBER
PERM (SOLOVIEV)	D-30F6	TURBOFAN	34,000 LBT	MIG-31, FIGHTER
PROGRESS	D-27	PROPFAN	24,700 LBT	IN DEVELOP. (AN-70T, YAK-46)
PROGRESS	D-36	TURBOFAN	14,500 LBT	AN-72, YAK-42
SOYUZ (TUMANSKY)	M701C500	TURBOJET	1,960 LBT	L-29 AERO, TRAINER
SOYUZ (TUMANSKY)	R-15BD-300	TURBOJET	24,700 LBT	MIG-25M
SOYUZ (TUMANSKY)	R-25	TURBOJET	24,000 LBT	MIG-21, SU-15 FIGHTERS
SOYUZ (TUMANSKY)	R-27-22A	TURBOJET	22,500 LBT	MIG-23M AND -23U
SOYUZ (TUMANSKY)	R-27V-300	TURBOFAN	23,000 LBT	IN DEV. YAK-38, VEC/THRUST
SOYUZ (TUMANSKY)	R-35F-300	TURBOJET	28,660 LBT	MIG-23ML
SOYUZ (TUMANSKY)	R-45	TURBOJET	5,000 LBT	MIG-15, TRAINER
SOYUZ (TUMANSKY)	R-79V-300	TURBOFAN	34,100 LBT	IN DEV. YAK-141 VEC/THRUST
SOYUZ (TUMANSKY)	R-195	TURBOJET	9,920 LBT	SU-25 ATTACK AIRCRAFT

GAS TURBINE ENGINES FOR AIRCRAFT

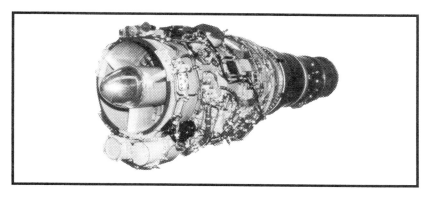 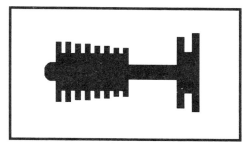

INTERNATIONAL TURBINE ENGINE COMPANIES

COUNTRY: YUGOSLAVIA

MANUFACTURER: ORAO AIR FORCE DEPOT
BELGRADE

DESIGNATION (military): Viper MK. 632-41.

BACKGROUND: Produced under Rolls-Royce License, Great Britain.
Date First Viper Engine Produced: 1980.
Date If Out-of-Production: N/A. Number of Licensed Viper Engines Produced to Date: Unknown.

ENGINE TYPE: Turbojet. Single-Shaft.

COMPRESSOR TYPE: 8-Stage Single-Spool, Axial Flow Compressor.

COMPRESSOR DATA: Compressor Pressure Ratio: 5.9: 1, Total Mass Airflow: 58.4 lb/sec (26,5 kg/sec).

TURBINE TYPE: 2-Stage Axial Flow.

COMBUSTOR TYPE: Through-Flow, Can-Annular with 7 liners.

POWER (take-off) RATING: 4,000 lbt (17,8 kN thrust). Rating Approved to: 59°F (15°C).

TAKE-OFF SPECIFIC FUEL CONSUMPTION (SFC): 0.9 lb/hr/lbt (0,25 mg/Ns).

BASIC DRY WEIGHT: 825 lbs (374 kg).

POWER/WEIGHT RATIO: 4.8 : 1 lbt/lb.

APPLICATION:
Soko "Orao 1/2" Attack Aircraft, Yugoslavia.
Cinair IAR-93 "Yurom" Trainer Aircraft, Rumania.
Afterburning Model in Orao-2 and Yurom-B.

OTHER ENGINES IN SERIES:
Viper MK.22 in Soko G-2A "Galeb" Trainer.
Viper MK.531 in J-1 "Jastreb" Attack/Recon Aircraft.
Viper MK.531 in TJ-1 "Jastreb" Attack/Recon Aircraft.
Viper Mk.632 in Soko G-4 "Super-Galeb" Trainer Aircraft.
Locate other Viper models under Rolls-Royce, Great Britain listing.

GAS TURBINE ENGINES FOR AIRCRAFT

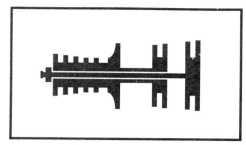

MULTI-NATIONAL TURBINE ENGINE COMPANIES

COUNTRY: GERMANY, ENGLAND

MANUFACTURER: BMW, ROLLS ROYCE GmbH (FORMERLY KHD) OBERURSEL, WEST GERMANY

DESIGNATION (military): T53L-13.

BACKGROUND: Manufactured under T-53 License, Textron-Lycoming Company (USA). Date First T53L-13 Produced in Germany: 1968.
Date If Out-of-Production: 1971. Number of T53L-13 Engines Produced to Date: 441.

ENGINE TYPE: Turboshaft. Dual-Shaft (Compressor Drive & Power Output Drive).

COMPRESSOR TYPE: Combination Compressor: 5-Stage Axial, 1-Stage Centrifugal.

COMPRESSOR DATA: Compressor Pressure Ratio: 7:1. Mass Airflow: 12.7 lb/sec (5,75 kg/sec).

TURBINE TYPE: 4-Stage Axial Flow Turbine including: 2-Stage Gas Producer (compressor drive) Turbine, 2-Stage Power (free) Turbine.

COMBUSTOR TYPE: Annular, Reverse Flow.

POWER (take-off) RATING: 1,400 shp (1,044 kw). Rating Approved to: 59°F (15°C).

TAKE-OFF SPECIFIC FUEL CONSUMPTION (SFC): 0.58 lb/hr/shp (0,353 kg/hr/kw).

BASIC DRY WEIGHT: 540 lbs (245 kg).

POWER/WEIGHT RATIO: 2.59 :1 shp/lb (4,26 :1 kg/kw).

APPLICATION:
Bell UH-1D "Iroquois" Helicopter.

OTHER ENGINES IN SERIES:
Locate other T-53 Models under Textron-Lycoming Company, USA.

GAS TURBINE ENGINES FOR AIRCRAFT

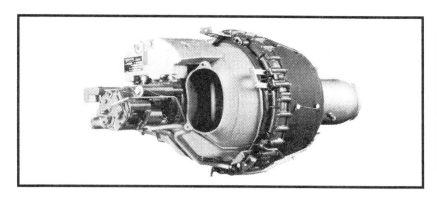 

MULTI-NATIONAL TURBINE ENGINE COMPANIES

COUNTRY: GERMANY, ENGLAND

MANUFACTURER: BMW, ROLLS ROYCE GmbH (FORMERLY "KHD") OBERURSEL, WEST GERMANY

DESIGNATION (civil & military)**:** T117.

BACKGROUND: Date First T117 Produced: 1978.
Date If Out-of-Production: N/A . Number of T117 Engines Produced to Date: 52.

ENGINE TYPE: Turbojet. Single-Shaft.

COMPRESSOR TYPE: Centrifugal Compressor: 1-Stage, Single Entry.

COMPRESSOR DATA: Compressor Pressure Ratio: 5.5:1, Total Mass Airflow: 3.53 lb/sec (1,6 kg/sec).

TURBINE TYPE: 1-Stage Axial Flow Turbine.

COMBUSTOR TYPE: Annular, Reverse Flow.

POWER (take-off) RATING: 236 lbt (1,05 kN thrust). Rating Approved to: 59°F (15°C).

TAKE-OFF SPECIFIC FUEL CONSUMPTION (SFC): 1.2 lb/hr/lbt (33,9 mg/Ns).

BASIC DRY WEIGHT: 50 lbs (22,8 kg).

POWER/WEIGHT RATIO: 4.8 : 1 lbt/lb.

APPLICATION:
Dornier/Canadair CL-289 Target Drone.
Future Aircraft Applications.

OTHER ENGINES IN SERIES:
None.

GAS TURBINE ENGINES FOR AIRCRAFT

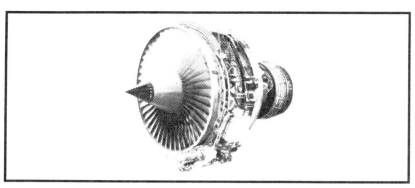
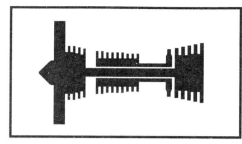

MULTI-NATIONAL TURBINE ENGINE COMPANIES

COUNTRY: USA, FRANCE

MANUFACTURER: CFM INTERNATIONAL: GENERAL ELECTRIC AND SNECMA CINCINNATI, OHIO, USA

DESIGNATION (civil): CFM56-5C3.

BACKGROUND: The original CFM 56 Utilized a General Electric, F101 Engine as its core.
Date First CFM56 Engine Produced: 1978.
Date If Out-of-Production: N/A. Number of CFM-56 Engines Produced to Date: 4,000.

ENGINE TYPE: Turbofan. Dual-Shaft, High-Bypass Front Fan.

COMPRESSOR TYPE: 14-Stage Dual-Spool, Axial Flow Compressor including: 1-Stage Front Fan and (4) additional Axial Stages in the Low Pressure Compressor, 9-Stage High Pressure Compressor.

COMPRESSOR DATA: Compressor Pressure Ratio: 37.5 : 1, Fan Pressure Ratio: 1 : 1, Fan Bypass Ratio: 6.5 : 1, Total Mass Airflow: 1,045 lb/sec (474 kg/sec).

TURBINE TYPE: 6-Stage Axial Flow Turbine including: 1-Stage High Pressure Turbine, 5-Stage Low Pressure Turbine.

COMBUSTOR TYPE: Annular, Through-Flow.

POWER (take-off) RATING: 32,500 lbt (144,56 kN thrust). Rating Approved to: 86°F (30°C).

TAKE-OFF SPECIFIC FUEL CONSUMPTION (SFC): 0.33 lb/hr/lbt (9,35 mg/Ns).

BASIC DRY WEIGHT: 5,645 lbs (2,561 kg).

POWER/WEIGHT RATIO: 5.75 : 1 lbt/lb.

APPLICATION:
CFM56-5C Series in Airbus A340 Airliner In Development.

OTHER ENGINES IN SERIES:
CFM56-2 series in McDonnell-Douglas DC-8 Super 71,72,73 and BAe "Trident" Upgrade.
CFM56-3 series in Boeing B737-300, -400, 500.
CFM56-5 in Airbus A320.
CFM56-5A1 In Airbus A340.
CFM56-5B in Airbus 320/321.
CFM56-2A Engine is Known as the F108-CF-100 in the military E-3C "Sentry", E-6A "Tacano" and upgraded KC-135R, In-Flight Refueling Aircraft.

GAS TURBINE ENGINES FOR AIRCRAFT

 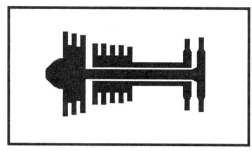

MULTI-NATIONAL TURBINE ENGINE COMPANIES

COUNTRY: GREAT BRITAIN, GERMANY, ITALY, SPAIN.

MANUFACTURER: EUROJET TURBO GmbH
(ROLLS-ROYCE, MTU, FIAT AVIO, ITP SPAIN)
Munich, Germany

DESIGNATION (military): EJ200.

BACKGROUND: Date First EJ200 Engine Produced: In Development.
Date If Out-of-Production: N/A. Number of EJ 200 Engines Produced to Date: N/A.

ENGINE TYPE: Turbofan. Dual-Shaft, Low-Bypass Front Fan with Afterburner.

COMPRESSOR TYPE: 8-Stage Dual-Spool, Axial Flow Compressor including: 3-Stage Fan Stages in the Low Pressure Compressor, 5-Stage High Pressure Compressor.

COMPRESSOR DATA: Compressor Pressure Ratio: 25 : 1, Fan Pressure Ratio: 4 : 1, Fan Bypass Ratio: 0.4 : 1, Total Mass Airflow: 170 lb/sec (77 kg/sec).

TURBINE TYPE: 2-Stage Axial Flow Turbine including: 1-Stage High Pressure Turbine, 1-Stage Low Pressure Turbine.

COMBUSTOR: Annular, Through Flow.

AFTERBURNER: Variable Stage, combined core and fan flow.

POWER (take-off) RATING: 13,500 lbt (60 kN thrust) [dry], 20,000 lbt (90 kN thrust) [A/B].
Rating Approved to: 59°F (15°C).

TAKE-OFF SPECIFIC FUEL CONSUMPTION (SFC): 0.8 lb/hr/lbt (Dry), 1.7 lb/hr/lbt (A/B).

BASIC DRY WEIGHT: 2,000 lbs (907 kg).

POWER/WEIGHT RATIO: 6.8 : 1 [dry], 10:1 [A/B] - lbt/lb.

APPLICATION:
EFA (European Fighter Aircraft), AlsoCalled "Eurofighter", under development.

OTHER ENGINES IN SERIES:
EJ-200 Dry Thrust Model being developed with 13,500 lbt., 1,800 lb Basic Dry Weight; with other data essentially same as above.

GAS TURBINE ENGINES FOR AIRCRAFT

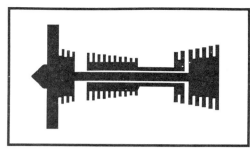

MULTI-NATIONAL TURBINE ENGINE COMPANIES

COUNTRY: USA, GREAT BRITAIN, JAPAN, GERMANY, ITALY

MANUFACTURER: INTERNATIONAL AERO ENGINES LTD.
(United Technologies, Rolls-Royce, Japanese Aero Engines, MTU, Fiat Aviazione).
Corporate Office East Hartford, CT, USA.

DESIGNATION (civil): V2500-A1.

BACKGROUND: Date First V2500 Engine Produced: 1988.
Date If Out-of-Production: N/A. Number of V2500 Series Engines Produced to Date: 125.

ENGINE TYPE: Turbofan. Dual-Shaft, High-Bypass Front Fan.

COMPRESSOR TYPE: 14-Stage Dual-Spool, Axial Flow Compressor including: 1-Stage Front Fan and (3) additional Axial Stages in the Low Pressure Compressor, 10-Stage High Pressure Compressor.

COMPRESSOR DATA: Compressor Pressure Ratio: 29.4 : 1, Fan Pressure Ratio: 1.7: 1, Fan Bypass Ratio: 5.4 : 1, Total Mass Airflow: 783 lb/sec (355 kg/sec).

TURBINE TYPE: 7-Stage Axial Flow Turbine including: 2-Stage High Pressure Turbine, 5-Stage Low Pressure Turbine.

COMBUSTOR TYPE: Annular, Through-Flow.

POWER (take-off) RATING: 25,000 lbt (111,2 kN thrust). Rating Approved to: 86°F (30°C).

TAKE-OFF SPECIFIC FUEL CONSUMPTION (SFC): 0.32 lb/hr/lbt (9,06 mg/Ns).

BASIC DRY WEIGHT: 4,942 lbs (2,242 kg).

POWER/WEIGHT RATIO: 5.06 : 1 lbt/lb.

APPLICATION:
Airbus 320-200.

OTHER ENGINES IN SERIES:
V2522-A5 in Airbus A319-100
V2522-D5 in McDonnell-Douglas MD-90-10, A321.
V2525-A5 in Airbus A320-200.
V2525-D5 in McDonnell-Douglas MD-90-30.
V2528-D5 in McDonnell-Douglas MD-90-40/50.
V2530-A5 in Airbus A321-100.
V2527-A5 in Airbus A330.

GAS TURBINE ENGINES FOR AIRCRAFT

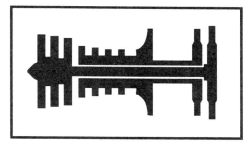

MULTI-NATIONAL TURBINE ENGINE COMPANIES

COUNTRY: USA, REPUBLIC OF CHINA-TAIWAN

MANUFACTURER: INTERNATIONAL TURBINE ENGINE CORPORATION.
(ALLIED SIGNAL COMPANY USA & AEROSPACE INDUSTRY DEV. CTR, TAIWAN)
PHOENIX, AR. USA

DESIGNATION (military): TFE 1042-70.

BACKGROUND: Date First TFE 1042 Engine Produced: In Development.
Date If Out-of-Production: N/A. Number of TFE 1042 Engines Produced to Date: N/A.

ENGINE TYPE: Turbofan. Low Bypass Fan, Dual-Shaft, with Afterburning.

COMPRESSOR TYPE: 8-Stage Compressor. A 3-Stage Fan acts as a Low Pressure Compressor. The High Pressure Compressor is Combination Axial and Centrifugal Flow design (5-stages total, 4-Stage Axial, 1-Stage Centrifugal).

COMPRESSOR DATA: Compressor Pressure Ratio: 22.5 : 1, Fan Pressure Ratio: 1.54 : 1, Fan Bypass Ratio: 0.4 : 1. Total Mass Airflow: 88 lb/sec (47,8 kg/sec).

TURBINE TYPE: 2-Stage Axial Flow Turbine including: 1-Stage High Pressure Turbine, 1-Stage Low Pressure Turbine.

COMBUSTOR TYPE: Annular, Through Flow.

POWER (take-off) RATING: 5,000 lbt (22,24 kN thrust) dry, 8,350 lbt (37,4 kN thrust) [A/B]. Rating Approved to: 59°F (15°C).

TAKE-OFF SPECIFIC FUEL CONSUMPTION (SFC): 0.78 lb/hr/lbt (22,1 mg/Ns) dry, 1.95 lb/hr/lbt (55,2 mg/Ns) [A/B].

BASIC DRY WEIGHT: 1,360 lbs (617 kg).

POWER/WEIGHT RATIO: 3.68 : 1 dry, 6.14 : 1 (A/B).

APPLICATION:
Republic of China-Taiwan A-3 "Lui-Ming" IDF (Indigenous Defensive Fighter).
Republic of China-Taiwan AT-3 "Tsu-Chiang" (Trainer Aircraft).

OTHER ENGINES IN SERIES:
F124-GA-100 Non-afterburner version, USA Future for Aircraft Applications.
F125-GA-100 afterburner version, USA Future for Aircraft Applications.

GAS TURBINE ENGINES FOR AIRCRAFT

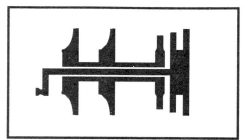

MULTI-NATIONAL TURBINE ENGINE COMPANIES

COUNTRY: GREAT BRITAIN, GERMANY, FRANCE

MANUFACTURER: MTU GmbH (MTU, TURBOMECA, ROLLS-ROYCE)
LONDON, ENGLAND

DESIGNATION (military): MTR 390.

BACKGROUND: Formerly designated MTM 385.
Date First MTR 390 Engine Produced: 1989.
Date If Out-of-Production: N/A. Number of MTR 390 Engines Produced to Date: 14.

ENGINE TYPE: Turboshaft. Two-Shaft (1-Compressor Drive & 1-Power Output Drive).

COMPRESSOR TYPE: Centrifugal Compressor: 2-Stage.

COMPRESSOR DATA: Compressor Pressure Ratio: 14:1. Mass Airflow: 7.05 lb/sec (3,2 kg/sec).

TURBINE TYPE: 3-Stage Axial Flow Turbine including: 1-Stage Gas Producer (compressor drive) Turbine, 2-Stage Power (free) Turbine.

COMBUSTOR TYPE: Annular, Reverse Flow.

POWER (take-off) RATING: 1,160 shp (865 kw). Rating Approved to: 59°F (15°C).

TAKE-OFF SPECIFIC FUEL CONSUMPTION (SFC): 0.456 lb/hr/shp (0,277 kg/hr/kw).

BASIC DRY WEIGHT: 372 lbs (169 kg).

POWER/WEIGHT RATIO: 3.12 : 1 shp/lb (5,12 : 1 kw/kg).

APPLICATION:
Eurocopter PAH-2 "Tiger" Helicopter (Aerospatiale, France).
Agusta A.129 "LAH", Helicopter, Italy.

OTHER ENGINES IN SERIES:
None.

GAS TURBINE ENGINES FOR AIRCRAFT

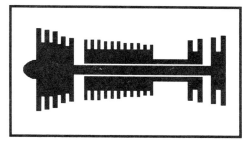

MULTI-NATIONAL TURBINE ENGINE COMPANIES

COUNTRY: GREAT BRITAIN, USA

MANUFACTURER: ROLLS-ROYCE , ALLISON TURBINE ENGINES (USA). LONDON, ENGLAND

DESIGNATION (military)**:** TF41-A-2.

BACKGROUND: No Civil Counterpart. Also known as Spey RB.168-66.
Date First TF41-A Engine Produced: 1968.
Date If Out-of-Production: N/A. Number of TF-41A Engines Produced to Date: 1414.

ENGINE TYPE: Turbofan. Dual-Shaft, Low-Bypass Front Fan (non-afterburning).

COMPRESSOR TYPE: 16-Stage Dual-Spool, Axial Flow Compressor including: 3-Stage Front Fan with (2) Additional Axial Stages in the Low Pressure Compressor, 11-Stage High Pressure Compressor.

COMPRESSOR DATA: Compressor Pressure Ratio: 21.4 : 1, Fan Pressure Ratio: 2.45 : 1, Fan Bypass Ratio: 0.75 : 1, Total Mass Airflow: 263 lb/sec (119,3 kg/sec).

TURBINE TYPE: 4-Stage Axial Flow Turbine including: 2-Stage High Pressure Turbine, 2-Stage Low Pressure Turbine.

COMBUSTOR TYPE: Can-Annular, Through-Flow, with 10 liners.

POWER (take-off) RATING: 15,000 lbt (66,72 kN thrust). Rating Approved to: 59°F (15°C).

TAKE-OFF SPECIFIC FUEL CONSUMPTION (SFC): 0.647 lb/hr/lbt (18,33 mg/Ns).

BASIC DRY WEIGHT: 2,980 lbs (1,353 kg).

POWER/WEIGHT RATIO: 5.03 : 1 lbt/lb.

APPLICATION:
LTV Vought A-7E & H "Corsair-II" Attack Aircraft.

OTHER ENGINES IN SERIES:
TF41-A-1 (Spey RB.168-62) in LTV Vought A-7D.
Spey Mk.101 in BAe "Buccaneer".
Spey Mk.202 in McDonnell-Douglas F-4M "Phantom".
Spey Mk.250 in BAe "Nimrod".
Spey Mk.807 (RB-169) in Proposed AMX Int'l. AM-X Fighter.

GAS TURBINE ENGINES FOR AIRCRAFT

 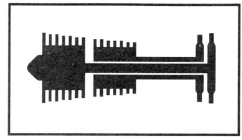

MULTI-NATIONAL TURBINE ENGINE COMPANIES

COUNTRY: GREAT BRITAIN, FRANCE

MANUFACTURER: ROLLS-ROYCE AND SNECMA
LONDON, ENGLAND

DESIGNATION (civil): Olympus 593 (Mk.610).

BACKGROUND: Developed from the Olympus 320 Turbojet in the TSR2 Bomber. The only commercial Gas Turbine Engine ever built for a supersonic aircraft other than the Russian equivalent in the TU-144.
Date First Olympus 593 Engine Produced: 1969.
Date If Out-of-Production: 1977. Number of Olympus 593 Engines Produced to Date: 136.

ENGINE TYPE: Turbojet, Dual-Shaft, with Afterburning.

COMPRESSOR TYPE: 14-Stage Dual-Spool, Axial Flow Compressor including: 7-Stage Low Pressure Compressor, 7-Stage High Pressure Compressor.

COMPRESSOR DATA: Compressor Pressure Ratio: 14.5 : 1, Mass Airflow: 420 lb/sec (186 kg/sec).

TURBINE TYPE: 2-Stage Axial Flow Turbine, including: 1-Stage High Pressure Turbine, 1-Stage Low Pressure Turbine.

COMBUSTOR TYPE: Annular, Through Flow.

AFTERBURNER: 1-Stage.

POWER (take-off) RATING: 38,000 lbt (169 kN thrust) [A/B]. 33,000 lbt dry thrust available. Rating Approved to: 59°F (15°C).

TAKE-OFF SPECIFIC FUEL CONSUMPTION (SFC): 1.19 lb/hr/lbt (33,7 mg/Ns) [A/B].

BASIC DRY WEIGHT: 7,588 lbs (3,449 kg).

POWER/WEIGHT RATIO: 5.0 :1 lbt/lb in afterburner.

APPLICATION:
British/French, "Concorde SST" Supersonic (Mach-2) Airliner.

OTHER ENGINES IN SERIES:
Locate other Models under Rolls-Royce Ltd. Great Britain.

GAS TURBINE ENGINES FOR AIRCRAFT

 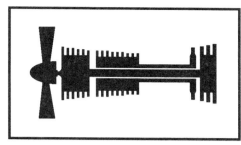

MULTI-NATIONAL TURBINE ENGINE COMPANIES

COUNTRY: GREAT BRITAIN, FRANCE, GERMANY

MANUFACTURER: ROLLS-ROYCE-SNECMA-MTU.
LONDON, ENGLAND

DESIGNATION (civil & military): Tyne-20 MK.21,22.

BACKGROUND: Date First Tyne-20 Engine Produced: 1963.
Date If Out-of-Production: N/A. Number of Tyne-20 20 Engines Produced to Date: 863.

ENGINE TYPE: Turboprop. Dual-Shaft (Compressor Drive & Power Output Drive).

COMPRESSOR TYPE: 15-Stage Dual-Spool, Axial Flow Compressor including: 6-Stage Low Pressure Compressor, 9-Stage High Pressure Compressor.

COMPRESSOR DATA: Compressor Pressure Ratio: 13.9 : 1, Total Mass Airflow: 46.5 lb/sec (21,1 kg/sec).

TURBINE TYPE: 4-Stage Axial Flow Turbine including: 1-Stage High Pressure Turbine, 3-Stage Low Pressure Turbine.

COMBUSTOR TYPE: Can-Annular, Through-Flow, with 10 liners.

POWER (take-off) RATING: 6,100 eshp (4,549 ekw). Includes 450 shp (335 kw) from 1,124 lbt (5.0 kN thrust). Rating Approved to: 86°F (30°C).

TAKE-OFF SPECIFIC FUEL CONSUMPTION (SFC): 0.47 lb/hr/eshp (0,29 kg/hr/ekw).

BASIC DRY WEIGHT: 2,489 lbs (1,129 kg).

POWER/WEIGHT RATIO: 2.45 :1 eshp/lb (4,03 :1 ekw/kg).

APPLICATION:
Tyne MK.21 in Updated Dassault-Breguet "Atlantic", Maritime Aircraft.
Tyne MK.22 in Franco-German Transall C-160 Transport Aircraft.

OTHER ENGINES IN SERIES:
Tyne-1 MK.506 in BAe-Vickers "Vanguard".
Tyne-11 MK.512 in BAe-Vickers "Vanguard".
Tyne-12 MK.515 in Canadair CL-44 "Yukon".
Tyne-12 MK.515-101 in Shorts "Belfast".

GAS TURBINE ENGINES FOR AIRCRAFT

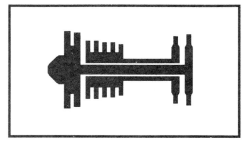

MULTI-NATIONAL TURBINE ENGINE COMPANIES

COUNTRY: GREAT BRITAIN, FRANCE

MANUFACTURER: ROLLS-ROYCE-TURBOMECA.
LONDON, ENGLAND

DESIGNATION (military)**:** Adour Mk.811/815.

BACKGROUND: No Civil Counterpart. USA Military Designation: F405-RR-401.
Date First Adour Mk. 811/815 Engine Produced: 1972.
Date If Out-of-Production: N/A. Number of Adour Mk.811/815 engines produced to date: Unknown.

ENGINE TYPE: Turbofan. Dual-Shaft, Low-Bypass Front Fan with Afterburning.

COMPRESSOR TYPE: 7-Stage Dual-Spool, Axial Flow Compressor including: 2-Stage Front Fan acting as a Low Pressure Compressor, 5-Stage High Pressure Compressor.

COMPRESSOR DATA: Compressor Pressure Ratio: 11.3 : 1, Fan Pressure Ratio: 1.9 : 1, Fan Bypass Ratio: 0.8 : 1, Total Mass Airflow: 94.8 lb/sec (43,1 kg/sec).

TURBINE TYPE: 2-Stage Axial Flow Turbine including: 1-Stage High Pressure Turbine, 1-Stage Low Pressure Turbine.

COMBUSTOR TYPE: Annular, Through-Flow.

AFTERBURNER: 1-Stage.

POWER (take-off) RATING: 5,700 lbt (25,35 kN thrust) dry, 8,400 lbt (37,36 kN thrust) A/B. Rating Approved to: 59°F. (15°C).

TAKE-OFF SPECIFIC FUEL CONSUMPTION (SFC): 0.78 lb/hr/lb (22,1 mg/Ns).

BASIC DRY WEIGHT: 1,720 lbs (780 kg).

POWER/WEIGHT RATIO: 3.3 : 1 lbt/lb (dry), 4.88 : 1 lbt/lb (A/B).

APPLICATION: Sepecat International Corp. as follows:
BAe "Jaguar" GR-1 Fighter, and T2 Trainer, Great Britain.
Dassault "Jaguar" Fighter, France.

OTHER ENGINES IN SERIES:
Adour MK.801 non-afterburning model in Mitsubishi T2 and F1, Japan.
Adour Mk.861, non-afterburning in BAe "Hawk" Trainer Aircraft.
Adour MK.861 and Mk.871, non-A/B models, in T-45A "Goshawk" Trainer Aircraft, USA.

GAS TURBINE ENGINES FOR AIRCRAFT

 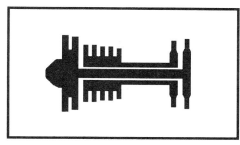

MULTI-NATIONAL TURBINE ENGINE COMPANIES

COUNTRY: GREAT BRITAIN, FRANCE

MANUFACTURER: ROLLS-ROYCE-TURBOMECA. LONDON, ENGLAND

DESIGNATION (military): Adour Mk.871.

BACKGROUND: No Civil Counterpart. (US Military Designation: F405-RR-401).
Date First Adour Mk.871 Engine Produced: 1980.
Date If Out-of-Production: N/A. Number of Adour Mk.871 engines produced to date: Unknown.

ENGINE TYPE: Turbofan. Dual-Shaft, Low-Bypass Front Fan, Non-Afterburning.

COMPRESSOR TYPE: 7-Stage Dual-Spool, Axial Flow Compressor including: 2-Stage Front Fan acting as a Low Pressure Compressor, 5-Stage High Pressure Compressor.

COMPRESSOR DATA: Compressor Pressure Ratio: 11.0: 1, Fan Pressure Ratio: 1.9 : 1, Fan Bypass Ratio: 0.8 : 1, Total Mass Airflow: 96.7 lb/sec (43,86 kg/sec).

TURBINE TYPE: 2-Stage Axial Flow Turbine including: 1-Stage High Pressure Turbine, 1-Stage Low Pressure Turbine.

COMBUSTOR TYPE: Annular, Through-Flow.

POWER (take-off) RATING: 5,845 lbt (25,99 kN thrust). Rating Approved to: 59°F (15°C).

TAKE-OFF SPECIFIC FUEL CONSUMPTION (SFC): 0.76 lb/hr/lbt (21,7 mg/Ns).

BASIC DRY WEIGHT: 1,303 lbs (591 kg).

POWER/WEIGHT RATIO: 4.49 : 1 lbt/lb.

APPLICATION:
BAe "Hawk" Mk.100 & 200 Trainer Aircraft.

OTHER ENGINES IN SERIES:
Adour Mk.811/815 Afterburning Model in Sepecat "Jaguar" Fighter.

GAS TURBINE ENGINES FOR AIRCRAFT

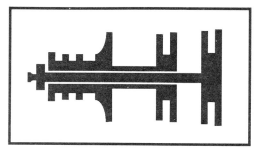

MULTI-NATIONAL TURBINE ENGINE COMPANIES

COUNTRY: GREAT BRITAIN, FRANCE

MANUFACTURER: ROLLS-ROYCE, TURBOMECA, RINALDO-PIAGGIO, MTU
LONDON, ENGLAND

DESIGNATION (military): RTM 322-01.

BACKGROUND: Date First RTM 322 Engine Produced: 1986.
Date If Out-of-Production: N/A. Number of RTM 322-01 engines produced to date: Unknown.

ENGINE TYPE: Turboshaft. Dual-Shaft (Compressor Drive & Power Output Drive).

COMPRESSOR TYPE: Combination Axial-Centrifugal Compressor: 3-Stage Axial, 1-Stage Centrifugal.

COMPRESSOR DATA: Compressor Pressure Ratio: 14.8:1. Mass Airflow: 17.4 lb/sec (7,9 kg/sec).

TURBINE TYPE: 4-Stage Axial Flow Turbine including: 2-Stage Gas Producer (compressor drive) Turbine, 2-Stage Power (free) Turbine.

COMBUSTOR TYPE: Annular, Reverse Flow.

POWER (take-off) RATING: 2,240 shp (1670 kw). Rating Approved to: 59°F (15°C).

TAKE-OFF SPECIFIC FUEL CONSUMPTION (SFC): 0.48 lb/hr/shp (0,29 kg/hr/kw).

BASIC DRY WEIGHT: 529 lbs (240 kg).

POWER/WEIGHT RATIO: 4.23 :1 shp/lb (6,96 :1 kw/kg).

APPLICATION:
EH Industries (Agusta-Westland), EH-101, "Merlin" Military Helicopter.
MBB GmbH, NH-90 Helicopter in Development (Germany, France, Italy).
Kamov, Ka-62R Civil Helicopter (CIS).

OTHER ENGINES IN SERIES:
RTM-321, 1700 SHP in development.

GAS TURBINE ENGINES FOR AIRCRAFT

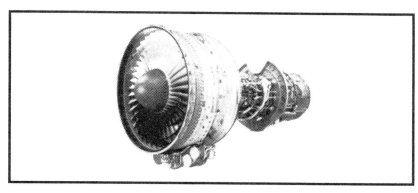
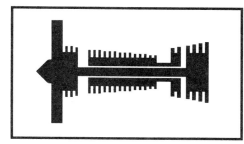

MULTI-NATIONAL TURBINE ENGINE COMPANIES

COUNTRY: USA, FRANCE

MANUFACTURER: SNECMA, GENERAL ELECTRIC (USA)

DESIGNATION (civil)**:** CF6-50E.

BACKGROUND: Date First CF6 Engine Produced: 1972.
Date If Out-of-Production: N/A. Number of CF6-50E engines produced to date: Unknown.
Military designation : F-103-GE-100. Manufacturing partners include: SNECMA and MTU.

ENGINE TYPE: Turbofan. Dual-Shaft, High-Bypass Front Fan.

COMPRESSOR TYPE: 18-Stage Dual-Spool, Axial Flow Compressor including: 1-Stage Front Fan and Three Axial Stages in the Low Pressure Compressor, 14-Stage High Pressure Compressor.

COMPRESSOR DATA: Compressor Pressure Ratio: 29.2 : 1, Fan Pressure Ratio: 1.5: 1, Fan Bypass Ratio: 4.4 : 1, Total Mass Airflow: 1,465 lb/sec (664,4 kg/sec).

TURBINE TYPE: 6-Stage Axial Flow Turbine including: 2-Stage High Pressure Turbine, 4-Stage Low Pressure Turbine.

COMBUSTOR TYPE: Annular, Through-Flow.

POWER (take-off) RATING: 51,000 lbt (226,8 kN thrust). Rating Approved to: 86°F (30°C).

TAKE-OFF SPECIFIC FUEL CONSUMPTION (SFC): 0.37 lb/hr/lbt (10,48 mg/Ns).

BASIC DRY WEIGHT: 8,731 lbs (3,960 kg).

POWER/WEIGHT RATIO: 5.84 : 1 lbt/lb.

APPLICATION:
Boeing B747-200 and Boeing B747-SP Airliners.

OTHER ENGINES IN SERIES:
Earlier models CF6-6 and CF6-50 in:
McDonnell-Douglas DC10-10, DC10-30, Boeing B-747, Airbus A300-B Airliners.
KC-10 Military Tanker.
Boeing E-4 Airborne Command Post.

GAS TURBINE ENGINES FOR AIRCRAFT

 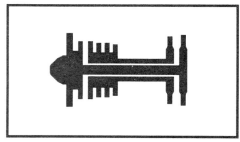

MULTI-NATIONAL TURBINE ENGINE COMPANIES

COUNTRY: FRANCE, GERMANY

MANUFACTURER: TURBOMECA & SNECMA-MTU Ltd.
PARIS, FRANCE

DESIGNATION (civil & military): LARZAC 04 -C20.

BACKGROUND: Date First Larzac 04-020 Engine Produced: 1988.
Date If Out-of-Production N/A. Number of Larzac 04-C20 Engines Produced to Date: 500.

ENGINE TYPE: Turbofan. Dual-Shaft, Low-Bypass Front Fan (non-afterburning).

COMPRESSOR TYPE: 6-Stage, Dual-Spool, Axial Flow Compressor including: 1-Stage Front Fan and (1) additional Axial Stage in the Low Pressure Compressor, 4-Stage High Pressure Compressor.

COMPRESSOR DATA: Compressor Pressure Ratio: 11.1 : 1, Fan Pressure Ratio: 2.2 : 1, Fan Bypass Ratio: 1.04 : 1, Total Mass Airflow: 63 lb/sec (28,6 kg/sec).

TURBINE TYPE: 2-Stage Axial Flow Turbine including: 1-Stage High Pressure Turbine, 1-Stage Low Pressure Turbine.

COMBUSTOR TYPE: Annular, Through-Flow.

POWER (take-off) RATING: 3,180 lbt (14,14 kN thrust). Rating Approved to: 59°F (15°C).

TAKE-OFF SPECIFIC FUEL CONSUMPTION (SFC): 0.71 lb/hr/lbt (20,1 mg/Ns)

BASIC DRY WEIGHT: 640 lbs (290 kg).

POWER/WEIGHT RATIO: 4.97 : 1 lbt/lb.

APPLICATION:
Dassault-Breguet/Dornier, "Alpha-Jet" Attack/Trainer Aircraft.

OTHER ENGINES IN SERIES:
Larzac 04 C6, also in the Alpha Jet.
Larzac 49 in Dassault Falcon-10 and Corvette Commuter/Business Jet Aircraft.

GAS TURBINE ENGINES FOR AIRCRAFT

 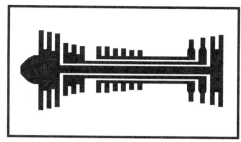

MULTI-NATIONAL TURBINE ENGINE COMPANIES

COUNTRY: ITALY, GERMANY, GREAT BRITAIN

MANUFACTURER: TURBO-UNION Ltd. (Fiat, MTU, Rolls-Royce).
BRISTOL, ENGLAND

DESIGNATION (military): RB-199-34 (MK.103).

BACKGROUND: No Civil Counterpart. An Original R-R Design.
Date First RB-199-34R Engine Produced: Unknown.
Date If Out-of-Production: N/A. Number of RB-199-34 engines produced to date: Unknown.

ENGINE TYPE: Turbofan. Triple-Shaft, Low-Bypass Front Fan with Afterburning.

COMPRESSOR TYPE: 12-Stage Triple-Spool, Axial Flow Compressor including: 3-Stage Front Fan acting also As a Low Pressure Compressor, 3-Stage Intermediate Pressure Compressor, 6-Stage High Pressure Compressor.

COMPRESSOR DATA: Compressor Pressure Ratio: 23.0 : 1, Fan Pressure Ratio: 3 : 1, Fan Bypass Ratio: 1.06 : 1, Total Mass Airflow: 190 lb/sec (86 kg/sec).

TURBINE TYPE: 4-Stage Axial Flow Turbine including: 1-Stage High Pressure Turbine, 1-Stage Intermediate Pressure Turbine, 2-Stage Low Pressure Turbine.

COMBUSTOR TYPE: Annular, Through-Flow.

AFTERBURNER: Variable Stage, combined Fan and Core flow.

POWER (take-off) RATING: 9,556 lbt (42,5 kN) Dry, 16,700 lbt (74,3 kN thrust) [A/B]. Rating Approved to: 59°F. (15°C).

TAKE-OFF SPECIFIC FUEL CONSUMPTION (SFC): 0.75 lb/hr/lbt (21,16 mg/Ns) dry, 1.95 lb/hr/lbt (55 mg/Ns) [A/B].

BASIC DRY WEIGHT: 2,387 lbs (1,083 kg).

POWER/WEIGHT RATIO: 4.0 : 1 lbt/lb (dry), 7.0 : 1 lbt/lb. (A/B).

APPLICATION:
Panavia Tornado GR-1 IDS "Interdictor-Strike" Fighter.
Panavia Tornado F-2 ADV "Air Defense Variant" Interceptor Aircraft.
Note: Panavia Corp. (Great Britain, Germany, Italy).

OTHER ENGINES IN SERIES:
RB-199-34 MK.104 and MK.105 in upgraded Tornado.

GAS TURBINE ENGINES FOR AIRCRAFT

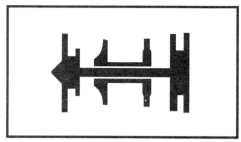

MULTI-NATIONAL TURBINE ENGINE COMPANIES

COUNTRY: USA, GREAT BRITAIN

MANUFACTURER: WILLIAMS INTERNATIONAL CORP.
(Williams & Rolls-Royce)
WALLED LAKE, MICHIGAN.

DESIGNATION (civil): FJ44.

BACKGROUND: Date First FJ44 Engine Produced: In Development (Due 1991).
Date If Out-of-Production N/A. Number of FJ44 Engines Produced to Date: N/A.

ENGINE TYPE: Turbofan. Dual-Shaft.

COMPRESSOR TYPE: Combination Axial-Centrifugal Flow: 1-Stage Fan followed by One-Stage of Axial Compression in the Low Pressure Compressor. I-Stage Centrifugal Compressor in the High Pressure Compressor.

COMPRESSOR DATA: Compressor Pressure Ratio: 12.8: 1, Fan Bypass Ratio: 3.28 : 1. Total Mass Airflow: 63.3 lb/sec (28,7 kg/sec).

TURBINE TYPE: 3-Stage Axial Flow Turbine including: 1-Stage High Pressure Turbine, 2-Stage Low Pressure Turbine.

COMBUSTOR TYPE: Annular, Through Flow.

POWER (take-off) RATING: 1,900 lbt (8,45 kN thrust). Rating Approved to: 72°F (21°C).

TAKE-OFF SPECIFIC FUEL CONSUMPTION (SFC): 0.47 lb/hr/lbt (13,31 mg/Ns).

BASIC DRY WEIGHT: 445 lbs (202 kg).

POWER/WEIGHT RATIO: 4.27 : 1 lbt/lb.

APPLICATION:
Swearingen SJ-30 Business Jet Aircraft. (originally known as SA-30),
Cessna Citation-Jet Business jet Aircraft. (Follow on aircraft in Citation Series)

OTHER ENGINES IN SERIES:
Williams International also manufactures F107-WR-series (600 lbt) Turbofan Engine in the Cruise Missile and several other engines for pilotless vehicles.

GAS TURBINE ENGINES FOR AIRCRAFT

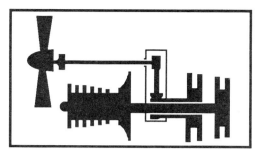

USA TURBINE ENGINE COMPANIES

MANUFACTURER: ALLISON GAS TURBINE DIVISION (General Motors). INDIANAPOLIS, INDIANA.

DESIGNATION (civil): 250-B17E.

BACKGROUND: Civil version of the military T-63 Turboshaft (first produced in 1963).
Date First Allison 250 Engine Produced: 1970.
Date If Out-of-Production: N/A. Number of "250" Engines Produced to Date: 25,000.

ENGINE TYPE: Turboprop. Dual-Shaft (Compressor Drive & Power Output Drive).

COMPRESSOR TYPE: Combination Compressor: 6-Stage Axial, 1-Stage Centrifugal.

COMPRESSOR DATA: Compressor Pressure Ratio: 7.2: 1, Total Mass Airflow: 3.6 lb/sec (1,63 kg/sec).

TURBINE TYPE: 4-Stage Axial Flow Turbine, including a 2-Stage Gas Producer (compressor drive) Turbine, 2-Stage Power (free) Turbine.

COMBUSTOR TYPE: Through Flow, Can-Type with a single Liner.

POWER (take-off) RATING: 420 eshp (313 ekw)., including 6.3 shp (4,7 kw) from 40 lbt (0,18 kN thrust). Rating Approved to: 59°F (15°C).

TAKE-OFF SPECIFIC FUEL CONSUMPTION (SFC): 0.66 lb/hr/eshp (0,40 kg/hr/ekw).

BASIC DRY WEIGHT: 195 lbs (88 kg).

POWER/WEIGHT RATIO: 2.15 : 1 eshp/lb (3,56 :1 ekw/kg).

APPLICATION:
Cessna 402 "Re-Engined", USA; Business Aircraft.
GAF N22 and 24 "Nomad" Australia; Civil/Military Aircraft.
Partenavia AP-68 "Sparticus" and "Viator", Italy; Business Aircraft.

OTHER ENGINES IN SERIES:
250-B17 in Beech Bonanza Re-Engined, USA.
250-B-17 in Maul M-7 Re-Engined, USA.
250-B17C in Cessna "Nomad" Business Aircraft, USA.
250-B17D in Arocet AT-9, USAF Trainer.
250-B17D in Fuji KM-2D "T-5" Re-Engined, Japan and RTAF-5, Thailand.
250-B17D in Hindustan HTT-34, Trainer Aircraft, India.
250-B17D/F in Siai-Marchetti SM-109E and SF-260, Civil/Military Aircraft, Italy.
250-B17D/F in Pilatus BN-2T Series, Civil/Military Aircraft, Great Britain.
250-B17F in Valmet L-90 "Redigo" Military Trainer Aircraft, Finland.

GAS TURBINE ENGINES FOR AIRCRAFT

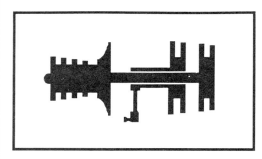

USA TURBINE ENGINE COMPANIES

MANUFACTURER: ALLISON GAS TURBINE DIVISION (General Motors). INDIANAPOLIS, INDIANA.

DESIGNATION (civil)**:** 250-C20R.

BACKGROUND: Civil version of the military T-63 Turboshaft.
Date First 250-C20R Engine Produced: 1970.
Date If Out-of-Production: N/A. Number of "250-C" Engines Produced to Date: 23,000.

ENGINE TYPE: Turboshaft. Dual-Shaft (Compressor Drive & Power Output Drive).

COMPRESSOR TYPE: Combination Compressor: 4-Stage Axial, 1-Stage Centrifugal.

COMPRESSOR DATA: Compressor Pressure Ratio: 7.9: 1, Total Mass Airflow: 3.8 lb/sec (1,72 kg/sec).

TURBINE TYPE: 4-Stage Axial Flow Turbine, including a 2-Stage Gas Producer (compressor drive) Turbine, 2-Stage Power (free) Turbine.

COMBUSTOR TYPE: Through Flow, Can-Type with a single Liner.

POWER (take-off) RATING: 450 shp (336 kw). Rating Approved to: 59°F (15°C).

TAKE-OFF SPECIFIC FUEL CONSUMPTION (SFC): 0.61 lb/hr/shp (0,37 kg/hr/kw).

BASIC DRY WEIGHT: 173 lbs (78,5 kg).

POWER/WEIGHT RATIO: 2.6 : 1 shp/lb (4,28 : 1 kw/kg).

APPLICATION:
Rhein-Flugbugbar FT-400 "FanTrainer" Ducted Fan type Helicopter.

OTHER ENGINES IN SERIES:
250-C20B, F, J; in: Bell 206B/L, and Hughes 500D/E, USA.
   250-C20B in: MBB BO.105, Germany.
               Brenda-Nardi 369D, India.
               Agusta AB 206, Italy.
               IPTN NBO.105, Indonesia.
               Kawasaki OH-6D, Japan.
               WSK-PZL Mi-2, Poland.
               Rogerson-Hiller UH-12/UN-12, USA.
   250-C20R in: MDHC MD-500/530, USA.
               Agusta A.109A, Italy.
   250-C20W in: Enstrom TH-28 and 480, USA.
               Schweizer-330, USA.

GAS TURBINE ENGINES FOR AIRCRAFT

 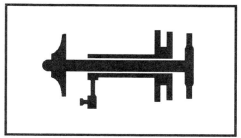

USA TURBINE ENGINE COMPANIES

MANUFACTURER: ALLISON GAS TURBINE DIVISION (General Motors). INDIANAPOLIS, INDIANA.

DESIGNATION (civil): 250-C34.

BACKGROUND: Civil version of military T703 Turboshaft.
Date First 250-C34 Engine Produced: In development.
Date If Out-of-Production: N/A. Number of 250-C Engines Produced to Date: 23,000.

ENGINE TYPE: Turboshaft. Dual-Shaft (Compressor Drive & Power Output Drive).

COMPRESSOR TYPE: Centrifugal Compressor: 1-Stage, Single Entry.

COMPRESSOR DATA: Compressor Pressure Ratio: 8.5:1, Total Mass Airflow: 5.5 lb/sec (2,49 kg/sec).

TURBINE TYPE: 3-Stage Axial Flow Turbine, including a 1-Stage Gas Producer (compressor drive) Turbine, 2-Stage Power (free) Turbine.

COMBUSTOR TYPE: Through Flow, Can-Type with a single Liner.

POWER (take-off) RATING: 735 shp (548 kw). Rating Approved to: 90°F (32°C).

TAKE-OFF SPECIFIC FUEL CONSUMPTION (SFC): 0.60 lb/hr/shp (0,37 kg/hr/kw).

BASIC DRY WEIGHT: 255 lbs (116 kg).

POWER/WEIGHT RATIO: 2.88 : 1 shp/lb (4,7 : 1 kw/kg).

APPLICATION:
McDonnell-Douglas MD-530 Notor™ Helicopter.

OTHER ENGINES IN SERIES:
250-C28B & C in Bell 206L, "LongRanger", USA; and Messerschmitt BO-105, Germany.
250-C30L,M,P,R in Hughes 500 "Defender" and Sikorsky S-76A "Spirit", USA.
250-C30G2 in Bell-230, USA.
250-C30L in Rhein-Flugbugbau FT-600 "Fantrainer", Germany.
250-C30S in MBB BK.117, Germany.
250-C30G in Bell 222 Re-Engined, USA.
250-C30R in Bell 406, USA.

GAS TURBINE ENGINES FOR AIRCRAFT

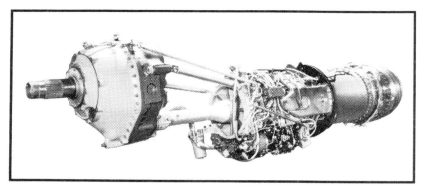 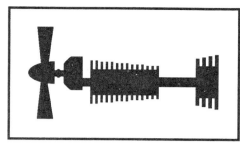

USA TURBINE ENGINE COMPANIES

MANUFACTURER: ALLISON GAS TURBINE DIVISION (General Motors). INDIANAPOLIS, INDIANA.

DESIGNATION (civil): 501-D22A & G.

BACKGROUND: Civil version of military T-56. This engine has the longest production history of any U.S. Gas Turbine Engine. It was the first U.S. engine to use cooled H.P. Turbine Blades. Date First Allison 501 Engine Produced: 1953.
Date If Out-of-Production: N/A. Number of "501" Engines Produced to Date: 14,600.

ENGINE TYPE: Turboprop. Single-Shaft Design.

COMPRESSOR TYPE: 14-Stage Single-Spool Axial Flow Compressor.

COMPRESSOR DATA: Compressor Pressure Ratio: 10.8:1, Total Mass Airflow: 32.4 lb/sec (14,7 kg/sec).

TURBINE TYPE: 4-Stage Axial Flow Turbine.

COMBUSTOR TYPE: Can-Annular, Through Flow, with 6 liners.

POWER (take-off) RATING: 4,680 eshp (3,490 ekw), includes 318 shp (237 kw) from 795 lbt (3,53 kN thrust). Rated at 1,049°C. Turbine Inlet Temperature.

TAKE-OFF SPECIFIC FUEL CONSUMPTION (SFC): 0.50 lb/hr/eshp (0,30 kg/hr/ekw).

BASIC DRY WEIGHT: 1,820 lbs (826 kg).

POWER/WEIGHT RATIO: 2.57:1 eshp/lb (4,2:1 ekw/kg).
Note: Turboprop models have high reduction gearbox weight and subsequent lower power/weight ratios than comparable Turboshaft models.

APPLICATION:
Lockheed L-100 "Hercules" Airliner/Transport.
Convair Super-580 and 600/660 Airliner/Transport.

OTHER ENGINES IN SERIES:
501-D-13 in Lockheed L-188 "Electra", Convair-540/580, USAF C-131.
501-D-22C in Super-Guppy Re-Engined.
Derivatives of the 501 include:
501-M78 (NASA Propfan Demonstrator).
501-M80C in Bell/Boeing V22, "Osprey", Tilt-Rotor Aircraft.

GAS TURBINE ENGINES FOR AIRCRAFT

 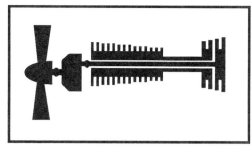

USA TURBINE ENGINE COMPANIES

MANUFACTURER: ALLISON GAS TURBINE DIVISION (General Motors). INDIANAPOLIS, INDIANA

DESIGNATION (civil): GMA-2100 (Re-rated version of military T406-AD-400).

BACKGROUND: Derived from T-56 and T701 Engines. Built under Rolls-Royce License.
Date First T406 Engine Produced: In Development.
Date If Out-of-Production: N/A. Number of T406 Engines Produced to Date: N/A.

ENGINE TYPE: Turboprop. Dual-Shaft (Compressor Drive & Power Output Drive).

COMPRESSOR TYPE: 14-Stage Single-Spool, Axial Flow Compressor.

COMPRESSOR DATA: Compressor Pressure Ratio: 14.1. Total Mass Airflow: 35.5 lb/sec (16,1 kg/sec).

TURBINE TYPE: 4-Stage Axial Flow Turbine, including a 2-Stage Gas Producer (compressor drive) Turbine, 2-Stage Power (free) Turbine.

COMBUSTOR TYPE: Annular, Through Flow.

POWER (take-off) RATING: 5,200 shp (3,877 kw). Rating Approved to: 90°F (32°C).

TAKE-OFF SPECIFIC FUEL CONSUMPTION (SFC): 0.426 lb/hr/shp (0,26 kg/hr/kw).

BASIC DRY WEIGHT: 971 lb (440 kg).

POWER/WEIGHT RATIO: 5.36 : 1 shp/lb (6,39 : 1 kw/kg).

APPLICATION:
Saab-Scania SAAB-2000, Business Aircraft, Sweden.
IPTN N-250 Transport, Indonesia.

OTHER ENGINES IN SERIES:
GMA-2100D in IPTN (N-250) and C-130J/L-100 in development.

GAS TURBINE ENGINES FOR AIRCRAFT

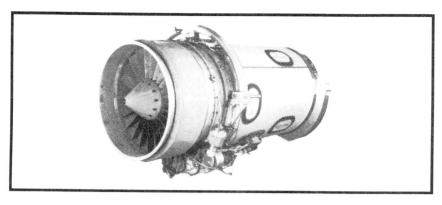 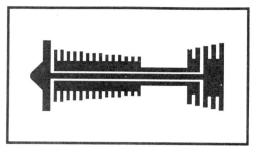

USA TURBINE ENGINE COMPANIES

MANUFACTURER: ALLISON GAS TURBINE DIVISION (General Motors). INDIANAPOLIS, INDIANA.

DESIGNATION (civil): GMA 3007.

BACKGROUND: This engine utilizes the T-406 Turboprop Core. Expected Date First GMA 3007 Engine Produced: In development.

ENGINE TYPE: Turbofan. Dual-Shaft Design.

COMPRESSOR TYPE: 15-Stage Axial Flow, including a 1-Stage Fan acting as a low pressure compressor and a 14-Stage High Pressure Compressor.

COMPRESSOR DATA: Compressor Pressure Ratio: 23 : 1. Mass Airflow: 145 lb/sec (66 kg/sec).

TURBINE TYPE: 5-Stage Axial Flow, including a 2-Stage High Pressure Turbine and a 3-Stage Low Pressure Turbine.

COMBUSTOR TYPE: Annular, Through Flow.

POWER (take-off) RATING: 7,000 lbt (31,1 kN). Rating Approved to: 59°F (15°C).

TAKE-OFF SPECIFIC FUEL CONSUMPTION (SFC): 0.49 lb/hr/lbt (estimated).

BASIC DRY WEIGHT: 1400 lbs (635 kg).

POWER/WEIGHT RATIO: 5.0 : 1 lbt/lb.

APPLICATION:
Embraer EMB-145, Brazil.
Cessna, Citation-X, USA.

OTHER ENGINES IN SERIES:
GAMA 3004 in development for Citation-X.

GAS TURBINE ENGINES FOR AIRCRAFT

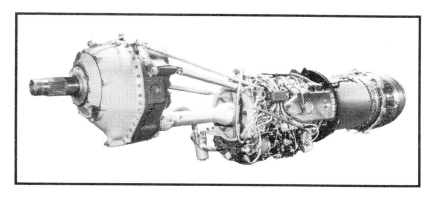
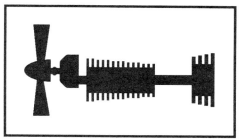

USA TURBINE ENGINE COMPANIES

MANUFACTURER: ALLISON GAS TURBINE DIVISION (General Motors). INDIANAPOLIS, INDIANA

DESIGNATION (military): T56-A-425.

BACKGROUND: Date First T56 Engine Produced: 1953.
Date If Out-of-Production: N/A . Number of Engines T56 Produced to Date: 14,600 .

ENGINE TYPE: Turboprop. Single-Shaft Design.

COMPRESSOR TYPE: 14-Stage Single-Spool Axial Flow Compressor.

COMPRESSOR DATA: Compressor Pressure Ratio: 9.6:1, Total Mass Airflow: 32.4 lb/sec (14.7 kg/sec).

TURBINE TYPE: 4-Stage Axial Flow Turbine.

COMBUSTOR TYPE: Through Flow, Can-Annular with 6 liners.

POWER (take-off) RATING: 4,591 eshp (3,424 ekw), includes 320 shp (239 kw) from 797 lbt (354 kN thrust). Rating Approved to: 59°F (15°C).

TAKE-OFF SPECIFIC FUEL CONSUMPTION (SFC): 0.54 lb/hr/eshp (0,33 kg/hr/ekw).

BASIC DRY WEIGHT: 1,899 lbs (861 kg).

POWER/WEIGHT RATIO: 2.42 : 1 eshp/lb (3.98 : 1 ekw/kg).

APPLICATION:
Grumman E-2C "Hawkeye" Military Early Warning Aircraft.

OTHER ENGINES IN SERIES:
T56-A-14 in Lockheed P-3C "Orion" USAF and CLP-140, Canada; Maritime Aircraft.
T56-A-7B in C-130 Military Transport.
T56-A-15 in Lockheed HC-130H, Hercules and AC-130 Gunship.
T56-A-101 in C-130 Super Hercules Military Transport.
T56-A-16 and -423 in C-130 USN Transport.
T56-A-427 (5,200 ESHP) in Grumman E-2C.

GAS TURBINE ENGINES FOR AIRCRAFT

 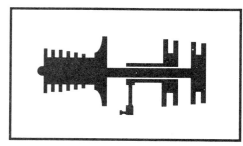

USA TURBINE ENGINE COMPANIES

MANUFACTURER: ALLISON GAS TURBINE DIVISION (General Motors). INDIANAPOLIS, INDIANA

DESIGNATION (military): T63-A-720.

BACKGROUND: Forerunner of the civil Allison 250 Turboshaft Series.
Date First T63 Engine Produced: 1959.
Date If Out-of-Production: N/A. Number of T-63 Engines Produced to Date: 4,850.
Total number of Allison 250 engines produced: 23,000.

ENGINE TYPE: Turboshaft. Dual-Shaft (Compressor Drive & Power Output Drive).

COMPRESSOR TYPE: Combination Compressor: 6-Stage Axial, 1-Stage Centrifugal.

COMPRESSOR DATA: Compressor Pressure Ratio: 7.3: 1, Total Mass Airflow: 3.6 lb/sec (1,63 kg/sec).

TURBINE TYPE: 4-Stage Axial Flow Turbine, including a 2-Stage Gas Producer (compressor drive) Turbine, 2-Stage Power (free) Turbine.

COMBUSTOR TYPE: Through Flow, Can-Type with a single Liner.

POWER (take-off) RATING: 420 shp (313 kw). Rating Approved to: 59°F (15°C).

TAKE-OFF SPECIFIC FUEL CONSUMPTION (SFC): 0.65 lb/hr/shp (0,40 kg/hr/kw).

BASIC DRY WEIGHT: 158 lbs (72 kg).

POWER/WEIGHT RATIO: 2.66 :1 shp/lb (4,35 :1 kw/kg).

APPLICATION:
Bell OH-58C "Kiowa" Helicopter.
Bell TH-57 "Jet Ranger" and "Sea Ranger" Helicopter.
MDHC MD-500E (OH-6) Helicopter.
Rogerson RH-1100 "Hornet" Helicopter.

OTHER ENGINES IN SERIES:
T63A-5 in early Rogerson RH-1100 (formerly Fairchild-Hiller) Helicopter.

GAS TURBINE ENGINES FOR AIRCRAFT

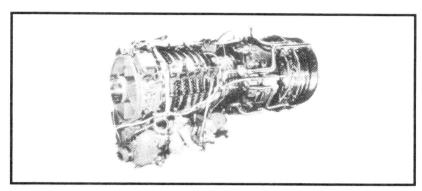 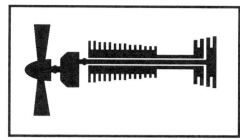

USA TURBINE ENGINE COMPANIES

MANUFACTURER: **ALLISON GAS TURBINE DIVISION (General Motors). INDIANAPOLIS, INDIANA**

DESIGNATION (military)**:** T406-AD-400. (civil version designated as GMA-2100).

BACKGROUND: Derived from T-56 and T701 Engines. Built under Rolls-Royce License. Date First T406 Engine Produced: In Development.

ENGINE TYPE: Turboprop. Dual-Shaft (Compressor Drive & Power Output Drive).

COMPRESSOR TYPE: 14-Stage Single-Spool, Axial Flow Compressor.

COMPRESSOR DATA: Compressor Pressure Ratio: 14.1. Total Mass Airflow: 35.5 lb/sec (16,1 kg/sec).

TURBINE TYPE: 4-Stage Axial Flow Turbine, including a 2-Stage Gas Producer (compressor drive) Turbine, 2-Stage Power (free) Turbine.

COMBUSTOR TYPE: Annular, Through Flow.

POWER (take-off) RATING: 6,150 shp (4,586 kw). Rating Approved to: 59°F (15°C).

TAKE-OFF SPECIFIC FUEL CONSUMPTION (SFC): 0.424 lb/hr/shp (0,26 Kg/hr/Kw).

BASIC DRY WEIGHT: 971 lb (440 kg).

POWER/WEIGHT RATIO: 6.3 : 1 shp/lb (10,4 : 1 kw/kg).

APPLICATION:
Bell-Boeing V-22 "Osprey", Tilt-Rotor Aircraft Military Transport.

OTHER ENGINES IN SERIES:
Turboprop Version in Development Designated as GMA-2100.

GAS TURBINE ENGINES FOR AIRCRAFT

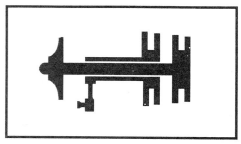

USA TURBINE ENGINE COMPANIES

MANUFACTURER: ALLISON GAS TURBINE DIVISION (General Motors)
INDIANAPOLIS, INDIANA

DESIGNATION (military): T703-A-700.

BACKGROUND: Military Version of Civil 250-C34.
Date First T703 Engine Produced: 1985.
Date If Out-of-Production: N/A. Number of 250 Engines Produced to Date: 23,000.

ENGINE TYPE: Turboshaft. Dual-Shaft (Compressor Drive & Power Output Drive).

COMPRESSOR TYPE: Centrifugal Compressor: 1-Stage, Single Entry.

COMPRESSOR DATA: Compressor Pressure Ratio: 8.6:1, Total Mass Airflow: 5.5 lb/sec (2,49 kg/sec).

TURBINE TYPE: 4-Stage Axial Flow Turbine, including a 2-Stage Gas Producer (compressor drive) Turbine, 2-Stage Power (free) Turbine.

COMBUSTOR TYPE: Through Flow, Can-Type with a single Liner.

POWER (take-off) RATING: 650 shp (485 kw). Rating Approved to: 59°F (15°C).

TAKE-OFF SPECIFIC FUEL CONSUMPTION (SFC): 0.59 lb/hr/shp (0,36 kg/hr/kw).

BASIC DRY WEIGHT: 240 lbs (109 kg).

POWER/WEIGHT RATIO: 2.70:1 shp/lb (4,45:1 kw/kg).

APPLICATION:
Bell OH-58D "Kiowa" Helicopter.
MDHC MD-530 Civil and H-6-530 Military NOTOR™ Helicopters.

OTHER ENGINES IN SERIES:
None.

GAS TURBINE ENGINES FOR AIRCRAFT

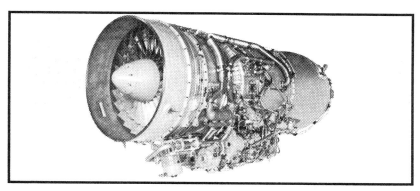 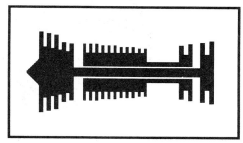

USA TURBINE ENGINE COMPANIES

MANUFACTURER: ALLISON GAS TURBINE DIVISION (General Motors). INDIANAPOLIS, INDIANA.

DESIGNATION (military): TF41-A-400.

BACKGROUND: A non-afterburning derivative of the RB.168, Rolls-Royce "Spey" Turbofan. The afterburning model designated as: 912-B23.
Date First TF41 Engine Produced: 1968.
Date If Out-of-Production: N/A. Number of TF41 Engines Produced to Date: 14,000.

ENGINE TYPE: Turbofan. Dual-Shaft, Low-Bypass, Front Fan (non-afterburning).

COMPRESSOR TYPE: 16-Stage Dual-Spool, including a 5-stage Low Pressure Compressor with 3 Fan stages and two Axial Flow Compressor stages and an 11-Stage High Pressure Compressor.

COMPRESSOR DATA: Compressor Pressure Ratio: 21.4 : 1, Fan Pressure Ratio: 2.45 : 1, Fan Bypass Ratio: 0.76: 1, Total Mass Airflow: 269 lb/sec (122 kg/sec).

TURBINE TYPE: 4-Stage Axial Flow Turbine including: 2-Stage High Pressure Turbine, 2-Stage Low Pressure Turbine.

COMBUSTOR TYPE: Can-Annular, Through-Flow, with 10 liners.

POWER (take-off) RATING: 15,000 lbt (66,7 kN thrust). Rating Approved to: 59°F (15°C)

TAKE-OFF SPECIFIC FUEL CONSUMPTION (SFC): 0.67 lb/hr/lbt (18,98 mg/Ns).

BASIC DRY WEIGHT: 3,024 lbs (1,372 kg).

POWER/WEIGHT RATIO: 4.96 : 1 lbt/lb.

APPLICATION:
LTV (Ling-Tempco-Vought) A-7E/H "Corsair-II" Attack Aircraft.

OTHER ENGINES IN SERIES:
TF-41-1B in LTV A-7D.
TF-41-A-1 in LTV A-7K.

GAS TURBINE ENGINES FOR AIRCRAFT

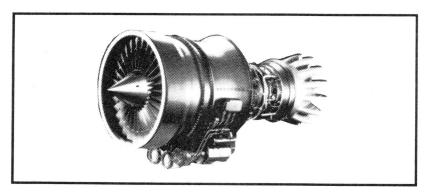
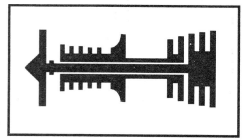

USA TURBINE ENGINE COMPANIES

MANUFACTURER: CFE COMPANY (General Electric and Garrett Div. of Allied Signal Aerospace Corp.)
PHOENIX, ARIZONA

DESIGNATION (civil)**:** CFE 738. Version of Military GE 38.

BACKGROUND: Developed from the GE27 Demonstrator Engine.
Expected Date First CFE 738 Engine Produced: 1992.
Date If Out-of-Production: N/A. Number of CFE 738 Engines Produced to Date: In Development.

ENGINE TYPE: Turbofan. Dual-Shaft, High Bypass Front Fan.

COMPRESSOR TYPE: 7-Stage Dual-Compressor with a single Front Fan acting as a Low Pressure Compressor and a Combination 5-Stage Axial Flow and 1-Stage Centrifugal Flow Compressor acting as a High Pressure Compressor.

COMPRESSOR DATA: Compressor Pressure Ratio: 25 : 1, Fan Pressure Ratio: 1.73 : 1, Fan Bypass Ratio: 5.4: 1, Total Mass Airflow: 254 lb/sec (115 kg/sec).

TURBINE TYPE: 5-Stage Axial Flow Turbine including: 2-Stage High Pressure Turbine coupled to the axial-centrifugal compressor, 3-Stage Low Pressure Turbine directly driving the single stage fan.

COMBUSTOR TYPE: Annular, Through Flow.

POWER (take-off) RATING: 5,990 lbt (26,64 kN thrust). Rating Approved to: 86°F (30°C).

TAKE-OFF SPECIFIC FUEL CONSUMPTION (SFC): 0.37 lb/hr/lbt (10,48 mg/Ns).

BASIC DRY WEIGHT: 1,230 lbs (558 kg).

POWER/WEIGHT RATIO: 4.87 : 1 lbt/lb.

APPLICATION:
Dassault-Falcon 2000, France; Commuter/Business Jet Aircraft.
IPTN N-250, Indonesia; Commuter/Business Jet Aircraft.
Lockheed P-3C "Orion" Maritime Aircraft, proposed.

OTHER ENGINES IN SERIES:
GE-38 turboprop P-3C Patrol Aircraft proposed.

GAS TURBINE ENGINES FOR AIRCRAFT

 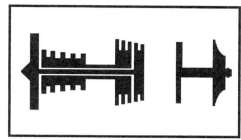

USA TURBINE ENGINE COMPANIES

MANUFACTURER: **GARRETT TURBINE ENGINE DIVISION (Allied Signal Aerospace) PHOENIX, ARIZONA.**

DESIGNATION (civil & military)**:** ATF3-6A.

BACKGROUND: Originally developed as the military F104-GA-100 for its unique side mounted exhaust and low infra-red signature design.
Date First ATF3-6A Engine Produced: 1975.
Date If Out-of-Production: N/A. Number of ATF3 Engines Produced to Date: 220.

ENGINE TYPE: Turbofan. Three-Shaft, Medium Bypass Front Fan.

COMPRESSOR TYPE: 7-Stage Triple Compressor, a Combination Compressor with both Axial Flow and Centrifugal Flow, including: Single Stage Fan acting as a Low Pressure Compressor, 5-Stage Axial Intermediate Pressure Compressor, 1-Stage Centrifugal High Pressure Compressor.

COMPRESSOR DATA: Compressor Pressure Ratio: 21.3 : 1, Fan Pressure Ratio: 1.5 : 1, Fan Bypass Ratio: 3 : 1, Total Mass Airflow: 188 lb/sec (85,3 kg/sec).

TURBINE TYPE: 6-Stage Axial Flow Turbine including: 1-Stage High Pressure Turbine, 3-Stage Intermediate Pressure Turbine, 2-Stage Low Pressure Turbine.

COMBUSTOR TYPE: Annular, Reverse Flow.

POWER (take-off) RATING: 5,440 lbt (24,2 kN thrust). Rating Approved to: 80°F (26.7°C).

TAKE-OFF SPECIFIC FUEL CONSUMPTION (SFC): 0.51 lb/hr/lbt (14,45 mg/Ns).

BASIC DRY WEIGHT: 1,125 lbs (510 kg).

POWER/WEIGHT RATIO: 4.84 : 1 lbt/lb.

APPLICATION:
Dassault-Breguet Falcon-200 Business Jet Aircraft, France.
Dassault-Breguet Falcon-Guardian Maritime Aircraft, France.

OTHER ENGINES IN SERIES:
F104-GA-100 (ATF3-6) model in the U.S. Coast Guard, HU-25A Patrol Aircraft.

GAS TURBINE ENGINES FOR AIRCRAFT

 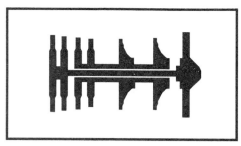

USA TURBINE ENGINE COMPANIES

MANUFACTURER: **GARRETT TURBINE ENGINE DIVISION (Allied Signal Aerospace) PHOENIX, ARIZONA.**

DESIGNATION (civil & military)**:** F109-GA-100

BACKGROUND: Civil Version known as TFE-109.
Date First F109 Engine Produced: 1990.
Date If Out-of-Production: N/A. Number of F109 Engines Produced to Date: 25.

ENGINE TYPE: Turbofan. Dual-Shaft, High-Bypass Front Fan.

COMPRESSOR TYPE: Combination Axial Flow and Centrifugal Flow Including: 1-Stage Fan Acting as a Low Pressure Compressor, followed by a 2-Stage Centrifugal Flow High Pressure Compressor.

COMPRESSOR DATA: Compressor Pressure Ratio: 14.0: 1, Fan Pressure Ratio: 1.5 : 1, Fan Bypass Ratio: 5 : 1, Total Mass Airflow: 52.3 lb/sec (23,7 kg/sec).

TURBINE TYPE: 4-Stage Axial Flow Turbine including: 2-Stage High Pressure Turbine, 2-Stage Low Pressure Turbine.

COMBUSTOR TYPE: Annular, Through-Flow.

POWER (take-off) RATING: 1,330 lbt (5,9 kN thrust). Rating Approved to: 59°F (15°C).

TAKE-OFF SPECIFIC FUEL CONSUMPTION (SFC): 0.39 lb/hr/lbt (11,05 mg/Ns).

BASIC DRY WEIGHT: 439 lbs (199 kg).

POWER/WEIGHT RATIO: 2.96 : 1 lbt/lb.

APPLICATION:
Promavia "Jet Squalus" Trainer Aircraft in Development, Belgium.
Promavia "ATTA-3000" Trainer Aircraft in Development, Belgium.
Fairchild T-46A "Goshawk" USAF Trainer Aircraft in Development, USA.

OTHER ENGINES IN SERIES:
TFE-109 Commercial model in Development.

GAS TURBINE ENGINES FOR AIRCRAFT

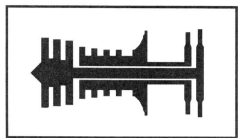

USA TURBINE ENGINE COMPANIES

MANUFACTURER: GARRETT TURBINE ENGINE DIVISION (Allied Signal Aerospace) PHOENIX, ARIZONA.

DESIGNATION (military): F124-GA-101.

BACKGROUND: F124 fighter engine derivative of Commercial TFE-731 Turbofan. Date First F124 Engine Produced: In Development.

ENGINE TYPE: Turbofan. Non-Afterburning, Dual-Shaft, Low Bypass Fan.

COMPRESSOR TYPE: Combination Axial Flow & Centrifugal Flow Compressor (8-stages total). The 3-Stage Low Pressure Axial Flow Compressor is in the form of a 3-stage Fan. The High Pressure Compressor is a Combination 4-stage Axial and Single-Stage Centrifugal Impeller.

COMPRESSOR DATA: Compressor Pressure Ratio: 22.5 : 1, Fan Pressure Ratio: 1.54 : 1, Fan Bypass Ratio: 0.3 : 1. Total Mass Airflow: 95.4 lb/sec (43,27 kg/sec).

TURBINE TYPE: 2-Stage Axial Flow Turbine including: 1-Stage High Pressure Turbine, 1-Stage Low Pressure Turbine.

COMBUSTOR TYPE: Annular, Through Flow.

POWER (take-off) RATING: 6,700 lbt (29,8 kN thrust). Rating Approved to: 59°F (15°C).

TAKE-OFF SPECIFIC FUEL CONSUMPTION (SFC): 0.84 lb/hr/lbt (23,79 mg/Ns) dry, unknown SFC in A/B.

BASIC DRY WEIGHT: 1,330 lbs (603,3 kg).

POWER/WEIGHT RATIO: 5.04 : 1 lbt/lb.

APPLICATION:
Proposed for Future Fighter Aircraft.

OTHER ENGINES IN SERIES:
TFE-1042 (Afterburning Export Model) in A-3 and AT-3, China-Taiwan.
See International Turbine Engine Company in Multi-National listing.

GAS TURBINE ENGINES FOR AIRCRAFT

USA TURBINE ENGINE COMPANIES

MANUFACTURER: GARRETT TURBINE ENGINE DIVISION (Allied Signal Aerospace) PHOENIX, ARIZONA

DESIGNATION (military)**:** T76-G10.

BACKGROUND: Date First T76 Engine Produced: 1965. Date If Out-of-Production: N/A. Number of T76 Engines Produced to Date: 1,040. Commercial versions designated TPE 331-10, -11, and -12.

ENGINE TYPE: Turboprop. Single Shaft Design.

COMPRESSOR TYPE: Centrifugal Flow, Two-Stage, Single Entry.

COMPRESSOR DATA: Compressor Pressure Ratio: 8.5 : 1. Mass Airflow: 6.17 lb/sec (2,8 kg/sec).

TURBINE TYPE: Three-Stage Axial Flow, Fixed Turbine.

COMBUSTOR TYPE: Annular, Reverse Flow.

POWER (take-off) RATING: 751 eshp (560 ekw), includes 36 shp (28,8 kw) from 90 lbt (0,4 kN thrust). Rating Approved to: 59°F (15°C).

TAKE-OFF SPECIFIC FUEL CONSUMPTION (SFC): 0.60 lb/hr/eshp (0,37 kg/hr/ekw).

BASIC DRY WEIGHT: 320 lbs (145 kg).

POWER/WEIGHT RATIO: 2.35 : 1 eshp/lb (3,86 : 1 ekw/kg).

APPLICATION:
Rockwell International OV-10A "Bronco" Attack Aircraft.

OTHER ENGINES IN SERIES:
T76-12 /-420/-421 Models in OV-10A/D.

GAS TURBINE ENGINES FOR AIRCRAFT

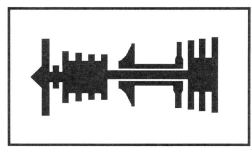

USA TURBINE ENGINE COMPANIES

MANUFACTURER: GARRETT TURBINE ENGINE DIVISION (Allied Signal Aerospace) PHOENIX, ARIZONA.

DESIGNATION (civil)**:** TFE 731-5A.

BACKGROUND: Date First TFE731 Engine Produced: 1971.
Date If Out-of-Production: N/A. Number of Engines Produced to Date: 6,100.

ENGINE TYPE: Turbofan. Dual-Shaft, High Bypass Front Fan.

COMPRESSOR TYPE: Combination Axial Flow & Centrifugal Flow Compressor (six stages total) with a Single Gear-Reduced Fan. The 5-Stage Low Pressure Axial Flow Compressor includes, the Fan and Four Axial Stages. The High Pressure Centrifugal Compressor is a Single-Stage, Single Entry Type.

COMPRESSOR DATA: Compressor Pressure Ratio: 15.5 : 1, Fan Pressure Ratio: 1.55 : 1, Fan Bypass Ratio: 3.61 : 1. Fan Gear Ratio: 0.55 : 1. Total Mass Airflow: 145.5 lb/sec (66 kg/sec).

TURBINE TYPE: 4-Stage Axial Flow Turbine including: 1-Stage High Pressure Turbine, 3-Stage Low Pressure Turbine.

COMBUSTOR TYPE: Annular, Reverse Flow.

POWER (take-off) RATING: 4,500 lbt (20 kN thrust). Rating Approved to: 73°F (23°C).

TAKE-OFF SPECIFIC FUEL CONSUMPTION (SFC): 0.469 lb/hr/lbt (13,3 mg/Ns).

BASIC DRY WEIGHT: 882 lbs (400 kg).

POWER/WEIGHT RATIO: 5.10 : 1 lbt/lb.

APPLICATION:
Business Jet Aircraft as Follows:
Cessna Citation-3, USA.   BAe 125-800, Great Britain.
CASA C101, Spain.   Dassault Falcon 900, France.
USAF C-23A and C-29A.

OTHER ENGINES IN SERIES:
TFE 731-1, -2, -3, -3A, -3B, -4, -5, -5A, -5B Installed as follows:

Learjets 31,35,36,55, USA.   BAe HS 125-700, 800, Gt. Britain.
C-21A and military Learjet, USA.   CASA C-101, Spain.
Cessna Citation-III,VI,VII, USA.   Dassault Falcon 10, 20, 50, 100, 900, France.
Rockwell JetStar and Sabreliner   FAMA IA-63A Trainer, Argentina.
Re-Engined, USA.   IAI Astra 1125, IAI Westwind-1124, Israel.

GAS TURBINE ENGINES FOR AIRCRAFT

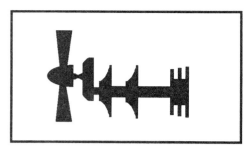

USA TURBINE ENGINE COMPANIES

MANUFACTURER: GARRETT TURBINE ENGINE DIVISION (Allied Signal Aerospace) PHOENIX, ARIZONA.

DESIGNATION (civil): TPE 331-14.

BACKGROUND: Originally designed as a military T76 Turboprop.
Date First TPE331 Engine Produced: 1959.
Date If Out-of-Production: N/A. Number of TPE331 Engines Produced to Date: 11,200.

ENGINE TYPE: Turboprop. Single Shaft Design.

COMPRESSOR TYPE: Centrifugal Flow, 2-Stage.

COMPRESSOR DATA: Compressor Pressure Ratio: 10.8 : 1. Mass Airflow: 11.4 lb/sec (5,2 kg/sec).

TURBINE TYPE: 3-Stage Axial Flow, Fixed Turbine.

COMBUSTOR TYPE: Annular, Reverse Flow.

POWER (take-off) RATING: 1,650 eshp (1,230 ekw), includes 67 shp (50 kw) from 168 lbt (0,75 kN thrust). Rating Approved to: 59°F (15°C).

TAKE-OFF SPECIFIC FUEL CONSUMPTION (SFC): 0.49 lb/hr/eshp (0,30 kg/hr/ekw).

BASIC DRY WEIGHT: 620 lbs (281 kg).

POWER/WEIGHT RATIO: 2.76 : 1 shp/lb (4,54 : 1 ekw/kg).

APPLICATION: Cessna Conquest II, USA.
Piper Cheyenne-400LS, USA. Mitsubishi MU-2, Japan

OTHER ENGINES IN SERIES: TPE 331-1 thru -15 ranging in power from 665 shp (488 kw) to 1,650 shp (932 kw). Installed in:

Beech Conquest-425, USA.
Beech KingAir B-100, USA.
C-212 and C-26A, USAF.
Fairchild Porter, USA.
Fairchild Metro-II,-III,-23, USA.
Fairchild Merlin-2,3,4, 300, USA.
Grumman S-2 Turbo-Tracker, USA.
Rockwell Commander JetProp Series, USA.

BAe JetStream 31/41, Great Britain.
CASA C-212, Spain.
Dornier DO-228, Germany.
Embraer EMB-312, Brazil.
FAMA IA-66, Argentina.
IPTN NC-212, Indonesia.
Shorts Skyliner, Great Britain.
Shorts Skyvan, Great Britain.
Shorts Tucano, Great Britain.

GAS TURBINE ENGINES FOR AIRCRAFT

 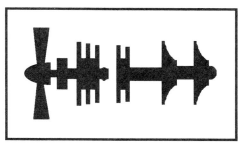

USA TURBINE ENGINE COMPANIES

MANUFACTURER: GARRETT TURBINE ENGINE DIVISION (Allied Signal Aerospace) PHOENIX, ARIZONA.

DESIGNATION (civil)**:** TPF 351-20.

BACKGROUND: A Derivative of the T800 Turboshaft Engine. Manufactured in both clockwise and counter-clockwise propeller rotation.
Date First TPF 351 Engine Produced: In Development.

ENGINE TYPE: Turboprop. Dual Shaft, Free Turbine Design with rear drive.

COMPRESSOR TYPE: Centrifugal Flow, 2-Stage.

COMPRESSOR DATA: Compressor Pressure Ratio: 13.3 : 1. Mass Airflow: 14 lb/sec (6,35 kg/sec).

TURBINE TYPE: 5-Stage Axial Flow, 2-Stage Gas Producer Compressor Drive Turbine, 3-Stage Power (free) Turbine.

COMBUSTOR TYPE: Annular, Reverse Flow.

POWER (take-off) RATING: 2,190 eshp (1,634 ekw), includes 91 shp (67,8 kw) from 227.5 lbt (1,01 kN thrust). Rating Approved to: 59°F (15°C).

TAKE-OFF SPECIFIC FUEL CONSUMPTION (SFC): 0.495 lb/hr/eshp (0,30 kg/hr/ekw).

BASIC DRY WEIGHT: 750 lbs (340 kg).

POWER/WEIGHT RATIO: 2.92 : 1 shp/lb (4,8 : 1 ekw/kg).

APPLICATION:
Embraer/FAMA, CBA-123 "Vector" Business Aircraft. Flat Rated to 1,300 ESHP.

OTHER ENGINES IN SERIES:
None.

GAS TURBINE ENGINES FOR AIRCRAFT

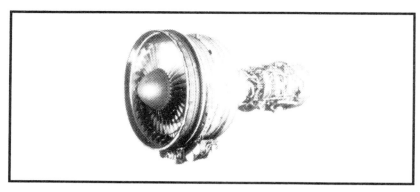 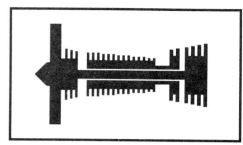

USA TURBINE ENGINE COMPANIES

MANUFACTURER: **GENERAL ELECTRIC AIRCRAFT ENGINES. LARGE ENGINE GROUP, EVANDALE, OHIO**

DESIGNATION (civil)**:** CF6-45A2.

BACKGROUND: Date First CF6 Engine Produced: 1972.
Date If Out-of-Production: N/A. Number of CF6 Engines Produced to Date: 4073.

ENGINE TYPE: Turbofan. Dual-Shaft, High-Bypass Front Fan.

COMPRESSOR TYPE: 18-Stage Dual-Spool, Axial Flow Compressor including: 1-Stage Front Fan and three Axial Stages in the Low Pressure Compressor, 14-Stage High Pressure Compressor.

COMPRESSOR DATA: Compressor Pressure Ratio: 26.3 : 1, Fan Pressure Ratio: 1.5 : 1, Fan Bypass Ratio: 4.64 : 1, Total Mass Airflow: 1,393 lb/sec (632 kg/sec).

TURBINE TYPE: 6-Stage Axial Flow Turbine including: 2-Stage High Pressure Turbine, 4-Stage Low Pressure Turbine.

COMBUSTOR TYPE: Annular, Through Flow.

POWER (take-off) RATING: 46,500 lbt (206,8 kN thrust). Rating Approved to: 86°F (30°C).

TAKE-OFF SPECIFIC FUEL CONSUMPTION (SFC): 0.35 lb/hr/lbt (9,91 mg/Ns).

BASIC DRY WEIGHT: 8,768 lbs (3,977 kg).

POWER/WEIGHT RATIO: 5.30 : 1 lbt/lb.

APPLICATION:
Boeing B747-100 and B747-SR Airliners.

OTHER ENGINES IN SERIES:
Earlier model: CF6-6 in DC-10 Airliner.
Locate later models, CF6-50 and CF6-80 in General Electric Listing.

GAS TURBINE ENGINES FOR AIRCRAFT

 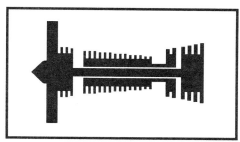

USA TURBINE ENGINE COMPANIES

MANUFACTURER: GENERAL ELECTRIC AIRCRAFT ENGINES. LARGE ENGINE GROUP, EVANDALE, OHIO

DESIGNATION (civil): CF6-50E.

BACKGROUND: Date First CF6 Engine Produced: 1972.
Date If Out-of-Production: N/A. Number of CF6-50 Engines Produced to Date: 2,224. Total number of CF6 model produced: 4073.
Military designation: F103-GE-100. Manufacturing partners include: SNECMA and MTU.

ENGINE TYPE: Turbofan. Dual-Shaft, High-Bypass Front Fan.

COMPRESSOR TYPE: 18-Stage Dual-Spool, Axial Flow Compressor including: 1-Stage Front Fan and Three Axial Stages in the Low Pressure Compressor, 14-Stage High Pressure Compressor.

COMPRESSOR DATA: Compressor Pressure Ratio: 29.2 : 1, Fan Pressure Ratio: 1.5: 1, Fan Bypass Ratio: 4.4 : 1, Total Mass Airflow: 1,465 lb/sec (664,4 kg/sec).

TURBINE TYPE: 6-Stage Axial Flow Turbine including: 2-Stage High Pressure Turbine, 4-Stage Low Pressure Turbine.

COMBUSTOR TYPE: Annular, Through-Flow.

POWER (take-off) RATING: 51,000 lbt (226,8 kN thrust). Rating Approved to: 86°F (30°C).

TAKE-OFF SPECIFIC FUEL CONSUMPTION (SFC): 0.37 lb/hr/lbt (10,48 mg/Ns).

BASIC DRY WEIGHT: 8,731 lbs (3,960 kg).

POWER/WEIGHT RATIO: 5.84 : 1 lbt/lb.

APPLICATION:
Boeing B747-200 and SP Airliners.
E-4A/B USAF Early Warning Aircraft.

OTHER ENGINES IN SERIES:
Earlier models include the CF6-50C, -50C1C2, -50 C2B in:
McDonnell-Douglas DC-10-10, DC10-30 and KC-10 Military Tanker.
Boeing B-747 Airliner and E-4 Military Aircraft.
Airbus Industries A300, Airliner.

GAS TURBINE ENGINES FOR AIRCRAFT

 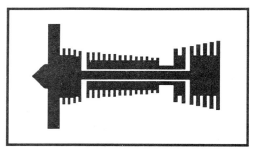

USA TURBINE ENGINE COMPANIES

MANUFACTURER: **GENERAL ELECTRIC AIRCRAFT ENGINES. LARGE ENGINE GROUP, EVANDALE, OHIO**

DESIGNATION (civil)**:** CF6-80C2A3.

BACKGROUND: Date First CF6-80 Engine Produced: 1985.
Date If Out-of-Production: N/A. Number of CF6-80 Engines Produced to Date: 1,362.
Total number of CF6 model produced: 4,073.
Manufacturing partners include: R-R, Fiat MTU, SNECMA, Volvo.

ENGINE TYPE: Turbofan. Dual-Shaft, High-Bypass Front Fan.

COMPRESSOR TYPE: 19-Stage Dual-Spool, Axial Flow Compressor including: 1-Stage Front Fan and four Axial Stages in the Low Pressure Compressor, 14-Stage High Pressure Compressor.

COMPRESSOR DATA: Compressor Pressure Ratio: 30.4 : 1, Fan Pressure Ratio: 1.7 : 1, Fan Bypass Ratio: 5.15 : 1, Total Mass Airflow: 1,754 lb/sec (795 kg/sec).

TURBINE TYPE: 7-Stage Axial Flow Turbine including: 2-Stage High Pressure Turbine, 5-Stage Low Pressure Turbine.

COMBUSTOR TYPE: Annular, Through-Flow.

POWER (take-off) RATING: 60,200 lbt (267,8 kN thrust). Rating Approved to: 86°F (30°C).

TAKE-OFF SPECIFIC FUEL CONSUMPTION (SFC): 0.33 lb/hr/lbt (9,35 mg/Ns).

BASIC DRY WEIGHT: 8,946 lbs (4,058 kg).

POWER/WEIGHT RATIO: 6.73 : 1 lbt/lb.

APPLICATION:
Airbus A300-600 and McDonnell-Douglas MD-11 Airliners.

OTHER ENGINES IN SERIES:
Earlier models include :
CF6-80A, -80A2, -80A3, -80B2, -80CA1, -80CA2, -80CA3, -80CB1, -80CB2, -80B4 in: Airbus A300, A310 and Boeing B767.
Later models include:
CF6-80C2A,B,C,D in Boeing B747-400, Boeing C-25A "AF-1", McDonnell-Douglas MD-11 and Airbus A300, A310.
CF6-80E (67,000 lbt) in Airbus A330.
GE-90 (90,000 lbt), in develolpment. Follow on Turbofan of CF6-80 Series.

GAS TURBINE ENGINES FOR AIRCRAFT

 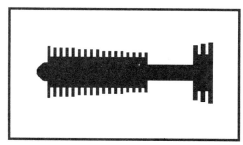

USA TURBINE ENGINE COMPANIES

MANUFACTURER: **GENERAL ELECTRIC AIRCRAFT ENGINES. LARGE ENGINE GROUP, EVANDALE, OHIO**

DESIGNATION (civil)**:** CJ805-3B.

BACKGROUND: Original model was the military J79, it was designed for the F-104 Starfigher (now out of service).
Date First CJ805 Engine Produced: 1960.
Date If Out-of-Production: 1963. Number of CJ-805 Engines Produced to Date: 638.

ENGINE TYPE: Turbojet. Single-Shaft.

COMPRESSOR TYPE: 17-Stage Single-Spool, Axial Flow Compressor.

COMPRESSOR DATA: Compressor Pressure Ratio: 13.5 : 1, Total Mass Airflow: 171 lb/sec (77,6 kg/sec).

TURBINE TYPE: 3-Stage Axial Flow Turbine.

COMBUSTOR TYPE: Can-Annular, Through Flow, with 10 liners.

POWER (take-off) RATING: 11,500 lbt (51,2 kN thrust). Rating Approved to: 59°F (15°C).

TAKE-OFF SPECIFIC FUEL CONSUMPTION (SFC): 0.81 lb/hr/lbt (22,95 mg/Ns).

BASIC DRY WEIGHT: 2,817 lbs (1,278 kg).

POWER/WEIGHT RATIO: 4.08 : 1 lbt/lb.

APPLICATION:
Convair CV-880, Airliner.

OTHER ENGINES IN SERIES:
CJ 805-23 Aft-Fan Model in Convair CV-990, Airliner.

GAS TURBINE ENGINES FOR AIRCRAFT

 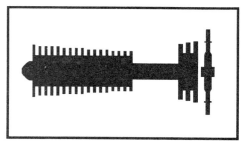

USA TURBINE ENGINE COMPANIES

MANUFACTURER: **GENERAL ELECTRIC AIRCRAFT ENGINES. LARGE ENGINE GROUP, EVANDALE, OHIO**

DESIGNATION (civil): CJ805-23.

BACKGROUND: A derivative of the military J79 turbojet designed for the F-104 aircraft. Also derivative of the CJ805-3 turbojet. The -23 model was the first Turbofan Engine in U.S. commercial service.
Date First CJ805 Engine Produced: 1962.
Date If Out-of-Production: 1963. Number of CJ805 Engines Produced to Date: 638.

ENGINE TYPE: Turbofan. Single compressor shaft with independent drive Aft-Fan.

COMPRESSOR TYPE: 17-Stage Single-Spool Axial Flow Compressor and an Independent Drive Aft Fan module mounted in the exhaust of the core turbine.

COMPRESSOR DATA: Compressor Pressure Ratio: 13.5 : 1. Fan Pressure Ratio: 1.5 : 1. Fan Bypass Ratio: 2.46 : 1. Total Mass Airflow: 422 lb/sec (191,4 kg/sec).

TURBINE TYPE: 3-Stage Axial Flow Compressor Drive Turbine, 1-Stage Axial Flow Fan Drive Turbine.

COMBUSTOR TYPE: Can-Annular, Through Flow, with 10 liners.

POWER (take-off) RATING: 16,100 lbt (71,6 kN thrust). Rating Approved to: 59°F (15°C).

TAKE-OFF SPECIFIC FUEL CONSUMPTION (SFC): 0.56 lb/hr/lbt (15,86 mg/Ns).

BASIC DRY WEIGHT: 3,766 lbs (1,708 kg).

POWER/WEIGHT RATIO: 4.28 : 1 lbt/lb.

APPLICATION:
Convair CV-990 Airliner.

OTHER ENGINES IN SERIES:
CJ805-3 Turbojet Model in Convair CV-880 Airliner.

GAS TURBINE ENGINES FOR AIRCRAFT

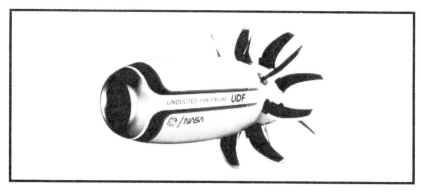 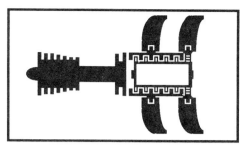

USA TURBINE ENGINE COMPANIES

MANUFACTURER: GENERAL ELECTRIC AIRCRAFT ENGINES. LARGE ENGINE GROUP, EVANDALE, OHIO.

DESIGNATION (civil): GE36-B22A.

BACKGROUND: Commonly referred to as the Unducted Fan, (UDF™) Engine. Based on the Military F404 Core.
Date First GE36 Engine Produced: 1986.
Date If Out-of-Production: N/A. Number of GE36 Engines Produced to Date: In Development.

ENGINE TYPE: Propfan. Aft-Fan with Ultra-High Bypass.

COMPRESSOR TYPE: 7-Stage Single-Spool Core Axial Flow Compressor.

COMPRESSOR DATA: Compressor Pressure Ratio: 8 : 1, Core Mass Airflow: 106 lb/sec (48 kg/sec), Propulsor (fan) Bypass Ratio: 36 : 1.

TURBINE TYPE: Two-Stage High Pressure Turbine and a 12-Stage Propulsor Unit acting as the Low Pressure Turbine.

COMBUSTOR TYPE: Annular, Through Flow.

POWER (take-off) RATING: 25,000 lbt (111.2 kN thrust). Rating Approved to: 59°F (15°C).

TAKE-OFF SPECIFIC FUEL CONSUMPTION (SFC): 0.24 lb/hr/lbt (6.79 mg/Ns).

BASIC DRY WEIGHT: Unknown.

POWER/WEIGHT RATIO: Unknown.

APPLICATION:
Projected for Boeing 7J7 and MD-92, Airliners.

OTHER ENGINES IN SERIES:
GE-38 Turboprop proposed for Lockheed P-7A, in Development.

NOTE: Missing information is confidential at this time.

GAS TURBINE ENGINES FOR AIRCRAFT

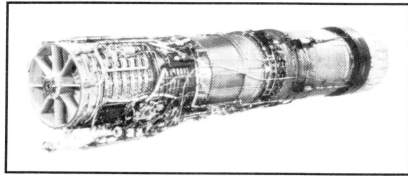
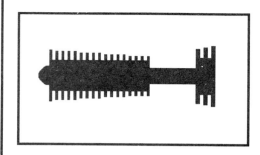

USA TURBINE ENGINE COMPANIES

MANUFACTURER: GENERAL ELECTRIC AIRCRAFT ENGINES. LARGE ENGINE GROUP, EVANDALE, OHIO.

DESIGNATION (military): J79-GE-119.

BACKGROUND: The First U.S. Engine designed with variable compressor stator vanes. The original J79 was designed for the F-104 aircraft. Commercial versions were designated as the CJ805-3 for the Convair 880 and a derivative CJ805-23 with an Aft-Fan for the Convair 990. Date First J79 Engine Produced: 1954.
Date If Out-of-Production: N/A. Number of J79 Engines Produced to Date: 13,686.

ENGINE TYPE: Turbojet. Single-Shaft with Afterburning.

COMPRESSOR TYPE: 17-Stage Single-Spool, Axial Flow Compressor.

COMPRESSOR DATA: Compressor Pressure Ratio: 13.4 : 1, Total Mass Airflow: 170 lb/sec (77,1 kg/sec).

TURBINE TYPE: 3-Stage Axial Flow Turbine.

COMBUSTOR TYPE: Can-Annular, Through Flow, with 10 liners.

AFTERBURNER: Four-Stage.

POWER (take-off) RATING: 11,870 lbt (52,80 kN thrust) [dry], 18,730 lbt (83,3 kN thrust) [A/B]. Rating Approved to: 59°F (15°C).

TAKE-OFF SPECIFIC FUEL CONSUMPTION (SFC): 0.84 lb/hr/lbt (23,79 mg/Ns) [dry], 2.05 lb/hr/lbt (58.07 mg/Ns) [A/B].

BASIC DRY WEIGHT: 3,855 lbs (1,749 kg).

POWER/WEIGHT RATIO: 3.08 : 1 [dry], 4.86 : 1 [A/B] - lbt/lb.

APPLICATION:
F-16 Series, "Fighting Falcon", Fighter Aircraft.

OTHER ENGINES IN SERIES:
J79-GE-5C in General Dynamics B-58 "Hustler", Bomber Aircraft.
J79-GE-8 in RF-4B "Phantom-II", Reconnaissance Aircraft.
J79-GE-10 in North American-Rockwell, RA-5C "Vigilante", Reconnaissance Aircraft.
J-79-GE-15 in RF-4C,D "Phantom-II" Reconnaissance Aircraft.
J-79-GE-15 in F-4C,D "Phantom-II" Fighter Aircraft.
J79-GE-17 in F-4E/G "Phantom-II" Fighter Aircraft.
J79-GE-19 in Lockheed F-l04 "StarFighter" series.
J-79-J1E in "Kfir" Fighter Aircraft, Israel: also called F-21A USAF, Fighter Aircraft.

GAS TURBINE ENGINES FOR AIRCRAFT

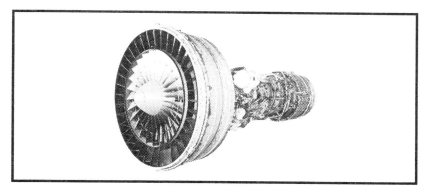 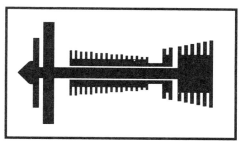

USA TURBINE ENGINE COMPANIES

MANUFACTURER: **GENERAL ELECTRIC AIRCRAFT ENGINES. LARGE ENGINE GROUP, EVANDALE, OHIO.**

DESIGNATION (military): TF39-GE-1C.

BACKGROUND: Developed for the military, Lockheed C5A Galaxy. Later to become the Commercial CF6 Turbofan. World's First Production High Bypass Turbofan Engine.
Date First TF39 Engine Produced: 1967.
Date If Out-of-Production: 1973. Number of TF39 Engines Produced to Date: 680.

ENGINE TYPE: Turbofan. Dual-Shaft, High-Bypass Front Fan.

COMPRESSOR TYPE: 18-Stage Dual-Spool, Axial Flow Compressor including: One and one-half-Stage Front Fan Providing 2-Axial Stages to the Core in the Low Pressure Compressor, 16-Stage High Pressure Compressor.

COMPRESSOR DATA: Compressor Pressure Ratio: 22.0: 1, Fan Pressure Ratio: 1.45 : 1, Fan Bypass Ratio: 7.88 : 1, Total Mass Airflow: 1,549 lb/sec (703 kg/sec).

TURBINE TYPE: 8-Stage Axial Flow Turbine including: 2-Stage High Pressure Turbine, 6-Stage Low Pressure Turbine.

COMBUSTOR TYPE: Annular, Through-Flow.

POWER (take-off) RATING: 43,000 lbt (191,3 kN thrust). Rating Approved to: 89.5°F, (32°C).

TAKE-OFF SPECIFIC FUEL CONSUMPTION (SFC): 0.32 lb/hr/lbt (9,06 mg/Ns).

BASIC DRY WEIGHT: 7,900 lbs (3,583 kg).

POWER/WEIGHT RATIO: 5.44 : 1 lbt/lb.

APPLICATION:
Lockheed C5A/B Military Transport.

OTHER ENGINES IN SERIES:
TF-39 Derivative Engines in:
McDonnell-Douglas KC-10 Tanker.
Boeing VC-25A Transport.
Boeing E-4 Early Warning Aircraft.

GAS TURBINE ENGINES FOR AIRCRAFT

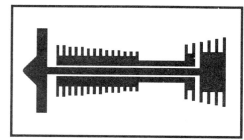

USA TURBINE ENGINE COMPANIES

MANUFACTURER: GENERAL ELECTRIC AIRCRAFT ENGINES
SMALL COMMERCIAL ENGINE DEPARTMENT, LYNN, MASS.

DESIGNATION (civil)**:** CF34-1A.

BACKGROUND: Originally Developed for the military as a TF34 Turbofan for the Lockheed S-3A and Fairchild A-10 aircraft.
Date First CF34 Engine Produced: 1982.
Date If Out-of-Production: N/A. Number of CF34 Engines Produced to Date: 300.

ENGINE TYPE: Turbofan. Dual-Shaft, High-Bypass Front Fan.

COMPRESSOR TYPE:: 15-stage dual spool axial flow compressor, including: 1-Stage Front Fan acting as the Low Pressure Compressor followed by a 14-Stage High Pressure Compressor.

COMPRESSOR DATA: Compressor Pressure Ratio: 21.0:1, Fan Pressure Ratio: 1.5:1, Fan Bypass Ratio: 6.2:1, Total Mass Airflow: 330 lb/sec (150 kg/sec).

TURBINE TYPE: 6-Stage Axial Flow Turbine including: 2-Stage High Pressure Turbine, 4-Stage Low Pressure Turbine.

COMBUSTOR TYPE: Annular, Through-Flow.

POWER (take-off) RATING: 9,140 lbt (40,7 kN thrust). Rating Approved to: 59°F (15°C).

TAKE-OFF SPECIFIC FUEL CONSUMPTION (SFC): 0.37 lb/hr/lbt (10,48 mg/Ns).

BASIC DRY WEIGHT: 1,625 lbs (737 kg).

POWER/WEIGHT RATIO: 5.62:1 lbt/lb.

APPLICATION:
Canadair Challenger-601 Commuter/Business Aircraft.
Canadair Challenger-RJ (Regional Jet) Commuter Aircraft.

OTHER ENGINES IN SERIES:
TF34-GE-100 in Fairchild Republic A-10 "Thunderbolt-II" Attack Aircraft.
TF34-GE-400A in Lockheed S-3A, "Viking" USN Patrol Aircraft.

GAS TURBINE ENGINES FOR AIRCRAFT

 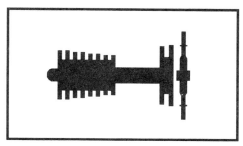

USA TURBINE ENGINE COMPANIES

MANUFACTURER: GENERAL ELECTRIC AIRCRAFT ENGINES
SMALL COMMERCIAL ENGINE DEPARTMENT, LYNN, MASS.

DESIGNATION (civil): CF700-2D2.

BACKGROUND: A derivative of the original CJ610 Turbojet in the LearJet aircraft.
Date First CF700 Engine Produced: 1965.
Date If Out-of-Production: 1981. Number of CF700 Engines Produced to Date: 1,170.

ENGINE TYPE: Turbofan. Single compressor shaft with independent drive Aft-Fan.

COMPRESSOR TYPE: 8-Stage Single-Spool, Axial Flow Compressor in the Core Engine, and an Independent Drive Aft-Fan Module mounted in the exhaust of the Core Turbine.

COMPRESSOR DATA: Core Compressor Pressure Ratio: 6.8 : 1, Fan Pressure Ratio: 1.6 : 1, Fan Bypass Ratio: 1.98 : 1, Total Mass Airflow: 130 lb/sec (59 kg/sec).

TURBINE TYPE: 2-Stage Axial Flow Compressor Drive Turbine, 1-Stage Axial Flow Aft-Fan Drive Turbine.

COMBUSTOR TYPE: Annular, Through-Flow.

POWER (take-off) RATING: 4,500 lbt (20 kN thrust). Rating Approved to: 59°F (15°C).

TAKE-OFF SPECIFIC FUEL CONSUMPTION (SFC): 0.65 lb/hr/lbt (18,41 mg/Ns).

BASIC DRY WEIGHT: 767 lbs (349 kg).

POWER/WEIGHT RATIO: 5.87 : 1 lbt/lb.

APPLICATION:
Rockwell Int'l Sabreliner-80, Business Aircraft (USA).
Dassault-Breguet Falcon D/E/F Business Jet Aircraft (France).

OTHER ENGINES IN SERIES:
CF700-2C in Dassault Falcon 20/200 Business Jet Aircraft.

GAS TURBINE ENGINES FOR AIRCRAFT

 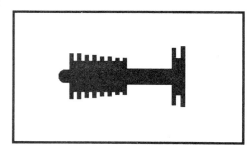

USA TURBINE ENGINE COMPANIES

MANUFACTURER: GENERAL ELECTRIC AIRCRAFT ENGINES. COMMERCIAL ENGINE DEPARTMENT, LYNN, MASS.

DESIGNATION (civil): CJ610-8A.

BACKGROUND: Civil version of Military J-85.
Date First CJ610 Engine Produced: 1964.
Date If Out-of-Production: 1978. Number of CJ610 Engines Produced to Date: 1,404.

ENGINE TYPE: Turbojet. Single-Shaft.

COMPRESSOR TYPE: 8-Stage Single-Spool, Axial Flow Compressor.

COMPRESSOR DATA: Compressor Pressure Ratio: 6.6 : 1, Total Mass Airflow: 44.0 lb/sec (20 kg/sec).

TURBINE TYPE: 2-stage axial flow turbine.

COMBUSTOR TYPE: Annular, Through-Flow.

POWER (take-off) RATING: 2,950 lbt (13,1 kN thrust). Rating Approved to: 59°F (15°C).

TAKE-OFF SPECIFIC FUEL CONSUMPTION (SFC): 0.97 lb/hr/lbt (27,48 mg/Ns).

BASIC DRY WEIGHT: 411 lbs (186 kg).

POWER/WEIGHT RATIO: 7.18 : 1 lbt/lb.

APPLICATION:
LearJet 24,25,28,29 Business Jet Aircraft.

OTHER ENGINES IN SERIES:
CJ610-1, -5, -8 in Hansa HBF-320, Germany.
CJ610-4 in LearJet 24 Business Jet Aircraft, USA.
CJ610-6 in LearJet 24B,C; 25,-25C Business Jet Aircraft, USA.
CJ610-8A in LearJet 24,25,28,29 Business Jet Aircraft, USA.
CJ610-8A also in early IAI Westwind Series, Business Jet Aircraft, Israel.
CJ601-8/9 in IAI, Jet Commander and Jet Commodore, Israel.

GAS TURBINE ENGINES FOR AIRCRAFT

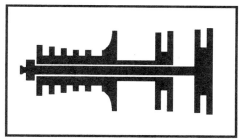

USA TURBINE ENGINE COMPANIES

MANUFACTURER: GENERAL ELECTRIC AIRCRAFT ENGINES. COMMERCIAL ENGINE DEPARTMENT, LYNN, MASS.

DESIGNATION (civil & military): CT7-6.

BACKGROUND: Civil version of Military T700.
Date First CT7 Engine Produced: 1978.
Date If Out-of-Production: N/A. Number of CT7/T700 Engines Produced to Date: 6,000.

ENGINE TYPE: Turboshaft. Dual-Shaft (Compressor Drive & Power Output Drive).

COMPRESSOR TYPE: 6-Stage Dual Compressor including a 5-Stage, Axial Flow Compressor, and a 1-Stage Centrifugal Flow Compressor.

COMPRESSOR DATA: Compressor Pressure Ratio: 18.6 : 1, Total Mass Airflow: 12 lb/sec (5,4 kg/sec).

TURBINE TYPE: 4-Stage Axial Flow Turbine including: 2-Stage Gas Producer (compressor drive) Turbine, 2-Stage Power (free) Turbine.

COMBUSTOR TYPE: Annular, Through Flow.

POWER (take-off) RATING: 2,000 shp (1,491 kw). Rating Approved to: 59°F (15°C).

TAKE-OFF SPECIFIC FUEL CONSUMPTION (SFC): 0.45 lb/hr/shp (0,27 kg/hr/kw).

BASIC DRY WEIGHT: 460 lbs (209 kg).

POWER/WEIGHT RATIO: 4.35 : 1 shp/lb (7,13 kw/kg).

APPLICATION:
EH Industries EH-101 "Merlin" military Helicopter. (Agusta-Westland)
EH Industries EH-101 "Heliliner" civil Helicopter. (Agusta-Westland)

OTHER ENGINES IN SERIES:
CT7-2A in Bell 214ST "Super Transport" Helicopter.
CT7-2B in Westland 30-200 and 300 Helicopter.
CT7-2C/2D in Sikorsky S-70C Helicopter.

GAS TURBINE ENGINES FOR AIRCRAFT

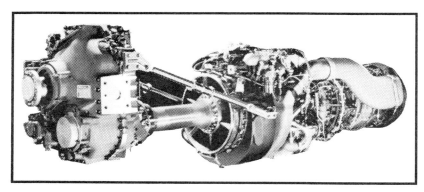
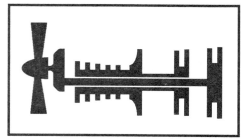

USA TURBINE ENGINE COMPANIES

MANUFACTURER: GENERAL ELECTRIC AIRCRAFT ENGINES. COMMERCIAL ENGINE DEPARTMENT, LYNN, MASS.

DESIGNATION (civil)**:** CT7-9.

BACKGROUND: Military version known as T700.
Date First CT7 Engine Produced: 1978.
Date If Out-of-Production: N/A. Number of CT7/T700 Engines Produced to Date: 6,000.

ENGINE TYPE: Turboprop. Dual-Shaft (Compressor Drive & Power Output Drive). Front Drive.

COMPRESSOR TYPE: 6-Stage Combination Compressor including a 5-Stage, Axial Flow Compressor, and a 1-Stage Centrifugal Flow Compressor.

COMPRESSOR DATA: Compressor Pressure Ratio: 18 : 1, Total Mass Airflow: 12 lb/sec (5,4 kg/sec).

TURBINE TYPE: 4-Stage Axial Flow Turbine including: 2-Stage Gas Producer (compressor drive) Turbine, 2-Stage Power (free) Turbine.

COMBUSTOR TYPE: Annular, Through Flow.

POWER (take-off) RATING: 1,870 shp (1,394, kw). Rating Approved to: 92°F (34°C).

TAKE-OFF SPECIFIC FUEL CONSUMPTION (SFC): 0.43 lb/hr/shp (0,26 kg/hr/ekw).

BASIC DRY WEIGHT: 795 lbs (360,6 kg).

POWER/WEIGHT RATIO: 2.35 : 1 eshp/lb (3,87 : 1 ekw/kg).

APPLICATION:
Saab Scania Saab-340, Business Aircraft, Sweden.
CASA 235-100, Business Aircraft, Spain.
Let L-610, Business Aircraft, Czechoslovakia.

OTHER ENGINES IN SERIES:
CT7-5 in Saab-340 Business Aircraft, Sweden.
CT7-7 in CASA CN-235 Business Aircraft, Spain.
CT-9C in IPTN CN-235, Indonesia
CT-9C in CASA CN-235 (T-19), Spain.

GAS TURBINE ENGINES FOR AIRCRAFT

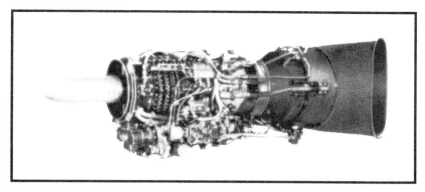
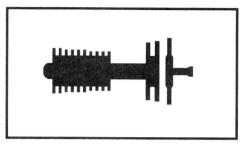

USA TURBINE ENGINE COMPANIES

MANUFACTURER: GENERAL ELECTRIC AIRCRAFT ENGINES. COMMERCIAL ENGINE DEPARTMENT, LYNN, MASS.

DESIGNATION (civil)**:** CT58-140.

BACKGROUND: Originally developed as a military T58-GE-10 for the Sikorsky CH-3.
Date First CT58-140 Engine Produced: 1965.
Date If Out-of-Production: 84. Number of CT58/T58 Engines Produced to Date: 585.

ENGINE TYPE: Turboshaft. Dual-Shaft (Compressor Drive & Power Output Drive). Rear Drive (output) Shaft.

COMPRESSOR TYPE: 10-Stage Single-Spool, Axial Flow Compressor.

COMPRESSOR DATA: Compressor Pressure Ratio: 8.4 : 1, Total Mass Airflow: 13.7 lb/sec (6,2 kg/sec).

TURBINE TYPE: 3-Stage Axial Flow Turbine including: 2-Stage Gas Producer (compressor drive) Turbine, 1-Stage Power (free) Turbine.

COMBUSTOR TYPE: Annular, Through Flow.

POWER (take-off) RATING: 1,500 shp (1,119 kw). Rating Approved to: 59°F (15°C).

TAKE-OFF SPECIFIC FUEL CONSUMPTION (SFC): 0.61 lb/hr/shp (0,37 kg/hr/kw).

BASIC DRY WEIGHT: 340 lbs (154 kg).

POWER/WEIGHT RATIO: 4.41 : 1 shp/lb (7,27 :1 kw/kg).

APPLICATION:
Sikorsky S-61 "Silver", Civil Helicopter, USA.
Agusta AS-61 "Silver" Civil Helicopter, Italy.
Sikorsky S-62, Civil Helicopter, USA.
Mitsubishi S-62B/C, Civil Helicopter, Japan.
Boeing-Vertol 107 Military Helicopter, USA.

OTHER ENGINES IN SERIES:
CT58-100-2 in Sikorsky S-61/62, Helicopters.
CT58-110-1/2 in S-61/62, Helicopters.
CT58-140-1/2 in S-61, Helicopter.
Note: The CT58-100, 110, and 140 also appear in the Boeing-Vertol 107 (CH-46).
Locate other T58 Models under T58-GE-16 in General Electric Aircraft Engines, USA.

GAS TURBINE ENGINES FOR AIRCRAFT

 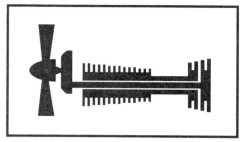

USA TURBINE ENGINE COMPANIES

MANUFACTURER: GENERAL ELECTRIC AIRCRAFT ENGINES. COMMERCIAL ENGINE DEPARTMENT, LYNN, MASS.

DESIGNATION (civil & military): CT64-820.

BACKGROUND: An FAA certified Turboprop derivative of the Military T64 Turboshaft Engine. Date First CT64-820 Engine Produced: 1975. Date If Out-of-Production: N/A. Number of T64 Turboprop and Turboshaft Engines Produced to Date: 3,000.

ENGINE TYPE: Turboprop. Dual-Shaft (Compressor Drive & Power Output Drive). Front Drive.

COMPRESSOR TYPE: 14-Stage Single-Spool, Axial Flow Compressor.

COMPRESSOR DATA: Compressor Pressure Ratio: 12.5 : 1, Total Mass Airflow: 27 lb/sec (12,2 kg/sec).

TURBINE TYPE: 4-Stage Axial Flow Turbine including: 2-Stage Gas Producer (compressor drive) Turbine, 2-Stage Power (free) Turbine.

COMBUSTOR TYPE: Annular, Through Flow.

POWER (take-off) RATING: 3,400 eshp (2,535 ekw) includes 95 eshp (71 ekw) from 210 lbt (93 kN thrust). Rating Approved to: 100°F (38°C).

TAKE-OFF SPECIFIC FUEL CONSUMPTION (SFC): 0.49 lb/hr/shp (0,31 kg/hr/kw).

BASIC DRY WEIGHT: 1,145 lbs (520 kg), includes propeller gearbox.

POWER/WEIGHT RATIO: 2.97 : 1 shp/lb (4,88 : 1 kw/kg).

APPLICATION:
DeHavilland DHC-5A "Caribou", and DHC-5D "Buffalo", Military Transports and DHC-5E "Transporter" commercial transport, Canada.
Alenia G.222 Transport, Italy.
Shin Meiwa PS-1 and US-1, Maritime Aircraft, Japan.

OTHER ENGINES IN SERIES (turboprop):
CT64-820-1/3 in Lockheed-Kawasaki P-2 Neptune, Maritime Aircraft, Japan.
Locate other T-64 models under T64-GE-419 in General Electric Aircraft Engines, USA.

GAS TURBINE ENGINES FOR AIRCRAFT

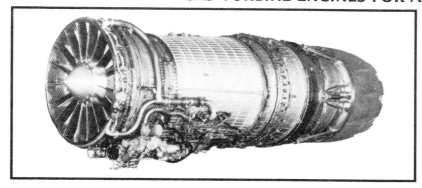 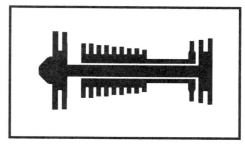

USA TURBINE ENGINE COMPANIES

MANUFACTURER: GENERAL ELECTRIC AIRCRAFT ENGINES. SMALL ENGINE GROUP, EVANDALE, OHIO.

DESIGNATION (military)**:** F101-GE-102.

BACKGROUND: Date First F101 Engine Produced: 1983.
Date If Out-of-Production: N/A. Number of F101 Engines Produced to Date: 470.

ENGINE TYPE: Turbofan. Dual-Shaft, Low-Bypass Front Fan with Afterburning.

COMPRESSOR TYPE: 11-Stage Dual-Spool, Axial Flow Compressor including: 2-Stage Front Fan in the Low Pressure Compressor, 9-Stage High Pressure Compressor.

COMPRESSOR DATA: Compressor Pressure Ratio: 26.5 : 1, Fan Pressure Ratio: 2.4 : 1, Fan Bypass Ratio: 1.8 : 1, Total Mass Airflow: 360 lb/sec (163 kg/sec).

TURBINE TYPE: 3-Stage Axial Flow Turbine including: 1-Stage High Pressure Turbine, 2-Stage Low Pressure Turbine.

COMBUSTOR TYPE: Annular, Through-Flow.

AFTERBURNER: 3-Stage, 1-Stage in Fan Bypass Exhaust, 2-Stages in the Core Exhaust.

POWER (take-off) RATING: 17,000 lbt (75,7 kN thrust) [dry], 30,000 lbt (133,4 kN thrust) [A/B]. Rating Approved to: 59°F (15°C).

TAKE-OFF SPECIFIC FUEL CONSUMPTION (SFC): 0.75 lb/hr/lbt (21 mg/Ns) [Dry] 2.3 lb/hr/lbt (65,16 mg/Ns) [A/B].

BASIC DRY WEIGHT: 4,400 lbs (1,996 kg).

POWER/WEIGHT RATIO: 3.9 : 1 [dry], 6.82 : 1 [A/B] - lbt/lb.

APPLICATION:
Rockwell International B-1B "Lancer" Bomber Aircraft.

OTHER ENGINES IN SERIES:
None.

GAS TURBINE ENGINES FOR AIRCRAFT

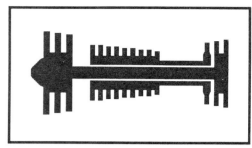

USA TURBINE ENGINE COMPANIES

MANUFACTURER: GENERAL ELECTRIC AIRCRAFT ENGINES. SMALL ENGINE GROUP, EVANDALE, OHIO.

DESIGNATION (military): F110-GE-400.

BACKGROUND: A derivative of the F101 Turbofan Engine.
Date First F110-GE-400 Engine Produced: 1987.
Date If Out-of-Production: N/A. Number of F110 Engines Produced to Date: 230.

ENGINE TYPE: Turbofan. Dual-Shaft, Low-Bypass Front Fan. With Afterburning.

COMPRESSOR TYPE: 12-Stage Dual-Spool, Axial Flow Compressor including: 3-Stage Front Fan acting as the Low Pressure Compressor, 9-Stage High Pressure Compressor.

COMPRESSOR DATA: Compressor Pressure Ratio: 29.9 : 1, Fan Pressure Ratio: 3.2 : 1, Fan Bypass Ratio: 0.76 : 1, Total Mass Airflow: 270 lb/sec (122 kg/sec).

TURBINE TYPE: 3-Stage Axial Flow Turbine including: 1-Stage High Pressure Turbine, 2-Stage Low Pressure Turbine.

COMBUSTOR TYPE: Annular, Through-Flow.

AFTERBURNER: 3-Stage, 1-Stage in Fan Bypass Exhaust, 2-Stages in the Core Exhaust.

POWER (take-off) RATING: 16,088 lbt (71,5 kN thrust) [dry], 29,000 lbt (129 kN thrust) [A/B]. Rating Approved to: 59°F (15°C).

TAKE-OFF SPECIFIC FUEL CONSUMPTION (SFC): .688 lb/hr/lbt (19,86 mg/Ns), 1.998 lb/hr/lbt (A/B).

BASIC DRY WEIGHT: 4,488 lbs (2,036 kg).

POWER/WEIGHT RATIO: 3.6 : 1 [dry], 6.0 : 1 lbt/lb (A/B).

APPLICATION:
F-14A+ and F-14 B/D Grumman "Tomcat", Fighter Aircraft.

OTHER ENGINES IN SERIES:
F110-GE-100 in F-16N "Fighting Falcon", Turkey.
F110-GE-129 in F-16C/D, USA.

GAS TURBINE ENGINES FOR AIRCRAFT

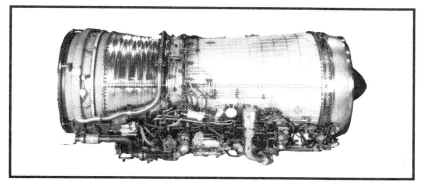 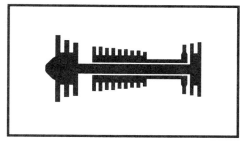

USA TURBINE ENGINE COMPANIES

MANUFACTURER: GENERAL ELECTRIC AIRCRAFT ENGINES. SMALL ENGINE GROUP, EVANDALE, OHIO.

DESIGNATION (military): F118-GE-100.

BACKGROUND: Date First F118 Engine Produced: 1988.
Date If Out-of-Production: N/A. Number of F118 Engines Produced to Date: 60.
Derivative Model of F110 Turbofan.

ENGINE TYPE: Turbofan. Dual-Shaft, Low-Bypass Front Fan. Non-Afterburning.

COMPRESSOR TYPE: 12 Stage Dual-Spool, Axial Flow Compressor including: 1 Stage Front Fan and two Axial Stages in the Low Pressure Compressor, 9 Stage High Pressure Compressor.

COMPRESSOR DATA: Compressor Pressure Ratio: 29.4 : 1, Fan Pressure Ratio: 1.4 : 1, Fan Bypass Ratio: 0.8 : 1, Total Mass Airflow: 285 lb/sec (129 kg/sec).

TURBINE TYPE: 3-Stage Axial Flow Turbine including: 1-Stage High Pressure Turbine, 2-Stage Low Pressure Turbine.

COMBUSTOR TYPE: Annular, Through-Flow.

POWER (take-off) RATING: 19,000 lbt (84,5 kN thrust). Rating Approved to: 59°F (15°C).

TAKE-OFF SPECIFIC FUEL CONSUMPTION (SFC): 0.673 lb/hr/lbt (19,06 mg/Ns).

BASIC DRY WEIGHT: 3,732 lbs (1,693 kg).

POWER/WEIGHT RATIO: 5.1 : 1 [dry] lbt/lb.

APPLICATION:
Northrop B-2, "Stealth Bomber".

OTHER ENGINES IN SERIES:
None.

GAS TURBINE ENGINES FOR AIRCRAFT

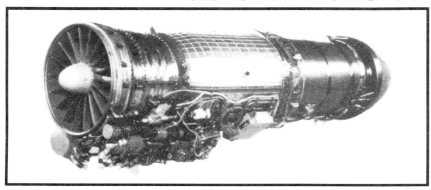
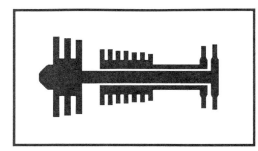

USA TURBINE ENGINE COMPANIES

MANUFACTURER: GENERAL ELECTRIC AIRCRAFT ENGINES. F404 PROJECT, LYNN, MASS.

DESIGNATION (military): F404-GE-400.

BACKGROUND: Date First F404 Engine Produced: 1979.
Date If Out-of-Production: N/A. Number of F404 Engines Produced to Date: 2,000.

ENGINE TYPE: Turbofan. Dual-Shaft, Low-Bypass Front Fan, with Afterburning.

COMPRESSOR TYPE: 10-Stage Dual-Spool, Axial Flow Compressor including: 3-Stage Fan Stages in the Low Pressure Compressor, 7-Stage High Pressure Compressor.

COMPRESSOR DATA: Compressor Pressure Ratio: 25 : 1, Fan Bypass Ratio: 0.34 : 1, Total Mass Airflow: 142 lb/sec (64,4 kg/sec).

TURBINE TYPE: 2-Stage Axial Flow Turbine including: 1-Stage High Pressure Turbine, 1-Stage Low Pressure Turbine.

COMBUSTOR TYPE: Annular, Through-Flow.

AFTERBURNER: Variable Stage, Combined Flow.

POWER (take-off) RATING: 10,800 lbt (48 kN thrust) [dry], 16,000 lbt (71,2 kN thrust) [A/B]. Rating Approved to: 59°F (15°C).

TAKE-OFF SPECIFIC FUEL CONSUMPTION (SFC): 0.8 lb/hr/lbt Dry, (1.85 lb/hr/lbt A/B).

BASIC DRY WEIGHT: 2,182 lbs (989 kg).

POWER/WEIGHT RATIO: 4.9 : 1 [dry], 7.33 : 1 [A/B] - lbt/lb.

APPLICATION:
McDonnell/Douglas F-18 "Hornet" Attack/Fighter Aircraft.
Grumman X-29 Experimental Fighter Aircraft.
Rockwell X-31A Experimental Fighter Aircraft.
Dassault-Breguet Rafale-A Fighter Aircraft in Development, France.

OTHER ENGINES IN SERIES:
F404-GE-100 in Northrop F-20 "Tigershark".
F404-GE-100D in Singapore A-4S "Super Seahawk".
F404-GE-400D in Grumman A-6F "Intruder".
F404-GE-402 in McDonnell-Douglas F-18 "Hornet".
F404-GE-F1D2 in Lockheed F-117 "Stealth Fighter".
F404/RM12 in JAS 39 Gripen, Sweden.
F404/F2J3 in LCA "Light Combat Aircraft", India.
F414-GE-400 I n "F/A18 E/F (22,000 Lbt in development).

GAS TURBINE ENGINES FOR AIRCRAFT

 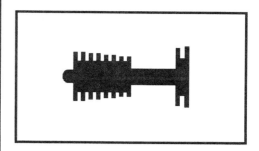

USA TURBINE ENGINE COMPANIES

MANUFACTURER: **GENERAL ELECTRIC ENGINES**
LYNN, MASS.

DESIGNATION (military)**:** J85-GE-17A.

BACKGROUND: Commercially this engine was designated the CJ610 and was installed in the early LearJet aircraft.
Date First J85 Engine Produced: 1966.
Date If Out-of-Production: 1978. Number of J85 Engines Produced to Date: 1,404.

ENGINE TYPE: Turbojet. Single-Shaft, Non-Afterburning.

COMPRESSOR TYPE: 8-Stage Single- Axial Flow Compressor.

COMPRESSOR DATA: Compressor Pressure Ratio: 6.9 : 1, Total Mass Airflow: 44.0 lb/sec (20 kg/sec).

TURBINE TYPE: 2-Stage Axial Flow Turbine.

COMBUSTOR TYPE: Annular, Through Flow.

POWER (take-off) RATING: 2,850 lbt (12,7 kN thrust). Rating Approved to: 59°F (15°C).

TAKE-OFF SPECIFIC FUEL CONSUMPTION (SFC): 0.99 lb/hr/lbt (28,04 mg/Ns).

BASIC DRY WEIGHT: 400 lbs (181 kg).

POWER/WEIGHT RATIO: 7.13 : 1 lbt/lb.

APPLICATION:
Cessna A-37A,B, "Dragonfly" Attack Aircraft.
Canadair CL-41G "Tutor" Trainer Aircraft.

OTHER ENGINES IN SERIES:
J85-GE-4 in Rockwell T-2, "Buckeye" Trainer Aircraft.
J85-GE-13/15 in Northrop F-5A/B Attack Aircraft and TF-5A Trainer Aircraft.
J85-GE-17 in Jet Pod AC-119G/K "Gunship"; Fairchild-Hiller "Packet-C" Transport.
J85-GE-17B in Saab-105, Attack/Trainer Aircraft.
J85-GE-19 in Lockheed XV-4B VTOL Aircraft.
J85-GE-LF1 in Ryan XV-5B VTOL Aircraft
J85-GE-CAN-15 in CF-5, NF-5 Fighter Aircraft.
J85-GE-CAN-40 in Canadair CL-41 Trainer Aircraft.
Locate other J85 Models under J85-GE-21 in General Electric Aircraft Engines, USA.

GAS TURBINE ENGINES FOR AIRCRAFT

 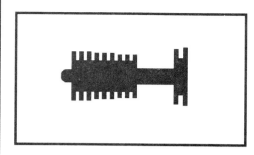

USA TURBINE ENGINE COMPANIES

MANUFACTURER: GENERAL ELECTRIC ENGINES
LYNN, MASS.

DESIGNATION (military): J85-GE-21.

BACKGROUND: Commercially this engine was designated the CJ610 and was installed in the early LearJet aircraft.
Date First J85 Engine Produced: 1972.
Date If Out-of-Production: 1989. Number of J85-GE-21 Engines Produced to Date: 3,314.

ENGINE TYPE: Turbojet. Single-Shaft with Afterburner.

COMPRESSOR TYPE: 9-Stage Single-Axial Flow Compressor.

COMPRESSOR DATA: Compressor Pressure Ratio: 8.3 : 1, Total Mass Airflow: 53 lb/sec (24 kg/sec).

TURBINE TYPE: 2-Stage Axial Flow Turbine.

COMBUSTOR TYPE: Annular, Through-Flow.

AFTERBURNER: One-Stage.

POWER (take-off) RATING: 3,500 lbt (15,6 kN thrust) [dry], 5,000 lbt (22,2 kN thrust) [A/B]. Rating Approved to: 59°F (15°C).

TAKE-OFF SPECIFIC FUEL CONSUMPTION (SFC): 1.24 lb/hr/lbt (35,13 mg/Ns) [dry], 2.13 lb/hr/lbt (60,34 mg/Ns) [A/B].

BASIC DRY WEIGHT: 684 lbs (310 kg).

POWER/WEIGHT RATIO: 5.12 : 1 [dry], 7.3 : 1 [A/B] - lbt/lb.

APPLICATION:
Northrop F-5E/F "Tiger-II" Fighter Aircraft.
RF-5E "TigerEye" Reconnaissance Aircraft.
F-5-21 (export model) for Fighter Aircraft.

OTHER ENGINES IN SERIES:
J85-GE-5 in T-38 Northrop, Talon" USA.
J85-GE-13 in Fiat G.91Y, Italy. and SF-5A, Spain.
J85-GE- Can-15 in CF-5, Canada.

GAS TURBINE ENGINES FOR AIRCRAFT

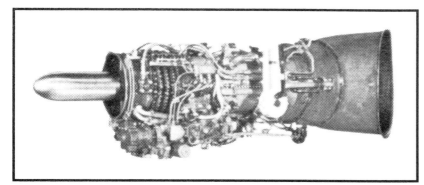 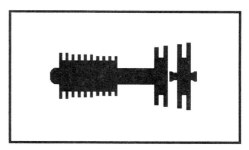

USA TURBINE ENGINE COMPANIES

MANUFACTURER: **GENERAL ELECTRIC ENGINES LYNN, MASS.**

DESIGNATION (military): T58-GE-16.

BACKGROUND: Date First T58 USA Military Engine Produced: 1974. Date If Out-of-Production: 1984. Number of T58 Engines Produced to Date: 742.

ENGINE TYPE: Turboshaft. Dual-Shaft (Compressor Drive & Power Output Drive).

COMPRESSOR TYPE: 10-Stage Single-Spool, Axial Flow Compressor.

COMPRESSOR DATA: Compressor Pressure Ratio: 8.6 : 1, Total Mass Airflow: 13.7 lb/sec (6,2 kg/sec).

TURBINE TYPE: 4-Stage Axial Flow Turbine including: 2-Stage Gas Producer (compressor drive) Turbine, 2-Stage Power (free) Turbine.

COMBUSTOR TYPE: Annular, Through Flow.

POWER (take-off) RATING: 1,870 shp (1,395 kw). Rating Approved to: 59°F (15°C).

TAKE-OFF SPECIFIC FUEL CONSUMPTION (SFC): 0.53 lb/hr/shp (0,32 kg/hr/kw).

BASIC DRY WEIGHT: 443 lbs (201 kg).

POWER/WEIGHT RATIO: 4.22 : 1 shp/lb (6,94 : 1 kw/kg).

APPLICATION:
Boeing AH-46 and CH-46 "Sea King" Helicopter, USA.
Sikorsky SH-3D "Jolly Green" Helicopter, USA.

OTHER ENGINES IN SERIES:
T58-GE-5 in Sikorsky CH-3, HH-3, S-61A4, S-67 Helicopters, USA.
T58-GE-5 in Kaman SH-42 Helicopter, USA.
T58-GE-5 in HH-52A USCG Helicopter, USA.
T58-GE-5 in Ishikawajima SH-3D/H Helicopter, Japan.
T58-GE-8F in Sikorsky CH-46A, SH-3G, SH-3F Helicopters, USA.
T58-GE-8F in Kaman H-2 "Seasprite" Helicopter, USA.
T58-GE-10 in Sikorsky CH-3B and SH-3B, Helicopters, USA.
T58-GE-10/-402 in C-113 and CH-118 Helicopters, Canada.
T58-GE-10M in Mitsubishi S-61 and SH-3B Helicopters, Japan.
T58-GE-10 in Agusta ASH-3H Helicopter, Italy.
T58-GE-100 in Sikorsky HH-3F (AS-61) and HH-3F "Pelican" Helicopters, USA.
Locate other T58 Models Under CT58 in General Electric Aircraft Engines, USA.

GAS TURBINE ENGINES FOR AIRCRAFT

 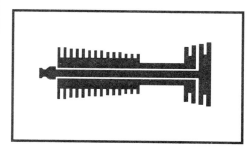

USA TURBINE ENGINE COMPANIES

MANUFACTURER: GENERAL ELECTRIC ENGINES
LYNN, MASS.

DESIGNATION (military): T64-GE-419.

BACKGROUND: Date First T64 Engine Produced: 1963 (T64-GE-1).
Date If Out-of-Production: N/A. Number of T64 Engines Produced to Date: In Development Expected 1993.

ENGINE TYPE: Turboshaft. Dual-Shaft (Compressor Drive & Power Output Drive).

COMPRESSOR TYPE: 14-Stage Single-Spool, Axial Flow Compressor.

COMPRESSOR DATA: Compressor Pressure Ratio: 14.8 : 1, Total Mass Airflow: 29 lb/sec (13,2 kg/sec).

TURBINE TYPE: 4-Stage Axial Flow Turbine including: 2-Stage Gas Producer (compressor drive) Turbine, 2-Stage Power (free) Turbine.

COMBUSTOR TYPE: Annular, Through Flow.

POWER (take-off) RATING: 4,750 shp (3,542 kw). Rating Approved to: 59°F (15°C).

TAKE-OFF SPECIFIC FUEL CONSUMPTION (SFC): 0.47 lb/hr/shp (0,29 kg/hr/kw).

BASIC DRY WEIGHT: 755 lbs (342 kg).

POWER/WEIGHT RATIO: 6.3 : 1 shp/lb (10,4 : 1 kw/kg).

APPLICATION:
Sikorsky CH-53D "Sea Stallion" Helicopter.
Sikorsky CH-53E "Super Stallion" Helicopter.
Sikorsky HH-53E "Sea Dragon" Helicopter.
Sikorsky MH-53 "Pave Low" Helicopter.

OTHER ENGINES IN SERIES:
T64-GE-3 in Sikorsky HH-53 Helicopter.
T64-GE-6 in Sikorsky CH-53A "Sea Stallion", MH-53 "Sea Dragon" and HH-53 "Super Jolly".
T64-GE-7 in Sikorsky CH-53 and HH-53D/E "Super Stallion" Helicopter.
T64-GE-100 in Sikorsky H-53 Helicopter.
T64-GE-413 in Sikorsky CH-53D Helicopter
T64-GE-415 and -416 in Sikorsky RH-53 Helicopter.
T64-GE-540 in Sikorsky S-64 "SkyCrane" Helicopter.
T64-GE-716 in Lockheed AH-56A "Cheyenne" Helicopter.
T64-P4D in Aleina/USAF C-27A Transport Aircraft.
Locate other T64 Models under Germany, Italy, Japan, USA listings.

GAS TURBINE ENGINES FOR AIRCRAFT

 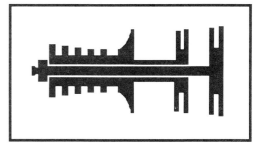

USA TURBINE ENGINE COMPANIES

MANUFACTURER: **GENERAL ELECTRIC ENGINES
LYNN, MASS.**

DESIGNATION (military): T700-GE-701A.

BACKGROUND: T700 is the military version of the CT7 commercial engine.
Date First T700 Engine Produced: 1978.
Date If Out-of-Production: N/A. Number of T700/CT7 Engines Produced to Date: 7,000.

ENGINE TYPE: Turboshaft. Dual-Shaft (Compressor Drive & Power Output Drive).

COMPRESSOR TYPE: 6-Stage Combination Compressor including 5-axial Stages and 1-Centrifugal stage.

COMPRESSOR DATA: Compressor Pressure Ratio: 25 : 1, Total Mass Airflow: 16 lb/sec (7,3 kg/sec).

TURBINE TYPE: 4-Stage Axial Flow Turbine including: 2-Stage Gas Producer (compressor drive) Turbine, 2-Stage Power (free) Turbine.

COMBUSTOR TYPE: Annular, Through Flow.

POWER (take-off) RATING: 1,715 shp (1,279 kw). Rating Approved to: 59°F (15°C).

TAKE-OFF SPECIFIC FUEL CONSUMPTION (SFC): 0.47 lb/hr/shp (0,29 kg/hr/kw).

BASIC DRY WEIGHT: 437 lbs (198 kg).

POWER/WEIGHT RATIO: 3.92 : 1 shp/lb (6,46 : 1 kw/kg).

APPLICATION:
Sikorsky UH-60 "Blackhawk" Helicopter.
Sikorsky S-70 Commercial Helicopter Version of UH-60.
Westland WS-1 Helicopter, Great Britain.

OTHER ENGINES IN SERIES:
T700-GE-401 in: Sikorsky SH-60B "Seahawk", HH-60D "Nighthawk", Bell AH-1W "Sea Cobra" and "Super Cobra", Kaman SF-2 "Super SeaSprite" Helicopters, USA.
T700-GE-401 in Westland WS-30 Helicopter, Great Britain.
T700-GE-401A in EH Industries EH101 "Merlin" Helicopter, Great Britain.
T700-GE-700 in Sikorsky UH-60A "Blackhawk", Boeing YUH-60A/61A/63/64 Helicopters, USA.
T700-GE-700 in Mitsubishi ISH-60J Helicopter, Japan.
T700-GE-701 series in McDonnell-Douglas AH-64A "Apache" Helicopter, USA.
T700/T6A (2,000 SHP) in EH-101 Helicopter.

GAS TURBINE ENGINES FOR AIRCRAFT

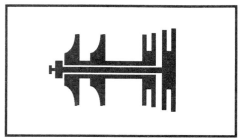

USA TURBINE ENGINE COMPANIES

MANUFACTURER: LIGHT HELICOPTER TURBINE ENGINE COMPANY-LHTEC (Allison and Garrett)
PHOENIX, ARIZONA.

DESIGNATION (civil)**:** CTS 800.

BACKGROUND: Date First CTS Engine Produced: In Development.
Date If Out-of-Production: N/A. Number of CTS 800 Engines Produced to Date: N/A.

ENGINE TYPE: Turboshaft. Three Shaft, (Dual Compressor Drive & Single Power Output Drive).

COMPRESSOR TYPE: 2 Independent Drive Centrifugal Flow Stages.

COMPRESSOR DATA: Compressor Pressure Ratio: 14 : 1, Total Mass Airflow: 7.56 lb/sec (3,43 kg/sec).

TURBINE TYPE: 4-Stage Axial Flow Turbine including: 2-Stage Gas Producer (compressor drive) Turbine, 2-Stage Power (free) Turbine.

COMBUSTOR TYPE: Annular, Reverse Flow.

POWER (take-off) RATING: 1,200 shp (900 kw). Rating Approved to: 59°F (15°C).

TAKE-OFF SPECIFIC FUEL CONSUMPTION (SFC): 0.46 lb/hr/shp (0,280 kg/hr/kw).

BASIC DRY WEIGHT: 315 lbs (142 kg).

POWER/WEIGHT RATIO: 3.8 : 1 shp/lb (6,3 :1 kw/kg).

APPLICATION:
Proposed for Agusta A-139 Helicopter, Italy. (In Development)

OTHER ENGINES IN SERIES:
T800-LHT-800 military version.
CTP-800 Turboprop in development.

GAS TURBINE ENGINES FOR AIRCRAFT

 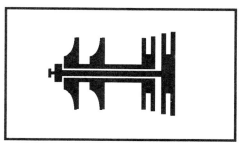

USA TURBINE ENGINE COMPANIES

MANUFACTURER: LIGHT HELICOPTER TURBINE ENGINE COMPANY-LHTEC (Allison and Garrett)
PHOENIX, ARIZONA.

DESIGNATION (military): T800-LHT-800.

BACKGROUND: Based on the Garret F109 Core.
Date First T800 Engine Produced: In Development.

ENGINE TYPE: Turboshaft. Three Shaft, (Dual Compressor Drive & Single Power Output Drive).

COMPRESSOR TYPE: 2 Independent Drive Centrifugal Flow Stages.

COMPRESSOR DATA: Compressor Pressure Ratio: 14 : 1, Total Mass Airflow: 7.56 lb/sec (3,43 kg/sec).

TURBINE TYPE: 4-Stage Axial Flow Turbine including: 2-Stage Gas Producer (compressor drive) Turbine, 2-Stage Power (free) Turbine.

COMBUSTOR TYPE: Annular, Reverse Flow.

POWER (take-off) RATING: 1,200 shp (900 kw). Rating Approved to: 59°F (15°C).

TAKE-OFF SPECIFIC FUEL CONSUMPTION (SFC): 0.46 lb/hr/shp (0,280 kg/hr/kw).

BASIC DRY WEIGHT: 315 lbs (142 kg).

POWER/WEIGHT RATIO: 3.8 : 1 shp/lb (6,3 : 1 kw/kg).

APPLICATION:
Agusta A-129 Helicopter, Italy.
Aerospatiale AS-565 "Panther" Helicopter, France.
Proposed for Boeing-Sikorsky RAH-66 Military Helicopter, USA.

OTHER ENGINES IN SERIES:
CTS-800 civil version.

GAS TURBINE ENGINES FOR AIRCRAFT

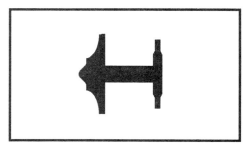

USA TURBINE ENGINE COMPANIES

MANUFACTURER: TELEDYNE CAE CORPORATION
TOLEDO, OHIO

DESIGNATION (military)**:** J69-T-25.

BACKGROUND: Originally built under license Turbomeca, France (Marbore II, Turbojet).
Date First J69 Engine Produced: 1954.
Date If Out-of-Production: N/A   Number of J69 Engines Produced to Date: 9,782.

ENGINE TYPE: Turbojet. Single-Shaft (non-afterburning).

COMPRESSOR TYPE: Centrifugal Compressor: 1-Stage, Single Entry.

COMPRESSOR DATA: Compressor Pressure Ratio: 3.9:1, Total Mass Airflow: 20 lb/sec (9 kg/sec).

TURBINE TYPE: 1-Stage Axial Flow Turbine.

COMBUSTOR TYPE: Annular, Through Flow.

POWER (take-off) RATING: 1,025 lbt (4,56 kN thrust). Rating Approved to: 59°F (15°C).

TAKE-OFF SPECIFIC FUEL CONSUMPTION (SFC): 1.14 lb/hr/lbt (32,29 mg/Ns).

BASIC DRY WEIGHT: 358 lbs (162,4 kg).

POWER/WEIGHT RATIO: 2.86 : 1 lbt/lb.

APPLICATION:
Cessna T-37B "Tweet" Trainer Aircraft.

OTHER ENGINES IN SERIES:
J69-2 and J69-9 in Early Trainer Aircraft.
J69-6,-17,-19 Early Target Drone Aircraft.
J69-29,-406 in Ryan Target Drones Aircraft.
J69-41A Military Special Purpose Aircraft.

GAS TURBINE ENGINES FOR AIRCRAFT

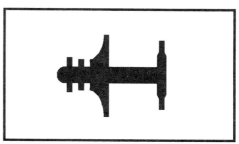

USA TURBINE ENGINE COMPANIES

MANUFACTURER: **TELEDYNE CAE CORPORATION**
TOLEDO, OHIO

DESIGNATION (military): J402-CA-702.

BACKGROUND: Civil designation 373-8B.
Date First J402 Engine Produced: 1985.
Date If Out-of-Production: N/A. Number of J402 Engines Produced to Date: 462.

ENGINE TYPE: Turbojet. Single-Shaft (non-afterburning).

COMPRESSOR TYPE: Combination Axial-Centrifugal Compressor: 2-Stage Axial and 1-Stage Centrifugal.

COMPRESSOR DATA: Compressor Pressure Ratio: 8.5 1, Mass Airflow: 13.7 lb/sec (6,2 kg/sec).

TURBINE TYPE: 1-Stage Axial Flow Turbine.

COMBUSTOR TYPE: Annular, Through Flow.

POWER (take-off) RATING: 960 lbt (4,27 kN thrust). Rating Approved to: 59°F (15°C).

TAKE-OFF SPECIFIC FUEL CONSUMPTION (SFC): 1.03 lb/hr/lbt (29,18 mg/Ns).

BASIC DRY WEIGHT: 138 lbs (62,6 kg).

POWER/WEIGHT RATIO: 6.96 : 1 lbt/lb.

APPLICATION:
Beech Remotely Piloted Vehicle and Ryan Reconnaissance Drone Aircraft.
Future Aircraft Applications.

OTHER ENGINES IN SERIES:
J402-CA-400 in McDonnell-Douglas Harpoon Missile.
J402-CA-700 in Beech Target Drone Aircraft.

GAS TURBINE ENGINES FOR AIRCRAFT

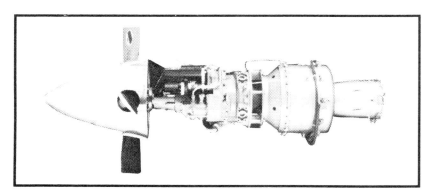 

USA TURBINE ENGINE COMPANIES

MANUFACTURER: TELEDYNE CONTINENTAL MOTORS
MOBILE, ALABAMA

DESIGNATION (civil): TP-500.

BACKGROUND: Date First TP-500 Engines Produced: 1988 (prototype).
Date Out-of-Production: N/A.   Number of TP-500 Engines Produced: 6.

ENGINE TYPE: Turboprop. Single Shaft.

COMPRESSOR TYPE: 1-Stage Centrifugal Flow Compressor.

COMPRESSOR DATA: Compressor Pressure Ratio: 8 : 1, Total Mass Airflow: 4.2 lb/sec (1,9 kg/sec).

TURBINE TYPE: 2-Stage Axial Flow.

COMBUSTOR TYPE: Annular, Through-Flow.

POWER (take-off) RATING: 447 eshp (333 ekw), including 22 shp (16,4 ekw) from 55 lbt (0,245 kN thrust). Rating Approved to: 75°F (24°C).

TAKE-OFF SPECIFIC FUEL CONSUMPTION (SFC): 0.675 lb/eshp/hr (0,41 kg/hr/ekw).

BASIC DRY WEIGHT: 340 lbs (155 kg).

POWER/WEIGHT RATIO: 1.3 : 1 eshp/lb (2,15 ekw/kw).

APPLICATION:
Proposed General Aviation aircraft.

OTHER ENGINES IN SERIES:
None

GAS TURBINE ENGINES FOR AIRCRAFT

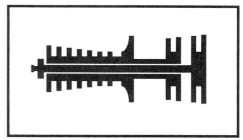

USA TURBINE ENGINE COMPANIES

MANUFACTURER: **TEXTRON-LYCOMING**
STRATFORD, CONN.

DESIGNATION (civil)**:** AL5512.

BACKGROUND: Date First AL5512 Engine Produced: 1981.
Date If Out-of-Production: N/A. Number of AL5512 Engines Produced to Date: 44.

ENGINE TYPE: Turboshaft. Dual-Shaft (Compressor Drive & Power Output Drive).

COMPRESSOR TYPE: Combination Compressor: 7-Stage Axial, 1-Stage Centrifugal.

COMPRESSOR DATA: Compressor Pressure Ratio: 8.2:1. Mass Airflow: 25.0 lb/sec (11,3 kg/sec).

TURBINE TYPE: 4-Stage Axial Flow including: 2-Stage Gas Producer (compressor drive) Turbine, 2-Stage Power (free) Turbine.

COMBUSTOR TYPE: Annular, Reverse Flow.

POWER (take-off) RATING: 4,075 shp (3,039 kw). Rating Approved to: 59°F (15°C).

TAKE-OFF SPECIFIC FUEL CONSUMPTION (SFC): 0.54 lb/hr/shp (0,32 kg/hr/kw).

BASIC DRY WEIGHT: 780 lbs (353 kg).

POWER/WEIGHT RATIO: 5.25 : 1 shp/lb (8,60 : 1 kw/kg).

APPLICATION:
Boeing Helicopter Model B-234 "Chinook International".

OTHER ENGINES IN SERIES:
Locate Military versions, T55-L-712 under Textron-Lycoming, USA.

GAS TURBINE ENGINES FOR AIRCRAFT

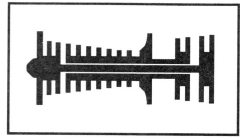

USA TURBINE ENGINE COMPANIES

MANUFACTURER: **TEXTRON-LYCOMING**
STRATFORD, CONN.

DESIGNATION (civil)**:** ALF 502L.

BACKGROUND: Date First ALF 502 Engine Produced: 1978.
Date If Out-of-Production: N/A. Number of ALF 502 Engines Produced to Date: 1,200. Military designation assigned to this engine is F102-LD-100, however, there is no present military application.

ENGINE TYPE: Turbofan. High Bypass, Dual-Shaft.

COMPRESSOR TYPE: 11-Stage Dual Compressor including a Geared Front Fan with 2-Booster Stages in the Low Pressure Compressor, followed by a Combination 7-Stage Axial and 1-Stage Centrifugal High Pressure Compressor.

COMPRESSOR DATA: Compressor Pressure Ratio: 13.3 : 1. Mass Airflow: 256 lb/sec (116,1 kg/sec). Fan Pressure Ratio: 1.5:1. Fan Bypass Ratio 5:2 : 1. Turbine/Fan Gear Ratio 2.3 :1.

TURBINE TYPE: 4-Stage Axial Flow including: 2-Stage High Pressure Turbine, 2-Stage Low Pressure Turbine.

COMBUSTOR TYPE: Annular, Reverse Flow.

POWER (take-off) RATING: 7,500 lbt (33,36 kN thrust). Rating Approved to: 59°F (15°C).

TAKE-OFF SPECIFIC FUEL CONSUMPTION (SFC): 0.41 lb/hr/lbt (11,61 mg/Ns).

BASIC DRY WEIGHT: 1,288 lbs (584 kg).

POWER/WEIGHT RATIO: 5.82 : 1 lbt/lb.

APPLICATION:
Canadair Challenger-600, Commuter/Airliner Aircraft.
Canadian Air Force CC-144 "Challenger" Transport.

OTHER ENGINES IN SERIES:
ALF 502L, L2, L2C, L3 in Challenger-600.
ALF 502R-3 in British Aerospace BAe-146 and BAe RJ70. (Regional Jet)

GAS TURBINE ENGINES FOR AIRCRAFT

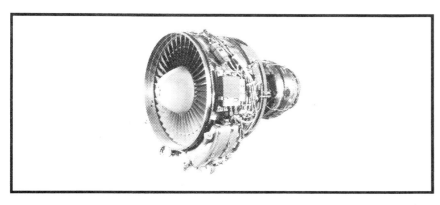
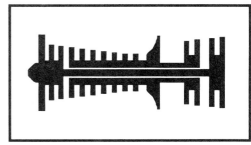

USA TURBINE ENGINE COMPANIES

MANUFACTURER: TEXTRON-LYCOMING
STRATFORD, CONN.

DESIGNATION (civil): LF-507-1F.

BACKGROUND: Date First LF-507 Engine Produced: In Development.

ENGINE TYPE: Turbofan. High Bypass, Dual-Shaft.

COMPRESSOR TYPE: 11-Stage Dual Compressor including a Front Fan with 2-Booster Stages in the Low Pressure Compressor, followed by a Combination 7-Stage Axial and 1-Stage Centrifugal High Pressure Compressor.

COMPRESSOR DATA: Compressor Pressure Ratio: 13 : 1. Mass Airflow: 252 lb/sec (114,3 kg/sec). Fan Pressure Ratio: 1.5:1. Fan Bypass Ratio 5 : 1.

TURBINE TYPE: 4-Stage Axial Flow including: 2-Stage High Pressure Turbine, 2-Stage Low Pressure Turbine.

COMBUSTOR TYPE: Annular, Reverse Flow.

POWER (take-off) RATING: 7,000 lbt (31,17 kN thrust). Rating Approved to: 74°F (23°C).

TAKE-OFF SPECIFIC FUEL CONSUMPTION (SFC): 0.406 lb/hr/lbt (11,5 mg/Ns).

BASIC DRY WEIGHT: 1,385 lbs (628 kg).

POWER/WEIGHT RATIO: 5.1 : 1 lbt/lb.

APPLICATION:
BAe-146 and BAe RJ, Commuter/Airliner Aircraft.

OTHER ENGINES IN SERIES:
None.

GAS TURBINE ENGINES FOR AIRCRAFT

USA TURBINE ENGINE COMPANIES

MANUFACTURER: TEXTRON-LYCOMING
STRATFORD, CT.

DESIGNATION (civil)**:** LTP101-700A-1.

BACKGROUND: Military designation T702-LD-700.
Date First LTP101 Engine Produced: 1976.
Date If Out-of-Production: N/A. Number of LTP 101 Engines Produced to Date: 156.

ENGINE TYPE: Turboprop. Dual-Shaft (Compressor Drive & Power Output Drive).

COMPRESSOR TYPE: Combination Compressor: 1-Stage Axial, 1-Stage Centrifugal.

COMPRESSOR DATA: Compressor Pressure Ratio: 8.6:1. Mass Airflow: 5.6 lb/sec (2,5 kg/sec).

TURBINE TYPE: 2-Stage Axial Flow including: 1-Stage Gas Producer (compressor drive) Turbine, 1-Stage Power (free) Turbine.

COMBUSTOR TYPE: Annular, Reverse Flow.

POWER (take-off) RATING: 700 eshp (522 ekw), includes 20 shp (14,9 kw) from 50 lbt (0,02 kN thrust). Rating Approved to: 59°F (15°C).

TAKE-OFF SPECIFIC FUEL CONSUMPTION (SFC): 0.55 lb/hr/eshp (0,33 kg/hr/ekw).

BASIC DRY WEIGHT: 335 lbs (152 kg).

POWER/WEIGHT RATIO: 2.09 : 1 eshp/lb (3,43 : 1 ekw/kg).

APPLICATION:
Riley-Cessna-421 "Turbine-Rocket" and "Turbine-Eagle" Agricultural Aircraft.

OTHER ENGINES IN SERIES:
No other Turboprop Models in Production.

GAS TURBINE ENGINES FOR AIRCRAFT

 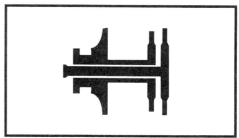

USA TURBINE ENGINE COMPANIES

MANUFACTURER: TEXTRON-LYCOMING
STRATFORD, CONN.

DESIGNATION (civil & military): LTS101-750C-1.

BACKGROUND: Date First LTS101 Engine Produced: 1975.
Date If Out-of-Production: N/A. Number of LTS 101 Engines Produced to Date: 1,881.

ENGINE TYPE: Turboshaft. Dual-Shaft (Compressor Drive & Power Output Drive).

COMPRESSOR TYPE: Combination Compressor: 1-Stage Axial, 1-Stage Centrifugal.

COMPRESSOR DATA: Compressor Pressure Ratio: 8.5:1. Mass Airflow: 5.9 lb/sec (2,68 kg/sec).

TURBINE TYPE: 2-Stage Axial Flow including: 1-Stage Gas Producer (compressor drive) Turbine, 1-Stage Power (free) Turbine.

COMBUSTOR TYPE: Annular, Reverse Flow.

POWER (take-off) RATING: 684 shp (510 kw). Rating Approved to: 59°F (15°C).

TAKE-OFF SPECIFIC FUEL CONSUMPTION (SFC): 0.57 lb/hr/shp (0,35 kg/hr/kw).

BASIC DRY WEIGHT: 244 lbs (111 kg).

POWER/WEIGHT RATIO: 2.8 : 1 shp/lb (4,59 : 1 kw/kg).

APPLICATION:
Bell 222 (A,B,UT) Model Helicopters.

OTHER ENGINES IN SERIES:
LTS101-600A-3 in Aerospatiale SA.350D Helicopter.
LTS101-650C-2 in Bell 222 Helicopter.
LTS101-650B-1 and -750 in Messerschmitt and Kawasaki BK 117 Helicopter
LTS101-750A/B in Aerospatiale HH-65A Military and SA.336, SA.366 Civil Model Helicopters.

GAS TURBINE ENGINES FOR AIRCRAFT

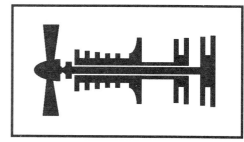

USA TURBINE ENGINE COMPANIES

MANUFACTURER: TEXTRON-LYCOMING
STRATFORD, CONN.

DESIGNATION (military): T53-L-701A.

BACKGROUND: Company designation LTC1F-5A.
Date First T53-L-701 Engine Produced: 1968.
Date If Out-of-Production: N/A. Number of T53 Turboprop Engines Produced to Date: 1,286.

ENGINE TYPE: Turboprop. Dual-Shaft (Compressor Drive & Power Output Drive).

COMPRESSOR TYPE: Combination Compressor: 5-Stage Axial, 1-Stage Centrifugal.

COMPRESSOR DATA: Compressor Pressure Ratio: 7:1. Mass Airflow: 12.7 lb/sec (5,62 kg/sec).

TURBINE TYPE: 4-Stage Axial Flow Turbine including: 2-Stage Gas Producer (compressor drive) Turbine, 2-Stage Power (free) Turbine.

COMBUSTOR TYPE: Annular, Reverse Flow.

POWER (take-off) RATING: 1,400 shp (1,044 kw). Rating approved to: 59°F (15°C).

TAKE-OFF SPECIFIC FUEL CONSUMPTION (SFC): 0.59 lb/hr/shp (0,36 kg/hr/kw).

BASIC DRY WEIGHT: 693 lbs (314 kg).

POWER/WEIGHT RATIO: 2.02.:1 shp/lb (3,32:1 kw/kg).

APPLICATION:
Grumman OV-1 "Mohawk" Observation Aircraft, USA.
TH-C-1A "Chung-Hsing" Attack/Trainer, Taiwan.

OTHER ENGINES IN SERIES:
No other Turboprop Engines in Production.

GAS TURBINE ENGINES FOR AIRCRAFT

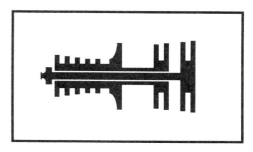

USA TURBINE ENGINE COMPANIES

MANUFACTURER: **TEXTRON-LYCOMING**
STRATFORD, CONN.

DESIGNATION (military & civil)**:** T53-L-703.

BACKGROUND: The original T53 was the first U.S. Gas Turbine Engine designed for Helicopter use. Commercial version designated as T5317A.
Date First T-53 Engine Produced: 1956.
Date If Out-of-Production: N/A. Number of T53 Turboshaft Engines Produced to Date: 18,961.

ENGINE TYPE: Turboshaft. Dual-Shaft (Compressor Drive & Power Output Drive).

COMPRESSOR TYPE: Combination Compressor: 5-Stage Axial, 1-Stage Centrifugal.

COMPRESSOR DATA: Compressor Pressure Ratio: 7.7 :1. Mass Airflow: 12.8.0 lb/sec (5,8 kg/sec).

TURBINE TYPE: 4-Stage Axial Flow Turbine including: 2-Stage Gas Producer (compressor drive) Turbine, 2-Stage Power (free) Turbine.

COMBUSTOR TYPE: Annular, Reverse Flow.

POWER (take-off) RATING: 1,485 shp (1,107 kw). Military Rating Approved to: 105°F. (40°C) Commercial Rating of 1500 shp for the 17A approved to 95°F (35°C).

TAKE-OFF SPECIFIC FUEL CONSUMPTION (SFC): 0.57 lb/hr/shp (0,35 kg/hr/kw).

BASIC DRY WEIGHT: 545 lbs (247 kg).

POWER/WEIGHT RATIO: 2.72 : 1 shp/lb (4,48 : 1 kw/kg).

APPLICATION:
Bell AH-1 (F,P,Q,S) "Cobra" Helicopter, USA.
Enhanced UH-1H "Iroquois-Huey" Helicopter, USA.
Updated Bell 205A1 "Iroquois" Helicopter, USA.
Dornier UH-1S Helicopter, Germany.

OTHER ENGINES IN SERIES:
T53-L-11 in Bell UH-1 B,C, D and Kaman HH-43 "Husky" Helicopter, USA.
T53-L-11A/B in Bell 204 Series "Iroquois" Helicopters, USA.
T53-L-11D in Agusta CH-47 "Chinook" Helicopter, Italy.
T53-L-13A in Bell-Japan-205 "Iroquois" Helicopter, Japan.
T53-L-13/-13B in Fuji AH-1S, UH-1S and FB-204 Helicopters, Japan.
T53-L-13B in Bell UH-1H "Iroquois-Huey" Helicopter, USA.
T5313B (commercial version) in Bell 205 Series Helicopters, USA.

GAS TURBINE ENGINES FOR AIRCRAFT

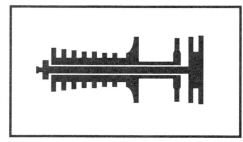

USA TURBINE ENGINE COMPANIES

MANUFACTURER: **TEXTRON-LYCOMING**
STRATFORD, CONN.

DESIGNATION (civil)**:** T5508D.

BACKGROUND: Company military designation LTC4B-8D.
Date First T5508D Engine Produced: 1974.
Date If Out-of-Production: N/A. Number of T5508/LTC4B-8D Engines Produced to Date: 521.

ENGINE TYPE: Turboshaft. Dual-Shaft (Compressor Drive & Power Output Drive).

COMPRESSOR TYPE: Combination Compressor: 7-Stage Axial, 1-Stage Centrifugal.

COMPRESSOR DATA: Compressor Pressure Ratio: 6.0:1. Mass Airflow: 25.0 lb/sec (11,3 kg/sec).

TURBINE TYPE: 3-Stage Axial Flow including: 1-Stage Gas Generator (compressor drive) Turbine, 2-Stage Power (free) Turbine.

COMBUSTOR TYPE: Annular, Reverse Flow.

POWER (take-off) RATING: 2,250 shp (1,676 kw). Rating Approved to: 118°F (48°C).

TAKE-OFF SPECIFIC FUEL CONSUMPTION (SFC): 0.63 lb/hr/shp (0,38 kg/hr/kw).

BASIC DRY WEIGHT: 618 lbs (280 kg).

POWER/WEIGHT RATIO: 3.64 : 1 shp/lb (5,99 : 1 kw/kg).

APPLICATION:
Bell 214 (A,B,C) "Biglifter" Helicopters.

OTHER ENGINES IN SERIES:
Military versions of the T55-L-7 Series.

GAS TURBINE ENGINES FOR AIRCRAFT

 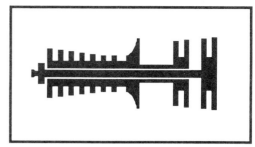

USA TURBINE ENGINE COMPANIES

MANUFACTURER: TEXTRON-LYCOMING
STRATFORD, CONN.

DESIGNATION (military)**:** T55-L-712.

BACKGROUND: Date First T55 Engine Produced: 1962, first -712 in 1981.
Date If Out-of-Production: N/A. Number of T55 Engines Produced to Date: 4,098, including 1,406 of the -712 Series.

ENGINE TYPE: Turboshaft. Dual-Shaft (Compressor Drive & Power Output Drive).

COMPRESSOR TYPE: Combination Compressor: 7-Stage Axial, 1-Stage Centrifugal.

COMPRESSOR DATA: Compressor Pressure Ratio: 8.2 :1. Mass Airflow: 25.0 lb/sec (11,3 kg/sec).

TURBINE TYPE: 4-Stage Axial Flow including: 2-Stage Gas Generator (compressor drive) Turbine, 2-Stage Power (free) Turbine.

COMBUSTOR TYPE: Annular, Reverse Flow.

POWER (take-off) RATING: 3,750 shp (2,796 kw). Rating Approved to: 92°F (33°C).

TAKE-OFF SPECIFIC FUEL CONSUMPTION (SFC): 0.518 lb/hr/shp (0,322 kg/hr/kw).

BASIC DRY WEIGHT: 750 lbs (340 kg).

POWER/WEIGHT RATIO: 5.0 : 1 shp/lb (8,22 : 1 kw/kg).

APPLICATION:
Boeing CH-47D and MH-47E "Chinook" Helicopters, USA.
Kawasaki CH-47C Helicopter, Japan.

OTHER ENGINES IN SERIES:
T55-L-11A/C/D/E In Boeing CH-46 and CH-47C Helicopter, USA.
T55-L-712E,F in Boeing 414 "Chinook Int'l" and British RAF CH-47 Helicopters.
T55-ACE (5,500 SHP) in development for Chinook Helicopter upgrade.

GAS TURBINE ENGINES FOR AIRCRAFT

 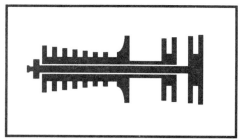

USA TURBINE ENGINE COMPANIES

MANUFACTURER: TEXTRON-LYCOMING
STRATFORD, CONN.

DESIGNATION (military)**:** T55-L-714.

BACKGROUND: Date First T55 Engine Produced: 1962, first -714 in 1990. Date If Out-of-Production: N/A. Number of T55 Engines Produced to Date: 4,098, including 12 of the -714 Series.

ENGINE TYPE: Turboshaft. Dual-Shaft (Compressor Drive & Power Output Drive).

COMPRESSOR TYPE: Combination Compressor: 7-Stage Axial, 1-Stage Centrifugal.

COMPRESSOR DATA: Compressor Pressure Ratio: 9.3 :1. Mass Airflow: 29.1 lb/sec (13,2 kg/sec).

TURBINE TYPE: 4-Stage Axial Flow including: 2-Stage Gas Generator (compressor drive) Turbine, 2-Stage Power (free) Turbine.

COMBUSTOR TYPE: Annular, Reverse Flow.

POWER (take-off) RATING: 4,777 shp (3,565 kw). Rating Approved to: 59°F (15°C).

TAKE-OFF SPECIFIC FUEL CONSUMPTION (SFC): 0.513 lb/hr/shp (0,315 kg/hr/kw).

BASIC DRY WEIGHT: 832 lbs (378 kg).

POWER/WEIGHT RATIO: 5.7 : 1 shp/lb (9,43 : 1 kw/kg).

APPLICATION:
Boeing MH-47E "Chinook" Helicopter.

OTHER ENGINES IN SERIES:
None.

GAS TURBINE ENGINES FOR AIRCRAFT

 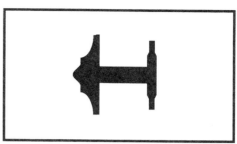

USA TURBINE ENGINES COMPANIES

MANUFACTURER: TURBINE TECHNOLOGIES LTD.
CHETEK, WISCONSIN

DESIGNATION (Civil): SR-30.

BACKGROUND: Date First SR-30 Engine Produced: 1990.
Date If Out-of-Production: N/A. Number of SR-30 Engines Produced to Date: Unknown.

ENGINE TYPE: Turbojet, Single-Shaft.

COMPRESSOR TYPE: 1-Stage Centrifugal Flow Compressor.

COMPRESSOR DATA: Compressor Pressure Ratio: 3.2 : 1, Total Mass Airflow: 0.75 lb/sec (0,34 kg/sec).

TURBINE TYPE: Single Stage Axial Flow.

COMBUSTOR TYPE: Annular, Reverse Flow.

POWER (take-off) RATING: 32 lbs Thrust (0,14 kN thrust). Rating Approved to: 59°F (15°C).

TAKE-OFF SPECIFIC FUEL CONSUMPTION (SFC): 1.555 lb/hr/lbt (43,9 mg/Ns).

BASIC DRY WEIGHT: 9.95 lbs (4,5 kg).

POWER/WEIGHT RATIO: 3.2 lbt/lb.

APPLICATION:
Proposed for Military and General Aviation.

OTHER ENGINES IN SERIES:
None.

GAS TURBINE ENGINES FOR AIRCRAFT

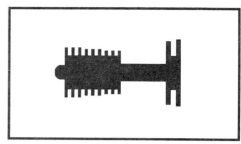

USA TURBINE ENGINE COMPANIES

MANUFACTURER: **UNITED TECHNOLOGIES INC.**
(PRATT & WHITNEY AIRCRAFT)
COMMERCIAL PRODUCTS DIVISION, EAST HARTFORD, CONN.

DESIGNATION (military): J60-P-6.

BACKGROUND: Civil Version is designated JT12A-5.
Date First J60 Engine Produced: 1958.
Date If Out-of-Production: 1962. Number of J-60 Engines Produced to Date: 1,341.
The J60 was also produced under license at Pratt & Whitney of Canada for the Royal Canadian Air Force.

ENGINE TYPE: Turbojet. Single-Shaft (non-afterburning).

COMPRESSOR TYPE: 9-Stage Single-Spool, Axial Flow Compressor.

COMPRESSOR DATA: Compressor Pressure Ratio: 6.4 : 1, Mass Airflow: 49.5 lb/sec (22,5 kg/sec).

TURBINE TYPE: 2-Stage Axial Flow Turbine.

COMBUSTOR TYPE: Can-Annular, Through Flow, with 8 liners.

POWER (take-off) RATING: 3,000 lbt (13,3 kN thrust). Rating Approved to: 59°F (15°C).

TAKE-OFF SPECIFIC FUEL CONSUMPTION (SFC): 0.96 lb/hr/lbt (27,19 mg/Ns).

BASIC DRY WEIGHT: 448 lbs (203 kg).

POWER/WEIGHT RATIO: 6.7 : 1 lbt/lb.

APPLICATION:
North American-Rockwell T-2B "Buckeye", Trainer Aircraft.
Rockwell CT-39 "Sabreliner" Military Light Transport and T-39 Trainer Aircraft.
Lockheed C-140 "JetStar", Military Light Transport.

OTHER ENGINES IN SERIES:
J60-P-5 in Gates-Learjet C-21A Military Light Transport.
J60-P-3,-3A,-5,-5A also in earlier Models of above Aircraft.

GAS TURBINE ENGINES FOR AIRCRAFT

 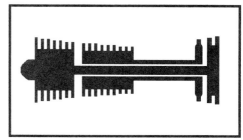

USA TURBINE ENGINE COMPANIES

MANUFACTURER: **UNITED TECHNOLOGIES INC.
(PRATT & WHITNEY AIRCRAFT GROUP)
COMMERCIAL PRODUCTS DIVISION, EAST HARTFORD, CONN.**

DESIGNATION (civil): JT3C-7.

BACKGROUND: Civil version of the military J-57 Turbojet.
Date First JT3C Engine Produced: 1951.
Date If Out-of-Production: 1972. Number of JT3C Engines Produced to Date: 20,640.

ENGINE TYPE: Turbojet. Dual-Shaft.

COMPRESSOR TYPE: 16-Stage Dual-Spool, Axial Flow Compressor including: 7-Stage Low Pressure Compressor, 9-Stage High Pressure Compressor.

COMPRESSOR DATA: Compressor Pressure Ratio: 13 : 1, Mass Airflow: 182 lb/sec (83 kg/sec).

TURBINE TYPE: 3-Stage Axial Flow Turbine, including: 1-Stage High Pressure Turbine, 2-Stage Low Pressure Turbine.

COMBUSTOR TYPE: Through Flow, Can-Annular with 8 liners.

POWER (take-off) RATING: 12,000 lbt (53,4 kN thrust). Rating Approved to: 59°F (15°C).

TAKE-OFF SPECIFIC FUEL CONSUMPTION (SFC): 0.77 lb/hr/lbt (21,81 mg/Ns).

BASIC DRY WEIGHT: 3,495 lbs (1,585 kg).

POWER/WEIGHT RATIO: 3.43 : 1 lbt/lb.

APPLICATION:
Boeing B-707 "Dash-80" and B-720 Airliners.

OTHER ENGINES IN SERIES:
JT3C-6 Water Injected Model in Boeing B707-120 & McDonnell-Douglas DC-8-10 Airliners.
JT3C-12 in Boeing B-720 Early Airliner.

GAS TURBINE ENGINES FOR AIRCRAFT

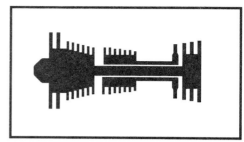

USA TURBINE ENGINE COMPANIES

MANUFACTURER: UNITED TECHNOLOGIES INC.
(PRATT & WHITNEY AIRCRAFT GROUP)
COMMERCIAL PRODUCTS DIVISION, EAST HARTFORD, CONN.

DESIGNATION (civil)**:** JT3D-7.

BACKGROUND: Military Version is designated TF-33. A derivative of the J-57 Turbojet. The first U.S. commercial Front Turbofan Engine to enter airline service.
Date First JT3D Engine Produced: 1959.
Date If Out-of-Production: 1983. Number of JT3D Engines Produced to Date: 8,325.

ENGINE TYPE: Turbofan. Dual-Shaft, Low-Bypass Front Fan. Water Injected.

COMPRESSOR TYPE: 15-Stage Dual-Spool, Axial Flow Compressor including: 2-Stage Front Fan and 6 additional Axial Stages in the Low Pressure Compressor, 7-Stage High Pressure Compressor.

COMPRESSOR DATA: Compressor Pressure Ratio: 13.4 : 1, Fan Pressure Ratio: 1.82 : 1, Fan Bypass Ratio: 1.43 : 1, Total Mass Airflow: 472 lb/sec (214 kg/sec).

TURBINE TYPE: 4-Stage Axial Flow Turbine including: 1-Stage High Pressure Turbine, 3-Stage Low Pressure Turbine.

COMBUSTOR TYPE: Can-Annular, Through-Flow, with 8 liners.

POWER (take-off) RATING: 19,000 lbt (85,5 kN thrust). Rating Approved to: 84°F (29 °C). [dry], 100°F (38°C). [wet].

TAKE-OFF SPECIFIC FUEL CONSUMPTION (SFC): 0.55 lb/hr/lbt (15,58 mg/Ns).

BASIC DRY WEIGHT: 4,340 lbs (1,969 kg).

POWER/WEIGHT RATIO: 4.38 : 1 lbt/lb.

APPLICATION:
Boeing B707-120B, -320B, -320C Airliners.
Boeing B-720B Early Airliner.
McDonnell-Douglas DC-8-50, -50F, -61F, -62F, -63 Airliners.
EC-18 (B707) Early Warning Aircraft.
EC-24 (DC-8) Early Warning Aircraft.

OTHER ENGINES IN SERIES:
JT3D-1, -3, -3B, -5A In B707 and DC-8 Airliners.
JT3D-3B In VC-137C, USAF VIP Model B-707 Transport.

GAS TURBINE ENGINES FOR AIRCRAFT

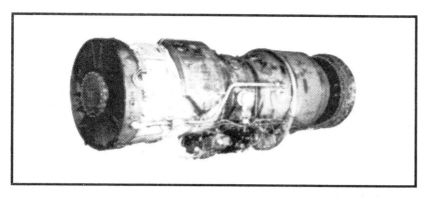 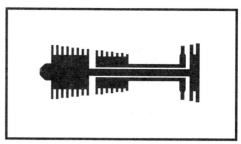

USA TURBINE ENGINE COMPANIES

MANUFACTURER: **UNITED TECHNOLOGIES INC.**
(PRATT & WHITNEY AIRCRAFT GROUP)
COMMERCIAL PRODUCTS DIVISION, EAST HARTFORD, CONN.

DESIGNATION (civil): JT4A-11.

BACKGROUND: Civil version of military J-75 Turbojet.
Date First J-75 Engine Produced: 1956.
Date First JT4D Engine Produced: 1957.
Date If Out-of-Production: 1967. Number of JT4 Engines Produced to Date: 2,579.

ENGINE TYPE: Turbojet. Dual-Shaft.

COMPRESSOR TYPE: 15-Stage Dual-Spool, Axial Flow Compressor including: 8-Stage Low Pressure Compressor, 7-Stage High Pressure Compressor.

COMPRESSOR DATA: Compressor Pressure Ratio: 12.5 : 1, Mass Airflow: 256 lb/sec (116 kg/sec).

TURBINE TYPE: 3-Stage Axial Flow Turbine, including: 1-Stage High Pressure Turbine, 2-Stage Low Pressure Turbine.

COMBUSTOR TYPE: Can-Annular, Through Flow, with 8 liners.

POWER (take-off) RATING: 17,500 lbt (77,8 kN thrust). Rating Approved to: 59°F (15°C).

TAKE-OFF SPECIFIC FUEL CONSUMPTION (SFC): 0.84 lb/hr/lbt (23,79 mg/Ns).

BASIC DRY WEIGHT: 5,100 lbs (2,313 kg).

POWER/WEIGHT RATIO: 3.43 : 1 lbt/lb.

APPLICATION:
Boeing B707-220 and -320 Airliners.
McDonnell-Douglas DC-8-20 and -30 Airliners.

OTHER ENGINES IN SERIES:
JT4A -3 and -5 in Boeing B707-320 and McDonnell-Douglas DC-8-20 Airliners.
JT4A-9 in Boeing B707-220/-320 and McDonnell-Douglas DC-8-20/-30 Airliners.

GAS TURBINE ENGINES FOR AIRCRAFT

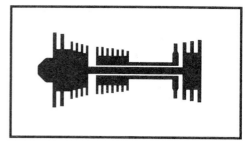

USA TURBINE ENGINE COMPANIES

MANUFACTURER: UNITED TECHNOLOGIES INC.
(PRATT & WHITNEY AIRCRAFT GROUP)
COMMERCIAL PRODUCTS DIVISION, EAST HARTFORD, CONN.

DESIGNATION (civil): JT8D-17AR.

BACKGROUND: A derivative of the military J-52 Turbojet. Originally developed for the Boeing B-727 Airliner.
Date First J-52 Engine Produced: 1958. Date First JT8D Engine Produced: 1963.
Date If Out-of-Production: N/A. Number of JT8D Engines through -17 model Produced to Date: 12,000

ENGINE TYPE: Turbofan. Dual-Shaft, Low-Bypass Front Fan.

COMPRESSOR TYPE: 13-Stage Dual-Spool, Axial Flow Compressor including: 2-Stage Front Fan and 4 Axial Stages in the Low Pressure Compressor, 7-Stage High Pressure Compressor.

COMPRESSOR DATA: Compressor Pressure Ratio: 17.5 : 1, Fan Pressure Ratio: 2.11 : 1, Fan Bypass Ratio: 1.0 : 1, Total Mass Airflow: 331 lb/sec (150 kg/sec).

TURBINE TYPE: 4-Stage Axial Flow Turbine including: 1-Stage High Pressure Turbine, 3-Stage Low Pressure Turbine.

COMBUSTOR TYPE: Can-Annular, Through-Flow, with 9 liners.

POWER (take-off) RATING: 17,400 lbt (77,4 kN thrust). Rating Approved to: 84°F (29°C).

TAKE-OFF SPECIFIC FUEL CONSUMPTION (SFC): 0.62 lb/hr/lbt (17,56 mg/Ns).

BASIC DRY WEIGHT: 3,500 lbs (1,588 kg).

POWER/WEIGHT RATIO: 4.97 : 1 lbt/lb.

APPLICATION: Boeing B727-100 and 200 Airliners.

OTHER ENGINES IN SERIES:
JT8D-1 in Boeing B-727 Airliner and McDonnell-Douglas DC-9-5 and DC9-10 Airliners.
JT8D-7 in B-727, DC-9, Aerospatiale Caravelle and Super Caravelle Airliners.
JT8D-9,-9A in B-727, B-737, DC-9, Caravelle Airliners.
JT8D-11 in DC-9, Dassault Mercure Airiners.
JT8D-15 in B-727, B-737, DC-9, Dassault Mercure-2 Airliners.
JT8D-15A in B-727-100/200, B-737, DC-9 Airliners.
JT8D-17,-17A in B-727, B-737, DC-9-30, -50 Airliners.
JT8D-17 also in Indonesian "Surveiller" Patrol Aircraft.
JT8D-17R in B-727 Airliner.
JT8D-9 In Kawasaki C-1 Transport Aircraft.

GAS TURBINE ENGINES FOR AIRCRAFT

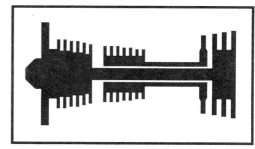

USA TURBINE ENGINE COMPANIES

MANUFACTURER: **UNITED TECHNOLOGIES INC. (PRATT & WHITNEY AIRCRAFT GROUP), COMMERCIAL PRODUCTS DIVISION, EAST HARTFORD, CONN.**

DESIGNATION (civil): JT8D-219.

BACKGROUND: A variation of the basic JT8D, developed for the DC-9 Super-80. A physically larger engine than the basic JT8D, with numerous changes. It could easily have been given a new designation.
Date First JT8D-200 Series Engine Produced: 1977.
Date If Out-of-Production: N/A. Number of JT8D-200 Series Engines Produced to Date: 1,350.

ENGINE TYPE: Turbofan. Dual-Shaft, Low-Bypass Front Fan.

COMPRESSOR TYPE: 14-Stage Dual-Spool, Axial Flow Compressor including: 1-Stage Front Fan and 6 Axial Stages in the Low Pressure Compressor, 7-Stage High Pressure Compressor.

COMPRESSOR DATA: Compressor Pressure Ratio: 17.4 : 1, Fan Pressure Ratio: 1.96 : 1, Fan Bypass Ratio: 1.82 : 1, Total Mass Airflow: 471 lb/sec (214 kg/sec).

TURBINE TYPE: 4-Stage Axial Flow Turbine including: 1-Stage High Pressure Turbine, 3-Stage Low Pressure Turbine.

COMBUSTOR TYPE: Can-Annular, Through-Flow, with 9 liners.

POWER (take-off) RATING: 19,250 lbt (85,6 kN thrust). Rating Approved to: 84°F (29°C).

TAKE-OFF SPECIFIC FUEL CONSUMPTION (SFC): 0.50 lb/hr/lbt (14,16 mg/Ns).

BASIC DRY WEIGHT: 4,410 lbs (2,000 kg).

POWER/WEIGHT RATIO: 4.37: 1 lbt/lb.

APPLICATION:
McDonnell-Douglas MD-83 Airliner.
(Formerly known as DC-9 Super 80 Series, now identified with an MD designation).

OTHER ENGINES IN SERIES:
JT8D-209 in McDonnell-Douglas MD-81 Airliner.
JT8D-216/218 in development for MD-90 series.
JT8D-217 and -217A in MD-81 and MD-82 Airliners.
JT8D-217C in MD-87 Airliner.
RTF-180, follow on Turbofan for the JT8D Series (PW & MTU, Germany)

GAS TURBINE ENGINES FOR AIRCRAFT

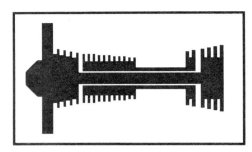

USA TURBINE ENGINE COMPANIES

MANUFACTURER: UNITED TECHNOLOGIES INC.
(PRATT & WHITNEY AIRCRAFT GROUP)
COMMERCIAL PRODUCTS DIVISION, EAST HARTFORD, CONN.

DESIGNATION (civil): JT9D-7R4G2.

BACKGROUND: Developed for Boeing B-747 aircraft. The early -3 and -7 models were water injected (wet); newer models are dry.
Date First JT9D Engine Produced: 1968.
Date If Out-of-Production: 1990. Number of JT9D Engines Produced to Date: 3,177.
Designation when the JT9D is installed in commercial aircraft for military use is F105-PW-100.

ENGINE TYPE: Turbofan. Dual-Shaft, High-Bypass Front Fan.

COMPRESSOR TYPE: 16-Stage Dual-Spool, Axial Flow Compressor including: 1-Stage Front Fan and 4 Axial Stages in the Low Pressure Compressor, 11-Stage High Pressure Compressor.

COMPRESSOR DATA: Compressor Pressure Ratio: 26.3 : 1, Fan Pressure Ratio: 1.7 : 1, Fan Bypass Ratio: 4.8 : 1, Total Mass Airflow: 1,695 lb/sec (769 kg/sec).

TURBINE TYPE: 6-Stage Axial Flow Turbine including: 2-Stage High Pressure Turbine, 4-Stage Low Pressure Turbine.

COMBUSTOR TYPE: Annular, Through-Flow.

POWER (take-off) RATING: 54,750 lbt (243,5 kN thrust). Rating Approved to: 86°F (30°C).

TAKE-OFF SPECIFIC FUEL CONSUMPTION (SFC): 0.36 lb/hr/lbt (10,2 mg/Ns).

BASIC DRY WEIGHT: 9,140 lbs (4,146 kg).

POWER/WEIGHT RATIO: 5.99 : 1 lbt/lb.

APPLICATION:
Boeing B747-200 (B,C,F and -300) Wide-Body Airliners.

OTHER ENGINES IN SERIES:
JT9D-3 in Boeing B747-100.
JT9D-7,7A,7F,7J in B-747, B-747SR.
JT9D-7Q in B747-200 (B,C,F).
JT9D-7R4D, -7R4E and E4 in B767-200.
JT9D-7R4D1, -7R4E1 and -7R4E3 in Airbus A310.
JT9D-7R4H1 in Airbus A310-600.
JT9D-20J in McDonnell-Douglas DC-10-40.
JT9D-59A/70A in DC-10-40, Airbus A300-70A and B747-200.

GAS TURBINE ENGINES FOR AIRCRAFT

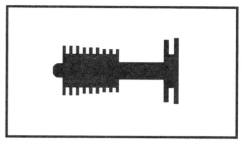

USA TURBINE ENGINE COMPANIES

MANUFACTURER: UNITED TECHNOLOGIES INC.
(PRATT & WHITNEY AIRCRAFT GROUP),
COMMERCIAL PRODUCTS DIVISION, EAST HARTFORD, CONN.

DESIGNATION (civil): JT12A-8.

BACKGROUND: Military version is designated J-60 in USAF T-39 and USN T2 Aircraft.
Date First JT12 Engine Produced: 1959.
Date If Out-of-Production: 1975. Number of JT12/J-60 Engines Produced to Date: 2,361.

ENGINE TYPE: Turbojet. Single-Shaft.

COMPRESSOR TYPE: 9-Stage Single-Spool, Axial Flow Compressor.

COMPRESSOR DATA: Compressor Pressure Ratio: 6.8 : 1, Mass Airflow: 50.5 lb/sec (22,9 kg/sec).

TURBINE TYPE: 2-Stage Axial Flow Turbine.

COMBUSTOR TYPE: Can-Annular, Through Flow, with 8 liners.

POWER (take-off) RATING: 3,300 lbt (14,68 kN thrust). Rating Approved to: 59°F (15°C).

TAKE-OFF SPECIFIC FUEL CONSUMPTION (SFC): 0.99 lb/hr/lbt (28,04 mg/Ns).

BASIC DRY WEIGHT: 468 lbs (212 kg).

POWER/WEIGHT RATIO: 7.05 : 1 lbt/lb.

APPLICATION:
Rockwell Sabreliner Business Jet Aircraft.
Lockheed JetStar Business Jet Aircraft.

OTHER ENGINES IN SERIES:
JT12A-6A also in early Sabreliner and JetStar.

GAS TURBINE ENGINES FOR AIRCRAFT

 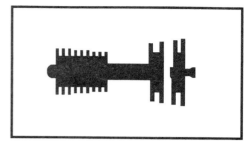

USA TURBINE ENGINE COMPANIES

MANUFACTURER: UNITED TECHNOLOGIES INC.
(PRATT & WHITNEY AIRCRAFT GROUP),
COMMERCIAL PRODUCTS DIVISION, EAST HARTFORD, CONN.

DESIGNATION (military): JTFD12A-5A.

BACKGROUND: A derivative of the JT12 Turbojet Engine. Its military application is designated T73-P-700.
Date First JTFD12A Engine Produced: 1961.
Date If Out-of-Production: 1970. Number of JTFD12 Engines Produced to Date: 354.

ENGINE TYPE: Turboshaft. Dual-Shaft (Compressor Drive & Power Output Drive).

COMPRESSOR TYPE: 9-Stage Axial Flow Compressor.

COMPRESSOR DATA: Compressor Pressure Ratio: 6.8 :1. Mass Airflow: 51.5 lb/sec (23,4 kg/sec).

TURBINE TYPE: 4-Stage Axial Flow Turbine including: 2-Stage Gas Producer (compressor drive) Turbine, 2-Stage Power (free) Turbine.

COMBUSTOR TYPE: Can-Annular, Through Flow, with 8 liners.

POWER (take-off) RATING: 4,800 shp (3,579 kw). Rating Approved to: 59 °F (15°C).

TAKE-OFF SPECIFIC FUEL CONSUMPTION (SFC): 0.65 lb/hr/shp (0,40 kg/hr/kw).

BASIC DRY WEIGHT: 935 lbs (424 kg).

POWER/WEIGHT RATIO: 5.13 : 1 shp/lb (8,44 : 1 kw/kg).

APPLICATION:
Sikorsky CH-54A "Tarhe" Military Helicopter and its counterpart the S-64E "Skycrane" Civil Helicopter.

OTHER ENGINES IN SERIES:
JTFD12A-4A also In early CH-54A "Tarhe" and S-64A,E; "Skycrane" Helicopter.

GAS TURBINE ENGINES FOR AIRCRAFT

 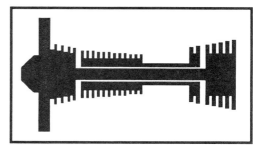

USA TURBINE ENGINE COMPANIES

MANUFACTURER: **UNITED TECHNOLOGIES INC.
(PRATT & WHITNEY AIRCRAFT GROUP)
COMMERCIAL PRODUCTS DIVISION, EAST HARTFORD, CONN.**

DESIGNATION (civil)**:** PW 2037.

BACKGROUND: Originally designated as JT10D. Developed for commercial aircraft as a joint effort with MTU in West Germany and Fiat in Italy. Not a variant of a former engine but a completely new design. Military designation assigned to this engine is F117-PW-100.
Date First PW2037 Engine Produced: 1984.
Date If Out-of-Production: N/A. Number of PW 2000 Series Engines Produced to Date: 340.

ENGINE TYPE: Turbofan. Dual-Shaft, High-Bypass Front Fan.

COMPRESSOR TYPE: 17-Stage Dual-Spool, Axial Flow Compressor including: 1-Stage Front Fan and 4 Axial Stages in the Low Pressure Compressor, 12-Stage High Pressure Compressor.

COMPRESSOR DATA: Compressor Pressure Ratio: 27.4 : 1, Fan Pressure Ratio: 1.69 : 1, Fan Bypass Ratio: 6.0 : 1, Total Mass Airflow: 1,210 lb/sec (549 kg/sec).

TURBINE TYPE: 7-Stage Axial Flow Turbine including: 2-Stage High Pressure Turbine, 5-Stage Low Pressure Turbine.

COMBUSTOR TYPE: Annular, Through-Flow.

POWER (take-off) RATING: 38,250 lbt (170,1 kN thrust). Rating Approved to: 87°F (31°C).

TAKE-OFF SPECIFIC FUEL CONSUMPTION (SFC): 0.32 lb/hr/lbt (9,07 mg/Ns).

BASIC DRY WEIGHT: 7,300 lbs (3,311 kg).

POWER/WEIGHT RATIO: 5.24 : 1 lbt/lb.

APPLICATION:
Boeing B-757 Airliner.
McDonnell-Douglas C-17A Military Transport.
Ilyushin Il-99M Transport.

OTHER ENGINES IN SERIES:
PW 2040 Uprated Model in Boeing B-757 and B-757F with 41,700 lbt.
ADP (Advanced Ducted PropFan) 50,000 lbt; follow on enigne of the PW 2000 Series.

GAS TURBINE ENGINES FOR AIRCRAFT

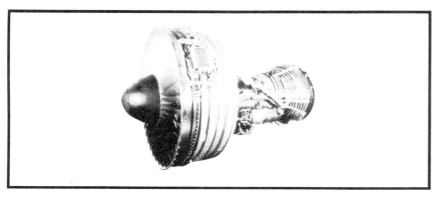
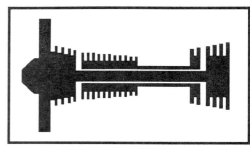

USA TURBINE ENGINE COMPANIES

MANUFACTURER: UNITED TECHNOLOGIES INC.
(PRATT & WHITNEY AIRCRAFT GROUP)
COMMERCIAL PRODUCTS DIVISION, EAST HARTFORD, CONN.

DESIGNATION (civil): PW 4152.

BACKGROUND: No military counterpart. Developed for commercial aircraft.
Date First PW 4152 Engine Produced: 1986.
Date If Out-of-Production: N/A. Number of PW 4000 Series Engines Produced to Date: 360.

ENGINE TYPE: Turbofan. Dual-Shaft, High-Bypass Front Fan.

COMPRESSOR TYPE: 16-Stage Dual-Spool, Axial Flow Compressor including: 1-Stage Front Fan and 4-Stages in the Low Pressure Compressor, 11-Stages in the High Pressure Compressor.

COMPRESSOR DATA: Compressor Pressure Ratio: 27.1 : 1, Fan Pressure Ratio: 1.66 : 1, Fan Bypass Ratio: 5.0 : 1, Total Mass Airflow: 1,645 lb/sec (746 kg/sec).

TURBINE TYPE: 6-Stage Axial Flow Turbine including: 2-Stage High Pressure Turbine, 4-Stage Low Pressure Turbine.

COMBUSTOR TYPE: Annular, Through-Flow.

POWER (take-off) RATING: 52,000 lbt (231 kN thrust). Rating Approved to: 108°F (42°C).

TAKE-OFF SPECIFIC FUEL CONSUMPTION (SFC): 0.31 lb/hr/lbt (9,06 mg/Ns).

BASIC DRY WEIGHT: 9,200 lbs (4,173 kg).

POWER/WEIGHT RATIO: 5.65 : 1 lbt/lb.

APPLICATION:
Airbus Industries A310 Wide-body Airliner.

OTHER ENGINES IN SERIES:
PW 4056 in Boeing B767-200/300 and Boeing B747-400 Airliners.
PW 4060 in Boeing B767-300 & B747-400 Airliners.
PW 4152/56/58 in Airbus A300-600 and A310-300 Airliners.
PW 4460 in McDonnell-Douglas MD-11 Airliner.
PW 4073/84 in develoment for Boeing B777 Airliner.
PW 4168 in Airbus A-330 and McDonnell/D MD12 Airliners

GAS TURBINE ENGINES FOR AIRCRAFT

 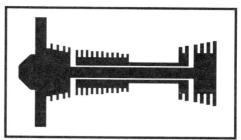

USA TURBINE ENGINE COMPANIES

MANUFACTURER: **UNITED TECHNOLOGIES INC.**
(PRATT & WHITNEY AIRCRAFT GROUP)
COMMERCIAL PRODUCTS DIVISION, EAST HARTFORD, CONN.

DESIGNATION (civil)**:** PW 4460.

BACKGROUND: No military counterpart. Developed for commercial aircraft.
Date First PW 4460 Engine Produced: 1988.
Date If Out-of-Production: N/A. Number of PW 4000 Series Engines Produced to Date: 360.

ENGINE TYPE: Turbofan. Dual-Shaft, High-Bypass Front Fan.

COMPRESSOR TYPE: 16-Stage Dual-Spool, Axial Flow Compressor including: 1-Stage Front Fan and 4-Stages in the Low Pressure Compressor, 11-Stages in the High Pressure Compressor.

COMPRESSOR DATA: Compressor Pressure Ratio: 32.3 : 1, Fan Pressure Ratio: 1.74: 1, Fan Bypass Ratio: 4.7 : 1, Total Mass Airflow: 1,766 lb/sec (801 kg/sec).

TURBINE TYPE: 6-Stage Axial Flow Turbine including: 2-Stage High Pressure Turbine, 4-Stage Low Pressure Turbine.

COMBUSTOR TYPE: Annular, Through-Flow.

POWER (take-off) RATING: 60,000 lbt (266,9 kN thrust). Rating Approved to: 86°F (30°C).

TAKE-OFF SPECIFIC FUEL CONSUMPTION (SFC): 0.33 lb/hr/lbt (9,35 mg/Ns).

BASIC DRY WEIGHT: 9,200 lbs (4,173 kg).

POWER/WEIGHT RATIO: 6.52 : 1 lbt/lb.

APPLICATION:
McDonnell Douglas MD-11 Wide-Body Airliner.

OTHER ENGINES IN SERIES:
Locate other PW 4000 Models under PW 4152 in United Technologies, USA.

GAS TURBINE ENGINES FOR AIRCRAFT

 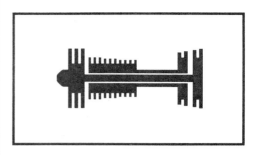

USA TURBINE ENGINE COMPANIES

MANUFACTURER: **UNITED TECHNOLOGIES INC. (PRATT & WHITNEY AIRCRAFT) GOVERNMENT PRODUCTS DIVISION), WEST PALM BEACH, FL.**

DESIGNATION (military)**:** F100-PW-220.

BACKGROUND: No Civil Counterpart. Original designation JTF-22.
Date First F100 Engine Produced: 1972.
Date If Out-of-Production: N/A. Number of F100 Engines Produced to Date: 767.

ENGINE TYPE: Turbofan. Dual-Shaft, Low-Bypass Front Fan with Afterburning.

COMPRESSOR TYPE: 13-Stage Dual-Spool, Axial Flow Compressor including: 3-Stage Front Fan with no other Axial Stages in the Low Pressure Compressor, 10-Stage High Pressure Compressor.

COMPRESSOR DATA: Compressor Pressure Ratio: 25 : 1, Fan Pressure Ratio: 2.60 : 1, Fan Bypass Ratio: 0.63 : 1, Total Mass Airflow: 228 lb/sec (103 kg/sec).

TURBINE TYPE: 4-Stage Axial Flow Turbine including: 2-Stage High Pressure Turbine, 2-Stage Low Pressure Turbine.

COMBUSTOR TYPE: Annular, Through-Flow.

AFTERBURNER: 5-Stage, 1-Stage in Fan Bypass Duct, 4-Stages in the Core Exhaust.

POWER (take-off) RATING: 14,670 lbt [dry], 23,830 [A/B]. Rating Approved to: 59°F (15°C).

TAKE-OFF SPECIFIC FUEL CONSUMPTION (SFC): 0.72 lb/hr/lbt (20,4 mg/Ns) [dry], 2.17 lb/hr/lbt (61,47 mg/Ns) [A/B].

BASIC DRY WEIGHT: 3,200 lbs (1,452 kg).

POWER/WEIGHT RATIO: 4.58 : 1 [dry], 7.45 : 1 [A/B] - lbt/lb.

APPLICATION:
McDonnell-Douglas Corp. F-15E "Eagle" Fighter Aircraft.
General Dynamics Corp. F-16C/D "Fighting Falcon" Fighter Aircraft.
LTV Corp. A-7F "Corsair-II" Upgrade Attack/Fighter Aircraft.

OTHER ENGINES IN SERIES:
F100-PW-100 in early F-15, locate under Ishikaeajima-Harima, Japan.
F100-PW-200 in early F-16 and TF-16N.
F100-PW-229 in Upgrade F-15 and F-16.
F119-PW-100 In Development for F-16AT and YF-22A and F-22A Fighter Aircraft, Data Classified.

GAS TURBINE ENGINES FOR AIRCRAFT

 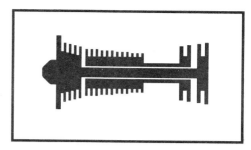

USA TURBINE ENGINE COMPANIES

MANUFACTURER: **UNITED TECHNOLOGIES INC. (PRATT & WHITNEY AIRCRAFT) GOVERNMENT PRODUCTS DIVISION), WEST PALM BEACH, FL.**

DESIGNATION (military)**:** F117-PW-100.

BACKGROUND: Derivative of the commercial PW 2037 Turbofan.
Date First F117 Engine Produced: 1990.
Date If Out-of-Production: N/A. Number of F117 Engines Produced to Date: Unknown.

ENGINE TYPE: Turbofan. Dual-Shaft, High-Bypass, Front Fan (non-afterburning).

COMPRESSOR TYPE: 17-Stage Dual-Spool, Axial Flow Compressor including: 1-Stage Front Fan and 4 Axial Stages in the Low Pressure Compressor, 12-Stage High Pressure Compressor.

COMPRESSOR DATA: Compressor Pressure Ratio: 31.8 : 1, Fan Pressure Ratio: 1.69 : 1, Fan Bypass Ratio: 6.0 : 1, Total Mass Airflow: 1,210 lb/sec (549 kg/sec).

TURBINE TYPE: 4-Stage Axial Flow Turbine including: 2-Stage High Pressure Turbine, 2-Stage Low Pressure Turbine.

COMBUSTOR TYPE: Annular, Through-Flow.

POWER (take-off) RATING: 41,700 lbt (185 kN thrust). Rating Approved to: 87°F (31 °C).

TAKE-OFF SPECIFIC FUEL CONSUMPTION (SFC): 0.34 lb/hr/lbt (9,6 mg/Ns).

BASIC DRY WEIGHT: 7,160 lbs (3,248 kg).

POWER/WEIGHT RATIO: 5.82 : 1 lbt/lb.

APPLICATION:
McDonnell-Douglas C-17 Military Transport.

OTHER ENGINES IN SERIES:
None.

GAS TURBINE ENGINES FOR AIRCRAFT

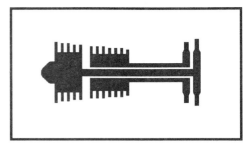

USA TURBINE ENGINE COMPANIES

MANUFACTURER: UNITED TECHNOLOGIES INC.
(PRATT & WHITNEY AIRCRAFT)
GOVERNMENT PRODUCTS DIVISION, WEST PALM BEACH, FL.

DESIGNATION (military): J52-P-408.

BACKGROUND: P&W designation JT8B-5. The J52 provides the core of the civil JT8D. Afterburning Model Designated as PW 1216.
Date First J52 Engine Produced: 1959.
Date If Out-of-Production: 1990. Number of J-52 Engines Produced to Date: 2,647.

ENGINE TYPE: Turbojet. Dual-Shaft (non-afterburning).

COMPRESSOR TYPE: 12-Stage Dual-Spool, Axial Flow Compressor, including: 5-Stage Low Pressure Compressor, 7-Stage High Pressure Compressor.

COMPRESSOR DATA: Compressor Pressure Ratio: 14.6 : 1, Mass Airflow: 143 lb/sec (64,9 kg/sec).

TURBINE TYPE: 2-Stage Axial Flow Turbine, including: 1-Stage High Pressure Turbine, 1-Stage Low Pressure Turbine.

COMBUSTOR TYPE: Can-Annular, Through Flow, with 9 liners.

POWER (take-off) RATING: 11,200 lbt (49,8 kN thrust). Rating Approved to: 59°F (15 °C).

TAKE-OFF SPECIFIC FUEL CONSUMPTION (SFC): 0.89 lb/hr/lbt (25,21 mg/Ns).

BASIC DRY WEIGHT: 2,318 lbs (1,051 kg).

POWER/WEIGHT RATIO: 4.83 : 1 lbt/lb.

APPLICATION:
McDonnell-Douglas A-4M "Skyhawk" Attack Aircraft.
Grumman EA-6B "Prowler" Electronic Warfare Aircraft.

OTHER ENGINES IN SERIES:
J52-P-6A/B in McDonnell-Douglas TA-4F/J "Skyhawk" Trainer Aircraft.
J52-P-8A/B in Grumman A-6E "Intruder" Attack Aircraft.
J52-P-8A/B in early Grumman EA-6A "Prowler" Electronic Warfare Aircraft.
J52-P-8A/B in Grumman KA-6D" "Intruder-Tanker" Inflight Refueling Aircraft.

GAS TURBINE ENGINES FOR AIRCRAFT

 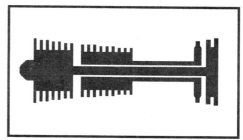

USA TURBINE ENGINE COMPANIES

MANUFACTURER: **UNITED TECHNOLOGIES INC.
(PRATT & WHITNEY AIRCRAFT)
GOVERNMENT PRODUCTS DIVISION, WEST PALM BEACH, FL.**

DESIGNATION (military): J57-P-59W.

BACKGROUND: Civil version is designated JT3C-8.
Date First J57-P-59W Engine Produced: 1958.
Date If Out-of-Production: 1964. Number of J-57 Water Injected Engines Produced to Date: 10,949.

ENGINE TYPE: Turbojet. Dual-Shaft, Water Injected.

COMPRESSOR TYPE: 16-Stage Dual-Spool, Axial Flow Compressor including: 7-Stage Low Pressure Compressor, 9-Stage High Pressure Compressor.

COMPRESSOR DATA: Compressor Pressure Ratio: 12.5 : 1, Mass Airflow: 185 lb/sec (83,9 kg/sec).

TURBINE TYPE: 3-Stage Axial Flow Turbine, including: 1-Stage High Pressure Turbine, 2-Stage Low Pressure Turbine.

COMBUSTOR TYPE: Can-Annular, Through Flow, with 8 liners.

POWER (take-off) RATING: 13,750 lbt (61,2 kN thrust) [wet], 11,200 lbt (49,8 kN thrust) [dry]. Rating Approved to: 100°F (38°C).

TAKE-OFF SPECIFIC FUEL CONSUMPTION (SFC): 0.95 lb/hr/lbt (26,91 mg/Ns).

BASIC DRY WEIGHT: 4,320 lbs (1,960 kg).

POWER/WEIGHT RATIO: 3.18 : 1 [wet], 2.59 : 1 [dry] - lbt/lb.

APPLICATION:
Boeing KC-135A "StratoTanker" Inflight Refueling Aircraft.
Boeing C-135C "StratoLifter" Military Transport Aircraft.
Boeing EC-135 "StratoLifter" Early Warning Aircraft.

OTHER ENGINES IN SERIES:
J-57-P-29W/WA in Boeing B-52 (B,C,D,E) "StratoFortress" Bomber Aircraft.
J-57-P-43W,WA/WB in Boeing B-52 (F,G) "StratoFortress" and KC-135A "StratoTanker".
Locate Afterburning J-57 Models under J-57-P-420 in United Technologies, USA.

GAS TURBINE ENGINES FOR AIRCRAFT

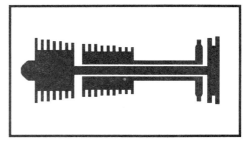

USA TURBINE ENGINE COMPANIES

MANUFACTURER: UNITED TECHNOLOGIES INC.
(PRATT & WHITNEY AIRCRAFT)
GOVERNMENT PRODUCTS DIVISION, WEST PALM BEACH, FL.

DESIGNATION (military): J57-P-420.

BACKGROUND: The original J-57 was the first U.S. built Turbojet with a Dual-Axial Flow Compressor. It was flight tested in the B-50 Aircraft. It was also the first Gas Turbine Engine to power an aircraft beyond the speed of sound in the YF-100 SuperSabre, in May, 1953. The J-57 Commercial version is designated the JT3C. The -420 is designated JT3C-31.
Date First J57 Engine Produced: 1951.
Date If Out-of-Production: 1964. Number of J57 Engines Produced to Date: 3,103.

ENGINE TYPE: Turbojet. Dual-Shaft, with Afterburning.

COMPRESSOR TYPE: 16-Stage Dual-Spool, Axial Flow Compressor including: 7-Stage Low Pressure Compressor, 9-Stage High Pressure Compressor.

COMPRESSOR DATA: Compressor Pressure Ratio: 12.3 : 1, Mass Airflow: 180 lb/sec (81,6 kg/sec).

TURBINE TYPE: 3-Stage Axial Flow Turbine, including: 1-Stage High Pressure Turbine, 2-Stage Low Pressure Turbine.

COMBUSTOR TYPE: Can-Annular, Through Flow, with 8 liners.

AFTERBURNER: 1-Stage.

POWER (take-off) RATING: 12,400 lbt (55,16 kN thrust) [dry], 19,600 lbt (87,2 kN thrust)[A/B]. Rating Approved to: 59°F (15°C).

TAKE-OFF SPECIFIC FUEL CONSUMPTION (SFC): 0.87 lb/hr/lbt (24,6 mg/Ns) [dry], 2.30 lb/hr/lbt (65,15 mg/Ns) [A/B].

BASIC DRY WEIGHT: 4,840 lbs (2,195 kg).

POWER/WEIGHT RATIO: 2.56 : 1 [dry], 4.05 : 1 [A/B] - lbt/lb.

APPLICATION: Chance-Vought F-8J "Crusader" Attack Aircraft and TF-8A Trainer.

OTHER ENGINES IN SERIES:
J57-P-4 in Vought F-8 and A-7B.
J57-P-10 in McDonnell A-3B "Sky Warrior".
J57-P-13 in McDonnell F101, RF-101 "Voodoo".
J57-P-16 in Vought F-8C "Crusader".
J57-P-20 in French Crusader (F-8E).

J57-P-21A in Rockwell F-100 "Super Sabre".
J57-P-23 in Gen/Dyn F-102A "Delta Dagger".
J57-P-27 in Martin RB-57D "Recon Aircraft".
J57-P-55 in McDonnell/D F-101 "Voodoo".

GAS TURBINE ENGINES FOR AIRCRAFT

 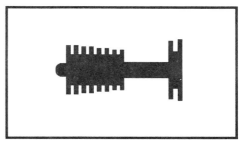

USA TURBINE ENGINE COMPANIES

MANUFACTURER: **UNITED TECHNOLOGIES INC.**
(PRATT & WHITNEY AIRCRAFT)
GOVERNMENT PRODUCTS DIVISION, WEST PALM BEACH, FL.

DESIGNATION (military)**:** J58-P-4.

BACKGROUND: Designed for the High Altitude Reconnaissance Aircraft YF-12A and later the SR-71, which holds several speed and altitude records. Many of its performance factors are military classified or not available. Sometimes referred to as A TurboRam Engine.
P&W designation is JT11D-20B
Date First J-58 Engine Produced: 1964. Number of J58-P-4 engines produced: Unknown.

ENGINE TYPE: Turbojet. Single-Shaft, with Afterburning.

COMPRESSOR TYPE: 8-Stage Single-Spool, Axial Flow Compressor. A portion of Compressor Discharge is diverted (bypassed) to the afterburner.

COMPRESSOR DATA: Compressor Pressure Ratio: 8.5 : 1, Mass Airflow: 327 lb/sec (148,6 kg/sec).

TURBINE TYPE: 2-Stage Axial Flow Turbine.

COMBUSTOR TYPE: Can-Annular, Through Flow, with 8 liners.

AFTERBURNER: 1-Stage.

POWER (take-off) RATING: 25,000 lbt (111,2 kN thrust) [dry], 34,000 lbt (151,2 kN thrust) [A/B]. Rating Approved to: 59°F (15°C).

TAKE-OFF SPECIFIC FUEL CONSUMPTION (SFC): 0.80 lb/hr/lbt (22,66 mg/Ns) [dry], 1.90 lb/hr/lbt (53,82 mg/Ns) [A/B].

BASIC DRY WEIGHT: 6,250 lbs (2,835 kg).

POWER/WEIGHT RATIO: 4.0 : 1 [dry], 5.44 : 1 [A/B] - lbt/lb.

APPLICATION:
Lockheed SR-71 "Blackbird" Reconnaissance Aircraft.
Lockheed SR-71B Trainer Aircraft.
Lockheed YF-12A Interceptor Aircraft. All now retired.

OTHER ENGINES IN SERIES:
None.

GAS TURBINE ENGINES FOR AIRCRAFT

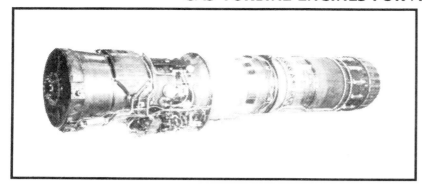 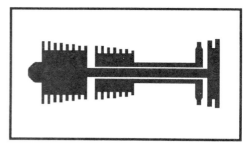

USA TURBINE ENGINE COMPANIES

MANUFACTURER: **UNITED TECHNOLOGIES INC. (PRATT & WHITNEY AIRCRAFT) GOVERNMENT PRODUCTS DIVISION), WEST PALM BEACH, FL.**

DESIGNATION (military): J75-P-19W.

BACKGROUND: Civil version is designated JT4A-29. One the few engines built with both afterburning and water injection.
Date First J75 Engine Produced: 1959.
Date If Out-of-Production: 1964. Number of J-75 Engines Produced to Date: 1,469.

ENGINE TYPE: Turbojet. Dual-Shaft, Afterburning & Water Injection.

COMPRESSOR TYPE: 15-Stage Dual-Spool, Axial Flow Compressor including: 8-Stage Low Pressure Compressor, 7-Stage High Pressure Compressor.

COMPRESSOR DATA: Compressor Pressure Ratio: 12.0 : 1, Mass Airflow: 253 lb/sec (115 kg/sec).

TURBINE TYPE: 3-Stage Axial Flow Turbine, including: 1-Stage High Pressure Turbine, 2-Stage Low Pressure Turbine.

COMBUSTOR TYPE: Can-Annular, Through Flow, with 8 liners.

AFTERBURNER: 1-Stage.

POWER (take-off) RATING: 16,100 lbt [dry], 26,500 lbt [Augmented]. Rating Approved to: 59°F (15°C).

TAKE-OFF SPECIFIC FUEL CONSUMPTION (SFC): 0.82 lb/hr/lbt (23,23 mg/Ns) [dry], 2.20 lb/hr/lbt (63,32 mg/Ns) [Augmented].

BASIC DRY WEIGHT: 5,960 lbs (2,703 kg).

POWER/WEIGHT RATIO: 2.70 : 1[dry], 4.45 : 1 [A/B] - lbt/lb.

APPLICATION:
Fairchild-Republic F-105 (D,F,G) "ThunderChief" Fighter Aircraft.
Lockheed U-2 and TR-1 High Altitude Reconnaissance Aircraft.

OTHER ENGINES IN SERIES:
J75-P-13B in U-2 and TR-1.
J75-P-17 in Convair-General Dynamics F-106 "Delta Dart".

GAS TURBINE ENGINES FOR AIRCRAFT

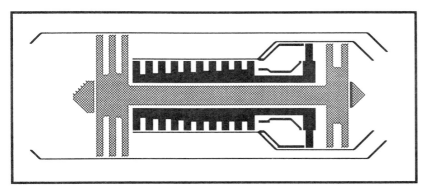
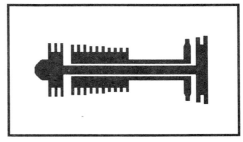

USA TURBINE ENGINE COMPANIES

MANUFACTURER: UNITED TECHNOLOGIES INC.
(PRATT & WHITNEY AIRCRAFT,
GOVERNMENT PRODUCTS DIVISION), WEST PALM BEACH, FL.

DESIGNATION (military): PW 1115.

BACKGROUND: No Civil Counterpart.
Date First PW 1115 Engine Produced: In development.

ENGINE TYPE: Turbofan. Dual-Shaft, Low-Bypass Front Fan, Non-Afterburning.

COMPRESSOR TYPE: 13-Stage Dual-Spool, Axial Flow Compressor including: 3-Stage Front Fan forming the Low Pressure Compressor, 10-Stage High Pressure Compressor.

COMPRESSOR DATA: Compressor Pressure Ratio: 28 : 1, Fan Pressure Ratio: 3.9 : 1, Fan Bypass Ratio: 0.2 : 1, Total Mass Airflow: 184 lb/sec (83,6 kg/sec).

TURBINE TYPE: 3-Stage Axial Flow Turbine including: 1-Stage High Pressure Turbine, 2-Stage Low Pressure Turbine.

COMBUSTOR TYPE: Annular, Through-Flow.

POWER (take-off) RATING: 14,551 lbt. (64,7 kN thrust) Rating Approved to: 59°F (15°C).

TAKE-OFF SPECIFIC FUEL CONSUMPTION (SFC): 0.79 lb/hr/lbt (22,36 mg/Ns).

BASIC DRY WEIGHT: 2,450 lbs (1,111 kg).

POWER/WEIGHT RATIO: 5.9 : 1 lbt/lb.

APPLICATION:
International Military Aircraft.

OTHER ENGINES IN SERIES:
None.

GAS TURBINE ENGINES FOR AIRCRAFT

 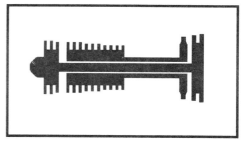

USA TURBINE ENGINE COMPANIES

MANUFACTURER: **UNITED TECHNOLOGIES INC. (PRATT & WHITNEY AIRCRAFT) GOVERNMENT PRODUCTS DIVISION), WEST PALM BEACH, FL.**

DESIGNATION (military)**:** PW 1120.

BACKGROUND: No Civil Counterpart. Derivative of F100 Turbofan Engine.
Date First PW 1120 Engine Produced: 1986.
Date If Out-of-Production: N/A. Number of PW 1120 Engines Produced to Date: 4.

ENGINE TYPE: Turbofan, Low Bypass. Dual-Shaft, with Afterburning.

COMPRESSOR TYPE: 13-Stage Dual-Spool, Axial Flow Compressor including: 3-Stage Front Fan forming the Low Pressure Compressor, 10-Stage High Pressure Compressor.

COMPRESSOR DATA: Compressor Pressure Ratio: 27 : 1, Total Mass Airflow: 178 lb/sec (80,7 kg/sec). Fan Bypass ratio 0.2 : 1.

TURBINE TYPE: 3-Stage Axial Flow Turbine including: 1-Stage High Pressure Turbine, 2-Stage Low Pressure Turbine.

COMBUSTOR TYPE: Annular, Through-Flow.

AFTERBURNER: 4-Stage.

POWER (take-off) RATING: 13,550 lbt [dry], 20,600 lbt [A/B]. Rating Approved to: 59°F (15°C).

TAKE-OFF SPECIFIC FUEL CONSUMPTION (SFC): 0.80 lb/hr/lbt (22,66 mg/Ns) [dry], 1.86 lb/hr/lbt (52,69 mg/Ns) [A/B].

BASIC DRY WEIGHT: 2,848 lbs (1,292 kg).

POWER/WEIGHT RATIO: 4.76 : 1 [dry], 7.23 : 1 [A/B] - lbt/lb.

APPLICATION:
McDonnell-Douglas F-4 "Phantom-II" Fighter Aircraft, Modification.
Israel Aircraft Industries (IAI) "Lavi" Fighter Aircraft.

OTHER ENGINES IN SERIES:
None.

GAS TURBINE ENGINES FOR AIRCRAFT

 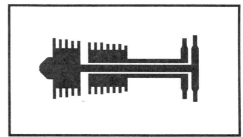

USA TURBINE ENGINE COMPANIES

MANUFACTURER: **UNITED TECHNOLOGIES INC. (PRATT & WHITNEY AIRCRAFT, GOVERNMENT PRODUCTS DIVISION), WEST PALM BEACH, FL.**

DESIGNATION (military): PW 1212.

BACKGROUND: No Civil Counterpart. Derivative of the J-52 Turbojet (military designation J52-P-409).
Date First PW 1212 Engine Produced: In development.

ENGINE TYPE: Turbojet. Dual-Shaft, Non Afterburning.

COMPRESSOR TYPE: 12-Stage Dual-Spool, Axial Flow Compressor including: 5 Axial Stages in the Low Pressure Compressor, 7-Stage High Pressure Compressor.

COMPRESSOR DATA: Compressor Pressure Ratio: 15.3 : 1, Total Mass Airflow: 147 lb/sec (66,8 kg/sec).

TURBINE TYPE: 2-Stage Axial Flow Turbine including: 1-Stage High Pressure Turbine, 1-Stage Low Pressure Turbine.

COMBUSTOR TYPE: Can-Annular, Through-Flow, with 9 liners.

POWER (take-off) RATING: 12,000 lbt (53,4 Kn). Rating Approved to: 59°F (15°C).

TAKE-OFF SPECIFIC FUEL CONSUMPTION (SFC): 0.88 lb/hr/lbt (24,9 mg/Ns).

BASIC DRY WEIGHT: 2,318 lbs (1,051 kg).

POWER/WEIGHT RATIO: 5.2 : 1 lbt/lb.

APPLICATION:
Proposed for Grumman EA-6B "Prowler" Electronic Warfare Aircraft.

OTHER ENGINES IN SERIES:
Locate other J52 Models under J52-P-408 in United Technologies, USA.

GAS TURBINE ENGINES FOR AIRCRAFT

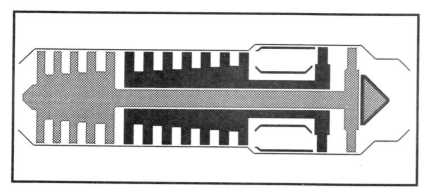 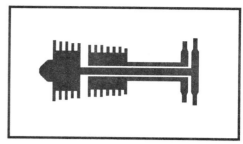

USA TURBINE ENGINE COMPANIES

MANUFACTURER: **UNITED TECHNOLOGIES INC. (PRATT & WHITNEY AIRCRAFT, GOVERNMENT PRODUCTS DIVISION), WEST PALM BEACH, FL.**

DESIGNATION (military)**:** PW 1216.

BACKGROUND: No Civil Counterpart. Derivative of the J-52 Turbojet. Date First PW 1216 Engine Produced: In development.

ENGINE TYPE: Turbojet. Dual-Shaft with Afterburning.

COMPRESSOR TYPE: 12-Stage Dual-Spool, Axial Flow Compressor including: 5 Axial Stages in the Low Pressure Compressor, 7-Stage High Pressure Compressor.

COMPRESSOR DATA: Compressor Pressure Ratio: 15.2 : 1, Total Mass Airflow: 147.4 lb/sec (66,8 kg/sec).

TURBINE TYPE: 2-Stage Axial Flow Turbine including: 1-Stage High Pressure Turbine, 1-Stage Low Pressure Turbine.

COMBUSTOR TYPE: Can-Annular, Through-Flow with 9 liners.

AFTERBURNER: 3-Stage

POWER (take-off) RATING: 11,480 lbt [dry], 16,500 lbt [A/B]. Rating Approved to: 59°F (15°C).

TAKE-OFF SPECIFIC FUEL CONSUMPTION (SFC): 0.91 lb/hr/lbt (25,8 mg/Ns) [dry], 1.99 lb/hr/lbt (56,4 mg/Ns) [A/B].

BASIC DRY WEIGHT: 2,800 lbs (1,270 kg).

POWER/WEIGHT RATIO: 4.1 : 1 [dry], 5.9 : 1 [A/B] - lbt/lb.

APPLICATION:
Proposed for future Attack/Fighter Aircraft.

OTHER ENGINES IN SERIES:
None.

GAS TURBINE ENGINES FOR AIRCRAFT

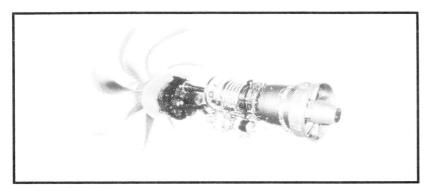 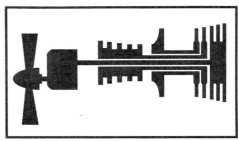

USA TURBINE ENGINE COMPANIES

MANUFACTURER: **UNITED TECHNOLOGIES INC. (PRATT & WHITNEY AIRCRAFT), GOVERNMENT PRODUCTS DIVISION), WEST PALM BEACH, FL.**

DESIGNATION (military): PW 3005.

BACKGROUND: Manufactured in Partnership with MTU.
Date First PW3005 Engine Produced: In development.

ENGINE TYPE: Turboprop. Three-Shaft (Dual Compressor Drive & Single Power Output Drive).

COMPRESSOR TYPE: Combination Axial-Centrifugal Compressor. Dual-Compressor of 6-Stages total compression including: 5-stages Low Pressure Axial Compressor followed by a 1-Stage independently rotating Centrifugal High Pressure Compressor.

COMPRESSOR DATA: Compressor Pressure Ratio: Unknown, Total Mass Airflow: Unknown.

TURBINE TYPE: 5-Stage Axial Flow Turbine includes: Gas Producer Section consisting of a 1-Stage High Pressure Turbine, 1-Stage Low Pressure Turbine and a Power Output Section consisting of a 3-Stage Power (free) Turbine.

COMBUSTOR TYPE: Annular, Through-Flow.

POWER (take-off) RATING: 4,750 shp (3,542 kw).   Rating Approved to: Unknown.

TAKE-OFF SPECIFIC FUEL CONSUMPTION (SFC): Unknown

BASIC DRY WEIGHT: 860 lbs (390 kg).

POWER/WEIGHT RATIO: 5.5 : 1 shp/lb (9,1 : 1 kw/kg).

APPLICATION:
Proposed for Advanced Lockheed P-3 "Orion" Maritime Aircraft.

OTHER ENGINES IN SERIES:
None.

GAS TURBINE ENGINES FOR AIRCRAFT

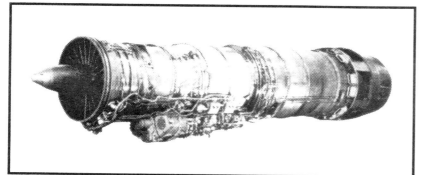 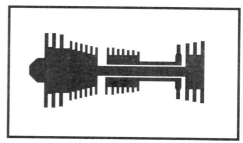

USA TURBINE ENGINE COMPANIES

MANUFACTURER: **UNITED TECHNOLOGIES INC. (PRATT & WHITNEY AIRCRAFT) GOVERNMENT PRODUCTS DIVISION), WEST PALM BEACH, FL.**

DESIGNATION (military): TF30-P-100.

BACKGROUND: Original designation the JTF10A-32C. The world's first Turbofan Engine with Afterburning. No Civil Counterpart.
Date First TF30 Engine Produced: 1965.
Date If Out-of-Production: 1987. Number of TF30 Engines Produced to Date: 3,483.

ENGINE TYPE: Turbofan. Dual-Shaft, Low-Bypass Front Fan with Afterburning.

COMPRESSOR TYPE: 16-Stage Dual-Spool, Axial Flow Compressor including: 3-Stage Front Fan and 6 Axial Stages in the Low Pressure Compressor, 7-Stage High Pressure Compressor.

COMPRESSOR DATA: Compressor Pressure Ratio: 22.0 : 1, Fan Pressure Ratio: 1.9 : 1, Fan Bypass Ratio: 0.7 : 1, Total Mass Airflow: 260 lb/sec (118 kg/sec).

TURBINE TYPE: 4-Stage Axial Flow Turbine including: 1-Stage High Pressure Turbine, 3-Stage Low Pressure Turbine.

COMBUSTOR TYPE: Can-Annular, Through-Flow, with 8 liners.

AFTERBURNER: 5-Stage, 3-Stages in Fan Bypass Exhaust, 2-Stages in the Core Exhaust.

POWER (take-off) RATING: 14,560 lbt [dry], 25,100 lbt [A/B]. Rating Approved to: 59°F (15°C).

TAKE-OFF SPECIFIC FUEL CONSUMPTION (SFC): 0.69 lb/hr/lbt (19,55 mg/Ns) [dry], 2.45 lb/hr/lbt (69,4 mg/Ns) [A/B].

BASIC DRY WEIGHT: 4,022 lbs (1,824 kg).

POWER/WEIGHT RATIO: 3.62 : 1 [dry], 6.24 : 1 [A/B] - lbt/lb.

APPLICATION:
General Dynamics F-111F "Raven" Swing-Wing Fighter/Bomber Aircraft.

OTHER ENGINES IN SERIES:
TF30-P-3/P-103 and -P7/P-107 in early F-111 Fighter/Bomber Aircraft.
TF30-P-6 in LTV Corp. A-7A "Corsair-II" Attack/Fighter Aircraft.
TF-30-P-109 in EF-111 "Raven" Electronic Warfare Aircraft.
TF30-P-408 in LTV Corp. A-7B/C/P Attack/Fighter and TA-7C Trainer Aircraft.
TF30-P-412 and P-414 in Grumman F-14A "Tomcat" Fighter Aircraft.

GAS TURBINE ENGINES FOR AIRCRAFT

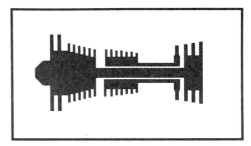

USA TURBINE ENGINE COMPANIES

MANUFACTURER: **UNITED TECHNOLOGIES INC. (PRATT & WHITNEY AIRCRAFT) GOVERNMENT PRODUCTS DIVISION), WEST PALM BEACH, FL.**

DESIGNATION (military)**:** TF33-PW-102.

BACKGROUND: Civil Version JT3D-3A. The first U.S. Military Turbofan Engine.
Date First TF33 Engine Produced: 1960.
Date If Out-of-Production: N/A. Number of TF33 Engines Produced to Date: 2,984.

ENGINE TYPE: Turbofan. Dual-Shaft, Low-Bypass Front Fan (non-afterburning).

COMPRESSOR TYPE: 15-Stage Dual-Spool, Axial Flow Compressor including: 2-Stage Front Fan and 6 Axial Stages in the Low Pressure Compressor, 7-Stage High Pressure Compressor.

COMPRESSOR DATA: Compressor Pressure Ratio: 13.6 : 1, Fan Pressure Ratio: 1.8 : 1, Fan Bypass Ratio: 1.4 : 1, Total Mass Airflow: 465 lb/sec (211 kg/sec).

TURBINE TYPE: 4-Stage Axial Flow Turbine including: 1-Stage High Pressure Turbine, 3-Stage Low Pressure Turbine.

COMBUSTOR TYPE: Can-Annular, Through-Flow with 8 liners.

POWER (take-off) RATING: 18,000 lbt (80,1 kN thrust).   Rating Approved to: 84°F (29°C).

TAKE-OFF SPECIFIC FUEL CONSUMPTION (SFC): 0.54 lb/hr/lbt (15,3 mg/Ns).

BASIC DRY WEIGHT: 4,340 lbs (1,970 kg).

POWER/WEIGHT RATIO: 4.18 : 1 lbt/lb.

APPLICATION:
Boeing KC-135 "StratoTanker" In-Flight Refueling Aircraft.

OTHER ENGINES IN SERIES:
TF33-PW-3 in Boeing B-52H "StratoFortress" Bomber Aircraft.
TF33-PW-5/9 in C-135B "StratoLifter" Transport and EC-135B Early Warning Aircraft.
TF33-PW-7/7A in Lockheed C-141 "StarLifter" Military Transport.
TF33-PW-9 in McDonnell-Douglas EC-24A (DC-8) Early Warning Aircraft.
TF33-PW-11 in Martin RB-57F "Canberra" Reconnaissance Aircraft.
TF33-PW-100A in Boeing E-3A/B "Sentry" Early Warning Aircraft.

GAS TURBINE ENGINES FOR AIRCRAFT

AUXILIARY POWER UNITS (USA)

MANUFACTURER: GARRETT TURBINE ENGINE COMPANY
DIVISION OF ALLIED SIGNAL CORP.
PHOENIX, Arizona 85034
TEL: (602) 231-1000

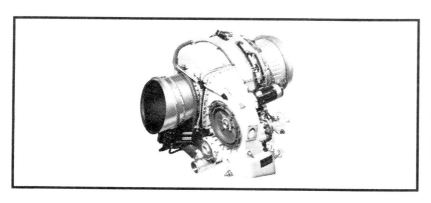

TSCP-700

COMBINATION LOADS AT 60°F INLET TEMPERATURE:

EXAMPLE: 0 (40) - 35 (10): INDICATES 0 SHP AT 35 LB./MIN. AIR BLEED OR 40 SHP AT 10 LB./MIN. AIR BLEED.

DESIGNATION	OUTPUT SHP	OUTPUT LB/MIN	APPLICATION
GTCP 30-9	40	35	AERO COMMANDER, FALCON 20
GTP 30-54	70	35	FAIRCHILD, F-27
GTP 30-95	85	35	CONVAIR "DART-600"
GTCP 30-92	0(40)	35(10)	BAE-125, JETSTAR-I/II, FALCON 20, JET COMM.
GTP 30-106	76	0	LOCKHEED, AH-56A
GTP 30-141	100	0	CONVAIR-600
GTP 30-142	0(50)	45(30)	BAe HS-125, USAF "C-140", HFB-320
GTCP 36-4A	0(50)	60(48)	FOKKER, F-28
GTCP 36-6	0(30)	65(55)	GULFSTREAM II
GTCP 36-16	0(50)	70(55)	NIHON YS-11 (Japan)
GTCP 36-16A	0(50)	70(55)	AERITALIA G-222
GTCP 36-17	0(50)	70(55)	FAIRCHILD-HILLER F-228
GTCP 36-28	0(50)	70(55)	VFW 614 (Germany)
GTCP 36-50	0(50)	55(40)	FAIRCHILD A-10

GAS TURBINE ENGINES FOR AIRCRAFT

GARRETT CONT'D.

DESIGNATION	OUTPUT SHP	OUTPUT LB/MIN	APPLICATION
GTCP 36-55 H	125	0	HUGHES, AH-64A (APACHE)
GTCP 36-100 A	0(50)	60(45)	DASSAULT "FALCON-50"
GTCP 36-100 C	0(50)	60(45)	DASSAULT "FALCON-20"
GTCP 36-100 E	0(50)	60(45)	CANADAIR CL-600/CL601
GTCP 36-100 G	0(50)	60(45)	GULFSTREAM II & III
GTCP 36-100 H	0(50)	60(45)	BAe/HS 125-700, JETSTAR I/II
GTCP 36-100 K	0(50)	60(45)	DHC-5, BUFFALO
GTCP 36-100 M	0(50)	60(45)	BAE-146
GTCP 36-150A	0(25)	50(45)	EMB-120, SAAB SF-340, BAE ATP/300 FALCON 20, CANADAIR CL-600
GTCP 36-150 C	0(25)	70(60)	AERITALIA, ATR-42
GTCP 36-150 E	0(25)	70(60)	SAAB-FAIRCHILD SF-340
GTCP 36-150 F	0(16)	62(58)	DASSAULT FALCON 900
GTCP 36-150 J	0(25)	50(45)	BAe "ATR"
GTCP 36-150R	0(25)	50(45)	FOKKER, F-100
GTC 36-200	0	120	NORTHROP, F-18
GTCP 36-201	80	120	LOCKHEED, S-3A
GTCP 36-201C	80	120	GRUMMAN, C-2A
GTCP 36-300	400	111	AIRBUS A320
GTCP 36-350	1,000	111	AIRBUS A330/340
GTC 85-37	200	130	GULFSTREAM I
GTC 85-71/86	200	130	LOCKHEED C-130
GTC 85-90F	0	120	CONVAIR-580, ELECTRA
GTCP 85-98 B	0(100)	110(0)	BOEING B-707 VIP
GTCP 85-98 C	0(100)	130(90)	BOEING B-727, DOUGLAS DC-9
GTCP 85-98 CK (A)	0(100)	130(90)	LOCKHEED C-130/L-282, BOEING VC-137

GAS TURBINE ENGINES FOR AIRCRAFT

GARRETT CONT'D.

DESIGNATION	OUTPUT SHP	OUTPUT LB/MIN	APPLICATION
GTCP 85-98 CK (B)	0(200)	180(90)	BOEING, B-707
GTCP 85-98 CK (D)	0(200)	180(90)	Mc/DOUGLAS, DC-8
GTCP 85-98 CU	0(100)	130(90)	B 727-100
GTCP 85-98 D	0(150)	170(100)	Mc/DOUGLAS, DC-9, USAF C-9
GTCP 85-98 DCK	0(150)	150(100)	Mc/DOUGLAS, DC-9-50
GTCP 85-98DHF	0(150)	160(100)	Mc/DOUGLAS, DC-9 SERIES 80
GTCP 85-98DJ	0(150)	140(80)	KAWASAKI, C-1
GTCP 85-98DK	0(150)	160(100)	Mc/DOUGLAS, DC-9
GTC 85-98 F	0(150)	160(100)	CONVAIR-580
GTCP 85-98W	0(150)	160(100)	DC-9 (TWA)
GTCP 85-99	0(150)	160(100)	SUD "CARAVELLE"
GTCP 85-100	0(150)	160(100)	BREGUET "ATLANTIQUE"
GTCP 85-106	0(150)	160(100)	LOCKHEED, C-141
GTCP 85-115	0(150)	120(40)	BAC-1-11
GTCP 85-115H	0(150)	160(100)	TRIDENT III
GTC 85-116	0	130	GRUMMAN, C2-A
GTCP 85-127A	75	0	LOCKHEED, AC-119 GUNSHIP
GTCP 85-129	0(200)	160(60)	BOEING, B-737, T-43, CONVAIR 990
GTCP 85-131J	0(150)	160(100)	SHIN MEIWA PS-1
GTCP 85-134	0(150)	140(80)	GULFSTREAM II, GRUMMAN TC-4C, VFW TRASALL-160
GTCP 85-139	0(150)	160(100)	TRIDENT I/II
GTCP 85-163CK	0(150)	150(90)	DASSAULT "MERCURE"
GTCP 85-180L	0(200)	160(100)	LOCKHEED, C-130 Commercial, USAF C-130H
GTCP 85-185	0(200)	150(80)	LOCKHEED, L100
GTCP 85-291C	190	0	CONVAIR-540

GAS TURBINE ENGINES FOR AIRCRAFT

GARRETT CONT'D.

DESIGNATION	OUTPUT SHP	OUTPUT LB/MIN	APPLICATION
GTCP 85-291E	190	0	LOCKHEED "ELECTRA"
GTCP 95-2	0(500)	140(100)	LOCKHEED, P3-A
GTCP 95-5	0(200)	150(80)	LOCKHEED C-121
GTCP 165-1	0(130)	160(120)	LOCKHEED, C-5, BOEING E3
GTCP 165-7	427	80	ROCKWELL, B-1A
GTCP 165-9	427	80	ROCKWELL, B-1B
GTCP 331-200A	450	260	BOEING B-757/767
GTCP 331-250F	375	65	AIRBUS A310/A300-600
GTCP 331-350	1200	0	B-777 IN DEVELOPMENT
GTCP 660-4	0(400)	600(500)	BOEING B-747, E4A
TSCP 700-4B	142	385	Mc/DOUGLAS DC-10
TSCP 700-5	142	385	AIRBUS 300B
TSCP 700-4E	142	385	Mc/DOUGLAS MD-11

MANUFACTURER: HAMILTON STANDARD CORP.
DIVISION OF UNITED TECHNOLOGIES CORP.
WINDSOR LOCKS, CN 06096
TEL: (203) 654-6000

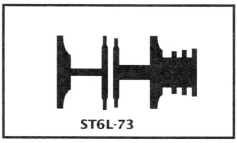

ST6L-73

DESIGNATION	OUTPUT SHP	OUTPUT LB/MIN	APPLICATION
ST6L-73	720	386	LOCKHEED TRISTAR, L-1011

GAS TURBINE ENGINES FOR AIRCRAFT

MANUFACTURER: PRATT & WHITNEY OF CANADA
DIVISION OF UNITED TECHNOLOGIES
LONGUEUIL, QUEBEC, CANADA J4G 1A1
TEL: (514) 677-9411

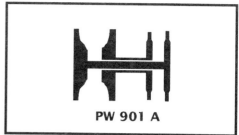

DESIGNATION	OUTPUT SHP	OUTPUT LB/MIN	APPLICATION
PW 901A	1265	551	BOEING, B747-400

MANUFACTURER: SUNDSTRAND CORP.
4400 RUFFIN ROAD
SAN DIEGO, CA 92318
TEL: (619) 569-4699

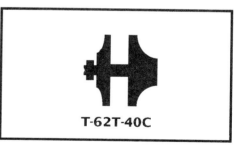

DESIGNATION	OUTPUT SHP	APPLICATION
T-20G-10C3/A	28	LEARJET, KINGAIR, CITATION II
T-62T-2A1	95	CH-47A HELICOPTER
T-62T-2B	95	CH-47D HELICOPTER
T-62T-2C	95	BOEING 234 HELICOPTER
T-62T-11	95	CH-46 HELICOPTER

GAS TURBINE ENGINES FOR AIRCRAFT

SUNDSTRAND CONT'D

DESIGNATION	OUTPUT SHP	APPLICATION
T-62T-16	71	VH-30 PRESIDENTIAL HELICOPTER
T-62T-16A2	95	CH-54 HELICOPTER
T-62T-16B1	95	CH-3C HELICOPTER
T-62T-25	88	FAIRCHILD-HILLER FH-227
T-62T-27	140	CH-53 HELICOPTER

DESIGNATION	OUTPUT SHP	OUTPUT LB/MIN	APPLICATION
APS 2000[1]	60	125	BOEING B-737
T-62T-29A	32(0)	0(20)	FALCON 20, JETSTAR, SABRELINER, HS-125
T-62T-39	40(0)	0(37)	FALCON, JETSTAR, SABRELINER, HS-125
T-62T-40-1	90(0)	0(72)	BLACKHAWK, SEAHAWK HELICOPTERS
T-62T-40-5	40(0)	0(61)	DeHAVILLAND BUFFALO
T-62T-40C	60(0)	0(72)	HS-125, FALCON 20, SABRELINER
T-62T-40C2	40(0)	0(60)	GULFSTREAM II, JETSTAR
T-62C-40C3	40(0)	0(49)	FALCON-200
T-62T-40C3A	60(0)	0(76)	FALCON-50
T-62T-40C4	40(0)	0(64)	FOKKER F-27
T-62T-40C7	40(0)	0(64)	DeHAVILLAND DASH 7
T-62T-40C7A	60(0)	0(76)	CESSNA CITATION III, DeHAVILLAND DASH 7
T-62T-40C7B	87(0)	0(72)	DeHAVILLAND DASH 8
T-62T-40C7D	50(0)	0(50)	DeHAVILLAND DASH 8
T-62T-40C7E1	95(0)	0(57)	EMBRAER EMB 120
T-62T-40C8D	40(0)	0(40)	BAe 125-800
T-62T-40LC-1	0	150	STARTING UNIT FOR USN
T-62T-40LC-2	40(0)	0(150)	KC-135R TANKER
T-62T-46-1	300(0)	0(52)	V-22 OSPREY

GAS TURBINE ENGINES FOR AIRCRAFT

SUNDSTRAND CONT'D

DESIGNATION	OUTPUT SHP	OUTPUT LB/MIN	APPLICATION
T-62T-46C1	150(0)	0(97)	FOKKER 50
T-62T-47-1	280(56)	0(150)	A-12
T-62T-47C1	95(0)	0(186)	B-737
T-62T-50C	110	N/A	CONVAIR SUPER 580

KEY: 1. SUNDTRAND-TURBOMECA (AUXILIARY POWER INT'L CORP. "APIC")

MANUFACTURER: WILLIAMS INTERNATIONAL
2280 WEST MAPLE ROAD
WALLED LAKE, MICHIGAN 48088
TEL: (313) 624-5200

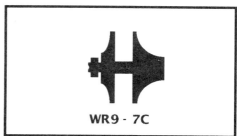

DESIGNATION	OUTPUT SHP	APPLICATION
WR9-7C	82	CC-115 BUFFALO (CANADA)
WR27-1	133	LOCKHEED, S-3 (USN)

SECTION II

LISTING OF

GAS TURBINE POWERED AIRCRAFT

CONTENTS OF SECTION II

		PAGE
1.	FINDER KEYS	252
2.	AIRCRAFT MANUFACTURERS ADDRESSES	253
3.	INTERNATIONAL - FIXED WING AIRCRAFT	255
4.	INTERNATIONAL ROTARY WING AIRCRAFT	261
5.	USA AGRICULTURAL AIRCRAFT	264
6.	USA BUSINESS - TURBOPROP AIRCRAFT	265
7.	USA BUSINESS - TURBOJET/TURBOFAN AIRCRAFT	266
8.	USA COMMERCIAL CARGO AIRCRAFT	267
9.	USA COMMERCIAL PASSENGER AIRCRAFT	267
10.	USA ROTARY WING AIRCRAFT (Civil & Military)	268
11.	USA MILITARY - TURBOJET, TURBOFAN, AND TURBOPROP AIRCRAFT	270
	ATTACK	270
	BOMBERS	270
	CARGO	270
	EARLY WARNING	271
	ELECTRONIC WARFARE	271
	FIGHTERS	272
	OBSERVATION AND PATROL	272
	RECONNAISSANCE	273
	TRAINING	273
	UTILITY	273
12.	USSR MILITARY AND COMMERCIAL - TURBINE POWERED AIRCRAFT	274
	BOMBERS	274
	MARITIME	274
	ATTACK AND FIGHTERS	274/5
	RECONNAISSANCE, ECM, EARLY WARNING	275
	TRANSPORTS	275
	TRAINERS	276
	HELICOPTERS	276
	CIVIL AIRCRAFT	277

FINDER KEYS:

ENGINE MANUFACTURER:

ALLISON	ALLISON-GMC (USA)
EUROJET	EUROJET TURBO (FIAT AVIO, IPT SPAIN, MTU, ROLLS-ROYCE)
CAE	TELEDYNE (USA)
FLYG	FLYGMOTOR (SWEDEN)
GARRETT	GARRETT (ALLIED SIGNAL CORP.) USA
G.E.	GENERAL ELECTRIC (USA)
GTRE	GOV. TECHNICAL RESEARCH (INDIA)
IAE	INT'L AERO ENGINES
IAI	ISRAELI AVIATION INDUSTRIES
IHI	ISHIKAWAJAMA-HARIMA (JAPAN)
LHTEC	LIGHT HELICOPTER TURBINE ENGINE CO. (USA)
LTV	LING-TEMCO-VOUGHT (USA)
M-T	MICROTURBO (FRANCE)
NAL	NATIONAL AEROSPACE (JAPAN)
P&W	PRATT & WHITNEY (USA)
PZL	POLSKIE ZAK LADY LOTNICZE (POLAND)
P&WC	PRATT & WHITNEY (CANADA)
R-R	ROLLS-ROYCE (GR. BRITAIN)
TEXTRON	TEXTRON-LYCOMING (USA)
T-M	TURBOMECA (FRANCE)
T-U	TURBO-UNION (GERMANY)
WILLIAMS	WILLIAMS INTERNATIONAL (USA)

AIRCRAFT MISSION:

AEW	AIRCRAFT EARLY EARNING AIRCRAFT
ASW	ANTI-SUBMARINE WARFARE AIRCRAFT
MARITIME	PATROL AIRCRAFT
MIL	MULTI-ROLE MILITARY USE AIRCRAFT
PAX	PASSENGER AIRCRAFT
PAX/CARGO	INCLUDES BUSINESS, COMMUTER, AIRLINER SIZED AIRCRAFT
RECON	RECONNAISSANCE AIRCRAFT
SAR	SEARCH AND RESCUE AIRCRAFT
STOL	SHORT TAKEOFF AND LANDING AIRCRAFT
VSTOL	VERTICAL-SHORT TAKEOFF AND LANDING AIRCRAFT
TRANSPORT	MILITARY PAX/CARGO AIRCRAFT

AIRCRAFT MANUFACTURERS

ADVANCED AERODYNAMICS & STRUCTURES INC., 10703 VANOWEN ST., N. HOLLYWOOD, CA 91605, USA. TEL: (818) 753-1888
AERO INDUSTRY DEVELOPMENT CTR., P.O.BOX 8676, 400 TAICHUNG, TAIWAN. TEL: 042-523051
AERO MODIFICATIONS INC., 2201 SCOTT AVE. FT. WORTH, TX 76103, USA. TEL: (817) 535-1936
AERITALIA - COMPANY NAME CHANGED TO ALENIA.
AERMACCHI SPA, VIA SAN VITO 80, 1-21100 VARESE, ITALY. TEL: 0332 25411ROS
AEROSPATIALE, 37 BLVD. DE MONTMORENCY, PARIS, CEDEX 16, F-75781, FRANCE. TELE: +(081) 725-2111
SUB. AEROSPATIALE AIRCRAFT (USA) CORP., SUITE 300, 1101 15TH STN.W. WASHINGTON, DC. 20005, USA. TEL: (202) 293-0650
SUB. AEROSPATIALE HELICOPTER (USA) CORP., 2701 FORUM DR. GRAND PRARIE, TX, 75053-4005, USA. TEL: (800) 873-0001
AEROSTAR INTERNATIONAL INC., P.O. BOX 5057 SIOUX FALLS, SD, USA, 57117. TEL: (605) 331-3500
AGUSTA AEROSPACE CORP.; 21 VIA CALDERA, MILAN, 20153, ITALY. TEL: 2 4525151-141-051
SUB. AGUSTA AEROSPACE CORP., (USA) CORP., P.O. BOX 16002, PHILADELPHIA, PA. 19154-0002, USA. TEL: (215) 281-1400
AIR TRACTOR INC., P.O. BOX 485, OLNEY, TX. 76374, USA. TELE: (817) 564-5616
AIRBUS INDUSTRIES NORTH AMERICA[1]**,** SUITE 300, 539 HERNDON PKWY, HERNDON, VA. 22070, USA. TEL: (703) 834-3400
AIRTECH USA (CASA & IPTN), SUITE 3810 CONCORDE PKWY, CHANTILLY, VA. 22021, USA. TEL: (703) 802-1000
ALENIA SPA, SAIAPA GRUPPOAEII, TIASPORTO, VIA DEL RIPPOSO ALLA, DOGANELLA, CAPOCICHINO, NAPLES, ITALY
AMX INTERNATIONAL, (AERMACCHI/AERITALIA/EMBRAER), BRAZIL. TEL: (55) 123-227178
ARCTIC AIRCRAFT COMPANY, P.O. BOX 6-141, ANCHORAGE, AK. 99502, USA. TEL: (907) 243-1580
ASTA (GOV. AIRCRAFT FACTORY), AVALON AIRFIELD, BEECH ROAD, LARA, VIC. AUSTRALIA 3212. TEL: 52-82-1188
ASTRA JET (IAI) CORPORATION, P.O. BOX 7517, PRINCETON, NJ. 08543, USA. TEL: (609) 987-1125
ATLAS AIRCRAFT, P.O.BOX 11, ATLAS ROAD, KEMPTON PARK, 1620, TRANSVAL, SO. AFRICA
ATR CORP. (AEROSPATIALE/AERITALIA), PIAZZALE VIA TECCHINO 51/A 80125 NAPLES, ITALY, TEL. 39 (81) 619522
AVALON CORP., AVALON AIRPORT, BEECH ROAD, LARA, VIC. AUSTRALIA 3212. TEL: 52-82-2988
AVTEC JET CORPORATION, 4680 CALLE CARGA CAMARILLO, CA. 93010, USA. TEL: (805) 482-2700
AYRES CORP, P.O. BOX 3090, ALBANY, GA. 31708, USA. TEL: (912) 883-1440
BASLER TURBO CONVERSIONS INC., OSHKOSH, WI. 54901 USA TEL: (414) 236-7820
BEDE CORP., P.O. BOX 706, NEWTON, KS. 67114
BEECH AIRCRAFT CORP., P.O. BOX 85, WICHITA, KS, 67201-0085, USA. TEL: (316) 681-7111
BELL HELICOPTER COMPANY, P.O. BOX 16858, PHILADELPHIA, PA, 19142, USA. TEL: (215) 591-2121
BELL HELICOPTER TEXTRON, P.O. BOX 482, FORT WORTH, TEXAS, 76101, USA. TEL: (817) 280-8200
BELL HELICOPTER TEXTRON CANADA, 12800 RUE DE L'AVENIR, ST. JANVIER, QUEBEC J0N 1L0 CANADA (514) 437-3400
BOEING COMMERCIAL AIRCRAFT COMPANY, P.O. BOX 3707, SEATTLE, WA. 98124-2207, USA. TEL: (206) 237-2121
BOEING-DEHAVILLAND CANADA INC., GARRETT BLVD. DOWNSVIEW, ONTARIO, CANADA MK3 1Y5. TEL: (416) 633-7301
BOMBARDIER (LEAR) CORP. P.O. BOX 11186, TUCSON, AZ, 85734, USA. TEL: (602) 746-5100
BRENDA NARDI CORP., AEROPORTO FORLANINI, 1-20090, MILAN, ITALY. TEL: 02-756-0241
BRITISH AEROSPACE, 11 STRAND, LONDON WC2N 5JT, ENGLAND. TEL: +(01) 930-1020
SUB. BRITISH AEROSPACE (USA) INC., P.O. BOX 17414, DULLES INT'L AIRPORT, WASHINGTON DC 20041, USA. TEL: (703) 435-9100
BROMON AIRCRAFT INC., 4085 NEVSO DRIVE, LAS VEGAS, NEVADA, 89103, USA. TEL: (702) 362-7121
CANADAIR GROUP, BOMBARDIER INC., P.O.BOX 6087 STATION-A, MONTREAL, H3C 3G9, CANADA. TEL: (514) 744-1511
SUB. CANADAIR CHALLENGER INC., 8 GRIFFIN ROAD, N. WINDSOR, CT 06095, USA. TEL: (203) 688-7767
CAPRONI VIZZPOLA (USA). MORRIS AVIATION LTD.P.O. BOX 718 STATESBORO AIRPORT, GA 30458, USA. TEL: (912) 489-8161
CASA, (CONSTUCCIONES AERONAUTICA S.A.) REY FRANCISCO 4, APARTAFO 193, MADRID E-28008, SPAIN. TEL: + 1-247-2500
SUB. CASA AIRCRAFT INC. (USA), 3810 CONCORDE PKY, SUITE 1000, CHANTILLY, VA. 22021, USA. TEL: (703) 378-CASA
CASA-IPTN CORP., REY FRANCISCO 4, 28008 MADRID, SPAIN, TEL. 34 (1) 247 00
CATIC (CHINA NATIONAL AERO-TECHNOLOGY IMPORT & EXPORT CORP.) 67 JIANO NAN ST. P.O.BOX 1671 BEIJING, PR. CHINA
CESSNA AIRCRAFT COMPANY, P.O. BOX 1521, WICHITA, KS 67201, USA. TEL: 1-(800) 4-CESSNA
CONAIR AVIATION, ABBOTSFORD BRITISH COLUMBIA, CANADA TEL: (604) 855-1171
CZEC AVIATION INDUSTRY, VODOCHODY-ODOLENA NODA, PRAGUE, CSSR-250-70, CZEC. TEL: 84-2551
DASSAULT AVIATION, 27 RUE DU PROFESSEUR PAUCHET, VAUCRESSAN F-92402, FRANCE. TEL: + (1) 479-58585
SUB. DASSAULT-BREGUET AVIATION (USA), FALCON JET CORP., TETERBORO, NJ 07608, USA. TEL: (201) 288-5300
DEHAVILLAND AIRCRAFT COMPANY OF CANADA, GARRETT BLVD. DOWNSVIEW, ONTARIO, CANADA, M3K 1Y5, TEL: (416) 633-7310
DEE HOWARD COMPANY, 9610 JOHN SAUNDERS RD., SAN ANTONIO, TX, 78216, USA. TEL: (512) 828-1341
DORNIER LUFTFARHT GMBH, POSTFACH 3, D-8031 WESSLING OPERPFAFFENHOFEN, GERMANY. TEL: + 08153-300
SUB. DORNIER AVIATION NORTH AMERICA INC., 22445 DAVIS DRIVE, SUITE 100 STERLING, VA. 22170, USA. TEL: (703) 769-7228
E.H. INDUSTRIES LTD., 500 CHISWICK HIGH ROAD, LONDON, ENGLAND W4 5RG. TEL: 01 995 8221
SUB. E.H. INDUSTRIES INC., 1735 JEFFERSON DAVIS HWY, SUITE 805, ARLINGTON, VA, 22202, USA. TEL: (703) 486-8000
EMBRAER-EMPRESA BRASILEIRA DE AERONAUTICA S.A., AV. BRIG. FARIA LIMA, 2170 SAS JOSE DOS CAMPOS, SP 12225, BRAZIL
SUB. EMBRAER AIRCRAFT (USA) CORP., 276 S.W. 34TH STREET, FORT LAUDERDALE, FL. 33315, USA. TEL: (305) 524-5755
ENSTROM HELICOPTER CORP., TWIN COUNTY AIRPORT, P.O. BOX 490, MENONINEE, MI 49858, USA. TEL: (906) 863-9971
EUROCOPTER CORP., CHEMIN DE LA PISTE, DONNEUIL-EN-FRANCE, 95500, GONESSE, FRANCE. TEL: (331) 49344374
EUROJET[2] **GMBH,** 2-20 AVE MARCEL CACHIN, 93126 LA COURERNEUVE, FRANCE. TEL: 1-49-34-40-00
FAIRCHILD AIRCRAFT CORP., P.O. BOX 32486, SAN ANTONIO, TX, 78279-0490, USA. TEL: (512) 824-9421
FAIRCHILD-REPUBLIC CORP., P.O. BOX 32486, SAN ANTONIO, TX, 78279-0490, USA. TEL: (512) 824-9421
FAMA CORP., AVENIDA FUERZA AEREA, ARGENTINA KM 5 1/2, 5103 GUARNICOIN AEREA, CORDOBA ARGENTINA. TEL: + (51) 606-661
FIAT AVIAZIONE SPA, VIA NIZZA 312, 10121, TORINO, ITALY. TEL: 011-6931241
FIRST AMERICAN AVIATION INC., 124 BROOKSTONE DR. PRINCETON, NJ, 08540, USA. TEL: (609) 921-2243
FOKKER AIRCRAFT B.V., P.O.BOX 12222, NL-1100 AE, AMSTERDAM-ZUIDOOST, THE NETHERLANDS. TEL: + (020) 605-9111
SUB. FOKKER AIRCRAFT (USA), 1199 N. FAIRFAX STREET, SUITE 500, ALEXANDRIA, VA. 22314, USA. TEL: (703) 838-0100
FRAKES, INC., ROUTE 3, BOX 229B, CLEBURNE, TX, 76031, USA (817) 645-9136
FUJI INDUSTRIES, 7-2 NISHI-SHINJUKU, 1-CHOME, SHIN JUKU-KU, TOKYO, JAPAN. TELE: 03-347-2525
GARRETT GENERAL AVIATION SERVICE, 6201 W. IMPERIAL HWY, LOS ANGELES, CA 90045, USA. TEL: (213) 568-3729
GATES LEARJET CORP. P.O. BOX 11186, TUCSON, AZ, 85734, USA. TEL: (602) 746-5100
GENERAL DYNAMICS CORP., PIERRE LACLEDE CTR. ST. LOUIS, MO 63105, USA. TEL: (314) 889-8200
GRUMMAN CORP. 1111 STEWART AVE, BETHPAGE, NY, 11714, USA. TEL: (516) 575-0574
GULFSTREAM AEROSPACE CORP., P.O. BOX 2206, SAVANNAH, GA. 31402-2206, USA. TEL: (912) 964-3000
HAWKER DEHAVILLAND VICTORIA LTD.; P.O. BOX 779H, GPO MELBOURNE, VICTORIA 3001, AUSTRALIA
HELIO AIRCRAFT CO., HANSCOM FIELD, BEDFORD, MA. 01730, USA. TEL: (617) 274-9130
HILLER AVIATION INC., 2075 W. SCRANTON AVE. PORTERVILLE, CA. 93257,USA. TEL: (209) 781-8000
HUNDISTAN AERONAUTICS LTD., BOX 5150, INDIAN EXPRESS BLDG. DR. AMBEDKDAR VEEDHI, BANGALORE, INDIA. 560-001. TEL: 76-901
HUGHES HELICOPTER CORP., CENTINELA AND TEALE STREETS, CULVER CITY, CA. 90230,USA. TEL: (213) 871-5212
IMP GROUP LTD. 2651 DUTCH VILLAGE ROAD, HALAFAX, NOVA SCOTIA, CANADA, B3L 4T1 (902) 453-2400
IPTN (INDUSTRI PESAWAT TERBANG NUSANTARA), P.O.BOX 563 JALAN PAJAJHRAN 154, BANDUNG, INDONESIA. TEL. 62 (022) 611081/2
ISHIDA AEROSPACE RESEARCH INC., 3908 SANDSHELL AVE., FORT WORTH, TX 76137, USA. TEL: (817) 847-5502
ISRAEL AIRCRAFT INDUSTRIES LTD.; BEN GURION INTERNATIONAL AIRPORT, 70100 TEL AVIV, ISRAEL. TEL: 973111
SUB. ISRAEL AIRCRAFT IND., USA: ATLANTIC AVIATION CORP., P.O. BOX 15000, WILMINGTON, DEL. 19850, USA. TEL: (302) 322-7223.

JAFFE HELICOPTER INC., 1770 SKYPLACE BLVD. INT'L AIRPORT, SAN ANTONIO, TX 78216, USA. TEL: (512) 821-6301
KAMAN CORP., BLUE HILLS AVE., BLOOMFIELD. CT 06002, USA. TEL: (203) 243-8311
KAWASAKI INDUSTRIES, 1-18 NAKAMACHI-DORI, 2-CHOME, CHOU-KU, KOBE, JAPAN. TEL: 078-341-7731
LAKE AIRCRAFT INC., LACONIA AIRPORT, LACONIA, NH, 03246, USA. TEL: (603) 524-5868
LEAR FAN CORP., P.O.BOX 60000, STEAD AIRPORT, RENO, NEV. 89506, USA. TEL: (701) 972-2600
LEARJET INC., P.O.BOX 7707 WICHITA MID-CONTINENT AIRPORT, WICHITA, KS, 67277, USA. TEL: (316) 946-2000
LEARJET-BOMBARDIER) P.O. BOX 11186, TUCSON, AZ, 85734, USA. TEL: (602) 746-5100
LET/OMNIPOL, LET NORODNI PODNIK, 686 04 UHERSKE HRADISTE KUNOVICE, CZECHOSLOVAKIA. TEL. 411111
LOCKHEED-GEORGIA CO., 86 SOUTH COBB DRIVE, MARIETTA, GA. 30060, USA. TEL: (404) 494-4411
LOCKHEED-CALIFORNIA CO., 4500 PARK GRANADA BLVD. CALABASAS, CA, USA. TEL: (818) 712-2000
L.T.V. AEROSPACE INC., P.O.BOX 655907, DALLAS, TX 75265, USA. TEL: (214) 266-2011
MAUL CORP., RT. 5, MOULTRIE, GA. 31768, USA. TEL: (912) 985-2045
MBB (MESSERSCHMITT-BOELKOW-BLOHM) HELICOPTER, P.O. BOX 80140, D-8000 MUNCHEN 80, GERMANY. TEL: (089) 6000-6488
SUB. MBB OF AMERICA INC. 900 AIRPORT ROAD, P.O.BOX 2343 WEST CHESTER, PA 19380, USA. TEL: (215) 431-4150
MCDONNELL DOUGLAS CORP., 3855 LAKEWOOD BLVD., LONG BEACH, CA. 90846, USA. TEL: (213) 593-9621
MCDONNELL DOUGLAS CORP., P.O. BOX 516, ST. LOUIS, MO. 63166, USA. TEL: (314) 232-0232
MCDONNELL DOUGLAS HELICOPTER COMPANY, 5000 E. MCDOWELL ROAD, MESA, AZ. 85205, USA. TEL: (602) 891-3000
MESSERSCHMITT-BOLKOW-BLOHM (MBB), 900 AIRPORT ROAD, P.O. BOX 2349, WEST CHESTER, PA. 19380, USA. TEL: (215) 436-9618
MICROTURBO CORP., CHEMIN DU PONT DO UPPE B.P. 2089-31019 TOULOUSE, CEDEX, FRANCE. TEL: (33) 61 70 11 27
SUB. MICROTURBO USA, 2707 FORUM DRIVE, GRAND PRARIE, TX 75051, USA. TEL: (214) 660-5545
MITSUBISHI AIRCRAFT INT'L., 1 LINCOLN CENTER, 5400 L.B.J. FREEWAY, SUITE 1500, DALLAS, TX. 75240, USA. TEL: (214) 387-5600
OMAC INC., P.O. BOX 3530, ALBANY, GA. 31708, USA. TEL: (912) 436-2425
NEW-CAL AVIATION, 14 RISER ROAD, LITTLE FERRY, NEW JERSEY, 07643, USA (201) 440-1990
NIHON AIRCRAFT (NOW MITSUBISHI AIRCRAFT INT'L)
NORMAN AEROSPACE LTD., CARDIFF WALES AIRPORT, RHOOSE ROAD, SO. GLAMORGAN, WALES, G.B. CF6-3BE. TEL: 0446-711884
NORTHROP CORP., 1890 CENTURY PART EAST, CENTURY CITY, LOS ANGELES, CA 90067, USA. TEL: (219) 553-6262
PANAVIA AIRCRAFT GMBH, MUNCHEN, ARABELLA STRASSE 16, F.R. GERMANY.
PARTENAVIA COSTRUZIONI AERONAUTICHE SPA, 80026 CASORIA NA, VIA CAVA, NAPLES, ITALY. TEL: 39-81-7596311
PEZETEL-MIELEC, ALEJA KRAKOWSKA, 110/114, 00-973 WARSZAWA, POLAND
PEZETEL MIELEC USA INC., 1200 FRONT STREET, SUITE 101, RALEIGH, NC. 27609, USA. TEL: (919) 828-7645
PILATUS FLUGZEUGWERKE AG, CH-6370 STANS/NW, SWITZERLAND. TEL.: 41 63 61 11
SUB. PILATUS AIRCRAFT, 1701 A1A VERO BEACH FL 32963, USA. (407) 234-5751
PILATUS BRITTEN-NORMAN LTD., BEMBRIDGE AIRPORT, BEMBRIDGE, ISLE OF WIGHT PQ35 5PR, ENGLAND. TEL: + 0983 872511
SUB. PILATUS BRITTEN-NORMAN (USA) LTD. (JONAS AIRCRAFT), 225 BROADWAY, NEW YORK, NY, 10007, USA. TEL: (212) 619-0330
PIPER AIRCRAFT CORP., P.O. BOX 1328, VERO BEACH, FL. 32961, USA. TEL: (305) 567-4361
PRECISION AIRMOTIVE CORP., EVERETT, WA, USA. TEL: (206) 353-8181
PROMAVIA SALES CORP., 2811 NIMITZ BLVD. SUITE-A, DAN DIEGO, CA, 96106, USA. TEL: (619) 222-0484
REIMS AVIATION, B.P.2745, F5012, REIMS CADEX, FRANCE. TEL: 26-484-6546
RHEIN-FLUGEUGBAU GMBH, P.O.BOX 408, FLUGPLATZ, MONCHENGLADBACH 1, GERMANY. TEL: 02-161-660231
RINALDO PIAGGIO CORP., VIA CIBRARIO 4, 1-16154 GENOA-SESTRI, ITALY. TEL: 010-60041
SUB. PIAGGIO (USA) CORP., 1802 W. 2ND STREET, WICHITA, KS, 67203, USA. TEL: (316) 262-3636
ROBINSON HELICOPTER COMPANY, 24747 CRENSHAW BLVD. TORRANCE, CA. 90505, USA. TEL: (213) 539-0508
ROCKWELL INT'L CORP., 2230 E. IMPERIAL HWY, EL SEGUNDO, CA 90245, USA. TEL: (213) 647 5000
ROGERSON-HILLER CORP., P.O.BOX 1425, 2140 W. 18TH STREET, PORT ANGELES, WA, 98362, USA. TEL: (206) 452-6891
SAAB-SCANIA AB, S-581 88, LINKOPING, SWEDEN. TEL: + (013) 180-000
SUB. SAAB AIRCRAFT OF AMERICA INC., 200 FAIRBROOK DRIVE, HEMDON, VA. 22070, USA. TEL: (703) 478-9720
SAAB-FAIRCHILD (USA), P. O. BOX 32486, SAN ANTONIO, TX. 78284, USA. TEL: (512) 824-9421
SABRELINER CORP., 18118 CHESTERFIELD AIRPORT. ROAD, CHESTERFIELD, MO 63005, USA. TEL: (314) 537-3660
SCHAFER AIRCRAFT MODIFICATIONS, ROUTE 10, BOX 301, WACO, TX, 76708, USA (817) 753-1551
SCHWEIZER AIRCRAFT CORP., P.O. BOX 147, ELMIRA NY. 14902, USA. TEL: (607) 739-3821
SEPECAT CORP. (DASSAULT-BAE), FALCON JET CORP., TETERBORO, NJ 07608, USA. TEL: (201) 288-5300
SHIN MEIWA INDUSTRIES, 5-25 KOSONE-CHO, 1 CHOME, NISHINOMIYA, HYOGO, JAPAN. TEL: 0798-47031
SHORTS BROTHERS PLC, P.O.BOX 341, AIRPORT ROAD, BELFAST BT3 9DZ, IRELAND. TEL: + (02) 325-8444
SUB. SHORTS BROTHERS USA INC., 2011 CRYSTAL DRIVE, SUITE 713, ARLINGTON, VA. 22202, USA. TEL: (703) 769-5555
SIAI MARCHETTI LTD., (USA), MUNICIPAL AIRPORT, RT. L, BOX 102, DENTON, TX. 76201, USA. TEL: (817) 383-1302
SIKORSKY AIRCRAFT COMPANY, 6900 NORTH MAIN STREET, STRATFORD, CT. 06601, USA. TEL: (203) 386-4000
SKYTRADER CORP., RICHARDS-GEBAUR AIR BASE, 15900 KENSINGTON, KANSAS CITY, MO. 64147, USA. TEL: (816) 322-2811
SMITH AERO CORP., P.O. BOX 430, STANTON COUNTY AIRPORT, JOHNSON CITY, KS. 67855, USA. TEL: (316) 492-6840
SNOW AVIATION INTERNATIONAL, P.O. BOX 307330, COLUMBUS, OHIO, 43230, USA. (614) 443-2711
SOKO AEROSPACE INDUSTRIES, RODOC B.B. YU-88000, MOSTAR, YUGOSLAVIA. TEL: 088-53749
SOLOY CONVERSIONS INC., 450 PAT KENNEDY WAY SW, OLYMPIA, WA 98502, USA. TEL: (206) 754-7000
STATE AIRCRAFT FACTORY, 67 JIAO NAN STREET, BEIJING, PEOPLES REP. OF CHINA. TEL: 44 2444
SCHWEIZER AIRCRAFT CORP., P.O. BOX 147, ELMIRA, N.Y. 14902, USA. TEL: (607) 796-2488
SWEARINGEN ENGINEERING INC., 1234 99TH STREET, SUITE-A SAN ANTONIO, TX 78214, USAL TEL: (512) 921-1208
TBM S.A., 12 RUE PASTEUR, SURENES F-92150, FRANCE. TEL: + 1-477-20934
SUB. TBM NORTH AMERICA, 8901 WESTMORE RD., SAN ANTONIO, TX 78216, USA. TEL: (512) 824-8283
TRIDAIR HELICOPTER INC., 3000 AIRWAY AVE., COSTA MESA, CA 92626, USA. TEL: (714) 540-3000
USSR AIRCRAFT, V/O /AVIA EXPORT TRUBNIKOVSKY PER. 19, MOSCOW G-69 1217 GSP, USSR. TEL. 7 290-01-71
VELMET (USA) CORP., 2520 RANDOLF AVE., AVERNEL NJ, 07001, USA. TEL: (201) 396-4160
VOPAR AIRCRAFT INC., 7945 WOODEY AVE. VAN NUYS, CA. 91406, USA. TEL: (818) 373-7990
WEATHERLY AVIATION CO., 2304 SAN FILEPE RD. HOLLISTER, CA. 95023, USA. TEL: (408) 637-5534
WESTLAND (USA) INC., 7135 JEFFERSON DAVIS HWY. SUITE 805, ARLINGTON, VA. 22202, USA. TEL: (703) 486-8000
WESTLAND AEROSPACE LTD., EAST COWES, ISLE OF WIGHT, PO32 6RH, ENGLAND. TEL: (0) 983-249101

[1] **AIRBUS INDUSTRIES,** (AEROSPATIALE, FRANCE AND DEUTSCHE AIRBUS, GERMANY)

[2] **EUROJET GMBH,** (MTU, GERMANY; FIAT AVIO, ITALY; ROLLS-ROYCE, GT. BRITAIN; INDUSTRIA DE TURBO PROPULSORES (ITP), SPAIN)

INTERNATIONAL - FIXED WING AIRCRAFT
(LISTING ALPHABETICAL BY COUNTRY)

AIRCRAFT MFR.	AIRCRAFT DESIGNATION	GROSS WEIGHT	ENGINE DESIG.	ENGINE TYPE	ENGINE POWER	ENGINE MFR.	AIRCRAFT MISSION
ARGENTINA							
FAMA CORP.[†]	IA 50 (GUARANI)	15,800	BASTAN-7	(2) TP	1,145 ESHP	T-M	TRAINER
	IA 58 (PACARA)	15,000	ASTAZOU-16G	(2) TP	1,021 ESHP	T-M	ATTACK
	IA 63 (PAMPA)	11,250	TFE731-2	(2) TF	3,500 LBT	GARRETT	TRAINER
	IA 66 (PACARA-C)	15,000	TFE331-11	(2) TP	1,100 ESHP	GARRETT	ATTACK
FAMA[5]	CBA-123 (VECTOR)	29,740	TPF-351-20	(2) PF	1,300 ESHP	GARRETT	PROPFAN (IN DEV.)
	[†] FABRICA MILITAR DE AVIONES						
AUSTRALIA							
GOV. ACFT.	N 22B SEARCH MASTER	8,950	250B-17C	(2) TP	420 ESHP	ALLISON	SAR
FACTORY	N 22C MISSION MASTER	8,950	250B-17C	(2) TP	420 ESHP	ALLISON	PAX/CARGO /MIL
(GAF)	N 24A (NOMAD)	9,400	250B-17C	(2) TP	420 ESHP	ALLISON	PAX/CARGO/MIL
BELGIUM							
PROMAVIA SA	F-1300 (JET SQUALUS) "SHARK"	5,290	TFE-109-1/3	(1) TF	1,600 LBT	GARRETT	TRAINER
	ATTA 3000	7,200	TFE-109-2	(2) TF	1,330 LBT	GARRETT	TRAINER (In Dev.)
	ARA 3600	7,200	TFE-109-2	(2) TF	1,330 LBT	GARRETT	ATTACK (In Dev.)
BRAZIL							
EMBRAER CORP.[†]	AT-26 (XAVANTE/MB-326)	11,500	VIPER 540	(1) TJ	3,401 LBT	R-R	TRAINER
	EMB-110 (BANDEIRANTE)	13,500	PT6A-34	(2) TP	783 ESHP	P&WC	PAX/CARGO/MARITIME
	EMB-111 (P-95)	15,400	PT6A-34	(2) TP	783 ESHP	P&WC	MARITIME
	EMB-120 (BRASILIA)	25,300	PW118	(2) TP	1,892 ESHP	P&WC	PAX/CARGO
	EMB-120 (BRASILIA)	25,300	PW118A	(2) TP	1,800 ESHP	P&WC	PAX/CARGO
	EMB-121 (XINGU-1)	13,000	PT6A-34	(2) TP	783 ESHP	P&WC	PAX/CARGO
	EMB-121A (XINGU-II)	13,000	PT6A-135	(2) TP	787 ESHP	P&WC	PAX/CARGO
	EMB-145	37,375	GMA-3007	(2) TF	7,200 LBT	ALLISON	IN DEVELOPMENT
	EMB-312 (TUCANO)	7,000	PT6A-25C	(1) TP	750 ESHP	P&WC	TRAINER
	EMB-312H (TUCANO)	7,500	PT6A-67R	(1) TP	1,600 ESHP	P&WC	TRAINER
	EMB-326	11,000	VIPER 20	(1) TJ	3,300 LBT	R-R	PAX/CARGO
	NE-821 (CARAJA)	8,000	PT6A-34	(1) TP	783 ESHP	P&WC	PAX/CARGO
EMBRAER[5]	CBA-123 (VECTOR)	29,740	TPF-351-20	(2) PF	2,081 ESHP	GARRETT	PROPFAN (IN DEV.)
EMBRAER[9]	AM-X (A-1)	28,300	SPEY MK.807	(1) TF	11,130 LBT	R-R	FIGHTER
	AM-XT	28,300	SPEY MK.807	(1) TF	11,130 LBT	R-R	TRAINER
	[†] EMPRESA BRASILEIRA DE AERONAUTICA						
CANADA							
BOEING	CF-5	20,000	J85-GE-13	(1) TJ	4,080 LBT	G.E.	FIGHTER
DeHAVILLAND	CF-104	31,000	J-79-11A	(1) TJ	15,800 LBT	G.E.	FIGHTER
CANADA	CT-142 (DHC-8)	34,500	PW120A	(2) TP	2,100 ESHP	P&WC	RCAF TRANSPORT
	CV-7	40,000	T64-GE-10	(2) TP	2,850 ESHP	G.E.	RCAF TRANSPORT
	DHC-2 (TURBO BEAVER)	5,370	PT6A-20	(2) TP	580 ESHP	P&WC	PAX/CARGO
	DHC-5A (CARIBOU)	41,000	CT64-820	(2) TP	3,133 ESHP	G.E.	PAX/CARGO
	DHC-5D (BUFFALO)	41,000	CT64-820	(2) TP	3,133 ESHP	G.E.	PAX/CARGO
	DHC-5E (TRANSPORTER)	41,000	CT64-820	(2) TP	3,133 ESHP	G.E.	PAX/CARGO
	DHC-6 (TWIN OTTER)	12,250	PT6A-27	(2) TP	620 ESHP	P&WC	PAX/CARGO
	DHC-7 (DASH-7)	44,000	PT6A-50	(4) TP	1,174 ESHP	P&WC	PAX/CARGO
	DHC-8 (DASH-8-100)	34,500	PW120/121	(2) TP	2,100 ESHP	P&WC	PAX/CARGO
	DHC-8 (DASH-8-300)	41,100	PW123	(2) TP	2,252 ESHP	P&WC	PAX/CARGO
	DHC-8 (TRITON-100)	34,500	PW120A	(2) TP	2,100 ESHP	P&WC	MARITIME
	DHC-8 (TRITON-300)	41,100	PW123	(2) TP	2,502 ESHP	P&WC	MARITIME
CANADA IMP. CO. (NOVA SCOTIA)	P-163 (TURBO-TRACKER S-2)	26,150	PT6A-67C	(2) TP	1,300 ESHP	P&WC	AWS (RE-ENGINE)
	CP-140 (ARCTURUS)	139,730	T56-A-14	(4) TP	4,591 ESHP	ALLISON	MODIFIED P-3
CANADAIR LTD.	CHALLENGER-600	41,250	ALF502-L3	(2) TF	7,800 LBT	TEXTRON	PAX/CARGO
	CHALLENGER-601	43,250	CF34-3A	(2) TF	9,220 LBT	G.E.	PAX/CARGO
	CANADAIR RJ-100	47,250	CF34-3A1	(2) TF	9,220 LBT	G.E.	PAX/CARGO
	CL-41A (CT-114/TUTOR)	7,500	J-85-CAN-40	(1) TJ	2,850 LBT	ORENDA	TRAINER
	CL-44 (YUKON)	210,000	TYNE 515	(4) TP	5,500 ESHP	R-R	PAX/CARGO
	CL-215T (AMPHIBIAN)	46,000	PW123AF	(2) TP	2,502 ESHP	P&WC	FIREFIGHTER
CONAIR AVIATION	TURBO-FIRECAT	26,150	PT6A-67AF	(2) TP	1,424 ESHP	P&WC	FIREFIGHTER (S-2 RE-ENGINE)
McKINNON-	G-21G (TURBO-GOOSE)	12,500	PT6A-27	(2) TP	715 ESHP	P&WC	PAX/CARGO (RE-ENGINE)
VIKING CORP.	TURBO BEAVER	13,500	PT6A-27	(2) TP	715 ESHP	P&WC	PAX/CARGO (RE-ENGINE)
SAUNDERS	ST-27	13,500	PT6A-34	(2) TP	783 ESHP	P&WC	PAX/CARGO

AIRCRAFT MFR.	AIRCRAFT DESIGNATION	GROSS WEIGHT	ENGINE DESIG.	ENGINE TYPE	ENGINE POWER	ENGINE MFR.	AIRCRAFT MISSION

CHINA (PEOPLES REPUBLIC)
NATIONAL AIRCRAFT FACTORIES (CATIC)

NAF (NANCHANG AIRCRAFT FACTORY)   SFA (SHANGHI FAR-EAST AERO)
SAC (SHENYANG AIRCRAFT CORP)   XAC (XIAN AIRCRAFT CORP)

AIRCRAFT MFR.	AIRCRAFT DESIGNATION	GROSS WEIGHT	ENGINE DESIG.	ENGINE TYPE	ENGINE POWER	ENGINE MFR.	AIRCRAFT MISSION
CHENGDU ACFT.	J-7 (MIG-21)[†]	16,600	WP-7B	(1) TJ	13,448 LBT	CHENGDU	FIGHTER
CHENGDU ACFT.	F-5A	UNKNOWN	UNKNOWN	------	---------------	CHENGDU	TRAINER
GUIZHOW ACFT.	FT-7 (MIG-21U)[†]	18,960	WP-7B	(1) TJ	13,448 LBT	LIYANG	FIGHTER
HARBIN ACFT.	H-5 (IL-28)[†]	46,700	KLIMOV VK-1	(2) TJ	5,950 LBT	KLIMOV	BOMBER
NAF	J-12	UNKNOWN	WP-6Z	(1) TJ	8,900 LBT	SAC	FIGHTER
NAF	K-8 (KAKORUM)	9,540	TFE-731-2A	(1) TF	3,500 LBT	GARRETT	TRAINER
NAF	Q-5 (MIG-19)[†] [A-5][#]	26,445	WP-6A	(1) TJ	8,260 LBT	SAC	ATTACK
NAF	Y-12 II	12,800	PT6A-27	(2) TP	680 ESHP	P&WC	UTILITY
SAC	J-5 (MIG 17) [F-5][#]	13,380	VK-1A	(1) TJ	7,600 LBT	KLIMOV	FIGHTER-BOMBER
SAC	J-6 (MIG-19)[†] (F-6, FT-6)[#]	22,040	WP-6	(2) TJ	7,165 LBT	SAC	FIGHTER
SAC	J-8 (F8-11)[*]	39,242	WP-13A	(2) TJ	14,800 LBT	LIYANG	FIGHTER
SFA	MD-82 [††]	130,000	JT8D-217A	(2) TF	20,850 LBT	P&W	PAX/CARGO
SHAANXI ACFT.	Y-8 (AN-12)[†]	134,500	WJ-6	(4) TP	4,250 ESHP	ZHUZHOU	PAX/CARGO/MIL
SHUISHANG ACFT.	SH-5 (PS-5)[*]	99,200	WJ-5A-1	(4) TP	3,150 ESHP	HARBIN	MARITIME
XAC	H-6 (TU-16)[†] [B-6][*]	160,000	WP-8	(2) TJ	21,000 LBT	XAC-WOPEN	BOMBER-MARITIME
XAC	H-7 (SU-16)[†]	60,600	WS-9 (SPEY 202)	(2) TF	20,500 LBT	XAC	ATTACK
XAC	JH-7	39,000	WP-13A	(2) TJ	14,800 LBT	XAC	FIGHTER (IN DEV.)
XAC	Y-7 (AN-24)[†]	48,000	WJ-5A	(2) TP	2,790 ESHP	HARBIN	PAX/CARGO
XAC	Y-7-200	32,500	PW124A	(2) TP	2,522 ESHP	XAC	PAX
XAC	Y-14 (AN-26)[†]	52,900	WJ-5A	(2) TP	2,790 ESHP	HARBIN	PAX/CARGO

BUILT UNDER LICENSE - [†] USSR, [††] USA, [*] WESTERN DESIGNATION, [#] EXPORT MODEL

CHINA (REPUBLIC OF TAIWAN)

AIRCRAFT MFR.	AIRCRAFT DESIGNATION	GROSS WEIGHT	ENGINE DESIG.	ENGINE TYPE	ENGINE POWER	ENGINE MFR.	AIRCRAFT MISSION
AERO INDUSTRIES DEV. CENTER (AIDC)	A-3 (LUI-MENG)	17,500	TFE 1042	(2) TF	3,500 LBT	GARRETT	ATTACK
	AT-3 (TSU-CHIANG)	17,500	TFE731-2	(2) TF	3,500 LBT	GARRETT	TRAINER
	T-HC-1 (CHUNG-HSING)	9,200	T53-L-701A	(1) TP	1,450 ESHP	TEXTRON	TRAINER
	XC-2	27,500	T53-L-701A	(2) TP	1,450 ESHP	TEXTRON	PAX/CARGO/MIL
	A-1 (CHING-KUO)	CLASSIFIED	TFE 1042	(2) TF	8,350 LBT	GARRETT	FIGHTER

CZECHOSLOVAKIA

AIRCRAFT MFR.	AIRCRAFT DESIGNATION	GROSS WEIGHT	ENGINE DESIG.	ENGINE TYPE	ENGINE POWER	ENGINE MFR.	AIRCRAFT MISSION
LET/OMNIPOL	LET L-410 UVP (TURBOLET)	12,800	M601D	(2) TP	780 ESHP	WALTER	PAX/CARGO/MIL
	LET L-410 UVP-E	14,100	M601E	(2) TP	809 ESHP	WALTER	PAX/CARGO/MIL
	LET L-410	14,100	PT6A-27	(2) TP	715 ESHP	P&WC	PAX/CARGO/MIL
	LET L-610	30,850	CT7-9B	(2) TP	1,870 ESHP	G.E.	PAX/CARGO/MIL
CZEC AVIATION INDUSTRY (AERO)[†]	L-29 DELFIN (MAYA)	7,800	M701C	(1) TJ	1,960 LBT	WALTER	TRAINER
	L-29A (AKROBAT)	5,700	M701C	(1) TJ	1,960 LBT	WALTER	ACROBATIC
	L-39 Z (ALBATROS)	12,300	AI-25TL	(2) TF	3,300 LBT	WALTER	ATTACK/TRAINER
	L-59 (FORMERLY L-39MS)	12,500	DV-2	(1) TF	4,850 LBT	LOZISK	TRAINER
	L-159	UNKNOWN	TFE-731-2	(1) TF	3,500 LBT	GARRETT	TRAINER
	ZLIN-Z 37T	5,600	M601Z	(1) TP	809 ESHP	WALTER	AGRICULTURE

[†] AERO VODOCHODY NATIONAL CORP.

FINLAND

AIRCRAFT MFR.	AIRCRAFT DESIGNATION	GROSS WEIGHT	ENGINE DESIG.	ENGINE TYPE	ENGINE POWER	ENGINE MFR.	AIRCRAFT MISSION
VALMET CORP.	L-80TP	4,200	250-B17D	(1) TP	420 ESHP	ALLISON	TRAINER
	L-90TP (REDIGO)	4,200	250-B17F	(1) TP	450 ESHP	ALLISON	TRAINER
	TBM-700[13]	6,007	PT6A-64	(1) TP	747 ESHP	P&WC	PAX/CARGO

FRANCE

AIRCRAFT MFR.	AIRCRAFT DESIGNATION	GROSS WEIGHT	ENGINE DESIG.	ENGINE TYPE	ENGINE POWER	ENGINE MFR.	AIRCRAFT MISSION
AEROSPATIALE	CARAVELLE	110,000	AVON 532	(2) TF	12,600 LBT	R-R	PAX/CARGO
	CARAVELLE (SUPER)	128,000	JT8D-7	(2) TF	14,500 LBT	P&W	PAX/CARGO
	CM-170 (MAGISTER)	7,200	MARBORE-VI	(2) TJ	1,058 LBT	T-M	TRAINER
	CORVETTE (SN-601)	14,000	JT15D-4	(2) TF	2,500 LBT	P&WC	PAX/CARGO
	N 262 (MOHAWK)	23,900	PT6A-45	(2) TP	1,254 ESHP	P&WC	PAX/CARGO
	N 262 (FREGATE)	23,900	BASTAN-7	(2) TP	1,145 ESHP	T-M	PAX/CARGO
	TB-31 OMEGA (SOCATA DIV.)	2,755	TM-319	(1) TP	488 ESHP	T-M	TRAINER
AEROSPATIALE[1]	ATR 42 (SUPER COMMUTER)	36,800	PW 120/121	(2) TP	2,100 ESHP	P&WC	PAX/CARGO
	ATR 72[17]	44,000	PW 124/127	(2) TP	2,522 ESHP	P&WC	PAX/CARGO
AEROSPATIALE[2]	CONCORDE	408,000	OLYMPUS 593	(4) TP	38,000 LBT	R-R	SUPERSONIC PAX
AEROSPATIALE[3]	TRANSALL C-160	112,350	TYNE MK.22	(2) TP	6,100 ESHP	R-R	PAX/CARGO/MIL
AIRBUS INDUSTRIES[4]	A 300B4/C4	363,700	CF6-50C2	(2) TF	52,500 LBT	G.E.	PAX/CARGO
	ALTERNATE ENGINE		JT9D-59A	(2) TF	53,000 LBT	P&W	
	A 300-600 & 600R	363,700	CF6-80C2A3	(2) TF	60,200 LBT	G.E.	PAX/CARGO
	ALTERNATE ENGINE		PW-4052	(2) TF	56,000 LBT	P&W	
	A 310-200 & 300	305,500	CF6-80C2A2	(2) TF	53,500 LBT	G.E.	PAX/CARGO
	ALTERNATE ENGINE		PW-4152	(2) TF	52,000 LBT	P&W	
	A 320-200	162,000	CFM56-5A1	(2) TF	25,000 LBT	G.E.	PAX/CARGO
	ALTERNATE ENGINE		V2500-A1	(2) TF	25,000 LBT	IAE	

AIRCRAFT MFR.	AIRCRAFT DESIGNATION	GROSS WEIGHT	ENGINE DESIG.	ENGINE TYPE	ENGINE POWER	ENGINE MFR.	AIRCRAFT MISSION
AIRBUS (CON'T)							
	A 321 (STRETCH	181,200	V2500-A5	(2) TF	30,000 LBT	IAE	PAX/CARGO
	A 330	456,000	CF6-80C2D1	(2) TF	65,500 LBT	G.E.,	PAX/CARGO
	A 330	458,500	TRENT-700	(2) TF	65,000 LBT	R-R	PAX/CARGO
	A 340	558,700	CFM-56-5C2	(4) TF	31,200 LBT	G.E.	PAX/CARGO
	SUPER-GUPPY [15]	180,000	501D-22	(4) TP	4,680 ESHP	ALLISON	PAX/CARGO
AVIONS	ATLANTIC (ATLANTIQUE)	100,000	TYNE R TY.20	(2) TP	6,100 ESHP	R-R/MULTI	MARITIME/AWS
MARCEL	BR-1050 (ALIZE)	18,078	DART MK.21	(1) TP	1,975 ESHP	R-R	PAX/CARGO
DASSAULT	ETENDARD (SUPER)	26,000	ATAR 8K-50	(1) TJ	9,700 LBT	SNECMA	FIGHTER
BREGUET	FALCON 10	18,740	TFE 731-2	(2) TF	3,500 LBT	GARRETT	PAX/CARGO
	FALCON 20F	28,600	CF700-2D2	(2) TF	4,500 LBT	G.E.	PAX/CARGO
	FALCON 20 (MODIFIED)	29,100	PW305-F2	(2) TF	5,225 LBT	P&WC	VOPAR (RE-ENGINE)
	731 FALCON (MODIFIED)	29,100	TFE731-5BR	(2) TF	4,500 LBT	GARRETT	GARRETT (RE-ENGINE)
	FALCON 50	38,800	TFE 731-3	(3) TF	3,700 LBT	GARRETT	PAX/CARGO
	FALCON (GUARDIAN HU-25)	32,000	ATF3-6A	(2) TF	5,440 LBT	GARRETT	MARITIME
	FALCON 100	18,730	TFE-731-2	(2) TF	3,230 LBT	GARRETT	PAX/CARGO
	FALCON 200	32,000	ATF3-6A	(2) TF	5,440 LBT	GARRETT	PAX/CARGO
	FALCON 900	45,700	TFE 731-5BR	(3) TF	4,750 LBT	GARRETT	PAX/CARGO
	FALCON-2000	35,000	CFE738	(2) TF	5,600 LBT	GARR'T/GE	In Development
	MERCURE	124,000	JT8D-15	(2) TF	14,500 LBT	P&W	PAX/CARGO
	MIRAGE 3B	29,800	ATAR 9C	(1) TJ	13,670 LBT	SNECMA	TRAINER
	MIRAGE 3E	29,800	ATAR 9C	(1) TJ	13,670 LBT	SNECMA	FIGHTER
	MIRAGE 3R	29,800	ATAR 9C	(1) TJ	13,670 LBT	SNECMA	RECON.
	MIRAGE 3NG	32,400	ATAR 9K-50	(1) TJ	15,870 LBT	SNECMA	FIGHTER
	MIRAGE 4	70,000	ATAR 9K-50	(2) TJ	15,870 LBT	SNECMA	BOMBER
	MIRAGE 5	30,200	ATAR 9C	(1) TJ	13,670 LBT	SNECMA	ATTACK
	MIRAGE 50	30,200	ATAR 9K-50	(1) TJ	15,870 LBT	SNECMA	FIGHTER
	MIRAGE F1	33,500	ATAR 9K-50	(1) TJ	15,870 LBT	SNECMA	FIGHTER
	MIRAGE 2000	37,375	M53-5	(1) TF	19,850 LBT	SNECMA	ATTACK/FIGHTER
	MIRAGE 2000B	37,375	M53-5	(1) TF	19,850 LBT	SNECMA	TRAINER
	MIRAGE 4000 (SUPER)	59,490	M53-P2	(2) TF	21,385 LBT	SNECMA	FIGHTER
	RAFAELE-D	57,000	M88-2	(2) TF	16,800 LBT	SNECMA	FIGHTER (In Dev.)
DASSAULT[7]	ALPHA JET	17,800	LARZAC O4-C20	(2) TF	3,180 LBT	SNECMA/T-M	ATTACK/TRAINER
DASSAULT[11]	JAGUAR	34,600	ADOUR MK.815	(2) TF	8,400 LBT	R-R/T-M	ATTACK
MICROJET	MICROJET 200B	2,866	TRS 18-2	(2) TJ	360 LBT	M-T	TRAINER
REIMS-CESSNA	CARAVAN-II	9,925	PT6A-112	(2) TP	528 ESHP	P&WC	PAX/CARGO
TBM INT'L.[13]	TBM-700	6,007	PT6A-64	(1) TP	747 ESHP	P&WC	PAX/CARGO

GERMANY

AIRCRAFT MFR.	AIRCRAFT DESIGNATION	GROSS WEIGHT	ENGINE DESIG.	ENGINE TYPE	ENGINE POWER	ENGINE MFR.	AIRCRAFT MISSION
DORNIER	DO 128-6 (SKYSERVANT)	9,600	PT6A-110	(2) TP	502 ESHP	P&WC	PAX/CARGO
GMBH.	DO 228-100,101,200,201	14,000	TPE 331-5	(2) TP	715 ESHP	GARRETT	PAX/CARGO
	D0 328	29,550	PW 119	(2) T	2,180 ESHP	P&WC	PAX/CARGO IN DEV.
DORNIER[7]	ALPHA JET	17,800	LARZAC O4-C20	(2) TF	3,180 LBT	SNECMA	TRAINER/ATTACK
DORNIER COMPOSITE GMBH.	SEASTAR CD-2	10,140	PT6A-135	(2) TP	728 ESHP	P&WC	PAX/CARGO
EQUATOR GmbH	P-550 (TURBO-EQUATOR)	4,400	PT6A-27	(1) TP	715 ESHP	P&WC	MIL STOL (AMPHIBIAN)
EUROFIGHTER[14]	EFA	37,500	EJ-200	(1) TF	20,000 LBT	EUROJET	FIGHTER
MBB GmbH	HANSAJET (HFB-320)	20,000	CJ-610-8	(2) TJ	3,100 LBT	G. E.	PAX/CARGO/RECON
MBB[3]	TRANSALL C-160	112,350	TYNE MK.22	(2) TP	6,100 ESHP	R-R	PAX/CARGO/MIL
RHEIN-FLUGBUG-BAU GmbH	FT-400 (FANTRAINER)	3,960	250-C20R	(1) TP	450 ESHP	ALLISON	TRAINER
	FT-600 (FANTRAINER)	5,070	250-C30L	(1) TP	650 ESHP	ALLISON	TRAINER
PANAVIA CORP.[10]	TORNADO	62,700	RB-199-34R	(2) TF	16,580 LBT	T-U	ATTACK/FIGHTER

GREAT BRITAIN

AIRCRAFT MFR.	AIRCRAFT DESIGNATION	GROSS WEIGHT	ENGINE DESIG.	ENGINE TYPE	ENGINE POWER	ENGINE MFR.	AIRCRAFT MISSION
BRITISH	ARGOSY AW-660	93,000	DART 526	(4) TP	2,020 ESHP	R-R	PAX/CARGO/MIL
AEROSPACE	ATP	50,500	PW 124A	(2) TP	2,522 ESHP	P&WC	PAX/CARGO
ENGINEERING	ATP	50,500	PW 126	(2) TP	2,786 ESHP	P&WC	PAX/CARGO
CORP. (BAe)	BAC 111-400	87,000	SPEY 512-14	(2) TF	12,000 LBT	R-R	PAX/CARGO
	BAC 111-400 (RE-ENGINE)	87,000	TAY 650-14	(2) TF	15,000 LBT	R-R	PAX/CARGO
	BAC 111-475	98,500	SPEY 512-14DW	(2) TF	12,500 LBT	R-R	PAX/CARGO
	BAC 111-500	104,500	SPEY 512-14SW	(2) TF	12,500 LBT	R-R	PAX/CARGO
	BAC 111 (RE-ENGINE)	104,500	TAY 650	(2) TF	15,100 LBT	R-R	DEE HOWARD CONV.
	BAC 145 (JET PROVOST)	8,150	VIPER 202	(1) TJ	2,500 LBT	R-R	TRAINER
	BAC 167 (STRIKEMASTER)	11,500	VIPER 535/540	(1) TJ	3,660 LBT	R-R	ATTACK
	BAe 125-400/600	25,500	VIPER 601	(2) TF	3,700 LBT	R-R	PAX/CARGO
	BAe 125-700	25,500	TFE 731-3R	(2) TF	3,700 LBT	GARRETT	PAX/CARGO
	BAe 125-800	27,520	TFE 731-5	(2) TF	4,300 LBT	GARRETT	PAX/CARGO
	BAe 125-1000	31,100	PW 305	(2) TF	5,200 LBT	P&WC	PAX/CARGO

AIRCRAFT MFR.	AIRCRAFT DESIGNATION	GROSS WEIGHT	ENGINE DESIG.	ENGINE TYPE	ENGINE POWER	ENGINE MFR.	AIRCRAFT MISSION
BAe (CONT'D)	BAe 146-100	84,000	ALF502R-5	(4) TF	6,970 LBT	R-R	PAX/CARGO
	BAe 146-200	84,000	LF-507-1F	(4) TF	7,000 LBT	R-R	PAX/CARGO
	BAe 146-300	97,000	ALF502R-5	(4) TF	6,970 LBT	R-R	PAX/CARGO
	BAe-146-NRA	97-000	CFM-56	(2) TF	22,000 LBT	G.E.	In Development
	BAe 748 (ANDOVER-C)	46,500	DART 532	(2) TP	2,020 ESHP	R-R	PAX/CARGO
	BAe 800	27,400	TFE731-5	(2) TF	4,300 LBT	GARRETT	PAX/CARGO
	BAe 801 (NIMROD)	177,500	SPEY RB.160	(4) TF	11,995 LBT	R-R	MARITIME
	BAe1000	28,000	PW 305B	(2) TF	5,200 LBT	P&WC	PAX/CARGO
	BAe JETSTREAM-31	15,300	TPE 331-10UG	(2) TP	1,000 ESHP	GARRETT	PAX/CARGO
	BAe SUPER-JET'M-31	16,200	TPE 331-12RU	(2) TP	1,100 ESHP	GARRETT	PAX/CARGO
	BAe JETSTREAM-41	23,000	TPE 331-14GR	(2) TP	1,500 ESHP	GARRETT	PAX/CARGO
	BAe RJ-70 (Mod. 146)	97,000	LF-507-1F	(4) TF	7,000 LBT	R-R	PAX/CARGO
	BAe RJ-85 (Mod. 146)	99,000	LF-507-1F	(4) TF	7,000 LBT	R-R	PAX/CARGO
	BRITANNIA	170,000	PROTEUS 765	(4) TP	4,250 ESHP	R-R	PAX/CARGO
	BUCCANEER	62,000	SPEY MK.250	(2) TF	11,995 LBT	R-R	ATTACK/RECON
	CANBERRA	57,000	AVON 206	(2) TJ	11,250 LBT	R-R	RECON
	HARRIER Mk.3	25,000	PEGASUS MK.103	(2) TF	21,500 LBT	R-R	ATTACK (VSTOL)
	HARRIER II Mk.5/7	31,000	PEGASUS MK.105	(2) TF	22,000 LBT	R-R	ATTACK (VSTOL)
	HARRIER-SEA	26,200	PEGASUS MK.104	(2) TF	21,500 LBT	R-R	ATTACK (VSTOL)
	HAWK 100/200	18,500	ADOUR MK.871	(1) TJ	6,600 LBT	R-R/T-M	TRAINER
	HERALD	43,000	DART 526	(2) TP	2,020 ESHP	R-R	PAX/CARGO/MIL
	JETSTREAM 1&2	12,500	ASTAZOU 16D	(2) TP	968 ESHP	T-M	PAX/CARGO
	P.135 (BAe 125)	27,520	TFE 731-5	(2) TF	4,300 LBT	GARRETT	MIL/AEW
	TRIDENT	150,000	SPEY 512	(3) TF	12,000 LBT	R-R	PAX/CARGO
	VANGUARD	146,000	TYNE 11	(2) TP	5,325 ESHP	R-R	PAX/CARGO
	VC-10	323,000	CONWAY 43	(4) TF	20,370 LBT	R-R	PAX/CARGO
	VC-10K & (SUPER VC-10)	323,000	CONWAY 43	(4) TF	21,800 LBT	R-R	TANKER/PAX/CARGO
	VISCOUNT	60,000	DART 526	(2) TP	1,910 ESHP	R-R	PAX/CARGO
BAe[2]	CONCORDE	408,000	OLYMPUS 593	(4) TP	38,000 LBT	R-R	PAX/SUPERSONIC
BAe[11]	JAGUAR	34,600	ADOUR MK.815	(2) TF	8,400 LBT	R-R/T-M	ATTACK
BAe-PANAVIA[10]	TORNADO	62,700	RB-199-34R	(2) TF	16,580 LBT	T-U	FIGHTER/BOMBER
HANLEY-PAGE	VICTOR-K	238,000	CONWAY Mk.201	(2) TF	20,600 LBT	R-R	TANKER
HAWKER	HUNTER FGA	17,750	AVON 207	(1) TJ	10,500 LBT	R-R	FIGHTER
NEW-CAL AVIATION	DHC-4 (CARIBOU)	31,000	PT6A-67R	(2) TP	1,509 ESHP	P&WC	PAX/CARGO/MIL
NORMAN AEROSPACE	NDN-1T (FIRECRACKER)	3,400	PT6A-25	(1) TP	580 ESHP	P&WC	TRAINER
	NDN-6 (FIELDMASTER)	10,000	PT6A-34	(1) TP	783 ESHP	P&WC	AGRICULTURE
PILATUS BRITTEN- NORMAN	BN-2T (ISLANDER)	7,000	250-B17D	(2) TP	420 ESHP	ALLISON	PAX/CARGO
	BN-2T (DEFENDER)	7,000	250-B17D	(2) TP	420 ESHP	ALLISON	PAX/CARGO
	BN-2TR (AEW DEFENDER)	8,500	250-B17F	(2) TP	450 ESHP	ALLISON	AEW
SHORTS BROTHERS	BELFAST	150,000	TYNE-12	(4) TP	5,200 ESHP	R-R	PAX/CARGO
	SD 330 (SHERPA)	22,900	PT6A-45R	(2) TP	1,254 ESHP	P&WC	PAX/CARGO
	SD 330 (SUPER SHERPA)	25,500	PT6A-65AR	(2) TP	1,509 ESHP	P&WC	PAX/CARGO
	SD 360 (ADVANCED)	27,100	PT6A-67R	(2) TP	1,509 ESHP	P&WC	PAX/CARGO
	SHORTS SKYLINER	13,700	TPE 331-2	(2) TP	715 ESHP	GARRETT	PAX/CARGO
	SHORTS SKYVAN	14,500	TPE 331-2	(2) TP	715 ESHP	GARRETT	PAX/CARGO
	SHORTS TUCANO S.312	7,716	TPE 331-12B	(1) TP	1,100 ESHP	GARRETT	TRAINER

INDIA

AIRCRAFT MFR.	AIRCRAFT DESIGNATION	GROSS WEIGHT	ENGINE DESIG.	ENGINE TYPE	ENGINE POWER	ENGINE MFR.	AIRCRAFT MISSION
HINDUSTAN AERONAUTICS LTD. (HAL)	JAGUAR	33,075	ADOUR MK.815	(2) TJ	8,000 LBT	R-R/T-M	FIGHTER
	HJT-16 (KIRAN)	10,900	ORPHEUS 701	(1) TJ	4,700 LBT	R-R	TRAINER
	AJEET (GNAT)	10,000	ORPHEUS 701	(1) TJ	4,700 LBT	R-R	TRAINER
	HTT-34	2,750	250-B17D	(1) TP	420 ESHP	ALLISON	TRAINER
	HS.748-2	44,400	DART-7	(2) TP	2,100 ESHP	R-R	PAX/CARGO/MIL
	HF-24 (MAHUT)	24,000	ORPHEUS 703	(2) TJ	4,850 LBT	R-R	PAX/CARGO/MIL
	LIGHT FIGHTER	36,000	GTX 35VS	(1) TF	18,000 LBT	GTRE	FIGHTER (IN DEV,)

INDONESIA

AIRCRAFT MFR.	AIRCRAFT DESIGNATION	GROSS WEIGHT	ENGINE DESIG.	ENGINE TYPE	ENGINE POWER	ENGINE MFR.	AIRCRAFT MISSION
IPTN[†]	N-250	42,500	GMA-2100	(2) TP	5,200 ESHP	ALLISON	PAX/CARGO/MIL
	NC-212 (AVIOCAR)	18,975	TPE 331-10R	(2) TP	1,000 ESHP	GARRETT	PAX/CARGO/MIL
IPTN[6]	CN 235 & 235M	31,740	CT7-9C	(2) TP	1,870 ESHP	G.E.	PAX/CARGO
	CN 235 (T.19)	33,300	CT7-9C	(2) TP	1,870 ESHP	G.E.	PAX/CARGO/MIL

[†] IPTN-INDUSTRI PESAWAT TERBANG NUSANTARA

ISRAEL

AIRCRAFT MFR.	AIRCRAFT DESIGNATION	GROSS WEIGHT	ENGINE DESIG.	ENGINE TYPE	ENGINE POWER	ENGINE MFR.	AIRCRAFT MISSION
ISRAEL AIRCRAFT INDUSTRIES (IAI)	LAVI	22,000	PW 1120	(1) TF	20,600 LBT	P&W	FIGHTER (PROTOTYPE)
	101/201 (ARAVA)	15,100	PT6A-36	(2) TP	786 ESHP	P&WC	PAX/CARGO
	202 (ARAVA)	15,000	PT6A-34	(2) TP	783 ESHP	P&WC	MIL (STOL)
	1121 (COMMANDER)	17,500	CJ-610-1	(2) TJ	3,000 LBT	G.E.	PAX/CARGO
	NAMMER	36,400	RM12	(1) TF	18,100 LBT	VOLVO	FIGHTER (PROTOTYPE)

AIRCRAFT MFR.	AIRCRAFT DESIGNATION	GROSS WEIGHT	ENGINE DESIG.	ENGINE TYPE	ENGINE POWER	ENGINE MFR.	AIRCRAFT MISSION
IAI (CONT'D)	NESHER	30,205	ATAR-9C	(1) TJ	13,670 LBT	SNECMA	ATTACK
	1123 (COMMODORE)	20,700	CJ-610-9	(2) TJ	3,000 LBT	G.E.	PAX/CARGO
	1124 (SEASCAN)	24,300	TFE-731-3	(2) TF	3,700 LBT	GARRETT	MARITIME
	KFIR (C2,C7,TC2,TC7)	36,300	J79-JIE	(1) TJ	18,750 LBT	G.E.	ATT/FTR/TNR
	WESTWIND 1/2 (1124)	23,600	TFE 731-3	(2) TF	3,700 LBT	GARRETT	PAX/CARGO
	WESTWIND-BEECH 18	15,000	PT6A-20/27	(2) TP	715 ESHP	P&WC	PAX/CARGO
ASTRA JET (USA SUBSIDIARY)	1125 (ASTRA)	24,600	TFE-731-3	(2) TF	3,700 LBT	GARRETT	PAX/CARGO

ITALY

AIRCRAFT MFR.	AIRCRAFT DESIGNATION	GROSS WEIGHT	ENGINE DESIG.	ENGINE TYPE	ENGINE POWER	ENGINE MFR.	AIRCRAFT MISSION
ALENIA SPA. (FORMERLY AERITALIA)	G.222	61,700	T64-GE-P4D	(2) TP	3,400 ESHP	G.E.	PAX/CARGO/MIL
	G.222T	61,700	TYNE-20	(2) TP	5,100 ESHP	G.E.	PAX/CARGO/MIL
	F-104S	28,700	J79-GE-19	(1) TJ	17,780 LBT	G.E.	FIGHTER
ALENIA[1]	ATR 42 (SUPER COMMUTER)	36,800	PW 120/121	(2) TP	2,100 ESHP	P&WC	PAX/CARGO
	ATR 72	44,000	PW 124	(2) TP	2,522 ESHP	P&WC	PAX/CARGO
ALENIA[9]	AM-X	28,300	SPEY MK.807	(1) TF	11,130 LBT	R-R	FIGHTER
	AM-XT	28,300	SPEY MK.807	(1) TF	11,130 LBT	R-R	TRAINER
AERMACCHI SPA.	MB 326K	11,500	VIPER 632-43	(1) TJ	3,970 LBT	R-R	ATTACK/TRAINER
	MB 329A	13,000	VIPER 632-43	(1) TJ	3,970 LBT	R-R	TRAINER
	MB 339K (VELTRO-2)	13,000	VIPER 680-43	(1) TJ	4,400 LBT	R-R	ATTACK/TRAINER
AERMACCHI[9]	AM-X	28,300	SPEY MK.807	(1) TF	11,130 LBT	R-R	FIGHTER
	AM-XT	28,300	SPEY MK.807	(1) TF	11,130 LBT	R-R	TRAINER
AGUSTA (FORMERLY SIAI-MARCHETTI)[16]	SF.260TP	2,900	250-B17F	(1) TP	450 ESHP	ALLISON	TRAINER
	S.211A	7,160	JT15D-4C	(1) TF	2,500 LBT	P&WC	TRAINER/ATTACK
	SF.260	8,050	250-B17D	(2) TP	420 ESHP	ALLISON	PAX/CARGO/MIL
	SF-600 (CANGURO)	7,495	250-B17F	(2) TP	450 ESHP	ALLISON	PAX/CARGO/MIL
	SM-109E	3,196	250-B17D	(1) TP	420 ESHP	ALLISON	OBSERVATION
AGUSTA CAPRONI	C22J (VANTRUA)	2,760	TRS 18-1	(2) TF	325 LBT	M-T	TRAINER (ACROBATIC)
CAPRONI SPA.	C22J	2,800	TRS 18-1	(2) TJ	326 LBT	M-T	TRAINER (ACROBATIC)
FIAT AVIO SPA.	G.91T/3	12,000	ORPHEUS 803	(1) TJ	5,000 LBT	R-R	TRAINER
	G.91Y	16,500	J85-GE-13A	(2) TJ	4,080 LBT	G.E.	FIGHTER/RECON
RINALDO PIAGGIO SPA.	P.166DL3	9,500	LTP101-700A	(2) TP	700 ESHP	TEXTRON	PAX/CARGO/MIL
	P.180 (AVANTI)	10,600	PT6A-66	(2) TP	905 ESHP	P&WC	PAX/CARGO
	PD.808	18,000	Viper Mk.526	(2) TJ	3,360 LBT	R-R	PAX/CARGO/MIL
PANAVIA CORP.[10]	TORNADO	62,700	RB-199-34R	(2) TF	16,580 LBT	T-U	ATTACK/FIGHTER
PARTENAVIA	AP 68TP-300 (SPARTICUS)	5,730	250-B17C/E	(2) TP	328 ESHP	ALLISON	PAX/CARGO
	AP 68TP-600 (VIATOR)	6,280	250-B17C/E	(2) TP	328 ESHP	ALLISON	PAX/CARGO
	P 68 (VIATOR)	4,385	250-B17C/E	(2) TP	328 ESHP	ALLISON	PAX/CARGO
COMMUTER AIRCRAFT CO.	CAC 100	37,500	PT6A-65R	(4) TP	1,276 ESHP	P&WC	PAX/CARGO

JAPAN

AIRCRAFT MFR.	AIRCRAFT DESIGNATION	GROSS WEIGHT	ENGINE DESIG.	ENGINE TYPE	ENGINE POWER	ENGINE MFR.	AIRCRAFT MISSION
FUJI INDUSTRIES	T-1 (SABER)	12,000	ORPHEUS	(1) TJ	4,000 LBT	R-R	TRAINER
	T-5 (RE-ENGINE)	3,490	250-B17D	(1) TP	420 ESHP	ALLISON	TRAINER
NIHON	YS-11A	54,000	DART 542-10	(2) TP	3,060 ESHP	R-R	PAX/CARGO
ISHIDA CORP.	TW-68	-----	PT6B-67	(2) TP	†1,960 ESHP	P&WC	TILT-WING (IN DEV.)

† TWIN-PAC ENGINE, 980 SHP EACH

AIRCRAFT MFR.	AIRCRAFT DESIGNATION	GROSS WEIGHT	ENGINE DESIG.	ENGINE TYPE	ENGINE POWER	ENGINE MFR.	AIRCRAFT MISSION
KAWASAKI INDUSTRIES	C-1 / EC-1	85,300	JT8D-M-9	(2) TF	14,500 LBT	P&W	PAX/CARGO/MIL
	P-2J (NEPTUNE)	75,000	T64-IHI-10E	(2) TP	2,850 ESHP	G.E.	MARITIME
	P-3C (ORION)	135,000	T56-IHI-14	(4) TP	4,910 ESHP	ALLISON	ASW
	T-4	16,100	F3-IHI-30	(2) TF	3,680 LBT	IHI	TRAINER
MITSUBISHI INDUSTRIES	DIAMOND-I (MU-300)	15,700	JT15D-4D	(2) TP	2,500 LBT	P&WC	PAX/CARGO
	DIAMOND-II (MU-300)	15,700	JT15D-5D	(2) TP	2,900 LBT	P&WC	PAX/CARGO
	F-15J (EAGLE)	41,500	F100-IHI-100	(2) TF	25,000 LBT	IHI	FIGHTER
	F-4EJ (PHANTOM)	46,600	J79-IHI-17	(2) TJ	17,820 LBT	IHI	FIGHTER
	F-1/T-2 (BUCKEYE)	30,200	TF40-IHI-801A†	(2) TF	7,300 LBT	IHI	ATTACK/TRAINER
	MU-2 (LR-1)	8,800	TPE 331-1	(2) TP	655 ESHP	GARRETT	RECON/SAR
	MU-2N (MARQUISE)	11,575	TPE 331-10/5	(2) TP	950 ESHP	GARRETT	PAX/CARGO
	MU-2P (SOLITAIRE)	10,700	TPE 331-10	(2) TP	950 ESHP	GARRETT	PAX/CARGO
	MU-2L	11,575	TPE-331-6	(2) TP	776 ESHP	GARRETT	PAX/CARGO

AIRCRAFT MFR.	AIRCRAFT DESIGNATION	GROSS WEIGHT	ENGINE DESIG.	ENGINE TYPE	ENGINE POWER	ENGINE MFR.	AIRCRAFT MISSION
NAL-KAWASAKI	C-1 (ASUKA)	85,200	FJR-710-600	(4) TF	14,330 LBT	NAL	STOL (RESEARCH)
SHIN MEIWA INDUSTRIES	PS-1 (FLYING BOAT)	86,860	T64-IHI-10J	(4) TP	3,490 ESHP	IHI	ASW (AMPHIBIAN)
	US-1A	99,180	T64-IHI-10J	(4) TP	3,490 ESHP	IHI	SAR (AMPHIBIAN)

† ROLLS-ROYCE, TURBOMECA (ADOUR) LICENSE.

NETHERLANDS

AIRCRAFT MFR.	AIRCRAFT DESIGNATION	GROSS WEIGHT	ENGINE DESIG.	ENGINE TYPE	ENGINE POWER	ENGINE MFR.	AIRCRAFT MISSION
FOKKER	F-27 (FRIENDSHIP)	45,900	DART 532/540	(2) TP	2,120 ESHP	R-R	PAX/CARGO
	F-27M	45,900	DART 532/540	(2) TP	2,120 ESHP	R-R	PAX/CARGO/MIL
	F-28 (FELLOWSHIP)	73,000	SPEY RB-183	(2) TF	9,900 LBT	R-R	PAX/CARGO
	F-50	45,900	PW 125A/B	(2) TP	2,699 ESHP	P&WC	PAX/CARGO
	F-100	98,000	TAY 650	(2) TF	15,100 LBT	R-R	PAX/CARGO
	F-27/50 (ENFORCER)	47,000	PW 124	(2) TP	2,400 ESHP	P&WC	MARITIME
	VFW-614	44,000	M-45H	(2) TF	7,780 LBT	R-R/SNECMA	PAX/CARGO

NEW ZEALAND

AIRCRAFT MFR.	AIRCRAFT DESIGNATION	GROSS WEIGHT	ENGINE DESIG.	ENGINE TYPE	ENGINE POWER	ENGINE MFR.	AIRCRAFT MISSION
PACIFIC AEROSPACE	CT4-CR	2,500	250-B17D	(1) TP	420 ESHP	ALLISON	TRAINER

POLAND

AIRCRAFT MFR.	AIRCRAFT DESIGNATION	GROSS WEIGHT	ENGINE DESIG.	ENGINE TYPE	ENGINE POWER	ENGINE MFR.	AIRCRAFT MISSION
WSK-PZL MIELEC	AN-28 (USSR LICENSE)	14,330	TVD-10B	(2) TP	990 ESHP	GLUSHENKOV	PAX/CARGO/MIL
	I-22 (IRYDA)	16,519	SO-3W22	(2) TJ	2,425 LBT	PZL	ATTACK/TRAINER
	I-22 (IRYDA-93)	16,519	K-15	(2) TJ	3,300 LBT	PZL	ATTACK/TRAINER
	PZL M15 (BELPHEGOR)	12,675	AI-25	(1) TF	3,000 LBT	IVCHENKO	TRAINER
	PZL 106 (TURBO KRUK)	7,715	PT6A-34AG	(1) TP	783 ESHP	P&WC	AGRICULTURAL
	TS-11 (ISKRA) "SPARK"	8,465	SO-3W	(1) TJ	2,425 LBT	PZL	TRAINER
	TS-11 (UPGRADE)	8,465	K-15	(1) TJ	3,300 LBT	PZL	TRAINER
PEZETEL	PZL M18 (TURBO DROMANDER)	12,500	PT6A-45AG	(1) TP	1,254 ESHP	P&WC	AGRICULTURAL
	PZL 130T (TURBO ORLIK)	4,750	PT6A-25	(1) TP	580 ESHP	P&WC	TRAINER
	PZL 130TB (TURBO ORLIK)†	5,750	M601E	(1) TP	750 ESHP	WALTER	TRAINER

† PZL-WARSZAWA-OKEICE

RUMANIA

AIRCRAFT MFR.	AIRCRAFT DESIGNATION	GROSS WEIGHT	ENGINE DESIG.	ENGINE TYPE	ENGINE POWER	ENGINE MFR.	AIRCRAFT MISSION
ICA CORP.	1-11 (BAC 1-11)	98,000	SPEY 512-14	(2) TF	12,550 LBT	R-R	PAX/CARGO
CINAIR CORP.	IAR.99 (SOIM) "HAWK"	12,250	VIPER MK.632	(1) TJ	4,000 LBT	R-R	ATTACK/TRAINER
	IAR-825 (TRIUMF)	3,042	PT6A-21	(1) TP	580 ESHP	P&WC	TRAINER
CINAIR [12]	IAR.93A/B (YUROM)	22,760	VIPER MK.632	(2) TJ	4,000 LBT	R-R	TRAINER

SOUTH AFRICA

AIRCRAFT MFR.	AIRCRAFT DESIGNATION	GROSS WEIGHT	ENGINE DESIG.	ENGINE TYPE	ENGINE POWER	ENGINE MFR.	AIRCRAFT MISSION
ATLAS ACFT.	MB 326M (IMPALA)	9,500	VIPER 540	(1) TJ	3,500 LBT	R-R	TRAINER
	CHEETAH (SA-315)	29,800	ATAR 09C	(1) TJ	13,670 LBT	SNECMA	TRAINER

SPAIN

AIRCRAFT MFR.	AIRCRAFT DESIGNATION	GROSS WEIGHT	ENGINE DESIG.	ENGINE TYPE	ENGINE POWER	ENGINE MFR.	AIRCRAFT MISSION
CONSTRUCCIONES AERONAUTICAS SA (CASA)	C 101 (AVIOJET)	13,880	TFE 731-5	(1) TF	4,300 LBT	GARRETT	ATTACK/TRAINER
	C 207A (AZOR)	36,400	HERCULES 730	(2) TP	2,000 ESHP	R-R	PAX/CARGO/MIL
	C 212 (AVIOCAR)	18,975	TPE 331-10R	(2) TP	1,000 ESHP	GARRETT	PAX/CARGO
	SF-5A (FREEDOM FIGHTER)	20,600	J85-GE-13	(2) TJ	4,080 LBT	G.E.	ATTACK/FIGHTER
	HA 220 (SUPER SAETA)	8,150	MARBORE-6	(2) TJ	1,060 LBT	T-M	ATTACK
CASA[6]	CN 235 & 235M	31,740	CT7-9C	(2) TP	1,870 ESHP	G.E.	PAX/CARGO
	CN 235 (T.19)	33,300	CT7-9C	(2) TP	1,870 ESHP	G.E.	PAX/CARGO/MIL

SWEDEN

AIRCRAFT MFR.	AIRCRAFT DESIGNATION	GROSS WEIGHT	ENGINE DESIG.	ENGINE TYPE	ENGINE POWER	ENGINE MFR.	AIRCRAFT MISSION
SAAB-SCANIA AB	SAAB-105 (SK60A,B,C)	9,800	AUBISQUE	(2) TF	1,635 LBT	T-M	ATTACK/TRAINER
	SAAB 105E	14,300	J85-GE-17B	(2) TJ	2,850 LBT	G.E.	EXPORT MODEL
	DRAKEN (SK 35C)	20,000	RM6B†	(1) TJ	14,500 LBT	FLYG	TRAINER
	DRAKEN (J 35D,F)	23,000	RM6C	(1) TJ	17,650 LBT	FLYG	FIGHTER
	DRAKEN (SK 35XD)	25,000	RM6C	(1) TJ	17,650 LBT	FLYG	ATTACK
	DRAKEN (SK 35X)	25,000	RM6C	(1) TJ	17,650 LBT	FLYG	TRAINER
	GRIPEN (JAS 39)	17,635	RM12	(1) TF	18,000 LBT	FLYG/G.E.	ATTACK/FIGHTER

AIRCRAFT MFR.	AIRCRAFT DESIGNATION	GROSS WEIGHT	ENGINE DESIG.	ENGINE TYPE	ENGINE POWER	ENGINE MFR.	AIRCRAFT MISSION
SAAB (CONT'D)	VIGGEN (AJ37)	35,000	RM8A††	(1) TF	25,990 LBT	FLYG	ATTACK
	VIGGEN (JA37)	37,500	RM8B	(1) TF	25,990 LBT	FLYG	FIGHTER
	VIGGEN (SK37, SH37)	35,000	RM8A	(1) TF	25,990 LBT	FLYG	TRAINER/RECON
	SAAB 340 (COMMUTER)	28,000	CT7-7E	(2) TP	1,700 ESHP	G.E.	PAX/CARGO
	SAAB 340B	28,000	CT7-9B	(2) TP	1,870 ESHP	G.E.	PAX/CARGO
	SAAB 2000	45,500	GMA-2100	(2) TP	4,500 ESHP	ALLISON	PAX/CARGO (IN DEV.)

† ROLLS-ROYCE AVON LICENSE,   †† P&W JT8D LICENSE.

SWITZERLAND

PILATUS	PC-6/B2-H4 (PORTER)	6,100	PT6A-27	(1) TP	715 ESHP	P&WC	PAX/CARGO (STOL)
AIRCRAFT	PC-7 (TURBO-TRAINER)	5,950	PT6A-25	(1) TP	580 ESHP	P&WC	TRAINER
LTD.	PC-9	7,050	PT6A-62	(1) TP	1,150 ESHP	P&WC	TRAINER
	PC-12	8,800	PT6A-67B	(1) TP	1,200 ESHP	P&WC	PAX/CARGO (IN DEV.)
	PC-X	9,500	PT6A-67B	(1) TP	1,200 ESHP	P&WC	PAX/CARGO (IN DEV.)

YUGOSLAVIA

SOKO	ORAO 1/2	22,760	VIPER MK.632-41	(2) TJ	4,000 LBT	R-R	ATTACK
AIRCRAFT	G-2A (GALEB)	9,480	VIPER MK.22-6	(1) TJ	2,500 LBT	R-R	TRAINER
INDUSTRIES †	G-4 (SUPER GALEB)	13,950	VIPER MK.632	(1) TJ	4,000 LBT	R-R	ATTACK
	J-1 (JASTREB)	11,025	VIPER MK.531	(1) TJ	3,120 LBT	R-R	ATTACK/RECON
	TJ-1 (JASTREB)	11,250	VIPER MK.531	(1) TJ	3,120 LBT	R-R	TRAINER
SOKO [12]	IAR.93A/B (YUROM)	22,760	VIPER MK.632	(2) TJ	4,000 LBT	R-R	TRAINER

† VAZDUHOPLOVNA INDUSTRIJA SOKO

MULTI-NATIONAL COMPANIES (SUPERSCRIPT DESIGNATORS):

1. AEROSPATIALE, FRANCE; ALENIA, ITALY.
2. AEROSPATIALE, FRANCE; BAe, GREAT BRITAIN.
3. AEROSPATIALE, FRANCE; MBB, GERMANY.
4. AEROSPATIALE, FRANCE; OTHER EUROPEAN COUNTRIES.
5. EMBRAER, BRAZIL; FAMA, ARGENTINA.
6. CASA, SPAIN; IPTN, INDONESIA, (AIRCRAFT TECHNOLOGIES INC. / AIRTECH CORP.).
7. DASSAULT, FRANCE; DORNIER, GERMANY.
8. PANAVIA CORP. (GERMANY, ITALY, GREAT BRITAIN).
9. AMX INT'L., (ALENIA & AERMACCHI, ITALY); EMBRAER, BRAZIL.
10. PANAVIA CORP. (GR. BRITAIN, GERMANY, ITALY).
11. SEPECAT CORP. (DASSAULT, FRANCE, BAe, GREAT BRITAIN).
12. SOKO-CINAIR (SOKO, YUGOSLAVIA, CINAIR, RUMANIA).
13. TBM INT'L. (AEROSPATIALE-SOCATA DIVISION, VALMET CORP., FINLAND).
14. GERMANY, FRANCE, SPAIN, ITALY.
15. AIRBUS CONVERSION (FORMERLY AERO-SPACELINES SUPER GUPPY).
16. AGUSTA-SESTO CALENDE WORKS (FORMERLY SIAI-MARCHETTI).
17. AVIONS TRANSPORT REGIONAL (ATR)

INTERNATIONAL ROTARY WING AIRCRAFT
(ALPHABETICAL BY COUNTRY)

AIRCRAFT MFR.	AIRCRAFT DESIGNATION	GROSS WEIGHT	ENGINE DESIG.	ENGINE TYPE	ENGINE POWER	ENGINE MFR.	AIRCRAFT MISSION

CANADA

BOEING VERTOL	CH-113 (LABRADOR)	21,000	T58-GE-8F	(2) TS	1,350 SHP	G.E.	SAR
CANADA BELL	CH-135 (BELL-212)	11,200	PT6T-3B	(2) TS	†1,800 SHP	P&WC	MIL (TWIN HUEY)
SIKORSKY	CH-118 (SH-3B)	21,000	T-58-GE-10	(2) TS	1,400 SHP	G.E.	MIL (SEA KING)

† TWIN-PAC ENGINE, 900 SHP EACH

CHINA (PEOPLES REPUBLIC OF) - NATIONAL AIRCRAFT FACTORIES (CATIC)

CHENGDU ACFT.	Z-8 (SA-321) †	28,600	WZ-6	(3) TS	1,550 SHP	CHENGDU	HELICOPTER
CHANGHE ACFT.	Z-9 (SA-365N) †	8,500	WZ-8	(2) TS	700 SHP	HARBIN	HELICOPTER

† BUILT UNDER LICENSE- FRANCE

AIRCRAFT MFR.	AIRCRAFT DESIGNATION	GROSS WEIGHT	ENGINE DESIG.	ENGINE TYPE	ENGINE POWER	ENGINE MFR.	AIRCRAFT MISSION
CHINA (REPUBLIC OF) TAIWAN							
BELL(TAIWAN)	205 (IROQUOIS)	9,500	T53-13A	(1) TS	1,400 SHP	TEXTRON	MILITARY
EGYPT							
GOVERNMENT (FRENCH LICENSE)	SA 342 (GAZELLE)	4,415	ASTAZOU 14M	(1) TS	590 SHP	T-M	ATTACK
FRANCE							
AEROSPATIALE[3] (EUROCOPTER INT'L)	AS 332B1 (SUPER PUMA)	19,840	MAKILA 1A1	(2) TS	1,878 SHP	T-M	MILITARY
	AS 332L1 (SUPER PUMA)	18,960	MAKILA 1A1	(2) TS	1,878 SHP	T-M	CIVIL
	AS 350B (ECUREUIL)	4,300	ARRIEL 1B	(1) TS	641 SHP	T-M	CIVIL
	AS 350D	4,300	LTS 101-600	(1) TS	615 SHP	TEXTRON	CIVIL
	AS 350G	4,300	250-C30M	(1) TS	650 SHP	ALLISON	MILITARY
	AS 355F (TWINSTAR)	5,600	250-C20F	(2) TS	420 SHP	ALLISON	CIVIL/MIL
	AS 355N (ECUREUIL-II)	5,500	ARRIUS (TM-319)	(2) TS	547 SHP	T-M	CIVIL/MIL
	AS 365N (DAUPHIN-II)	9,040	ARRIEL 1C1	(2) TS	724 SHP	T-M	CIVIL
	AS 365M (PANTHER)	9,040	TM 3331A	(2) TS	912 SHP	T-M	CIVIL
	AS 366G1 (DAUPHIN-II)	8,900	LTS 101-750A/B	(2) TS	742 SHP	T-M	MIL (HH65A)
	AS 532 (COUGAR)	19,840	MAKILA 1A1	(2) TS	1,878 SHP	T-M	MIL (AS-332)
	AS 550 (FENNECS)	4,300	ARRIEL 1B	(1) TS	1,200 SHP	T-M	MIL (AS 365)
	AS 565 (PANTHER-800)	9,340	T800-LHT-800	(2) TS	1,200 SHP	LHTEC	MIL (AS 365)
	AS 565 (PANTHER)	9,340	ARRIEL 1C1	(2) TS	724 SHP	T-M	MILITARY
	NORD N.262	10,600	BASTAN IV	(2) TS	1,080 SHP	T-M	CIVIL
	SA 315 (LAMA)	5,070	ARTOUSTE 3B	(1) TS	562 SHP	T-M	CIVIL
	SA 318 (ALOUETTE-II)	3,500	ARTOUSE II	(1) TS	360 SHP	T-M	CIVIL
	SA 316/319 (ALOUETTE-III)	4,900	ASTAZOU II	(1) TS	660 SHP	T-M	CIVIL
	SA 321 (SUPER FRELON)	28,660	TURMO 3C7	(3) TS	1,610 SHP	T-M	MILITARY
	SA 332M1 (SUPER PUMA)	19,840	MAKILA 1A1	(2) TS	1,878 SHP	T-M	CIVIL/MIL
	SA 341/342 (GAZELLE)	4,415	ASTAZOU 14M	(1) TS	590 SHP	T-M	MILITARY
	SA 360/361 (DAUPHIN)	6,600	ASTAZOU 16	(1) TS	1,021 SHP	T-M	CIVIL/MIL
	TIGER (HAC) / (HAP)	13,227	MTR-390	(2) TS	1,285 SHP	T-M/R-R	MILITARY
AEROSPATIALE[1]	NH-90	19,000	RTM 322	(2) TS	2,300 SHP	R-R/T-M	In Dev.
AEROSPATIALE[2]	AS 330 (PUMA)	14,100	TURMO IV	(2) TS	1,450 SHP	T-M	CIVIL/SAR
	HCC MK.4 (GAZELLE)	3,970	ASTAZOU IIIA	(1) TS	590 SHP	T-M	MILITARY
	HC MK.2	13,500	GNOME MK.10	(2) TS	1,350 SHP	R-R	MILITARY
GERMANY							
DORNIER GMBH	UH-1S (UPGRADE)	9,500	T53L-703	(1) TS	1,485 SHP	TEXTRON	CIVIL/MIL
MESSERSCHMITT-BOELKOW-BLOHM (MBB)[3] (EUROCOPTER INT'L)	BO.105CB	5,500	250-C20B	(2) TS	420 SHP	ALLISON	CIVIL/MIL
	BO.105LS	5,750	250-C28C	(2) TS	550 SHP	ALLISON	CIVIL/MIL
	BO.105LS-B1	5,750	PW 205B	(2) TS	509 SHP	P&WC	CIVIL/MIL
	EC-135 (formerly BO.108)	5,500	ARRIUS (TM-319)	(2) TS	742 SHP	T-M	CIVIL/MIL
	ALTERNATE ENGINE	5,500	PW 206B	(2) TS	750 SHP	P&WC	CIVIL/MIL
	PAH-1 (BO-105P)	5,500	250-C28C	(2) TS	550 SHP	ALLISON	MILITARY
	TIGER (PAH-2)	13,227	MTR-390	(2) TS	1,285 SHP	T-M/R-R	MILITARY
MBB[4]	BK.117	7,050	LTS101-750	(2) TS	742 SHP	TEXTRON	CIVIL/MIL
	BK.117C-1	7,050	ARRIEL 1E	(2) TS	750 SHP	T-M	CIVIL/MIL
	ALTERNATE ENGINE	7,050	250-C30S	(2) TS	750 SHP	ALLISON	CIVIL/MIL
MBB[1]	NH-90	19,000	RTM 322	(2) TS	2,300 SHP	R-R/T-M	In Dev.
GREAT BRITAIN (UK)							
EH INDUSTRIES[5]	EH 101 (MERLIN)	31,500	T700-GE-401	(3) TS	1,690 SHP	G.E.	MILITARY
	ALTERNATE ENGINE	31,500	RTM 322	(3) TS	2,240 SHP	T-M	MILITARY
	EH 101 (HELILINER)	31,500	CT7-6	(3) TS	2,250 SHP	G.E.	CIVIL
WESTLAND HELO LTD.	LYNX (AH-1)	9,600	GEM SERIES	(2) TS	900 SHP	R-R	MILITARY (UK/FR)
	LYNX-II (SUPER)	11,000	GEM 60-3	(2) TS	1,257 SHP	R-R	MILITARY (UK/FR)
	LYNX-SUPER MK.99	11,300	GEM 42-1	(2) TS	1,250 SHP	R-R	MILITARY
	COMMANDO	21,000	GNOME H 1400	(2) TS	1,680 SHP	R-R	MIL (USA LICENSE)

AIRCRAFT MFR.	AIRCRAFT DESIGNATION	GROSS WEIGHT	ENGINE DESIG.	ENGINE TYPE	ENGINE POWER	ENGINE MFR.	AIRCRAFT MISSION
WESTLAND HELO CONT.							
	HAS MK.2/3/4	9,750	GEM 2	(2) TS	900 SHP	R-R	NAVAL
	SEA KING (S-61)	21,000	GNOME H 1400	(2) TS	1,480 SHP	R-R	ASW,SAR (USA LICENSE)
	W30-100	12,800	GEM 60-3	(2) TS	1,257 SHP	R-R	MILITARY
	W30-140 (WESTLAND 30)	12,250	GEM 41-1	(2) TS	1,120 SHP	R-R	PAX/CARGO
	W30-160	12,800	GEM 60-3	(2) TS	1,257 SHP	R-R	NAVAL
	W30-200	12,800	CT7-2B	(2) TS	1,725 SHP	G.E.	CIVIL/MIL
	W30-300	15,500	T700-GE-401	(2) TS	1,690 SHP	G.E.	MILITARY
	WS-70 (UH-60 BLACKHAWK)	27,000	T700-GE-700	(2) TS	1,622 SHP	G.E.	MIL (USA LICENCE)
	WASP-SCOUT	6,000	NIMBUS MK.503	(1) TS	710 SHP	R-R	MILITARY
	WESSEX	13,500	GNOME MK.110	(2) TS	1,350 SHP	R-R	CIVIL/MIL
	WESSEX	13,500	GAZELLE	(2) TS	1,600 SHP	R-R	CIVIL/MIL
WESTLAND[2]	HCC MK.4 (GAZELLE)	3,970	ASTAZOU IIIA	(1) TS	590 SHP	T-M	MILITARY
	HC MK.2	13,500	GNOME MK.10	(2) TS	1,350 SHP	R-R	MILITARY
	SA 330 (PUMA)	14,100	TURMO IV	(2) TS	1,450 SHP	T-M	CIVIL/SAR
INDIA							
BREDA NARDI	369D (NH500D)	3,000	250-20B	(1) TS	420 SHP	ALLISON	MILITARY
HINDUSTAN	ALH	11,000	TM 333-2B	(2) TS	1,000 SHP	T-M	IN DEVELOPMENT
	CHEETAH (SA-315)	3,860	ARTOUSE 3B	(1) TS	562 SHP	T-M	MIL (FRENCH LICENSE)
	CHETAK (SA-316)	4,850	ARTOUSE 3B	(1) TS	562 SHP	T-M	MIL (FRENCH LICENSE)
	LAMA	4,600	ARTOUSE 3B	(1) TS	562 SHP	T-M	MIL (FRENCH LICENSE)
INDONESIA							
IPTN	NBO-105 (MBB LICENSE)	5,500	250-C20B	(2) TP	420 SHP	ALLISON	MILITARY
ITALY							
AGUSTA	A-109A	5,930	250-C20R-1	(2) TS	420 SHP	ALLISON	CIVIL/MIL
	A-109K	5,730	ARRIEL 1C	(2) TS	700 SHP	T-M	CIVIL/MIL
	A129 (MANGUSTA)	9,039	GEM 2 MK.1004	(2) TS	900 SHP	R-R	CIVIL/MIL
	AB 205 (IROQUOIS)	9,500	T53L-13B	(1) TS	1,400 SHP	TEXTRON	MILITARY
	AB 206B (JET RANGER)	3,200	250-C20B	(1) TS	420 SHP	ALLISON	MILITARY
	AB 206BL (LONG RANGER)	4,050	250-C28B	(1) TS	550 SHP	ALLISON	MILITARY
	AB 212	11,200	PT6T-3B	(2) TS	†1,800 SHP	P&WC	MILITARY/AWS
	AB 412	11,500	PT6T-3B	(2) TS	†1,800 SHP	P&WC	MILITARY
	ASH-3H & AS-61	21,000	T58-GE-10	(2) TS	1,400 SHP	G.E.	SAR
	CH-47C (CHINOOK)	46,000	T55L-11D	(2) TS	3,750 SHP	TEXTRON	MILITARY
AGUSTA[6]	HH-3F (AS 61N1)	22,000	T58-GE-100	(2) TS	1,500 SHP	G.E.	VIP
	HH-3F (PELICAN)	22,000	T58-GE-100	(2) TS	1,500 SHP	G.E.	SAR
AGUSTA[1]	NH-90	19,000	RTM 322	(2) TS	2,300 SHP	R-R/T-M	In Dev.
AGUSTA[7]	BELL-204 (IROQUOIS)	9,500	T53L-13B	(1) TS	1,400 SHP	TEXTRON	MIL
EH INDUSTRIES[5]	EH 101	31,500	T700-GE-401	(3) TS	1,690 SHP	G.E.	MILITARY
	ALTERNATE ENGINE		CT7-6	(3) TS	2,250 SHP	G.E.	CIVIL/MIL
	ALTERNATE ENGINE		RTM 322	(3) TS	2,240 SHP	T-M	MILITARY

† TWIN-PAC ENGINE, 900 SHP EACH

AIRCRAFT MFR.	AIRCRAFT DESIGNATION	GROSS WEIGHT	ENGINE DESIG.	ENGINE TYPE	ENGINE POWER	ENGINE MFR.	AIRCRAFT MISSION
JAPAN							
BELL (JAPAN)	205 (BELL)	9,500	T53-13A	(1) TS	1,400 SHP	TEXTRON	MILITARY
FUJI HEAVY INDUSTRIES	AH-1S (BELL)	10,000	T53K-13B	(1) TS	1,400 SHP	TEXTRON	ATTACK
	FB-204B2	8,500	T53K-13B	(1) TS	1,400 SHP	TEXTRON	CIVIL
	UH-IH (BELL)	9,500	T53K-13B	(1) TS	1,400 SHP	TEXTRON	ATTACK
KAWASAKI HEAVY INDUSTRIES	BK-117 (WITH MBB)	6,280	LTS101-650B	(2) TS	650 SHP	TEXTRON	MILITARY UTILITY
	CH-47C	54,000	T55K-712	(2) TS	3,750 SHP	TEXTRON	MILITARY (USA LICENSE)
	HK-500 (369HS)	2,550	CT63-M-5A	(1) TS	420 SHP	ALLISON	CIVIL (USA LICENSE)
	KV107-2A5 (UH-46)	21,400	CT58-IHI-140	(2) TS	1,500 SHP	G.E.	RESCUE (USA LICENSE)
	OH-6D (500D)	3,000	C250-C20B	(1) TS	420 SHP	ALLISON	MIL/CIVIL (USA LICENSE)
	414-100 (BOEING)	54,000	T55L-712S/SB	(2) TS	3,750 SHP	TEXTRON	CIVIL/INT'L MIL
MITSUBISHI HEAVY INDUSTRIES	S-61A (HSS-2)	22,000	T58-IHI-10M1	(2) TS	1,400 SHP	IHI	SAR (USA LICENSE)
	S-62	7,900	CT-58-140	(1) TS	1,500 SHP	G.E.	CIVIL/MIL
	SH-3B	20,500	T58-IHI-10M2	(2) TS	1,500 SHP	IHI	ASW (USA LICENSE)
	ISH-60J (S-70)	24,000	T700-GE-700	(2) TS	1,622 SHP	IHI	ASW (USA LICENSE)

AIRCRAFT MFR.	AIRCRAFT DESIGNATION	GROSS WEIGHT	ENGINE DESIG.	ENGINE TYPE	ENGINE POWER	ENGINE MFR.	AIRCRAFT MISSION
NETHERLANDS							
FOKKER[1]	NH-90	19,000	RTM 322	(2) TS	2,300 SHP	R-R/T-M	In Development
POLAND							
WSK-PZL	MI-2	7,800	GTD-350	(2) TS	396 SHP	KLIMOV	MILITARY
SWIDNIK	MI-2 (KUNIA)	7,800	C250-C20B	(2) TS	420 SHP	ALLISON	MILITARY
	MI-2M	7,800	GTD-350	(2) TS	396 SHP	KLIMOV	MILITARY
	W-3 SOKOL (FALCON)	15,000	PZL-10W (TWIN)[†]	(2) TS	1,780 SHP	PZL	CIVIL
	[†] 890 SHP EACH						
RUMANIA							
GOVERNMENT (FRENCH LICENSE)	SA 319 (ALOUETTE-III)	4,900	ASTAZOU II	(1) TS	860 SHP	T-M	CIVIL/MIL
SOUTH AFRICA							
ATLAS	ORYX (PUMA)	14,100	TURMO IV	(2) TS	1,450 SHP	T-M	CIVIL/SAR
AIRCRAFT	ROOIVALK CHS-2 (RED FALCON)	20,065	TOPAZ	(2) TS	1,600 SHP	ATLAS	MIL (In Dev.)
TURKEY							
BELL (TURKEY)	205	9,500	T53-13A	(1) TS	1,400 SHP	TEXTRON	CIVIL/MIL
YUGOSLAVIA							
GOVERNMENT (FRENCH LICENSE)	SA 342 (GAZELLE)	4,415	ASTAZOU 14M	(1) TS	590 SHP	T-M	ATTACK

MULTI-NATIONAL COMPANIES (SUPERSCRIPT DESIGNATORS):

1. AEROSPATIALE, FRANCE; AGUSTA, ITALY; MBB, GERMANY; FOKKER, NETHERLANDS.
2. AEROSPATIALE, FRANCE; WESTLAND, GREAT BRITAIN.
3. EUROCOPTER GMBH: MBB DIVISION OF DEUTSCHE AEROSPACE, GERMANY; AND AEROSPATIALE, FRANCE.
4. MBB, GERMANY; JAPAN.
5. WESTLAND, GREAT BRITAIN; AGUSTA, ITALY.
6. AGUSTA, ITALY; SIKORSKY, USA.
7. AGUSTA, ITALY; BELL USA.

USA AGRICULTURAL AIRCRAFT
(ALPHABETICAL BY MANUFACTURER)

AIRCRAFT MFR.	AIRCRAFT DESIGNATION	GROSS WEIGHT	ENGINE DESIG.	ENGINE TYPE	ENGINE POWER	ENGINE MFR.
AIR TRACTOR INC.	AT 400	7,900	PT6A-15AG	(1) TP	715 ESHP	P&WC
	AT 402	8,200	PT6A-34AG	(1) TP	783 ESHP	P&WC
	AT 502	8,500	PT6A-11AG	(1) TP	680 ESHP	P&WC
	AT 503	10,500	PT6A-45R	(1) TP	1100 ESHP	P&WC
AYRES CORP. (TURBO THRUSH)	S2R (T-11)	6,200	PT6A-11AG	(1) TP	528 ESHP	P&WC
	S2R (T-15)	6,200	PT6A-15AG	(1) TP	680 ESHP	P&WC
	S2R (T-34)	6,200	PT6A-34AG	(1) TP	783 ESHP	P&WC
	S2R (T-65)	10,500	PT6A-65AG	(1) TP	1,230 ESHP	P&WC
FRAKES INC.	TURBO-CAT	8,200	PT6A-34	(1) TP	783 ESHP	P&WC (RE-ENGINE)
	MOHAWK-298	23,370	PT6A-45	(2) TP	1,254 ESHP	P&WC
GRUMMAN & SCHWEIZER	G-164B (AG-CAT)	7,000	PT6-11AG	(1) TP	528 ESHP	P&WC
RILEY- CESSNA	421 TURBINE ROCKET	7,650	LTP 101-700	(1) TP	700 ESHP	TEXTRON (RE-ENGINE)
	421 TURBINE EAGLE	7,700	LTP 101-700	(1) TP	700 ESHP	TEXTRON (RE-ENGINE)
WEATHERLY CORP.	620-TP	6,300	PT6-11AG	(1) TP	528 ESHP	P&WC
	620A-TP	6,300	PT6-15AG	(1) TP	680 ESHP	P&WC

USA BUSINESS-TURBOPROP AIRCRAFT
(ALPHABETICAL BY MANUFACTURER)

AIRCRAFT MFR.	AIRCRAFT DESIGNATION	GROSS WEIGHT	ENGINE DESIG.	ENGINE TYPE	ENGINE POWER	ENGINE MFR.
ADVANCED AERODYNAMICS	JET CRUZER (In Dev.)	4,500	PT6A-27	(2) TP	750 ESHP	P&WC
AERO MODS INC.	TURBO DC-3 (RE-ENGINE)	28,750	PT6A-67R	(2) TP	1,509 ESHP	P&WC
AVTEK INC.	AVTEK 400	6,500	PT6A-135A	(2) TP	750 ESHP	P&WC
BASLER INC.	TURBO DC-3 (RE-ENGINE)	28,750	PT6A-67R	(2) TP	1,509 ESHP	P&WC
BEECH CORP.	B-99 AIRLINER	10,900	PT6A-28	(2) TP	680 ESHP	P&WC
	BE-1900	20,000	PT6A-65/67	(2) TP	1,280 ESHP	P&WC
	BONANZA (RE-ENGINE)	3,500	250-B17	(1) TP	420 ESHP	ALLISON
	EXEC-LINER	16,600	PT6A-65B	(2) TP	1,174 ESHP	P&WC
	KING AIR A,B,C-90	10,100	PT6A-21	(2) TP	610 ESHP	P&WC
	KING AIR F-90	11,100	PT6A-135	(2) TP	780 ESHP	P&WC
	KING AIR A-100	10,100	PT6A-28	(2) TP	715 ESHP	P&WC
	KING AIR BE-100	10,100	TPE331-6	(2) TP	715 ESHP	GARRETT
	STARSHIP BE-2000	14,900	PT6A-67A	(2) TP	1,272 ESHP	P&WC
	SUPER KING BE-200	12,500	PT6A-42	(2) TP	788 ESHP	P&WC
	SUPER KING BE-300	14,000	PT6A-60A	(2) TP	1,050 ESHP	P&WC
	SUPER KING BE-350	15,100	PT6A-60A	(2) TP	1,050 ESHP	P&WC
BROMON ACFT	BR-2004	31,750	CT7-9	(2) TP	1,225 ESHP	G.E.
CESSNA CORP.	205B CARGOMASTER	6,000	PT6A-114	(1) TP	632 ESHP	P&WC
	208 CARAVAN-I	8,035	PT6A-114	(1) TP	632 ESHP	P&WC
	208B GRAND CARAVAN	8,750	PT6A-114	(1) TP	632 ESHP	P&W
	425 CONQUEST-I	8,850	PT6A-112	(2) TP	500 ESHP	P&WC
	441 CONQUEST-II	9,950	TPE331-8/10	(1) TP	950 ESHP	GARRETT
	N-22 NOMAD	8,950	250-B17C	(2) TP	420 ESHP	ALLISON
CESSNA-REIMS	CARAVAN-II	9,435	PT6A-112	(2) TP	500 ESHP	P&WC
COX AVIATION	TURBO OTTER	8,000	PT6A-135	(2) TP	787 ESHP	P&WC
FAIRCHILD AIRCRAFT	MERLIN-II & III	12,500	TPE331-10	(2) TP	810 ESHP	GARRETT
	MERLIN-IV	14,500	TPE331-11	(2) TP	1,100 ESHP	GARRETT
	METRO-II	12,500	TPE331-11/12	(2) TP	1,100 ESHP	GARRETT
	METRO-III	15,500	TPE331-12	(2) TP	1,100 ESHP	GARRETT
	METRO-VI	17,000	TPE331-14	(2) TP	1,250 ESHP	GARRETT
	METRO-23	14,500	TPE331-11/12	(2) TP	1,100 ESHP	GARRETT
	METRO-23EF (IN DEV.)	16,500	TPE331-12UAR	(2) TP	1,100 ESHP	GARRETT
FAIRCHILD-FOKKER	FH-227 (F-27 Upgrade)	45,000	DART 532	(2) TP	2,200 ESHP	R-R
FAIRCHILD-SAAB	SAAB-340	28,000	CT7-7/9	(2) TP	1,800 ESHP	G.E.
FRAKES INC.	MALLARD	11,000	PT6A-45	(1) TP	1,254 ESHP	P&WC
GULFSTREAM CORP.	G-1 & 1C	60,000	DART 529	(2) TP	2,200 ESHP	R-R
	COMMANDER 840	10,800	TPE331-5-2	(2) TP	715 ESHP	GARRETT
	COMMANDER 980	10,325	TPE331-10	(2) TP	950 ESHP	GARRETT
	COMMANDER 1000	11,200	TPE331-10	(2) TP	950 ESHP	GARRETT
HELIO ACFT.	STALLION H-600	5,100	PT6A-27	(1) TP	680 ESHP	P&WC
IMP CORP.	TURBO TRACKER	23,000	PT6A-67A	(2) TP	1,283 ESHP	P&WC
INTERCEPTOR CORP	INTERCEPTOR 400	4,030	TPE331	(1) TP	400 ESHP	GARRETT
JAMES AVIATION	FLETCHER 1060	5,400	PT6A-34	(1) TP	750 ESHP	P&WC
LEAR CORP.	LEARFAN 2100	7,350	PT6B-35F	(2) TP	715 ESHP	P&WC
MARSH-GRUMMAN (RE-ENGINE)	S-2T (TURBO-TRACKER)	26,147	TPE331-15	(2) TP	1,645 ESHP	GARRETT
MAUL CORP.	M7-420 (RE-ENGINE)	2,750	250-B17	(1) TP	420 ESHP	ALLISON
	MX7-420 (RE-ENGINE)	2,500	250-B17	(1) TP	420 ESHP	ALLISON
MOONEY INC.[13]	TBM-700	6,007	PT6A-64	(1) TP	747 ESHP	P&WC
OMAC INC.	LASER 300	8,300	PT6A-135A	(2) TP	750 ESHP	P&WC
OMNI AVIATION	TURBO TITAN	8,400	PT6A-34	(2) TP	750 ESHP	P&WC

AIRCRAFT MFR.	AIRCRAFT DESIGNATION	GROSS WEIGHT	ENGINE DESIG.	ENGINE TYPE	ENGINE POWER	ENGINE MFR.
PIPER CORP.	CHEYENNE I	8,700	PT6A-11	(2) TP	528 ESHP	P&WC
	CHEYENNE II	9,050	PT6A-28	(2) TP	715 ESHP	P&WC
	CHEYENNE II XL	9,400	PT6A-135	(2) TP	780 ESHP	P&WC
	CHEYENNE III	11,200	PT6A-61	(2) TP	720 ESHP	P&WC
	CHEYENNE 400	12,050	TPE331-14A	(2) TP	1,250 ESHP	GARRETT
	TP-400 (TURBINE MALIBU)	5,300	PT6A-61	(2) TP	600 ESHP	P&WC
PIAGGIO INC.	AVANTI	10,800	PT6A-66	(2) TP	905 ESHP	P&WC
POTEZ INC.	841	9,000	PT6A-6	(2) TP	520 ESHP	P&WC
PRECISION AIRMOTIVE (FORMERLY ROCKWELL)	COMMANDER 680	10,300	TPE331-1	(2) TP	665 ESHP	GARRETT
	COMMANDER 690	10,300	TPE331-1	(2) TP	665 ESHP	GARRETT
	COMM. 840/900/980	10,700	TPE-331-5	(2) TP	715 ESHP	GARRETT
	COMMANDER 1000	11,250	TPE331-10	(2) TP	950 ESHP	GARRETT
	AERO-COMMANDER	10,300	PT6A-21	(2) TP	580 ESHP	P&WC
SCHAFER INC.	COMANCHERO 500	8,000	PT6A-20	(2) TP	610 ESHP	P&WC
	COMANCHERO 750	8,500	PT6A-135	(2) TP	787 ESHP	P&WC
	DC-3-65TP	26,900	PT6A-65AR	(2) TP	1,230 ESHP	P&WC
SKY TRADER INC.	ST-1400	13,000	TPE331-5	(2) TP	715 ESHP	GARRETT
	ST-1700	13,400	PT6A-45R	(2) TP	850 ESHP	P&WC
SMITH AERO INC.	PROPJET XP99	5,182	PT6A-42	(1) TP	550 ESHP	P&WC
SNOW AVIATION	SA-204C	51,830	PW-126A	(2) TP	2,652 ESHP	P&WC
SOLOY CORP.	CESSNA-206 (T-PROP CONV.)	3,800	250-C20S	(1) TP	420 SHP	ALLISON
	CESSNA 207 (T-PROP CONV.)	4,000	250-C20S	(1) TP	420 SHP	ALLISON
VOPAR INC.	TURBO-18	10,300	TPE331-1	(2) TP	715 ESHP	GARRETT
	CARGO LINER	11,500	TPE331-1	(2) TP	715 ESHP	GARRETT

USA BUSINESS-JET AIRCRAFT
(ALPHABETICAL BY MANUFACTURER)

AIRCRAFT MFR.	AIRCRAFT DESIGNATION	GROSS WEIGHT	ENGINE DESIG.	ENGINE TYPE	ENGINE POWER	ENGINE MFR.
ASTRA JET CORP.	ASTRA 1125	24,650	TFE-731-3	(2) TF	3,700 LBT	GARRETT
BEDE CORP.	BD-5J	970	TRS-18	(1) TJ	200 LBT	M-T
BEECH CORP.	BEECHJET 400 (DIAMOND-II)	16,300	JT15D-5	(2) TF	2,900 LBT	P&WC
CESSNA CORP.	CITATION-I	13,300	JT15D-1	(2) TF	2,500 LBT	P&WC
	CITATION-II	14,300	JT15D-4	(2) TF	2,500 LBT	P&WC
	CITATION-SII	15,300	JT15D-4B	(2) TF	2,500 LBT	P&WC
	CITATION-III, VI	22,200	TFE731-3B-100	(2) TF	3,650 LBT	GARRETT
	CITATION-IV, VII	24,200	TFE731-4R2S	(2) TF	4,000 LBT	GARRETT
	CITATION-V	16,100	JT15D-5A	(2) TF	2,900 LBT	P&WC
	CITATION-JET 525	10,400	FJ44	(2) TF	1,900 LBT	WILLIAMS
	CITATION-X	31,000	GMA-3007A	(2) TF	6,000 LBT	ALLISON (In Dev.)
GARRETT AVIATION	FALCON-20 (MOD.)	29,100	TFE731-5BR	(2) TF	4,500 LBT	GARRETT
GULFSTREAM CORP.	G-II	62,100	SPEY MK.511-8	(2) TF	11,400 LBT	R-R
	G-III	70,200	SPEY MK.511-8	(2) TF	11,400 LBT	R-R
	G-II/III (FIRST AMERICAN INC)	70,200	TAY-611	(2) TF	13,800 LBT	R-R (RE-ENGINE)
	G-IV	73,600	TAY-650	(2) TF	13,850 LBT	R-R
LEARJET (BOMBARDIER) CORP.	LEAR 23/24	14,000	CJ610-6	(2) TJ	2,950 LBT	G.E.
	LEAR 25/28/29	15,000	CJ610-8	(2) TJ	3,100 LBT	G.E.
	LEAR 31	16,700	TFE731-2	(2) TF	3,500 LBT	GARRETT
	LEAR 35A	18,250	TFE731-2	(2) TF	3,500 LBT	GARRETT
	LEAR 36A	18,500	TFE731-2	(2) TF	3,500 LBT	GARRETT
	LEAR 28/55C (LONGHORN)	21,250	TFE731-3A	(2) TF	3,700 LBT	GARRETT
	LEAR 60	23,000	PW 305	(2) TF	4,400 LBT	P&WC (IN DEV.)
LOCKHEED CORP.	JETSTAR I	20,372	JT12A-5	(4) TJ	3,000 LBT	P&W
	JETSTAR II	20,372	TFE731-3	(4) TF	3,700 LBT	GARRETT (RE-ENGINE)
	JETSTAR 731	20,372	TFE731-3	(4) TF	3,700 LBT	GARRETT (RE-ENGINE)
SABRELINER CORP.	40	19,900	TFE731-3R	(2) TF	3,700 LBT	GARRETT
	60	20,300	JT12A-8	(2) TF	3,300 LBT	P&W
	65	24,000	TFE731-3R	(2) TF	3,700 LBT	GARRETT
	75A	23,500	CF700-2D2	(2) TF	4,500 LBT	G.E.
	80	33,000	CF700-2D2	(2) TF	4,500 LBT	G.E.
	85	33,000	CF700-2D2	(2) TF	4,500 LBT	G.E.
SWEARINGEN	SJ 30 (FANJET)	9,250	FJ 44	(2) TF	1,900 LBT	WILLIAMS (In Dev.)
VOPAR CORP.	FALCON-20 (RE-ENGINE)	29,100	PW305-F2	(2) TF	5,225 LBT	P&WC

USA COMMERCIAL CARGO-JET AIRCRAFT
(ALPHABETICAL BY MANUFACTURER)

AIRCRAFT MFR.	AIRCRAFT DESIGNATION	GROSS WEIGHT	ENGINE DESIG.	ENGINE TYPE	ENGINE POWER	ENGINE MFR.
BOEING	B707-320	336,000	JT3D-7	(4) TF	19,000 LBT	P&W
	B707-320	336,000	JT3D-7	(4) TF	19,000 LBT	P&W
	B727-100	166,000	TAY-650	(3) TF	15,000 LBT	R-R
	B727-100C	235,000	JT8D-7	(4) TF	14,500 LBT	P&W
	B727-200F	166,000	JT8D-17A	(3) TF	16,000 LBT	P&W
	B737-200	103,000	JT8D-15A	(2) TF	15,500 LBT	P&W
	B737-200C	107,000	JT8D-17A	(2) TF	16,000 LBT	P&W
	B737-300	139,000	CFM56-3B1	(2) TF	20,000 LBT	G.E.
	B737-400	139,000	CFM56-3B2	(2) TF	22,000 LBT	G.E.
	B737-500	134,000	CFM56-3B1	(2) TF	22,000 LBT	G.E.
	B747-200F	630,000	JT9D-7R	(4) TF	48,000 LBT	P&W
	B747-200C	585,000	JT9D-7R	(4) TF	48,000 LBT	P&W
	B747-300	605,000	JT9D-7R	(4) TF	48,000 LBT	P&W
	B747-400	605,000	CF6-80C2B1F	(4) TF	57,900 LBT	G.E.
	B757-200F	210,000	PW2037	(2) TF	38,250 LBT	P&W
	B757-200F	210,000	RB211-535E4	(2) TF	43,100 LBT	R-R
CESSNA	CARAVAN-1	8,350	PT6A-114	(2) TF	632 ESHP	P&WC
CONVAIR	CV-540/580	54,000	501D-13	(2) TP	3,750 ESHP	ALLISON
	CV-600/640	50,000	DART 542	(2) TP	3,750 ESHP	R-R
FAIRCHILD	SA227-AT (EXPEDITER)	16,500	TPE331-11	(2) TP	1,100 ESHP	GARRETT
LOCKHEED	L100-30 (SUPER HERCULES)	135,000	501D-22A	(4) TP	4,500 ESHP	ALLISON
	L188 (ELECTRA)	116,000	501D-22A	(2) TP	4,500 ESHP	ALLISON
McDONNELL DOUGLAS	DC-8F (JET TRADER)	258,000	JT3D-3	(4) TF	18,000 LBT	P&W
	DC-8-61 (JET TRADER)	258,000	JT3D-3B	(4) TF	18,000 LBT	P&W
	DC-8-62 (JET TRADER)	250,000	JT3D-7	(4) TF	19,000 LBT	P&W
	DC-8-63 (JET TRADER)	275,000	JT3D-7	(4) TF	19,000 LBT	P&W
	DC-9-30CF	110,000	JT8D-11	(2) TF	15,000 LBT	P&W
	DC-10 (SERIES 10)	363,500	CF6-6D	(3) TF	40,000 LBT	G.E.
	DC-10 (SERIES 30)	436,000	CF6-50C2B	(3) TF	54,000 LBT	G.E.
	MD-11	458,000	PW 4360	(3) TF	60,000 LBT	P&W
	MD-11F	471,500	CF6-80C2-D1	(3) TF	61,500 LBT	G.E.

USA COMMERCIAL PASSENGER-JET AIRCRAFT
(ALPHABETICAL BY MANUFACTURER)

AIRCRAFT MFR.	AIRCRAFT DESIGNATION	GROSS WEIGHT	ENGINE DESIG.	ENGINE TYPE	ENGINE POWER	ENGINE MFR.
BEECH	1900	16,900	PT6A-65B	(2) TP	1,174 ESHP	P&WC
BOEING	B707-120B	258,000	JT3D-3	(4) TF	19,000 LBT	P&W
	B707-320C	336,000	JT3D-7	(4) TF	19,000 LBT	P&W
	B727-100	170,500	JT8D-1,7,9	(3) TF	14,500 LBT	P&W
	B727-200	204,500	JT8D-15A	(3) TF	15,000 LBT	P&W
	B727 (RE-ENGINE)	204,500	TAY-650	(3) TF	15,100 LBT	R-R
	B737-100	101,000	JT8D-7,9	(2) TF	14,500 LBT	P&W
	B737-200	116,000	JT8D-15A	(2) TF	15,500 LBT	P&W
	B737-300	139,000	CFM56-3-B1	(2) TF	20,000 LBT	CFM INT'L
	B737-400	150,000	CFM56-3C	(2) TF	23,500 LBT	CFM INT'L
	B737-500	134,000	CFM56-3B-1	(2) TF	22,000 LBT	CFM INT'L
	B747-100B (SUPERJET)	830,000	CF6-45A	(4) TF	46,500 LBT	G.E.
	B747-SR (SUPERJET)	830,000	CF6-45A	(4) TF	46,500 LBT	G.E.
	B747-200B (SUPERJET)	830,000	JT9D-7R4E3	(4) TF	50,000 LBT	P&W
	B747-200 (CONVERTIBLE)	833,000	JT9D-7R4E3	(4) TF	50,000 LBT	P&W
	B747-300B (STRETCH)	830,000	JT9D-7R4E3	(4) TF	50,000 LBT	P&W
	B747-SP (SUPERJET)	730,000	RB211-524D4	(4) TF	53,000 LBT	R-R
	B747-400 (ADV. SUPERJET)	870,000	CF6-80C2B1A	(4) TF	57,900 LBT	G.E.
	B747-400	870,000	RB211-524H	(4) TF	60,600 LBT	R-R
	B757-200	250,000	PW 2037	(2) TF	38,250 LBT	P&W
	B757-200	250,000	RB211-535E4	(2) TF	43,100 LBT	R-R
	B767-200	315,000	JT9D-7R4E	(2) TF	50,000 LBT	P&W
	B767-200ER	351,000	JT9D-7R4E	(2) TF	50,000 LBT	P&W
	B767-200ER	351,000	CF6-80A	(2) TF	48,000 LBT	G.E.

AIRCRAFT MFR.	AIRCRAFT DESIGNATION	GROSS WEIGHT	ENGINE DESIG.	ENGINE TYPE	ENGINE POWER	ENGINE MFR.
BOEING (CONT'D)	B767-300	351,000	JT9D-7R4E	(2) TF	50,000 LBT	P&W
	B767-300	351,000	CF6-80A	(2) TF	48,000 LBT	G.E.
	B767-300ER	400,000	PW4052	(2) TF	57,900 LBT	P&W
	ALTERNATE ENGINE	400,000	CF6-80C2	(2) TF	52,700 LBT	G.E.
	B777	590,000	GE-90	(2) TF	80,000 LBT	G.E. (IN DEV.)
	B777	590,000	TRENT-800	(2) TF	80,000 LBT	R-R (IN DEV.)
FAIRCHILD	SA227-AT (EXPEDITER)	16,500	TPE331-11	(2) TP	1,100 ESHP	GARRETT
CONVAIR	CV-540/580	53,000	501D-13D	(2) TP	3,750 ESHP	ALLISON
	CV-600/640	54,000	DART-542	(2) TP	3,750 ESHP	R-R
	CV-990	253,000	CJ805-23	(4) TF	16,500 LBT	G.E.
LOCKHEED CALIF.	L100-30 (SUPER HERCULES)	135,000	501D-22A	(4) TP	4,500 ESHP	ALLISON
	L188 (ELECTRA)	116,000	501D-22A	(4) TP	4,500 ESHP	ALLISON
	L1011-1 (TRISTAR)	430,000	RB211-22B	(3) TF	42,000 LBT	R-R
	L1011-100 (TRISTAR)	466,000	RB211-22B	(3) TF	42,000 LBT	R-R
	L1011-200 (TRISTAR)	466,000	RB211-524B42	(3) TF	50,000 LBT	R-R
	L1011-250 (TRISTAR)	496,000	RB211-524B42	(3) TF	50,000 LBT	R-R
	L1011-500 (TRISTAR)	504,000	RB211-524B42	(3) TF	50,000 LBT	R-R
McDONNELL DOUGLAS	DC-8-30	315,000	JT4A-9,11	(4) TJ	17,500 LBT	P&W
	DC-8-40	315,000	CONWAY-12	(4) TF	17,500 LBT	R-R
	DC-8-50	325,000	JT3D-3B	(4) TF	18,000 LBT	P&W
	DC-8-61	325,000	JT3D-3B	(4) TF	18,000 LBT	P&W
	DC-8-62	350,000	JT3D-7	(4) TF	19,000 LBT	P&W
	DC-8-63	350,000	JT3D-7	(4) TF	19,000 LBT	P&W
	DC-9-10	90,100	JT8D-1/7/209	(2) TF	14,500 LBT	P&W
	DC-9-20	98,300	JT8D-9/209	(2) TF	14,500 LBT	P&W
	DC-9-30	121,100	JT8D-9/209	(2) TF	14,500 LBT	P&W
	DC-9-40	121,000	JT8D-11/15/209	(2) TF	15,000 LBT	P&W
	DC-9-50	140,000	JT8D-15,17/209	(2) TF	16,000 LBT	P&W
	DC-10-10	440,500	CF6-6D	(3) TF	40,000 LBT	G.E.
	DC-10-15	440,500	CF6-50C2F	(3) TF	54,000 LBT	G.E.
	DC-10-30	580,000	CF6-50C2	(3) TF	54,000 LBT	G.E.
	DC-10-40	580,000	JT9D-59A	(3) TF	53,000 LBT	P&W
	MD-11	602,500	PW4460	(3) TF	60,000 LBT	P&W
	MD-11	602,500	CF6-80C2-CF	(3) TF	61,500 LBT	G.E.
	MD-11ER	602,000	CF6-80C2	(3) TF	60,000 LBT	G.E.
	MD-12	670,000	CF6-80C2	(4) TF	60,000 LBT	G.E. (IN DEV.)
	MD-81	140,000	JT8D-217219	(2) TF	19,250 LBT	P&W
	MD-82	150,000	JT8D-217/219	(2) TF	20,850 LBT	P&W
	MD-83	163,500	JT8D-219	(2) TF	21,700 LBT	P&W
	MD-87	147,000	JT8D-217/219	(2) TF	20,850 LBT	P&W
	MD-88	150,000	JT8D-219	(2) TF	20,850 LBT	P&W
	MD-90-30	156,000	V2500-D1	(2) TF	25,000 LBT	IAE
	MD-90-40	163,000	V2500-D5	(2) TF	25,000 LBT	IAE

USA ROTARY WING AIRCRAFT
(ALPHABETICAL BY MANUFACTURER-CIVIL AND MILITARY)

AIRCRAFT MFR.	AIRCRAFT DESIGNATION	GROSS WEIGHT	ENGINE DESIG.	ENGINE TYPE	ENGINE POWER	ENGINE MFR.	POPULAR NAME	TYPE SERVICE
AEROSPATIALE	HH-65A (AS565)	8,900	ARRIEL 1C1	(2) TS	724 SHP	T-M	DOLPHIN	USCG
	ALTERNATE ENGINE	8,900	T800-LHT-800	(2) TS	IN DEV.	LHTEC	DOLPHIN	USCG
	ALTERNATE ENGINE	8,900	LTS101-750	(2) TS	684 SHP	TEXTRON	DOLPHIN	USCG
BELL-TEXTRON HELICOPTER	204 (UH-1H / HH-1H)	9,500	T53L-13B	(1) TS	1,400 SHP	TEXTRON	IROQUOIS-HUEY	CIVIL/MIL
	205 (UH-1 SERIES)	9,500	T53L-13A	(1) TS	1,400 SHP	TEXTRON	IROQUOIS	CIVIL/MIL
	206 (OH-58)	3,200	250-C20B	(1) TS	420 SHP	ALLISON	KIOWA	USAR
	206 (OH-58A)	3,000	T703-A-700	(1) TS	650 SHP	ALLISON	KIOWA	USAF
	206 (OH-58D)	3,200	250-C30R	(1) TS	650 SHP	ALLISON	KIOWA WARRIOR	USAF
	206A	3,200	T63A-700	(1) TS	420 SHP	ALLISON	JET RANGER	CIVIL
	206B (TH-57)	3,200	250-C20J	(1) TS	420 SHP	ALLISON	SEA/JET RANGER	USAR/USN
	206L-1	4,150	250-C28	(1) TS	550 SHP	ALLISON	LONG RANGER 1	CIVIL
	206L-3	4,150	250-C30	(1) TS	650 SHP	ALLISON	LONG RANGER 3	CIVIL
	206L-3 (TWIN RE-ENGINE)	4,150	250-C30P	(2) TS	420 SHP	ALLISON	GEMINI	In Dev.
	209 (AH-1G)	9,500	T53L-13B	(1) TS	1,400 SHP	TEXTRON	HUEY COBRA	USAR
	209 (AH-1J)	10,000	T400-CP-400	(2) TS	†1,800 SHP	P&WC	SEA COBRA	USN/USMC
	209 (AH-1S)	10,000	T53L-703	(1) TS	1,485 SHP	TEXTRON	UPGUN COBRA	USAR
	209 (AH-1S)	10,000	T53L-703	(1) TS	1,485 SHP	TEXTRON	MOD. COBRA	USAR
	209 (AH-1T)	14,000	T400-CP-402	(2) TS	††1,850 SHP	P&WC	SEA COBRA MOD.	USN
	209 (AH-1W)	14,700	T700-GE-401	(2) TS	1,690 SHP	G.E.	SUPER COBRA	USMC
	212 (UH-1N)	10,500	T400-CP-400	(2) TS	†1,800 SHP	P&WC	TWIN HUEY	CIVIL/MIL
	214A/B	17,500	T5508D	(2) TS	2,250 SHP	TEXTRON	BIG LIFTER	MIL
	214ST	17,500	CT7-2A	(2) TS	1,725 SHP	G.E.	SUPER TRANS'PT	CIVIL/MIL
	222	7,850	LTS101-650	(2) TS	650 SHP	TEXTRON	NONE	CIVIL
	222B	8,250	LTS101-750C	(2) TS	735 SHP	TEXTRON	NONE	CIVIL

† TWIN-PAC ENGINE, 900 SHP EACH - †† 925 SHP EACH

AIRCRAFT MFR.	AIRCRAFT DESIGNATION	GROSS WEIGHT	ENGINE DESIG.	ENGINE TYPE	ENGINE POWER	ENGINE MFR.	POPULAR NAME	TYPE SERVICE
BELL (CON'T)	222UT	8,250	LTS101-750C	(2) TS	735 SHP	TEXTRON	UTILITY TWIN	CIVIL
	222 (RE-ENGINE)	8,250	250-C30G	(2) TS	650 SHP	ALLISON	NONE	CIVIL
	230 (UPGRADE 222)	8,250	250-C30G2	(2) TS	700 SHP	ALLISON	NONE	CIVIL
	406 (OH-58D)	4,500	250-C30R	(1) TS	650 SHP	ALLISON	AERO SCOUT	USAF
	406 (OH-58D)	5,000	T703-A-700	(1) TS	650 SHP	ALLISON	COMBAT SCOUT	USAR
	412 SP	11,900	PT6T-3B/6	(2) TS	†1,800 SHP	P&WC	NONE	CIVIL
	901 (V-22)	47,000	T406-AD-400	(2) TS	2,650 SHP	ALLISON	OSPREY	In Dev.
BOEING HELICOPTERS	CH-46/UH-46	23,000	T55L-11	(2) TS	3,700 SHP	TEXTRON	CHINOOK	USMC,USN
	H-46	23,000	T55L-11	(2) TS	3,700 SHP	TEXTRON	SEA KNIGHT	USMC
	AH-47E	24,300	T58-GE-16	(2) TS	1,870 SHP	G.E.	SEA KING	USN/USMC
	107 (CH-46)	23,000	T58-GE-16	(2) TS	1,870 SHP	G.E.	SEA KING	USN/USMC
	145 (CH-47D)	50,000	T55L-712	(2) TS	3,750 SHP	TEXTRON	CHINOOK	USAR
	145 (MH-47E)	54,000	T55L-712	(2) TS	3,750 SHP	TEXTRON	CHINOOK	USAF
	234	48,500	AL5512	(2) TS	4,075 SHP	TEXTRON	CHINOOK-234	CIVIL
	414-100	54,000	T55L-712S/SB	(2) TS	3,750 SHP	TEXTRON	CHINOOK INT'L	CIVIL/ MIL
BOEING-SIKORSKY	RAH-66	CLASSIFIED	T800-LHT-800	(1) TS	1,200 SHP	LHTEC	COMANCHE	IN DEV.
ENSTROM HELICOPTER	TH-28A	2,700	250-C20W	(1) TS	225 SHP	ALLISON	NONE	USAR
	480 (TH-28)	2,700	250-C20W	(1) TS	285 SHP	ALLISON	NONE	In Dev.
JAFFE CORP.	BELL-222 SP	7,850	250-C30	(2) TS	650 SHP	ALLISON	RE-ENGINE	CIVIL
KAMAN HELICOPTER	HUSKY (HH-43)	8,400	T53L-11	(1) TS	1,100 SHP	TEXTRON	HUSKY	USAF
	K-860 (H-2)	12,800	T58-GE-8F	(2) TS	1,350 SHP	G.E.	SEASPRITE	SAR/ASW
	K-888 (SH-2G)	13,500	T700-GE-401	(2) TS	1,622 SHP	G.E.	SUPER SEASPRITE	USN
	K-894 (SH-2F)	13,500	T700-GE-401	(2) TS	1,622 SHP	G.E.	SUPER SEASPRITE	USN
LOCKHEED HELICOPTER	MODEL 86	11,000	PT6T-6	(2) TS	†1,800 SHP	P&WC	NONE	USAR
	AH-56A	18,300	T64-GE-716	(1) TS	3,400 SHP	G.E.	CHEYENNE	USAR
McDONNELL-DOUGLAS HELICOPTER (MDHC)	AH-64A	21,000	T700-GE-701C	(2) TS	1,800 SHP	G.E.	APACHE	USAR
	MD-500E (OH-6)	3,000	T-63A-720	(1) TS	420 SHP	ALLISON	CAYUSE-NOTAR	CIVIL/MIL
	MD-500F (AH-6)	3,500	250-C30	(1) TS	425 SHP	ALLISON	LITTLE BIRD-NOTAR	USAR
	MD-520N	3,300	250-C20R	(1) TS	450 SHP	ALLISON	BLACK TIGER-NOTAR	CIVIL/MIL
	MD-530 (H-6-530)	3,100	T703-A-700	(1) TS	650 SHP	ALLISON	DEFENDER-NOTAR	USAR
	MD-530-MG	3,100	250-C30	(1) TS	425 SHP	ALLISON	NIGHTFOX-NOTAR	USAR
	MD-900 (MDX)	5,600	PW206A	(2) TS	603 SHP	P&WC	EXPLORER-NOTAR	IN.DEV.
ROGERSON-HILLER CORP. (FORMERLY FAIRCHILD-HILLER)	TWINJET	5,000	TSE 231	(2) TS	474 SHP	GARRETT	HORNET	CIVIL
	H-1100	2,750	250-C20B	(1) TS	420 SHP	ALLISON	NONE	CIVIL/AG
	RH-1100 (OH5A)	2,750	T-63A-720	(1) TS	425 SHP	ALLISON	HORNET	CIVIL/MIL
	RH-1100S	3,300	250-C20B	(1) TS	420 SHP	ALLISON	NONE	MIL
	1099	11,000	PT6B-3	(1) TS	900 SHP	P&WC	NONE	USMC
SCHWEIZER	MODEL-330 (TH-330)	2,050	250-C20W	(1) TS	225 SHP	ALLISON	In Development	USAR/CIVIL
SIKORSKY	CH-3B (S-61R)	21,000	T58-GE-10	(2) TS	1,400 SHP	G.E.	NONE	USAF
	SH-3B (S-61)	21,000	T58-GE-10	(2) TS	1,400 SHP	G.E.	SEA KING	USN
	SH-3D/G/F (S-61)	21,000	T58-GE-10/16	(2) TS	1,400 SHP	G.E.	JOLLY GREEN	USAR
	CH-53D (SH-65A)	42,000	T64-GE-419	(3) TS	4,750 SHP	G.E.	SEA STALLION	USAR
	CH-53E (SH-65A)	69,000	T64-GE-419	(3) TS	4,750 SHP	G.E.	SUPER STALLION	USAF
	HH-53A (S-62)	7,900	CT-58-140	(2) TS	1,500 SHP	G.E.	NONE	USAR
	HH-53B	42,000	T64-GE-7A	(3) TS	4,350 SHP	G.E.	SUPER JOLLY	USAF
	HH-53 (SH-65A)	42,000	T64-GE-419	(3) TS	4,750 SHP	G.E.	PAVE LOW	USAF
	MH-53E (SH-65A)	42,000	T64-GE-419	(3) TS	4,750 SHP	G.E.	SEA DRAGON	USN/USMC
	CH-54B (S-64)	42,000	JTFD-12A	(3) TS	2,830 SHP	P&W	SKYCRANE-TARHE	USAR
	HH-60 (S-70)	20,200	T700-GE-701	(2) TS	1,698 SHP	G.E.	NIGHT HAWK	USN/USCG
	HH-60 (S-70)	20,200	T700-GE-701	(2) TS	1,698 SHP	G.E.	PAVE HAWK	USAF
	HH-60J (S-70)	20,200	T700-GE-701	(2) TS	1,698 SHP	G.E.	JAYHAWK	USCG
	SH-60J (S-70B)	21,000	T700-GE-401	(2) TS	1,622 SHP	G.E.	SEAHAWK	USN/USMC
	SH-60F (S-70B)	21,000	T700-GE-401C	(2) TS	1,900 SHP	G.E.	OCEANHAWK	USN
	UH-60A (S-70A)	22,000	T700-GE-700	(2) TS	1,690 SHP	G.E.	BLACKHAWK	USAR
	VH-60 (S-70)	22,000	T700-GE-700	(2) TS	1,622 SHP	G.E.	BLACKHAWK	VIP-USMC
	S-61	22,000	CT-58-140	(2) TS	1,500 SHP	G.E.	SILVER	CIVIL
	S-62	7,900	CT-58-140	(1) TS	1,500 SHP	G.E.	NONE	USCG
	S-64	47,000	T64-GE-540	(3) TS	3,750 SHP	G.E.	SKYCRANE	CIVIL
	S-70C (HH-60)	22,000	CT-7-2C	(2) TS	1,622 SHP	G.E.	NONE	CIVIL
	AUH-76 (S-76B)	10,500	PT6B-36	(2) TS	981 SHP	P&WC	EAGLE	USAF
	S-76A	10,500	250-C30	(2) TS	650 SHP	ALLISON	SPIRIT	CIVIL/MIL
	S-76B	11,700	PT6B-36	(2) TS	980 SHP	P&WC	SPIRIT	CIVIL
	S-76A+/C	11,500	ARRIEL 1S	(2) TS	750 SHP	T-M	EAGLE (H-76)	CIVIL
	S-76D	11,700	ARRIEL 2S	(2) TS	850 SHP	T-M	EAGLE	MIL
	S-80M (MH-53)	69,000	T64-GE-419	(3) TS	4,750 SHP	G.E.	EXPORT	MIL
SOLOY CORP.	BELL-47 (RE-ENGINED)	3,200	250-C20/20B	(1) TS	420 SHP	ALLISON	RANGER/SIOUX	MIL
	UH-12 (RE-ENGINED)	3,100	250-C20/20B	(1) TS	420 SHP	ALLISON	HILLER UH-12	CIVIL/AG

† TWIN-PAC ENGINE, 900 SHP EACH

USA MILITARY - TURBINE POWERED AIRCRAFT
USA MILITARY AIRCRAFT (ATTACK)
(ALPHABETICAL BY MANUFACTURER)

AIRCRAFT MFR.	AIRCRAFT DESIGNATION	GROSS WEIGHT	ENGINE DESIG.	ENGINE TYPE	ENGINE POWER	ENGINE MFR.	POPULAR NAME	TYPE SERVICE
BRITISH AEROSPACE	AV-8A	25,000	PEGASUS 11	(1) TF	21,500 LBT	R-R	HARRIER	USMC
CESSNA	A-37 A/B	14,000	J85-GE-17	(2) TJ	2,850 LBT	G.E.	DRAGONFLY	USAF
FAIRCHILD-REPUBLIC	A-10A	48,000	TF34-GE-100	(2) TF	9,275 LBT	G.E.	THUNDERBOLT-II	USAF
GRUMMAN	A-6E	60,400	J52-P-8B	(2) TJ	9,300 LBT	P&W	INTRUDER	USN/USMC
	A-6F	60,400	F404-GE-400	(2) TF	10,800 LBT	G.E.	INTRUDER	USN/USMC
	S.211A	7,160	JT15D-4C	(1) TF	2,500 LBT	P&WC	NONE	USN
LOCKHEED	AC-130 A,H,U	155,000	T56-A-15	(4) TP	4,591 ESHP	ALLISON	SPECTRE-GUNSHIP	USAF
LTV	A-7D	42,000	TF41-A-1B	(1) TF	14,500 LBT	ALLISON	CORSAIR-II	ANC
	A-7E	42,000	TF41-A-400	(1) TF	15,000 LBT	ALLISON	CORSAIR-II	USN
	A-7F	46,000	F100-PW-220	(1) TF	13,400 LBT	P&W	A-7 PLUS	USAF
	A-7H	42,000	TF30-P-408	(1) TF	15,000 LBT	P&W	A-7 PLUS	USAF
	A-7K	42,000	TF41-A-1	(1) TF	14,500 LBT	ALLISON	CORSAIR-II	ANC
	F-8	27,550	J-57-P-420	(1) TJ	19,000 LBT	P&W	CRUSADER	USN
McDONNELL DOUGLAS	A-4F	24,500	J52-P-8A	(1) TJ	9,300 LBT	P&W	SKYHAWK	USMC
	A-4M	24,500	J52-P-408	(1) TJ	11,200 LBT	P&W	SKYHAWK	USMC
	A-4S	24,500	F404-GE-100D	(1) TF	10,800 LBT	G.E.	SUPER SKYHAWK	USMC
McD-D/BR. AERO.	AV-8B	29,750	F402-RR-406	(1) TF	22,000 LBT	R-R	HARRIER-II	USMC
McD-D/NORTHROP	F/A-18	51,900	F404-GE-402	(2) TF	17,700 LBT	G.E.	HORNET	USN
	F-/A18 E,F	51,900	F414-GE-400	(2) TF	22,000 LBT	G.E.	HORNET	USN
NORTHROP	F-5A, B	20,670	J85-GE-13,15	(2) TJ	5,000 LBT	G.E.	TIGER-II	USAF
ROCKWELL INT'L.	AC-130U	175,000	T56-A-15	(4) TP	4,591 ESHP	ALLISON	GUNSHIP	USAF
SABRELINER	A/T-33	17,250	J33A-35	(1) TJ	4,600 LBT	ALLISON	SHOOTING STAR	ANC

USA MILITARY AIRCRAFT (BOMBERS)
(ALPHABETICAL BY MANUFACTURER)

AIRCRAFT MFR.	AIRCRAFT DESIGNATION	GROSS WEIGHT	ENGINE DESIG.	ENGINE TYPE	ENGINE POWER	ENGINE MFR.	POPULAR NAME	TYPE SERVICE
BOEING CORP.	B-52 C,F,G	488,000	J57-P-43W	(8) TJ	13,750 LBT	P&W	STRATO-FORTRESS	USAF
	B-52H	488,000	TF33-P-3	(8) TF	17,000 LBT	P&W	STRATO-FORTRESS	USAF
GENERAL DYNAMICS	FB-111	114,000	TF30-P-7	(2) TF	20,350 LBT	P&W	RAVEN	USAF
NORTHROP	B-2A	400,000	F118-GE-100	(4) TF	19,000 LBT	G.E.	STEALTH BOMBER	USAF
ROCKWELL	B-1B	477,000	F101-GE-102	(4) TF	30,800 LBT	G.E.	LANCER	USAF

USA MILITARY AIRCRAFT (CARGO-TRANSPORT)
(ALPHABETICAL BY MANUFACTURER)

AIRCRAFT MFR.	AIRCRAFT DESIGNATION	GROSS WEIGHT	ENGINE DESIG.	ENGINE TYPE	ENGINE POWER	ENGINE MFR.	POPULAR NAME	TYPE SERVICE
ALENIA	C-27A	61,728	T64-GE-P4D	(2) TP	3,400 ESHP	G.E.	G.222	USAF
BEECH	VC-6B	9,650	PT6A-20	(2) TP	610 ESHP	P&WC	BEECH-90, VIP	USAF
	C-12F	12,500	PT6A-42	(2) TP	786 ESHP	P&WC	HURON/KINGAIR	USAF/USN
	C-12J	16,600	PT6A-65B	(2) TP	1,174 ESHP	P&WC	HURON-II/KINGAIR	USAF
BOEING	C-135A	275,000	J57-59W	(4) TJ	13,750 LBT	P&W	STRATO-LIFTER	USAF
	C-135B	275,000	TF33-P-3	(4) TF	17,000 LBT	P&W	STRATO-LIFTER	USAF
	C-135C	319,000	CFM-56-2	(4) TF	22,000 LBT	CFM INT'L	STRATO-LIFTER	USAF
	KC-135A	297,000	J57-9-59W	(4) TJ	13,750 LBT	P&W	STRATO-TANKER	USAF
	KC-135B	275,000	TF33-P-3	(4) TF	17,000 LBT	P&W	STRATO-TANKER	USAF
	KC-135R (CFM-56)	322,500	F108-CF-100	(4) TF	22,000 LBT	CFM INT'L	STRATO-TANKER	USAF
	C-22B	154,500	JT8D-15A	(3) TF	15,000 LBT	P&W	B-727	USAF
	VC-25A	833,000	CF6-80C2B1	(4) TF	56,700 LBT	G.E.	B-747-200B	AF-1
	VC-137A	328,000	JT3D-3B	(4) TF	18,000 LBT	P&W	STRATOLINER-707	USAF

AIRCRAFT MFR.	AIRCRAFT DESIGNATION	GROSS WEIGHT	ENGINE DESIG.	ENGINE TYPE	ENGINE POWER	ENGINE MFR.	POPULAR NAME	TYPE SERVICE
BRITISH AEROSPACE (BAE)	C-29A	27,500	TFE731-5	(2) TF	4,300 LBT	GARRETT	BAe 125	USAF
	C-550	13,500	JT15D-4	(2) TF	2,900 LBT	P&WC	CITATION	USAF
CASA	C-212	17,600	TPE331-10R	(2) TP	900 ESHP	GARRETT	AVIOCAR	USAR
CONVAIR	C-131	54,500	501-D13H	(2) TP	4,591 ESHP	ALLISON	CONVAIR 340	USN
DEHAVILLAND	C-8A	41,000	CT-64-820	(2) TP	3,133 ESHP	G.E.	CARIBOU	USN
	UV-18A	12,500	PT6A-27	(2)TP	715 ESHP	P&WC	TWIN OTTER	USAF
FAIRCHILD	C-26A	16,500	TPE-331-11U	(2) TP	1,100 ESHP	GARRETT	METRO III	USAF
FOKKER	F-27	5,900	DART MK.536-7R	(2) TP	2,140 ESHP	R-R	FRIENDSHIP	USAR
GATES	C-21A	18,300	J60-P-5	(2) TJ	3,000 LBT	P&W	LEARJET	USAF
	C-21A	18,300	TFE-731-2A	(2) TJ	3,500 LBT	GARRETT	LEARJET	USAF
GRUMMAN	KA-6E	60,400	J52-P-8B	(2) TJ	9,300 LBT	P&W	INTRUDER TANKER	USN/USMC
GULFSTREAM	C-4C	36,000	DART-542	(2) TP	3,025 ESHP	R-R	GULFSTREAM G-1	USN
	C20-A,B,D,E	69,700	F113-RR-100	(2) TF	12,500 LBT	R-R	GULF. G-III (SPEY)	USAF
	C-20F	73,600	TAY	(2) TF	13,850 LBT	R-R	GULFSTREAM G-IV	USAF
GRUMMAN	C-2A	55,740	T56-A-425	(2) TP	4,591 ESHP	ALLISON	GREYHOUND	USN
LOCKHEED GEORGIA	C-130H	155,000	T56-A-15	(4) TP	4,591 ESHP	ALLISON	HERCULES	USAF
	C-130H-30	155,000	T56-A-15	(4) TP	4,591 ESHP	ALLISON	SUPER HERCULES	USAF
	C-140A	40,920	J60-P-6	(2) TJ	3,000 LBT	P&W	JETSTAR	USAF
	C-141B	343,000	TF33-P-7	(4) TF	20,350 LBT	P&W	STARLIFTER	USAF
	C-5A,B	837,000	TF39-GE-1C	(4) TF	43,050 LBT	G.E.	GALAXY	USAF
	KE-130	175,000	T56-A-423	(4) TP	4,900 ESHP	ALLISON	TANKER	USAF/USN
	KS-3	52,530	TF34-GE-400A,B	(2) TF	9,295 LBT	G.E.	VIKING TANKER	USN
McDONNELL-DOUGLAS	C-9A	108,000	JT8D-9	(2) TF	14,500 LBT	P&W	NIGHTINGALE	USAF
	C-9B	110,000	JT8D-9	(2) TF	14,500 LBT	P&W	SKYTRAIN-II	USN
	C-17A	580,000	F117-PW-100	(4) TF	41,600 LBT	P&W	NONE	USAF
	KC-10A	590,000	CF6-50C2	(4) TF	54,000 LBT	G.E.	EXTENDER	USAF
	VC-9C	110,000	JT8D-9	(2) TF	14,500 LBT	P&W	NONE	USAF
ROCKWELL	C-140	17,760	J60-P3	(2) TJ	3,000 LBT	P&W	JETSTAR	USAF
	CT-39A	17,760	J60-P3	(2) TJ	3,000 LBT	P&W	SABRELINER	USAF
SHORT BROTHERS	C-23A	26,450	TFE731-5	(2) TF	4,300 LBT	GARRETT	SHERPA	USAF
	C-33B	25,500	PT6A-65	(2) TP	1,100 ESHP	P&WC	SHORTS 330	USAR

USA MILITARY AIRCRAFT (EARLY WARNING)
(ALPHABETICAL BY MANUFACTURER)

AIRCRAFT MFR.	AIRCRAFT DESIGNATION	GROSS WEIGHT	ENGINE DESIG.	ENGINE TYPE	ENGINE POWER	ENGINE MFR.	POPULAR NAME	TYPE SERVICE
BOEING	E-3A,B (B-707)	325,000	TF33-P-100A	(4) TF	21,000 LBT	P&W	SENTRY	USAF
	E-3C (B-707)	325,000	F108-CF-100	(4) TF	24,000 LBT	CFM INT'L	SENTRY (CFM-56)	USAF
	E-4 A,B (B-747)	800,000	CF6-50E	(4) TF	52,500 LBT	G.E.	NEACP	USAF
	E-6A (B-707)	342,000	F108-CF-100	(4) TF	24,000 LBT	CFM INT'L	TACANO (CFM-56)	USN
	EC-18A	47,000	JT3D-7	(4) TF	19,000 LBT	P&W	B-707 (ARIA)	USAF
	EC-135	275,000	J57-59W	(4) TJ	13,750 LBT	P&W	AIR/COMM/POST	USAF
	EC-135C	275,000	TF-33-P-9	(4) TF	18,000 LBT	P&W	AIR/COMM/POST	USAF
	NKC-135	275,000	TF33-P-9	(4) TF	18,000 LBT	P&W	BIG CROW	USAF
GRUMMAN	E-2C	51,900	T56-A-427	(2) TP	5,250 ESHP	ALLISON	HAWKEYE	USN
G/DYNAMICS	EF-111	114,000	TF30-P-109	(2) TF	20,350 LBT	P&W	RAVEN	USAF
LOCKHEED	EC-130	155,000	T56-A-15	(4) TP	4,591 ESHP	ALLISON	HERCULES	USAF
LTV	EA-7D	42,000	TF41-A-1B	(1) TF	14,500 LBT	ALLISON	CORSAIR-II	ANG
MC/DOUGLAS	EC-24	315,000	JT3D-3B	(4) TF	18,000 LBT	P&W	DC-8	ANC
	EC-24A	350,000	JT3D-7	(4) TF	19,000 LBT	P&W	DC-8	USN

USA MILITARY AIRCRAFT (ELECTRONIC WARFARE)
(ALPHABETICAL BY MANUFACTURER)

AIRCRAFT MFR.	AIRCRAFT DESIGNATION	GROSS WEIGHT	ENGINE DESIG.	ENGINE TYPE	ENGINE POWER	ENGINE MFR.	POPULAR NAME	TYPE SERVICE
BEECH	CU-21	10,200	PT6A-41	(2)NTP	900 ESHP	P&WC	SUPER KINGAIR	USAF
BOEING	E-8A	47,000	JT3D-7	(4) TF	19,000 LBT	P&W	B-707	USAF
BOEING CANADA	E-9A	34,500	PW-120A	(2) TP	2,100 ESHP	P&WC	DASH-8	USAF
BOEING USA	NKC-135C	275,000	TF33-P-9	(4) TF	18,000 LBT	P&W	BIG CROW	USAF

AIRCRAFT MFR.	AIRCRAFT DESIGNATION	GROSS WEIGHT	ENGINE DESIG.	ENGINE TYPE	ENGINE POWER	ENGINE MFR.	POPULAR NAME	TYPE SERVICE
GEN. DYNAMICS	EF-111A	88,950	TF30-P-3	(2) TF	18,500 LBT	P&W	RAVEN	USAF
GRUMMAN	EA-6A	54,500	J52-P-8	(2) TJ	9,300 LBT	P&W	INTRUDER	USMC-USN
	EA-6B	60,000	J52-P-408	(2) TJ	11,200 LBT	P&W	PROWLER	USMC-USN
McD/DOUGLAS	EC-24A	350,000	TF33-P-9	(4) TF	19,000 LBT	P&W	DC-8	USN

USA MILITARY AIRCRAFT (FIGHTERS)
(ALPHABETICAL BY MANUFACTURER)

AIRCRAFT MFR.	AIRCRAFT DESIGNATION	GROSS WEIGHT	ENGINE DESIG.	ENGINE TYPE	ENGINE POWER	ENGINE MFR.	POPULAR NAME	TYPE SERVICE
GENERAL DYNAMICS	F-102	22,000	J-57-P-23	(2) TJ	16,000 LBT	P&W	DELTA DAGGER	USAF
	F-106A	36,000	J75-P-17	(1) TJ	24,500 LBT	P&W	DELTA DART	USAF
	F-111 A,D,E,F	47,000	TF30-P-100	(2) TF	25,100 LBT	P&W	NONE	USAF
	F-16 A,B	33,360	F100-PW-200	(1) TF	23,830 LBT	P&W	FIGHTING FALCON	USAF
	F-16 C,D	34,537	F110-GE-100	(1) TF	28,000 LBT	G.E.	FIGHTING FALCON	USAF
	F-16 C,D (RE-ENGINE)	34,537	F100-PW-229	(1) TF	29,000 LBT	P&W	FIGHTING FALCON	USAF
	F-16N	35,470	F110-GE-100/129	(1) TF	28,000 LBT	G.E.	FIGHTING FALCON	USN
	F-16AT	35,470	F119-PW-100	(1) TF	CLASSIFIED	P&W	FALCON-21 (IN Dev.)	USAF
GRUMMAN	F-14A	59,700	TF30-P-414	(2) TF	20,900 LBT	P&W	TOMCAT	USN
	F-14B	70,000	F110-GE-400	(2) TF	27,000 LBT	G.E.	SUPER TOMCAT	USN
	X-29A	17,800	F404-GE-400	(1) TF	16,000 LBT	G.E.	FWD SWEPT WING	USAF
IAI	F-21A	22,000	J79-GE-J1E	(1) TJ	18,750 LBT	G.E.	KFIR	USN
LOCKHEED CA.	F-22A	65,000	F119-PW-100	(2) TF	35,000 LBT	P&W	IN DEVELOPMENT	USAF
	F-104	31,000	J-79-GE-19	(1) TJ	18,000 LBT	G.E.	STARFIGHTER	USAF
	F-117A	52,500	F404-GE-F1D2	(2) TF	10,800 LBT	G.E.	STEALTH FIGHTER	USAF
McDONNELL DOUGLAS	F-4D	56,000	J79-GE-10,-15	(2) TJ	17,820 LBT	G.E.	PHANTOM-II	USAF
	F-4 E,G	58,000	J79-GE-17	(2) TJ	17,820 LBT	G.E.	PHANTOM-II	USAF
	F-4S	58,000	J79-GE-17	(2) TJ	17,820 LBT	G.E.	PHANTOM-II	USN
	F-4S	58,000	SPEY-202	(2) TJ	17,820 LBT	R-R	LICENSE BUILT	USN
	F-15 C,D	68,000	F100-PW-100	(2) TF	23,830 LBT	P&W	EAGLE	USAF
	F-15 C,D	68,000	F100-PW-200	(2) TF	23,830 LBT	P&W	EAGLE	USAF
	F-15E	61,000	F100-PW-220	(2) TF	23,830 LBT	P&W	EAGLE	USAF
	F-101	49,000	J-57-P-13	(2) TJ	15,000 LBT	P&W	VOODOO	USAF
McDONNELL-D & NORTHROP	F/A-18 A,B,C,D	51,900	F404-GE-400	(2) TF	16,000 LBT	G.E.	HORNET	USMC-USN
	F-/A18 E,F	51,900	F414-GE-400	(2) TF	22,000 LBT	G.E.	HORNET	USMC-USN
NORTHROP	F-5 A,B	20,670	J85-GE-13,-15	(2) TJ	5,000 LBT	G.E.	FREEDOM FIGHTER	USAF/USN
	F-5 E,F	24,720	J85-GE-21	(2) TJ	5,000 LBT	G.E.	TIGER-II	USAF-USN
	F-20	26,290	F404-GE-400	(1) TF	16,000 LBT	G.E.	TIGERSHARK	USAF
REPUBLIC	F-105F	54,580	J75-P-19W	(1) TJ	26,500 LBT	P&W	THUNDERCHIEF	USAF
ROCKWELL	X-31A	15,935	F404-GE-400	(1) TF	16,000 LBT	G.E.	NONE	USAF

USA MILITARY AIRCRAFT (OBSERVATION)
(ALPHABETICAL BY MANUFACTURER)

AIRCRAFT MFR.	AIRCRAFT DESIGNATION	GROSS WEIGHT	ENGINE DESIG.	ENGINE TYPE	ENGINE POWER	ENGINE MFR.	POPULAR NAME	TYPE SERVICE
CESSNA	OA-37 A,B	14,000	J85-GE-17	(2) TJ	2,850 LBT	G.E.	DRAGONFLY	USAF
GRUMMAN	OV-1 B,C,D	18,300	T53L-701A	(2) TP	1,400 ESHP	TEXTRON	MOHAWK	ARMY
ROCKWELL INT'L	OV-10A	13,850	T76G-421	(2) TP	1,040 ESHP	GARRETT	BRONCO	UAF-USMC
	OV-10D	14,450	T76G-420	(2) TP	1,040 ESHP	GARRETT	NOS	USMC

USA MILITARY AIRCRAFT (PATROL)
(ALPHABETICAL BY MANUFACTURER)

AIRCRAFT MFR.	AIRCRAFT DESIGNATION	GROSS WEIGHT	ENGINE DESIG.	ENGINE TYPE	ENGINE POWER	ENGINE MFR.	POPULAR NAME	TYPE SERVICE
FAIRCHILD	U-23	6,100	TPE331-1-101	(1) TP	650 ESHP	GARRETT	PORTER	EXPORT
LOCKHEED CA.	HU-25A	32,000	ATF3-6	(2) TF	5,440 LBT	GARRETT	GUARDIAN (FALCON)	USCG
	P-3C	139,760	T56-A-14	(4) TP	4,591 ESHP	ALLISON	ORION	USN
	P-7A	171,000	GE-38	(4) TP	5,150 ESHP	G.E./TEX.	In Development	USN
	S-3 A,B	52,530	TF34-GE-400A,B	(2) TF	9,295 LBT	G.E.	VIKING	USN
BEECH	RC-12 D,G,H,K	16,000	PT6A-67A	(2) TP	1,272 ESHP	P&WC	GUARDRAIL (KINGAIR)	ARMY
	RU-21 A,E,G,H	10,200	T74-CP-700	(2) TP	618 ESHP	P&WC	UTE (KINGAIR)	ARMY

USA MILITARY AIRCRAFT (RECONNAISSANCE)
(ALPHABETICAL BY MANUFACTURER)

AIRCRAFT MFR.	AIRCRAFT DESIGNATION	GROSS WEIGHT	ENGINE DESIGNATION	ENGINE TYPE	ENGINE POWER	ENGINE MFR.	POPULAR NAME	TYPE SERVICE
LOCKHEED CA.	SR-71	172,000	J-58-P-4	(2) TJ	34,000 LBT	P&W	BLACKBIRD	USAF
	TR-1A	40,600	J75-P-13B	(1) TJ	26,500 LBT	P&W	NONE	USAF
	TR-1A (RE-ENGINE)	40,600	F118-GE-100	(1) TJ	26,500 LBT	G.E.	NONE	USAF
	U-2R	40,000	J-75-P-13B	(1) TJ	26,500 LBT	P&W	NONE	USAF
	WC-130	155,000	T56-A-425	(4) TP	4,591 ESHP	ALLISON	HERCULES	USAF
MARTIN	RB-57F	70,000	TF33-P-11	(2) TF	18,000 LBT	P&W	CANBERRA	USAF
McDONNELL	RF-4B,C	58,000	J79-10B,15	(2) TJ	17,820 LBT	G.E.	PHANTOM-II	USN/ANG
NORTHROP	RF-5E	25,200	J85-GE-21	(2) TJ	5,000 LBT	G.E.	TIGEREYE	USAF
	TR-3A	CLASSIFIED	---------------	-------	---------------	-----	BLACK MAMBA	USAR
NORTH AMERICAN	RA-5	66,000	J-79-GE-10	(2) TJ	17,820 LBT	G.E.	VIGILANTE	USN

USA MILITARY AIRCRAFT (TRAINING)

AIRCRAFT MFR.	AIRCRAFT DESIGNATION	GROSS WEIGHT	ENGINE DESIGNATION	ENGINE TYPE	ENGINE POWER	ENGINE MFR.	POPULAR NAME	TYPE SERVICE
AROCET INC.	AT-9	2,700	250-B17D	(1) TP	420 SHP	ALLISON	NONE	USAF
BEECH	T-1A (Beechjet-400A)	16,080	JT15D-5	(2) TF	2,900 LBT	P&WC	JAYHAWK	USAF
	T-34A	4,300	PT6A-25	(1) TP	580 ESHP	P&WC	TURBO MENTOR	USAF
	T-34 B/C	4,300	PT6A-25	(1) TP	580 ESHP	P&WC	TURBO MENTOR	USN
	T-44A	10,100	PT6A-34B	(2) TP	783 ESHP	P&WC	KING AIR	USN
BOEING	T-43A	109,000	JT8D-9	(2) TF	14,500 LBT	P&W	B-737	USAF
CESSNA	T-37 B,C	7,500	J69T-25 †	(2) TJ	1,025 LBT	CAE	TWEET	USAF
	T-37A	6,600	MARBORE VI	(2) TJ	925 LBT	T-M	TWEET	EXPORT
	T-47A	15,500	JT15D-5	(2) TF	2,900 LBT	P&WC	CITATION	USN
	† Marbore License, France							
FAIRCHILD	T-46A	7,295	F109-GA-100	(2) TF	1,330 LBT	GARRETT	IN DEVELOPMENT	USAF
G/DYNAMICS	TF-16N	23,360	F100-PW-200	(1) TF	23,830 LBT	P&W	FIGHTING FALCON	USAF
GULFSTREAM	TC-4	36,000	DART MK.529-8X	(2) TP	2,185 ESHP	R-R	ACADEME/G-1	USN
JAFFE CORP.	SA-32T	2,600	250-B17D	(1) TF	420 SHP	ALLISON	TRAINER-PROPOSED	USAF
LOCKHEED	T-33A	14,400	J33A-35	(1) TJ	4,650 LBT	ALLISON	SHOOTING STAR	EXPORT
	T-33V	14,400	PW 305	(1) TJ	5,250 LBT	P&WC	SHOOTING STAR	EXPORT
	TP-3	139,760	T56-A-14	(4) TP	4,591 ESHP	ALLISON	ORION	USN
LTV	TA-7D	42,000	TF41-A-1B	(1) TF	14,500 LBT	ALLISON	CORSAIR-II	ANG
Mc/DOUGLAS	TA-4	24,500	J52-P-8A	(1) TJ	9,300 LBT	P&W	SKYHAWK	USMC
McD/BAe	T-45A	12,750	F405-RR-401	(1) TF	5,800 LBT	R-R	GOSHAWK (ADOUR)	USN
	TAV-8B	29,750	F402-RR-406	(1) TF	22,000 LBT	R-R	HARRIER-II	USMC
McD/NORTHROP	TF-18	51,900	F404-GE-400	(2) TF	10.800 LBT	P&W	HORNET	USMC/USN
NORTHROP	TF-5 A, B	20,670	J85-GE-13,15	(2) TJ	5,000 LBT	G.E.	FREEDOM FIGHTER	USAF/USN
	T-38B	12,093	J85-GE-5	(2) TJ	3,850 LBT	G.E.	TALON	USAF/USN
ROCKWELL	T-2B	13,190	J60-P3	(2) TJ	3,000 LBT	P&W	BUCKEYE	USAF
	T-2C	13,190	J85-GE-4	(2) TJ	2,950 LBT	G.E.	BUCKEYE	USN
SABRELINER	T-39 A,B	10,000	J60-P-3	(2) TJ	3,000 LBT	P&W	SABRELINER	USAF
	T-39N	10,000	JT12A-8A	(2) TJ	3,000 LBT	P&W	SABRELINER	USN

USA MILITARY AIRCRAFT (UTILITY)

AIRCRAFT MFR.	AIRCRAFT DESIGNATION	GROSS WEIGHT	ENGINE DESIGNATION	ENGINE TYPE	ENGINE POWER	ENGINE MFR.	POPULAR NAME	TYPE SERVICE
BEECH	AU-24	5,100	PT6A-27	(1) TP	680 ESHP	P&WC	STALLION	ARMY
	UC-12	12,500	PT6A-114	(2) TP	632 ESHP	P&WC	SUPER KINGAIR	USN/USMC
	U-21 A,G	9,650	T74-CP-700	(2) TP	618 ESHP	P&WC	UTE (KINGAIR)	ARMY
	U-21F	11,500	PT6A-28	(2) TP	715 ESHP	P&WC	UTE (KINGAIR)	ARMY
	U-21J	15,000	PT6A-41	(2) TP	632 ESHP	P&WC	UTE (KINGAIR)	ARMY
CESSNA	U-27A	8,035	PT6A-114	(2) TP	632 ESHP	P&WC	CARAVAN	ARMY
DeHAVILLAND	UV-18A	12,500	PT6A-27	(2) TP	715 ESHP	P&WC	UTE	ARMY
PILATUS	UV-20A	6,100	PT6A-27	(1) TP	715 ESHP	P&WC	PORTER PC-6	ARMY

USSR MILITARY AND CIVIL JET AIRCRAFT
(USSR IS PRESENTLY REFERRED TO AS THE COMMONWEALTH OF INDEPENDENT STATES)
LISTING ALPHABETICAL BY MANUFACTURER

BOMBERS AND MARITIME MILITARY AIRCRAFT
(B) - BOMBER (M) - MARITIME

AIRCRAFT MFR.	AIRCRAFT DESIGNATION	POPULAR NAME	GROSS WEIGHT	ENGINE DESIGNATION	ENGINE TYPE	ENGINE RATING
BERIEV BUREAU	BE-12 (M)	MAIL (TCHAIKA)	68,000	PROGRESS-IVCHENKO AI-20D	(2) TP	4,190 ESHP (3,124 EKW).
	BE-42 [A-40] (AMPHIBIAN)	MERMAID-ALBATROSS (M)	190,000	PERM-SOLOVIEV D-30KV	(2) TF	24,452 LBT (117,6 KN THRUST). Plus (2) RD-60 Boost-Turbojets.
ILYUSHIN BUREAU	IL-28 (B)	BEAGLE	46,700	KLIMOV VK-1	(2) TP	5,950 LBT (26,5 KN THRUST).
	IL-38 (M)	MAY	140,000	PROGRESS-IVCHENKO AI-20M	(4) TP	4,250 ESHP (3,169 EKW).
MYASISHCHEV BUREAU	M-4 (B)	BISON	350,000	MIKULIN AM-3D KOLIESOV VD-7	(4) TJ (4) TJ	21,000 LBT (93,4 KN THRUST), NON-A/B. 23,150 LBT (103 KN THRUST), NON-A/B.
	M-50 (B)	BOUNDER	500,000	KOLIESOV VD-7	(4) TJ	30,865 LBT (137.3 KN THRUST) WITH A/B.
TUPOLEV BUREAU	TU-16 (B)	BADGER	165,000	MIKULIN RD-3M	(2) TJ	20,950 LBT (93,2 KN THRUST) NON A/B.
	TU-22 (TU-105)	BLINDER (B)	185,000	KOLIESOV VD-7	(2) TJ	30,900 LBT (137,4 KN THRUST) WITH A/B.
	TU-26 (TU-22M)	BACKFIRE (B/M)	286,000	KUZNETSOV NK-144	(2) TF	44,090 LBT (196,1 KN THRUST) WITH A/B.
	TU-95 (B)	BEAR	414,000	KUZNETSOV NK-12MV	(4) TP	14,795 ESHP (11,033 EKW).
	TU-104 (B)	CAMEL	165,000	MIKULIN RD-3M	(2) TJ	20,950 LBT (93,2 KN THRUST) NON A/B.
	TU-142 (B/M)	BEAR-F	414,000	KUZNETSOV NK-12	(4) TP	14,795 ESHP (11,032 EKW) [UPGRADE TU-95].
	TU-160 (B)	BLACKJACK	606,000	KOLIESOV VD-57	(4) TF	44,100 LBT (196 KN THRUST) WITH A/B.

ATTACK & FIGHTER MILITARY AIRCRAFT
(ALPHABETICAL BY MANUFACTURER)

AIRCRAFT MFR.	AIRCRAFT DESIGNATION	POPULAR NAME	GROSS WEIGHT	ENGINE DESIGNATION	ENGINE TYPE	ENGINE RATING
MIKOYAN BUREAU	MIG-21	FISHBED	21,600	TUMANSKY R-13-300	(1) TJ	11,240 LBT DRY (50 KN), AND 14,550 LBT (64,5 KN THRUST) WITH A/B.
	MIG-23	FLOGGER-B	41,000	TUMANSKY R-29B	(1) TJ	17,635 LBT DRY (78,5 KN) AND 27,500 LBT (122,32 KN THRUST) WITH A/B.
		FLOGGER-C	41,000	TUMANSKY R-27	(1) TJ	22,485 LBT (100 KN THRUST) WITH A/B.
		FLOGGER-G	41,000	TUMANSKY R-35F-300	(1) TJ	28,660 LBT (127 KN THRUST) WITH A/B.
	MIG-25	FOXBAT	82,500	TUMANSKY R-31	(2) TJ	27,100 LBT (120,5 KN THRUST) WITH A/B.
	MIG-27	FLOGGER D/J	44,300	TUMANSKY R-29B	(1) TJ	25,350 LBT (113 KN THRUST) WITH A/B.
	MIG-29	FULCRUM	39,700	KLIMOV RD-33	(2) TF	18,300 LBT (81,4 KN THRUST) WITH A/B.
	MIG-31	FOXHOUND	90,700	PERM-SOLOVIEV D-30F6	(2) TF	34,162 LBT (152 KN THRUST) WITH A/B.
SUKHOI BUREAU	SU-7	FITTER-A	30,865	LYULKA AL-7F-L00	(1) TJ	21,150 LBT (93,6 KN THRUST) WITH A/B.
	S-7U	MOUJIK	28,000	LYULKA AF-7F-100	(1) TJ	21,150 LBT (93,6 KN THRUST).
	SU-9/11	FISHPOT	35,250	LYULKA AL-7F	(1) TJ	22,046 LBT (98,1 KN THRUST) WITH A/B.
	SU-15/21	FLAGON	39,600	TUMANSKY R13F2-300	(2) TJ	15,875 LBT (70,6 KN THRUST) WITH A/B.
	SU-17K/20	FITTER-K	42,700	LYULKA AL-21F-3	(1) TJ	24,700 LBT (110 KN THRUST) WITH A/B.
	SU-19/22	FITTER-F	30,865	TUMANSKY R-29-B-300	(1) TJ	23,350 LBT (103 KN THRUST) WITH A/B.
	SU-24	FENCER	87,000	LYULKA AL-21F-3	(2) TJ	24,700 LBT (110 KN THRUST) WITH A/B.
	SU-25/28	FROGFOOT	38,800	TUMANSKY R-195	(2) TJ	9,921 LBT (44 KN THRUST) NON A/B.

AIRCRAFT MFR.	AIRCRAFT DESIGNATION	POPULAR NAME	GROSS WEIGHT	ENGINE DESIGNATION	ENGINE TYPE	ENGINE RATING
SUKHOI CONT.	SU-27	FLANKER	66,135	LYULKA AL-31F	(2) TF	27,550 LBT (122.5 KN THRUST) WITH A/B.
	SU-35	----------	----------	LYULKA AL-31SM	(2) TF	IN DEVELOPMENT.
TUPOLEV BUREAU	TU-28P (TU-128)	FIDDLER	100,000	LYULKA AL-21F-3	(2) TJ	24,700 LBT (110 KN THRUST) WITH A/B.
YAKOVLEV BUREAU	YAK-28P	FIREBAR	44,000	TUMANSKY R-11	(2) TJ	13,120 LBT (58,4 KN THRUST) WITH A/B.
	YAK-38 ALTERNATE	FORGER-VSTOL	25,795 23,100 LBT	LYULKA AL-21 TUMANSKY R-27V-300	(1) TJ (1) TF	17,700 LBT(110 KN THRUST) NON A/B, PLUS (2) KOLIESOV 7,165 LBT, RD-38 LIFT-FANS.
	YAK-41	VSTOL	26,000	LYULKA AL-21	(1) TJ	YAK-38 UPGRADE, (IN DEVELOPMENT).
	YAK-141	FREESTYLE	34,100	TUMANSKY R-79	(1) TF	34,170 LBT - VSTOL WITH (2) RD-41 SOYUZ LIFTFANS OF 9,000 LBT (IN DEVELOPMENT).

RECONNAISSANCE, ECM AND EARLY WARNING MILITARY AIRCRAFT
(ALPHABETICAL BY MANUFACTURER)

AIRCRAFT MFR.	AIRCRAFT DESIGNATION	POPULAR NAME	GROSS WEIGHT	ENGINE DESIGNATION	ENGINE TYPE	ENGINE RATING
ANTONOV BUREAU	AN-12	CUB	134,000	PROGRESS-IVCHENKO AI-20K	(4) TP	4,250 ESHP (3,169 KW).
	AN-74	MADCAP	70,000	PROGRESS-LOTAREV D-36	(2) TF	14,350 LBT (63,8 KN), UPGRADE OF AN-72, WITH WHEEL-SKI LANDING GEAR.
ILYUSHIN BUREAU	IL-20	COOT-A	134,600	PROGRESS-IVCHENKO AI-20K	(4) TP	4,000 ESHP (2,983 KW).
	IL-38	MAY	140,000	PROGRESS-IVCHENKO AI-20M	(4) TP	4,250 ESHP (3,169 KW).
	IL-76 (A-50)	MAINSTAY	374,700	PERM-SOLOVIEV D-30KP	(4) TF	24,455 LBT (117,7 KN THRUST).
MIKOYAN BUREAU	MIG-21	FISHBED	12,600	TUMANSKY R-13-300	(1) TJ	14,500 LBT (64,5 KN THRUST) WITH A/B.
	MIG-25	FOXBAT-B/D	73,000	TUMANSKY R-31	(2) TJ	27,100 LBT (120,5 KN THRUST) WITH A/B.
MYASISHCHEV BUREAU	M-17	MYSTIC-A	48,500	KOLIESOV RD-36	(1) TF	15,400 LBT (68,5 KN THRUST).
	M-55	MYSTIC-B	Unknown	PERM-SOLOVIEV-?	(2) TF	ESTIMATED 11,000 LB THRUST).
SUKHOI BUREAU	SU-17	FITTER	30,800	LYULKA AL-21F-3	(1) TJ	24,700 LBT (109,8 KN THRUST) WITH A/B.
	SU-24	FENCER-E	87,000	LYULKA AL-21F-3	(2) TJ	24,700 LBT (110 KN THRUST) WITH A/B.
TUPOLEV BUREAU	TU-16	BADGER H/J	165,000	MIKULIN RD-3M	(2) TJ	20,950 LBT(93,19 KN THRUST) NON A/B.
	TU-22	BLINDER	185,000	KOLIESOV VD-7	(2) TJ	30,900 LBT (137,4 KN THRUST) WITH A/B.
	TU-126	MOSS	361,000	KUZNETSOV NK-12M	(4) TP	14,795 ESHP (11,033 EKW).
	TU-142	BEAR-F	414,000	KUZNETSOV NK-12	(4) TP	14,795 ESHP (11,030 KW).
YAKOVLEV BUREAU	YAK-28	BREWER	44,000	TUMANSKY R-11	(2) TJ	13,120 LBT (58,36 KN THRUST) WITH A/B.

TRANSPORT MILITARY AIRCRAFT
(ALPHABETICAL BY MANUFACTURER)

AIRCRAFT MFR.	AIRCRAFT DESIGNATION	POPULAR NAME	GROSS WEIGHT	ENGINE DESIGNATION	ENGINE TYPE	ENGINE RATING
ANTONOV BUREAU	AN-12	CUB	131,000	PROGRESS-IVCHENKO AI-20	(4) TP	4,000 ESHP (2,983 KW).
	AN-22	COCK	551,000	KUZNETSOV NK-12M	(4) TP	15,000 ESHP (11,186 EKW).
	AN-26	CURL	52,900	PROGRESS-IVCHENKO AI-24VT	(2) TP	2,820 ESHP (2,103 EKW). PLUS ONE RU-19A (TJ) AUXILIARY ENGINE FOR TAKE-OFF.
	AN-30	CLANK	59,700	IVCHENKO AI-24	(2) TP	2,820 ESHP (2,013 KW).
	AN-32	CLINE	59,000	IVCHENKO AI-20DM	(2) TP	5,180 ESHP (3,683 EKW).
	AN-72	COALER	70,000	PROGRESS-LOTAREV D-436T	(2) TF	16,575 LBT (73,7 KN THRUST), UPGRADE D-36.
	AN-74	COALER-B	70,000	PROGRESS-LOTAREV D-36	(2) TF	14,350 LBT (63,8 KN), UPGRADE OF AN-72, WITH WHEEL-SKI LANDING GEAR.
	AN-124	RUSLAN (CONDOR)	892,800	PROGRESS-LOTAREV D-18T	(4) TF	51,650 LBT (229,3 KN THRUST).
	AN-128	----------	120,000	IVCHENKO AI-20K	(4) TP	4,000 ESHP (2,983 EKW).
	AN-225	MRIYA (DREAM)	1,322,750	PROGRESS-LOTAREV D-18T	(6) TF	51,650 LBT (229,3 KN THRUST).
ILYUSHIN BUREAU	IL-62	CLASSIC	368,000	KUZNETSOV NK-8-4	(4) TF	22,273 LBT (99,1 KN).
	IL-76	CANDID	374,785	PERM-SOLOVIEV D-30KP	(4) TF	26,455 LBT (117,67 KN THRUST).
	IL-78	MIDAS	374,700	PERM-SOLOVIEV D-30KP	(4) TF	24,455 LBT (TANKER VERSION OF IL-76).
MYASISHCHEV BUREAU	M-4	BISON TANKER	355,000	MIKULIN AM-30	(4) TJ	19,180 LBT (85,3 KN THRUST).
TUPOLEV BUREAU	TU-95	BEAR-H	414,000	KUZNETSOV NK-12	(4) TP	14,795 ESHP (11,033 EKW).
	TU-126	MOSS/CLEAT	361,000	KUZNETSOV NK-12	(4) TP	14,795 ESHP (11,030 KW).
	TU-142	BEAR-F	414,000	KUZNETSOV NK-12	(4) TP	14,795 ESHP (11,030 KW).
	TU-144	CHARGER SST	396,000	KUZNETSOV NK-144	(4) TF	44,090 LBT (196,1 KN).
	TU-144D	CHARGER SST	396,000	RD36-51A	(4) TF	44,090 LBT (IN DEVELOPMENT).
	TU-334T	----------	91,000	PROGRESS-LOTAREV D-236	(4) TP	10,000 ESHP (IN DEVELOPMENT).

AIRCRAFT MFR.	AIRCRAFT DESIGNATION	POPULAR NAME	GROSS WEIGHT	ENGINE DESIGNATION	ENGINE TYPE	ENGINE RATING

TRAINER MILITARY AIRCRAFT (TWO-PLACE COCKPIT)
(ALPHABETICAL BY MANUFACTURER-CIVIL AND MILITARY)

AIRCRAFT MFR.	AIRCRAFT DESIGNATION	POPULAR NAME	GROSS WEIGHT	ENGINE DESIGNATION	ENGINE TYPE	ENGINE RATING
AERO BUREAU	L-29	DELFIN	7,800	TUMANSKY M701C500	(1) TJ	1,960 LBT (8,72 KN THRUST).
	L-39	ALBATROS	10,300	PROGRESS-IVCHENKO AI-25-TL	(1) TF	3,750 LBT (17,1 KN THRUST).
MIKOYAN BUREAU	MIG-15UT	MIDGET	15,000	KLIMOV VK-1	(1) TJ	5,800 LBT (22,2 KN THRUST).
	MIG-15UT	FAGOT	15,000	TUMANSKY RD-45F	(1) TJ	5,000 LBT (22,2 KN THRUST).
	MIG-21U	MOGOL	20,500	TUMANSKY R-29B	(1) TJ	17,635 LBT DRY, (78,5 KN) AND 27,500 LBT (122,3 KN THRUST) WITH A/B.
SUKHOI BUREAU	SU-7	FITTER-A	30,865	LYULKA AL-7F-L00	(1) TJ	21,150 LBT (93,6 KN THRUST) WITH A/B.
	SU-15	FLAGON	39,600	TUMANSKY R13F2-300	(2) TJ	15,875 LBT (70,6 KN THRUST) WITH A/B.
	SU-17	FITTER-C	30,865	LYULKA AL-21F-3	(1) TJ	24,700 LBT (110 KN THRUST) WITH A/B.
TUPOLEV BUREAU	TU-22U	BLINDER	185,000	KOLIESOV VD-7	(2) TF	30,900 LBT (137,4 KN THRUST) WITH A/B.
YAKOVLEV BUREAU	YAK-28P	MAESTRO	44,000	TUMANSKY R-11	(2) TJ	13,120 LBT (58,4 KN THRUST) WITH A/B.
	YAK-38	FORGER-P	25,795	LYULKA AL-21F-3	(1) TJ	24,700 LBT(110 KN THRUST) NON A/B. PLUS (2) KOLIESOV RD-38, 7,165 LBT LIFT-FANS.

ROTARY WING AIRCRAFT
(ALPHABETICAL BY MANUFACTURER-CIVIL AND MILITARY)

AIRCRAFT MFR.	AIRCRAFT DESIGNATION	POPULAR NAME	GROSS WEIGHT	ENGINE DESIGNATION	ENGINE TYPE	ENGINE RATING	AIRCRAFT MISSION
KAMOV BUREAU	KA-25	HORMONE	16,300	GLUSHENKOV GTD-3BM	(2) TS	990 SHP (738 KW)	ASW/CIVIL
	KA-26	HOODLUM	9,000	ISOTOV TV-0-100	(1) TS	720 SHP (537 KW)	MIL
	KA-27	HELIX A/B/D	24,455	KLIMOV TV3-117V	(2) TS	2,225 SHP (1,644 KW)	ASW/SAR
	KA-28/29	EXPORT HELIX	24,455	KLIMOV TV3-117V	(2) TS	2,225 SHP (1,644 KW)	ASW/SAR
	KA-32	HELIX-C	24,455	KLIMOV TV3-117V	(2) TS	2,225 SHP (1,644 KW)	MIL/CIVIL
	KA-41	HOKUM	16,500	KLIMOV TV3-117V	(2) TS	2,205 SHP (1,644 KW)	ATTACK
	KA-50	----------	23,700	KLIMOV TV3-117V	(2) TS	2,205 SHP (1,644 KW)	ATTACK
	KA-62	----------	25,500	KOLIESOV TVD-1500V	(2) TS	1,500 SHP (1,120 KW)	MIL
	KA-62R	----------	25,500	R-R/TM RTM-322	(2) TS	2,200 SHP (1,641 KW)	CIVIL
	KA-126	HOODLUM	9,000	SOYUZ (ISOTOV) TV-0-100	(1) TS	720 SHP (537 KW)	MIL
	KA-126	--------	3,200	250-C20B	(1) TS	420 SHP (313 KW)	CIVIL
	KA-226	--------	UNKNOWN	250-C20B	(1) TS	420 SHP (313 KW)	CIVIL
MIL DESIGN BUREAU	MIL MI-2	HOPLITE	8,157	KLIMOV GTD-350 (POLISH BUILT)	(2) TS	400 SHP (298 KW)	MIL/CIVIL
	MIL MI-6	HOOK	93,700	PERM-SOLOVIEV D-25V	(2) TS	5,500 SHP (4,101 KW)	MIL/CIVIL
	MIL MI-8	HIP/HAZE	26,455	KLIMOV TV2-117A	(2) TS	1,700 SHP (1,268 KW)	MIL/CIVIL
	MIL MI-10	HARKE	95,800	PERM-SOLOVIEV D-25V	(2) TS	5,500 SHP (4,100 KW)	MIL/CIVIL SKYCRANE
	MIL MI-12	HOMER	231,000	PERM-SOLOVIEV D-25V	(4) TS	6,500 SHP (4,847 KW)	MIL TRANSPORT
	MIL MI-14	HAZE	30,800	KLIMOV TV3-117	(2) TS	2,200 SHP (1,641 KW)	MIL/CIVIL
	MIL MI-17	HIP-H	28,660	KLIMOV TV3-117MT	(2) TS	1,900 SHP (1,417 KW)	MIL TRANSPORT
	MIL MI-24/25/35	HIND	26,455	KLIMOV TV3-117	(2) TS	2,200 SHP (1,641 KW)	ATTACK/TRANS
	MIL MI-26	HALO	123,400	PROGRESS-LOTAREV D-136	(2) TS	11,400 SHP (8,501 KW)	MIL/CIVIL
	MIL MI-28	HAVOC	25,130	KLIMOV TV3-117	(2) TS	2,225 SHP (1,659 KW)	MIL TRANSPORT
	MIL-MI-38	----------	28,000	KLIMOV TV7-117TG	(2) TS	2,370 SHP (1,120 KW)	MIL/CIVIL

AIRCRAFT MFR.	AIRCRAFT DESIGNATION	POPULAR NAME	GROSS WEIGHT	ENGINE DESIGNATION	ENGINE TYPE	ENGINE RATING

CIVIL AIRCRAFT
(ALPHABETICAL BY MANUFACTURER-CIVIL AND MILITARY)

AIRCRAFT MFR.	AIRCRAFT DESIGNATION	POPULAR NAME	GROSS WEIGHT	ENGINE DESIGNATION	ENGINE TYPE	ENGINE RATING
ANTONOV BUREAU	AN-22	COCK	551,000	KUZNETSOV NK-12M	(4) TP	15,000 ESHP (11,186 EKW).
	AN-24	COKE	46,500	PROGRESS-IVCHENKO AI-24VT	(2) TP	2,820 ESHP (2,103 EKW).
	AN-26	CURL	52,900	PROGRESS-IVCHENKO AI-24VT	(2) TP	2,820 ESHP (2,103 EKW), PLUS ONE RU-19A AUXILIARY (TJ) ENGINE FOR TAKE-OFF.
	AN-28	CASH	20,000	GLUSHENKOV TVD-10	(2) TP	990 ESHP (738 KW).
	AN-30	CLANK	50,700	PROGRESS-IVCHENKO AI-20	(2) TP	2,820 ESHP (2,301 KW).
	AN-32	CLINE	59,000	PROGRESS-IVCHENKO AI-20DM	(2) TP	5,180 ESHP (3,683 EKW).
	AN-38	----------	IN DEV.	KOLIESOV TVD-1500B	(2) TP	1725 ESHP (1,301 EKW).
	AN-72	COALER	70,000	PROGRESS-LOTAREV D-436T	(2) TF	16,575 LBT (73,73 KN THRUST), UPGRADE D-36.
	AN-74	COALER-B	70,000	PROGRESS-LOTAREV D-36	(2) TF	14,350 LBT (63,8 KN). UPGRADE OF AN-72 WITH WHEEL-SKI LANDING GEAR.
	AN-120	CAT	121,500	PROGRESS-IVCHENKO AI-20	(4) TP	4,000 ESHP (2,983 KW).
	AN-124	CONDOR/RUSLAN	892,800	PROGRESS-LOTAREV D-18T	(4) TF	51,700 LBT (230 KN THRUST).
	AN-180	PROPFAN	IN DEV.	PROGRESS-LOTAREV D-27	(2) TP	14,000 SHP (10,340 KW).
	AN-218	WIDE-BODY	IN DEV.	PROGRESS-LOTAREV D-18T	(2) TF	51,700 LBT (230 KN THRUST).
	AN-225	MRIYA	1,322,750	PROGRESS-LOTAREV D-18T	(6) TF	51,700 LBT (230 KN THRUST).
BERIEV BUREAU	BE-200	----------	95,000	PROGRESS-LOTAREV D-436T	(2) TF	16,575 LBT (73,73 KN THRUST) UPGRADE D-36.
ILYUSHIN BUREAU	IL-18	COOT	141,000	PROGRESS-IVCHENKO AI-20	(4) TP	4,000 ESHP (2,983 KW).
	IL-62	CLASSIC	368,000	KUZNETSOV NK-8-4	(4) TF	22,273 LBT (99,1 KN THRUST).
	IL-62M	CLASSIC	368,000	PERM-SOLOVIEV D-30KU	(4) TF	24,250 LBT (107,9 KN THRUST).
	IL-76	CANDID	374,000	PERM-SOLOVIEV D-30KP	(4) TF	26,455 LBT (117,7 KN THRUST).
	IL-86	CAMBER	450,000	KUZNETSOV NK-86	(4) TF	28,600 LBT (127 KN) WIDE-BODY.
		ALTERNATE ENGINE		CFM-56-5	(4) TF	26,000 LBT (115,6 KN THRUST).
	IL-96-300	WIDE-BODY	476,200	PERM PS-90A	(4) TF	35,000 LBT (155,7 KN THRUST).
	IL-96M	WIDE-BODY	476,200	PW-2337	(4) TF	35,000 LBT (155,7 KN THRUST).
	IL-108	BUSINESS JET	UNKNOWN	PROGRESS DV-2	(2) TF	4,852 LBT (21,6 KN THRUST).
	IL-114	----------	46,300	KLIMOV TV7-117V	(2) TP	2,500 ESHP (1,862 KW).
MIL BUREAU	MI-6A & P	HELICOPTER	93,700	PERM-SOLOVIEV D-25	(2) TS	5,500 SHP (4,101 KW).
	MI-8	HELICOPTER	26,400	KLIMOV TV2-117	(2) TS	1,500 SHP (1,119 KW).
	MI-14	HELICOPTER	30,800	KLIMOV TV3-117	(2) TS	2,200 SHP (1,641 KW).
	MI-26	HELICOPTER	123,000	PROGRESS D-136	(2) TS	11,400 SHP (8,501 KW).
ROS-AEROPROGRESS	T-101	PAX/CARGO	11,500	UNKNOWN	(1) TP	IN DEVELOPMENT.
SUKHOI BUREAU	SU-51	PAX/CARGO	68,000	LYULKA AL-36	(2) TF	SUPERSONIC (IN DEVELOPMENT).
SUKHOI/NIZHY	A.90	SUPER-SEAPLANE/ EKRANOPLANE	IN DEV.	KUZNETSOV NK-12	(1) TP	15,000 ESHP, PLUS (2) NK-8 (TJ).

AIRCRAFT MFR.	AIRCRAFT DESIGNATION	POPULAR NAME	GROSS WEIGHT	ENGINE DESIGNATION	ENGINE TYPE	ENGINE RATING
CIVIL AIRCRAFT (CON'T)						
TUPOLEV BUREAU	TU-104	AIRLINER	IN DEV.	MIKULIN AM-3M	(3) TJ	20,950 LBT (93,2 KN THRUST).
	TU-114	CLEAT	361,100	KUZNETSOV NK-12	(4) TP	14,795 ESHP (11,030 KW).
	TU-124	COOKPOT	83,700	PERM-SOLOVIEV D-20P	(2) TF	11,905 LBT (52,93 KN THRUST).
	TU-126	CLEAT	361,000	KUZNETSOV NK-12	(4) TP	14,795 ESHP (11,030 KW).
	TU-134	CRUSTY	98,000	PERM-SOLOVIEV D-30	(2) TF	15,000 LBT (66,6 KN THRUST).
	TU-144	CHARGER SST ALTERNATE ENGINE	396,000	KUZNETSOV NK-144 RYBINSK RD-36-51	(4) TF 4 (TJ)	44,090 LBT (196,1 KN THRUST). 44,100 (196,2 KN THRUST).
	TU-154	CARELESS	220,400	KUZNETSOV NK-8-2	(3) TF	22,273 LBT (99 KN THRUST).
	TU-154M	CARELESS	220,400	PERM-SOLOVIEV D-30KU	(3) TF	23,380 LBT (104 KN THRUST).
	TU-204	AIRLINER	206,100	PERM PS-90A	(2) TF	35,000 LBT (155,7 KN THRUST).
	TU-204	AIRLINER	206,100	RB-211-535ER	(2) TF	35,000 LBT (155,7 KN THRUST).
	TU-334	----------	91,000	PROGRESS-LOTAREV D-436T	(2) TP	16,575 LBT (IN DEVELOPMENT).
YAKOVLEV BUREAU	YAK-40	CODING	39,200	LYULKA AI-25 LYCOMING LF-507	(3) TF	3,850 LBT (17,1 KN THRUST). ALTERNATE ENGINE
	YAK-42	CLOBBER	110,000	PROGRESS-LOTAREV D-36	(3) TF	14,350 LBT (63,8 KN THRUST).
	YAK-42M	CLOBBER	110,000	PROGRESS-LOTAREV D-436T	(3) TF	16,575 LBT (73,73 KN THRUST), UPGRADE D-36.

SECTION III

DICTIONARY OF
THE GAS TURBINE ENGINE

A

A-CHECK
Locate under Letter Checks.

A-WEIGHTED DECIBELS (DBA)
A unit of noise measurement in general use for non-aircraft purposes by the civil community. A-Weighted Decibels are often converted to units of aircraft engine noise measurement for airport noise control. Aircraft units are referred to as Effective Perceived Noise Decibels (epndB). *See also: Decibel (dB).*

ABORTED START
An unscheduled termination of the engine starting cycle after engine light-off occurs. The abort procedure is initiated either by the operator or by an automatic engine shut-down device. Reasons for an aborted start might include: Out-of-limits readings on engine instruments, unusual noises within the engine, fuel or oil leaks etc.

ABRADABLE SEAL
An engine component part. The generic description of a removable shroud ring seal or knife-edged seal designed to: 1) wear away clean and smooth, or 2) become slightly trenched, should contact with a rotor blade occur. After contact, a sufficient amount of the seal surface will remain intact to preserve the tip-sealing function, and the blade length will be preserved. *See also: Abradable Shroud.*

ABRADABLE SHROUD (fan & compressor)
An engine component part. A shroud ring constructed of a plastic type material that will wear (abrade) away should a rotating engine part such as a fan blade tip or compressor blade tip contact the shroud. The abradable shroud is designed to lose its material, leaving a clean smooth surface. This arrangement minimizes airflow disturbance, blade length reduction, and blade efficiency losses. Also called Attrition Lining.

ABRADABLE SHROUD (turbine)
An engine component part. A shroud ring constructed of sintered metal or honeycomb metal that will wear (abrade) away upon contact with a turbine blade tip. The abradable shroud protects against blade length reduction and reduced blade efficiency.
See also: 1) Abrasive Tip, 2) Shroud-Ring.

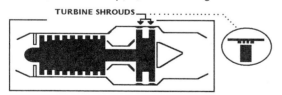

ABRASION
Roughened surface, varying from light to severe. The result of fine foreign material present between moving parts. Locate also under Bearing Inspection and Distress Terms and Engine Inspection and Distress Terms in Appendix.

ABRASIVE TIP (turbine blade)
An engine component part. A turbine blade tip with a silicone carbide insert capable of cutting into its turbine shroud ring should high blade loading or case warpage cause momentary contact. *See also: Abradable Shroud.*

ABSOLUTE PRESSURE
Pressure above absolute zero pressure as read on a barometer type instrument. For example, at International Standard Day (ISD) condition, ambient barometric pressure is 14.7 psia (29.92 in.hg. absolute).

ABSOLUTE PRESSURE GAUGE
A gauge that indicates ambient pressure when no other pressure is applied. This gauge is used where influence of ambient pressure in the readout is required. Such a gauge is included in an Engine Trim Kit to measure Turbine Discharge Pressure. *See also: Relative Pressure Gauge.*

AC STANDARD
Locate under Aircraft Standards.

ACC
Abbreviation for Active Clearance Control.

ACCEL-DECEL TEST (engine)
An engine stall-performance test based on acceleration and deceleration characteristics of an engine as an evaluation criteria. This test requires the Power Lever be moved in one second or less from idle speed to a specified high power setting near take-off power, after which the Power Lever is rapidly moved back to the idle stop. The engine is expected to exhibit relatively stall-free performance on both power changes. *See also: 1) Acceleration Check, 2) Interrupted Decel Test.*

ACCELERATION
A physics term. It represents the change in velocity divided by time. Mathematically defined as final velocity, minus initial velocity, divided by time, or:
$A = [V_2 - V_1] \div t$. Where: "A" is acceleration, V_2 is final velocity, V_1 is initial velocity and "t" is time.

$$A = [V_2 - V_1] \div t$$

ACCELERATION (engine)
Refers to a change from a low engine power setting to a high power setting, such as would be required during aircraft take-off. *See also: 1) Accel-Decel Test, 2) Cold Acceleration.*

ACCELERATION (of gravity)
A physics term. It is the acceleration of a free falling body due to the attraction of gravity expressed as the rate of increase in velocity per unit of time. At sea level, acceleration of gravity is expressed as 32.17 ft./sec.^{2} (9.80665 m/sec.^{2}). Acceleration due to the influence of gravity decreases with increased altitude, dropping to zero upon leaving the earth's gravitational field. *See also: Gravity Constant as a factor in the turbine engine thrust formula, Thrust (gross).*

ACCELERATION BLEED VALVE (or Band)
Locate under Compressor Bleed Valve.

ACCELERATION BLEED SYSTEM
Locate under Anti-Stall Systems.

ACCELERATION CHECK
An engine operational check to measure spool-up time, generally from idle speed to take off power setting. A typical engine spool-up time as required for FAA Type Certification, is approximately 6 to 10 seconds.

ACCELERATION CONTROL UNIT (ACU)
A component of a Fuel Control which ensures correct fuel scheduling for smooth, stall-free engine acceleration.

ACCELERATION THERMOSTAT
A small accessory. A bimetallic probe protruding into the exhaust stream of an auxiliary power unit. When heated to its cracking temperature, the thermostat expands to dump air pressure and create a signal to the fuel control unit, which in turn reacts to bypass fuel and prevent over-temperature within the engine.

ACCELEROMETER
A small engine accessory. A type of Vibration Pickup (transducer). It is engine mounted and contains piezo-electrical crystals which produce an electrical current proportional to the squeezing action on the crystals induced by engine vibration. *See also: Vibration Pickup.*

ACCESSORY
Most widely used to mean definable units, such as mechanisms with moveable or removable parts, attached to an engine. These units can be either essential or non-essential to engine operation. This dictionary identifies all such mechanisms as Accessories and will not distinguish between essential Basic Engine accessories and non-essential Quick Engine Change (QEC) Accessories. *See also: 1) Accessory (basic engine), 2) Accessory (QEC).*

ACCESSORY (basic engine)
Mechanisms supplied by the engine manufacturer in addition to the core portion of an engine. Basic Engine Accessories are considered essential to engine operation and include fuel and oil pumps, fuel controls, valves, solenoids, control mechanisms, etc. Non-essential accessories are generally called QEC Accessories.

ACCESSORY (QEC)
Mechanisms supplied by the aircraft manufacturer and attached during installation of the Quick Engine Change (QEC) kit to a basic engine. These accessories are considered non-essential to the operation of the engine and include such items as hydraulic pumps, electrical generators, fluid pressure transmitters, electrical harnesses, etc.

ACCESSORY DRIVE GEARBOX (ADG)
An engine component. The main unit of the Accessory Section. The gearbox provides mounting points for Accessories, both QEC and Basic Engine types. The mounting pad drive points are gear reduced from compressor speed via internal reduction gearing. Most accessories drive at speeds of 3,000 to 5,000 RPM. The ADG is in most cases also a return point for scavenged oil before it is pumped back to the main oil tank. Also called 1) Main Gearbox, 2) Power Transmission. *See also: Airframe Mounted Accessory Drive.*

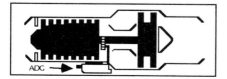

ACCESSORY DRIVE SECTION
Another name for Accessory Section.

ACCESSORY DRIVE SHAFT
An engine component part. A radial shaft that drives an Accessory Drive Gearbox or Auxiliary Drive Gearbox from a bevel gear connected to the compressor rotor shaft.

ACCESSORY GEARBOX
Another name for Accessory Drive Gearbox.

ACCESSORY GEARBOX MODULE
An engine component. The Accessory Drive Gearbox and all of its basic engine accessories. It is designed to be removed and replaced as a single modular unit. Modules are generally tracked independently of engine operating time using module operating hours and/or cycles.

ACCESSORY SECTION
The Accessory Section includes the Accessory Drive Gearbox with all of its accessories mounted [both Basic Engine accessories and Quick Engine Change accessories]. *See also: Engine Sections.*

ACE PROGRAM
Locate under All Cast Engine Program.

ACFE
Abbreviation for Aircraft Company Furnished Equipment.

ACOUSTIC LINERS
Sound absorbing metal liners used in an engine cowling or a tailpipe to reduce noise emissions.

ACTIVATED DIFFUSION HEALING (ADH)
A process for repair of turbine engine parts; especially suited to repair of damaged turbine nozzle guide vanes. ADH is a braze repair method stated to be an improvement over traditional TIG Welding in that it is a batch-type repair process rather than a one part at a time repair. The process begins with a chemical cleaning, followed by a slurry bath in superalloy power and a vacuum furnace brazing.

ACTIVE CLEARANCE CONTROL SYSTEM
An engine system. A recent development in tip clearance control to improve turbine efficiency. Active Clearance Control systems are similar to Passive Clearance Control but with a means of varying the amount of cooling air applied at various points on the engine to match growth factors and maintain very close rotor blade to case clearances. This system is often controlled via an electronic fuel control system. *See also: Passive Clearance Control (turbine).*

ACU
Abbreviation for Acceleration Control Unit.

AD NOTE
Short for Airworthiness Directive, issued by the Federal Aviation Administration.

ADDITIVES (fuel)
Locate under 1) Anti-Misting, 2) Prist.

ADDITIVES (oil)
Locate under Oil Additives.

ADEL CLAMP™
A small expendable part. A type of rubber cushioned loop-shaped clamp. Locate under Loop-Cushioned Clamp.

ADH
Abbreviation for Activated Diffusion Healing.

ADIABATIC
A physics term. Refers to airflow being compressed without loss of heat or gain of heat from outside influences. This would be the case with the ideal compression process.

ADIABATIC (process)
A physics term. The change of state of a gas in which no gain or loss of heat occurs. The large volumetric capacity of a Gas Turbine Engine makes heat gain or loss through its casings negligible. The process of compression and expansion through the Gas Turbine Engine is, therefore, essentially adiabatic. The process, however, is not Isentropic because turbulence (entropy) is present. *See also:* 1) *Diabatic (process)*, 2) *Isentropic Compression*.

ADIABATIC EFFICIENCY (compressor)
A physics term. The actual compressor efficiency expressed in percent. A ratio of the energy input required to achieve a given compressor pressure ratio Isentropically to the energy input actually required. The modern compressor is in the range of 90% to 95% efficient. *See also: Compressor Efficiency.*

ADIABATIC EFFICIENCY (turbine)
A physics term. The actual turbine efficiency expressed in percent. A ratio of the Isentropic temperature drop across the turbine to the actual temperature drop. The modern turbine is in the range of 90% to 95% efficient.

ADIABATIC LAPSE RATE
Locate under Temperature Lapse Rate.

ADP
Abbreviation for Advanced Ducted Propfan.

ADVANCED DUCTED PROPFAN (ADP)
A type of Ultra High Bypass Turboprop engine. One such ADP features a variable pitch, reversible, ducted fan driven by a free turbine and an output reduction gearbox. The bypass ratio is expected to be in the 10:1 to 15:1 range.

ADVANCED DUCTED PROPULSOR
The propeller unit installed on an Advanced Ducted Propfan.

ADVANCED INTEGRATED MANUFACTURING SYSTEM (AIMS)
A highly computerized and automated engine parts production system. For example, one which can take a rough forging, such as a turbine disc, and machine, clean, treat, and inspect it all in one process, with no human interaction.

ADVANCED MECHANICAL ENGINE DEMONSTRATOR (AMED)
A research and development prototype engine built specifically to provide test run data, to prove feasibility of production, and to meet the requirements of FAA-Type Certification.

ADVANCED TURBOPROP (ATP)
Another name for Propfan.

ADVISORY CIRCULAR (AC)
Information pamphlets published by the FAA for the purpose of advising flight personnel, maintenance personnel, and the general public on aviation matters.

ADZE TRIMMER
An altitude compensator used to automatically lean out a fuel/air mixture.

AERATING FUEL NOZZLE
Locate under Fuel Nozzle (aerating).

AERO-GAS TURBINE
Refers to a Gas Turbine Engine for aircraft use, as opposed to an Industrial Gas Turbine.

AERODYNAMIC BLOCKAGE (thrust reverser)
An engine component. A reverser in which a blocker door and a series of cascade vanes are employed to redirect exhaust gas flow to a forward direction. This type reverser is commonly used on the core (hot stream), the fan (cold stream), or both. Also called Cascade Reverser. Often called a Pre-Exit Reverser because the Aerodynamic type reverser redirects the gases prior to their reaching the exhaust nozzle opening. *See also: Thrust Reverser.*

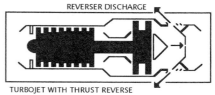

TURBOJET WITH THRUST REVERSE

AERODYNAMIC COUPLING
Refers to the method in which a low pressure turbine rotor is rotated by the spill-over gas energy from the high pressure turbine. There is no direct mechanical coupling between these two units.

AERODYNAMIC DESIGN POINT (compressor)
Point on a compressor stall margin chart where most efficient compression will occur when other factors, such as mass airflow and engine speed, are in match.

AERODYNAMIC DRAG
Form Drag resulting from the shape of airfoils such as compressor blades and fan blades. The imposed resistance on an airfoil by the oncoming airstream which must be overcome by energy extracted within the engine. *See also: Drag.*

AERODYNAMIC EFFICIENCY
The amount of lift as compared to the amount of drag that an airfoil experiences at a given speed. *See also: Lift/Drag Ratio.*

AERODYNAMIC ENGINE STATIONS
A numbering system starting with station (1) at the engine inlet and with progressively higher numbers along the gas path to the exhaust nozzle. Generally the number is accompanied by a prefix F (fan), P (pressure) or T (temperature). For example, P_1 and T_1 stations are in the inlet, P_5 and T_5 are further down

the gaspath. *See also: Engine Stations.*

AEROSPACE GROUND EQUIPMENT (AGE)
All non-airborne equipment required to supply both air and electrical power to aircraft systems when the main engines are not operating. AGE units are also needed to inspect, repair, and test engines and aircraft.

AERO-THERMODYNAMIC DUCT
Locate under Athodyd and Ramjet.

AERO-THERMODYNAMIC ENGINE STATIONS
Numbered reference points along the gaspath where significant changes in gas temperature take place. *See also: Engine Stations.*

AFBMA
Abbreviation for Anti-Friction Bearing Manufacturer's Association.

AFT-FAN (turbofan engine)
An older Turbofan engine design in which the fan is an extension of the power turbine blades rather than an extension of the compressor blades as is the case with forward fan turbofan engines. The Aft-Turbofan was never widely accepted due to exhaust seal leakage problems and resultant loss of efficiency as compared to front fan engines. *See also: 1) Ducted Aft Turbofan, 2) Turbofan Engine, 3) Unducted Fan Engine (UDF).*

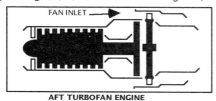
AFT TURBOFAN ENGINE

AFT-THRUST REVERSER
An aircraft manufacturer's supplied component. The hot exhaust reverser on a turbofan engine where the fan exhaust and the core exhaust are unmixed.

AFTERBODY
A nacelle component part. The rear-most portion of the engine nacelle surrounding the engine tailpipe.

AFTERBURNER (A/B)
An engine component. A type of exhaust duct in which exhaust gases leaving a Turbojet or Turbofan engine are reignited to provide maximum gas velocity and thrust. Afterburners provide thrust at the expense of fuel consumption because fuel is introduced after compression has dropped to only a small percentage of what it was in the combustor. The final exit of an afterburner is designed as a variable flow orifice, and referred to as an Eyelid Exhaust Nozzle, a Flap Exhaust Nozzle, or an Iris Exhaust Nozzle. Locate under above terms in this Listing.

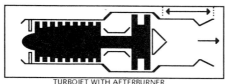
TURBOJET WITH AFTERBURNER

See also: 1) Augmentor, 2) Convergent-Divergent Ehaust Duct 3) Fan-Duct Burner, 4) Plenum Chamber Burner, 4) Variable Area Exhaust Nozzle, 5) Screech Liner.

AFTERBURNER (hard light)
An afterburner which has only one fuel flow mode. When fuel is introduced, an immediate increase in power and noise level occurs. The two-position afterburner is typical of the hard light type afterburner, wherein the full power increase occurs when the nozzle actuates to the open position.

AFTERBURNER (smooth light)
An afterburner with a variable fuel flow capability. The afterburner lights off in the first stage and minimum thrust increase occurs, after which additional stages may be selected using the cockpit power lever. A computer operated, controllable area nozzle, is normally employed in a modern smooth light afterburner. *See also: Variable Area Exhaust Nozzle.*

AFTERBURNER (two-position)
Locate under Eyelid Exhaust Nozzle.

AFTERBURNER THRUST (calculation)
Additional thrust via use of an afterburner is the result of temperature and velocity change in the exhaust duct and is calculated as follows: When afterburner temperature ratio (afterburner outlet temperature divided by its inlet temperature in degrees Kelvin) is known, the exhaust velocity increase is calculated as the square root of the A/B temperature ratio. Percent of thrust increase is proportional to the percentage increase in exhaust velocity. For example: If Temperature Ratio (TR) = $1,542 K. \div 913 K. = 1.69$. Then Velocity Increase = $\sqrt{1.69} = 1.3$ or 30%. Therefore, thrust is increased by 30% of the non-afterburning thrust level. Due to the abundance of oxygen in a turbofan exhaust, the basic engine thrust level can be increased up to approximately 75 percent by the afterburning process.

AFTER-FIRE
A fire occurring in the combustor after engine shutdown. This condition most often is the result of a malfunctioning fuel manifold drain valve or combustion case drain valve which leaves some fuel in the combustor to be reignited by residual heat after a normal shutdown has been accomplished. *See also: 1) Pressurizing and Drain Valve, 2) Combustor Drain Valve.*

AGB
Abbreviation for Accessory Gearbox.

AIAA
Abbreviation for American Institute of Aeronautics and Astronautics.

AIDS
Abbreviation for Airborne Instrument Data System or Aircraft Integrated Data System.

AIMS
Abbreviation for Advanced Integrated Manufacturing System.

AIR
Locate under specific usage such as ambient air, primary air, ram air pressure, secondary air, static air pressure, etc.

AIR (in fuel or oil)
Locate under Entrained Air.

AIR ADAPTER
An engine component. An external duct used to direct airflow. The Air Adapter is located between the diffuser and the combustor in a centrifugal flow engine of the multiple-can combustor type (rarely seen today).

AIR BEARING
Locate under Bearing (air).

AIR-BLAST FUEL NOZZLE
Locate under Fuel Nozzle.

AIR-BLEED SYSTEMS
Locate under Bleed Air (systems).

AIR-BLOWN LABYRINTH SEAL
Another name for Labyrinth Air-Oil Seal.

AIR-BUFFERED OIL SEAL (GTCP Units)
Another name for Labyrinth Air-Oil seal.

AIR-COOLED BLADES AND VANES
Locate under Turbine Cooling.

AIR DENSITY
Locate under Density (air).

AIR DIFFUSER HOUSING
Another name for Diffuser Case. A main structural part of a turbine engine.

AIR EXTRACTION PAD
Locate under Extraction Pad.

AIR-FUEL RATIO
Locate under Stoichiometric Mixture.

AIR-GAP TUBE (igniter system)
A component in the ignition exciter unit secondary circuit that acts as an electronic switch. Each time the storage capacitor has sufficient electron build up to overcome the air-gap, current flows through this specially designed glass tube to the igniter plug.

AIR IMPINGEMENT STARTING
Locate under Impingement Starter.

AIR INLET CONTROLLER
An aircraft computer system that schedules the position of airflow control devices in the flight inlet of supersonic aircraft. *See also: Convergent-Divergent Inlet.*

AIR INLET SEPARATOR
Locate under Inertial Ice and Sand Separator.

AIR-MAZE OIL SEAL
Another name for Labyrinth Air-Oil Seal.

AIR-MIST FUEL NOZZLE
Locate under Fuel Nozzle.

AIR- MIST OIL JET
Locate under Oil Jet.

AIR-OIL COOLER
An engine accessory. An air cooled, radiator-type device with oil passing through its cores and ambient, ram air passing over the exterior of the cores. The function of an oil cooler is to maintain a predetermined engine oil temperature. A thermostatic bypass valve in the inlet of the cooler allows oil to bypass the cooling chamber on start up when the oil is cold. Not as widely used as the fuel cooled oil cooler.
See also: Oil Cooler.

AIR-OIL SEPARATOR (oil tank)
Another name for Deaerator (oil tank).

AIR-OIL SEPARATOR (rotary)
A small engine accessory. A lubrication system device in the vent subsystem. It is a centrifuge or slinger located within the Accessory Drive Gearbox and into which oil laden air is directed. The oil is centrifuged back into the gearbox sump and relatively clean air passes out to the atmosphere. Also called 1) Breather Impeller, 2) Gearbox Deaerator, 3) Rotary Deaerator, 4) Rotary Breather, 5) Rotary Air-Oil Separator. Sometimes additional small rotary separators may be located in bearing sumps to reduce aeration of scavenge subsystem oil. *See also: Center Vent System.*

AIR SEAL
An engine component part. Air Seals are typically in the form of thin metal rims commonly called knife-edge seals, positioned between stationary and rotating parts. Air seals are designed to act as air dams to control air leakage. These seals can be seen on turbine blade tips and compressor and turbine shafts. Often air seals also form a metering orifice to supply a precise amount of air to internal locations of the engine for such purposes as cooling turbines blades, turbine vanes, turbine discs, etc.

AIR SEAL (flange)
An engine component part. Generally a seal of rubberized material serving as a gasket to prevent air leakage through mating flanges. In higher heat areas the air seal may be a gas-filled metal o-ring type seal.

AIR SPLITTER
Another name for Splitter Fairing.

AIR START
Engine re-starting after in-flight shutdown or flameout. This procedure is normally accomplished by initiating engine ignition and employing the dynamic forces of the airstream to windmill the engine until combustion occurs. When combustion does not occur within approximately 30 seconds, fuel is stop-cocked and the procedure is repeated. The starter is not normally needed in this procedure. When airstream forces do not provide sufficient engine rotation (approximately 20% N2 Speed), a starter assisted Air Start procedure is used. Also called In-Flight Relight.

AIR STARTER OR AIR START MOTOR
Additional names for Pneumatic Starter.

AIR STREAM (engine primary)
Locate under Primary Airstream (turbofan).

AIR STREAM (engine secondary)
Locate under Secondary Airstream (turbofan).

AIR TAKE-OFF
British term for extraction pad (customer bleed air).

AIR TRANSPORT ASSOCIATION OF AMERICA

An association made up of aerospace professionals, that publishes technical standards for the aviation industry. For example: ATA NO.100 establishes a standard numbering system for the published technical data used to maintain all commercial aircraft, aircraft engines and aircraft accessories. The ATA method employs a six-part number, the first two digits identify the Chapter containing a particular system, the second two digits identify a Section of that system, and the last two digits identify Individual Units. For example the reference 72-55-02 in any aircraft maintenance manual published under ATA No. 100 can be understood to mean: 72 identifies the Engine, 55 identifies the Turbine Section within the engine, and 02 identifies the Gas Producer Turbine unit within the turbine section. Location 72-55-02 in the aircraft maintenance manual, therefore, contains information about the engine's Gas Producer Turbine.

AIR TURBINE STARTER

Another name for Pneumatic Starter.

AIR TURBO-RAMJET (ATR)

Locate under 1) Turbo-Ramjet, 2) Variable Cycle Engine.

AIRACOMET

Locate under Bell Airacomet.

AIRBORNE INSTRUMENT DATA SYSTEM (AIDS)

An automated data acquisition system which records in-flight engine and aircraft parameters. The AIDS is used to plot trend analyses for detection of early part failures and for planning of maintenance actions. Also called Aircraft Instrument Data System. *See also: Ground-Based Engine Monitoring (GEM).*

AIRBORNE VIBRATION MONITORING SYSTEM (AVM)

A cockpit system used to monitor the vibration levels at various points on an engine where vibration pickups (transducers) are located. Pickups are small generators containing a crystalline ceramic element which generates an electrical signal when vibrated. The signal is sent to a cockpit indicator which displays vibration level in AVM Units which cam be converted to thousandths of inches (MILS).
See also: AVM Unit.

AIRCRAFT COMPANY FURNISHED EQUIPMENT (ACFE)

Accessories that configure an engine to a particular aircraft. Often the same as Accessories (QEC).

AIRCRAFT INLET

Locate under Flight Inlet.

AIRCRAFT INSTRUMENT DATA SYSTEMS (AIDS)

Another name for Airborne Instrument Data Systems.

AIRCRAFT INTEGRATED DATA SYSTEM (AIDS)

Another name for Airborne Instrument Data System.

AIRCRAFT OPERATING CYCLE

Completed take-off and landing sequence. Touch and go landings are counted as Aircraft Operating Cycles.

AIRCRAFT SPECIFICATIONS

A detailed description of aircraft materials, products and services. For Example: Military Specifications (Mil Specs).

AIRCRAFT STANDARDS

A rule or basis of comparison of quality, content or value established by law or general usage. In the aviation industry, Aircraft Standards are commonly used for such items as hardware, tubing, oils and fuels. The most common Standards accepted by the aircraft industry are as follows:

AC	Air Corps (rarely used today)
AIAA	American Institute of Aeronautics and Astronautics
AISI	American Iron and Steel Institute
AMS	Aeronautical Material Standard
AN	Army-Navy or Air Force-Navy
API	American Petroleum Institute
AS	Aeronautical Standard
ASA	American Standards Association
ASTM	American Society for Testing Materials
MS	Military Standard
NAS	National Aircraft Standard

AIRFLOW MODULATOR RING (engine inlet)

A blocking ring that moves into the inlet at low RPM to reduce flow area excluding unwanted airflow. Used as an anti-stall device. Similar in function to a variable inlet guide vane system.

AIRFLOW STATIONS

Locate under Aerodynamic Engine Stations.

AIRFOIL (3-D)

Airfoils such as compressor and turbine vanes that have a noticeable backwards bend from base to tip.

AIRFRAME BLEED AIR

Another name for Customer Bleed Air.

AIRFRAME MOUNTED ACCESSORY DRIVE (AMAD)

A new concept in Accessory Drive Gearbox design. The AMAD is mounted in the nacelle rather than on the engine and is engine driven via an external power take-off shaft. It is generally a wet sump type gearbox containing the bulk of the engine oil supply.

AIRPLANE MAINTENANCE ENGINEERING TIME STANDARD (AMETS)

A Boeing Company system of calculating time requirements for completing maintenance tasks on airplanes. The items typically considered are: man-hours, crew size, elapsed time, tooling requirements, etc.

AIRWORTHY (airworthiness)

In terms of the Federal Aviation Administration, the condition of an aircraft, engine or component that meets all of the original requirements for its FAA Certification.

AIRWORTHINESS DIRECTIVE

A written notice published under FAR Part 39, sent to registered owners of FAA Certificated aircraft and engines to inform them of a problem area that must be corrected if the equipment is to remain in an airworthy status. Sometimes referred to as an AD or an AD Note.

ALFA RANGE

Locate under Alpha Range.

ALFA ROMERO COMPANY (ITALY)
Gas Turbine Engine manufacturer. Locate in section entitled "Gas Turbine Engines for Aircraft".

ALL CAST ENGINE PROGRAM (ACE)
An industry-wide effort to cast in one piece, parts that were formerly assemblies of forged or cast parts. Casting in this way reduces cost of parts, reduces assembly time, and reduces future maintenance costs. Newer procedures for thin wall casting have made the ACE Program more feasible.

ALLISON TURBINE ENGINE COMPANY
Locate in section "Gas Turbine Engines for Aircraft".

ALLOYS-SUPER
Locate under Super alloys.

ALPAK COATING
A type of aluminum thermal barrier coating used especially on turbine blades and vanes. *See also: Thermal Barrier Coating.*

ALPHA RANGE
Flight operating regime of Turboprop and Turboshaft engines.

ALTERNATE THRUST RATING
Locate under Reserve Thrust Rating.

ALTITUDE SENSING UNIT (ASU)
A component part of a Fuel Control. Typically it is a bellows unit which sends a pressure signal to the control for fuel scheduling purposes.

ALUMINA OXIDE
Another name for Aluminum Oxide.

ALUMINIDE COATING
A surface coating designed to provide resistance to oxidation and sulfidation. The part, packed in aluminum powder, is placed in a furnace where aluminum chloride gas is present. The process leaves an aluminum deposit on the surface. Further heat treating creates a diffusion bond of the aluminum into the base metal 0.002 to 0.004 inch thick.
See also: 1) Hot Corrosion, 2) Plasma Spray Coating, 3) Thermal Barrier Coating.

ALUMINIZING
A metal plasma spray coating process that bonds either a corrosion resistant or a wear resistant outer surface to a base metal. *See also: Plasma Spray Coating.*

ALUMINUM OXIDE (igniter plug)
A white colored material used as an insulator at the firing end of many High Voltage Type turbine igniter plugs. This material has a heat limit of approximately 1,300°F. *See also: Beryllium Oxide.*

AMAD
Abbreviation for Airframe Mounted Accessory Drive.

AMBIENT AIR
Air under atmospheric pressure moving over or encompassing all sides of an engine (of an aircraft).

AMBIENT PRESSURE
Air at prevailing atmospheric pressure. For example, at International Standard Day (ISD) conditions, ambient pressure is 14.7 psia (29.92 in.hg. absolute).

AMBIENT TEMPERATURE
Air at prevailing atmospheric temperature. For example, at International Standard Day (ISD) conditions, ambient temperature is 59°F (15°C).

AMED
Abbreviation for Advanced Mechanical Engine Demonstrator.

AMERICAN INSTITUTE OF AERONAUTICS AND ASTRONAUTICS (AIAA)
An association of aerospace professionals who establish standards for the aerospace industry. *See also: Aircraft Standards.*

AMERICAN SOCIETY OF TESTING MATERIALS (ASTM)
An association of aviation professionals who publish standards for aerospace and other industries. Fuels and oils are commonly designed by ASTM Standards.

AMETS
Abbreviation for Airplane Maintenance Engineering Time Standard.

AMS
Abbreviation for Aviation Material Specification. *See also: Aircraft Standards.*

AN STANDARD
Locate under Aircraft Standards.

ANECHOIC CHAMBER
A test chamber free of echoes and reverberations in which sound transmissions are measured. For example, tests to determine the level of engine noise intrusion into an aircraft cabin are completed is an Anechoic Chamber.

ANGLE-OF-ATTACK (compressor blades)
The resultant of two airflow vector forces, 1) compressor blade RPM effect on airflow, and 2) inlet airflow effect. Compressor stalls occur when the limits of this relationship are exceeded, similar to the way in which an aircraft wing stalls. *See also: Effective Angle-of-Attack.*

ANGLE-OF-INCIDENCE (compressor blades)
Angle at which blades are set into the compressor disc. It is a fixed angle in all cases except in a variable pitch fan. Angles are set for optimum airflow at altitude-cruise flight condition. *See also: Effective Angle-of-Attack.*

ANGLED GEARBOX
An engine component. A type of auxiliary gearbox. This component receives its name because its drive shaft is angled rather than perpendicular to the rotor shaft bevel gear system.

ANGLED DRIVE SHAFT
An engine component part. A shaft that drives an Angled Gearbox.

ANNULAR COMBUSTOR
An engine component. Annular meaning ring-shaped, this combustor contains a cylindrical or ring-shaped

liner. It is the most widely used combustor design for both small and large engines. Its liner has less wall surface to cool and it is shorter in length for a given volume of combustion space than other combustor liners. See below for different designs of annular combustors. *See also: Combustor.*

ANNULAR COMBUSTOR (dual-stage)
A combustor with a single compartment liner, but one utilizing two fuel nozzles (one downstream of the other), and two flame zones. This arrangement provides for improved emission control over the conventional combustor designs. The forward nozzle operates continuously, the downstream nozzle operates only at higher power settings of flight idle to take-off power. Similar in operation to the Annular Combustor (dual-zone). Also called (dual-dome).

ANNULAR COMBUSTOR (dual-step)
An older design. Whereas the annular combustor has only one combustion area, sometimes referred to as a one-basket combustor, the dual-step has two combustion areas. The inner combustion area is coaxially located within the outer, each section with its own fuel nozzles. Both sections are in operation at all engine power settings. Also called (double annular).

ANNULAR COMBUSTOR (dual-zone)
A design with a two compartment liner. One compartment is used as a pilot flame zone for start-up to idle speed; the other is used as a main flame zone during higher power operation. At higher power settings both zones are in operation. This design was developed for improved emission control. This combustor requires two fuel manifolds, one for each flame zone. The dual-zone combustor is similar in function to the two-stage combustor. Also called 1) Dual-Annular Combustor, 2) Double-Annular Combustor. *See also: Annular Combustor (dual-stage).*

ANNULAR COMBUSTOR (machined ring)
A combustor containing a liner constructed of separate machined rings welded together to form a single unit. It is of heavier construction than the traditional rolled sheet metal liner. Also called 1) Ring Type Combustor, 2) Roll Ring Combustor.

ANNULAR COMBUSTOR (reverse-flow)
A combustor design with an S-shaped path the air follows as it flows from the diffuser section to the turbine section. This design shortens engine length because the liner is coaxial to the turbine wheel(s), rather than in tandem to the turbine wheel(s), as in a conventional annular combustor. Another advantage is that reverse-flow keeps heavier than air particles such as water, sand, etc., from entering the primary combustion zone. This combustor is the most popular design for Turboprop and Turboshaft engines. However, its width prohibits use in larger Turbojet and Turbofan engines. Also called 1) External Annular Combustor, 2) Folded or Foldback Annular Combustor.

ANNULAR COMBUSTOR (scroll-type)
A combustor that utilizes scroll carburetion to atomize fuel instead of a high pressure atomizing spray. Locate under Scroll Carburetion.

ANNULAR DIFFUSER
Another name for Diffuser. Most compressor diffusers are annular in shape.

ANNULUS
The opening between two concentric rings. For example, the flow area between a turbine shaft housing and the combustor outer casing could be referred to as the combustor annulus.

ANNUNCIATION PANEL
A cockpit display containing warning lights, information lights, and worded messages to assist the flight crew.

ANTI-FRICTION BEARING
Locate under Bearings.

ANTI-FRICTION BEARING MANUFACTURER'S ASSOCIATION (AFBMA)
An association which sets standards for U.S. industry regarding bearings, bearing terminology, bearing dimensions, tolerances, gauging practices, identification coding, load ratings, etc.

ANTI-ICER SYSTEM (engine)
An engine system. A hot air system that ducts Engine Bleed Air into the inlet area to prevent the formation of ice. Some Turbofan engines do not need anti-ice in the engine inlet because the slinging action of the fan during ground runup and the high inlet stagnation temperatures in flight are sufficient to prevent ice build-up. The inlet cowling, however, is ice protected by bleed air or by electrical heat strips. On some smaller engines, heated air discharging from the engine Air-Oil Cooler is used to supplement Engine Bleed Air.

ANTI-ICING (fluid)
Locate under Freezing Point Dispersant.

ANTI-ICING (fuel)
Locate under 1) Fuel Heater, 2) Fuel Icing, 3) Prist.

ANTI-LEAK CHECK VALVE (lubrication system)
Locate under Static Anti-Leak Check Valve.

ANTI-MISTING FUEL ADDITIVE
A safety additive which changes the molecular structure of jet fuels to provide a jel-type viscosity with low misting characteristics when fuel is released in air. When misting is minimized, danger of post accident fuel fires is reduced. This additive is in an experimental stage and not in general use.

ANTI-ROTATION PIN
A small metal pin that prevents rotation of stationary engine components. A convenient, light-weight means of fastening without a nut or bolt when compression and tension loads are not required.

ANTI-SCREECH LINER
Another name for Screech Liner.

ANTI-SEIZE COMPOUND
A thread lubricant used to prevent hot corrosion on threaded fasteners in the hot section of turbine engines.

ANTI-SIPHON LOOP (lubrication system)
A Lubrication System scavenge return line with a loop positioned higher than the oil tank inlet. The loop breaks the suction effect and prevents the oil tank from siphoning back down to the sumps after engine shutdown, especially when the oil level in the tank is slightly high. Also called Anti-Siphon Tube.

ANTI-STALL SYSTEMS
Engine systems designed to aid in preventing compressor stalls and surges. Locate under 1) Compressor Bleed Band System, 2) Compressor Bleed Valve System, 3) Variable-Angle Vane System.

ANTI-STATIC LEAK CHECK VALVE (lubrication system)
Locate under Static Anti-Leak Check Valve.

ANTI-SURGE BLEED VALVE
Another name for Compressor Bleed Valve.

ANTI-WINDMILLING BRAKE
An engine accessory. A friction brake fitted as an accessory to the Accessory Drive Gearbox of some older Turbojet engines. The brake was in place to prevent engine rotation during in-flight shutdowns. This device is rarely seen today. Some Turboshaft and Turboprop engines include a braking system in their reduction gearbox mechanism, but this is not the friction brake described here. *See also Brake (propeller or rotor).*

API
Abbreviation for American Petroleum Institute. *See also: Aircraft Standards.*

APR
Abbreviation for Automatic Power Reserve.

APPROACH IDLE SPEED
Another name for Flight Idle Speed.

AROMATIC FUEL
Aromatic hydrocarbon additives are blended into reciprocating engine fuels to increase rich-mixture performance ratings. Aromatics are not used with Jet Fuels.

ARP
Abbreviation for Automatic Reserve Power. Same as Automatic Power Reserve (APR).

ARRESTED PROPELLER SYSTEM
Locate under Propeller Brake.

ART
Abbreviation for Automatic Reserve Thrust. Another name for Automatic Reserve Power.

ARTICULATED GUIDE VANE
A component part of a GTCP auxiliary engine. A variable angle guide vane in a load compressor inlet. Its function is to partially block the air inlet when aircraft demand for pressurized air is at a minimum level. *See also: 1) Gas Turbine Compressor-Power Unit, 2) Load Compressor.*

AS STANDARD
Locate under Aircraft Standards.

ASA
Locate under Aircraft Standards.

ASPECT RATIO (compressor blade)
Ratio of blade length to mean chord width. A high aspect ratio means increased blade length and decreased chord width. High bypass Turbofan blades are generally of this design to produce less drag and improve airflow. The wider, low aspect ratio blade is used where increased mechanical strength is required. *See also: Low Aspect Ratio Blading.*

A.S.T.M.
Locate under American Society of Testing Materials. *See also: Aircraft Standards.*

ASU
Abbreviation for Altitude Sensing Unit.

ATHODYD
A type of jet engine. Acronym for Aero-thermo-dynamic duct, better known as a ramjet. *See also: Ramjet.*

ATOMIC ABSORPTION SPECTROMETER
A portable spectrometer device used to measure contamination levels in engine oil. *See also: Atomic Emission Spectrometer.*

ATOMIC EMISSION SPECTROMETER
A larger unit than an Atomic Absorption Spectrometer. It will more quickly analyze oil samples than a portable unit. *See also: Spectrometric Oil Analysis Program.*

ATOMIZATION (fuel)
Refers to fine droplets of fuel entering the engine combustor. The droplets are of a size such that five to ten are needed to cover an area the size of the head of a pin. *See also: Atomizing Spray Fuel Nozzle.*

ATOMIZER (fuel)
Refers to Atomizing Spray Fuel Nozzle.

ATOMIZING SPRAY FUEL NOZZLE
Fuel nozzle that creates a very fine fuel mist in a cone shaped pattern by forcing large quantities of fuel under high pressure through a small metering orifice. *See also: Fuel Nozzles.*

ATP
Abbreviation for Advanced Turboprop.

ATR
Abbreviation for Air-Turbo-Ramjet.

ATTRITION LINING
Another name for Abradable Seal.

AUGMENTATION SYSTEMS (engine)
Systems used to boost engine thrust, mainly during take-off. Most common types of augmentation are engine afterburning and water injection. Less common today is the use of Jet Assisted Takeoff (JATO), a liquid rocket assist system. *See also: 1) Afterburner, 2) Water Injection.*

AUGMENTED TURBOFAN
A Turbofan Engine configured with an afterburner or a water injection system. Also applies to Turbojet Engines.

AUGMENTED WING
Another name for Blown Flap Augmentation.

AUGMENTOR
A engine component. A recent name in reference to a type of Afterburner which employs both core discharge and fan discharge to support combustion. *See also: Afterburner.*

AUGMENTOR (rotary)
An engine component. Early name given to the fan portion of a Turbofan engine.

AUSTENITIC METAL
A non-magnetic, corrosive-resistant carbon steel with high nickel content, referred to as stainless steel. Stainless Steels have many applications in the Turbine Engine.

AUTO-FEATHERING
Refers to a system which quickly feathers a propeller in the event of engine failure to reduce aerodynamic drag. *See also: Thrust Sensitive System.*

AUTO-FEATHERING SYSTEM
Locate under Thrust Sensitive System.

AUTO-IGNITION
Locate under Ignition Systems.

AUTO-ROTATION
Refers to airstream induced rotation of a helicopter rotor when engine power is lost to the system. Helicopters are capable of landing successfully in this condition because airflow is in the reverse direction upwards; slowing the aircraft sink rate. Could also mean propeller windmilling.

AUTOMATIC CLEARANCE CONTROL SYSTEM
Another name for Active Clearance Control (British).

AUTOMATIC FUEL CONTROL UNIT
Another name for Fuel Control Unit.

AUTOMATIC PERFORMANCE RESERVE (APR)
Another name for Automatic Reserve Power.

AUTOMATIC PRESSURE RELIEF VALVE (filters)
Another name for By-Pass Relief Valve.

AUTOMATIC PULSED CHIP DETECTOR
Locate under Chip Detector (capacitive discharge).

AUTOMATIC RESERVE POWER (ARP)
An aircraft system capable of monitoring engine speeds on various types of turbine engines. When the speed of one engine drops below a predetermined value, the remaining engine(s) receive a thrust boost signal through the engine Reserve Take-off Thrust (RTT) system. *See also: Reserve Thrust Rating.*

AUTONOMOUS ELECTRO-MECHANICAL PITCH CHANGE UNIT
A propeller pitch changing system located within the propeller dome. This unit contains no slip rings, as with other electric pitch change mechanisms. It is designed for large Prop-Fans of the future.

AUTOSYN TRANSMITTER™
A trademark of the Bendix Corporation. An A.C. electric transmitter that powers an aircraft cockpit instrument. This type of transmitter provides inputs to such indicators as oil pressure, fuel pressure, and fuel flow. *See also: Selsyn Transmitter.*

AUXILIARY ACCESSORY GEARBOX
An engine component part. A small gearbox, other than the Accessory Drive (main) Gearbox, on which accessories are mounted. This gearbox is often located on the front of the engine and driven by the Low Pressure Compressor. Also called Inlet or Low Speed Gearbox when located on the front of the engine.

AUXILIARY DRIVE GEARBOX
An engine component part. A small gearbox that is driven by the main rotor shaft by a beveled gear system and a radial shaft. This gearbox is used to drive the Accessory Drive Gearbox via a horizontal drive shaft. Used where a direct drive to the Accessory Drive Gearbox is not desired. Also called an Intermediate Drive Gearbox.

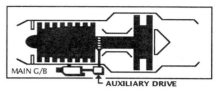

AUXILIARY ENGINE
An aircraft installed engine that is not a prime engine. In years past many reciprocating engine powered aircraft also had one or more small Turbojet Auxiliary Engines to assist in Take-Off at highly loaded conditions. Modern Helicopters designed for high speed flight are also sometimes fitted with small Turbojets to assist in high speed, forward flight.

AUXILIARY POWER UNIT (APU)
Aircraft installed small gas turbine engine used primarily to supply electrical power, bleed air, or both to aircraft systems when the main engines are not operating. This unit can also be operated as an emergency electrical power source in flight. APU's that drive an electrical generator only are classed as Turboshaft engines. When the APU also provides bleed air it is classed as a Turbo-Compressor. The typical APU can supply air for engine starting up to 25,000 feet and electrical power up to 35,000 feet altitude. *See also: 1) Gas Turbine Compressor-Power Unit, 2) Gas Turbine Power Unit, 3) Twin-Spool Compressor-Power Unit.*

AVCO CORPORATION (Lycoming Division)
Former name of Textron-Lycoming Company. Manufacturer of Gas Turbine Engines.

AvFuel-2M™
A type of Fuel System Icing Inhibitor.

AVIATION STANDARDS
Locate under Aircraft Standards.

AVM
Abbreviation for Airborne Vibration Monitoring System.

AVM UNIT
The units of engine vibration observed on a cockpit display. AVM units do not usually read out in Mils as do engine Vibration Meters in engine test cells. AVM units provide for ease of establishing limits. For example; one Fan AVM unit may equal 1.75 mils, whereas one Turbine AVM unit may equal 2.0 mils.

AXIAL
Along an axis or around an axis as in axial airflow or axial flow compressor.

AXIAL-CENTRIFUGAL COMPRESSOR
Locate under Combination Flow Compressor.

AXI-CENTRIFUGAL COMPRESSOR
Another name for Combination Flow Compressor.

AXIAL CLEARANCE
Generally thought of as clearances measured between rotating and stationary engine components along the length of the engine. For example, the clearance between the inlet guide vane assembly and the first stage compressor disc.

AXIAL FACE-TYPE CARBON SEAL
Another name for Carbon Oil Seal (face type).

AXIAL FLOW COMPRESSOR
A major engine component. A component which utilizes a rotor blade and stator vane arrangement to increase air pressure at each of its several stages. In larger commercial engines, Axial Flow Compressors traditionally have from 15 to 20 stages. A Fan in a Turbofan Engine is also considered an Axial Flow Compressor and it can be configured with one or more stages. Locate types of axial compressors under: 1) Single-Spool Compressor, 2) Dual-Spool Compressor, 3) Triple-Spool Compressor. *See also: Compressor.*

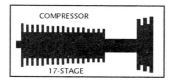

AXIAL FLOW ENGINE
An Engine configured with an axial flow compressor.

AXIAL FLOW TURBINE
Locate under Turbine (axial flow).

AXIAL LOADING
Forces created along the axis of the engine; induced by internal engine pressures. These forces try to move the compressor forward and the turbine rearward and are controlled by ball (thrust) bearings located on the main rotor shaft.

AXIAL RUNOUT CHECK
A dial micrometer measurement of the flatness (fore and aft movement) of a part when rotated 360° about its axis. For example, a runout measurement on the rear face of a turbine disc. *See also: Radial Runout Check.*

B

B-CHECK
Locate under Letter Checks.

BACKLASH CHECK
A dial micrometer measurement to ensure the correctness of mesh (clearance) between mating gears. For example, the clearance between the compressor driven bevel gear, which drives the Accessory Drive Gearbox, and its pinion gear.

BALANCE CHAMBER (thrust)
An air chamber within the engine. A chamber pressurized with air extracted from the compressor section. Its function is to absorb a portion of the axial loading created by the compressor and turbine and thereby assist the main thrust bearings. Also called 1) Balance Piston, 2) Thrust Balance Chamber.

BALANCE CHECKS (compressor & turbine rotors)
Static and dynamic (spin) checks performed on rotating components such as compressor rotors, fan rotors and turbine rotors. Balance checks are performed during the component final assembly during engine build and also on rotors in service during specified maintenance checks. Balance Checks fall into three categories; Locate under the following: 1) Single-Plane Balancing, 2) Moment Weight Balancing, 3) Dynamic Balancing. *See also: 1) Geometral Multiplane Balancing Procedure, 2) Fan Trim Balance.*

BALANCE PISTON
Another name for Balance Chamber.

BALANCE WASHER
Another name for Balance Weight.

BALANCE WEIGHTS
Small component parts. Metallic weights fitted to rotating assemblies such as compressor and turbine rotors to reduce their tendencies to vibrate during operation. Also called Counterweights.

BALANCED-PISTON TORQUEMETER
Refers to the traditional Engine Oil Pressure Operated, hydromechanical torque measuring system seen on many Turboprop and Turboshaft Engines. *See also: Torquemeter.*

BALL BEARING
Locate under Bearing.

BANDING (bearings)
Locate under Bearing Inspection and Distress Terms in Appendix.

BARBER, JOHN
Eighteenth century scientist who, in 1791, obtained a patent for a machine that operated on a thermodynamic cycle of intake, compression, expansion and exhaust. This cycle was later to become the principle of operation of the internal combustion engines of that era and the modern Gas Turbine Engine.

BAROMETRIC PRESSURE
The absolute static pressure in Pounds per Square Inch absolute (PSIA) or Inches of Mercury (In. Hg. absolute). The weight of a column of air from point of measurement to the top of the atmosphere in balance with a column of mercury. For example, on a Standard Day (ISD) at sea level altitude, atmospheric air will be in perfect balance with a one square inch column of mercury 29.92 inches in height.

BAROMETRIC TEMPERATURE
In U.S. Units, the absolute temperature in degrees Fahrenheit plus 460, also called degrees Rankine. In Metric Units, degrees Celsius plus 273, also called degrees Kelvin.

BARRIER COATING
Locate under Thermal Barrier Coating.

BARRIER PAPER
A moisture proof paper used for wrapping engine parts or whole engines, for protection against corrosion or contamination during storage.

BARRIER TYPE FILTER (fuel or oil)
A traditional full flow filter wherein pressurized fluid passes through a barrier (filtering element) on its way to other units in the system. The barrier traps debris entrained in the fluids and prevents further movement through the system. *See also: Filters.*

BASIC ENGINE
An engine with all the accessories, lines, components, and fittings necessary to operate, but without a Quick Engine Change Kit installed. Quite often the basic engine is all that is supplied by the engine manufacturer. The remaining accessories, lines, fixtures, fasteners etc. are supplied by the aircraft manufacturer to adapt the engine to a particular aircraft. *See also: 1) Built-Up Engine, 2) Power Pack, 2) Quick Engine Change Kit.*

BAYONET OIL LEVEL GAUGE
Another name for Dip Stick.

BEARING (air)
Type of plain bearing with a ceramic outer shell separated from a rotating shaft by a high pressure film of air and with no fluid lubrication. Presently under development as a low friction type bearing in locations where oil film strength tends to break down at high temperatures.

BEARING (anti-friction)
Refers to ball bearings and roller bearings which utilize the principles of rolling friction. These bearings provide much less friction than the sliding friction produced by a plain bearing. *See also: Bearing (engine).*

BEARING (ball)
A type of anti-friction bearing used as a main bearing in a Gas Turbine Engine. A caged anti-friction bearing designed to absorb both axial (thrust) loads and radial (centrifugal) loads. A Ball Bearing is caged and rides in grooves in both its inner and outer races. Ball bearings are of two common types: 1) Non-Separable Type (installed and removed as one unit), 2) Split Inner Race Type (with a two-piece split inner race). *See also: Thrust Bearing.*

BEARING (checks)
Bearings are critical engine parts, and therefore have very specific inspection criteria. Locate under Bearing Inspection and Distress terms in Appendix.

BEARING (engine)
Small engine components. Locate under Bearing: 1) air, 2) anti-friction, 3) ball, 4) oil damped, 5) roller.

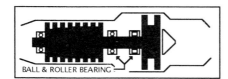

BEARING (oil damped)
A bearing with a vibration damping design. This bearing has pressurized engine lubrication oil between the outside surface of its outer race and the bearing housing. The oil film dissipates energy from unbalanced motion in rotating parts and prevents dynamic loads from being transmitted into the bearing housing. Also called Squeeze-Film Damper Bearing.

BEARING (plain)
A bearing constructed of flat metal and without balls or rollers. For example, a bushing used to support a shaft in an Accessory Drive Gearbox. A type of bearing not used as a main bearing in a Gas Turbine Engine. Also called Friction Bearing or Sleeve Bearing.

BEARING (roller)
A type of anti-friction bearing used as a main bearing in a Gas Turbine Engine. Roller bearings are fitted into a groove on only one race, either the inner or the outer. This design allows the rollers to shift axially on one race as heat expansion changes dimensions of both rotating and stationary parts. The Roller Bearing is used to absorb radial loading in compressor and turbine rotors.

BEARING (squeeze film)
Another name for Bearing (oil damped).

BEARING CAGE
A component part of a bearing assembly. The part of a bearing assembly that houses the balls or rollers and maintains uniform spacing.

BEARING CAVITY
Locate under Bearing Housing.

BEARING DEFECTS
Locate under Bearing Inspection and Distress Terms in Appendix.

BEARING DEGAUSSER
A unit of support equipment. A device that removes magnetism which accumulates in bearings and other rotating ferrous metal parts of turbine engines. *See also: Bearing Field Detector.*

BEARING DISTRESS TERMS
Locate in Appendix.

BEARING FIELD DETECTOR
A small tool. A device employed in detecting the presence of magnetism in ferrous engine parts. This magnetism is induced by constant rotation of engine parts such as bearings and journals. *See also: Bearing Degausser.*

BEARING HEATER TANK
A unit of support equipment. An oil bath type heater used to expand bearing components so they can be hand fitted over shafts or into bearing housings.

BEARING HOUSING
The inner portion of a bearing support which houses a main bearing. Generally a sealed compartment designed to contain the oil mist created by motion of the rotating parts. Also called 1) Bearing Cavity, 2) Bearing Chamber, 3) Bearing Sump. *See also: Bearing Support.*

BEARING INSPECTION TERMS
Locate in Appendix.

BEARING NOISE AND FEEL TEST
A vibration and sound test requiring special equipment and a sound-proof room. Excessive vibration or sound emanating from a spinning bearing under controlled conditions is a cause for rejection.

BEARING PINCH
The squeezing force applied to a main bearing race, either inner or outer, so that it is retained in place and does not rotate independently of its shaft or in its housing.

BEARING SCRATCH DETECTOR (manual)
A small tool. A hand held ball-bearing-tipped tool of a prescribed diameter used to pass over bearing surfaces to discriminate between surface marks and scratches. When the tool finds a depression in the surface, the bearing is generally rejected.

BEARING SEAL
Locate under Brush Air Seal, Carbon or Labyrinth Seal.

BEARING SUMP
The lower portion of a Bearing Housing. The point to which oil gravity flows before being scavenged back to the oil reservoir.

BEARING SUPPORT
Often the inner hub of a major engine case. The hub is supported by radial struts attached to an outer engine case. *See also: Bearing Housing.*

BELL AIRACOMET (XP-59A)
In 1942, the first American purely jet propelled aircraft to fly. It was powered by a General Electric GE I-A Turbojet engine. *See also: 1) Gloster E28, 2) Heinkel 178.*

BELL INLET
Another name for Bellmouth Inlet.

BELLCRANK
Attaching point of many linkages on the turbine engine and its interface with the aircraft. It is an angled arm used to move all component parts is unison, to change direction of mechanical movement, or to change rate of movement. For example, the bellcrank on a variable compressor vane linkage.

BELLMOUTH INLET
A convergent shaped air inlet duct designed to provide the optimum aerodynamic effect and maximum massflow for air entering the engine. Used primarily on ground installed gas turbines and slow moving aircraft such as rotorcraft. The Bellmouth's high drag diameter prevents its use on faster aircraft.

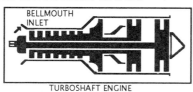

BELLVILLE SPRING OR WASHER
Locate under Spring Washer.

BELTING-IN (engine)
Motor over procedure of engine (no ignition) for the purpose of pre-oiling newly overhauled engines prior to initial runup. Also procedure for purging fuel system of preservatives after engine removal from storage.

BELVALVE™
A small part of an accessory. A poppet type relief valve as seen in fuel and oil filter bowls. This valve functions without springs and allows bypass when the differential pressure acting on its inner and outer surfaces reaches a predetermined value.

BENCH CHECK
A check performed on accessories and components to determine their condition. Bench Checks are typically completed during engine overhaul or during engine troubleshooting.

BENCH TEST (engine)
Refers to the test running of a Turbine Engine for checkout after maintenance has been performed. *See also: Test Cell.*

BEND
Locate under Engine Inspection and Distress Terms in Appendix.

BERNOULLI'S THEOREM
A physics principle stating static pressure and velocity (dynamic) pressure of a fluid are inversely proportional when passing through a duct at constant, subsonic flowrate. That is, total pressure and density will not change. This is the case in a subsonic aircraft inlet, where velocity decreases and static pressure increases as the airsteam approaches the engine face. Also known as Bernoulli's Principle.

BERYLLIUM OXIDE (igniter plug)
An insulator material used at the firing end of some high voltage turbine igniter plugs. This material has a high heat limit of approximately 1,700° F. It is used where heat strength of other materials for igniter tips would be exceeded. Beryllium Oxide is a toxic substance and is listed by the EPA as a hazardous material requiring special handling. *See also: Aluminum Oxide.*

BET-SHEMESH ENGINES LTD (Israel)
Manufacturer of Gas Turbine Engines. Locate in section entitled "Gas Turbine Engines for Aircraft".

BETA MODE
Locate under Beta Range.

BETA RANGE
Propeller ground operating range and propeller non-governing mode of operation for a Turboshaft or Turboprop engine. Blade angle in this range is controlled manually by a cockpit lever.

BEVEL GEAR (spiral type)
A pair of spiral cut gears set at right angles to each other. The drive gear mates with its pinion gear to change direction of motion 90 degrees. This arrangement provides higher strength than straight cut gears by its greater contact surface. The spiral cut is also said to be quieter in operation. *See also: Gear Designs.*

BEVEL GEAR (spur type)
A pair of straight cut gears set at right angles to each other. The drive gear mates with its pinion gear to change direction of motion 90 degrees.

BEVEL GEAR SYSTEM (gearbox drive)
A pair of gears set at right angles to each other, often arranged with one large driven gear and a smaller pinion gear. It is a means of extracting power from the compressor shaft to drive the Auxiliary Gearbox

or an Accessory Drive Gearbox. Also called: 1) Inlet Bevel Gearbox, 2) Power Takeoff Gearbox.

BEVEL PINION GEAR
The smaller of the two gears in a Bevel Gear.

BIFURCATED DUCT
A nacelle component used to direct air. The two common types are: 1) A split air inlet duct, 2) A split exhaust duct used on lift-fan engines. *See also: Inlet (configurations).*

BITE PANEL
Locate under Built In Troubleshooting Equipment Panel.

BLACK LIGHT INSPECTION
A non-destructive inspection technique which involves inspection with an ultra-violet light after a part has been immersed in a penetrating dye solution. The light illuminates cracks where the dye has penetrated the surface of the part. Green Dye is the more common in this procedure.

BLACK OXIDE COATING
A coating applied to iron and steel, generally by a dip method in an oxidizing bath. The coating improves the anti-chafing and anti-friction qualities of the metal.

BLADE
An engine component part. A rotating airfoil, as in fan blade. Blades are used to create lift, as is the case with compressor blades and fan blades, or to extract energy, as is the case with turbine blades. Some manufacturers also use the word "blade" to describe stationary airfoils such as compressor stator vanes. *See also: 1) Blade Base, 2) Blade Fillet, 3) Blade Root.*

BLADE ANGLE (Propeller, Variable Pitch Fan,)
Angle between the chord of an airfoil and a plane perpendicular to the axis of rotation. For example, a blade with its chord at right angles to its hub is at zero degrees. If the leading edge is forward, a positive angle is present. If the leading edge is rear of center, the pitch is negative.

BLADE ASPECT RATIO
Locate under Aspect Ratio.

BLADE BACK
The low pressure side of the blade. Also called the convex side or cambered side.

BLADE BASE
Portion of the blade between the contoured (airfoil) section and the root section. Also called blade platform. *See also: Blade.*

BLADE CAMBER
Refers to the curvature of the Blade Back.

BLADE CHORD
The imaginary straight line between the blade leading and trailing edge.

BLADE COOLING
Locate under Turbine Cooling.

BLADE-DISC ASSEMBLY
An engine component. Forged one-piece disc and blade unit, either compressor or turbine. Also referred to as a Blisk. *See also: 1) Blisk, 2) Disc and Blade Assembly 3) Integrated Blade and Rotor.*

BLADE FACE
The concave side, or the side opposite the Blade Back. Also called the Pressure Side.

BLADE FILLET
The portion of the blade where the airfoil section meets the blade base. It is a critical stress area where the least blade damage is allowed. *See also: Blade.*

BLADE FLUTTER
A self-excited vibration in which the alternating force sustaining the motion is created by the motion itself. When the motion stops, the alternating force disappears. Blade Flutter is opposite to Forced Vibration in that the sustaining alternate force exists independently of the motion. *See also: Forced Vibration.*

BLADE LOADING
Forces exerted on fan blades, compressor blades, or turbine blades by pressure differentials across the chord of the blades. Blade Loading also refers to the total horsepower absorbed by a fan, compressor, or turbine rotor assembly divided by the number of blades in the rotor. Newer high strength metals allow for higher blade loading and fewer blades.
See also: 1) Disc Loading, 2) Stage Loading.

BLADE NOISE
The three categories of blade noise are generally thought of as:
1) Aerodynamic Loading (unsteady noise) - The noise resulting from lift and drag forces experienced by the blade and emitted at the Blade Passing Frequency and Harmonics.
2) Aerodynamic Loading (steady noise) - The noise resulting from the interaction of blades within a non-uniform airflow.
3) Blade Volume (noise at blade passing frequency) - Noise caused by blades with sections approaching or exceeding the speed of sound and displacement of air by the blades.

BLADE PASSING FREQUENCY (BPF)
The speed (frequency) in seconds at which blades of an airfoil section such as a fan pass a given point.

$$BPF = Blade\ count \times RPM \div 60$$

BLADE PLATFORM
Another name for Blade Base.

BLADE PROTRUSION
Permanent axial shift of a compressor or turbine blade from its original flush position with the disc. When the shift reaches a predetermined limit, blade replacement and/or a new blade retainer is required.

BLADE RETAINER
A small expendable part. A device that prevents axial shift of either compressor or turbine blades; usually metal locktabs, rivets, or roll pins.

BLADE ROOT
Portion of the blade that fits into the disc. It is usually dovetail shaped in compressor blades and firtree shaped in turbine blades. *See also: Blade.*

BLADE SCATTER
The position compressor blades take relative to their root mountings when the rotor is operating below its designed speed.

BLADE SHIFT
Another name for blade protrusion.

BLADE SHANK
Another name for blade root.

BLADE SING
Jargon for vibration.

BLADE SPAN SHROUD
Locate under Span Shroud.

BLADE STAGGER ANGLE
Locate under Blade Twist.

BLADE STRESS
The three classifications of stress acting on rotating blades are stated to be: 1) Centrifugal Tensile Stress (radial or stretching loads), 2) Centrifugal Bending Stress (bending force in reaction to rotation), 3) Gas Bending Stress (axial loads due to air pressure forces).

BLADE TIP
Portion of blade opposite to blade root. Locate under 1) End-Bend Compressor Blade, 2) Flat-Machine Tip, 3) Open-Tip Blade, 4) Shrouded Tip Blade, 5) Squeeler-Tip Blade, 6) Vortex-Tip.

BLADE TIP VORTEX
A disturbance to normal airflow. A vortex occurs when the high pressure on a blade's lower surface moves outward and upward through the tip clearance to the upper, low pressure side of the blade. *See also:* 1) *Tip Clearance,* 2) *Vortex Tip.*

BLADE TUNING
Locate under Frequency Tuning.

BLADE TWIST
The change in angle-of-incidence of compressor and turbine blades from base to tip. This angle represents the angle between the engine centerline and the blade leading edge. The change in angle provides a uniform axial discharge velocity along the blade trailing edge from base to tip. Also called the blade Stagger Angle. *See also: Pitch Distribution.*

BLAST DEFLECTOR
A structure placed in back of an operating turbine engine to reduce a ground safety hazard. The deflector turns the jet exhaust (blast) upwards to diffuse its high velocity. Also called Blast Fence.

BLAST TUBE
A tube that directs air (usually ram air) to cool an AC or DC Generator.

BLEED AIR (systems)
Engine systems which use pressurized air bled away from the compressor. *See also:* 1) *Anti-Icer System,* 2) *Anti-Stall Systems,* 3) *Bleed Air (uses).*

BLEED AIR (uses)
Air taken from the engine compressor for various purposes. Locate under 1) Compressor Bleed Air, 2) Customer Bleed Air, 3) Engine Bleed Air.

BLEED AIR HOGGING
A condition of Customer Service Air being over-supplied by one engine rather than a shared load from all engines. If not corrected, this condition can cause over-temperature and stall problems on the over-worked engine. Also called Bleed Load Hogging. *See also: Customer Bleed Air.*

BLEED BAND SYSTEM
Locate under Compressor Bleed Band.

BLEED LOAD HOGGING
Locate under Bleed Air Hogging.

BLEED-OFF VALVE
Locate under Compressor Bleed Valve.

BLEED PORT (engine)
Points on the outer compressor case where a Bleed Band or Bleed Valve discharges Compressor Bleed Air back to the atmosphere. For Customer Bleed Air ports see Extraction Pads.

BLEED STRAP
Another name for Bleed Band.

BLEED VALVE SYSTEM
Locate under Compressor Bleed Valve.

BLEEDER RESISTOR (ignition system)
A component part of an ignition Exciter Unit. A permanently connected resistor in the secondary circuit which bleeds off the capacitor charge after the unit is shut off. The bleed resistor removes an electrical shock hazard for maintenance personnel when disconnecting the ignitor plug lead. It also protects the Ignition Exciter Unit from overheating if operated with an open igniter plug circuit. Also called 1) Discharge Resistor, 2) Safety Resistor.

BLENDING
General term for a metal filing method to re-contour damaged compressor and turbine airfoils back to an aerodynamic shape after incurring erosion or impact damage to their surface. A typical blend will have a length to depth ratio of approximately four to one. Blending may be thought of as completed in the following steps:
1) *Dressing* - by cropping, which involves cutting off material if necessary, or scalloping, which involves filing to a saddle or dished-out shape.
2) *Fine Filing* - to smooth out dressing scores and radiusing of edges.
3) *Polishing* - to restore the original surface.

BLING
An engine component. A cast, one piece bladed ring. One of several used to form an axial flow compressor.

BLISK
An engine component. A cast, one-piece bladed disk, one or more of which are used to form an axial flow compressor or turbine assembly. Blisk can also refer to a turbine disk and blade unit which utilizes single-crystal blades that are diffusion bonded to a powder-metal alloy disc. By eliminating the attachment points, the Blisk has a significant weight advantage over an assembly-design. *See also: Integrally Bladed Rotor.*

BLISK FAN
An engine component. A single stage fan for a Turbofan engine that is formed as one unit, with no removable blading. Not as widely used as a fan with separate disc and blade components.

BLISTER
Locate under Engine Inspection and Distress Terms in Appendix.

BLOCK TEST
The process of test running turbine engines out of the aircraft and in a test facility, typically called a Test Cell or Test Stand. Test runs are generally for troubleshooting that cannot be accomplished in the aircraft or test running after engine overhaul. *See also: Test Cell.*

BLOCKER DOOR (thrust reverser)
A thrust reverser door used to block airflow from its normal rearward direction of flow.

BLOCKING DIODE (ignition system)
A component part of an ignition exciter unit. The diode is located in the secondary circuit to block one-half the cycle. This action converts the power transformer's alternating current to pulsating direct current being delivered to the capacitor, which fires the igniter plug.

BLOOM
An engine instrument overshoot (below the maximum limit) which has an initial high (within limits) transient value and which returns to normal, as in EGT Bloom or N1 Bloom.

BLOW-DOWN LINE (scavenge oil system)
A scavenge oil line that takes both air and oil back to the oil tank from bearing sumps. Generally, when this term is used, there are no separate air venting lines for the oil system.

BLOW-IN DOORS
A nacelle component part. Small doors in the flight inlet of some Turbofan engines that move inward from suction at high engine power and low airspeeds to increase the effective inlet flow area.

BLOW-OFF (compressor)
Refers to compressor air released to the atmosphere through a Bleed Valve. *See also: Bleed Valve System.*

BLOW-OFF VALVE (compressor)
Locate under Compressor Bleed Valve.

BLOW-OUT
Generic name for Flame-Out. Can also specifically mean Rich Blowout.

BLOWN FLAP
A recent design concept for Turbofan powered aircraft wherein the fan exhaust, the core exhaust, or both, discharge through a slotted landing flap. The primary purpose of this system is to increase aircraft lift for short runway takeoff and for landing (STOL) operations. *See also: 1) Blown Flap Augmentation, 2) Over Wing Blowing.*

BLOWN FLAP AUGMENTATION
A thrust augmentation system under development. A derivative of the Blown Flap design enhanced by special discharge nozzles. The engine Low Pressure (LP) compressor will release some of its air discharge through the pylon and out of a series of wing trailing edge discharge nozzles. The primary purpose of this system is to increase short runway takeoff and landing (STOL) capabilities.

BLOWN WING (turbofan engine)
Locate under Over-Wing Blowing.

BLUCKET
An engine component part. A combination turbine blade and fan blade used in a ducted Aft-fan type Turbofan engine. For example, the General Electric CJ-805-23 and the TF-700 turbofan engines.

BLOWOUT
Another name for Flameout.

BODY-BOUND BOLT
A small expendable part. A bolt that fits into a close tolerance hole for alignment purposes as well as retention.

BOOST PUMP (fuel)
A small aircraft or engine accessory. A type of pump used to give an initial small increase in fuel pressure to help prevent vapor lock. Most common are the impeller or convolute pumps which are located in aircraft fuel tanks and also mounted as small accessories on engines.

BOOSTER APU
An auxiliary power unit capable of producing emergency aircraft propulsion.

BOOSTER COMPRESSOR
Another name for Low Pressure Compressor of a dual-spool axial flow turbine engine.

BOOSTER FAN
Another name for the forward fan of a dual-spool axial flow Turbofan engine.

BOOSTER STAGES (compressor)
Locate under Compressor Booster Stages.

BOOSTER VANE (compressor)
A vane in the Booster Stage of a Low Pressure Compressor.

BOSS
A raised or reinforced portion of a casting to which a smaller part is fastened.

BORESCOPE
A special tool. A rigid or flexible probe type viewing device used to inspect internal parts in the gaspath of an engine. It often has both an illuminating and a magnifying capability. *See also: Fiber Optics Borescope.*

BORESCOPE CRANKING PAD
An access pad on an engine. A means of inserting a small cranking handle into the engine (usually at the accessory gearbox) to rotate a compressor and facilitate borescope inspections.

BORESCOPE PLUG
A small plug that covers a borescope access port into which a viewing probe can be inserted.

BORON SLURRY FUEL
A dense, high energy fuel under study as an alternate for presently used Jet Fuels. *See also: Fuel Types-Jet.*

BOUNDARY LAYER AIR
A thin turbulent (slow moving) layer of air adjacent to an airfoil surface which extends up to the free airstream. The thinner the layer, the less disturbance to the free airstream flow and the greater the effective flow area. With a laminar flow, no boundary layer is present. *See also: 1) Laminar Flow, 2) Vortex Generator.*

BOUNDARY LAYER DIVERTER
A nacelle component. The movable portion of a Variable Geometry type convergent-divergent inlet duct which controls inlet flow area. Sometimes this unit is called a Vari-Ramp or Movable Wedge. *See also: Convergent-Divergent Inlet Duct.*

B.O.V.
Abbreviation for Bleed-Off Valve.

BOYLE'S LAW
A physics term. A physical law which states that in an ideal gas, the volume (V) of a confined body of gas varies inversely with its absolute pressure (P) when the temperature remains constant. Mathematically stated as ($P_1V_1 = P_2V_2$). *See also: Charles' Law.*

$$P_1V_1 = P_2V_2$$

BRAKE (propeller or rotor)
Locate under 1) Anti-Windmilling Brake, 2) Propeller Brake, 3) Rotor Brake.

BRAKE HORSEPOWER (turboprop & turboshaft)
The older term for actual horsepower delivered to the output side of the power output gearbox, after overcoming the braking force of the rotor system and gearbox. For example, the propeller shaft of a Turboprop engine. *See also: 1) Shaft Horsepower, 2) Friction Horsepower.*

BRANCA, GIOVANNI
Seventeenth century inventor of an industrial gas turbine system of impulse wheels driven by steam. A system that was the forerunner of the modern industrial gas turbine.

BRAYTON CYCLE
Named for George Brayton, an American scientist, who first described the thermodynamic pressure-volume relationship of an open-cycle, continuous combustion type internal combustion engine. The modern Gas Turbine Engine is such an engine with a cycle of events including: Inlet, Compression, Expansion and Exhaust. The four events are on-going simultaneously with the working gases finally returning back to the atmosphere. Also called a Constant Pressure Cycle. *See also: 1) Open-Cycle Engine, 2) Recuperated Brayton Cycle.*

BREAK (materials)
Locate under Engine Inspection and Distress Terms in Appendix.

BREAK-AWAY POINT
Another name for Shear Point.

BREAK-OUT BOX
Jargon for a test equipment kit used for operationally checking engine systems.

BREATHER IMPELLER
Another name for Air-Oil Separator (rotary).

BREATHER PRESSURIZING AND VENT VALVE
A small engine accessory. A valve in the Vent Subsystem of the engine lubrication system that acts as a vent to the atmosphere. At low altitude the aneroid operated valve is open. At higher altitudes it closes to lock in a sea level air pressure condition at bearing sumps. Sump pressurization is utilized to provide a controlled back-pressure on oil flow from the oil jets. This unit also contains a relief valve to prevent excessive pressure buildup. Also called 1) Breather Pressurizing Valve, 2) Vent Pressurizing Valve.

BREATHER SUBSYSTEM
Another name for Oil Vent Subsystem.

BRINELL HARDNESS
A unit of hardness determined by measuring an indentation pressed into a solid part.

BRINELLING (bearings)
There are two classifications of Brinelling: 1) True Brinelling, 2) False Brinelling. Locate under Bearing Inspection and Distress Terms in Appendix.

BRITISH ENGINES Locate by manufacturer: Rolls-Royce Ltd. or Noel Penny Turbines Ltd. in section entitled "Gas Turbine Engines for Aircraft".

BRITISH THERMAL UNIT (BTU)
A unit of heat measurement. The heat energy required to raise one pound of water one degree Fahrenheit. For example, one pound of Jet Fuel Type-A contains approximately 18,700 BTU's of heat energy.

BRITTLENESS (bearings)
Locate under Bearing Inspection and Distress Terms in Appendix.

BROOM-STICK CLIP
A small expendable part. A spring tension type clamp used to position electrical leads.

BRUNSBOND™
A Brunswick Corporation plasma spray process in which a heat barrier type coating of zirconia is applied to combustor and turbine parts. *See also: Thermal Barrier Coating.*

BRUNSALOY™ SHROUD RING
A Brunswick Corporation sintered metal alloy. It is mainly used to construct abradable turbine-shroud rings. Sintered metals are light-weight, porous alloys that are formed without melting. *See also: Shroud Ring.*

BRUSH AIR SEAL
A small engine component part. An improved design of a main bearing shaft air seal that fits a bristled metal seal in a stationary housing around the rotating part of the shaft. Brush Air Seals, being a full contact

seal, are stated to be more effective than the traditional labyrinth or knife edge air seals used to control turbine engine internal airflow.

BTU
Abbreviation for British Thermal Unit.

BUCKET (turbine)
Jargon for turbine blade.

BUCKLING
Locate under Engine Inspection and Distress Terms in Appendix.

BUDDY START
Starting of a gas turbine engine with air supplied by a second aircraft. This procedure would be used when no auxiliary or ground air starting unit is available. The procedure involves an air hose being connected between the ground air service points of both aircraft, with the operating aircraft supplying starting air to the inoperative aircraft.

BUILD-UP KIT
Another name for Quick Engine Change Kit.

BUILD-UP STAND
A maintenance stand in which a turbine engine is assembled (built up).

BUILT-UP ENGINE
A Basic Engine with its Quick Engine Change Kit installed. See also: 1) Basic Engine, 2) Power Pack, 3) Quick Engine Change Kit.

BUILT-IN TEST EQUIPMENT
Another name for Built-In Troubleshooting Equipment.

BUILT-IN TROUBLE-SHOOTING EQUIPMENT (BITE)
A cockpit fault display panel that is capable of identifying an engine or aircraft system fault and also the faulty unit. Also called Bite-Panel.

BULB-FIT (blade)
A design for compressor blade attachment to the disc in which the blade base and its slot in the compressor are shaped in a bulb fashion. Also an early design for turbine blades not seen today.

BURN
Locate under Engine Inspection and Distress Terms in Appendix.

BURNER
Jargon for Turbine Engine combustor or afterburner.

BURNER CAN
Jargon for combustion liner.

BURNER CAN PRESSURE
Locate under Burner Pressure.

BURNER EFFICIENCY
Another name for Combustor Efficiency.

BURNER PRESSURE (SYMBOL Pb)
Static pressure sensed via a probe within a selected combustion liner (burner can). This pressure provides a signal used as a measure of mass airflow through an engine against which a fuel schedule can be established.

BURNING
Locate under Bearing Inspection and Distress Terms in Appendix.

BURNISHING
Locate under Bearing Inspection and Distress Terms in Appendix.

BURR
Locate under Engine Inspection and Distress Terms in Appendix.

BURST MARGIN (metals)
Another name for Ultimate Strength of a material.

BUSHING
Locate under Bearing (plain).

BUSHING (pump shaft)
A small part of an accessory. A soft metal bushing, typically of bronze or copper, that serves as an end-bearing. Common to oil pumps and fuel pumps where, in most cases, the fluid being pumped lubricates the bushings.

BUTT LINE (BL)
A reference point on a gas turbine engine. The distance from the vertical center-line, to either right or left, as viewed from the rear of the engine looking forward. Example: When the basic reference is BL100. BL 99.0 is one inch to the left of center, BL 101.0 is one inch to the right of center. Also called Buttress Line. See also: 1) Water Line, 2) Engine Station.

BUTTERFLY CLAMP
Locate under Loop-Cushion Clamp.

BUZZ BOMB
The WWII German V-1 flying bomb. Its engine was not a Gas Turbine Engine but rather a Pulse-Jet.

BUZZ SAW NOISE
The sound the emits from the inlet of some Turbofan engines which employ supersonic fan tip speeds.

BYPASS DUCT
The cold airstream duct of a Turbofan engine. Also called 1) Fan Exhaust Duct, 2) Secondary Exhaust Duct.

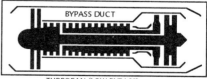
TURBOFAN (LOW BYPASS)

BYPASS DUCT (Nozzle Exit)
Locate under Exhaust Nozzle Exit.

BYPASS DUCT (Nozzle Throat)
Locate under Exhaust Nozzle Throat.

BYPASS INDICATOR
Locate under Filter Bypass Indicator.

BYPASS FAN
Another name for the Fan of a Turbofan engine. Also

called Bypass Engine.

BYPASS JET
A form of Turbojet Engine in which a portion of the compressor air is by-passed around the combustor and into the tailpipe. *See also: 1) Turbo-Ram Engine, 2) Turbo-Ramjet Engine.*

BYPASS RATIO (Turbofan Engine)
Ratio of the fan weight of airflow to core engine weight of airflow. For example, an engine with a 5 to 1 Fan Bypass Ratio would have: for a total mass airflow of 1500 lb./sec., a core flow of 250 lb./sec. and a fan flow of 1250 lb./sec. Also locate in section entitled "Gas Turbine Engines for Aircraft" for specific engine applications.

BYPASS REGULATOR VALVE
A small component part of a Fuel Control. Another name for a Differential Pressure Regulating Valve.

BY-PASS RELIEF VALVE
A small part of an accessory. A pop-off valve used to relieve excessive pressure at oil filters, oil coolers, fuel filters, etc. These valves are set to crack open at a predetermined differential pressure (PSID), and are generally not adjustable. *See also: Filters.*

BYPASS TURBOJET
An engine that diverts some of its compressor discharge air away from the combustor to a separate exhaust duct outside of the engine. This usually occurs at high ram conditions and adds to the total engine thrust. The bypass turbojet is more of a theoretical than an operational engine design and is the forerunner of the bypass Turbofan engine of today.

C

C-CHECKS
Locate under Letter Checks.

CABIN BLEED AIR
Another name for Customer bleed Air.

C.A.D.
Abbreviation for Computer Aided Design. A process widely used in the design of Gas Turbine Engines.

C.A.M.
Abbreviation for Computer Aided Manufacture.

CAGE FAILURE (bearings)
Locate under Bearing Inspection and Distress Terms in Appendix.

CALCIUM SILICATE
Locate under Thermal Barrier Coatings.

CAN
Jargon for Combustion Can or Combustion Liner.

CAN-ANNULAR COMBUSTOR
Type of combustion chamber with multiple liners fitted coaxially around the turbine shaft housing. The annulus is formed at its inner diameter by the turbine shaft housing and at its outer diameter by the combustor case. Also called Turbo-Annular Combustor. *See also: Combustor.*

CAN-TYPE COMBUSTOR
Another name for Multiple-Can Combustor.

CANADIAN MANUFACTURED ENGINES
Locate by manufacturer: Hawker Siddeley Canada Ltd., Pratt & Whitney of Canada, in section entitled "Gas Turbine Engines for Aircraft".

CANNULAR COMBUSTOR
Another name for Can-Annular Combustor.

CANTILEVERED ANNULAR COMBUSTION LINER
Locate under Combustion Liner (cantilevered).

CANTILEVERED PROPELLER GEARBOX
A propeller reduction gearbox on a Turboprop engine that is separated from the engine by support struts. The propeller shaft is usually on a different plane to the axis of the engine. *See also: Propeller Reduction Gearbox.*

CAPACITOR DISCHARGE IGNITION
An engine system. High energy (wattage) electrical ignition which delivers capacitor stored energy to an igniter plug. Turbine engine ignition is used for engine starting and for protection against flameout during certain flight operations. Also Called High Energy Ignition (HEI). *See also: Ignition System.*

CARBO-BLAST™
An abrasive grit used to clean turbine engine compressors This material is a lignocellulose substance consisting of granulated walnut shells and apricot pits. *See also: Field Cleaning.*

CARBON-CARBON COMPOSITE
A multi-layer, low weight material consisting of high-strength graphite, cloth-like fiber treated with a phenolic resin. The phenolic is converted to carbon by heat treating. This material has a predicted heat strength up to 4,000° F. and only one-half the weight of metallic materials currently used in Gas Turbines. This graphite/phenolic composite, however, is a brittle material and currently is used in non-structural areas, mostly as a heat shield. Because of its high specific strength, its future projected use is for hot section parts such as turbine vanes and blades. *See also: Composite Materials.*

CARBON FIBER MATERIALS
A family of carbon-fiber and glass-fiber reinforced epoxy resin materials blended in various ways to enhance their mechanical properties. At present, composites are used only in low to moderate heat areas in Gas Turbine Engines but are useful due to their high strength to weight characteristics. Projections are that heat strength of many composites may be raised considerably in the near future. Often called Composite Fiber Materials. *See also: 1) Carbon-Carbon Composite, 2) Ceramic-Matrix Composite, 3) Kevlar™, 4) Nomex™.*

CARBON OIL SEAL
A fluid seal commonly used at main bearing and drive shaft locations.

CARBON OIL SEAL (bearing)
An engine component part. Single or multiple rings of carbon material that are spring-loaded onto a highly polished metal surface. Carbon seals are found at main bearing locations to prevent leakage of engine

lubricating oil into the gaspath. The two common types of Carbon Oil Seals are: 1) face type, 2) ring type. *See also: Carbon Seal Pressurization.*

CARBON OIL SEAL (face type)
A main bearing shaft oil seal. The seal assembly consists typically of a single carbon ring and a highly polished metallic rub ring. The carbon ring remains stationary and its metallic rub ring moves with the shaft to which it is attached. The seal and rub ring are placed to face each other axially on the shaft, under spring tension, to prevent oil leakage past the seal location. Also called Axial-Face-Type Carbon Seal.

CARBON OIL SEAL (ring type)
A main bearing shaft oil seal. The seal assembly consists of single or multiple carbon rings which ride on a highly polished metallic rub ring in the shape of a shaft journal. This seal is arranged such that the carbon rings are preloaded to their rub rings via spring tension. Also called a 1) Centrifugal Shaft-Rubbing Carbon Oil Seal, 2) Radial Carbon Oil Seal. *See also: Carbon Oil Seal (split ring).*

CARBON OIL SEAL (shaft)
A component part of a fuel, oil, or hydraulic pump. Typically a single carbon ring that is spring-loaded onto a highly polished metal surface. This seal prevent pump fluids from escaping via the external drive shaft.

CARBON OIL SEAL (split ring)
A carbon seal of the ring type with more than one carbon sealing ring.

CARBON OIL SEAL PRESSURIZATION
Refers to air pressure which assists spring pressure in preloading the carbon sealing rings onto their running surfaces or to provide a form of air seal back-up to the carbon seal.

CARBON OIL SEAL RUB RING
Locate under Carbon Oil Seal.

CARBURIZING FUEL NOZZLE
Locate under Fuel nozzle.

CART START
Refers to an engine start by means of the cartridge mode of a Cartridge-Pneumatic starter. Could also mean engine start using a power cart rather than aircraft power.

CARTRIDGE HANG-FIRE
Refers to a Cartridge-Pneumatic Starter. The cartridge fires but with too slow a burning rate to rotate the engine. A condition resulting from moisture or other contamination within the starter breech.

CARTRIDGE MISFIRE
Refers to a Cartridge-Pneumatic Starter. The cartridge does not ignite when the starter switch in the cockpit is actuated to supply power to the cartridge electrode.

CARTRIDGE-PNEUMATIC STARTER
An engine accessory. A combination pneumatic starter and cartridge starter. It can be operated by either engine bleed air or by an explosive charge. In both methods, the gases exhaust through a turbine wheel and reduction gearbox assembly attached to an output shaft. This starter is usually mounted on the engine accessory drive gearbox.

CARTRIDGE TYPE FILTER
An engine accessory. A filter that utilizes a disposable element of paper, cellulose or the like, which collects contaminants as the fluid or gas passes through. Commonly used as oil, fuel and air filters on turbine engines.

CAPTIVE NUT
A small expendable part. A nut permanently attached to a component part of an engine and which is quite often riveted to a single or multiple retainer. The retainers are generally replaceable.

CASCADE THRUST REVERSER
Another name for Aerodynamic Blockage type thrust reverser.

CASCADE VANE
A general term for an air turning vane. One common use is in the thrust reverser in the exhaust section.

CASING (engine)
A shell-type structure such as a compressor, combustor, or turbine case, spaced between engine frames. Often constructed in two halves to facilitate removal.

CASTING (metal forming process)
Traditional method of molten metal poured into a mold and allowed to cool until hardened. Used especially in forming aluminum or magnesium cases of turbine engines. *See also: Investment Casting.*

CASTING (thin wall)
A process of casting to a very low thickness of 20 mils (0.020 inch). Referred to as casting down to a sheet metal thickness.

CATENARY THERMAL SHIELD
An engine component part. A curved sheet metal section located between multiple drum-type turbine wheels to serve as a heat barrier between the gaspath and the inner portions of the wheel assembly.

CAVITATION
Locate under Pump Cavitation.

CBO
Abbreviation for Cycles Between Overhaul.

C-D (inlet or exhaust)
Locate under Convergent Divergent Inlet or Convergent Divergent Exhaust.

C.D.P.
Abbreviation for Compressor Discharge Pressure.

CEILING (engine)
Maximum altitude at which engine thrust is capable of overcoming drag for straight and level flight but not sufficient for climb or acceleration. That is, the altitude at maximum lift/drag ratio where thrust is sufficient only to overcome drag.

CELLULAR COMBUSTOR
Another name for Can-Type Combustor.

CENTERBODY (Exhaust)
Another name for Tailcone.

CENTERBODY (Inlet)
Locate under Inlet Centerbody.

CENTER-VENT SYSTEM
An engine lubricating system venting arrangement, wherein the main rotor shaft acts an air-oil centrifuge. After separation, oil is scavenged back to the oil tank and the air is vented via the rotor shaft into the exhaust duct.

CENTIPOISE (cP)
One-hundredth of a poise. A metric unit of dynamic viscosity, Dyne sec./cm^2. Centistoke is the more common unit of viscosity measurement for Turbine Oils.

CENTISTOKE (cSt)
One-hundredth of a stoke. A measurement unit of kinematic viscosity (resistance to flow), to describe the thickness of synthetic turbine oils. Viscosity is referred to as Kinematic Viscosity due to the measuring process employed. A stoke represents the force it takes to move a flat surface of one square centimeter across another flat surface with a test fluid of one centimeter width in between, at a rate of one centimeter per second. The average turbine oil viscosity at normal engine operating oil temperature of 210° F. is approximately 5-centistokes. By comparison, the kinematic viscosity of water at 68° is 1-centistoke. *See also: 1) Kinematic Viscosity, 2) Synthetic Lubricating Oils.*

CENTRIFUGAL BREATHER
Another name for Air-Oil Separator.

CENTRIFUGAL FILTER (fuel)
Locate under Fuel Filter (centrifugal).

CENTRIFUGAL FLOW COMPRESSOR (single-stage)
A major engine component. An impeller rotor and housing designed to take air in at its center, called the Inducer Section, and accelerate that air outward into a Diffuser Manifold for the purpose of increasing air pressure. Also called Radial Outflow Compressor. *See also: 1) Compressor, 2) Diffuser Vanes, 3) Inducer.*

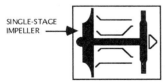

CENTRIFUGAL FLOW COMPRESSOR (dual entry)
An engine component with a back to back impeller design. This arrangement effectively doubles the mass airflow of the engine as it takes in air simultaneously at both impeller faces. This arrangement does not, however, increase the compressor pressure ratio. *See also: Centrifugal Flow Compressor (two stage).*

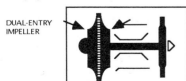

CENTRIFUGAL FLOW COMPRESSOR (two stage)
An engine component. A compressor containing two single sided impellers in series, for the purpose of enhancing compressor pressure ratio. The newer types contain dual, independently rotating impellers. The older types were fixed, to rotate as one unit. *See also: Combination Compressor.*

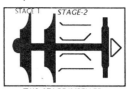

TWO-STAGE IMPELLER

CENTRIFUGAL FORCE
Action of an object moving outward when no centripetal (inward) force exists to restrain it. Current usage substitutes such words as reaction, action, or loading for "force", due to the fact that no actual force in pounds or kilograms exists on the object, only the absence of centripetal force.

CENTRIFUGAL PUMP
Locate under Pump (centrifugal).

CENTRIFUGAL SHAFT-RUBBING CARBON SEAL
Another name for Carbon Oil Seal (ring type).

CENTRIFUGAL TWISTING MOMENT (CTM)
Air loading on a rotating airfoil that causes the blade to twist in the direction of decreased pitch.

CENTRIFUGE (Air-Oil)
Locate under Air-Oil Separator.

CENTRIPETAL FORCE
An inward pulling force on an object in circular motion. For example, the radial force on a blade at its root as the blade travels around in a circular orbit.

CERABOREX™
A Ceramel with iron powder compounds added to a base of zirconium dibromide to increase material strength.

CERAMEL™
A combined ceramic and metal alloy. Chromium is one such alloy that accepts bonding of ceramic crystals. Chromium gives mixture strength and shock resistance and ceramics add high temperature strength. Also called Cermet. *See also: Ceramic Materials.*

CERAMET™
Another name for Ceramel.

CERAMIC (materials)
A broad classification of hard, brittle, heat-resistant and corrosion-resistant materials made by firing clay. For many years the high heat strength of ceramics has attracted its use to the Gas Turbine Engine. But problems of brittleness have prevented wide-spread applications of solid ceramic parts or even for barrier coatings. Two of the more common ceramics used today are silicon carbide and silicon nitride. The following few definitions describe some of these materials and uses. *See also: Ceramic-Matrix Composite.*

CERAMIC BLADES (rotor and stator)
A recent development in the use of silicone nitride, a high temperature strength material for turbine parts. Its use is projected to improve performance by allowing higher internal engine temperatures without adding to the cooling air requirements.

CERAMIC COATING (hot section parts)
A technique of coating hot section metallic parts by a dip method. The coating provides a heat barrier over a base metal. A newer method is to apply by plasma stray. Where mechanical load is low, such as combustor liners, a zirconious material is often used; where loading is higher, such as turbine vanes and blades, a silicon carbide and silicon nitride material is used. See also: Thermal Barrier Coating.

CERAMIC-MATRIX COMPOSITE (CMC)
A composite with a layered silicon carbide fiber formed within a solid silicon carbide material. The fibers reinforce the solid, more brittle material. This blend of ceramic materials has a predicted temperature strength up to 4,000° F. and are much less brittle than some earlier solid ceramic materials

CERAMIC SINGLE-CRYSTAL MATERIALS
Ceramics produced in single crystals with no grain boundaries. See also: Combustion Liner (floatwall).

CERMET™
Another name for Ceramel.

CESIUM- BARIUM 137
A radioactive substance used to coat the air-gap points (electronic switch) of some ignition system transformer units. The air-gap points in the Discharge Tube act as a synchronizing switch for current flow to the igniter plugs.

CFD
Abbreviation for Computational Fluid Dynamics.

CHAFING (bearings)
Locate under Bearing Inspection and Distress Terms in Appendix.

CHAMBER-SCOPE
A type of Borescope. This unit provides a means of visually inspecting inside the combustion chamber of an assembled engine.

CHAMFER
Term referring to a beveled edge often used to provide alignment between mating parts.

CHANNEL NUT
A small expendable part. A type of captive nut, usually one of several held in place in a sheet metal channel plate. Also called Gang Nut.

CHARLES' LAW
A physics term. A physical law that states that the volume (V) of a gas varies in direct proportion to the absolute temperature (T). Mathematically stated as: $V_1/V_2 = T_1/T_2$ at constant pressure (P), and $P_1/P_2 = T_1/T_2$ at constant volume. The basis of the General Gas Law. See also: 1) Boyle's Law, 2) General Gas Law.

$$\boxed{V_1/V_2 = T_1/T_2} \quad \boxed{P_1/P_2 = T_1/T_2}$$

CHEMICAL VAPOR DISPOSITION PROCESS (CVD)
A metal plating process to increase resistance to wear and corrosion of turbine engine parts.

CHINE TIRES
An aircraft tire with a ridge on its outer casing called a chine. The Chine suppresses the upward movement of ground water that would otherwise go into engine inlets.

CHIP DETECTOR
Locate under Chip Detector, 1) Capacitive Discharge Type, 2) Continuity Type, 3) Indicating Type, 4) Magnetic Type. See also: Quantative Debris Monitoring.

CHIP DETECTOR (capacitive discharge type)
A small engine accessory. A Chip Detector, sometimes of the indicating type, with a special pulsing feature. By use of a cockpit switch, the operator can discharge an electrical pulse capable of burning off small ferrous or non-ferrous particles at the gap-end that could be the cause of a false warning light. Some Chip Detectors have a self-clearing capability, and are called Automatic Pulsed Chip Detectors. See also: Chip Detector (indicating type).

CHIP DETECTOR (continuity type)
A small engine accessory. A magnetic chip detector that can be tested for presence of electrical continuity with an ohmmeter. Should the gap-end become bridged with metallic debris, continuity will show on the meter connected between the center electrode and the case-ground electrode. The check is accomplished in lieu of visual inspection.

CHIP DETECTOR (indicating type)
A small engine accessory. A magnetic plug similar to a Chip Detector of the magnetic type except that the magnet is connected to an electrical circuit and a cockpit warning light. If the gap-end becomes bridged, a warning light illuminates in the cockpit.

CHIP DETECTOR (magnetic type)
A small engine accessory. A magnetic plug, generally located in the scavenge portion of the lube system to attract ferrous particles, the size and number of which indicate engine condition. This unit requires removal before visual inspection can be performed. See also: Fines.

CHIP DETECTOR (self-closing)
A Chip Detector designed with a check valve which closes off oil flow when the probe is removed from an oil line or oil sump for inspection.

CHIPPING
Locate under Bearing Inspection and Distress Terms and Engine Inspection Terms in Appendix.

CHOKED (airflow)
A condition in air (in compressible flow condition) where flow from a convergent shaped nozzle is at Mach-one in relation to the temperature of the flowing air. In this condition, air cannot be further accelerated to a higher Mach number, regardless of additional pressure that could be applied to the flow. In the turbine engine, a choked airflow is designed to be present during high power engine operation at the turbine nozzle and the exhaust nozzle. This condition, however, can cause shock stall turbulence in the com-

pressor. *See also: Choked Exhaust Nozzle.*

CHOKED EXHAUST NOZZLE
Flow condition in which the exhaust nozzle gasses are flowing at Mach-one in relation to the internal gas temperature. The pressure across the nozzle exit at this time is raised above ambient pressure to provide a Pressure-Thrust in addition to the Reactive-Thrust of the flowing gasses. Gas flow exiting the nozzle is, however, higher than Mach-one in relation to ambient temperature. *See also: Thrust (choked nozzle).*

CHOKED FLOW
Locate under Choked.

CHOKED INLET
An aircraft inlet in which shock waves cause the airflow to choke. *See also: Convergent-Divergent Inlet Duct.*

CHOO-CHOO
A name to describe a mild, audible compressor stall condition caused by insufficient pressure ratio across the compressor section. It is also called chugging or coughing. *See also: Compressor Stall.*

CHOPPER (OIL)
A small component part of a labyrinth air-oil seal. A mechanical slinger ring with a notched surface which converts oil molecules in vent air to a liquid state and slings the oil back into a sump. Often used in conjunction with a labyrinth type air-oil seal to reduce the amount of lubricating oil in the area of the seal. *See also: 1) Labyrinth Air-Oil Seal, 2) Windback Seal.*

CHORD
Locate under Blade Chord.

CHROME ALUMINIDE DIFFUSION COATING
Locate under Aluminide Coating.

CHROMEL-ALUMEL
The bimetallic electrical conductors used in the exhaust gas temperature indicating system. Alumel contains an excess of free electrons which when heated move into the chromel portion of the system via a cockpit gauge. The current flow is read as an indication of engine exhaust gas temperature. *See also: Exhaust Gas Temperature.*

CHUGGING
Locate under Choo Choo.

CIP
Abbreviation for Compressor Inlet Pressure.

CIT
Abbreviation for Compressor Inlet Temperature.

CIT SENSOR
A small accessory. Compressor Inlet Temperature (CIT) Sensor. A device which sends an engine inlet duct temperature signal to a fuel scheduling device such as the main fuel control. It is a heat sensitive coil, often helium filled, connected to a capillary tube that carries a pressure signal to a servo mechanism in the fuel control. Often Called Tt_2 Sensor.

CIVIL AVIATION KEROSENE
The type fuel used by most commercial carriers and general aviation operators in gas turbine powered aircraft. Also called 1) Commercial Aviation Kerosene, 2) Jet Kerosene. 3) Jet Fuel. *See also: Fuel Types-Jet.*

CIVV
Abbreviation for Compressor Inlet Variable Vanes. Locate under Variable Inlet Guide Vane System.

CLAMSHELL THRUST REVERSER
Locate under Mechanical Blockage (thrust reverser).

CLAPPER (fan blade)
Another name for Span-Shroud or Snubber.

CLASH ENGAGEMENT (starter)
Locate under Crash Engagement.

CLEAR ICE
Locate under Icing (clear).

CLEARING THE ENGINE PROCEDURE
The process of motoring an engine over with the starter after an aborted start. Motoring allows the compressor discharge airflow to help vaporize residual liquid fuel and to purge fuel vapors from the combustor. Also called Purge Cycle.

CLOGGING INDICATOR
Another name for Filter Bypass Indicator.

CLOSED THERMODYNAMIC CYCLE
Referring to an engine, one which can transfer energy but cannot transfer matter between the boundaries of its systems.

CMC
Abbreviation for Ceramic Matrix Composite.

CO-ANNULAR EXHAUST DUCT
Locate under Exhaust Duct (configurations).

CO-ANNULAR TURBINE SHAFT
Another name for Coaxial Turbine Shaft.

COALESCER FILTER (fuel)
An engine accessory. A special filtering unit that filters out solid particulate matter such as algae or paraffin, but also allows water to form into droplets and sink to the bottom to be drained away. Used mainly in large turbine fuel storage tanks.

COANDA EFFECT
The tendency of air moving over a curved surface to continue bending after it has left the curved surface. A principle used to duct air downward in some Short Take-Off and Landing (STOL) aircraft.

COASTDOWN CHECK
The time an engine takes to rotate to a complete stop from idle speed after fuel is terminated. Generally a maintenance function during test cell operational checks.

COATINGS (metal)
Locate under 1) Aluminide Coating, 2) Thermal Barrier Coatings.

COAXIAL TURBINE SHAFT
An arrangement in which the rear turbine shaft is fitted concentrically through the front turbine shaft to drive the front (low pressure) compressor of a dual

axial flow engine. This is also the arrangement for driving the output gearbox of a Turboshaft or Turboprop engine. Also called Concentric Turbine Shaft.

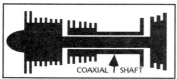

TURBOFAN (DUAL-SPOOL)

COBALT-BASE ALLOY
A cobalt, tungsten, molybdenum alloy of extremely high temperature strength, in a family of turbine engine super alloys. A very expensive material and often used only in the stator vanes of the engine hot section. Generally thought of as being too heavy for rotating parts such as turbine blades. *See also: Super Alloys.*

COBBING
A failure condition of compressors and turbines where a disc sheds most or all of its blading. The causes include: 1) foreign object induced damage, 2) blade failure which causes progressive damage down the gaspath.

COCKPIT DISPLAYS
Displays of flight and engine instrumentation other than the traditional instruments. Displays typically include: 1) CRT Displays (cathode ray tube), indicators with a changeable format, 2) LED Displays (light emitting diode), indicators having a fixed format. *See also: 1) Electronic Cockpit, 2) Electronic Flight Instrument System.*

COCKPIT GAUGES
Another name for Cockpit Indicators or Cockpit Instruments.

COCKPIT INDICATORS (engine)
Primary and Optional gauges in aircraft cockpits and described in FAR 33 and AC 20-88. The two general categories of indicators are as follows: 1) Condition Indicators, 2) Performance Indicators. Also locate under individual names such as N1 Speed, N2 Speed, EGT, EPR, Fuelflow, Oil Pressure, etc.

COCKPIT VOICE RECORDER
Part of a Flight Data Recorder system. A voice tape recording system which, at selected flight times, continuously records voice communications within and from the cockpit to other communication points. The voice tape is an effective means of reconstructing aircraft, engine and other events during post accident investigations. *See also: Flight Data Recorder.*

COKING (fuel)
Formation of (black) carbon deposits from unburned fuel. These deposits can be seen most visibly on the head of fuel nozzles. Coke will also move along the gaspath and impinge on turbine section surfaces causing erosion. This erosion is a starting point for a condition called Hot Corrosion.

COKING (oil)
Formation of (black) carbon deposits from the decomposition of oil under heat loads. Coke is more prone to occur where engine lubricating oil is stagnating under high heat or oil-laden vent air is subjected to high heat. Coke accumulations are a common cause of restricted vent airflow or sticking carbon oil seals. *See also: Lacquer (oil).*

COLD DAY ENRICHMENT (engine)
Abbreviation (CDE). Refers to providing the engine with additional fuel during cold day starts to match the high density air condition of the flight inlet. *See also: Stoichiometric Mixture.*

COLD ACCELERATION (engine)
A procedure where an engine is started and brought to idle speed, then with no heat stabilization time, accelerated to full power to check for stall tendencies. This check must be done within strict guidelines because blade tip clearances are at a maximum and compressor inefficiencies may result in an overtemperature. *See also: Re-Slam Acceleration.*

COLD FLOW REVERSER
Another name for Fan Thrust Reverser.

COLD HANG-UP
A starting cycle condition wherein an excessive time interval occurs between the point of initiating the starter and engine acceleration to ground idle speed.

COLD NOZZLE
Refers to the Secondary Exhaust Duct (fan exhaust) of a Turbofan Engine.

COLD OIL RELIEF VALVE
Another name for Cold Starting Oil Relief Valve.

COLD PROP
A propeller of a Turboprop or Ultra High Bypass Engine in which the core hot exhaust does not pass through the propeller. *See also: Hot Prop.*

COLD SECTION (turbine engine)
A typical reference to the air inlet and compression sections of the engine. The Cold Section starts at the engine inlet case and ends at the rear flange of the diffuser case. A borescope inspection performed on the engine Cold Section would be an example of usage of this term. *See also: 1) Fire Seal, 2) Hot Section.*

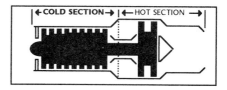

COLD SECTION MODULE
Locate under Module.

COLD SHIFT (VSV system)
A shift towards the open position. It is either a compressor Variable Stator Vane (VSV) performance shift which occurs normally at a decreased inlet temperature or a malfunction condition in which the VSV system actuates to a greater degree of vane opening than desirable.

COLD SOAKED
Refers to any mechanism that is subjected to cold ambient temperatures for a period of time and has

thereby acquired the same temperature.

COLD SOAK TEST (engine)
Term applied to engine test running after remaining inoperative in a low ambient temperature environment for a predetermined period of time. Cold Soak Testing is a study of mechanical failure tendencies and of engine starting characteristics under high lubricating oil viscosities and fuel filter icing conditions.

COLD STALL (compressor)
Very slight stalling condition in flight resulting from disrupted inlet airflow that is generally inaudible and undetectable on cockpit instruments. A Cold Stall will most often clear itself as inlet conditions return to normal. See also: Compressor Stall.

COLD-START BYPASS OIL VALVE
Another name for Cold Starting Oil Relief Valve.

COLD-STARTING OIL RELIEF VALVE
A small engine accessory. A bypass relief valve that acts as an emergency pop-off valve when cold oil causes excessive system back-pressures during start up. Typically used in systems having no regulating-type oil pressure relief valve.

COLD STREAM (Turbofan)
Another name for Secondary Airstream (Turbofan).

COLD STREAM REVERSER (Turbofan)
Another name for Fan Thrust Reverser.

COLD-TANK OIL SYSTEM
An engine lubrication system in which the oil cooler is located in the scavenge subsystem. This arrangement allows the oil to return to the tank in a cooled condition. See also: 1) Hot-Tank Oil system, 2) Lubrication System.

COLD THRUST
Refers to maximum engine thrust output, without afterburning.

COLD-WORKING (metals)
The process of hand working parts with hand tools such as files, hammers, etc. For example, the reworking of a damaged compressor blade by a blending (filing) technique.

COLLANT
An agent used to absorb heat.

COLLECTIVE PITCH LEVER
The rotorcraft cockpit lever that controls rotor blade angle and simultaneously controls fuel metering to the combustor. An up and down movement of the lever controls rotor blade pitch and a twisting motion controls fuel flow. See also: Speed Selector Lever.

COLORBRITE PENCIL™
A special pencil, free of carbon or zinc, used to mark hot section parts at areas of distress during inspection, or for part identification.

COMBINATION FLOW COMPRESSOR (engine)
A major engine component. A compressor design that utilizes both an axial and a centrifugal compressor in combination, with the axial section serving as the front stages and the centrifugal section as the last stage. In most engines, the two sections are fitted together to form one rotor. In some newer engines, the axial and centrifugal compressors are split and rotate independently. Also called 1) Axi-Centrifugal Compressor, 2) Compound Compressor.

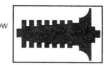
COMBINATION FLOW COMPRESSOR

COMBINED EXHAUST PROBES
A small engine accessory. A Combination Exhaust Gas Temperature probe and Turbine Discharge Pressure probe.

COMBINING GEARBOX
An engine component. An output reduction gearbox on a Twin-Pac™, Turboshaft Engine that receives power from two separate engines and produces one gear reduced output to a propeller or to a rotorcraft transmission.

COMBUSTION
The ignition of air and hydrocarbon fuels in a Gas Turbine Engine. Locate under Combustor Types.

COMBUSTION CAN
Jargon for Combustion Liner.

COMBUSTION CHAMBER
A major engine section. The point in the engine where fuel in introduced and combustion of fuel and air occurs. The combustion chamber consists of an inner liner and an outer case. Also called Combustor. See also: 1) Combustion Liner, 2) Combustor Types, 3) Engine Sections.

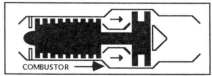
TURBOJET (SINGLE-SPOOL)

COMBUSTION EFFICIENCY
Locate under Combustor Efficiency.

COMBUSTION FLAME ZONE
Another name for Combustion Zone.

COMBUSTION LINER
An engine component. A perforated metal cylinder within the combustion chamber in which combustion occurs and fuel burning is controlled. The traditional liner is constructed of rolled and stamped sheet stainless steel or nickel alloy. Many of the modern liners are constructed of weld-assembled, machined rings. Because liner cooling requirements are directly proportional to flame-front length, newer, rapid fuel-air mixing techniques have been developed to provide shorter flame zones and therefore physically shorter liners. See also: 1) Combustion Liner (float-wall), 2) Combustion Liner (machined ring), 3) Combustion Liner (transply), 4) Combustion Liner (variable geometry).

COMBUSTION LINER (cantilevered)
An annular (one piece) combustion liner with its flow

chamber angled out from the center-line of the engine (front to back) for the purpose of reducing engine length. *See also: Combustion Liner.*

COMBUSTION LINER (dual dome)
Another name for Annular Combustor (dual-stage).

COMBUSTION LINER (float-wall)
A recent development in annular combustor design which utilizes a double wall liner of weld-assembled, cast rings. Its inner surface is made up of individual, movable shingles which attach to the outer shell. The outer surface is constructed of rolled metal. This design allows thermal expansion both axially and circumferential to suppress hoop-stress and low cycle fatigue. The shingles are often of single-crystal ceramic materials. *See also: Combustors Types.*

COMBUSTION LINER (machined ring)
An assembly of recent design and constructed of a series of cast metallic segments (machined rings) which are welded together. This liner is stated to have three to four times the expected service life of the stamped metal liner.

COMBUSTION LINER (transply™)
A development of Rolls-Royce Ltd. A liner stated to require less cooling air than traditional combustion liners. Transply is a multi-layer metal matrix with air cooling channels between each layer that more effectively utilize available cooling air.

COMBUSTION LINER (variable geometry)
A liner undergoing development. One which utilizes adjustable swirl vanes that open at high temperature operating modes and close to restrict primary airflow for combustion at low operating temperatures. Primary air can be as high as 55 percent of total airflow with the swirl vanes actuated fully open, whereas fixed swirl vanes are relegated to approximately 25 to 40 percent. This liner is said to produce higher thermal efficiencies and lower smoke emissions than traditional liners. *See also: Combustion Zone.*

COMBUSTION LINER LOUVER
Small slots in the liner to direct cooling airflow and provide the inner walls of the liner with a cooling air blanket.

COMBUSTION NOISE
Locate under Combustor Instability.

COMBUSTION RATIO
Locate under Stoichiometric Mixture.

COMBUSTION STARTER
Another name for Fuel-Air Combustion Starter.

COMBUSTION TRANSITION DUCT
Locate under Combustor Transition Duct.

COMBUSTION ZONE
The area at the fuel nozzle flame front. Approximately 25 to 40 percent of total airflow is utilized as Primary Air (for combustion). Swirl Vanes in the head end of the combustion liner introduce approximately 15 to 20 % of the incoming air to the area of the fuel nozzle. Radial holes in the liner head-end introduce another 10 to 20 % to create a region of low velocity air which helps to promote complete combustion. The Stoichiometric Temperature in the flame zone is approximately 3,500 to 4000° Fahrenheit. The remainder of total airflow is known as Secondary Air and is used to cool the liner and reduce gas temperature to an acceptable level before entering the turbine section.

COMBUSTOR
Another name for Combustion Chamber.

COMBUSTOR (flow direction)
The two flow categories typically seen in Turbine Engine combustors are: 1) Through-Flow Combustor, in which gases flow in an axial path, 2) Reverse-Flow Combustor in which gases flow in a reverse-S path. Through-Flow combustors are more common to larger engines and Reverse-Flow combustors are in wider use for smaller engines. *See also: 1) Through-Flow Combustor, 2) Annular Combustor (reverse-flow).*

COMBUSTOR (types)
Locate under Major Classifications: 1) Annular Combustor, 2) Can-Annular Combustor, 3) Can-Type Combustor, or Minor Classifications: 1) Finwall Combustor, 2) Pre-Chamber Combustor, 3) Ram Induction Combustor, 4) Scroll Combustor, 5) Combustor (variable geometry). *See also: Combustion Liner.*

COMBUSTOR CASE
The outer case of a Combustion Chamber.

COMBUSTOR DRAIN VALVE
A small component part. A small gas pressure closed and spring opened valve at the low point of a combustor outer case. This valve reduces the risk of after-fire or hot starts by draining away fuel which might remain after an aborted start.

COMBUSTOR EFFICIENCY
A physics term. The actual change in combustor gas temperature divided by the theoretical temperature change. In other words, the BTU value of heat attained versus the BTU potential of the fuel introduced into the combustor (typically 98 to 99%).

COMBUSTOR INSTABILITY (rumble)
Rumble is heard as a low frequency vibrating type noise when a mismatch occurs between fuel-flow and airflow. This condition results from fluctuating fuel pressure caused by dynamic interaction of fuel system components. Rumble most often occurs at low engine speeds and is usually not an airworthiness concern. But when instability persists at higher power settings, changing a worn fuel pump, fuel control or pressurizing and dump valve often removes the rumble.

COMBUSTOR LINER
Locate under Combustion Liner.

COMBUSTOR MECHANICAL DRAIN
Another name for Combustor Drain Valve.

COMBUSTOR OUTLET DUCT
Another name for Combustor Transition Duct.

COMBUSTOR TRANSITION DUCT
An engine component. A duct positioned such that it guides gases from the rear of multiple-can and can-annular combustion liners to the first stage turbine nozzle. This duct also provides physical support for

the rear of the combustion liners.

COMBUSTOR TURBINE
Another name for Gas Generator Turbine.

COMMERCIAL AVIATION KEROSENE
Another name for Civil Aviation Kerosene.

COMMERCIAL ENGINE POWER RATINGS
Locate under Engine Ratings.

COMMON-FLOW AFTERBURNER
Refers to a Turbofan Engine in which both the fan and core discharge are mixed in the afterburning process.

COMMON GEARBOX
Locate under Combining Gearbox.

COMPARATOR (EPR system)
Another name for Transmitter. Locate under Engine Pressure Ratio system.

COMPONENT (engine manufacturer supplied)
Generally thought of as major assemblies of an engine. An essential part that must be present in order to operate an engine. Components of this type include compressor rotors, compressor stators, turbine rotors, turbine stators, major casings, etc. *See also: Accessories.*

COMPONENT (aircraft manufacturer supplied)
A major assembly of an engine. A part not essential to the operation of the basic engine but in actuality essential to an engine in an aircraft. Components of this type include afterburners, thrust reversers, flight inlets, tailpipes etc.

COMPONENT LIFE
The established service life of a component as given by the engine manufacturer and after which the unit is overhauled or discarded.

COMPONENT PART (aircraft)
For the purposes of this listing, a sub-assembly of the aircraft.

COMPONENT PART (engine)
For the purposes of this listing, a sub-assembly or smaller part of a Component.

COMPOSITE MATERIALS
A generic term. Locate under the following: 1) Carbon-Fiber, 2) Ceramic-Matrix 3) Metal-Matrix Materials 4) Thermo-Plastic Composites.

COMPOSITE FAN BLADE
A new blade design of epoxy-resin and graphite-fiber materials. It is said to be 20% to 30% lighter than metal blades of the same mechanical strength.

COMPOSITOR
Locate under Ignition Compositor.

COMPOUND CYCLE GAS TURBINE
An engine design not yet in production that utilizes a high pressure, high speed, two-stroke diesel engine in place of a traditional combustor and turbine to drive the compressor.

COMPOUND COMPRESSOR
Another name for Combination Flow Compressor.

COMPRESSED AIR STARTER MOTOR
Another name for Pneumatic Starter.

COMPRESSIBILITY (air)
The aerodynamic effect of air moving at the speed of sound, where it is no longer incompressible and where its density is changing.

COMPRESSION (air)
The result of forces acting towards each other on the same line to cause a reduction in airstream volume and an increase in pressure. Below 300 M.P.H., air in movement has so little force acting on it that air acts as if it were incompressible, with a pressure rise but no density increase. *See also: Compressibility Effect.*

COMPRESSIBILITY EFFECT (air)
Relates to the ideas concerning compression of air, in that air acts as an incompressible fluid with no change in density as pressure increases while moving at subsonic speeds within a duct. Then at high speeds, near the speed of sound, the compressibility effect causes an increase in density as pressure rises.

COMPRESSION FAILURE (material)
Locate under Stress Failure in "Engine Inspection and Distress Terms" of Appendix.

COMPRESSION LIFT VEHICLE
Another name for Waverider Vehicle.

COMPRESSION RATIO (engine or fan)
Often used to describe compressor pressure ratio, but not quite technically correct due to the fact that compression ratio is a comparison of compressor discharge density to inlet density. Whereas compressor pressure ratio compares compressor discharge pressure to compressor inlet pressure. On ground runup, compression ratio is essentially the same as compressor pressure ratio, but in flight the two would be very different due to ram effect. *See also: Compressor Pressure Ratio. Note: In a reciprocating engine, compression ratio is a ratio of volumes within the cylinder and not a ratio of actual pressure changes.*

COMPRESSOR (blisk)
Locate under Blisk.

COMPRESSOR (engine)
A major engine component. A Compressor of either the Centrifugal Flow (impeller) Type or the Axial Flow Type. Compressors that are driven at high speeds by a turbine rotor for the purpose of moving a mass airflow into a reducing volume and thereby increasing air pressure. The pressurized air is then delivered to the combustor to facilitate the combustion process and also to provide air for cooling. *See also specific compressor types: 1) Centrifugal Flow, 2) Axial Flow, 3) Combination, 4) Dual-Spool, 5) Low Pressure, 6) High Pressure, 7) Triple-Spool.*

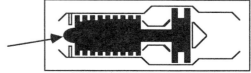

TURBOJET (SINGLE AXIAL COMPRESSOR)

COMPRESSOR ADIABATIC EFFICIENCY
Locate under Adiabatic Efficiency.

COMPRESSOR BLADE
A component part of a compressor. A rotating airfoil designed to increase air velocity and to raise static pressure in air. *See also: 1) Axial Flow Compressor, 2) Centrifugal Flow Compressor, 3) End-Bend Compressor Blade.*

COMPRESSOR BLADE (root)
Locate under: 1) Dovetail Root, 2) Bulb Root. *See also: Blade Tips.*

COMPRESSOR BLADE (shapes)
There are two common blade designs utilized in Gas Turbine Engine compressors: 1) Circular-Arc shape, wherein the leading and trailing edge are sharp cut, 2) Controlled-Diffusion shape, wherein the leading edge is blunt and the trailing edge is sharp cut or slightly blunted. *See also: 1) Controlled-Diffusion Airfoils, 2) Orthogonal Compressor Design, 3) Swept-Back Compressor Blade, 4) Wide Chord Blade.*

COMPRESSOR BLADE (tips)
Locate under Blade Tips.

COMPRESSOR BLADE (wing)
A stiffener across the chord of a compressor blade designed to add resistance to blade distortion; especially after impact damage by foreign objects.

COMPRESSOR BLEED AIR
Air that is dumped overboard from a compressor bleed band, bleed valve or other anti-stall bleed air device. Compressor bleed is accomplished for the purpose of relieving drag on the compressor at low to intermediate engine speeds. *See also: 1) Compressor Bleed Band, 2) Compressor Bleed Valve.*

COMPRESSOR BLEED BAND
A compressor anti-stall system. A means of releasing a portion of the pressurized air from the compressor for anti-stall/surge control. The band is located around the outer compressor case. It slackens at low RPM to release air into the fan duct or directly to the atmosphere. This air is not reused by the engine. The aerodynamic function of the Bleed Band is similar to that of the Compressor Bleed Valve. *See also: 1) Compressor Bleed Air, 2) Compressor Bleed Valve.*

COMPRESSOR BLEED VALVE
A compressor anti-stall system. A system with one large air flapper valve or several small poppet valves which provide a means of unloading the compressor. The older types have a two position valve, either fully open or fully closed, the newer types have a variable flow capability. At low and intermediate speeds the valves are scheduled open, allowing a portion of compressor air to escape. This reduces the pressure ratio and increases the air velocity to a point necessary to aerodynamically maintain the correct angle-of-attack within the compressor. Sometimes the valve itself is called a Blow-Off Valve. *See also: 1) Compressor Bleed Air, 2) Variable Bleed Valve.*

COMPRESSOR BLISK
Locate under Blisk.

COMPRESSOR BOOSTER STAGES
The axial compressor stages in addition to the fan in the low pressure compressor of a Turbofan engine. Booster Stages are located downstream of the fan to raise the pressure delivered to the High Pressure Compressor. Also called Compressor Super-Charger.

COMPRESSOR CASE
An engine component. The outer compressor housing and a major structural part of the engine. The compressor case also provides mounting space on its inner walls for the numerous stator vanes.

COMPRESSOR CASE (constant radius)
A case with no change in its internal flow area from front to back. It is utilized in conjunction with a conical shaped compressor rotor of either the disc or drum type.

COMPRESSOR CASE (conical radius)
A case with a decreasing flow area front to back. It is used with a constant radius compressor of the disc or drum type.

COMPRESSOR CHARACTERISTICS
Refers to the compressor efficiency for a given pressure ratio, mass airflow, and compressor speed. *See also: Compressor Stall Margin Graph.*

COMPRESSOR CLEARANCE CONTROL
Locate under Active Clearance Control.

COMPRESSOR DIFFUSER
Another name for Diffuser (engine).

COMPRESSOR DISCHARGE PRESSURE (CDP)
A pressure signal that is taken at the compressor exit and sent to the fuel control for scheduling the air to fuel ratios. On many engines, the CDP signal is referred to as Ps_4 (pressure static, station four).

COMPRESSOR DISC
An engine component part. The circular, inner part of a compressor rotor assembly to which the blading is attached. Disc can mean a single casting to accommodate all of the blading in the rotor, but generally it refers to one of a set of discs fitted together with long through-bolts to form a one-piece rotating assembly. *See also: 1) Blisk, 2) Compressor Drum, 3) Integrally Bladed Rotor.*

COMPRESSOR DISC AND BLADE ASSEMBLY
An engine component part. A compressor disc including its removable blading. The rotating portion of the compressor assembly, often called the compressor rotor. *See also: Blisk.*

COMPRESSOR DRUM
An engine component. A hollow, cylindrically-shaped metal drum to which the compressor blades attach to form a compressor rotor assembly. Used in place of a Compressor Disc design. *See also: 1) Compressor Disc, 2) Torque Cone.*

COMPRESSOR EFFICIENCY
A physics term. A measure of the aerodynamic efficiency of a compressor. A major factor present here is the compressor's ability to compress air to the maximum with the minimum air turbulence and temperature rise. The modern compressor is in the range of

85 to 90% efficient. *See also: 1) Adiabatic Efficiency, 2) Polytropic Efficiency, 3) Isentropic Efficiency, 4) Isothermal Efficiency.*

COMPRESSOR EXIT GUIDE VANES
Locate under Exit Guide Vanes (compressor).

COMPRESSOR FIELD CLEANING
Locate under Field Cleaning.

COMPRESSOR FRONT FRAME
Another name for Inlet Case (engine).

COMPRESSOR HANDLING QUALITY
Refers to a compressor's ability to remain stall and surge free when engine power is changed from one setting to another.

COMPRESSOR HUB
The front or rear portion of the compressor rotor to which the compressor shafts attach.

COMPRESSOR HUB/TIP RADIUS RATIO
Locate under Hub to Tip Radius Ratio.

COMPRESSOR INLET GUIDE VANES
Locate under Inlet Guide Vanes (compressor).

COMPRESSOR INLET PRESSURE (CIP)
Pressure Total (Pt) taken in the engine inlet. This parameter provides an indication of inlet conditions for such functions as fuel scheduling and variable vane scheduling. It is also one of the Engine Pressure Ratio (EPR) sensors along with Turbine Discharge Pressure. On many engines this parameter is labeled Pressure Total Station Two (Pt_2). *See also: Inlet Pressure (Pt_2).*

COMPRESSOR INLET TEMPERATURE (CIT)
Temperature Total (Tt) taken in the engine inlet. This parameter provides an indication of inlet conditions for such function as fuel scheduling and variable vane scheduling. On many engines this parameter is labeled Temperature Total Station Two (Tt_2). *See also: Inlet Temperature (Tt_2).*

COMPRESSOR INLET VARIABLE VANES
Another name for Variable Inlet Guide Vanes.

COMPRESSOR INTERMEDIATE CASE
Locate under intermediate case.

COMPRESSOR MAP
Another name for Compressor Stall Margin Graph.

COMPRESSOR PRESSURE RATIO (compressor)
The ratio of compressor discharge pressure to compressor inlet pressure. For example, with a compressor discharge pressure of 147 psia and a compressor inlet pressure of 14.7 psia, the ratio would be stated as 10 to 1. *See also: Cycle Pressure Ratio. See section entitled "Gas Turbine Engines for Aircraft" for specific engine aplications.*

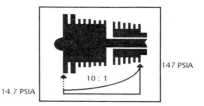

COMPRESSOR PRESSURE RATIO (per stage)
A means of calculating the average compression rise per stage. If a compressor were to produce a 10:1 pressure ratio over 12 stages, it could be stated that the pressure ratio per stage is 1.21:1, this figure is arrived at by calculating the 12th root of 10.

COMPRESSOR REAR FRAME
An engine frame positioned between the Compressor and Combustor sections. Another name for Diffuser (engine). *See also: Frames.*

COMPRESSOR RECOUP VALVE
Another name for Compressor Bleed Valve

COMPRESSOR RICE HULL CLEANING
A dry grit material used to field clean engine compressors. *See also: Field Cleaning.*

COMPRESSOR SECTION
The portion of a turbine engine in which air is compressed before entering the Diffuser Section. A complete compressor assembly consisting of all rotating parts, stationary parts, and casings. *See also: 1) Engine Sections, 2) Module.*

COMPRESSOR STAGE (axial compressor)
A rotor blade set followed by a stator vane set. Simply stated, the rotating airfoils (blades) create increased air velocity which changes to a static pressure rise in the diverging passageways formed by the stator vanes. Stages are numbered front to back. Stage one is generally the first fan of a Turbofan Engine or the first compressor stage of a Turbojet, Turboprop and Turboshaft engine.

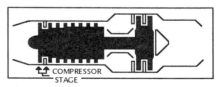

COMPRESSOR STAGE (centrifugal compressor)
A single impeller and housing. For example, the first impeller of a two stage centrifugal compressor is referred to as the first stage.

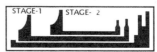

COMPRESSOR STAGE EFFICIENCY
Locate under Compressor Polytropic Efficiency.

COMPRESSOR STALL
A disruption of normal air flow through a compressor. A localized discontinuity (low frequency oscillation in air) that can be heard as a pulsating sound. There are many causes for this condition, one of which is an

over rich mixture, causing over pressure in the combustor and a slow down of compressor gas flow. Normally the fuel scheduling system automatically corrects this problem or the operator will retard the power lever to correct the stall. A second cause of stall is aircraft maneuvering at high angle-of-attack or sideslip angle creating airflow inlet distortion. A third cause of stall is Shock Stall of airflow at the blade tips.
See also: 1) Cold Stall, 2) Compressor Surge, 3) Hot Stall, 4) Off-Idle Stall, 5) Sub-Idle Stall.

COMPRESSOR STALL-MARGIN GRAPH
A graph showing the relationship between the compressor pressure ratio and mass airflow that must be maintained for any particular engine RPM. If one of the three variables goes out of limits, a compressor stall will likely result. The Stall Margin itself is defined as a measure of the separation between the stall line on the graph and the normal operating line. Also called 1) Compressor Map, 2) Compressor Surge-Margin Graph.

COMPRESSOR STATOR VANE
An engine component part. Stationary airfoils placed to the rear of the rotor blades to form a stage of compression. See also: 1) Compressor Stage, 2) End-Bend Compressor Vane.

COMPRESSOR SUPER-CHARGER
Name sometimes given to the axial compressor stages in back of a fan in a Turbofan engine. Also known as Compressor Booster.

COMPRESSOR SURGE
An engine malfunction. A progressive compressor stall across the entire compressor that can result in engine flame-out or even engine damage if not quickly corrected. This condition occurs from a complete stoppage of compressor airflow or reversal of compressor airflow. Also called "Hot Stall". See also: 1) Compressor Stall, 2) Compressor Stall-Margin Graph.

COMPRESSOR TIP CLEARANCE (cold)
The cold running clearance between the blade tips and their stationary shrouds. This clearance is taken as a maintenance inspection procedure. When the cold clearance is correct, it is expected that the hot running clearance will also be correct. Cold clearance is typically in the range of 20 to 30 thousandths of an inch, reducing to approximately 5 to 10 thousandths of an inch at hot running condition. Tip clearances are affected primarily by the following: 1) Thermal distortion of outer cases, 2) Mechanical distortion of outer cases, 3) Effects of coefficients of thermal expansion. See also: 1) Active Clearance Control, 2) Tip Clearance.

COMPRESSOR TIP/HUB RATIO
Locate under Hub to Tip Ratio.

COMPRESSOR TURBINE (CT)
Another name for Gas Generator Turbine.

COMPRESSOR VARIABLE GEOMETRY SYSTEM
Another name for Variable Angle Stator Vane System.

COMPRESSOR ZONE INSPECTION (CZI)
A major inspection of the core portion of the engine including specified measurement checks of the disks, blades, and vanes.

COMPUTATIONAL FLUID DYNAMICS (CFD)
The study of three-dimensional gas flow patterns over compressor blading, through combustors, over turbine blading, etc. This technique allows examination of many possible configurations before a final design is decided upon.

COMPUTED TOMOGRAPHY (CT)
An inspection process used on metals. An X-ray technique similar to a medical Cat-Scan. See also: Non-Destructive Inspection.

COMPUTER AIDED DESIGN (CAD)
Newer technique using computers to quickly produce multiple design drawings, which then undergo consideration towards final designs. Other terms that spin off from this concept are Computer Aided Manufacturing (CAM) and Computer Integrated Manufacture (CIM).

CONCENTRIC TURBINE SHAFT
Another name for Co-Axial Turbine Shaft.

CONCENTRICITY CHECK
A dial micrometer check to determine out-of-roundness of a circular base such as a bearing race, a compressor case, or a turbine case.

CON-DI NOZZLE
Another name for convergent-divergent exhaust duct.

CONDITION CODE
Locate under Part Condition Code.

CONDITION INDICATOR (engine)
A category of cockpit indicators. Includes most engine gauges such as EGT, Percent RPM, Fuelflow, Oil Pressure, etc. If these indicators show abnormally high readings, it generally means wear, malfunction, or failure of an associated part of the engine. See also: 1) Cockpit Gauges, 2) Performance Indicator (engine).

CONDITION LEVER
A cockpit lever in a Turboprop aircraft which provides inputs to the fuel control. In some installations it serves as a speed control lever for low and high idle speed range, in others it serves only as a fuel On-Off lever. Sometimes called the RPM Lever or Speed Lever.

CONDITION MONITORING SYSTEM
A system of physical inspections performed on aircraft and engines, along with electronic monitoring of selected operating parameters. The resultant data is later used to determine engine condition and to plan work schedules. See also: Engine Condition Monitoring System (ECM).

CONE BOLT (engine mount)
A tapered main mount bolt that ensures correct alignment and minimum wear on the airframe mount when correctly fastened. In many pylon mountings, the Cone Bolt also provides a structural fuse that will allow the engine to break away cleanly in the presence of extreme shear forces during a sudden engine seizure. See also: Engine Mount.

CONICAL VORTEX CHAMBER
Another name for Spin Chamber in a fuel nozzle.

CONSTANT DELIVERY PUMP
Another name for Positive Displacement Pump.

CONSTANT DYE PENETRANT INSPECTION
Another name for Dye-Penetrant method of inspection.

CONSTANT PRESSURE CYCLE
The name in physics for the Brayton Cycle. The Constant Pressure Cycle is so named because combustion takes place at nearly constant pressure in the Gas Turbine Engine. Also called the Continuous Thermodynamic Cycle of the Gas Turbine Engine.

CONSTANT PRESSURE, DUCTED JET
Older name for Ram-Jet engine.

CONSTANT PRESSURE PUMP
Another name for Positive Displacement Pump.

CONSTANT SPEED DRIVE (CSD)
An engine accessory. A hydromechanical drive unit used to power an A.C. Generator. Generally the CSD is mounted on the engine Accessory Drive Gearbox and the generator is mounted on the CSD.

CONSTANT SPEED PROPELLER
A propeller that automatically maintains a constant, preset speed. If the engine speed tends to increase, the blade angle is increased by the propeller governor to keep the propeller on speed.

CONTAINMENT RING (fan)
An engine component part. A strengthened shroud ring positioned outside of the fan case. It often consists of honeycomb material covered with an epoxy resin such as Kevlar™. The ring will disintegrate in stages should contact with a dislodged blade occur, preventing a complete fan case rupture. *See also: Fan Failure (uncontained).*

CONTAINMENT RING (turbine)
An engine component part. A heavy metal shroud placed outside the engine at the plane of tip rotation of a turbine wheel during test runs. The ring is designed to contain radial movement of blades in the event of a turbine failure.

CONTINGENCY POWER RATING
Locate under Engine Power Ratings.

CONTINUOUS FLOW CYCLE
Another name for Brayton Cycle.

CONTINUOUS COMBUSTION CYCLE
Refers to the Brayton Cycle.

CONTINUOUS IGNITION SYSTEM
Locate under Ignition System (continuous duty).

CONTINUOUS LOOP FIRE DETECTOR (electrical)
An aircraft fire detection system having one detecting loop per fire zone as opposed to several spot detectors. The loop contains a core wire with electrical potential. An overheat condition or fire acting on the loop causes the insulation material in the loop to break down and the potential to find ground. When this occurs, a warning light will illuminate in the cockpit. *See also: Fire Detector.*

CONTINUOUS LOOP FIRE DETECTION (gas)
Locate under Systron Donner System.

CONTINUOUS PRIMER SYSTEM
Locate under Primer System.

CONTINUOUS THERMODYNAMIC CYCLE
Refers to the Brayton Cycle.

CONTOUR VALVE (fuel control)
Another name for Main Metering Valve.

CONTRA-ROTATING PROPELLER
Another name for Counter-Rotating Propeller.

CONTRA-ROTATING TURBINES
Another name for Counter-Rotating Turbines.

CONTROL ALTERNATOR (engine)
Locate under Permanent Magnet Generator.

CONTROL LEVER (fuel)
Another name for Power Lever.

CONTROLLED-DIFFUSION AIRFOILS
Airfoils of recent design with thicker leading and trailing edges and wider chord which provide less tendency to shock-stall at supersonic airflow conditions. A compressor configured with controlled-diffusion blades is allowed to rotate faster than previous designs, resulting in an improved compressor pressure ratio. This design is said to also have the following advantages: 1) improved airflow pressure distribution, 2) reduced shock wave formation in airfoil passageways, 3) less flow separation and higher tip speeds, 4) higher efficiency over a wider rotational range, 5) more erosion resistant and tolerant of Foreign Object Damage. *See also: Compressor Blade (shapes).*

CONVECTION COOLING (turbine blades & vanes)
Blades and vanes with internal through-flow cooling air passageways from base to tip. As air leaves the tip area, it carries away heat by convection and allows the blades and vanes to reside in a temperature environment much higher than the heat stress limits of the metal. Sometimes referred to as Single-Pass Cooling or Internal Convection Cooling. *See also: Turbine Cooling.*

CONVERGENT-DIVERGENT EXHAUST DUCT (subsonic aircraft)
An engine component part. A recent design concept employed in the construction of secondary (mixed flow) exhaust nozzles of Turbofan engines. A slight convergent-divergent shape is introduced to promote an acceleration in the exhaust efflux and optimize engine performance at take-off power setting.

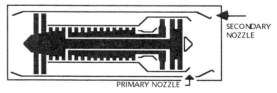

The secondary nozzle, in this case, is of fixed geometry. *See also: Convergent-Divergent Exhaust Duct (supersonic aircraft).*

CONVERGENT-DIVERGENT EXHAUST DUCT (supersonic aircraft)
A supersonic aircraft exhaust duct (afterburner). Its forward section is convergent to increase gas velocity

to sonic speed, and the aft section is divergent to increase gas velocity still further to supersonic speeds. In this manner the aircraft can attain supersonic flight speeds. The jet nozzle in this arrangement is of variable geometry, actuated by a hydraulic or pneumatic system. *See also: Afterburner.*

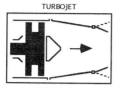

C-D EXHAUST DUCT ARRANGEMENTS

CONVERGENT-DIVERGENT INLET DUCT (C-D)
A nacelle component. A supersonic aircraft inlet duct of either fixed (C-D) geometry or of variable (C-D) geometry. Its forward inlet section is convergent to reduce air velocity from supersonic speed to sonic speed (Mach 1.0) and thereby increase static pressure. The aft section is divergent (acting as a subsonic diffuser) to increase air pressure still further and slow airflow to approximately Mach 0.5 before it enters the engine. Also called 1) Mixed Compression Inlet, 2) Variable Geometry Inlet. *See also: 1) Air Inlet Controller 2) Inlet Spike, 3) Vari-Ramp.*

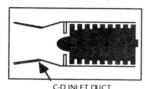

C-D INLET DUCT

CONVERGENT DUCT
A duct that decreases in cross sectional flow area from front to back. As air passes through the duct at constant subsonic flowrate (friction losses negligible), total pressure in the air will remain constant but ram pressure (velocity) will increase and static pressure will decrease. An aircraft tailpipe is an example of a convergent duct with resulting air velocity increase and static pressure decrease.

CONVERGENT EXHAUST DUCT (TAILPIPE)

CONVERTIBLE GAS TURBINE ENGINE
A prototype engine designed with multiple functions. It can produce fan thrust or shaft power, or any combination of the two, as power outputs.

CONVOLUTE PUMP
A pump with a spiral shaped impeller. It is mainly used as a fuel pump where only a small fuelflow is required.

COOKIE-CUTER (exhaust mixer)
Another name for Exhaust Mixer Nozzle.

COOLING
Locate under Turbine Cooling.

CORE or CORE ENGINE
The Gas Generator portion of a Gas Turbine Engine, sometimes referred to as a Simple Turbojet portion of an engine or the Core Module. *See also: Gas Generator.*

CORE MIXER
Refers to an Exhaust Mixer located at the core engine exhaust.

CORE SPEED INDICATION
High Pressure Compressor (HPC) speed of a dual-spool engine.

CORE SPEED SENSOR
A small engine accessory. Generally an electrical tachometer generator on a dual-spool engine, that is driven by the Accessory Drive Gearbox. The voltage it generates powers a cockpit Percent RPM Indicator. *See also: Tachometer Generator.*

CORN-COBBING
Locate under Cobbing.

CORRECTED SHP
Locate under Shaft Horsepower (corrected).

CORRECTED THRUST
Locate under Thrust (corrected).

CORROSION
Locate under Bearing Distress Terms and Inspection Terms in Appendix. *See also: Hot Corrosion.*

CORRUGATED PERIMETER NOISE SUPPRESSOR
An older type noise suppressor unit fitted in place of a conventional tailpipe. The final opening is a scalloped (corrugated) shape providing several openings for the escaping gas. In this manner jet efflux is broken up into several smaller airstreams, eliminating the large eddies that produce the loudest noises. This unit also has a means of mixing the ambient, free airstream with hot gases to reduce the shear-noise effect. *See also: Noise Suppressor Unit.*

COUNTER-FLOW COMBUSTOR
An older name for Reverse Flow Combustor. There is only one commonly used Reverse Flow Combustor, the Annular, Reverse Flow. *See also: Annular Combustor (reverse flow).*

COUNTER-ROTATING INTEGRATED SHROUDED PROPFAN (CRISP)
One of the new generation of Ultra High Bypass (UHB) turbine engines configured with dual shrouded propellers. CRISP engines are predicted to perform up to Mach 0.9 airspeeds, whereas, engines with open tipped prop-fans will be limited to approximately Mach 0.8 airspeeds. *See also: Counter-Rotating Propellers.*

COUNTER-ROTATING COMPRESSOR
A design in which one compressor in a dual-spool arrangement rotates in the opposite direction to the other. This is not a common design, but some high pressure compressors are designed to counter-rotate to reduce gyroscopic motion of the rotor system. *See also: Counter-Rotating Turbine.*

COUNTER-ROTATING PROPELLERS (CRP)
Newer Propfan Engines are being designed with two propellers mounted in tandem, which counter-rotate

on a common axis. The advantage of this design is that the swirl energy from the first propeller can be effectively angled into the second propeller, essentially eliminating swirl energy losses. For a given mass airflow, the added air velocity into the second propeller provides additional thrust that more than overcomes the weight and drag penalty of two propellers. In a single-rotating propeller, swirl energy in the prop wake is considered to be a loss of energy to the system. *See also: 1) Counter Rotating Integrated Shrouded Propfan, 2) Propellers.*

COUNTER-ROTATING TURBINES
High Pressure (HP) and Low Pressure (LP) turbines of Thrust producing Engines, or Gas Producer and Free Turbines of torque producing engines, which rotate in opposite directions. This turbine design lessens vibration and torsional stresses within the engine. However, counter-rotation is not a popular design. *See also: Counter-Rotating Compressor.*

COUNTERWEIGHT (balancing)
Another name for Balance Weight.

COUPLING (shaft)
A sleeve that fits over two mating quill-shaft splines to form one rotating shaft. For example, the compressor rear shaft and the turbine front shaft.

COWL or COWLING (engine)
Detachable parts of an engine nacelle. The term Cowling is often used to mean the engine nacelle itself.

CRACK
Often a minute material defect which has grown larger under cyclic and vibratory loads. Also locate under Engine Inspection and Distress Terms in Appendix.

CRACKING PRESSURE (relief valve)
The pressure at which a relief valve moves (cracks) off its seat to create a bypass condition. *See also: Relief Valve.*

CRASH ENGAGEMENT (starter)
Starter engagement into an operating engine, above the recommended engine speed limit. Crash engagement is allowed only in emergencies because the starter can be damaged as its drive shaft engages into the moving drive point on the Accessory Drive Gearbox. *See also: Starter Clash Re-engagement.*

CRASH RESISTANT FUEL SYSTEM (CRFS)
A fuel system in which an anti-misting additive is used in the fuel tanks. *See also: Anti-Misting Fuel Additive.*

CRAZING
Locate under Engine Inspection and Distress Terms in Appendix.

CREEP
A slow stretching process. A condition of permanent elongation in rotating airfoils after experiencing thermal stress and centrifugal loading. Creep is a normal occurrence in Turbine Engines and one that requires periodic measurement of certain rotating parts such as compressor and turbine blades.

CREEP STRENGTH
Refers to metal strength and resistance to creep under centrifugal and heat loading. *See also: Single-Crystal (turbine parts).*

CRFS
Abbreviation for Crash Resistant Fuel System.

CRISP
Abbreviation for Counter-Rotating Integrated Shrouded Propfan.

CRITICAL COMPONENT (engine)
Any failed component which could contribute to a condition of reduced flight safety.

CRITICAL ENGINE
Refers to any (one) engine of a multi-engine aircraft that is shut down in flight.

CRITICAL MACH NUMBER
An airfoil is said to be at its Critical Mach Number when airflow over the top (cambered) side reaches Mach One. The airfoil's leading edge can be traveling at speeds as low as Mach 0.7 when Critical Mach Number is reached. Sweep-back is one method of delaying the shock wave formation. *See also: Super-Critical Airfoil.*

CRITICAL PRESSURE (compressor)
The pressure ratio at which an axial flow compressor experiences a choked airflow along with flow breakdown and stalling.

CRITICAL PRESSURE (exhaust nozzle)
The gas pressure at which an exhaust nozzle airflow will choke. A choked condition is defined as the point at which flow velocity reaches the speed of sound.

CRITICAL PRESSURE RATIO (exhaust nozzle)
The pressure ratio of Turbine Discharge Pressure to Ambient Pressure at which exhaust nozzle airflow will choke. Choking occurs when Turbine Discharge Pressure is 1.893 times ambient. With an aircraft operating on the ground, and the Engine Pressure Ratio (EPR) gauge reading 1.893 or over, the exhaust nozzle is said to be choked.

CRITICAL RESONANCE
Locate under Resonance.

CRITICAL REYNOLDS NUMBER
The point at which air becomes unlaminar on an airfoil, due to the effect of viscosity and density changes in air. *See also: Reynolds Number Effect.*

CRITICAL ROTOR SPEED
The speed at which a rotating component such as a compressor or turbine produces hazardous vibration. A speed which must be avoided.

CRITICAL VELOCITY (air)
The speed at which flow becomes sonic.

CROPPED FAN ENGINE
A Turbofan Engine in which the fan has been reduced in diameter.

CROPPING (blades)
Meaning to cut off. Locate under Blending.

CROSS BLEED AIR SYSTEM
Refers to the engine starting air manifold in an aircraft. This manifold interconnects with bleed air points on the engine and the auxiliary power unit, to deliver air to engine Pneumatic Starter units.

CROSS IGNITION TUBE
Another name for Flame Propagation Tube.

CROSS SHAFT TRANSMISSION
A single transmission powered by two engines. This unit is installed in Tandem-Rotor Helicopters and also Rotary-Wing and Tilt-Rotor type aircraft. During a One-Engine-Inoperative (OEI) condition, the transmission is capable of driving both rotors from power supplied by the one remaining engine. *See also: 1) Power Output Gearbox, 2) Prop Rotor, 3) Transmission.*

CROSSOVER DUCT
A centrifugal impeller discharge duct designed with both diffuser and de-swirler capabilities.

CRT DISPLAY (cockpit)
Locate under Cockpit Displays.

CRUDE OIL
Petroleum oil in its natural state. There are two distinct crude oil bases: 1) Napthenic base also called Asphaltic, and 2) Paraffinic base. Turbine lubricants do not use crude oils in their natural state but use a synthesized product as a base stock. *See also: Synthetic Lubricating Oils.*

CRUISE POWER SETTING
An Engine Power Rating as set by FAA Standards. No time limit is imposed on this rating. *See also: Engine Power Ratings.*

CRYOGENIC FUELS
Fuels under consideration as alternates for the hydrocarbon jet-fuels presently in use. Fuels stored as liquids at very low temperatures. When heated by ambient air in a heat exchanger, the liquid turns to a gas before being fed to the engine. Locate under: 1) Hydrogen Fuel, 2) Liquefied Natural Gas. *See also: Fuel Types-Jet.*

Cs (symbol)
Symbol for the local speed of sound. The formula for computing Cs is as follows: 49.022 times the square root of the absolute temperature in degrees Rankine. The answer obtained will be in units of feet per second. Cs can also be calculated by the formula $\sqrt{G \times R \times K \times °R(TAM)}$. In this formula, $G = 32.2$ ft./sec.^{2}, $R = 53.3$ ft.lbs., derived from [778(0.24-0.1715)], $K =$ specific heat ratio 1.4, derived from (0.24 ÷ 0.1715). The value 0.24 is the constant pressure factor, 0.1715 is the constant volume factor, and (TAM) is temperature ambient.

$$Cs = 49.022 \sqrt{°R}$$

CSR
Abbreviation for Cycles Since Restoration.

CUPWASHER
A small expendable part. A safety locking washer that is depressed into slots in a spanner type nut or similar fastener to prevent loosening. Similar in function to a tablock or lockwasher.

CURL (combustor)
Component part of an Annular Reverse Flow combustion liner. Its function is to turn expanding combustor gases 180° into the direction of the turbine nozzle.

CURVIC COUPLING
A circular set of radial, gear-like teeth on each of two mating flanges. When meshed together and bolted in place, the curvic coupling provides a positive engagement with little likelihood of slippage. It is used primarily to attach a series of compressor discs to each other and to attach turbine shafts to turbine discs.

CUSTOMER BLEED AIR
Air extracted from the engine (usually at the diffuser) to supply pressurized air to the aircraft for such purposes as air conditioning, fuel tank pressurization, engine starting etc. *See also: 1) Bleed Air Hogging, 2) Extraction Pad.*

CUSTOMER SERVICE AIR
Another name for Customer Bleed Air.

CUT-OFF TURBINE
A type of turbine blade and vane design. Refers to the proper selection (cut off) of numbers of airfoils in a particular unit to reduce noise emissions.

CVG SYSTEM
Abbreviation for Compressor Variable Geometry System. Locate under Variable Angle Stator Vanes.

CVV SYSTEM
Abbreviation for Compressor Variable Vane System. Locate under Variable Angle Vane System.

CYCLE
Locate under Engine Cycle.

CYCLE, AIRCRAFT OPERATING
See Aircraft Operating Cycle.

CYCLE BETWEEN OVERHAUL
Locate under Time Between Overhaul.

CYCLE COUNTER
A cockpit instrument designed to count engine starts for purposes of managing a life-cycle-limited parts program. *See also: Cyclic Life Limit.*

CYCLE EFFICIENCY (engine)
Another name for Thermal Efficiency.

CYCLE, ENGINE OPERATING
See Engine Operating Cycle.

CYCLE PRESSURE RATIO
The result of multiplying Flight Inlet Pressure Ratio and Compressor Pressure Ratio. For example, the Concord SST aircraft, at a flight speed of Mach 2, has an aircraft Inlet Pressure Ratio of 7.37 : 1 and a Compressor Pressure Ratio of 12.07 : 1 The multiple of these two values results in a Cycle Pressure Ratio of 88.96 to 1. The combustor will then receive pressurized air at 88.96 times ambient value.

CYCLES BETWEEN OVERHAUL (CBO)
Locate under Time Between Overhaul.

CYCLES SINCE RESTORATION (CSR)
Locate under Time Since Restoration.

CYCLIC LIFE LIMIT
An engine part that has a manufacturer's suggested life limit based on engine cycles rather than operating hours. *See also: Life Limited Part.*

CYCLONE BREATHER
Another name for Air-Oil Separator (rotary).

CZECHOSLOVAKIA OMNIPOL TRADE GROUP
Engine manufacturer of Gas Turbine Engines. Locate in section entitled "Gas Turbine Engines for Aircraft".

CZI
Abbreviation for Compressor Zone Inspection.

D

D-CHECK
Locate under Letter Checks.

DAISY TUNER NOZZLE
Another name for Exhaust Mixer Nozzle.

DAMPER
Broad term for any number of vibration reducing devices attached to rotating engine components.

DARK COCKPIT
A cockpit fault indicating display. A push button system that is functionally dark unless indicating a fault or when the system is turned off.

DASHPOT
A fluid scheduling metering valve device seen particularly in fuel systems. Dashpots employ viscosity of the fluid to dampen motion of the metering valve.

DATA PLATE SPEED CHECK
An engine performance run used to compare the present engine condition to the " as new" performance data stamped on the engine data plate.

DBA
Abbreviation for Decibels, A-weighted.

DE-AERATOR
A type of air-oil separator in a turbine engine lubrication system.

DE-AERATOR (main gearbox)
Another name for Air-Oil Separator (rotary).

DE-AERATOR (oil tank)
A component part of an oil tank. An air-oil separator located at the scavenge oil return point in the main oil tank. Often the Deaerator is a cylindrical hopper tank which vents air to the expansion space at the top of the tank while allowing oil to seep through openings in the bottom. The return oil then seeks its own level along with the oil supply in the tank. *See also: Deaerator Tray.*

DE-AERATOR TRAY (oil tank)
A Deaerator in the shape of a tray within an oil tank. The tray allows scavenge return oil to spread out in a thin layer to promote air separation before mixing with the oil supply in the tank. *See also: Swirler (oil tank).*

DEBRIS TESTER (oil chip detector)
A device used to inspect metallic contaminants found on chip detectors. The data collected in this inspection is used to form a trend analysis of wear-metals generated by the oil wetted components of the engine.

DECIBEL (dB)
A measurement of sound equal to one-tenth Bell. One dB of sound is defined as 0.0002 dynes/cm^2, and is considered the threshold of hearing in humans. By comparison, sounds over 120 dB cause pain to humans and sounds over 160 dB are considered lethal. Turbine engines at take-off power emit sounds in the range of 90 to 120 dB. Aircraft noises are measured in Effective Perceived Noise Decibels. *See also: 1) A-Weighted Decibels, 2) Effective Perceived Noise Decibels.*

DECS
Abbreviation for Digital Engine Control System.

DECU
Abbreviation for Digital Electronic Control Unit.

DEEP CHUTE MIXER
A primary exhaust nozzle with a deep scalloped or corrugated shape at its exit. Another name for Exhaust Mixer Nozzle.

DEER
Abbreviation for Digital Electronic Engine Control.

DEFECT, CRITICAL
A defect that constitutes a hazardous or unsafe condition, or as determined by experience and judgement, could conceivably become so relative to its deleterious effect on the prime intended function or mission capability of the aircraft or its operating personnel.

DEFECT, MAJOR
A defect, other than critical, that could result in failure or materially reduce the usability of the unit or part for its intended purpose.

DEFECT, MINOR
A defect that does not materially reduce the usability of the unit or part for its intended purpose, or is a departure from standards which have no significant bearing on the effective use or operation of the unit or part.

DEFERRED MAINTENANCE ITEM
Engine defects of a non-critical nature for which maintenance actions are delayed until a part is available or until a maintenance scheduled is established.

DE-ICING (engine)
Removal of ice by melting with heat. This procedure is accomplished at fuel filters of turbine engines by use of an accessory called a Fuel Heater. *See also: 1) Anti-Icing, 2) Fuel Heater.*

DeHAVILLAND, GEOFFREY JR.
Early pioneer in supersonic flight, killed in the late 1940's while test flying a British DH-108 experimental Mach-1, pulse jet aircraft.

DeLAVAL NOZZLE
Named after its original designer, it refers to a convergent-divergent exhaust duct. Originally used in Steam Turbines.

DeLAVAN FUEL NOZZLE
Named after its original designer, it refers to a Simplex Atomizing Fuel Nozzle.

DELTA-P (ΔP)
Shortened form for Differential Pressure. The difference between two pressure values or a pressure change between one point and another.

DELTA-P INDICATOR (oil or fuel filters)
Either a pop-out indicator on the body of a filter or a warning light in the cockpit. The Delta-P indication serves as a warning that a dangerous condition of pressure drop has occurred, most probably from a partially clogged filter.

DELTA-P SWITCH (oil or fuel filter)
A pressure differential switch that actuates an electrical microswitch. The microswitch turns on a cockpit Delta-P warning light when a dangerous pressure loss condition exists. For example, when a filter is clogging the light warns the operator of an impending bypass condition.

DENSIOMETER (fuel)
An instrument used to measure fuel density in an aircraft fuel system.

DENSITY (of air)
The ratio of a mass of air to its volume, at a given temperature and pressure. Also called mass density. In U.S. units, density is expressed in lb./cu.ft.

DENSITY-AIR (formula)
Density at International Standard Day (ISD) condition defined mathematically as Specific Weight of air divided by the Gravity Constant. That is, 0.076474 lb./ft^3 ÷ 32.1714 ft./sec.^{2}) = 0.0023769 lb. sec^2/ft^4. Also to determine Density in dry air where Density is expressed in lb./cu.ft. (d = 1.325 x Po ÷ T), where Po is barometric (static) pressure (in.hg.) and T is absolute temperature in degrees Rankine. *See also: 1) Temperature (To), 2) Pressure (Po), 3) Specific Weight.*

$$d = 1.325 \times Po \div T$$

DENSITY ALTITUDE
The pressure altitude corrected for non-standard day temperature. The altitude based on both Pressure Altitude and Ambient Temperature of air. Density altitude computed by using a chart. The altitude that corresponds with an air density value in standard atmosphere. Example: at a density of 0.65896 lb./ft^3, the altitude is 5,000 feet. *See also: Pressure Altitude.*

DENT
Locate under Engine Inspection and Distress Terms in Appendix.

DE-OILER
Another name for Air-Oil Separator (rotary).

DE-OILER VALVE (APU)
A lubrication system relief valve which protects the system from high oil pressure build-up during cold weather starts.

DERATED ENGINE
An engine in which the originally intended power rating has been reduced for one of the following reasons: 1) Because full power is not required in a particular aircraft installation, 2) Because of poor or unreliable performance or poor service life at the higher value, 3) To accommodate a limitation placed, not on the engine but on another component such as rotorcraft transmission. *See also: 1) Engine Down-Rating, 2) Rerated Engine.*

DESALINATION WASH (compressor)
Procedure of motoring the engine with the starter while injecting a wash solution into the compressor, followed by a pure water rinse to remove salt deposits. Under light corrosion conditions, only the water rinse may be employed. Also called "motoring wash". *See also: Performance Recovery Wash.*

DESIGN POINT (compressor)
Refers to the optimum operating mode where the best compression results.

DE-SWIRLER VANES
A turning vane used to straighten airflow. For example, the exit guide vanes in the diffuser of a centrifugal flow compressor.

DETENTE (throttle)
A notch on the cockpit power lever quadrant that locates a desired position by feel rather than sight. For example, the Idle Detente or Afterburner Detente.

DETONATION GUN COATING
Another name for Plasma Spray coating.

DETROIT DIESEL ALLISON CORP.
Former name of Allison Turbine Engine Company.

DIABATIC PROCESS
A thermodynamic process in which a transfer of heat occurs across boundaries of the system. *See also: Adiabatic (process).*

DIAMETRAL RUNOUT
The total indicator reading (TIR) of a dial micrometer on a perfectly round shaft that is rotated 360° about its axis. Also called Radial Runout.

DICHRONATE TREATMENT (metals)
Typically a dip method of sodium dichromate to provide an anti-corrosion treatment to a metal surface.

DIFFERENTIAL PRESSURE (ΔP or psid)
Usually described as a pressure drop between two points. For example, on an oil system schematic diagram one might see Oil Filter (8 to 10 psid), indicating the maximum allowable pressure drop across the filtering element is 8 to 10 lb./sq. inch. *See also: Delta-P.*

DIFFERENTIAL PRESSURE REGULATING VALVE (fuel control)
A fuel bypass regulating valve which establishes a predetermined pressure drop (ΔP) across the fuel Main Metering Valve. The constant pressure differential ensures that fuel scheduling by weight takes place across the Main Metering Valve. *See also: 1) Main Metering Valve, 2) Weight of Fuel.*

DIFFUSER
In terms of the Gas Turbine Engine, a component in which gas velocity energy is converted to pressure energy. *See also: ; 1) Diffuser (engine), 2) Convergent-Divergent Inlet Duct, 3) Exhaust Diffuser, 4) Flight Inlet, 5) Pipe Diffuser, 6) Vane Diffuser.*

DIFFUSER (engine)
A major engine section located between the compressor and the combustor. A Diffuser in this location refers to an increasing flow area allowing air to spread out, which results in a pressure increase and a velocity decrease. The air is then considered to be of the correct pressure and velocity for delivery to the combustor. *See also: Divergent Duct.*

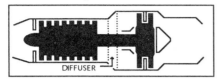

DIFFUSER CASE
A major engine component. Typically refers to an outer, mid-engine supporting case. Sometimes the diffuser assembly is a separate removable component, part of a mid-engine support case. Also called Mid-Frame Case.

DIFFUSER CASE (split-duct type)
A main engine Diffuser used with a Dual-Annular Combustor. *See also: Annular Combustor (dual-zone).*

DIFFUSER CONE (exhaust)
Another name for Tailcone or Exhaust Plug.

DIFFUSER RAMP
A nacelle component. The rear-most part of a vari-ramp type supersonic flight inlet. This section forms a divergent subsonic section. *See also: Convergent-Divergent Inlet Duct.*

DIFFUSER RATIO
The ratio of outlet to inlet flow area.

DIFFUSER SECTION
The portion of a turbine engine in which air is allowed to spread out and raise in pressure before entering the combustor. *See also: 1) Diffuser, 2) Engine Sections.*

DIFFUSER SCROLL
An engine component part. A type of engine diffuser which collects compressor discharge air and then acts as a flow divider to distribute air to the combustor in two or more airstreams.

DIFFUSER VANES
An engine component part. Vanes which receive air from the tips of an impeller type compressor. The vanes form divergent passageways which raise air pressure by the process of diffusion. The air then proceeds through Swirl Vanes to the Combustor.

DIFFUSION BONDING PROCESS (fan blades)
A fan blade manufacturing process of bonding laminated sheets of titanium. A process in which atoms of one sheet move into the lattice of adjoining sheets. A recent development for Wide Chord Fan Blade manufacture is referred to as Superplastically Formed, Diffusion Bonding (SPF/DB). This process involves an inert gas inflation of a titanium sandwich in a vacuum furnace, to provide the required shape. *See also: Wide Chord Blades.*

DIGITAL ELECTRONIC ENGINE CONTROL (DEEC)
An engine accessory. A full authority electronic fuel scheduling system with several advantages over traditional Electro-Hydromechanical systems. The advantages include: 1) its ability to give full power at approximately the same power lever angle (PLA) regardless of ambient temperature, 2) its unrestricted allowable movement (excursion time and PLA) throughout the flight envelope without over-scheduling fuel, 3) being digital, the DEEC utilizes more computer software programs than the hardware one would see in analog systems. Also called a Digital Electronic Control Unit (DECU). *See also: Fuel Control Unit.*

DIGITAL ENGINE CONTROL SYSTEM (DECS)
Engine fuel controlling system which uses a Digital Electronic Engine Control (DEEC) as its main fuel scheduling device.

DILUTION AIR (combustor)
The portion of combustor Secondary Airflow that is directed into the combustion liner to control the temperature of the gases just prior to entering the turbine nozzle vanes. The remainder of secondary airflow is used to provide a cooling air barrier over the inner and outer surfaces of the liner. *See also: Combustor.*

DILUTION CHUTE (combustor)
An opening in a combustion liner which directs dilution air into the hot gas path.

DILUTION ZONE (combustor)
The rear part of a combustion liner into which dilution air is directed. It is approximately 50% of the total compressor discharge. Also called Tertiary Combustion Zone.

DIP STICK (oil)
A small component part. A thin metal bar, usually calibrated with markings to indicate the number of quarts needed to fill the oil reservoir to the approved level.

DIRECT DRIVE TURBINE
Another name for Fixed Turbine.

DIRECT PRESSURE GAUGE
A Bourdon tube or bellows type non-electric gauge, primarily used in engine test equipment or for troubleshooting. *See also: Relative Pressure Gauge.*

DIRECTED FLOW THRUST REVERSER
Locate under Thrust Reverser.

DIRECTIONAL REFERENCE (engine)
An industry agreed upon means of describing turbine engine locations. The correct orientation is to stand at the rear of the engine looking forward and to use a standard o'clock reference. Right, left, up, and down are also obtained in this manner.

DIRECTIONAL SOLIDIFICATION
A metal casting process in which the grain boundaries are made to run axially to improve the metal's mechanical and temperature strength. In this process,

transverse grain boundaries are virtually eliminated, leaving long columns of crystals aligned along the line of stress. Refers to Directional Solidified Eutectics.

DIRECTIONAL SOLIDIFIED EUTECTICS
A directional solidification casting process for producing super-alloys used in Hot Sections of turbine engines. A nickel alloy manufacturing process that involves setting inter-metallic fibers into a nickel base. The result is a material of high metal strength with tolerance to heat and centrifugal loading. *See also: Super Alloys.*

DIRT SEPARATOR (bearing seal)
A small accessory. A filter in a line directing air to a labyrinth type main bearing oil seal.

DISC (compressor or turbine)
Locate under Compressor Disc or Turbine Disc.

DISC AND BLADE ASSEMBLY (compressor or turbine)
An engine component. Name given to either a compressor rotor or a turbine rotor when it is designed as an assembly of parts (discs and blades). *See also: Blisk.*

DISC LOADING
The horsepower absorbed at the blade attaching points on compressor, fan or turbine discs. *See also: Blade Loading.*

DISCHARGE RESISTOR (ignition system)
Another name for Bleeder Resistor.

DISK
Alternate spelling for Disc.

DISTILLATE FUELS
Crude oil that is processed by distillation into a range of fuels from very heavy, such as diesel oil, to light volatile gasoline type fuels. For example, gas turbine fuel is a middle distillate in the kerosene to gasoline range. *See also: Fuel Types-Jet.*

DISTORTION
Locate under Engine Inspection and Distress Terms in Appendix.

DISTRESS (engine)
A broad term to mean the various structural deteriorations, damage and failures that engine parts experience during their life time. Locate list of Engine and Bearing Distress Terms in Appendix.

DIVERGENT DUCT
A duct in which the cross sectional area increases in the direction of flow, resulting in an air pressure increase and a velocity decrease (with subsonic flow). *See also: Diffuser (engine).*

D.O.D.
Abbreviation for Domestic Object Damage.

DOGBONE MOUNT (engine)
A mount named as such because of its ball-shaped ends that allow for movement in a socket. Usually it is a front lower engine mount designed to support minimal loads. *See also: Engine Mount.*

DOMESTIC OBJECT DAMAGE (DOD)
Damage which occurs in the engine gaspath from aircraft material failure or part failure within the engine, rather than ingestion of foreign objects from outside the aircraft. *See also: Foreign Object Damage.*

DOODLE-BUG
Another name given to the German V-1, Pulse-Jet powered "Buzz Bomb".

DOUBLE ANNULAR COMBUSTOR
Locate under Annular Combustor (dual-zone).

DOUBLE ENGINE
British term for Dual-Pac Engine.

DOUBLE ENTRY COMPRESSOR
British term for Dual-Entry Centrifugal Flow Compressor.

DOUGHNUT SEAL
A small expendable part. A type of compressible synthetic rubber packing used to seal air, fuel or oil lines. The doughnut seal fits over a metal tube and under a collar-type compression nut.

DOVETAIL FIT (compressor blade)
Shape of the blade root similar to a cabinet maker's dovetail cut. The dovetail cut is primarily used to fit compressor blades to the compressor disc. This design provides resistance to radial blade movement when the disc is in motion.

DOVETAIL FIT (turbine blade)
Another name for Dovetail Serration.

DOVETAIL SERRATION (turbine blade root)
A modified dovetail shape with two or three steps instead of one, as in a traditional dovetail cut. Similar to a fir-tree fit but with fewer serrations.

DOWN-RATED ENGINE
Locate under Engine Down-Rating.

DRAG
The force that opposes an airfoil's movement in air. For example, the force which opposes propulsion induced thrust of an airplane. Drag goes up from increasing air resistance as an airfoil moves faster. Airfoils traveling in air experience three common types of drag: 1) Drag (induced), 2) Drag (parasite), 3) Drag (shock). *See also: 1) Kinetic Heating, 2) Lift/Drag Ratio.*

DRAG (induced)
Drag created by the lift of an airfoil when the downward displacement of air by the airfoil interferes with the oncoming airstream. This drag increases and decreases along with angle-of-attack changes and changes in airfoil speed.

DRAG (parasite)
Drag due to surface shape. Aerodynamic resistance to movement through air, relatively independent of angle-of-attack, and one which increases with airfoil speed. Four types of parasite drag are present on an airfoil in motion. 1) Form drag, due to the shape of the airfoil, 2) Interference drag, due to junctions at sections of connecting airfoil parts, 3) Scrubbing Drag caused by fan and core exhaust gases washing over

surfaces such as pylons and gas generator cowlings, 4) Skin Friction Drag created by viscosity of air passing over airfoils.

DRAG (shock)
Drag occurring at supersonic speeds as a result of shock wave formation on airfoil surfaces that reduce the approach velocities of the oncoming airstream. *See also: Transonic Drag.*

DRAG (wave)
Another name for Drag (shock).

DRAIN CANNISTER (fuel)
Locate under Ecology Drain.

DRAIN MAST (nacelle)
An overboard drain. The point on lower engine cowls to which the engine and accessories drain vapors and fluids which leak past barrier seals.

DRAIN VALVE (combustor)
A small component part. A spring opened and combustor pressure closed, mechanical valve located in the lower portion of the combustor outer case. This valve is present to prevent fuel puddling after an aborted start.

DRAIN VALVE (fuel manifold)
Another name for Dump Valve.

DRESSING (blades)
Locate under Blending.

DRIP VALVE (fuel manifold)
Another name for Dump Valve.

DRIVE CONE (turbine wheel)
The front (drive) end of a Turbine Drum.

DROOP (engine speed)
In a fuel system with a mechanical governor, droop refers to a slight change in engine speed when a load is applied, such as when increasing pitch of a propeller. This results is a spring tension change on the flyweight speed governor in the main fuel control. When the flyweight speeder spring is extended during this underspeed condition, less flyweight force (speed) is required to bring about equilibrium of the two forces, weight and RPM. With a weaker spring force present, less RPM results. The loss of speed in this case is called negative droop. Conversely, when a load is removed and engine speed input to the governor allows the engine to stabilize at a higher speed, it is called positive droop.

DRUM (compressor)
Locate under Compressor Drum.

DRY FILM LUBRICANT
A lubricant of graphite base sprayed on wear surfaces, such as accessory spine drives, to act as an anti-galling and anti-wear agent.

DRY MOTORING
Engine rotation via the starter (no fuel input). Another name for Motor-Over (engine).

DRY SPLINE (accessory drive gearbox)
Refers to accessory drive pad points which are coated with an anti-wear type grease rather than oil from the engine lubrication system. *See also: Wet Spline.*

DRY SUMP (lubrication system)
A system in which the main oil supply is stored in a tank rather than in the sump area of the engine Accessory Drive Gearbox. *See also: 1) External Dry Sump, 2) Integral Dry Sump, 3) Lubrication System.*

DRY TAKEOFF
Locate under Take-Off (dry).

DRY THRUST
Thrust produced when the engine is not assisted by augmentation systems such as Water Injection or Afterburning. *See also: Take-Off Dry.*

DUAL ANNULAR (dual-dome) COMBUSTOR
Locate under Annular Combustor (dual-zone).

DUAL-AXIAL FLOW COMPRESSOR
British term for Dual-Spool Axial Flow Compressor.

DUAL-CENTRIFUGAL FLOW COMPRESSOR
An engine component. A compressor design with two independently rotating radial-outflow impellers. *See also: Compressor.*

DUAL-ENTRY CENTRIFUGAL FLOW COMPRESSOR
An engine component. A single compressor rotor with back to back impellers. This design essentially doubles the engine massflow. It is rarely seen today.

DUAL FUEL PUMP
Locate under Fuel Pump (multi-stage).

DUAL-FLOW FUEL NOZZLE
Another name for Fuel Nozzle (duplex).

DUAL IMPELLER
Locate under Dual-Centrifugal Flow Compressor.

DUAL-LINE DUPLEX FUEL MANIFOLD
A component part of an engine. A fuel manifold that contains two separate fuel lines to the primary and secondary fuel orifices of duplex fuel nozzles. Normally fuel is distributed to the two lines via an accessory called a Pressurizing and Dump Valve. *See also: 1) Primary Fuel Flow, 2) Secondary Fuel Flow.*

DUAL-ORIFICE, PRESSURE ATOMIZING FUEL NOZZLES
Another name for Fuel Nozzles (duplex).

DUAL PAC (engine)
A pair of engines mounted side by side and connected to one central gearbox. *See also: Twin Pac™.*

DUAL-ROTOR TURBINE ENGINE
A Turbine Engine which contains a Dual-Spool Compressor and Turbine arrangement.

DUAL-SPOOL COMPRESSOR
An engine component. Refers to an axial-flow compressor design with two independently rotating compressors, each driven by its own turbine wheel(s). The front compressor is referred to as the Low Pressure Compressor (LPC) and the rear as the High Pressure Compressor (HPC). Not generally used to describe a Dual-Centrifugal Flow Compressor. *See also: 1) Com-*

pressor, 2) Dual Centrifugal Flow Compressor.

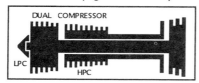

DUAL THERMOCOUPLE
Locate under Thermocouple.

DUCT
A passage or part of a passage used to direct gases.

DUCT BURNER
Locate under Fan-Duct Burning.

DUCT BUZZ (inlet)
Another name for Inlet Buzz.

DUCT HEATER
Another name for Fan Duct Afterburning.

DUCT LOSS (inlet)
Refers to difference between potential mass airflow through the inlet and the actual mass received at the engine face. Loss occurs from surface friction, flow separation, turbulence, etc.

DUCT RECOVERY POINT
Another name for Ram Recovery Point.

DUCT SHAPES
Locate under Convergent, Divergent, Convergent-Divergent.

DUCTED AFT TURBOFAN
A proposed new engine design. This engine will be configured with dual, ducted fans attached to a contra-rotating power turbine. Other new engines of similar design are unducted. It is stated that the Ducted Aft Turbofan will have better under-wing capabilities and thereby less noise than the Unducted Fan Type engines.

DUCTED-FAN ENGINE
Another name for Turbofan in reference to a Gas Turbine Engine.

DUCTED JET (constant pressure)
Older name for Ram-Jet engine.

DUCTED PROPULSOR
Locate under Ultra-High Bypass Engine.

DUMP VALVE (fuel system)
An engine accessory. A fuel pressure operated or solenoid operated valve. The valve is located at the low portion of the combustor fuel manifold and opens at each engine shutdown. The Dump Valve's function is to prevent fuel boiling from residual combustor heat which can cause carbonizing of fuel nozzles.

DUPEL FUEL NOZZLE
Locate under Fuel Nozzle (duplex).

DUPLEX BEARING
Generally refers to two matched, main bearings installed as one in areas of high thrust loading.

DUPLEX FUEL NOZZLE
Locate under Fuel Nozzle (duplex).

DUROMETER SEAL
Another name for Doughnut Seal.

DWELL CHAMBER (oil tank)
Another name for Deaerator (oil tank).

DYE PENETRANT INSPECTIONS
Fluid penetrant, Non-Destructive-Inspections of turbine engine parts. The two commonly used methods are as follows: 1) the red-dye method, 2) the fluorescent penetrant method.

DYE PENETRANT (red-dye)
A process involving a sprayed on application of a red penetrating dye to a metal surface. If cracked, the surface will absorb the dye by capillary action. Then when a developer is applied, the crack will show as a red trace in the white developer. The Red-Dye method is especially suitable for small area inspection on non-ferrous parts such as gearbox housings, cases etc. *See also: Non-Destructive Inspection.*

DYE PENETRANT (fluorescent)
A process involving application of a green penetrating dye to a metal surface which, if cracked, will absorb the dye by capillary action. Then when viewed under a black light, cracks will show up as a green line. The Green-Dye method is especially suitable for wide area inspection of larger engine parts because the dye can be applied by immersion as well as spraying. *See also: Non-Destructive Inspection.*

DYNAMIC BALANCING (compressor & turbine rotors)
A method of adding or removing weight during spin testing in order to obtain vibration-free rotation of engine parts such as compressor and turbine rotors. The rotating mass will remain free of vibration during normal operation, as long as all centrifugal forces created by the various parts of the mass remain balanced by opposing forces. This check is generally done in a horizontal balance unit only after the disk has been statically balanced. Also called the Influence Coefficient Method. *See also: 1) Balance Checks, 2) Balance Weights.*

DYNAMIC FILTER (fuel or oil)
Another name for Centrifugal Filter.

DYNAMIC ICE SHEDDER
The front portion of the Spinner on a Turboprop or Turbofan Engine. The Ice Shedder is pointed and designed to shed ice by centrifugal action before heavy build-up occurs. *See also: Spinner (turbofan).*

DYNAMIC PRESSURE
Another name for Ram Air Pressure.

DYNAMIC THRUST
In reference to the Turbine Engine, another name for Thrust.

DYNAMOMETER
A unit of support equipment. A device used to measure shaft horsepower (torque) output of Turboshaft and Turboprop Engines against an electrical or fluid resistance. For example, on engines being test run after overhaul. *See also: Water Brake Unit.*

E

EAD
An abbreviation for Engine and Alert Display.

EAR MUFF (defender)
A hearing protector worn over the ears. Used by flightline personnel when in the vicinity of operating Gas Turbine Engines.

EAR PLUG
A plastic or wax insert which fits into the ear canal for hearing protection for persons working in the vicinity of operating turbine engines. Ear plugs are not considered as effective a hearing protector as the Ear Muff.

EBU KIT
Abbreviation for Engine Build-Up Kit.

E.C.M.
Abbreviation for Engine Condition Monitoring.

ECOLOGY DRAIN TANK
A component part of an Engine Nacelle. A drain cannister into which fuel from the Pressurizing and Drain Valve collects after engine shut-down. On some installations, a jet (ejector) pump returns the drained fuel to the fuel supply tank. Often called an Ecology Kit which includes the tank, valves and associated hardware. *See also: 1) Ecology Valve, 2) Jet Pump.*

ECOLOGY VALVE (fuel system)
An accessory, either aircraft or engine. A valve assembly which allows fuel from the Pressurizing and Drain Valve to enter the Ecology drain tank.

EDDY CURRENT INSPECTION
A method of passing electrical current through a part to locate a discontinuity in its subsurface. Eddy Current inspection is especially accurate on compressor and turbine blades and other small parts. This inspection utilizes a primary current flow through a hand held probe which induces a secondary current flow through the test piece. Cracks are detected by observing an imbalance between the primary and secondary current paths on a CRT type screen. *See also: Non-Destructive Inspection.*

EDGE-TYPE OIL FILTER
A filter which utilizes a cylindrical filtering element. The element consisting of numerous flat disks and screens arranged such that they allow oil to enter radially from the outer surface (edges). When disassembled, debris trapped on the screens becomes accessible for inspection. A Stacked Disc Oil Filter is of this type.

EDUCTOR
Another name for Jet Pump.

EFFECTIVE ANGLE OF ATTACK (compressor blade)
The angle between the blade chord line and the resultant airflow vector at a blade's leading edge. The resultant airflow vector represents the change in airflow brought about by two interacting vectors: 1) airflow direction and velocity into the engine from the flight inlet, 2) airflow direction and velocity due to blade rotation. Note: The angle between the chord line and the plane of rotation of the blade is referred to as the "apparent angle of attack". *See also: Angle of Attack.*

EFFECTIVE HORSEPOWER
The horsepower delivered to a propeller.

EFFECTIVE PERCEIVED NOISE DECIBEL (EPNdB)
A unit of noise measurement employed by the Federal Aviation Administration as described in FAA Regulation Part 36. This measurement system accounts for both noise volume and frequencies and is used to compute maximum allowable noise levels of all civil aircraft. EPNdB is a blend of atmospheric noise generated at various engine locations such as the fan, the turbine, and the jet exhaust. *See also: Decibel.*

EFFECTIVE PERCEIVED NOISE LEVEL (EPNL)
A noise level measured in EPNdB's.

EFFECTIVE PITCH
The distance an aircraft advances in its flightpath for one revolution of the propeller.

EFFECTIVE PROPELLER THRUST
The net propulsive force after subtracting airstream drag from propeller thrust.

EFFICIENCIES (engine)
Locate under 1) Combustor Efficiency, 2) Compressor Efficiency, 3) Overall Efficiency, 4) Propeller Efficiency, 5) Propulsive Efficiency, 6) Thermal Efficiency, 7) Turbine Efficiency.

EFFLUX
The process of flowing outward. Often used to mean Jet Efflux (the hot gas discharge from a turbine engine), or the Fan Efflux of a Turbofan engine.

EFIS
Abbreviation for Electronic Flight Instrument System.

EGT
Abbreviation for Exhaust Gas Temperature.

EGT BLOOM
Locate under Bloom.

EGT GAUGE
A d'Arsonval electrical indicator in a conventional cockpit or a CRT display in a Glass Cockpit, used to indicate the temperature at the turbine discharge area of a turbine engine. Locate also under Exhaust Gas Temperature (EGT). *See also: 1) ITT Gauge, 2) TIT Gauge, 3) TOT Gauge.*

EGT MARGIN
The difference between the "red line value" (the maximum allowable EGT) and the actual EGT indication being observed.

EGT PROBE
Locate under Probes.

EHP
Abbreviation for Equivalent Horsepower. Another name for Equivalent Shaft Horsepower.

EHR
Abbreviation for Engine History Recorder.

EkW
Abbreviation for Equivalent Kilowatt.

EICAS
Abbreviation for Engine Instrument Crew Alerting (or Advisory) System.

EIS
Abbreviation for Engine Instrument System.

EJECTOR EXHAUST SUPPRESSOR
An aircraft manufacturer's supplied component. A new type of noise suppression unit that attaches to the exhaust of a turbine engine. It has the appearance of an early Multi-Tube Sound Suppressor. The ejector unit is designed with an opening in front which brings ambient air into a mixing chamber. The cold-hot gases mix to dampen the shear (noise) effect which occurs when hot exhaust gases meet cold atmospheric air at the jet nozzle. *See also: Noise Suppressor.*

EJECTOR PUMP (fuel boost)
A small aircraft accessory with no moving parts. A type of fuel tank booster pump which utilizes a high pressure fuel stream though a venturi-like device. The venturi fuel produces a motive force which carries fuel surrounding the venturi, now slightly pressurized, along with venturi fuel to the engine.

ELASTO-METRIC O-RING SEAL
A type of synthetic rubber packing similar to a common O-Ring.

ELECTRICAL DISCHARGE MACHINING (EDM)
A precision metal removal process employing an electrode in a cutting tool capable of eroding metal to very exact dimensions. EDM is used as a final process in forming many hardened turbine engine super-alloy parts.

ELECTRIC ENGINE TEST SET (ESTS)
Typically a portable troubleshooting instrument used to perform checks on electronic fuel controls. These test units generally have two modes of operation: 1) Dynamic Mode, which is a real time microprocessor mode analyzing engine operating parameters as a means of fuel scheduling system check out, 2) Static Mode, which is a test unit supplied set of parameters to the microprocessor as a means of system check out.

ELECTRICAL STARTER
An engine accessory. An electrical motor-operated starter used on many smaller turbine engines of all types. It is generally mounted on the Accessory Drive Gearbox. Electrical starters engage the engine by a retractable jaw or by an overrunning clutch mechanism. A high weight to power ratio prevents the use of electrical starters on larger engines. *See also: Pneumatic Starter.*

ELECTRO-HYDROMECHANICAL FUEL CONTROL
An engine accessory. A type of fuel control that combines the older hydromechanical designs with limited electronic controlling features. Generally the electronic component's only function is to provide over-temperature protection. *See also: Fuel Control Unit (main).*

ELECTRO-MAGNETIC IMAGING (EMI)
A materials flaw detecting technique employed in Kirlian Photography.

ELECTRO-MECHANICAL FUEL CONTROL
A fuel control utilizing both electric and mechanical forces to operate its mechanisms.

ELECTRON BEAM WELDING
A fusion welding process performed in a vacuum chamber and commonly used to produce and to repair many gas turbine metallic components. *See also: Welding.*

ELECTRONIC COCKPIT
Refers to the new generation of aircraft control panels in which aircraft and engine parameters are shown on CRT or LED displays as opposed to traditional cockpit indicators. Many gauges are continuously displayed, such as EPR, N1 Speed, or Torque. Other gauges are displayed only if selected or if an over-limits condition occurs. Electronic cockpits are often described as being made up of several displays, as follows: 1) primary flight displays (PFD), 2) navigational displays (ND), 3) flight management system displays, 4) engine and alert displays (EAD), and 5) system displays (SD). Also called Glass Cockpit. *See also: Cockpit Displays.*

ELECTRONIC ENGINE CONTROL (EEC)
One of a variety of new electronic engine accessories. A part-time or sometimes full-authority electronic fuel controlling device. The EEC schedules fuel in accordance with engine parameters such as power lever angle (PLA), percent RPM, bleed valve position, variable vane position, engine pressures and temperatures, etc. The EEC supplements the traditional hydromechanical fuel control unit on some newer aircraft to give more precise fuel scheduling and prevent over-temperatures and over-speeds throughout the entire flight envelope. *See also: Fuel Control.*

ELECTRONIC FLIGHT INSTRUMENT SYSTEM (EFIS)
A primary flight instrument and weather display system often referred to as a Glass Cockpit. The EFIS employs cathode ray tube (CRT) displays and LED displays in place of traditional instruments. *See also: 1) Cockpit Displays, 2) Engine Instrument Crew Alerting System.*

ELECTRONIC PROPULSION CONTROL SYSTEM (EPCS)
An engine control system that can interface with other engine systems such as: 1) Electronic Engine Control (EEC) systems, 2) Active Clearance Control (ACC) systems, 3) Variable Stator Vane (VSV) systems, 4) Variable Bleed Valve (VBV) systems. *See also: Fuel Control.*

ELECTRONIC TACHOMETER SYSTEM
Locate under Tachometer System.

ELECTROLYTIC ACTION
Locate under Engine Inspection and Distress Terms in Appendix.

ELOXED HOLES
Elongated cooling air holes as seen in turbine blades and vanes for film cooling of surfaces exposed to hot gases. The holes appear in fish-gill, tear-drop, or rectangular shapes. *See also: Turbine Cooling.*

EMERGENCY FUEL SYSTEM
A fuel back up system that is activated in the event of a main fuel system failure. Seen especially on single engine military aircraft.

EMERGENCY POWER RATING (engine)
Locate under: 1) Maximum Contingency Power, 2) Super Contingency Power.

EMERGENCY OIL SYSTEM
A small oil supply carried in an air pressurized accumulator. Should the main oil system fail, emergency oil is piped to an air-oil mist type oil nozzle. The emergency system is designed to provide a short term protection only. The engine will eventually have to be shut down if the main oil supply remains interrupted.

EMINENT BYPASS INDICATOR (fuel and oil filters)
Locate under Filter Bypass Indicator.

EMISSIONS (engines)
Pollutants created in the atmosphere caused by combustor inefficiencies. Gas Turbine Engine emissions classed as pollutants by the Environmental Protection Agency are: 1) Smoke-carbon particles, 2) Oxides of nitrogen (NOx), 3) Carbon monoxide (CO), 4) Unburned hydrocarbons (HC). Of particular concern is Nitric Oxide from the hot exhaust because it reduces atmospheric Ozone when forming Nitrogen Dioxide.

EMS
Abbreviation for, 1) Engine Management System, 2) Engine Monitoring System.

END-BEND COMPRESSOR BLADE
A recent development in blade design which employs a twist at both the base and tip. The aerodynamic effect of the twist counteracts slow or stagnant inner and outer boundary layer effect on air flow. The blade tip area is referred to as a dog-ear shape because it has the appearance of a dog-ear on the page of a book.

END-BEND COMPRESSOR VANE
A recent development in vane design. Compressor stator vanes with a noticeable twist in their outer section. The twist gives greater swirl to the air and a resulting higher compression per stage.

END-WALL CLEARANCE
Clearances between rotating and stationary components. For example, the radial clearance between blade tips and shrouds.

END-WALL EFFECT (rotor blade)
The negative aerodynamic effect on airflow at blade tips due to tip clearance induced air turbulence.

ENDOTHERMIC HYDROCARBON FUEL
A fuel for very high speed aircraft. Fuels that are routed through portions of an aircraft to act as a heat sink and reduce aircraft surface temperatures. *See also: Fuel Types-Jet.*

ENDURANCE LIMIT (of materials)
The predicted failure point due to cyclic stress over time. Repeated stresses at or near the Limit Load which can fatigue fail a part as a result of numerous cycles of operation. *See also: Material Strength.*

ENERGY
In terms of physics, the capacity of doing work. *See also: 1) Kinetic Energy, 2) Potential Energy.*

ENGINE (turbine)
A Gas Turbine Engine. An internal combustion device designed to produce mechanical work. An engine made up of various sections, components and accessories to form a working assembly. *See also: Gas Turbine Engine.*

ENGINE (types)
Locate under Gas Turbine Engines.

ENGINE AERODYNAMIC STATION
Locate under Engine Stations.

ENGINE ALERT DISPLAY
A CRT display in an electronic cockpit that pertains to the operation and performance of the gas turbine engine. *See also Glass Cockpit.*

ENGINE BLEED AIR
Air extracted from the engine at various points along the compressor section. Engine Bleed Air is used to service engine systems such as Anti-Ice, Balance Chamber, Sump Pressurization, and Turbine Cooling. *See also: 1) Compressor Bleed Air, 2) Customer Bleed Air.*

ENGINE BUILD-UP KIT
A kit which includes all the necessary lines, electrical harnesses, accessories and fittings to adapt a basic engine for installation into a particular aircraft. Also called 1) EBU Kit, 2) QEC Kit, 3) Quick Engine Change Kit.

ENGINE CAN
Generic term for a metal engine shipping container. The can is pressurized with dry air and contains desiccants to prevent corrosion.

ENGINE COMPONENT
Locate under Component.

ENGINE COMPONENT PART
Locate under Component Part.

ENGINE CONDITION MONITORING SYSTEM (ECM)
An electronic system that automatically records engine condition and performance data such as Vibration, EGT, RPM, EPR, etc. These data are later used to plot trend analysis of engine operational wear characteristics for maintenance scheduling purposes. *See also: 1) Engine Monitoring System (EMS), 2) Ground-based Engine Monitoring system (GEM).*

ENGINE CYCLE
Most commercial operators record one cycle as: 1) one take-off and landing which, in most instances, will involve only one engine start, one full power operation, and one shutdown. 2) An air start relight after shutdown in flight. 3) A touch-and-go landing. 4) A flight go-around. In addition to cycles, most aircraft also have a means of recording engine hourly

operating time using a Hobbs Meter type device. Some aircraft also utilize Event Counters to record the numbers of engine starts. Also called Load Cycle. *See also: 1) Event Counter, 2) Hobbs Meter, 3) High Cycle Life. Note: For military aircraft in which numerous power changes are required, one cycle is typically counted for each engine acceleration to full power.*

ENGINE DERATING
Locate under Derating.

ENGINE DESIGNATIONS
Manufacturer's alpha-numeric means of identifying different engine types and models. Locate the various engine types in the section titled "Gas Turbine Engines for Aircraft".

ENGINE DOWN-RATING
Refers to the manufacture of a smaller, less powerful model of an existing engine. *See also: Derated Engine.*

ENGINE ELECTRONIC CONTROL (EEC)
Another name for Electronic Engine Control.

ENGINE EMISSIONS
Locate under Emissions.

ENGINE HISTORY RECORDER (EHR)
An electronic recording system located in the aircraft to record engine operating hours and to record any over-limits conditions sensed by cockpit instruments such as EGT, RPM, Oil Pressure, etc. The hard copy data are later used for Condition Monitoring purposes. *See also: 1) Condition Monitoring, 2) Engine Condition Monitoring System.*

ENGINE INLET
Refers to the area directly in front of the first stage of compression. *See also: Flight Inlet.*

ENGINE INSTRUMENT AND CREW ALERTING SYSTEM (EICAS)
Part of an Electronic Flight Instrument System (EFIS). An electronic cockpit engine parameter display system. EICAS displays CRT readouts of EGT, RPM, Oil Pressure, etc., and color-coded alpha-numeric messages and symbols to advise flight and ground crews of core engine and engine system malfunctions. The three most common categories of messages are: 1) Warnings (red), 2) Cautions (amber), 3) Advisories (amber). *See also: Electronic Cockpit.*

ENGINE INSTRUMENT SYSTEM (EIS)
Name given to the newer "Glass Cockpit" type of aircraft instrument panel with LED and CRT displays. *See also: Electronic Cockpit.*

ENGINE INSPECTION AND DISTRESS TERMS
Words to clarify, describe and identify distressed areas of the engine whether from normal wear or early failure. The listing is located in Appendix.

ENGINE MAINTENANCE UNIT (EMU)
Another name for Module. *See also: Modular Maintenance.*

ENGINE MAJOR SECTIONS
The largest sections of an engine. The basic engine sections are considered to be the: 1) Compressor, 2) Combustor, 3) Turbine. The three major sections through which gases flow can be expanded to six sections as follows: 1) Inlet, 2) Compressor, 3) Diffuser 4) Combustor, 5) Turbine, 6) Exhaust. *See also: Engine Sections.*

ENGINE MODULE
Locate under 1) Modular Maintenance, 2) Module.

ENGINE MONITORING SYSTEM (EMS)
General term given to a common engine maintenance program used by airlines. The program typically includes: 1) Automatic and/or manual recording of engine performance data to be used in later trend analysis, 2) Scheduling of On-Condition Maintenance to include inspections, parts repairs and replacements as necessary, 3) Scheduling Hard Time replacement of parts at certain time intervals, no matter the condition. *See also: 1) Engine Condition Monitoring (ECM), 2) Ground-Based Engine Monitoring (GEM).*

ENGINE MOUNT
A static engine structural part which supports the engine in the aircraft. Mounts suppress gyroscopic moments, inertial loads during maneuvers, and also axial, vertical and side loads. *See also: 1) Cone Bolt, 2) Dogbone Mount, 3) Hangar Mount, 4) Trunnion Mount.*

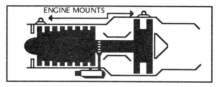

ENGINE OILER SYSTEM
An aircraft mounted oil tank which is capable of replenishing the oil reservoirs on individual engines via an electrical pumping and oil distribution system. It is used during ground servicing only.

ENGINE OPERATING CYCLE
A completed engine thermal cycle including the application of takeoff power.

ENGINE POWER RATINGS
Engine power parameters that are certificated by the FAA. These ratings list the thrust or shaft horsepower that must be available when selected by the pilot. Locate under headings of: 1) Take-Off (dry), 2) Take-Off (wet), 3) Super Contingency Power, 4) Maximum Contingency Power, 5) Maximum Continuous, 6) Maximum Cruise, 7) Cruise, 8) Idle. Ratings are given on Type Certificate Data Sheets and are listed typically in values of EPR, Percent N1, or Torque.

ENGINE PRESERVATION
A process of introducing corrosion inhibitors to oil and fuel systems and to wrapping of the engine in vapor barrier materials. Often the engine is placed in a pressurized container. Sometimes called Pickling of the engine. *See also: Engine Can.*

ENGINE PRESSURE RATIO (EPR)
The ratio of two engine parameters. The ratio of Turbine Discharge Total Pressure to Compressor Inlet Total Pressure. An "EPR" gauge in the cockpit is used by the flight crew as an indication of engine thrust. In many aircraft, Take-Off, Cruise and all other Engine

Power Ratings are set in the cockpit using the EPR gauge. This system evolved because thrust is a function of certain pressure values within the engine. The acronym (E-PER) is used in reference to this system. *See also: 1) Engine Power Ratings, 2) Integrated Engine Pressure Ratio.*

ENGINE PRESSURE RATIO (EPR) INDICATING SYSTEM
An aircraft and engine system consisting of: 1) Cockpit Indicator, 2) Compressor Inlet Total Pressure Probe, 3) EPR Transmitter, 4) Turbine Discharge Total Pressure Probes and Manifold. Signals generated by the engine are delivered to the cockpit gauge to indicate the thrust being produced by the engine. The indicator is called an EPR Gauge or Indicator.

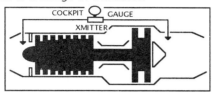

ENGINE PRODUCTION & OVERHAUL CONTROL SYSTEM (EPOCS)
A computerized maintenance data control system which provides overhaul information on modules, components, component parts, and accessories. This data follows an engine and all of its associated parts and hardware through the various stages of overhaul.

ENGINE RATING PLUG
A plug in unit and part to an electronic fuel control. This plug inputs the engine take-off thrust rating into the controlling system. *See also: Electronic Engine Control.*

ENGINE RATINGS
Locate under Engine Power Ratings.

ENGINE RERATING
Locate under Rerating.

ENGINE ROLL-BACK
An engine malfunction. Loss of engine power with Engine Percent RPM decreasing (rolling back). Frequently EGT will rise when roll-back occurs as the direct result of an engine stall. *See also: Stall.*

ENGINE SECTIONS
Portions of a turbine engine which have major aerodynamic and thermodynamic functions. Sections are made up of various components. Sections are not generally thought of as being particular pieces of turbomachinery but rather locations on an engine. For example, 1) Cold Section and Hot Section, or 2) Accessory Section, Compressor Section, Diffuser Section Combustor Section, Turbine Section, Exhaust Section.

ENGINE SPEED
Refers to the R.P.M. of the gas producer rotor (compressor-turbine) system.

ENGINE SPEED SENSOR (electric)
A small engine accessory. A single phase generator with a rotor in the form of a soft iron gear revolving within a stator housing. This unit produces an electrical frequency signal proportional to engine speed and sends it to a cockpit gauge. Similar to a traditional tachometer generator. *See also: 1) Tachometer Generator.*

ENGINE SPEED SENSOR (ELECTRONIC)
Locate under Tachometer System (Electronic).

ENGINE STATIONS (physical)
Reference points on an engine along its horizontal center-line. Two commonly used systems are: 1) Aerodynamic Engine Station numbering to identify locations along the engine gaspath from station-1 at the inlet, to station (last) at the exhaust. 2) Inch Station Numbering to identify inch locations along the engine's physical horizontal length. *See also: 1) Aero-Thermodynamic Engine Station, 2) Butt Line, 3) Water Line.*

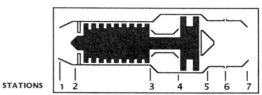

ENGINE STATIONS (thermodynamic)
Locate under Aero-Thermodynamic Engine Stations.

ENGINE SYSTEMS
Systems placed on a turbine engine to support its operation. Locate under 1) Anti-Icer System, 2) Anti-Stall System 3) Fuel System, 4) Ignition System, 5) Inlet Guide Vane System, 6) Lubrication System, 7) Starter System, etc.

ENGINE TEST RUN
Locate under Test Cell.

ENGINE TOPPING
A rotorcraft engine full power check. A performance check accomplished in flight under full load conditions.

ENGINE TRIM KIT
Locate under Trim Kit.

ENGINE TRIMMING
Locate under Part-Power Trimming.

ENGINE VANE CONTROL SYSTEM
Another name for Variable Angle Stator Vane System.

ENGINE VIBRATION MONITORING (EVM)
Another name for Airborne Vibration Monitoring.

ENGINEERING CHANGE PROPOSAL (ECP)
Another name for Service Bulletin.

ENGINEERING ORDER (E.O.)
A manufacturer's in-house type of Service Bulletin not always issued to the general public.

ENGOBE COATING™ (igniter plug)
A Champion Company product name. A baked on semi-conductor coating at the firing end of a low-voltage igniter plug. The coating bridges the electrical gap between the center electrode and the shell. *See also: Igniter Plug (low voltage).*

ENHANCED STOP DRILL REPAIR™ (ESDR)
A procedure that involves conventional stop drilling and then pulling a special mandrel through a split sleeve placed in the stop-drill hole. The hole is cold expanded approximately five-percent and this imparts an improved fatigue life in the material surrounding the hole. *See also: Stop Drill Repair.*

ENRICHMENT
Locate under Fuel Enrichment Valve.

ENRICHMENT BURST
Another name for Torching.

ENTHALPY (engine)
A physics term. A thermodynamic measurement of an engine's total heat value when its internal energy is added to the product of pressure and volume. Mathematically defined as:

$$H = E + (PV)$$

Where H = enthalpy, E = entropy, P = absolute pressure, V = volume. *See also: Entropy.*

ENTRAINED AIR (fuel)
Air which mixes with fuel in the fuel tanks and which flows through the entire aircraft and engine fuel systems. In small quantities this air is of no concern but in large quantities it can cause vapor lock. *See also: Vapor Lock.*

ENTRAINED AIR (oil)
Mainly foam or bubbles in the scavenged oil caused by heat and the centrifugal action of the oil wetted parts of the engine. Oil with large quantities of air entrained is a poor lubricant, and for this reason it is removed by mechanisms within the lubrication system. *See also: Deaerator.*

ENTRAINED WATER (FUEL)
Locate under Water (in fuel).

ENTROPY
A physics term. A thermodynamic measurement of the engine's heat energy which is present but which is not capable of doing work. A large amount of entropy means air in the engine has motion in the disordered form of turbulence and is not in useful axial motion. Mathematically defined as:

$$E = C_v \times T$$

where E = entropy, Cv = specific heat at constant volume and T = absolute temperature in degrees Rankine. A system with constant entropy is said to be isentropic. A Turbine Engine is Adiabatic but not isentropic. *See also: Enthalpy.*

EPA COLLECTOR TANK
Another name for Ecology Drain Tank.

EPCS
Abbreviation for Electronic Propulsion Control System.

EPNdB
Abbreviation for Effective Perceived Noise Decibels.

EPNL
Abbreviation for Effective Perceived Noise Level.

EPOCS
Abbreviation for Engine Production & Overhaul Control System.

EPOXY MATERIALS
Locate under Composite Materials.

EPR
Abbreviation for Engine Pressure Ratio.

EPR GAUGE
An engine performance indicator in the cockpit which displays engine thrust. For example; take-off thrust, climb thrust, and cruise thrust are set by the pilot while observing this gauge. *See also: Engine Pressure Ratio.*

EPR PROBE
Locate under Probes.

EPR RATED ENGINE
A turbine engine in which Engine Power Ratings are guaranteed to occur at certain Engine Pressure Ratio (EPR) settings on the cockpit EPR indicator. *See also: 1) Engine Power Ratings, 2) Speed Rated Engine.*

EQUIVALENT HORSEPOWER (EHP)
Another name for Equivalent Shaft Horsepower.

EQUIVALENT KILOWATT (EkW)
Metric Unit of power output of Turboprop Engines which combines both the Kilowatts (kW) of measured power delivered to the propeller and the Gross (static) Thrust being produced at the hot exhaust. *See also: Equivalent Horsepower.*

EQUIVALENT SPECIFIC FUEL CONSUMPTION (ESFC)
A physics term. A ratio of Weight of Fuel (Wf) to ESHP. A means of comparison for Turboprop Engines and some Turboshaft Engines where: ESFC = Fuelflow (Wf) ÷ Equivalent Shaft Horsepower (ESHP).

$$ESFC = (Wf) \div (ESHP)$$

EQUIVALENT SHAFT HORSEPOWER (ESHP)
U.S. unit of power output of Turboprop Engines which combines both the Shaft Horsepower (SHP) delivered to the propeller and Gross (static) Thrust being produced at the hot exhaust. Mathematically expressed as: ESHP = SHP + HP(Jet). Where SHP is the Shaft Horsepower as measured on a dynamometer and HP(Jet) is defined as Gross Thrust ÷ 2.5. The conversion factor 2.5 is an industry accepted standard.

$$ESHP = SHP + HP_{(Jet)} \quad HP_{(Jet)} = Fg \div 2.5$$

EROSION (hot section)
A condition in which the hot section materials erode away from impingement of hot gases and airborne contaminants on heated surfaces. Also locate under Engine Inspection and Distress Terms in Appendix.

ESDR
Abbreviation for Enhanced Stop Drill Repair.

ESHP
Abbreviation for Equivalent Shaft Horsepower.

ESTS
Abbreviation for Electric System Test Set.

EVENT COUNTER
A device which automatically records numbers of engine starts. The data collected in this manner is later

used to plan engine maintenance schedules. *See also: Engine Cycle.*

EXCITER UNIT OR BOX (ignition system)
Locate under Ignition Exciter.

EXDUCER (turbine)
The exit (rearmost) portion of a radial outflow turbine wheel.

EXHAUST
Locate under Exhaust Duct.

EXHAUST CENTERBODY
A fixed fairing within the exhaust stream, the more modern name for Tailcone. This term was also used to refer to the early design of a movable plug located within the exhaust stream. *See also: 1) Plug Nozzle, 2) Tailcone.*

EXHAUST COLLECTOR
Another name for Exhaust Cone (outer) or Turbine Exhaust Case.

EXHAUST CONE (outer)
An engine component. Generally refers to the outer casing located between the turbine case and the exhaust duct (tailpipe). This outer case is also called an 1) Exhaust Collector, 2) Turbine Exhaust Case. Some manufacturers also refer to the Exhaust Cone (outer) as the Tailcone, but generally the Tailcone is thought of as the Exhaust Cone (inner).

EXHAUST CONE (inner)
Another name for Exhaust Centerbody.

EXHAUST DIFFUSER
An aircraft manufacturer's supplied component. A type of exhaust duct (tailpipe) commonly used on Turboshaft Engines; it is divergent in shape to nullify the little remaining thrust typical of this type engine and facilitate a stable hover condition of the aircraft.

EXHAUST DUCT
Most often an aircraft manufacturer's supplied component. In recent times also supplied by engine manufacturers. A sheet metal duct attached to the rear of

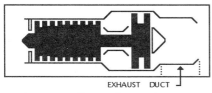

the basic engine to configure the engine to a particular aircraft. Same as tailpipe. A convergent shaped tailpipe is used on subsonic airplanes and a convergent-divergent shaped tailpipe is used on supersonic airplanes. *See also: 1) Exhaust Diffuser, 2) Exhaust Nozzle, 3) Internal-Flow Nozzle, 4) Mixed Exhaust Duct, 5) Plug Nozzle, 6) Primary Exhaust Duct, 7) Secondary Exhaust Duct.*

EXHAUST DUCT CONFIGURATIONS (turbofans)
There as three basic Exhaust Duct configurations currently in use on Turbofan Engines, as follows: 1) Co-Annular Exhaust Duct-where the hot stream duct is centered within the cold stream duct and both exit independently without mixing, 2) Plain Compound or Mixed Exhaust Duct-where the hot stream duct has the traditional plain round opening which feeds into a common tailpipe along with the cold stream duct, 3) Mixed Compound Exhaust Duct-where the hot and cold streams mix in a common tailpipe in the manner of a Plain Compound Duct but one with a cookie-cutter shaped hot stream nozzle called a Exhaust Mixer Nozzle.

EXHAUST FRAME
Another name for Turbine Rear Frame.

EXHAUST GAS TEMPERATURE (EGT)
A basic engine parameter. Total temperature generally taken at the turbine exit by means of a thermocouple system which transmits an electrical signal to a cockpit EGT Indicator. The total temperature probe absorbs both heat of ram (friction) and static heat from the flowing gases and produces a current flow proportional to the heat applied. Some newer systems employ laser light or infrared sensors as temperature measuring methods in place of the traditional thermocouple system. *See also: 1) EGT Gauge, 2) Thermocouple (exhaust temperature).*

EXHAUST GAS TEMPERATURE (gauge or indicator)
Locate under EGT Gauge.

EXHAUST GAS TEMPERATURE LIMITS
Two categories of temperature limits are most often given in flight manuals, as follows: 1) Intermediate Limits-engines experiencing this limit typically require inspection or possibly overhaul of the Hot Section depending on the time duration of the over temperature condition., 2) Maximum Limit-engines operated over this limit, regardless of the time interval, generally require immediate Hot Section overhaul. *See also: Over-Temperature.*

EXHAUST MIXER NOZZLE
An engine component. A recent development in noise suppressor designs for fully ducted Turbofan Engines. At high power settings, in the combined exhaust of a Turbofan Engine, the core discharge velocity is sonic and the fan discharge velocity is in the range of 0.4 to 0.6 Mach. The mixer nozzle smooths the shear-flow between the two streams in such a way that noise peaks are reduced and a blended noise of lower volume results. In some installations the combined exhaust results in improved thrust due to a more uniform velocity. Sometimes called: 1) Cookie-Cutter Mixer, 2) Daisy Tuner Mixer, 3) Deep-Chute Mixer, 4) Forced Exhaust Mixer. *See also: Noise Suppressor Unit.*

EXHAUST NOZZLE
In a strict sense, the final opening at the rear of a Core or Fan exhaust duct (tailpipe), but commonly used to mean the exhaust duct itself. Also called Jet Nozzle. *See also: 1) Exhaust Duct, 2) Fixed Area Exhaust Nozzle, 3) Variable Area Exhaust Nozzle, 4) Subsonic Exhaust Nozzle, 5) Supersonic Nozzle.*

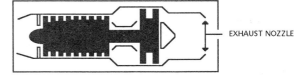

325

EXHAUST NOZZLE (2-D)
A recent development in low drag design, a nozzle rectangular in shape. 2-D refers to the fact that the nozzle opening has both length and width rather than the more traditional round nozzle shape. Also known as a Non-axisymmetric (off-center) Exhaust Nozzle or a 2D-CD Nozzle. May also be used to refer to an Exhaust Nozzle (vectored thrust).

EXHAUST NOZZLE (bypass duct)
Refers to the final opening for fan discharge airflow. Also called Fan Exhaust Nozzle. *See also: Secondary Exhaust Duct (turbofan).*

EXHAUST NOZZLE (fan or secondary)
Another name for Exhaust Nozzle (bypass duct).

EXHAUST NOZZLE (primary)
The final opening to the atmosphere of the Core Engine (hot) discharge airflow of a Turbofan Engine.

EXHAUST NOZZLE (vectored thrust)
An engine component. An exhaust duct containing a nozzle which has the capability of angling its flow in an up and down direction. Vectoring aids in short field take-off and landing, and in aircraft maneuverability at low speeds. For example, when the nozzle is angled up, the exhaust gases being directed upwards cause the nose of the aircraft to also move up. Vectoring can be accomplished in both Dry and Afterburning mode. Also referred to as a 2D-CD Nozzle. *See also: 1) Exhaust Nozzle 2-D, 2) Viffing.*

EXHAUST NOZZLE EXIT
A physics term. The point slightly downstream of the Nozzle Throat (opening to the atmosphere) of a convergent exhaust duct where complete expansion of gases takes place.

EXHAUST NOZZLE THROAT
The point on an exhaust duct most constricting to exhaust gases. The throat is located just at the lip of the final opening to the atmosphere, in a convergent exhaust duct. In a Convergent-Divergent Exhaust duct the throat is the narrowest part of a variable flow area mechanism. *See also: Convergent-Divergent Exhaust Duct (supersonic aircraft).*

EXHAUST PRESSURE RATIO
Locate under Nozzle Pressure Ratio.

EXHAUST REHEATER
British term for Afterburner.

EXHAUST SECTION
The portion of a turbine engine through which gases return to the atmosphere. *See also: Engine Sections.*

EXHAUST SILENCER
Another name for Noise Suppressor Unit.

EXHAUST STATOR VANE
British term for Turbine Exit Guide Vane.

EXHAUST PIPE
Another name for Exhaust Duct.

EXHAUST PLUG
Another name for Exhaust Centerbody.

EXHAUST STRUT
An engine component part. A spoke-like, stationary support located at the rear of the turbine wheels, and one which extends inward from the exhaust cone casing to support the tailcone. The Exhaust Struts may also at times support a rear bearing housing. *See also: Turbine Exhaust Fairing.*

EXIT GUIDE VANE (compressor or turbine)
An engine component part employed to remove whirl from flowing gasses. Stationary airfoils that straighten (turn) airflow back to an axial direction after it leaves the compressor or turbine rotors. Also called Outlet Guide Vanes. *See also 1) Guide Vanes, 2) Whirl.*

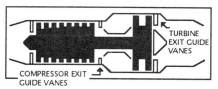

EXIT SILENCER
Another name for Noise Suppressor Unit.

EXPANDER TURBINE NOZZLE
An engine component. A flow orifice arrangement that accelerates cooling air over turbine nozzle vane surfaces to provide a thermal barrier.

EXPANSION RATIO (exhaust duct)
The ratio of inlet to outlet pressure. The Exhaust Duct airflow will choke at a ratio of 1.89.

EXPANSION SPACE (oil tank)
Space above the oil level in an oil supply tank. An FAA requirement of Parts 23, 25 and 33 which states that an expansion space of 10 percent or 0.5 gallons, whichever is greater, must be provided.

EXTENDED-ROOT BLADE (turbine)
A blade configured with a long root length allowing for a narrow disc diameter.

EXTERNAL ANNULAR COMBUSTOR
Another name for Annular Combustor (reverse flow).

EXTERNAL COMBUSTION ENGINE
An engine in which the working fluid is not used in the combustion process. In years past, the Gas Turbine Engine was classed as an external combustion engine because combustion occurred in a different location from the compression of air. Today the turbine is classed as an internal combustion engine because it uses the same working fluid for compression and combustion within one outer engine casing. *See also: Internal Combustion Engine.*

EXTERNAL COMPRESSION INLET
A type of flight inlet in which all of the ram pressure rise (compression) occurs outside the forward lip of the inlet. Some of the compression is used by the engine and the remainder spills over the lip. *See also: Flight Inlet.*

EXTERNAL DRY SUMP (lubrication system)
A lubrication system in a Gas Turbine Engine in which the main supply of oil is located in an externally mounted oil tank. *See also: Lubrication System.*

EXTERNAL GEARBOX
British for Accessory Drive Gearbox.

EXTERNAL TOOTH LOCKWASHER
Locate under Shakeproof Washer.

EXTRACTION PAD (customer bleed air)
Points on the compressor case or diffuser case where air is extracted from the engine and directed into aircraft air systems. Also called Service Bleeds. *See also: Customer Bleed Air.*

EXTRACTION PAD (engine bleed air)
Points on the compressor or diffuser case where air is extracted from the engine and directed back into engine air systems such as anti-icing, turbine cooling air, etc.

EYELID EXHAUST NOZZLE
An engine component part. An afterburner exhaust nozzle consisting of two movable segments. Usually a two-position type unit, either fully open at maximum flow area or fully closed at minimum flow area. The smallest flow area is used for non-afterburning engine operation and the largest flow area for engine operation in afterburner. The two-position exhaust nozzle system generally includes no capability for modulating power in the afterburner mode. *See also: 1) Afterburner (hard light), 2) Exhaust Nozzle.*

F

FAA
Abbreviation for Federal Aviation Administration.

FAB SWITCH
Locate under: 1) Fuel Filter Impending Bypass Switch, 2) Oil Filter Impending Bypass Switch.

FACE RUNOUT CHECK
Another name for Axial Runout Check.

FACE-TYPE SEAL (fuel or oil)
A shaft seal in which a hard rotating surface and a carbon or soft metal stationary part are in sliding contact. The stationary part is spring loaded onto the highly polished mating surface to form a fluid seal.

FADEC
Abbreviation for Full Authority Digital *Electronic* Control, or Full Authority Digital *Engine* Control.

FAFC
Abbreviation for Full Authority Fuel Controller.

FAIRED FAN TIP
Locate under Fan Tip (faired).

FALSE START
An unsuccessful engine start, wherein idle speed is not achieved and the engine start cycle is terminated.

FAN
A component of a Turbofan Engine. The typical Fan is mounted on the front of the Low Pressure Compressor. However, some Fans independently rotate, being driven by a separate turbine system. Also a limited number of Fans were mounted at the rear of the engine as extensions to the turbine blading. Fans act as compressors, to pressurize air and to move a large mass airflow. *See also: 1) Aft-Fan, 2) Turbofan.*

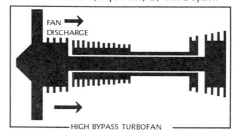

FAN AND SUPERCHARGER
The fan and following stages that make up a Low Pressure Compressor in a Turbofan Engine.

FAN ABRADABLE SHROUD RING
A component part of the fan case. A ring of an abradable material positioned just outside of the plane of rotation of the fan blade tips. *See also: Abradable Shroud*

FAN BALANCE CHECK
Locate under Fan Trim Balance Check.

FAN BLADE
A component part of a fan assembly. A rotating airfoil which increases the velocity of a mass of air and facilitates an increase in static air pressure. *See also: Turbofan.*

FAN BLADE REMOVAL PORT
A covered opening in the fan outer case which facilitates individual fan blade replacement. The blades, in this case, generally have a pin type attachment rather than a dovetail attachment, which allows the blade to move radially out of the disc.

FAN BLADE SHINGLING
A fan blade wear condition. A condition of overstress on fan blades which allow the span-shrouds to overlap (shingle-fashion) rather than support each other face to face. Also locate in Engine Inspection and Distress Terms in Appendix. *See also: Span-Shrouds.*

FAN BLADE SHROUD
Locate under Span-Shroud (fan blade).

FAN BLISK
An engine component. A disc and blade unit which is cast as one solid piece. The blades in this unit are not removable. *See also: Fan Disc and Blade assembly.*

FAN BOOSTER
An engine component. Name given to the compressor stages behind the fan in the low pressure (LP) compressor of a Turbofan engine. This section further boosts static pressure before the air is delivered to the High Pressure Compressor (HPC).

FAN BOOSTER AIR
Another name for Low Pressure Compressor discharge air.

FAN BYPASS RATIO
Locate under Bypass Ratio (fan).

FAN CASE
An engine component. Locate under Fan Containment.

FAN CONTAINMENT CASE
An engine component. The outer fan case. Named as such because it is designed to protect against radial blade movement in the event of a fan failure. *See also: Containment Ring.*

FAN CONTAINMENT RING
Locate under Containment Ring.

FAN COWL
A nacelle component part. The portion of a Turbofan Engine nacelle that surrounds the fan. *See also: Translating Cowl.*

FAN DISC
An engine component part. The solid center section of a fan rotor to which the blades are attached.

FAN DISC AND BLADE ASSEMBLY
An engine component. The fan disc with all of its replaceable blades attached. Also called Fan Rotor.

FAN DRIVE SHAFT
An engine component part. The low pressure turbine front shaft which connects to the low pressure compressor (fan). The Fan Drive Shaft usually runs concentrically through the high pressure compressor drive shaft.

FAN DUCT BURNER
An afterburner system placed within the fan exhaust as a means of augmenting thrust. Most often, fan duct afterburning is only a portion of the total afterburning capability, the remainder being in the traditional location of the core engine exhaust. Also called 1) Plenum Chamber Burner, 2) Secondary Power System. *See also: Afterburner.*

FAN ENGINE
Another name for Turbofan Engine.

FAN EXHAUST DUCT
Locate under Secondary Exhaust Duct (turbofan).

FAN FAILURE (contained)
A fan break-up during engine operation which is completely contained by its casings, with all of the debris created moving axially along the gas path. *See also: 1) Containment Ring (fan), 2) Uncontained Fan Failure.*

FAN FAILURE (uncontained)
A failure which results in the escape of fan fragments through the outer fan case. Engines are FAA Type Certificated to standards which in most cases will prevent this type of failure. However, industry experience indicates that fragments weighing three pounds or more traveling at speeds of near 1,000 ft./sec. during a fan break up, often result in unpredicted failures not contained by the fan cases. *See also: 1) Containment Ring (fan), 2) Fragment Spread Angle.*

FAN FRAME
Another name for Fan Containment Case.

FAN GEAR-RATIO
Locate under Geared Turbofan.

FAN HUB
Refers to the inner portion of the fan of a Turbofan Engine that discharges its air into the core.

FAN JET
Generic name for an aircraft powered by a Turbofan Engine.

FAN NOZZLE
Another name for Fan Exhaust Nozzle.

FAN PRESSURE RATIO
The ratio of fan discharge pressure to fan inlet pressure. Approximately 1.5 to 1 for a single stage fan and 2.0 to 1 for a two stage fan. *See section entitled "Gas Turbine Engines for Aircraft" for specific engine applications*

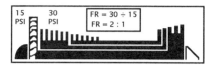

FAN ROTOR
An engine component. The fan disc and blade assembly with the front spinner and rear shaft attached. *See also: Fan Disc and Blade Assembly.*

FAN SPEED
Speed of the fan rotor in a turbofan engine, given in percent RPM. Often used as a thrust parameter in the cockpit and identified as N1 Fan Speed. Take-off and all other power ratings are set in terms of N1 Fan Speed when it is used as the primary power indication. This instrumentation is stated to have an accuracy of ± 1 percent of the actual thrust being produced.

FAN SPEED PICKUP
Another name for Fan Speed Sensor.

FAN SPEED SENSOR
A small engine accessory. A unit in an Electronic Tachometer System. One such sensor is placed outside the plane of rotation of the fan blade tips. Another similar system uses a sensor positioned at the plane of rotation of a gear-type wheel pressed onto the fan shaft. As the blades or the gear teeth cut through an eddy current generated by the sensor, an electronic circuit calculates the speed in percent RPM. These systems are stated to have an accuracy of ± 0.2 Percent of the actual speed. Also called 1) Fan Speed Pickup, 2) Magnetic Pickup, 3) Signal Conditioner. *See also: Tachometer System (electronic).*

FAN STATOR
An engine component. The stationary vane system located to the rear of the fan rotor. The stators diffuse air to increase static pressure and angle the airstream back to an axial direction.

FAN THRUST REVERSER
Typically a Blocker-Door type. Locate under Aerodynamic Thrust Reverser.

FAN TIP (faired)
A notch-shaped cutout along the chord line of an airfoil near its tip. The Faired Tip is an aerodynamic design to increase the stall margin at that point on the blade. *See also: Lenticular Profile.*

FAN TRIM BALANCE CHECK (airborne)
A check on the balance condition of the fan in a Turbofan Engine. As the fan becomes imbalanced

through normal wear in service, an on-board vibration monitor predicts the placement of fan balance weights that will bring the unit back to original specifications. *See also: Ground-Based Engine Monitoring (GEM) system.*

FAN TRIM BALANCE CHECK (ground)
A maintenance check to determine whether the fan is in balance. A typical check is as follows: 1) find the point of highest vibration on the N1 tachometer indicator, while slowly accelerating the engine from ground idle speed to take-off power. 2) If out of limits, plot the data on standard graphs to calculate the weight to be used to bring the fan into balance. 3) Shut down the engine and add balance weights to the fan hub. This check is often required after fan blade blending repairs or fan module change. *See also: Balance Checks.*

FAN TURBINE INLET TEMPERATURE
Refers to a temperature indication taken in front of the low pressure (fan) turbine.

FAN VANE
The vanes which make up the Fan Stator.

FANJET AIRCRAFT
An aircraft that is powered by a Turbofan Engine.

FANJET ENGINE
Another name for Turbofan Engine.

FANWHEEL
Another name for Fan Disc and Blade Assembly.

FAR'S
Federal Aviation Administration Regulations. The most common which affect operation and maintenance of Gas Turbine Engines and Gas Turbines Engine powered aircraft are:

FAR Part 23	Airworthiness standards, normal, utility, acrobatic aircraft.
FAR Part 25	Airworthiness standards, transport category aircraft.
FAR Part 33	Airworthiness standards, Aircraft Engines.
FAR Part 35	Airworthiness standards, Propellers.
FAR Part 36	Noise standards, aircraft.
FAR Part 39	Airworthiness Directives.
FAR Part 43	Maintenance, Preventive Maintenance, Rebuilding, Alteration.
FAR Part 65	Certification of Airmen Other Than Flight Crewmembers.
FAR Part 91	General Operating and Flight Rules.
FAR Part 120	Air carrier operations.
FAR Part 135	Air Taxi Operations.

FATIGUE FAILURE
A phenomenon which occurs in most materials after repeated cycles of loading and unloading, even though the stresses never reach the yield strength or elastic limit of the material. In turbine engines, fatigue failure is usually the result of vibration, G-forces, and heat. *See also:* 1) *High Cycle Fatigue,* 2) *Low Cycle Fatigue,* 3) *Material Strength,* 4) *Thermal Fatigue.* Fatigue is also listed under Engine Inspection and Distress Terms in Appendix.

FATIGUE LIFE
A service life factor published by a manufacturer. A minimum time expressed in engine operating hours or cycles that an engine part is expected to accumulate without failure.

FATIGUE LIMIT
The maximum stress that a part can withstand for a specified number of engine operating hours or cycles without failure.

FAULT INDICATORS (engine)
A traditional reference to all cockpit instruments that monitor engine performance. The newer cockpits also include visual, pictorial, and worded warnings of engine faults.

FAULT-TREE
A systematic troubleshooting chart used by maintenance crews to locate the source of malfunctions in turbine engines. Also called logic tree.

F.C.O.C.
Abbreviation for Fuel-cooled Oil Cooler. Locate under Oil cooler.

FEDERAL AVIATION ADMINISTRATION (FAA)
The Federal agency, under the Department of Transportation, which issues directives and regulates aircraft operation, safety, and certification of personnel and equipment. The FAA was formerly known as the Federal Aviation Agency.

FEEDBACK CABLE
A cable that connects two or more engine accessories or components. The feedback cable provides a mechanical sense of position between units in a system. *See also: Teleflex Cable.*

FEEL TEST
An engine bearing check whereby the technician hand rotates the bearing and then compares the feel it produces to the feel of a new bearing when rotated. *See also: Bearing Noise and Feel Test.*

FEFI SYSTEM
Abbreviation for Flight Environment Fault Isolation System.

FELT-METAL™
A porous metal used in the manufacture of abradable air seals for turbine engines by the Brunswick Corporation. These seals form shroud rings placed outside of the plane of rotation of the blade tips. Should contact loading occur between the seal and the tip, a slight wearing-in occurs to regain the required clearance. The seals, being of soft metal, also prevent reduction in blade length. Over time, the seals that exhibit excessive wear are replaced. Felt-Metal Seals are used in place of traditional honeycomb seals. *See also: Shroud Ring.*

FENWALL FIRE DETECTOR™
A Fenwal Company thermoswitch type detector, or a continuous loop type detector which completes an electrical cockpit warning circuit in the presence of an overheat condition. The detectors are located in the nacelle area surrounding the exterior of a turbine engine. *See also: Fire Detector.*

FFR
Abbreviation for Fuel Flow Regulator.

FIAT AVIAZIONE COMPANY (Italy)
Manufacturer of Gas Turbine Engines. Locate in section entitled "Gas Turbine Engines for Aircraft".

FIB SWITCH
Locate under Fuel Filter Impending Bypass Switch.

FIBERGLASS MATERIALS
Locate under Composite Materials.

FIBERSCOPE
Refers to a Fliber Optics Borescope.

FIBER OPTICS BORESCOPE
A unit of test equipment. A flexible borescope that utilizes light transmitting fibers and is used in place of a traditional rigid borescope. The probe of a flexible borescope is capable of being bent by a hand control in order to view at various angles within the engine. See also: Borescope.

FIELD CLEANING (compressor)
The process of injecting a dry grit blast cleaning agent or liquid emulsion cleaner into the compressor section of turbine engines. The objective of Field Cleaning is to clean airfoil surfaces and restore their aerodynamic shape. This procedure is accomplished when engine performance starts to degrade. Cleaning with the abrasive grit is normally accomplished while the engine is being run at alternating power settings. Cleaning with emulsion is generally accomplished during a motor-over with the starter or during an idle speed operation. See also: 1) Carbo-Blast, 2) Compressor Rice Hull Cleaning.

FIELD CLEANING (turbine)
Turbine cleaning is accomplished by way of compressor field cleaning when certain types of emulsion cleaners are used.

FIELD DETECTOR
Locate under Bearing Field Detector.

FILLET AREA (compressor or turbine blades)
Locate under Blade Fillet.

FILM AIR COOLING (turbine blades and vanes)
Air that flows from small holes in the airfoil surfaces to form an insulating layer of cooling air. See also: 1) Turbine Cooling.

FILTERS
Small engine accessories that remove contaminants from air, fuel, and oil systems. Locate under 1) Barrier Type Filter, 2) Fuel Filter, 3) Oil Filter.

FILTER BACKFLUSH (oil system)
A bench check procedure to flow fluid in the reverse direction through a filter and to collect the contaminant particles for analysis.

FILTER BYPASS INDICATOR (fuel or oil)
A cockpit warning light that indicates: 1) An impending bypass condition exists due to a partially blocked filter. Should filter clogging worsen, filter bypassing ultimately occurs delivering unfiltered fluid to mechanisms down stream of the filter. 2) A filter bypass condition actually exists. Also called clogging indicator, impending bypass indicator, eminent bypass indicator, flag indicator. See also: 1) Fuel Filter Impending Bypass Switch, 2) Oil Filter Impending Bypass Switch.

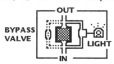

FILTER BYPASS INDICATOR (pop-out)
A mechanically operated pop-out (or pop-up) button on the filter body which serves as a visual warning of a clogging filter. Generally, the button can only be reset by removal of the filter bowl. This indicator is typical of fuel, hydraulic, and oil systems and is used in place of a more expensive cockpit warning light system.

FILTER DEBRIS ANALYSIS
Locate under Oil Filter Debris Analysis.

FILTER MAINTENANCE INDICATOR
Another name for a Filter Bypass Indicator (Pop-Out).

FILTER RATINGS (micronic)
Generally two ratings are given to turbine engine filters, the nominal and the absolute ratings. The nominal rating is defined as the ability to trap 95% to 98%, by weight, of the particles at or larger than the filter rating size. The absolute rating defines the largest size particles that will pass through the pores of the filter. If only one rating is given, it is most likely the absolute rating. Micronic ratings describe the diameter of the particles to be trapped, such that a 15 micron filter (absolute) will trap all particles over 15 microns in diameter.

FILTER RATINGS (U.S. sieve number)
A mesh per linear inch system of describing filtration values. For example, a 323 mesh fuel filter has 323 sieve openings per linear inch. The U.S. Sieve Rating is more common to turbine engine fuel filters. A conversion table is available to convert from the U.S. Sieve Number to the more popular Micron rating. Locate U.S. Sieve Ratings in Appendix.

FIN
British term for Knife-Edge Tip.

FINENESS RATIO
Ratio of an engine's length to its diameter. A factor involved in streamlining an engine for a pod and pylon mounting.

FINE PITCH STOP
A small component part. A stop which sets a limit on propeller blade rotation. Fine Pitch is a low blade angle employed during engine start-up, and opposite to Coarse Pitch.

FINES
Name given to small metallic particles that appear as gray granules or powder on magnetic chip detectors. A condition often caused by a thrust loaded main bearing experiencing skidding failure or bearing cage wear. See also: (skidding) in Bearing Inspection and Distress Terms in Appendix.

FINGER WASHER
Another name for Spring Washer.

FINWALL COMBUSTOR™
An engine component. A type of double wall combustion liner with cooling air layered in between. *See also: Combustors Types.*

FIR-TREE FIT (turbine blade)
The most common turbine blade root design. A blade root in a tapered (V-shape) with multiple, axial serrations. This design provides a maximum contact surface to resist radial movement under high centrifugal and heat loading. The fir-tree is sometimes described as: 1) multiple serration fit, 2) multiple dovetail fit.

FIRE CAN
Jargon for combustor liner.

FIRE DETECTOR
An aircraft accessory typically mounted in the engine nacelle. An electrical heat sensor which alarms a fire warning light in the cockpit. This detector is also classed as an overheat detector because it will alarm if acted upon by hot air as well as from a fire fed heat source. Also locate under the following fire detector systems: 1) Continuous Loop, 2) Fenwall, 3) Kidde, 4) Pneumatic, 5) Thermocouple, 6) Thermoswitch.

FIRE RETARDANT FUEL
JET-A fuel which includes the anti-misting agent called "AMK", a substance containing long chain molecules that surround kerosene molecules to suppress misting. *See also: Fuel Types.*

FIRE POINT (fuel and oil)
A point on a temperature scale. The temperature at which fuel and oil vapors will ignite and continue to burn. *See also: Flash Point.*

FIRE SEAL
A fireproof bulkhead that separates the turbine engine Cold Section from the Hot Section when the engine is installed in its nacelle. Generally part of a QEC Kit installation.

FIREWALL
A fire barrier between an aircraft engine and the airframe.

FIREWALL POWER
Jargon for maximum or emergency power setting. *See also: Engine Power Ratings.*

FIRE WARNING SYSTEM
Locate under Fire Detector.

FIT
Locate under Engine Inspection and Distress Terms in Appendix.

FIXED AREA EXHAUST NOZZLE
An exhaust duct with a non-adjustable exit flow area. The duct is generally constructed of rolled sheet metal. The fixed area nozzle is the most common type of exhaust ducting system used on turbine engines. *See also: 1) Internal-Flow Nozzle, 2) Plug Nozzle.*

FIXED DISPLACEMENT PUMP
Another name for Positive Displacement Pump.

FIXED GEOMETRY INLET DUCT (supersonic aircraft)
An inlet of convergent-divergent flow chamber design with no movable or adjustable parts. *See also: Convergent-Divergent Inlet Duct.*

FIXED GEOMETRY INLET DUCT (subsonic aircraft)
An inlet of divergent flow chamber design with no movable or adjustable parts. *See also: Flight Inlet.*

FIXED TURBINE
An engine component. A turbine rotor in a Turboprop or Turboshaft Engine that is mechanically attached to drive both a compressor rotor and an output reduction gearbox. The Fixed Turbine in this sense is directly attached to the propeller of a Turboprop Engine or to the transmission drive of a Turboshaft Engine. Also called Direct Drive Turbine. *See also: Free Power Turbine.*

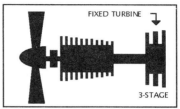

TURBOPROP (FIXED TURBINE DESIGN)

FIXED TURBINE ENGINE
A classification of torque producing turbine engines. A Turboprop or Turboshaft Engine in which the turbine rotor(s) are mechanically connected to the compressor, the output reduction gearbox, and the propelling shaft. *See also: Free Turbine Engine.*

FLADE
Another name for Fan Blade. Not commonly used.

FLAG OIL FILTER
Another name for Filter Bypass Indicator.

FLAKING
Locate under Engine Inspection and Distress Terms in Appendix.

FLAME ARRESTOR (afterburner)
Another name for Flame-holder.

FLAME IGNITER
Another name for Hot Streak Igniter.

FLAME HOLDER (afterburner)
An afterburner component. A stationary ring-shaped fixture located to the rear of the afterburner fuel spraybars. Flame Holders are designed to create turbulence as the flowing gases contact their surfaces. This turbulence reduces the gas velocity and increases air/fuel mixing time in the afterburner flame zone. Flame Holders are also utilized in Ram-Jets and Pulse-Jets. In British terminology also called a, 1) Flame Stabilizer, 2) Vapor Gutter.

FLAME-OUT
The unintentional extinction of combustion, typically caused by: 1) interruption of fuel flow by such conditions as fuel filter icing or fuel vapor lock, 2) interruption to normal inlet airflow and resulting compressor stall

while operating in turbulent weather.
See also: 1) Lean Die-out, 2) Rich Blow-out.

FLAME TUBE
British term for Combustion Liner.

FLAME PROPAGATION TUBE (combustor)
A small component part of a combustion liner. Connectors fitted between individual liners of multiple-can or can-annular type combustors. During engine start, the tubes provide for flame propagation from the liners which contain the electrical igniters to the remaining liners. Also called 1) flameover tubes, 2) interconnector tubes, 3) crossover tubes.

FLAME SENSOR (afterburner)
An afterburner component. An ultra-violet radiation sensing device which sends a signal to the cockpit to indicate the afterburner is functioning properly.

FLAME STABILIZER (afterburner)
Another name for Flame Holder.

FLANGE (engine)
Main separating points on an engine. Most flanges are located between major engine sections. Many flanges also have alpha or numeric code designations to facilitate maintenance instructions. *See also: Exhaust Nozzle.*

FLAP EXHAUST NOZZLE (single)
A component part of an afterburner. A single ring of movable flaps that form a final exit for the exhaust gases. The smallest flow area occurs when the flaps are closed in non-afterburning engine operation. The largest flow area occurs when in afterburner.

FLAP EXHAUST NOZZLE (dual)
A component part of an afterburner. A double set of movable flaps, one set with a larger flow area than the other. In combination they form a controllable final exit for the exhaust gases. Both sets of flaps open fully when the engine is in afterburning operation. Flap nozzles can be seen on both older two-position afterburners and the newer variable flow area afterburners. *See also: 1) Primary Exhaust Nozzle (afterburner), 2) Secondary Exhaust Nozzle (afterburner).*

FLASH POINT (fuel or oil)
A point on a temperature scale. The temperature at which a volatile fluid, such as fuels or oils, give off vapors which if in the presence of a flame will ignite and burn for one to five seconds. *See also: Fire Point.*

FLAT-MACHINE BLADE TIP
Compressor or turbine blade tips that have a constant cross-section at the tip. *See also: Blade Tip.*

FLAT-RATE TEMPERATURE (FRT)
Refers to Flat-Rating (engine).

FLAT-RATING (engine)
A manufacturer's method of referring to engine Rated Thrust or Rated SHP which is guaranteed to be available up to a specified ambient temperature, known as the Flat-Rate Temperature. For example: At Sea Level Barometric Pressure, the Pratt & Whitney PW-2037 Turbofan is flat-rated at 37,000 lb. thrust to 86° F. The term flat-rating also refers to the flat shape of an ambient temperature vs. thrust curve. A PW-2037 flat-rating curve would show that from the lowest operating temperature possible up to 86° F., the engine is capable of producing the full 37,000 lbs. of thrust. Above 86° F., full power lever and full thrust will not be permitted due to combustor temperature limitations. Knowing the flat-rating of an engine is a means of comparing its thrust capability to that of other similar engines. Flat-Rated thrust also applies to power settings such as Max Continuous, Max Climb, Max Cruise, etc., where the manufacturer would also have to specify the thrust, ambient temperature and altitude. Also called Rerated Engine. *See also: 1) Full Throttle Engine, 2) ISA Rating, 3) Thermodynamic Power Rating.*

FLEXIBLE TAKE-OFF (FTO)
A take-off procedure at less than full power when the aircraft weight is less than Maximum Take-Off Weight (MTOW).

FLIGHT CYCLE
Another name for Engine Cycle.

FLIGHT DATA RECORDER (FDR)
An aircraft accessory. A recorder that continuously monitors selected vital operating aircraft and engine parameters. Typically the recorder erases and repeats its cycle every thirty minutes. The recording tape is later used to schedule maintenance actions or, in the case of an accident, to reveal operating conditions before or during the accident to FAA and NTSB accident investigators. All turbine powered aircraft operating under FAR 121 are required to have FDR equipment. *See also: Cockpit Voice Recorder.*

FLIGHT ENGINE
A prime mover aircraft engine as opposed to a turbine engine used as an auxiliary power unit.

FLIGHT ENVELOPE
Refers to the allowable flight speed and altitude of a particular aircraft. All aircraft have a specified Flight Envelope as part of their FAA Type Certification.

FLIGHT ENVELOPE GRAPH (air carrier)
A graph showing the maximum altitude permissible per the flight Mach Number and also the absolute ceiling limitation of an aircraft. For example, Mach 0.5 to 0.85 airspeed is often a permissible speed range at maximum flight ceiling in a Turbine Powered Aircraft.

FLIGHT ENVIRONMENT FAULT ISOLATION (FEFI)
A flight crew troubleshooting procedure. A systematic method by which flight crews, using specific check procedures and instrument indications, record cockpit data. The data is later used by maintenance crews to complete the troubleshooting of aircraft and engine malfunctions.

FLIGHT IDLE SPEED
The lowest engine speed allowable in flight. Used primarily for final descent, approach, and landing. Flight Idle may occur at the same full rearward position of the power lever as ground idle, but more likely it will be approximately 10% RPM higher than ground idle speed. Some cockpits use a mechanical gate to block ground idle position. Setting flight idle speed higher than ground idle speed gives greater stall protection and quicker response on power up. The cockpit

may also be configured with an automatic reset for flight idle speed, the higher the altitude the higher the idle speed setting. Also called Approach Idle Speed. *See also: Ground Idle.*

FLIGHT INLET
A component part of the engine nacelle. The air inlet portion of the nacelle which supplies air to the engine inlet. It is divergent in shape to promote inlet compression. The nacelle is generally considered an aircraft component rather than an engine component. However, in recent times engine manufacturers have also started producing engine nacelles. Also called Inlet Cowl because the nacelle is made up of a series of cowl panels or cowlings which encompass the engine. *See also: Inlet (configurations).*

FLIGHT MANAGEMENT SYSTEM (FMS)
A computer based system that monitors engine fuel consumption vs. aircraft range. The objective is to select the correct airspeed which results in the best fuel economy. Also known as Performance Management System (PMS). *See also: Maximum Range Cruise Speed.*

FLIGHT NOZZLE
Another name for Exhaust Duct.

FLIGHT REGIMES
Flight speeds expressed in Mach Numbers as follows: 1) Subsonic, below Mach number 0.75, 2) Transonic, between Mach 0.75 and Mach 1.2, 3) Supersonic, between Mach 1.2 and 5.0, 4) Hypersonic, above Mach 5.0. *See also: Mach Number.*

FLOAT-WALL COMBUSTOR
Locate under Combustor Liner (float-wall).

FLOW AUGMENTOR
Another name for Jet Pump.

FLOW DENSITY (oil)
Defined as the parts per million (PPM) of oil flow as per the flow area. For example, as a filter screen begins to clog, the flow area decreases causing a flow density increase. This is a case where essentially the same PPM of oil is trying to flow through a smaller surface area.

FLOW DIVIDER (fuel)
A device that directs fuel into two or more distinct flow paths. Flow Dividers are typically used at: 1) Split fuel manifolds that connect to the combustor fuel nozzles, 2) Dual manifolds feeding into duplex fuel nozzles.

FLOWMETER
A shortened form for Flowmeter Transmitter.

FLOWMETER INDICATOR
A cockpit instrument that shows the fuel being consumed in pounds per hour (PPH) by the engine. The indicator receives its operating data from an engine mounted transmitter. Also called Fuelflow Indicator.

FLOWMETER TRANSMITTER (motor driven massflow type)
A QEC Accessory. An electrical transmitter through which fuel flows to the combustor. It is an older type and contains a small motor driven turbine to swirl the fuel and give a massflow reference against which actual flow can be measured in pounds per hour. This transmitter sends continuous operating data to a cockpit fuel Flowmeter Indicator.

FLOWMETER TRANSMITTER (coroilis type)
A QEC Accessory. An electronic motorless transmitter through which fuel flows to the combustor. It is the most popular type and contains magnetic pickups that twist under influence of fuel flow to give a massflow signal. A time factor is used as a reference against which actual fuel flow can be measured. This transmitter sends continuous operating data to a cockpit Flowmeter Indicator. Also called Motorless Massflow Fuelflow Transmitter.

FLOW REVERSER
Another name for Thrust Reverser.

FLUIDIC AMPLIFIER
A tailpipe design under development for small engines. One in which engine thrust is amplified by gases exhausting through several small discharge nozzles as opposed to the traditional single opening.

FLYBACK LIMIT
The allowable aircraft/engine malfunction or damage condition in which an aircraft can be ferried back to its maintenance base. Aircraft operation under Flyback Limits requires that flight-crew personnel only be aboard the aircraft.

FLY-BY-LIGHT
Refers to aircraft and engine systems that are controlled or powered by light emissions carried in fiber optic bundles throughout the aircraft. This system is being developed for use in high performance aircraft for its weight saving advantage over fly-by-wire systems. *See also: Fly-By-Wire.*

FLY-BY-WIRE
Refers to aircraft and engine systems that are controlled by or powered by electrical/electronic means rather than by traditional methods such as hydraulics, pneumatics, or cable systems. Many newer aircraft are fly-by-wire controlled and cannot operate with an electrical power failure. *See also: Fly-By-Light.*

FLYWEIGHT GOVERNOR
A small component part of a fuel control unit. A system of weights on a rotating table plate that move outward when under a centrifugal load as engine speed increases. The weights push against a slider and speeder spring on a movable shaft that connects to a mechanical fuel scheduling linkage. In this manner, a flyweight governor on a turbine engine schedules fuel in reference to engine speed. This type governor allows for quick engine spool-up until the flyweight force moves into equilibrium with the opposing (spring) force, allowing the engine to seek an on-speed operating condition.

FLYWEIGHT SHOE
The points on a rotating weight in a flyweight governor that make contact to move the slider.

FMS
Abbreviation for Flight Management System.

F.O.D.
Abbreviation for Foreign Object Damage.

F.O.D PROTECTOR (inlet)
Another name for Inertial Ice and Sand Separator.

FOGGER OIL JET
Another name for Oil Jet (air-mist).

FOLDED ANNULAR COMBUSTOR
Another name for Annular Combustor (reverse flow).

FORCE
A physics term. Force is described as energy exerted on an object, or weight pull of gravity on an object. Expressed in units of Pounds in the U.S. System and Newtons in the Metric System. Mathematically expressed as:

$$F = P \times A$$

Where "F" is Force, "P" is pressure and "A" is area. The thrust produced by a Turbojet and Turbofan Engine is a type of force. For example, a large Turbofan Engine with a thrust rating of 56,000 lbs. (249 kN).

FORCED EXHAUST MIXER
An engine component. Locate under Exhaust Mixer Nozzle.

FORCED VIBRATION
A vibration that occurs when a variable force on an engine component causes it to vibrate at the frequency of the applied force rather than at its own natural vibration frequency. *See also: Vibration.*

FOREIGN OBJECT DAMAGE (F.O.D)
Damage primarily to blades and vanes in the gaspath of Gas Turbine Engines. F.O.D. results from ingestion of objects not originally part of the aircraft or engine; items such as ground debris, birds, and other objects in the air. F.O.D also occurs from hardware dropped into internal engine parts such as gearboxes and pumps. *See also: Domestic Object Damage (D.O.D.).*

FORGING
Process of hot pressing or hammering of a piece of metal to a desired shape. Formerly, the sole process suitable for manufacture of higher strength metallic parts. Now isothermal forging and investment casting also produce the same high quality metals for turbine engines. *See also: Isothermal Forging.*

FORM DRAG
Locate under Drag (parasite).

FORWARD FAN ENGINE
A Turbofan with its fan section located at the front of the compressor. It may be mechanically attached to the low pressure compressor or it may be a separate rotor as is the case with a three-spool Turbofan engine. *See also: 1) Aft-Fan, 2) Turbofan.*

FRACTURE
Locate under Engine Inspection and Terms in Appendix.

FRAGMENT SPREAD ANGLE (fan)
The angle measured fore and aft from the center of rotation of a fan which catastrophically fails, and the fragmented pieces move radially through the fan containment case. *See also: Fan Failure (uncontained).*

FRAME (engine)
A stationary main engine component configured with radial spoke-like struts. Frames are spaced along the engine to act as support members for rotating components at its inner diameter and to provide mounting points for major cases at its outer diameter. The struts usually perform an aerodynamic function of straightening airflow as gas path air passes over their surfaces. *See also: Casings.*

FRANZ, ANSELM, DR.
The German designer of the Junkers Jumo-004, the world's first axial flow turbojet engine. It was installed in the Messerschmitt Me-262 fighter in the early 1940's.

FREE FAN ENGINE
Another name for Propfan.

FREE POWER TURBINE An engine component. A turbine wheel that drives an output reduction gearbox of a Turboshaft or Turboprop Engine rather than an engine compressor. The Free Turbine engine design permits: 1) Variable gearbox speed output at constant core engine speed, 2) constant gearbox speed output over a range of core engine speeds. Also called Free Turbine or Power Turbine. *See also: 1) Free Turbine Engine, 2) Gas Generator Turbine.*

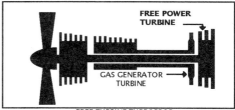

FREE TURBINE TURBOPROP

FREE STREAM AIRFLOW
Air in motion not affected by the proximity of an airfoil.

FREE TURBINE
Locate under Free Power Turbine.

FREE TURBINE ENGINE
An engine that contains a free power turbine. When the Free Turbine drives a propeller, it is called a Turboprop engine. When it powers a transmission drive it is called a Turboshaft engine. The flexibility in the Free Turbine design is that by varying core engine speed (Ng) with load changes: 1) Variable power output can be achieved at constant Free Power Turbine (Nf) speeds, 2) Constant power output can be achieved over a range of Free Power Turbine (Nf) speeds. *See also: Fixed Turbine Engine.*

FREE VIBRATION
A vibration that occurs after a vibratory force has been exerted on a component and is then removed. The vibration continues until the energy remaining dissipates. *See also: Vibration.*

FREEZING POINT DEPRESSANT (FPD)
A solution used to control ice formation on aircraft surfaces. This fluid can be applied: 1) unmixed for ice removal, 2) mixed with warm water for ice removal, 3) as an ice suppressant after ice removal with warm water.

FRENCH ENGINES
Locate by manufacturer, Microturbo, SNECMA, Turbomeca, in section entitled "Gas Turbine Engines for Aircraft".

FRETTING
Locate under Bearing Inspection And Distress Terms in Appendix.

FREQUENCY TUNING
In reference to compressor and turbine blades, an aerodynamic design that increases the blades' natural vibration frequency beyond the frequency of engine rotation. Tuning prevents initial resonance that may over time increase to a point of blade failure. A Squeeler Tip is one such design feature. *See also: 1) Resonance Of Blades (critical, 2) Resonant Frequency, 3) Vortex Tip.*

FRICTION
A force that opposes motion.

FRICTION (fluid)
A force that opposes flow of liquids or gases. This force opposes rolling and sliding friction in the molecular layer of liquids or gases between two bodies bearing on each other.

FRICTION (rolling)
A force that resists relative motion between two solid bodies when one or both roll over the other.

FRICTION (sliding)
A force that resists relative motion between sliding solid bodies.

FRICTION BEARING
Locate under Bearing (plain).

FRICTION CLUTCH
Another name for Slip-Clutch.

FRICTION HORSEPOWER (turboprop and turboshaft)
The horsepower lost between the power section and the propelling shaft of a turbine engine. That is, the power given up to turn the output reduction gearbox of a Turboshaft Engine or the propeller reduction gearbox of a Turboprop Engine. *See also: 1) Brake Horsepower, 2) Shaft Horsepower.*

FRONT ACCESSORY CASE
An accessory case located on the inlet case of a Turbojet or Turbofan Engine. Small driven accessories are sometimes attached to this case.

FRONT FRAME
Another name for Inlet Case.

FROSTING (bearings)
Locate under Bearing Inspection and Distress Terms in Appendix.

FROUDE EFFICIENCY
Another name for Propulsive Efficiency.

FSII
Abbreviation for Fuel System Icing Inhibitor.

FRT
Abbreviation for Flat Rate Temperature.

FTIT GAUGE
Abbreviation for Fan-Turbine Inlet Temperature.

FUEL (jet)
Locate under Fuel Types.

FUEL (metered)
Fuel leaving the fuel control unit (metered by weight) on its way to the combustor.

FUEL (primary)
Locate under Primary Fuel (combustor).

FUEL (secondary)
Locate under Secondary Fuel (combustor).

FUEL (unmetered)
Fuel discharged from the fuel pump being delivered to the fuel control unit.

FUEL (wide-cut)
A hydro-carbon turbine fuel consisting mainly of gasoline. A blend of approximately three parts gasoline and one part kerosene. A volatile fuel for improved cold weather starting and high altitude performance. Civil Aviation Fuel type Jet-B and Military Fuel type JP-4 are examples of wide-cut fuels. *See also: Fuel Types.*

FUEL ADDITIVES
Locate under 1) Anti-Misting Additive, 2) Fuel System Icing Inhibitor, 3) Prist.

FUEL-AIR COMBUSTION STARTER
A engine accessory. A unit containing a combustion section similar to an engine combustor with its expanding gases exhausted through a turbine wheel; the wheel being connected to an output reduction gearbox fitted with an external drive shaft. The drive shaft splines into the engine Accessory Drive Gearbox to provide starting torque sufficient to rotate the engine during its starting cycle. This starter contains its own fuel scheduling system, but it must be supplied with high pressure air from an aircraft storage bottle or ground support compressor. Also called *Combustion Starter*. *See also: Jet Fuel Starter.*

FUEL-AIR MIXTURE RATIO
Locate under Stoichiometric Mixture.

FUEL ATOMIZER
Another name for Atomizing Spray Fuel Nozzle.

FUEL BOOST PUMP
Locate under Fuel Pump.

FUEL CONDITION ACTUATOR (cockpit)
An engine start control lever that functions as a fuel on-off lever, and which may have one or more additional functions: 1) Rich Position, (for cold starting), 2) Ground Idle Position, 3) Flight Idle Position.

FUEL CONTAMINATION
Locate under 1) Prist, 2) Water (in fuel).

FUEL CONTROL PARAMETERS
Locate under Signals (fuel control).

FUEL CONTROL UNIT (abbreviations)
The most common abbreviations are: 1) DEEC for Digital Electronic Engine Control, 2) EEC for Electronic Engine Control, 3) FCU for Fuel Control Unit, 4) JFC for Jet Fuel Control, 5) HMU for Hydromechanical (control) Unit, 6) MEC for Main Engine Control, 7) MFC for Main Fuel Control.

FUEL CONTROL UNIT (emergency)
A small accessory or part of a main fuel control unit. This unit can be activated in the event of a main fuel control failure. The engine in emergency fuel mode is directly responsive to fuel inputs from the power lever, with few if any automatic signals governing fuel flow. *See also: Emergency Fuel System.*

FUEL CONTROL UNIT-FCU (main)
An engine accessory. The main fuel scheduling mechanism of an engine. This unit receives a power lever command from the cockpit and also various pressure and temperature signals from specific locations on the engine. These signals provide for automatic scheduling of fuel at all ambient conditions of ground and flight operation. See also specific fuel control types as follows: 1) Electro-Hydromechanical, 2) Electro-Mechanical, 3) Digital Electronic Engine Control, 4) Electronic Engine Control, 5) Full Authority Digital Electronic Control, 6) Hydromechanical, 7) Pneumatic-Mechanical.

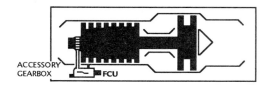

FUEL CONTROL UNIT (starting)
A small accessory used in conjunction with a main fuel control unit. The starting fuel control supplies fuel to start the engine and then its operation terminates. It is not common to see this unit today.

FUEL DEICER (engine)
Locate under Fuel Heater.

FUEL DUMP VALVE
Locate under Dump valve.

FUEL DRAINAGE CHECK
A check on the amount of fuel issuing from a seal drain point on an engine. The limits are most often given in drops of fuel per minute or cubic centimeters per minute. *See also: 1) Gang Drain, 2) Fuel Seepage Check.*

FUEL ENRICHMENT VALVE
A component part of a fuel metering system. A means of providing a rich mixture to the combustor during cold weather engine starting. Enrichment is generally initiated by a cockpit switch and required when ambient temperature is 32°F and below. *See also: Torching.*

FUEL FILTER (centrifugal)
A small accessory. An engine filter assembly containing a rotary filtering element to throw contaminants outward into a cleanable sediment trap.
See also: 1) Filters, 2) Fuel Purifier.

FUEL FILTER (engine)
A small engine accessory. Porous paper, cellulose, or stainless steel screen filtration units. Fuel filters are placed at various locations within the fuel system to provide protection against contamination to system mechanisms. Fuel filters have either Micronic Ratings or U.S. Sieve Ratings. A conversion table is required to convert from one system to the other. Locate table in Appendix. *See also: 1) Fuel Filters listed below, 2) Filters, 3) Wash Filter.*

FUEL FILTER (high pressure)
Typically a filter located downstream of the main fuel pump. Unmetered Fuel on its way to the main fuel control passes through this filter. *See also: 1) Fuel Filter Bypass Indicator, 2) Unmetered Fuel.*

FUEL FILTER (last chance)
A small filter located at the most downstream point in the fuel system, typically located in the fuel nozzle.

FUEL FILTER (low pressure)
Typically a filter located downstream of the engine boost pump and upstream of the main fuel pump. This filter will have a larger surface area and higher micronic rating than a high pressure fuel filter.

FUEL FILTER (self-relieving)
A filter arrangement which allows the filter screen to lift off its seat and bypass fuel when clogging of the filtering element reaches a predetermined level. *See also: Filters.*

FUEL FILTER (servo)
A small filter with a very fine mesh, used to filter only a small portion of the total fuelflow. Fuel passing through this filter is typically going into the Fuel Control to operate servo mechanisms and for lubrication.

FUEL FILTER IMPENDING BYPASS SWITCH
A fuel pressure operated switch that actuates a cockpit warning light should the main filter screen clog sufficiently to cause a pressure drop equal to the alarm setting of the switch. Often called FIB Switch. If a further pressure drop occurs, the filter will go into a bypass condition, and this may be indicated by a Fuel Filter Actual Bypass Switch (FAB Switch and Warning Light). *See also: Filters.*

FUEL FLOW INDICATOR
Locate under Flowmeter Indicator.

FUEL FLOW REGULATOR (FFR)
Another name for Fuel Control Unit.

FUEL FLOW TRANSMITTER
Locate under Flowmeter Transmitter.

FUEL FOGGING
A condition in which fuel vapor is being issued from the tailpipe during the engine start cycle when the engine fails to light off.

FUEL HEATER
An engine accessory. Typically a device which heats fuel by command of a cockpit switch. The heater is turned on when the temperature of fuel in the tanks drops to 32°F. Fuel heat prevents entrained water in the fuel from forming ice crystals at fuel filter screens. If ice has already formed, this unit acts as a fuel

deicer. Sometimes called a Fuel Deicer.

FUEL HEATER (air type)
A type of air to fuel heat exchanger in which a transfer of heat from hot engine bleed air into the engine fuel takes place. This heater often supplements a Fuel-Oil Cooler that also transfers some of its oil heat into the fuel. Also called Fuel Pre-Heater. *See also: 1) Fuel Heater (oil type), 2) Fuel-Oil Cooler.*

FUEL HEATER (oil type)
A type of oil to fuel heat exchanger in which a transfer of heat from the hot engine lubricating oil into the engine fuel takes place. Both the air type heater and the oil type heater operate in a similar fashion. Both utilize a Soda-Straw type heat transfer chamber to carry fuel through the heater. *See also: Fuel Heater (air type).*

FUEL ICING
Ice formed when entrained water contacts supercooled parts such as filter screens. *See also: 1) Fuel Heater, 2) Fuel System Icing Inhibitor, 3) Prist, 4) Water (in fuel).*

FUEL INJECTOR
A small engine component part. Generally means a fuel distribution device other than the more common atomizing type fuel nozzle. As with other fuel distributors, the injector delivers fuel directly into the engine combustor. *See also: 1) Fuel Nozzles.*

FUEL INJECTOR (carburizing)
An injector tube which utilizes low fluid pressure over a large flow area to deliver a fuel spray directly into a scroll shaped combustion chamber. The scroll is designed to impart a swirling motion to the air. The scroll, in combination with the injector, maximizes atomization of fuel. This design is said to provide good resistance to carbon build-up on the head of the injector. *See also: Scroll Carburetor.*

FUEL INJECTOR (centrifugal)
A type of rotating fuel nozzle that atomizes by a slinging action. This nozzle is not in wide use but can be found in a some single annular type combustors.

FUEL INJECTOR (scroll cup)
A type of spray nozzle that delivers fuel to a small scroll (snail) shaped mixing chamber located in the head of the combustion liner. The scroll causes counter-rotation of air and the turbulence created helps the atomization process.

FUEL LEAKAGE CHECK
Locate under 1) Fuel Drainage Check, 2) Fuel Seepage Check.

FUEL MANAGEMENT SYSTEM
Locate under Flight Management System.

FUEL MANIFOLD
An engine component part. A series of fuel lines manifolded together which carry fuel to the fuel nozzles.

FUEL MANIFOLD (simplex)
A fuel distribution system to a series of simplex fuel nozzles.

FUEL MANIFOLD (duplex)
A fuel distribution system to a series of duplex fuel nozzles. Also called Dupel Fuel Manifold (British).

FUEL NOZZLE
A small component part. The unit that introduces fuel into the combustion liner. The three most common fuel nozzle types are the: 1) Aerating Type, which creates a fine spray under low delivery pressure, 2) Pressure Atomizing Type, which creates a fine spray under high delivery pressure, 3) Fuel Vaporizing type, which creates a fine spray under a high heat condition. See Fuel Nozzle entries below. *See also: 1) Fuel Injector, 2) Primer Fuel System, 3) Fuel Vaporizing Tube.*

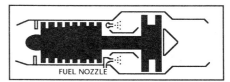

FUEL NOZZLE (aerating)
A recent development of several distinct designs. A simplex fuel nozzle or duplex fuel nozzle which creates a finely atomized spray by introducing both fuel under pressure and a portion of the primary air for combustion into the fuel flow orifice. This process allows a large, carbon resistant fuel orifice to be used with fuel under relatively low pressure. The use of high velocity air instead of high velocity fuel, as in the pressure atomizing type nozzle, is stated to provide better atomization at low fuel flow rates and avoid rich concentrations of fuel. Also called, 1) Air-Blast Fuel Nozzle), 2) Air-Mist Fuel Nozzle. *See also: Hybrid Air-Blast Nozzle.*

FUEL NOZZLE (duplex)
A dual orifice, pressure atomizing fuel distributor, that directs fuel into the combustor at two distinct flow rates through two separate flow orifices. One orifice flows at low engine power settings and both flow in a combined spray pattern at power settings above idle speed. The spray angle can be the same from each orifice or there can be two distinct spray angles. Often the low flow orifice, for engine starting, has the wider spray angle to facilitate good fuel-air mixing. Also called a Dupel Fuel Nozzle.

FUEL NOZZLE (Lubbock)
An older style, variable area atomizing nozzle and forerunner to the modern duplex fuel nozzle.

FUEL NOZZLE (pressure atomizing)
A general classification. A simplex or duplex fuel nozzle which creates a fine atomized spray by forcing a fluid under high pressure and velocity, first through a spin chamber, and then through a small spray orifice into the combustor. *See also: 1) Fuel Nozzle (duplex), 2) Fuel Nozzle (simplex).*

FUEL NOZZLE (simplex)
A pressure atomizing fuel distributor with only one spray orifice and one spray pattern. Fuel is delivered to a single, fixed-area orifice, through a spin chamber which induces a swirl in the fuel. The swirl creates a finely atomized spray cone. This type nozzle is most efficient at higher engine power settings and less effective at low power settings. *See also: Fuel Nozzle (duplex).*

FUEL NOZZLE (single stage)
Another name for Fuel Nozzle (simplex).

FUEL NOZZLE (spill-spray)
A type of simplex fuel nozzle in which fuel is supplied to the flow orifice spin chamber under high pressure even at low engine speeds, to provide good atomization. In this condition, a portion of the fuel is continually being bled away. An old style seldom seen today.

FUEL NOZZLE (vaporizing type)
Locate under Fuel Vaporizing Tube.

FUEL NOZZLE FERRULE
A small component part. A receptacle in the head of the combustion liner into which the fuel nozzle tip is inserted.

FUEL-OIL COOLER
An engine accessory. A heat exchange device that cools the oil by transferring its heat into the fuel. It is a radiator-like unit with fuel passing though internal cores and oil passing around the cores. The oil flow is generally controlled by a thermostatic valve to route the oil through the cooling chamber only after a predetermined oil temperature is reached. Until oil reaches its operating temperature, the oil bypasses the cooler. On some engines no separate Fuel Heater is required due to the high heat exchange rate of this cooler. Also called Fuel-Oil Heat Exchanger. *See also: Oil Cooler.*

FUEL-OIL HEAT EXCHANGER
Another name for Fuel-Oil Cooler.

FUEL PRESSURE GAUGE
A cockpit instrument. An electrical engine condition gauge showing metered fuel pressure being delivered to the combustor. *See also: Fuel (metered).*

FUEL PRESSURE TRANSMITTER
Locate under Transmitter (fuel pressure).

FUEL PRESSURIZING VALVE
Locate under Pressurizing Valve.

FUEL-PRIMARY
Locate under Primary Fuel (combustor).

FUEL PUMP (aircraft boost)
A small aircraft accessory. Typically a small centrifugal pump submerged in a fuel tank. This pump places a pressure head of approximately 10 to 25 psig on fuel in route to the engine. Its principle function is to reduce the possibility of vapor lock by keeping entrained air in motion. *See also: Pump Centrifugal.*

FUEL PUMP (engine boost)
A small engine accessory. Generally a centrifugal type pump, either mounted on the main fuel pump or mounted separately on or near the engine This pump increases fuel system pressure beyond fuel tank boost pressure to provide a fuel pressure head of approximately 50 to 60 psig. Its principle function, like the aircraft boost pump, is to reduce the possibility of vapor lock by keeping entrained air in motion. *See also: Pump (centrifugal).*

FUEL PUMP (engine main)
An engine accessory generally mounted on the Accessory Drive Gearbox. This pump is most often of the spur gear or vane type and its function is to deliver fuel under pressure to the Fuel Control Unit. Most gear and vane pumps delivers a constant volume of fuel per revolution and are classed as positive displacement pumps. *See also: 1) Fuel Pump (variable displacement), 2) Fuel Pump (parallel-stage), 3) Pump (positive displacement).*

FUEL PUMP (multi-stage)
A main engine fuel pump of positive displacement with two or more pumping stages in series. For example, a three stage pump might include a centrifugal impeller boost stage, a primary stage gear pump to supply fuel at normal pressure to the fuel control, and a secondary gear stage of higher pressure output. The secondary stage, in this case, provides motive force for hydraulic mechanisms such as inlet guide vane actuators and for other fuel pressure operated devices.

FUEL PUMP (parallel-stage)
A main engine fuel pump of positive displacement with two gear elements in parallel, often called a primary element and a secondary element. A built-in safety feature provides a shear-type shaft arrangement which allows for a one-element failure. Should this occur, the remaining element retains its pumping capacity. Single element operation is sufficient for most flight conditions except take-off.

FUEL PUMP (piston)
An engine accessory. A type of pump seldom used on the Gas Turbine Engine. This pump is capable of delivering fuel at higher pressures than other type pumps and can also function as a variable displacement unit by varying its output per revolution. Also called a Plunger-Type Pump.

FUEL PUMP (variable displacement)
A main engine fuel pump of the piston type or a spur gear type with two gear elements in series. This pump can vary its displacement per pump revolution to meet varying needs of the fuel control unit. *See also: Piston Fuel Pump.*

FUEL PUMP METERING UNIT (FPMU)
A combined main fuel pump and fuel control unit.

FUEL PURIFIER (engine)
A dynamic fuel filter which by centrifugal action forces contaminants into a cleanable trap. It is used mainly on extended duty stationary ground turbine engines and seldom used in flight engines. Also called Centrifugal Filter.

FUEL-SECONDARY
Locate under Secondary Fuel (combustor).

FUEL SEEPAGE CHECK
A determination of the severity of a slight amount of fuel that is seeping through a sealed surface. Four categories of seepage are recognized: 1) Slow Seep, 2) Seep, 3) Heavy Seep, 4) Running Seep. The Running Seep is also called a Fuel Leak and generally requires immediate attention. Whether the other less severe seepage problems require attention is a matter of interpretation by trained maintenance personnel. *See also: Leakage limits.*

FUEL SYSTEMS (engine)
The three common types of fuel systems employed to support engine operation are: 1) emergency fuel system, 2) main fuel system, 3) primer (starting) fuel system.

FUEL SYSTEM (main)
Locate under Main Fuel System.

FUEL SYSTEM ICING INHIBITOR (FSII)
An additive to prevent water in fuel from forming microscopic ice crystals which can block fuel filters and restrict flow. *See also: 1) AvFuel-2M, 2) Prist.*

FUEL TEMPERATURE SENSOR
Most often a Resistance Bulb type unit that is placed in a fuel tank. The sensor is connected electrically to a cockpit Fuel Temperature Gauge, and when fuel temperature drops to 32°F., the operator is advised to turn on the Fuel Heater to prevent water in fuel from freezing. *See also: 1) Fuel Heater, 2) Resistance Temperature Detector.*

FUEL TYPES-JET (civil-commercial)
Fuels in common use are: 1) Turbo Fuel-A, also called Jet-A, 2) Turbo Fuel-A1, also called Jet-A1, 3) Turbo Fuel-B, also called Jet-B. Jet-A and A1 are described as civil aviation kerosene of high flash point, with a weight of 6.75 lb./gal. Jet-B is described as a wide-cut fuel of relatively low flash point, with a weight of 6.50 lb./gal. These fuels are used in all types of Gas Turbine Engines and can, in general, be mixed or matched in any quantities as servicing requires. Jet Fuels conform to Mil-T-5624, Mil-J-5624, ASTM 1655. *See also: 1) Fuel (wide-cut), 2) Civil Aviation Kerosene, 3) Middle Distillate Fuel, 4) Naphtha Base Fuel, 5) Synjet-A.*

FUEL TYPES-JET (dirty fuel)
Name given to ground turbine engine fuels for power utility companies and the like. It is not as highly filtered for removal of impurities as are fuels for flight engines.

FUEL TYPES-JET (emergency alternate)
Many Gas Turbine Engines are authorized to use fuels with a Mil. Spec. MIL-G-5572 (aviation fuel grades 115/145 and lower) for a limited number of operating hours. However, when gasoline fuels are used in turbine engines, a specified type of hydrocarbon oil at a rate of approximately 3% is normally added to ensure correct lubricity to fuel system components.

FUEL TYPES-JET (military)
Military fuels are designated by Jet Petroleum (JP) numbers, such as JP-1 through JP-8. JP-1, JP2, and JP3 are rarely seen today. JP-4 fuels are similar to Commercial Jet-B, and JP-5 fuels are similar to Commercial Jet-A & A1. JP-6 and JP-7 are special purpose airplane fuels. JP-8 is similar to Commercial Jet-A and is being widely distributed within the military. It may one day completely replace JP-4.

FUEL TYPES-JET (under development)
Locate under: 1) Boron Slurry Fuel, 2) Cryogenic Fuels, 3) Endothermic Hydrocarbon Fuel, 4) Fire Retardant Fuel, 5) Synjet-A Fuel.

FUEL VAPORIZING TUBE
A fuel distributor typically of L-shape or T-shape design, that injects a vaporized fuel and air mixture into the combustor. The vaporizer tube, located in a hot zone of the combustor, receives both fuel and air under low pressure. The high heat of combustion outside of the mixing chamber facilitates vaporization of the mixture. *See also: 1) Fuel Nozzle, 2) Fuel Injector.*

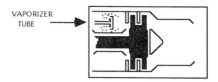

FUELFLOW INDICATOR OR TRANSMITTER
Locate under Flowmeter.

FULL ANNULAR FAN DUCT
Another name for Co-annular Exhaust Duct.

FULL AUTHORITY DIGITAL ELECTRONIC CONTROL (FADEC)
An integrated engine and aircraft fuel controlling system which includes a flight management system and an Electronic Engine Control (EEC). The FADEC is an improvement over the traditional hydromechanical control unit assisted by an analog electronic control. The FADEC unit contains a digital electronic control and operates on a full time basis to prevent overshoots of internal engine temperatures. FADEC is said to improve response time, decrease fuel consumption, and enhance engine service life. The FADEC system communicates with an Engine Instrument and Crew Alerting System (EICAS) to compute its display parameters.

FULL AUTHORITY FUEL CONTROL (FAFC)
An Electronic Fuel Control that is on line during all engine operation regimes. The older types are analog units and the newer units are digital. *See also: Fuel Control Unit.*

FULL INDICATOR READING
Another name for Total Indicator Reading.

FULL INTERMEDIATE POWER
A recent military designation for what was previously called Military Power. I.E. Full take-off power without augmentation. *See also: Takeoff (dry).*

FULL POWER TRIMMING
Locate under Trimming.

FULL THROTTLE ENGINE
A Gas Turbine Engine which receives its take-off thrust at maximum travel of the power lever (throttle). The power lever is advanced to a fixed stop on the throttle quadrant at each take-off. Various limiters in the fuel system prevent the engine from over-boosting. Many fighter aircraft are fitted with this throttle configuration. *See also: Part Throttle Engine.*

FULLY-DUCTED TURBOFAN
A Turbofan Engine with a fan duct covering the entire length of the core engine and in which the fan and core engine exhausts are mixed. Also called 1) Fully-Shrouded, 2) Long-Ducted. *See also: Turbofan Engine.*

FULLY-SHROUDED TURBOFAN
Another name for fully ducted Turbofan.

G

G-ROTOR PUMP
Locate under Oil Pump.

GALLERY (oil)
An oil passageway within the engine.

GALLING
Locate under Bearing Inspection and Distress Terms in Appendix.

GANG DRAIN
A small engine accessory. A holder for a series of overboard fluid seal drains located at the bottom of an engine. The Gang Drain connects to the Nacelle Drain Mast to carry away fluids emanating from Accessory Drive Gearbox shaft seals, the oil tank Scupper Drain, and the Fuel Dump Valve. The individual drains are often called Witness Drains. *See also: 1) Leakage Limits, 2) Seepage Checks.*

GANG NUT
A nut plate often used to fasten internal component parts together. A series of nuts attached loosely to a metal strip. Gang Nut strips are designed for ease of maintenance while assembling or disassembling an engine. Also called Channel Nut.

GANTRY
A component of an engine test cell. A large metal frame into which a Gas Turbine Engine is mounted for test running.

GARLOCK SEAL™
A type of rubber seal formed around a drive shaft of an engine mounted accessory or within an Accessory Gearbox drive point. This seal is in place to prevent fluid leakage via the shaft.

GARRETT TURBINE ENGINE COMPANY
Manufacturer of Gas Turbine Engines. Locate in section entitled "Gas Turbine Engines for Aircraft".

GAS BEARING
Locate under Bearing (air).

GAS BENDING STRESS
The stress that gaspath forces apply to compressor and turbine blades. This stress tends to concentrate at the fillet area where the airfoil section adjoins to its base.

GAS GENERATOR
Refers to the basic power producing (hot gas generating) portion of a turbine engine. Includes the compressor, combustor and turbine and excludes the free power turbine of a Turboshaft or Turboprop engine. Also excluded are the aircraft inlet and tailpipe. Some manufacturers also exclude the fan of a Turbofan engine when referring to the gas generator. Often referred to as the Core Engine of a Turbofan engine.

GAS GENERATOR TURBINE (engine)
An engine component. The high pressure turbine wheel(s) that drive the compressor of a Turboprop or Turboshaft engine. Not generally used to refer to the compressor-drive or fan-drive turbines of Turbofan engines. Also Called 1) Compressor Turbine, 2) Gas Producer Turbine. *See also: Free Power Turbine.*

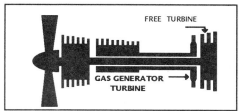

FREE TURBINE TURBOPROP

GAS PRODUCER TURBINE
Another name for Gas Generator Turbine.

GAS TUNGSTEN ARC WELDING (GTAW)
Another name for Tungsten Inert Gas Welding.

GAS TURBINE
Short name for Gas Turbine Engine.

GAS TURBINE COMPRESSOR UNIT (GTC)
A Ground Power Unit or Auxiliary Power Unit powered by a gas turbine engine. The GTC provides compressed air energy to aircraft systems. *See also: 1) Gas Turbine Power Unit (GTP), 2) Gas Turbine Compressor-Power Unit (GTCP).*

GAS TURBINE COMPRESSOR-POWER UNIT (GTPC)
A type of Ground Power Unit or Auxiliary Power Unit powered by a turbine engine. The GTPC provides both electrical power and compressed air energy to aircraft systems. *See also: 1) Gas Turbine Compressor Unit (GTC), 2) Gas Turbine Power Unit (GTP), 3) Twin-Spool Compressor-Power Unit.*

GAS TURBINE ENGINE
An open cycle family of engines in which the shaft power to drive the compressor(s) comes from a single or multiple stage turbine rotor. The four types of Gas Turbine Engines presently used in aviation include: 1) Turbojet, 2) Turbofan, 3) Turboprop, 4) Turboshaft. The Gas Turbine Engine is also considered to be a type of Jet Engine, the original Gas Turbine Engine design being the Turbojet. *See also: 1) Basic Engine, 2) Built-Up Engine, 3) Power Pack.*

GAS TURBINE ENGINES (under development)
Locate under: 1) Convertible, 2) Hypersonic, 3) Turbo-Compound, 4) Turbo-Ram, 5) Turbo-Rocket.

GAS TURBINE POWER UNIT (GTP)
A type of Ground Power Unit or Auxiliary Power Unit powered by a turbine engine. The primary function of a GTP is to provide electrical power to aircraft systems. It can also supply electrical power to other flightline support facilities and equipment. *See also: 1) Gas Turbine Compressor-Power Unit (GTCP), 2) Gas Turbine Compressor Unit (GTC).*

GAS TURBINE STARTER
Another name for Jet Fuel Starter.

GASIFIER TURBINE
Another name for Gas Generator Turbine.

GASPATH
The airflow portion of a turbine engine. From front to back the gaspath includes the following engine sections: 1) Air Intake, 2) Compressor, 3) Combustor,

4) turbine, 5) Exhaust.

GASPATH ANALYSIS (GPA)
A computer analysis of engine gas flow parameters designed to predict engine component wear. GPA is used primarily in making management decisions for engine repairs, especially module changes.

GATOR-GARD™
A United Technologies Corporation trademark. A procedure for metal plasma-spraying of engine parts to increase their resistance to wear, oxidation, erosion, and sulfidation. For example, anti-wear coatings on fan blade shrouds, or turbine blade shrouds, and thermal barrier coatings on combustion liners and turbine nozzle vanes. *See also: Thermal Barrier Coatings.*

GATORIZING™
An isothermal forging process developed by United Technologies Corp., wherein difficult to forge metals such as superalloys are heated to a plastic state before forging. *See also: Isothermal Forging.*

GAUGE MARKINGS (engine)
Colored lines placed on the face of certain gauges as range marking or limits. The FAA lists the suggested colors and types of markings in Advisory Circular 20-88.

GAUGE PRESSURE
Refers to the readout of a Relative Pressure Gauge, expressed as Pounds per square inch gauge (PSIG).

G. E.
Abbreviation for General Electric Corporation.

GEAE
Abbreviation for General Electric Aircraft Engines.

GEAR DESIGNS
In the Gas Turbine Engine, gears are often designed by what it referred to as the "hunting tooth method". Mating gears do not engage between the same set of opposing gears on each revolution, and wear in this way is spread out more evenly on all the teeth. *See also: 1) Bevel Gear, 2) Helical Gear, 3) Spiral Gear.*

GEARBOX
An engine component. Locate under: 1) Accessory Drive Gearbox, 2) Auxiliary Drive Gearbox, 3) Combined Gearbox, 4) Inlet Gearbox, 5) Output Reduction Gearbox, 6) Propeller Reduction Gearbox.

GEAR PUMP
Locate under Fuel Pump or Oil Pump.

GEARBOX DEAERATOR
Another name for Air-Oil Separator (rotary).

GEARED TURBOFAN
A Turbofan Engine with a geared-down fan rotor. A reduction gearing is interposed between the fan and the compressor to allow the compressor to rotate at a higher RPM than the fan. This provides higher compressor tip speeds, greater pressure ratio of the compressor, and improved Cycle Efficiency. The Geared Fan is also said to reduce the number of turbine wheels required and thus provides for a shorter engine length than the ungeared type Turbofan engine. *See also: Turbofan Engine.*

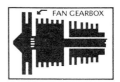

GEM
Abbreviation for Ground-Based Engine Monitoring.

GENERAL GAS LAW
A physics term. A physical law based on Boyle's Law and Charles' Law which figures significantly in the operation of the turbine engine. Mathematically stated as:

$$P_1 V_1 / T_1 = P_2 V_2 / T_2$$

GENERAL ELECTRIC AIRCRAFT ENGINES
Manufacturer of Gas Turbine Engines. The first U.S. company to manufacture Turbojet Engines. G.E. produced the GE-IA model for the Bell Airacomet aircraft in the early 1940's. Locate current listings in the section entitled "Gas Turbine Engines for Aircraft".

GENERAL MOTORS CORPORATION
Locate under Allison Turbine Engine Company.

GEOMETRAL MULTIPLANE BALANCING PROCEDURE (GMBP)
A manufacturing concept employed in producing rotating parts with critical diameters, runouts, and wall thickness. Data introduced into a computer program directs machine tools to remove material to obtain tight balancing tolerances. Also called Super-Balancing. *See also: Balance Checks.*

GEOMETRIC FLOW AREA (turbine stator)
Abbreviation GFA. The effective flow area of a turbine nozzle stator vane assembly. When the GFA is reduced, the constricted flow increases velocity and more energy is made available to the turbine wheel(s). Serious limits govern this process because as GFA is reduced, Turbine Inlet Temperature increases. Turbine nozzles are produced in "classes' that indicate their GFA.

GERMAN ENGINES
Locate under 1) Kloeckner Humbolt-Deutz, 2) Motoren-Und Turbinen-Union, in section entitled "Gas Turbine Engines for Aircraft".

GEROTOR OIL PUMP
Locate under Oil Pump.

GIDLE
Acronym for "Ground Idle" engine operating speed. Not widely used.

GILL HOLE (turbine blades and vanes)
Refers to elongated shape of turbine blade and nozzle vane air cooling exit holes. The shape of the hole aids in laminar flow of air over the surfaces. *See also: 1) Laser Beam Drilling, 2) Shower Head Cooling, 3) Turbine Cooling.*

GLASS COCKPIT
Locate under Electronic Cockpit.

GLOSTER-WHITTLE, E-28/39 AIRCRAFT
An experimental prototype aircraft which flew in 1941, making the first purely jet propelled flight in Britain. It was powered by a Whittle (W-1) Turbojet Engine of approximately 1,000 lb. thrust. *See also: 1) Bell Airacomet, 2) Heinkel 178, 3) Whittle, Sir Frank, 4) Appendix "Early Engines".*

GLOW PLUG (relight system)
A system rarely seen today which includes a metallic probe that is heated to a red-hot condition by combustion gases. In the event of an engine flameout, the probe will retain its heat sufficiently to automatically reignite the fuel/air mixture.

GLOW PLUG (ignition system)
A small engine accessory. An electrical igniter plug in the form of a heater coil, with the appearance of an automobile cigarette lighter. Unlike igniter plugs with an intermittent spark, the glow plug maintains a continuous yellow hot condition that provides a torching effect when fuel is dripped onto its surface. The glow plug is found on some small turbine engines to provide an improved cold weather starting capability.

GOUGING
Locate under Bearing Inspection and Distress Terms in Appendix.

GPU
Abbreviation for Ground Power Unit.

GRAIN BOUNDARY OXIDATION
Locate under Engine Inspection and Distress Terms in Appendix.

GRAPHITE MATERIALS
Locate under Composite Materials.

GRAVITATIONAL ACCELERATION
A constant value in the turbine engine thrust formula represented as, 32.174 ft./sec.^{2}. In a vacuum or when neglecting friction in the atmosphere, gravitational acceleration is the constant acceleration rate of a falling object due to the pull of gravity towards the earth.

GRAVITY FILL POINT (oil tank)
The opening in an oil tank under the oil filler cap. Typically oil is added directly from a hand held oil container into the gravity fill point. *See also: Oil Tank.*

GRE
Abbreviation for Ground Runup Enclosure.

GREAT BRITAIN (engines)
Locate under British Manufactured Engines in section entitled "Gas Turbine Engines for Aircraft".

GRIFFITH, DR. A.
British engineer and member of the Royal Aircraft Establishment. Early developer of turbine engine theory and theory of airflow over airfoils.

GROOVING (bearings)
Locate under Bearing Inspection and Distress Terms in Appendix.

GROSS THRUST
Locate under Thrust (gross).

GROSS MOMENTUM THRUST
A physics term. Another name for Gross Thrust. Gross Momentum Thrust minus Inlet Momentum Drag is equal to Net Thrust.

GROUND-BASED ENGINE MONITORING (GEM)
An on-board data recording system which feeds data from engine mounted sensors into aircraft computers. Data such as internal engine temperatures, rotor speeds, vibration levels of rotating systems, etc., are fed into Propulsion Multiplexer (PMUX) units for analysis and storage. Ground personnel later use this data to adjust and repair engine modules and systems as the engine gains service life. GEM is part of an Airborne Instrument Data System (AIDS). *See also: Turbine Engine Module Performance Estimation Routine (TEMPER).*

GROUND HANDLING PAD
A pad on the exterior of the engine to which mounting fixtures can be attached. The fixtures support the engine in maintenance stands.

GROUND IDLE (engine)
The lowest allowable engine speed for ground operation. Ground Idle is typically in the range of 50% to 70% RPM. It is set with the power lever in conjunction with the: 1) N1 Percent Indicator on a single compressor engine, 2) N2 Percent Indicator on a dual-spool engine, 3) N3 Percent Indicator on a triple-spool engine. *See also: Flight Idle.*

GROUND IDLE (constant speed)
A ground idle speed which is preset regardless of inlet conditions of pressure and temperature. By comparison, Idle thrust may also be allowed to vary slightly with ambient conditions. *See also: Ground Idle (variable speed).*

GROUND IDLE (variable speed)
A ground idle speed which is reset electronically in order to hold idle thrust constant. This control system performs as follows: 1) Limits idle thrust on a cold day, 2) Resets idle speed higher on a warm day. Note: warm day reset keeps idle thrust up and ensures stall-free engine acceleration.

GROUND IDLE DWELL TIME
Generally a required engine ground idle operation time before engine shutdown. This cool down period is designed to prevent carbonizing of oil in hot-section areas which results from high residual engine heat after shut-down.

GROUND POWER TROLLEY
British for Ground Power Unit.

GROUND POWER UNIT (GPU)
A small conventional Gas Turbine Engine powered ground support unit. The GPU is used to support aircraft during ground operation. The turbine engine powered GPU often provides both electrical power and also bleed air from its compressor section to the Starting Air Manifold in the aircraft. The GPU is similar in design to an Auxiliary Power Unit (APU). *See also: 1) Gas Turbine Compressor Unit (GTC), 2) Gas Turbine Compressor-Power Unit (GTCP), 3) Gas Turbine Power Unit (GTP), 4) Twin-Spool Compressor-Power Unit (TSCP).*

GROUND RUNUP ENCLOSURE (GRE)
A new concept in noise suppressed jet aircraft runup facilities. A specially constructed, hangar-type structure lined with acoustic materials which substantially reduce the noise footprint of operating aircraft. Not as expensive as a Hush House design, which practically eliminates all noise outside of the structure. *See also: Hush House.*

GROUND STARTING UNIT
Another name for Ground Power Unit.

GROUND SUPPORT EQUIPMENT (GSE)
Any of the pieces of equipment required for completing maintenance tasks on aircraft and engines. Equipment such as Ground Power Units, Hydraulic Units, Electronic Test Units, Personnel Maintenance Stands, Aircraft Jacks, etc. GSE may also include equipment required for loading and unloading of aircraft.

GROWN TURBINE BLADE
Jargon for Single Crystal Turbine Blade.

GSE
Abbreviation for Ground Support Equipment.

GTAW
Abbreviation for Gas Tungsten Arc Welding.

GTC
Abbreviation for Gas Turbine Compressor unit.

GTCP
Abbreviation for Gas Turbine Compressor-Power unit.

GTP
Abbreviation for Gas Turbine Power unit.

GUIDE SPRING FINGER
An engine component part. A curved sheet metal part configured with split edges which act as a guide when the fingers are slipped over another section.

GUIDE VANE
An engine component part. A circular spoke-shaped airfoil section placed radially in the engine gas path to direct airflow. Locate under: 1) Inlet Guide Vane, 2) Exit Guide Vane 3) Turbine Inlet Guide Vane.

GUIDED-VANE PUMP
A small engine accessory. Another name for Vane Pump or Sliding Vane pump. A pump commonly used as an engine oil supply pump or scavenge oil pump.

H

HALF-LIFE INSPECTION
An inspection required on many engines at half the Time Between Overhaul (TBO). This inspection is primarily a "hot section" disassembly, inspection, and repair as necessary. Also called a Hot Section Inspection (HSI), although an HSI may be scheduled at times other than at half-life. *See also: Heavy Maintenance Inspection.*

HANGAR MOUNT (engine)
A main engine mount. Typically one of two such mounts that fit aft of the center of gravity and connect between the turbine casing and the airframe structure. The major portion of engine weight hangs on this mount. The engine will also be supported by a smaller inlet case hangar mount or a lower support strut type mount. *See also: Engine Mount.*

HANDS-ON-THROTTLE & STICK (HOTAS)
A recent development in turbine engine powered fighter aircraft controls. This design employs a combination stick-type flight controller and throttle lever.

HARD FACING (turbine blades)
A laser beam or electron beam process by which a hard metal powder is bonded or fused to the blade leading edge. This process allows blades to be produced of more workable materials and later hardened to reduce leading edge erosion.

HARD-LIFE TIMES
Refers to overhaul requirements of certain engine accessories, components, component parts, and modules, at specific accrued hourly time or accrued cycles. *See also: Soft-Life Times.*

HARD-TIME REPLACEMENT (or limit)
Refers to replacement of certain engine accessories, components, component parts and modules at specific hourly or cycle intervals regardless of their condition. Many parts are scrapped when their hard-time life is reached. Locate under 1) Engine Monitoring System, 2) Life-Limited Parts. *See also: Hard-Life Times.*

HARMONICS
In regards to a turbine engine, refers to vibration that accumulates in multiples of a base frequency. If not interrupted, harmonics induced vibration can completely fail an engine. *See also: 1) Resonant Frequency, 2) Vibration.*

HASTELLOY™
A trademark of the Union Carbide Corporation. Hastelloy is a nickel-base alloy in the family of turbine engine super-alloys of very high temperature strength. This alloy is widely used in manufacture of thin wall components such as combustion liners. *See also: Super-alloys.*

HAWKER SIDDELEY OF CANADA
Gas Turbine Engine manufacturer. Locate in section entitled "Gas Turbine Engines for Aircraft".

HEAT
The interaction that occurs when two systems at different temperatures are brought together. In the turbine engine the systems are: 1) energy of compressed air, 2) energy of hydrocarbon Jet Fuel.

HEAT EXCHANGER
An engine accessory. A device that transfers some of the heat in one medium into another medium. For examples of heat exchangers, see: 1) Fuel Heater, 2) Fuel-Oil Cooler, 3) Recuperator, 4) Regenerator.

HEAT OXIDIZING
Locate under Engine Inspection and Distress Terms in Appendix.

HEAT RECOVERY COMPONENTS
In reference to the Turbine Engine, locate under 1) Recuperator, 2) Regenerator.

HEAT SHIELD
An engine component part. An insulating or refracting barrier that protects an engine component part from high heat.

HEAT SOAKING (engine)
Refers to turbine engine operation for a sufficient period of time for heat absorption to take place and raise to internal operating temperature, certain critical engine components. Heat soaking ensures optimum clearances are attained for most efficient engine operation.

HEAVY MAINTENANCE
Aircraft or engine maintenance beyond the scope of Line Maintenance. Similar to Major Repair in FAA terms and includes such repairs as a complete engine overhaul or module disassembly for repair.

HEAVY MAINTENANCE INSPECTION (HMI)
A major engine inspection such as an engine Half-Life Inspection. *See also: Half-Life Inspection.*

HEAVY REPAIR
Another name for Heavy Maintenance.

HEI
Abbreviation for High Energy Ignition.

HEINKEL-178 (Germany)
In 1939, the aircraft to perform the first purely jet propelled flight in the history of aviation. The HE-178 was powered by a Heinkel He3B, Turbojet Engine. *See also: 1) Bell Airacomet, 2) Gloster-Whittle E-28, 3) Appendix "Early Engines".*

HELI-ARC WELDING™
A type of Tungsten Inert Gas (TIG) welding. A welding process often employed in Gas Turbine Engine metal crack repairs because it concentrates its heat in a very narrow area. TIG welding is also easily accomplished under atmospheric conditions. *See also: Welding.*

HELICAL GEAR
A type of spur gear with its teeth cut diagonally. Helical Gears mesh starting at one end of a tooth and in rotating contact, ride to the opposite end, as opposed to meshing at right angles as seen in Spiral Gears. Helical Gears are used mainly in areas of heavy gear loading such as propeller reduction gearboxes and in bevel gears that drive Accessory Gearboxes. *See also: Gear Designs.*

HELI-COIL INSERT™
A spring-steel coil in the form of a thread that is inserted into a soft metal part. Heli-Coils give a greater fastening strength than the base metal could provide. Commonly found in aluminum and magnesium casings in which bolts or studs are inserted as fasteners.

HEXOLOY SA™
A sintered alpha-silicon-carbide material under development for use in hot sections. *See also: Super Alloys.*

HYBRID AIR-BLAST FUEL NOZZLE
A type of air-blast fuel nozzle employing higher pressures than traditional air-blast fuel nozzles. Primarily-used on smaller engines where flow is low. *See also: Fuel Nozzle (aerating).*

HIDEK
Abbreviation for Highly Integrated Digital Electronic Control.

HIGH ASPECT RATIO BLADING
Locate under Aspect Ratio.

HIGH BYPASS TURBOFAN ENGINE
Turbofan engines with approximately 4:1 to 8:1 bypass ratio (BPR) fall into this category. BPR is described as a ratio of fan airflow versus core engine airflow. For example, an engine with 1,500 lb./sec. total mass airflow, in which 1,250 lb./sec. flows through the fan and 250 lb./sec. simultaneously flows through the core is described as having a BPR of 5:1. *See also: Turbofan.*

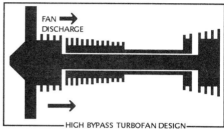
HIGH BYPASS TURBOFAN DESIGN

HIGH BYPASS RATIO TURBOFAN ENGINE
Another name for High Bypass Turbofan Engine.

HIGH COMPRESSOR TURBINE
Another name for High Pressure Turbine.

HIGH CYCLE FATIGUE (HCF)
Fatigue induced cracking or fracture that takes place in rotating airfoils from vibratory and thermal stresses. HCF results from forces created by the airstream when engines are operated at high speeds. In a time sense, generally thought of as stresses occurring several times per second. High Cycle Fatigue can also mean failure after a high number of Engine Cycles have accumulated during the engine aging process. *See also: Fatigue Failure.*

HIGH CYCLE LIFE
Refers to the high number of operating cycles expected during the in-service life span of a part or of an engine. *See also: Engine Cycle.*

HIGH ENERGY IGNITION (HEI)
Another name for Capacitor Discharge Ignition.

HIGH FREQUENCY FATIGUE
Another name for High Cycle Fatigue when referring to failure from vibratory and thermal stress.

HIGH-LOW PRESSURE STARTER
Locate under Pneumatic Starter (High-Low).

HIGH PERFORMANCE TURBINE ENGINE INITIATIVE (HPTEI)
A U.S. Defense Department initiative for development of future fighter aircraft engines. The stated objective of this program is for industry to develop engines with one-half the present fuel consumption while doubling the power-to-weight ratio.

HIGH PRESSURE AIR START VALVE
An engine accessory. A valve that connects a High

Pressure Compressor bleed air source to the Cross-Bleed starting air manifold. If the Low Pressure Air Start Valve is supplying too little air pressure, the High Pressure Air Start Valve is manually opened to maintain the required manifold pressure. *See also: 1) Cross Bleed Air System, 2) Low Pressure Air Start Valve.*

HIGH PRESSURE COMPRESSOR (HPC)
A major engine component. The rear section of a dual-spool, also called split-spool, compressor. The HPC is used in conjunction with a Low Pressure Compressor (LPC) to deliver air under pressure to the diffuser and the combustor. The high pressure compressor is also referred to as the 1) HP Compressor 2) N2 Compressor, 3) High Speed Compressor. *See also: Low Pressure Compressor.*

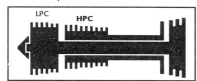
TURBOJET (DUAL SPOOL DESIGN)

HIGH PRESSURE COMPRESSOR (TRIPLE-SPOOL)
The rear section of a three spool compressor. Also called the N_3 Compressor.

HIGH PRESSURE ROTOR SYSTEM
An assembly consisting of a High Pressure Compressor connected to its High Pressure Turbine.

HIGH TEMPERATURE OXIDATION (engines)
Locate under Hot Corrosion.

HIGH PRESSURE TURBINE (HPT)
A major engine component. The forward most turbine wheel(s) of a dual or triple spool axial flow turbine engine that are mechanically attached to and drive the high pressure compressor. The HPT is named as such because it receives the highest gas pressure of

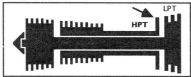
TURBOJET (DUAL SPOOL DESIGN)

all the turbines. Also called 1) HP Turbine 2) N2 Turbine in a dual spool, 3) N3 Turbine in a triple spool, 4) High Speed Turbine. *See also: Low Pressure Turbine.*

HIGH SPEED CIVIL TRANSPORT (HSCT)
A proposed high speed aircraft in the Mach 2.5 to Mach 3.0 speed range. It is expected to carry approximately 250 passengers and operate between flight levels 55 to 60.

HIGH SPEED ROTOR
Refers to the following : 1) High Pressure Rotor system of a dual-spool Turbofan or Turbojet Engine where the Low Pressure Rotor has a wider diameter and lower rotational speed, 2) Gas Generator Rotor system of a Turboprop or Turboshaft engine where the Free Power Turbine is of wider diameter and turns at a slower speed in relation to the Gas Producer Turbine. *See also: 1) Low Speed Rotor, 2) Tip Speed.*

HIGH SPEED SPIN TEST
Another name for Dynamic Balancing.

HIGH SPOOL SPEED
Refers to the speed of the High Pressure Compressor.

HIGH TEMPERATURE OXIDATION
Another name for Hot Corrosion.

HIGH TENSION LEAD (ignition system)
A small component part. A shielded electrical lead that carries the high voltage discharge from the ignition exciter to the igniter plug.

HIGH VELOCITY OXYGEN & FUEL DIAMOND JET
A combustion spraying process of aluminum-silicon and polyester which serves an a abradable coating inside compressor and fan cases. The coating provides compressor and fan blade tip clearance control.

HIGH VOLTAGE IGNITER PLUG
Locate under Igniter Plug (high voltage).

HIGH VOLTAGE IGNITION SYSTEM
Locate under Ignition System (high voltage).

HIGHLY INTEGRATED DIGITAL ELECTRONIC CONTROL (HIDEC)
An integration between fight and propulsion controlling systems for improved low and high speed maneuverability of advanced military aircraft. For example, the HIDEC computer system is capable of changing the Engine Pressure Ratio without pilot input to provide optimum engine performance during flight maneuvers. Adjustment to the Variable Geometry Inlet and the Variable Exhaust Nozzle flow area are also automatically accomplished as a function of aircraft angle of attack and Mach number in the HIDEC system. Control of Vectored Exhaust Nozzles is another proposed feature of this system.

HIP PROCESS
Locate under Hot Isostatic Pressing.

HMI
Locate under Heavy Maintenance Inspection.

HMU
Abbreviation for Hydomechanical Control Unit.

HOBBS METER
A counter which automatically records engine operating time. The data is used to plan engine maintenance. *See also: Engine Cycle.*

HONEYCOMB RUBSTRIP
Another name for Honeycomb Shroud Ring.

HONEYCOMB SHROUD-RING
An engine component part. A circular ring that fits outside the area of blade tip rotation of turbine wheels to minimize tip leakage. Should the ring warp inward or a blade stretch outward and contact occurs, the honeycomb can withstand a certain amount of abrasion and still remain functional. The use of abradable tip seals also prevents loss of blade length and thus, loss of turbine wheel efficiency. Also called an Abradable Seal. *See also: Shroud Ring.*

HOOP STRESS
The distortion associated with ring-shaped metallic parts under radial stresses that act to expand or distort the diameter of the part. For example, the disk or drum of a compressor, fan or turbine.

HOPPER TANK (lubrication system)
Another name for Deaerator (oil tank).

HORSEPOWER (HP)
A physics term. Power measurement in U.S. Units equal to 33,000 ft.lb./min., 550 ft.lb./sec., or 375 mi.lb./hr. See also: 1) Isentropic Gas Horsepower, 2) Shaft Horsepower, 3) Thermodynamic SHP, 4) Thrust Horsepower.

HORSEPOWER (fuel)
The potential for horsepower in jet fuel is: Hp in 1 lb. burned in 1 minute = BTU/min. X 778 ÷ 33,000; where 778 is the foot pounds of work in 1 BTU and 33,000 is the foot pounds per minute in 1 horsepower. Formula:

$$HP_{(FUELFLOW)} = lb/min \times 18,300\ btu/lb \times 778\ ft.lb./btu \div 550\ ft.lb./min$$

HORSEPOWER FORMULA (general)
A measurement system given as the power produced by a horse of average strength. Mathematically stated as: 1) [F x D ÷ t] ÷ 33,000 when power is expressed in ft.lb./min., 2) [F x D ÷ t] ÷ 550 when power is expressed in ft.lbs./sec., 3) [F x D ÷ t] ÷ 375 when power is expressed in mile-lbs./hr. Where: F = force in lbs., D = distance in feet, t = time in minutes, seconds, or hours. One horsepower is also equal to 746 watts.

$$HP = [F \times D \div t] \div 33,000$$

HORSEPOWER FORMULA (to drive the compressor)
The horsepower required to drive the compressor of a turbine engine is mathematically expressed as:

$$HP = 0.24 \times Tr \times Ms \times 778 \div 550$$

Where: 0.24 is the Specific Heat Ratio at constant pressure, Tr is air temperature (°F) rise above ambient temperature as a result of compression, Ms is mass airflow in lbs./sec., 778 is the ft.lbs. of work available per BTU, 550 is a conversion from power to horsepower.

HOT CORROSION
Corrosion that occurs in hot sections of turbine engines from a chemical reaction between sulfur in jet fuel and salt in the airstream. Hot Corrosion is more pronounced when turbine engines are operated predominantly in a salt laden environment. Hot Corrosion occurs readily at 1,400°F and above when sodium sulfate interacts with heated metal surfaces. Sodium sulfate forms when sodium chloride in air mixes with sulfates in fuel. Metal coatings are available that resist Hot Corrosion. Also called 1) High Temperature Oxidation, 2) Sulfidation. See also: 1) Aluminide Coating, 2) Coking.

HOT DAY THRUST
Refers to the maximum ambient temperature at which an engine can obtain its Rated Thrust. See also Flat Rating.

HOT END (engine)
Jargon for engine Hot Section.

HOT GAS FACILITY
A research and development facility for testing Gas Turbine Engine major components such as Compressors, Combustors, and Turbines. This testing is performed on pre-production components under expected operational loads.

HOT ISOSTATIC PRESSING (HIP)
A recently developed Powder Metallurgy process used to produce many types of alloys for the Gas Turbine Engine. This process is apart from traditional casting or forging. See also: 1) Powder Metallurgy, 2) Rapid Solidification Rate, 3) Super Alloys.

HOT PROP
A name given to the Turboprop engine pusher propeller or a Rear-Propfan, where some of the engine hot exhaust gases pass over the prop blading. See also: Cold Prop.

HOT SECTION (engine)
A typical reference to the combustion through exhaust sections of an engine. The Hot Section starts at the combustor front flange and ends at the exhaust nozzle. A Hot Section Inspection (HSI) performed on a turbine engine is an example of usage of this term. Also called Hot End. See also: Cold Section.

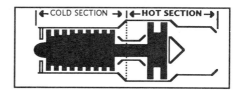

HOT SECTION DISTRESS
An umbrella term in reference to any of the metal deterioration conditions found in the "hot section" of an engine. For example: 1) Buckling, 2) Burn Through, 3) Cracking, 4) Hot Spots, 5) Warping. Also locate in this listing: 1) Creep, 2) Erosion, 3) High and Low Cycle Fatigue, 4) Hot Corrosion, 5) Inter-granular Oxidation, 6) Stress Rupture. See also: Appendix entitled "Engine Inspection and Distress Terminology".

HOT SECTION INSPECTION (HSI)
A major inspection required on some engines between overhauls or for investigation of engine performance loss attributable to hot section components. See also: Half-Life Inspection.

HOT SPOTS
A localized dark brown, gray or black discoloration on hot section component parts indicating a breakdown of cooling air or a harmful concentration of fuel at that point. Hot spots are often the result of a malfunctioning fuel nozzle that is hot streaking directly at an inner engine surface. See also: 1) Hot Section Distress, 2) Hot Streaking (fuel nozzle).

HOT STALL (compressor)
A severe compressor stall accompanied by combustor cooling air loss, a rise in exhaust gas temperature (EGT) and a loss of RPM. The operator must reduce fuelflow to the engine to remedy this condition or a

flame-out will most likely result. *See also: Compressor Stall.*

HOT STREAKING (fuel nozzle)
A concentration of fuel that has not been atomized properly at the nozzle orifice. In this condition, a stream of fuel has sufficient force to cut through the protective cooling air film and impinge directly on combustor and turbine surfaces causing serious Hot Section Distress. *See also: Hot Spots.*

HOT STREAK IGNITION (afterburner)
An open flame ignition system used in many afterburners to initiate combustion. The hot streak has a flame pattern similar to a blow-torch. It is ignited spontaneously from residual heat or via an electrical spark. Also called Flame Ignition. *See also: Ignition System (afterburner).*

HOT START
An engine start in which internal engine temperatures exceed the prescribed limits. This condition will warrant either a special hot section inspection or, in severe cases, engine overhaul. The maintenance action required often depends on both the Exhaust Gas Temperature (EGT) experienced and the time interval of the over-temperature condition.

HOT STREAM NOZZLE
Refers to Primary Air exhaust nozzle on a Turbofan Engine. *See also: Primary Air.*

HOT TANK OIL SYSTEM
An engine main lubrication system wherein the oil cooler is located in the Pressure Oil Subsystem, cooling the oil before it enters the engine. The scavenge (return) subsystem returns oil to the tank uncooled (hot). *See also: 1) Cold Tank Oil System, 2) Lubrication System.*

HOTAS
Abbreviation for Hands-on-Throttle and Stick.

HOTOL
Abbreviation for Horizontal Take-Off and Landing Aircraft.

HP
Abbreviation for Horsepower or for High Pressure.

HP COMPRESSOR (HPC)
Another name for High Pressure Compressor.

HPTEI
Abbreviation for High Performance Turbine Engine Initiative.

HP THRUST BEARING
Refers to a main bearing of the ball-type which prevents axial movement of the High Pressure Compressor. Main ball bearings absorb thrust (axial) loads when high gas pressure forces at the rear exert a push forward on the compressor rotor. *See also: Bearing Listing.*

HP SPEED SENSOR
A small engine accessory. A unit in an Electronic Tachometer System. One such sensor is placed outside the plane of rotation of a gear-type wheel pressed onto the HP rotor shaft. Another uses a sensor positioned near a gear within the Accessory Gearbox. As the gears cut through an eddy current generated by the sensor, an electronic circuit calculates the speed in percent RPM of the HP rotor. *See also: Fan Speed Sensor.*

HP TURBINE (HPT)
Another name for High Pressure Turbine.

HSI
Locate under Hot Section Inspection.

HSCT
Abbreviation for High Speed Civil Transport.

HUB TO TIP RADIUS RATIO (compressor and turbine blades)
The compressor or turbine disc diameter versus the unit diameter blade tip to blade tip. For example, an H/T ratio of (0.8) in reference to an 8 inch hub diameter and a 10 inch tip to tip diameter, indicates a short blade and a resultant low massflow design. A 12-inch tip diameter would indicate H/T = (8 ÷ 12), H/T = 0.67; a longer blade with a larger flow design capability.

HUMMING NOISE (engine)
A gaspath generated noise caused by an imbalance in a rotating component; a sound unlike the pulsating noise associated with a compressor stall condition.

HUNG ACCELERATION
An engine malfunction wherein an engine hangs up, failing to reach take-off thrust with the power lever at the take-off setting.

HUNG ENGINE
Minimal or no power response as the engine power lever is advanced above the idle power setting.

HUNG START
A condition wherein engine light-off is normal but the engine speed stabilizes at a low value and does not self-accelerate to idle speed. Also called Stagnated Start.

HUNTING TOOTH RATIO
Locate under Gear Design.

HUSH HOUSE
A special hangar in which jet aircraft runup is performed. The walls are acoustically treated to absorb substantially all sound that would otherwise create unwanted noise outside of the run facility. *See also: Ground Runup Enclosure.*

HUSH KIT
An engine and nacelle modification kit. A mod kit placed on older engines and engine nacelles to reduce noise emissions to meet new more stringent FAA Noise Abatement Standards of FAR Part 36. The kit includes acoustic linings for fan inlets and exhausts and for hot exhaust flow areas. This modification basically changes a short fan bypass duct to a full duct, providing increased surface area available for acoustical treatment of air. *See also: Noise Suppression.*

HVOF
Abbreviation for High Velocity Oxygen and Fuel spray process.

HYDRAULIC STARTER
An engine Accessory. The hydraulic starter is used to initiate starting of turbine engines in place of the more traditional electrical or pneumatic starters. The motive force for this starter is provided by aircraft hydraulic fluid under pressure from an aircraft accumulator or a ground power unit. One important feature of this starter is that an accumulator can be hand pumped to operational pressure to give the aircraft a self-starting capability, without auxiliary power support systems. This starter is seen mainly on military helicopters.

HYDRAULIC THRUST ROTOR CONTROL
A hydraulic piston supplied with engine oil pressure to act as a balance chamber and maintain a constant axial loading on engine thrust bearings during power changes. *See also: Balance Chamber (thrust).*

HYDROCARBON FUELS
A wide variety of petroleum fuels used in all types of internal combustion engines, including the gas turbine. Locate under Fuel Types - Jet.

HYDROGEN FUEL
A jet fuel proposed for future gas turbine powered aircraft when hydrocarbon fuels become scarce, or as the fuel of choice for future hypersonic aircraft. This fuel can be stored as a gas or as a cryogenic liquid. Present high cost and storage problems prevent its use today, except in experimental applications. *See also: Cryogenic Fuels.*

HYDROMECHANICAL FUEL CONTROL
One of several types of fuel controls used on Gas Turbine Engines. A primary fuel scheduling device on an engine; one which utilizes hydraulic servos in combination with mechanical linkages to operate its mechanisms. *See also: Fuel Controls Units.*

HYDROMECHANICAL CONTROL UNIT (HMU)
Another name for a Hydromechanical Fuel Control.

HYFIL FAN BLADE™
An engine component part. A newer, lightweight, hollow titanium blade filled with a honeycomb composite material of carbon fiber and resin. Not presently in wide use.

HYPERSONIC AIRCRAFT
An aircraft capable of speeds in excess of Mach 5.

HYPERSONIC ENGINE
A proposed powerplant for aircraft of Mach 5 and above. One design is composed of a Turbojet core for sonic to transonic flight, after which the inlet to the Turbojet will close off and a Supersonic Combustion Ramjet (SCRAMJET) in the form of an outer duct around the Turbojet will function for supersonic to hypersonic flight. Another proposal replaces the Scramjet with a Rocket-jet. *See also: 1) Scramjet, 2) Turbo-Ramjet Engine, 3) Variable-Cycle Engine.*

HYPERSONIC SPEED
Speeds over Mach 5. *See also: Flight Regimes.*

I

IBIS
Abbreviation for Integrated Blade Inspection System.

IBR
Abbreviation for Integrally Bladed Rotor.

I.C.A.O.
Abbreviation for International Civil Aviation Organization.

ICE ACCRETION TEST
A visual or measurement test for growth of ice on an aircraft surface, such as the surfaces of a Flight Inlet or an Engine Inlet. An ice buildup that could disrupt engine airflow. *See also: Icing.*

ICE DETECTOR (anti-icer system)
A small engine accessory that detects the presence of ice in the engine inlet. The detector circuit is capable of illuminating a cockpit warning light, and in some cases powering open the engine anti-ice system.

ICE SHEDDER
Locate under Dynamic Ice Shedder.

ICING (clear)
Clear colored ice which forms most readily between 0° and 15° F. Clear Ice often forms in layers. It is relatively transparent because of its homogeneous structure and small number of air spaces. Clear ice can build up in great mass and cause loss of aerodynamics to the Flight Inlet or Engine Inlet, resulting in compressor stalls. Also when dislodged, this ice can enter the engine causing serious impact damage. Also called Glaze Icing.

ICING (flight inlet)
Ice which forms on surfaces when moisture and temperature conditions reach a critical value. *See also: 1) Icing (clear), 2) Icing (rime), 3) Fuel Icing.*

ICING (rime)
A heavy frost of non-cohesive ice. A milky granular deposit of ice formed by rapid freezing of super-cooled water droplets as they impinge on surfaces such as the Flight Inlet or Engine Inlet. Rime ice weighs less than clear ice, but may cause more serious distortion to airfoil shapes, which can quickly diminish aerodynamic efficiency to a point where compressor stalls occur.

IDG
Abbreviation for Integrated Drive Generator.

IDEAL CYCLE (combustor)
One in which no pressure loss occurs across the combustion section. Not a practical condition for the Gas Turbine Engine, where a slight (approximately 3%) pressure drop is needed to promote uniform airflow from inlet to outlet, and for proper combustor cooling.

IDLE (engine)
Locate under 1) Flight Idle, 2) Ground Idle.

IDLE APPROACH SETTING
Another name for Flight Idle setting.

IDLE MINIMUM SETTING
Another name for Ground Idle setting.

IDLE RESET
Refers to a fuel control mechanism capable of automatically resetting engine idle speed in the increase direction to match an increase in compressor inlet temperature. Idle Reset prevents possible compressor stalls on acceleration during low inlet density conditions. *See also: Ground Idle Speed (variable).*

IEPR
Locate under Integrated Engine Pressure Ratio.

IFSD
Abbreviation for In-Flight-Shut-Down.

IGNITER
Short for Igniter Plug. Also spelled Ignitor.

IGNITER BOX
Another name for Ignition Exciter Box.

IGNITER LEAD
A small component part of an ignition system. A high voltage output lead that provides a path for electrical energy from the Exciter Box to the Igniter plug. The lead is shielded on its outer surface by a braided, stainless steel covering designed to prevent high altitude insulation breakdown and shield against Radio Frequency Interference (RFI).

IGNITER PLUG
A small component part of an Ignition System. The igniter plug produces a high intensity flashover to ignite the air-fuel mixture in the combustor liner during engine starting. The tip of the plug is most often located within the combustion liner at the perimeter of the fuel nozzle spray pattern. Igniter Plugs are of two general types as follows: 1) High Voltage Type also called Air-Gap type, 2) Low Voltage Type also called Surface-Gap Type. The typical plug will fire 60 to 100 times per minute with a flash-over of 1/2 to 3/4 inch in length. Also spelled Ignitor Plug. *See also: Ignition System.*

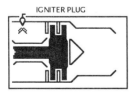

IGNITER PLUG (high voltage type)
An engine spark igniter used in a High Voltage Ignition system. An air gap separates the high voltage center electrode from the igniter plug shell. When current crosses the gap, a high intensity electrical flashover occurs. *See also: Ignition System (high voltage).*

IGNITER PLUG (low voltage type)
An engine spark igniter used in a Low Voltage Ignition system. A gap bridged by a semi-conductor material separates the voltage carrying center electrode from the igniter plug shell. When current crosses the gap a high intensity electrical flashover occurs. Two types of Low Voltage Plugs in use are the Coated Type and the Pellet Type. *See also: 1) Igniter Plug (coated), 2) Igniter Plug (pellet), 3) Ignition System (low voltage).*

IGNITER PLUG (coated, low voltage)
A low range plug which includes a semi-conductor material of Silicon Carbide (SiC) between the voltage carrying center electrode and shell electrode. A low voltage range plug of approximately 1000 to 5,000 volts. Also called an Engobe Coated™ Igniter Plug.

IGNITER PLUG (pellet, low voltage)
A high heat range igniter plug with a solid pellet type semi-conductor material of Silicon Carbide (SiC) between the center electrode and the shell electrode. A high voltage range plug of approximately 5,000 to 9,000 volts. This igniter plug has a longer expected service life than the coated type plug, it is more expensive, and it is more suited to Extended Duty ignition systems. Also called a homogeneous Semi-Conductor Igniter Plug.

IGNITER PLUG (semi-conductor type)
Another name for Igniter Plug (low voltage type).

IGNITER PLUG (shunted-gap type)
Another name for Igniter plug (low voltage type).

IGNITER PLUG TRACKING
A normal, saw-tooth wear pattern of erosion seen on the firing end of a low voltage Igniter Plug.

IGNITION COMPOSITOR
A step-up transformer located between the Ignition Exciter and the Igniter Plug. Today it is more common to place this unit inside the Ignition Exciter Box.

IGNITION EXCITER BOX
An engine accessory. The main component of the ignition system. The Exciter contains a transformer to step up low input voltage of 24/28 VDC or 115 VAC to several thousand volts and create a high intensity electrical flashover at a combustor mounted igniter plug. The D. C. input voltage units utilize an actuating point system, sometimes called a trembler unit, to make and break the primary circuit. The A. C. Input voltage units do not need the trembler, in that polarity reversal occurs automatically with the use of an A.C. voltage source. This unit also houses storage capacitors which power the igniter plugs. Also called Exciter Box or Transformer Box. *See also: 1) Ignition Exciter Box (dual), 2) Ignition Exciter Box (single).*

IGNITION EXCITER BOX (dual)
An Ignition Exciter Box containing two complete exciter units within one housing. It can be seen to have one input lead and two output leads.

IGNITION EXCITER BOX (single)
A Single Ignition Exciter Box containing a circuit which powers only one igniter plug. It can be seen to have one input lead and one output lead.

IGNITION GENERATOR
Another name for Ignition Exciter Box or sometimes in reference to the entire ignition system.

IGNITION POINT CAPACITOR
A component part of an Ignition Exciter Box. A small capacitor, electrically placed across the primary circuit trembler point system in an Ignition Exciter unit. The capacitor prevents point arcing as the points start to open by absorbing the initial current flow. *See also: Ignition Exciter Box.*

IGNITION STORAGE CAPACITOR
A component part of an Ignition Exciter Box. The main energy storage capacitor in the secondary circuit of an Ignition Exciter Box. The stored energy is used to create a flash-over at the Igniter Plug.

IGNITION SYSTEM
An engine mounted electrical capacitor discharge system. The typical system contains two Ignition Exciters, two High Voltage Output Leads, and two Igniter Plugs. Ignition systems range from 2 to 20 joules of stored electrical energy. Approximately one-third of the stored energy will reach the igniter plug to create flashover. The remainder will be lost in the discharge system. The principle function of this system is to ground start turbine engines and to provide in-flight relight in the event of a flameout. Other purposes for this system are to provide automatic relight in the event of flameout during engine operation, when the aircraft is being supplied with large amounts of Customer Bleed Air, while flying in adverse weather, or during take-off and landing. The following two systems are most common: 1) The Ignition System (low voltage type) and 2) The Ignition System (high voltage type). *See also: Joule Rating.*

IGNITION SYSTEM (A.C. input)
An engine ignition system that receives 115 VAC from the aircraft bus to power its circuits.

IGNITION SYSTEM (afterburner)
An electrical capacitor discharge ignition system. Afterburner combustion is most often initiated via an ignition system similar to a main engine system, but one which initiates combustion of a single stream of fuel called a Hot Streak. The fuel hot streak, in turn, ignites the afterburner main fuel. *See also: Hot Streak Ignition.*

IGNITION SYSTEM (automatic relight)
A means of automatically initiating engine ignition when a loss of engine power is sensed. The typical ignition alert systems contain devices sensitive to: 1) loss of engine torque, 2) loss of compressor discharge pressure, or 3) abnormally fast engine deceleration (found in electronic fuel scheduling systems).

IGNITION SYSTEM (continuous)
A secondary ignition system that can be operated with no duty cycle limitations. Typically a system of low output rating, approximately 4 to 6 Joules. Continuous ignition is installed along with a main system of much higher output. The continuous system is activated during takeoff, landing, or adverse weather operation to provide rapid relight in the event of flameout. Also called On-Demand Ignition. *See also: Ignition System (intermittent duty cycle).*

IGNITION SYSTEM (D.C. input)
An engine system that receives its power to operate from the aircraft D.C. bus; either the aircraft battery (24 VDC) or a generator (28 VDC).

IGNITION SYSTEM (extended duty cycle)
A main engine ignition system of recent design. One in which there is a long duty cycle or perhaps no time restriction. This newer, solid state type system, produces much less heat in comparison to the Intermittent Duty Cycle systems and therefore can be operated for longer periods.

IGNITION SYSTEM (high voltage type)
An engine ignition system that falls in a range of 15,000 VDC to 30,000 VDC output and 15 to 20 joules of stored energy. The High Voltage System is classed as a two-spark system because it utilizes a high voltage discharge from a trigger transformer and capacitor to create initial ionization and flashover at the igniter plug firing tip. After the igniter plug air-gap becomes ionized by the first flashover, a storage capacitor discharges to produce a second high intensity spark at the igniter plug. High Voltage Ignition systems are required on some engines to promote good high altitude starting characteristics. The large spark plume created by the high wattage available to this plug enhances engine relight capabilities after flameout or after in-flight shutdowns. Also called High Tension Ignition System. *See also: 1) Ignition System, 2) Ignition Trigger Capacitor, 3) Ionization.*

IGNITION SYSTEM (intermittent duty cycle)
A main engine ignition system which has been available for many years in turbine engines. A system that has a limit on its operating time to prevent heat damage to its Ignition Exciter Box components. A typical restricted cycle time is one minute on, two minutes off, one minute on followed by a 20 minute cooling period. Also called Restricted Duty Cycle.

IGNITION SYSTEM (laser)
A turbine ignition system under development which uses the heat of radiation from a laser source to ignite combustor fuel, as opposed to a traditional electrical power source.

IGNITION SYSTEM (low voltage type)
An ignition system that falls in a range between 1,000 VDC and 9,000 VDC output with a 2 to 8 joule rating. It is classed as a single-spark system because it does not contain a trigger spark system to ionize the gap at the firing tip as does the High Voltage Ignition System. The low voltage system instead uses a semiconductor material between the center electrode and the metal shell at the igniter plug firing tip. The semiconductor radiates enough heat to ionize the air-gap. After ionization, the storage capacitor discharges across the ionized gap to create a high intensity flashover. Also called Low Tension Ignition System. *See also: 1) Ignition System, 2) Ionization.*

IGNITION TRANSFORMER BOX
Another name for Ignition Exciter Box.

IGNITION TRIGGER CAPACITOR
A capacitor used to store energy for the trigger transformer in High Voltage Ignition Systems. *See also: Ignition System (high voltage type).*

IGNITION TRIGGER TRANSFORMER
Locate under Ignition System (high voltage type).

IGNITION TUBE
Another name for Flame Propagation Tube.

IGNITOR PLUG
Alternate spelling for Igniter Plug.

IGV
Abbreviation for Inlet Guide Vane.

IGV SCHEDULE CHECK
A maintenance check-run on a Turbine Engine to determine whether the compressor variable inlet guide vanes are opening and closing at the correct Percent RPM. Also called VIGV Schedule Check. *See also: Variable Inlet Guide Vanes.*

IHPTET
Abbreviation for Integrated High-Performance Turbine Engine Technology. Also used as an acronym pronounced "Ip-Tet".

IMPACT-REACTION TURBINE
Another name for Impulse-Reaction Turbine.

IMPACT PRESSURE
Another name for Ram Pressure.

IMPELLER
Locate under Centrifugal Flow Compressor.

IMPELLER CASE
Another name for Impeller Shroud.

IMPELLER HUB
Another name for Inducer.

IMPELLER SHROUD
An engine component part. The stationary outer casing of a centrifugal (impeller) compressor. This casing typically contains some type of Diffuser Vanes and Swirl Vanes to increase air pressure and turn airflow back to an axial direction.

IMPELLER FACE SHROUD
Another name for Impeller Shroud.

IMPELLER STAGE
An impeller rotor and an impeller shroud in which an air pressure rise is produced before entry into a second stage or into the combustor.

IMPENDING BYPASS INDICATOR (engine filters)
Locate under Filter Bypass Indicator.

IMPINGEMENT STARTING
A starting process requiring no engine mounted starter unit. Impingement starting was developed primarily for naval, carrier based aircraft. Air from a ground power source is directed through a special engine connection onto the engine turbine wheel(s) to cause engine rotation for starting, after which the air source is removed.

IMPULSE DUCT
Older name for a Pulse-Jet.

IMPULSE TURBINE SYSTEM
A turbine system in which the gas enters at low pressure and maximum velocity. A stator vane and rotor blade arrangement whereby the turbine nozzle vanes, one to the other, form convergent gas flow paths while the turbine blades form straight through paths. The turbine rotor is turned by impingement of gases on its surfaces (impulse). Not a common design for flight engine systems, but can be seen in turbine driven accessories such as the Pneumatic Starter and the Fuel-Air Combustion Starter. For comparison see also: 1) Impulse-Reaction Turbine, 2) Reaction Turbine.

IMPULSE-REACTION TURBINE SYSTEM
The type of turbine rotor system common to all gas turbine engines. Impulse-Reaction implies a combination of a 50% Impulse Turbine design and 50% Reaction Turbine design. A stator vane and rotor blade arrangement whereby the turbine vanes and blades have a slight twist from base to tip. The twist is called the stagger angle. The blade base area is an impulse design, twisting toward the blade tip to become a reaction design. *See also: 1) Impulse Turbine System, 2) Reaction Turbine, 3) Turbine Rotor Assembly.*

INCREASED PERFORMANCE ENGINE (IPE)
An engine in production which is raised to a higher performance level and issued as a derivative model.

IN-FLIGHT SHUT-DOWN (IFSD)
Generally an unscheduled engine shut-down, except during training flights.

IN-FLIGHT RELIGHT
Locate under Air Start.

IFSD
Abbreviation for In-Flight Shut-Down.

IN-LINE COMBUSTOR
Another name for Through-Flow Combustor.

INCLUSION
Locate under Bearing Inspection and Distress Terms in Appendix.

INCONEL™
A trademark of the International Nickel Company. A high nickel content alloy. Locate under Nickel-Base Alloy. *See also: 1) ODS Super-Alloy, 2) Super Alloys.*

INDENTATION
Locate under Bearing Inspection and Distress Terms and Engine Inspection and Distress Terms in Appendix.

INDEPENDENT TURBINE
Another name for Free Power Turbine.

INDICATED AIRSPEED
Airspeed indicated on the cockpit display as opposed to true airspeed. For example, when the Concorde SST's indicated airspeed is 575 MPH, the true airspeed is approximately 1,350 MPH. This results from the low density at cruising altitude of 55,000 ft. The pilot uses this information for control purposes, realizing that the aircraft is experiencing the force of only 575 MPH on its surfaces.

INDUCED DRAG
Locate under Drag (induced).

INDUCER (centrifugal impeller)
The center (inlet) portion of a Centrifugal Flow Compressor impeller. The Inducer contains curved air-turning vanes that induce air to flow at the best angle onto the impeller blading. Also called 1) Impeller Hub, 2) Rotating Guide Vane.

INDUCTION SYSTEM
Locate under Inlet.

INDUSTRE AERONATUICHE COMPANY (Italy)
Manufacturer of Gas Turbine Engines. Locate under

section entitled "Gas Turbine Engines for Aircraft".

INERTIA BONDING
Locate under Inertia Welding.

INERTIA WELDING
A recent development in welding, by heat of rubbing friction. A rotating piece is forced against a fast moving stationary piece and the friction heat generated causes the pieces to bond without melting. Inertia Welding was developed to join super-alloys that are difficult to weld by traditional methods. Also called Inertia Bonding.

INERT GAS WELDING
Another name for Heli-Arc™ welding.

INERTIAL ICE AND SAND SEPARATOR
A nacelle component. A mechanism that removes heavy particles such as ice, sand, and other F.O.D causing debris from Turboprop and Turboshaft powered aircraft inlets. A sharp turn in the incoming air is provided to allow only uncontaminated air to enter the engine while applying an inertial load to heavier than air particles. The particles, in this manner, are delivered to a cleanable trap or are sent overboard of the aircraft. In some installations, a small impeller blower is used to create a suction to aid in the process of particle separation. Also called an Inlet Particle Separator.

INERTIAL ICE AND SAND SEPARATOR (types)
The two most common types of separators are as follows: 1) Vane Type-uses vanes to rotate engine inlet air to the outer walls of the unit to carry heavier than air particles away from engine face. Also called Swirl-Type. 2) Vaneless Type-utilizes contoured walls to direct heavy particles away from the engine face.

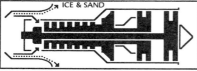

INERTIAL SEPARATOR (VANELESS TYPE)

INFLUENCE COEFFICIENT BALANCING
Locate under Dynamic Balancing.

INFRA-RED SUPPRESSOR
A device fitted to the tailpipe of some military aircraft to suppress the infra-red signature emitted by the hot exhaust and reduce the aircraft's visibility to heat seeking missiles.

INFRA-RED THERMOMETRY
An infra-red means of measuring exhaust gas temperatures in engines as opposed to the traditional thermocouple sensors. By aiming infra-red sensors at the heat source, the temperature value can be determined and transmitted to a cockpit gauge. This method can be employed in areas too hot for measurement by conventional metallic probes.

INLET (aircraft)
There are three common inlet configurations in aircraft, as follows: 1) Nacelle - An engine cowling which encompasses the engine and has a front facing inlet. 2) Bifurcated Duct - A front fuselage arrangement with a dual air inlet into a single engine. 3) S-Duct Inlet - A curved fuselage inlet feeding into an engine mounted on the center-line in the aft fuselage. See also: Flight Inlet.

INLET (engine)
The point at which air enters the engine compressor. In most Gas Turbine Engines, air moves parallel to the aircraft flight path through the inlet and directly into the engine. It is less common to see a Plenum arrangement interposed between the flight inlet and the engine inlet, except for some turboprop engines and older military aircraft. See also: 1) Inlet Case, 2) Inlet Plenum (GTCP), 3) Plenum Chamber, 4) Radial Air Inlet.

INLET (flight)
Locate under Flight Inlet.

INLET (external compression)
Locate under External Compression Inlet.

INLET ANTI-ICING SYSTEM
A means of preventing heavy ice formation in aircraft flight inlets.
1) Electro-Expulsion System - Electric pulses in a neoprene strip bonded to the flight inlet. Pulses cause a slight movement in the strip to break up ice as it forms.
2) Electro-Thermal System - A resistance type heat-strip in neoprene material, bonded to the flight inlet to prevent the start of ice formations.
3) Pneumatic System - A neoprene boot which removes early ice formations when inflated.

INLET AREA HAZARD MARKER
A painted line or mark on an engine nacelle to indicate a hazard. The indicating mark is placed near the flight inlet to alert personnel to the danger of inlet suction. See also: Turbine Area Hazard Marker.

Inlet Bevel Gearbox
Locate under Inlet Gearbox.

INLET BOUNDARY LAYER DIVERTER
Locate under Boundary Layer Diverter.

INLET BUZZ
An audible sound resulting from inlet pressure variations occurring in inlets of supersonic aircraft when shock waves alternately move in and out. This condition results when designed aircraft speeds are being exceeded.

INLET CASE (engine)
An engine component. The front compressor supporting member to which the flight inlet attaches in front and the compressor outer case attaches at the rear. The Inlet Case most often houses the front compressor support bearing. Also called Front Frame. The older term is Intake Case. See also: Inlet (engine).

INLET CENTERBODY
Newer term to mean Nose Dome.

INLET CONE
Forward most part of a Gas Turbine Engine. Another name for Nose Dome. See also: 1) Inlet Spike, 2) Spinner.

INLET COWL
Another name for Flight Inlet.

INLET DUCT
Another name for Flight Inlet.

INLET FAIRING
Another name for Nose Dome.

INLET FRAME
Another name for Inlet Case.

INLET GEARBOX
1) An engine component used to drive small accessories. A gearbox located at the engine inlet case and driven by the front shaft of the compressor. Not all engines are configured with this gearbox, due to the safety problems it creates by being located within the inlet gaspath. On a dual-spool engine, the Inlet Gearbox will be driven at Low Pressure Compressor speed.
2) An internal gearbox located at the front of a high pressure compressor and used as a power take-off to drive the Accessory Drive Gearbox. Also called Inlet Bevel Gearbox. *See also: Bevel Gear System.*

INLET GUIDE VANE SYSTEM (engine)
An engine system typically located within the Inlet Case. A set of stator vanes positioned in front of the compressor first stage rotor blades to direct airflow at the optimum angle into the blades, thereby reducing aerodynamic drag. Air is angled in the blade direction of rotation and the approach speed is reduced, creating a desirable angle-of-attack. The guide vanes can be of fixed angle or have a variable angle capability. *See also: 1) Guide Vanes, 2) Variable Angle Stator Vanes.*

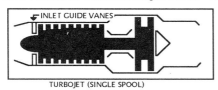
TURBOJET (SINGLE SPOOL)

INLET HAZARD MARKER
Locate under Inlet Area Hazard Marker.

INLET HUB FAIRING
British term for nose dome.

INLET MOMENTUM DRAG
Also called Ram Drag.

INLET PARTICLE SEPARATOR (IPS)
Another name for Inertial Ice and Sand Separator.

INLET PLENUM (GTCP)
An ambient air inlet duct from which air is simultaneously distributed to: 1) the engine compressor Bellmouth Inlet, 2) the Load Compressor inlet.

INLET PRESSURE (Pt_2)
Total pressure taken in the flight inlet just forward of the engine compressor. Pressure at this location is often a good indicator of air density against which fuel can be scheduled to the combustor. The symbol Pt_2 represents Pressure-total at engine station two. The symbol $P2$ is also used when total pressure is understood to be the value. If static pressure is read in the inlet, the symbol changes to Ps_2.

INLET PRESSURE RATIO
The pressure difference, front to back, in a flight inlet (i.e. pressure within the flight inlet divided by ambient pressure). For example, in flight at Mach 0.85 the average aircraft will produce a pressure ratio of approximately 1.6 : 1. Also called Inlet Ram Compression.

INLET RAM COMPRESSION
Locate under Inlet Pressure Ratio.

INLET SAND AND ICE SEPARATOR
Another name for Inertial Sand and Ice Separator.

INLET SCREEN
A foreign object damage (F.O.D) prevention device used on Turboprop, Turboshaft and stationary turbine engines. Not used on Turbojet or Turbofan installations because screens cause drag, flow blockage, and icing problems.

INLET SECTION (aircraft)
The portion of an aircraft in which ambient air is allowed to decrease in velocity and increase in pressure before entering the engine inlet. Also called Flight Inlet. *See also: 1) Inlet Section (engine), 2) Engine Sections.*

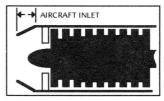

INLET SECTION (engine)
The portion of a turbine engine where air enters from the flight inlet. This section often houses inlet guide vanes, which optimize the flow angle of air entering the first stage of compression. On large Turbofan Engines, inlet guide vanes are not used and inlet section air flows directly into the fan rotor.

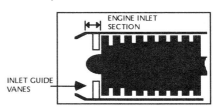

INLET SPIKE
A nacelle component for supersonic aircraft. A movable device used to control inlet geometry and shock wave formation in the inlet gaspath. The fore and aft moving spike diffuses supersonic airflow (increases pressure) and reduces velocity of air to subsonic speed before entry into the engine. This movement is an automatic function within the authority of a computerized engine scheduling system. Also known as Shock Spike. *See also: Convergent-Divergent Inlet Duct.*

INLET STRUT
A component part of an inlet case. Radially placed airfoils set in a spoke-fashion between a front bearing hub and the outer casing. The struts also provide a means of routing air lines, oil lines, electrical conduits,

etc., from the exterior to the interior of the engine. Inlet struts have been eliminated in many engines today, with the inlet guide vanes providing support for the front compressor bearing.

INLET TEMPERATURE (Tt$_2$)
Total temperature taken in the flight inlet just forward of the engine compressor. Temperature at this location is often a good indicator of air density against which fuel can be scheduled to the combustor. The symbol Tt$_2$ represents Temperature total at engine station two. The symbol T2 is also used when total temperature is understood to be the value. If static temperature is read in the inlet, the symbol changes to Ts$_2$. A signal similar in location to Inlet Pressure (Pt$_2$). Also known as Compressor Inlet Temperature (CIT).

INLET UNSTART
Locate under Unstart (inlet).

IN-LINE COMBUSTOR
Another name for Through-Flow Combustor.

INNER COMBUSTION CASING
An engine component. The portion of the combustor assembly which forms an outer housing for the turbine shaft.

INNER EXHAUST CONE
Another name for: 1) Exhaust Centerbody, 2) Exhaust Plug, 3) Tailcone.

INNER LINER (combustor)
Refers to the inner wall of an annular combustion liner nearest to the turbine shaft housing.

IPE
Abbreviation for Increased Performance Engine.

INSERT (tailpipe)
Locate under Tailpipe Insert.

INSPECTION
There are many inspection procedures which apply to the Gas Turbine Engine. Two basic categories of inspections are Physical Inspections and Special Inspections. Physical Inspection methods detect damaged or worn parts visually or with measuring instruments. Special Inspections require more sophisticated equipment such as Eddy Current, Ultra-Sound, and X-ray, used in Non-Destructive Inspection procedures. *See also: 1) Letter Checks, 2) Non-Destructive Inspection.*

INSPECTION AND DISTRESS TERMS
Listing is located in Appendix.

INSULATION BLANKET
An external engine covering. The blanket prevents heat rejected by the engine from overheating the engine nacelle area.

INTAKE
Another name for Inlet.

INTAKE BULLET
Another name for Spinner.

INTAKE CASE
Another name for Inlet Case.

INTAKE MOMENTUM DRAG
Another name for Inlet Momentum Drag. *See also: Ram Drag.*

INTAKE PRESSURE RECOVERY
Another name for Ram Recovery.

INTEGRAL DRY SUMP (lubrication system)
A lubrication system for a Gas Turbine Engine in which the main supply of oil is contained in a compartment located within the confines of the engine. Generally it is a space between two major engine cases. The Pratt and Whitney of Canada PT6 Turboprop is such an engine. *See also: Dry Sump (lubrication system).*

INTEGRAL OIL SYSTEM
Another name for Wet Sump.

INTEGRAL OIL TANK
Locate under Integral Dry Sump.

INTEGRAL PARTICLE SEPARATOR
Another name for Inertial Ice and Sand Separator.

INTEGRALLY BLADED DISC
Another name for Blisk.

INTEGRALLY BLADED ROTOR (IBR)
A one-piece, multi-stage compressor rotor constructed in the manner of a Blisk. Both axial-flow and mixed-flow compressors are under development using the IBR concept. *See also: Blisk.*

INTEGRATED AIR-OIL SEPARATOR (oil tank)
Another name for Deaerator (oil tank).

INTEGRATED BLADE AND ROTOR
Locate under Integrally Bladed Rotor.

INTEGRATED BLADE INSPECTION SYSTEM (IBIS)
An automated compressor and turbine blade inspection process performed in place of a manual inspection process. This inspection method includes use of penetrating dyes, xray and other non-destructive tests performed on newly manufactured parts.

INTEGRATED CIRCUIT TEMPERATURE TRANSDUCER
An engine test device. A semi-conductor type temperature transmitting device similar to a Thermistor, but very linear in output with temperature change. Unlike a thermocouple it requires external power. *See also: 1) Thermistor, 2) Thermocouple.*

INTEGRATED DRIVE GENERATOR (IDG)
A combined Constant Speed Drive Generator (CSD) unit and its electrical generator located within the same housing. *See also: Constant Speed Drive.*

INTEGRATED ENGINE PRESSURE RATIO (indicating system)
Similar to the aircraft Engine Pressure Ratio indicating system with the addition of a Fan Pressure Probe to integrate fan discharge pressure with the traditional turbine discharge pressure and compressor inlet pressure. Not a widely used cockpit indication. Mathematically expressed as:

$$IEPR = [Pf \times Af + (Pe \times Ae) \div (Af + Ae)] \div Pt_2$$

Where: Ae = Area of hot exhaust, Af = Area of fan exhaust, Pe = Pressure at hot exhaust, Pf = Pressure at fan exhaust, Pt_2 = Pressure total in the engine inlet. *See also: Engine Pressure Ratio.*

INTEGRATED EXHAUST DUCT (or nozzle)
Locate under Mixed Exhaust Duct.

INTEGRATED FRONT FAN
An engine component. Refers to a traditional design, a combination Front Fan and Low Pressure Compressor.

INTEGRATED ROTOR AND BLADE
Locate under Integrally Bladed Rotor.

INTER-CONNECTOR (combustor)
Another name for Flame Propagation Tube.

INTERFERENCE DRAG
Locate under Drag (parasite).

INTERFERENCE FIT
A zero-clearance or negative clearance condition between parts causing very tight fits when the parts are mated. To assemble, the mating parts are forced together by pressing, or they can be heated or chilled to facilitate the fit.

INTERGRANULAR OXIDATION
Locate under Engine Inspection and Distress Terms in Appendix.

INTERMEDIATE CASE
An engine component. The outer-most housing of a high pressure compressor. Also called N2 Case. May also be the housing around an Intermediate Pressure Compressor.

INTERMEDIATE DRIVE GEARBOX
Another name for Auxiliary Drive Gearbox.

INTERMEDIATE PRESSURE COMPRESSOR (IPC)
A major engine component. The middle (IPC) compressor of a triple-spool axial flow engine. Its function is to receive air under pressure from the LP compressor and deliver an even higher pressure to the HP

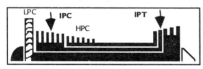

compressor. Also called 1) IP compressor, 2) N2 compressor, when the high pressure compressor is referred to as the N3 compressor. *See also: Intermediate Pressure Turbine.*

INTERMEDIATE PRESSURE TURBINE (IPT)
A major engine component. The center turbine of a triple-spool axial flow engine. The IPT mechanically attaches to and drives the Intermediate Pressure Compressor. *See also: Intermediate Pressure Compressor.*

INTERMEDIATE TURBINE TEMPERATURE (ITT)
Temperature taken between the high and low pressure turbine systems of a Turbojet or Turbofan engine. Also temperature taken between the gas producer turbine and free power turbine of a Turboprop or Turboshaft engine. Similar to the Exhaust Gas Temperature cockpit indicating system except for location of the temperature probes. Also called Interstage Turbine Temperature. *See also: Temperature (exhaust).*

INTERNAL COMBUSTION ENGINE
An engine in which fuel is mixed with its working fluid (air), and combustion occurs within the confines of the engine. A Gas Turbine Engine is an internal combustion engine in this sense. By comparison, a steam engine is an external combustion engine because its combustor utilizes atmospheric air and not its working fluid (steam). *See also: External Combustion Engine.*

INTERNAL CONVECTION COOLING (turbine)
Locate under Convection Cooling.

INTERNAL-FLOW NOZZLE
Refers to the exhaust nozzle at the rear of a traditional exhaust duct (tailpipe) and the location of its inner tailcone. The tailcone fairing within the duct in this case is shorter in length than the exhaust duct itself. *See also: Plug Nozzle.*

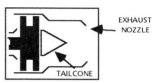

INTERNAL TOOTH LOCKWASHER
Locate under Shakeproof Washer.

INTERNATIONAL AERO ENGINES LTD.
Manufacturer of Gas Turbine Engines. Locate under section entitled "Gas Turbine Engines for Aircraft".

INTERNATIONAL CIVIL AVIATION ORGANIZATION (ICAO)
An international organization that sets and also regulates many aeronautical standards. For example, ICAO produces tables for the International Standard Atmosphere (ISA) and also sets the parameters for the International Standard Day (ISD).

INTERNATIONAL STANDARD ATMOSPHERE (ISA)
A table showing the various altitudes with corresponding ambient temperature, ambient pressure, speed of sound, and other atmospheric parameters as set forth by the International Civil Aviation Organization. *See also: ISA Rating (engine).*

INTERNATIONAL STANDARD DAY (ISD)
Internationally agreed upon atmospheric conditions, set forth by the International Civil Aviation Organization, as follows: 14.7 psia (29.92 in.hg), 59°F. (15°C.), 0% relative humidity, at 40° latitude gravity effect.

INTERPRETER (anti-icer system)
Another name for Ice Detector.

INTERRUPTED DECEL TEST (engine)
A rapid engine deceleration toward the idle stop followed by a rapid acceleration to the initial high power setting, after the engine speed passes through a specified turn-around point. This procedure is a test of engine stall tendencies.

INTERSTAGE TURBINE TEMPERATURE (ITT)
Another name for Intermediate Turbine Temperature.

INTRASCOPE
British term for a borescope.

INVERSE FLOW ANNULAR COMBUSTOR
Another name for Annular Combustor (reverse flow).

INVERSE VARIABLE CYCLE ENGINE
A Gas Turbine Engine in which inlet airflow separates into two distinct airstreams; one into the core and the other into a ram duct surrounding the core. The core airflow compresses initially by diffusion, after which it exhausts through a heat exchanger to cool and increase in density and then flows into an axial flow compressor and combustor. Both the core and ram duct later mix in a second combustor located in the tailpipe, before being discharged into the atmosphere. This design is under study and stated to be best suited for very high speed flight.

INVERTED FLOW ANNULAR COMBUSTOR
Another name for Annular Combustor (reverse-flow).

INVESTMENT CASTING PROCESS
A metal casting process called the lost wax method. After sand is packed (invested) around a wax pattern, molten metal is introduced and melts the wax, taking its place within a mold. The casting process is accomplished within a vacuum furnace spin chamber, to pull out entrained gases and produce a more dense and higher quality metal. This process is commonly used to cast small complex parts, such as turbine blades, to near-net-shape. Investment casting enables the use of harder materials than can be produced in a forging process. The Investment Casting process also reduces machining costs associated with forged parts of hard alloys. Note: Shell Molding remains the best method for casting large parts where temperature of the molten metal is not as great as in the investment cast method.

INWARD RADIAL FLOW TURBINE
Another name for Radial Inflow Turbine.

IONIZATION (ignition system)
The process of ionizing (making conductive) an air gap. Ionization takes place across igniter plug gaps and discharge tube gaps in the Ignition Exciter Box by utilizing the high electron discharge from capacitors. The process is described as the stripping of electrons from nitrogen and hydrogen molecules by using heat. An ion is a molecule stripped of electrons.

IP COMPRESSOR (IPC)
Another name for Intermediate Pressure Compressor.

IP THRUST BEARING
Refers to a main bearing of the ball-type which prevents axial movement of the Intermediate Compressor. Main ball bearings absorb thrust loads when high pressure forces at the rear exert a push forward on the rotor. *See also: Bearings.*

IP TURBINE (IPT)
Another name for Intermediate Pressure Turbine.

IRIS EXHAUST NOZZLE
A component part and final opening of an afterburner tailpipe. A design similar to a camera shutter with either a two-position flow area (partially open or fully open) or fully variable-flow area. The widest opening in each type of nozzle is the full afterburner position. The smallest opening is used for full power operation without afterburner in the variable-flow type and for all non-afterburner power settings in the two-position type. *See also: Exhaust Nozzle.*

IRON-CONSTANTAN
Locate under Thermocouple (exhaust temperature).

ISA
Abbreviation for International Standard Atmosphere.

ISA RATING (engine)
Refers to the ambient conditions of 14.7 psia (29.92 in.hg.) and 59 degrees F. (15 degrees C.) and the power rating of an engine. For example: The Garrett Companies TFE731-5B turbofan is rated at 4,750 lbt to ISA + 10° C., meaning that this thrust level is available up to 25° C. at sea level condition or to any runway condition corresponding to ISA + 10° C. Also called flat Rating. *See also: Flat Rating.*

ISD
Abbreviation for International Standard Day. Standard atmospheric conditions as established by the International Civil Aviation Organization (ICAO).

ISENTROPIC COMPRESSION
The ideal compression process wherein entropy is held constant as pressure rises. For example, in the engine inlet, pressure increases isentropically from ram effect when pressure losses due to friction are negligible. The compressor, however, is a non-isentropic compression process because entropy increases greatly due to friction heating of air. *See also: Entropy.*

ISENTROPIC GAS HORSEPOWER (IGHP)
Power rating applied to a dual-compressor auxiliary gas turbine. IGHP represents the potential work available in the gases leaving the Gas Generator Turbine and entering the Power Turbine. The Isentropic Horsepower is arrived at by dividing Shaft Horsepower by Power Turbine efficiency.

ISENTROPIC EFFICIENCY (compressor)
A physics term meaning without entropy. A measure of friction and turbulence losses to the compression process. The ideal compression process is one in which pressure increases without entropy increase. The Gas Turbine compressor is not isentropic because it has an unavoidable air friction and turbulence level, meaning an entropy increase. The compressor is, therefore, less than ideal and is non-isentropic. *See also: 1) Compressor Efficiency, 2) Isothermal Efficiency.*

ISHIKAWAJIMA-HARIMA COMPANY (Japan)
Manufacturer of Gas Turbine Engines. Locate in section entitled "Gas Turbine Engines for Aircraft".

ISOCHRONOUS SPEED GOVERNOR
Another name for Flyweight Governor. Isochronous means Actual Speed becoming equal-over-time to Desired Speed.

ISO-PROPYL-NITRATE STARTER
An engine accessory. A starter unit which utilizes Iso-Propyl-Nitrate rather than Jet-Fuel. The fluid is sprayed into a Fuel-Air Combustion type starter unit for rapid self-starting capability. Not in common use.

ISOSTATIC FORGING
Another name for Isothermal Forging.

ISOTHERMAL ALTITUDE
The boundary between atmospheric layers of constant thermal gradient. The lowest is 36,089 ft., after which pressure continues to drop but ambient temperature remains the same up to approximately 66,000 ft. (In accordance with U.S Standard Atmosphere).

ISOTHERMAL EFFICIENCY (compressor)
A physics term meaning without heat. High isothermal efficiency refers to a process of compression without heat rise. The turbine engine compressor, by comparison, has an unavoidable heat rise due to pressure rise, friction, and turbulence, and a corresponding loss of isothermal efficiency. However, if by some means during compression the temperature could be held constant by extracting all of the heat as it occurs, and pressure and volume were changing normally, 100% Isothermal Efficiency would result. *See also: Compressor Efficiency.*

ISOTHERMAL FORGING
A forging process used in super-alloy production. One such process involves hot pressing of alloy powder to form a solid metallic piece such as a turbine blade. It is referred to as a near-net-shape process because less waste is incurred during final machining. Also called Isostatic Forging.

ISOTHERMAL PRESSURE
Pressure at boundary altitudes between atmospheric layers. For example, the ambient pressure at an altitude of 36,089 ft. is 3.282 psia. *See also: Isothermal Altitude.*

ISOTHERMAL TEMPERATURE
Temperature at boundary altitudes between atmospheric layers. For example, the ambient temperature at 36,089 ft. is -69.7° F. *See also: Isothermal Altitude.*

ISRAELI MANUFACTURED ENGINES
Locate by manufacturer, Bet Shemesh Engines Ltd. in section entitled "Gas Turbine Engines for Aircraft".

ITALIAN MANUFACTURED ENGINES
Locate by manufacturer, Alfa Romero, Fiat Aviazione Spa, Industrie Aeronautiche, Meccaniche Rinaldo, Piaggio Spa., in section entitled "Gas Turbine Engines for Aircraft".

ITT GAUGE
Abbreviation for Intermediate Turbine Temperature gauge.

J

JAM ACCELERATION
Rapid movement of the cockpit Power Lever to cause engine spool-up (acceleration) in the shortest possible time. Also called 1) Ram or Slam Acceleration, 2) Throttle Burst.

JATO
Acronym for Jet Assisted Take-off. A liquid rocket-pac attached to an aircraft to provide additional takeoff thrust. Mainly used by military aircraft operating at maximum gross takeoff weight.

JAPANESE MANUFACTURED ENGINES
Locate under Ishikawajima-Harima Company, Kawasaki Company, Mitsubishi Company, National Aerospace Laboratory, Technical R & D Institute. Locate in section entitled "Gas Turbine Engines for Aircraft".

JET EFFLUX
Refers to the hot, high velocity gas flow from the Exhaust Nozzle of a Gas Turbine Engine. Also called Jet Exhaust or Jet Wake.

JET ENGINE
A family of engines which produce reactive thrust. The four common types are: 1) rocket-jet, 2) pulse-jet, 3) ram-jet, 4) turbo-jet. The original aircraft turbo-jet has grown into its own family of prime movers called Aircraft Gas Turbine Engines. The term "jet engine" is also widely used to mean "Gas Turbine Engine".

JET FUEL
Common name for all types of civil and military aviation fuels. Locate under Fuel Types-Jet.

JET FUEL STARTER (JFS)
An engine accessory. A complete small turbine engine used as a starter unit for a main engine. This unit is typically mounted on the accessory section of the engine or a gearbox attached to the engine accessory section. Some of these units have a direct drive, others utilize a free turbine drive. Differs from a Fuel-Air Combustion Starter which has no compressor and must be supplied with air from an external source. Also called a Turboshaft Starter.

JET KEROSENE
Another name for Civil Aviation Kerosene.

JET NOZZLE
Another name for Exhaust Nozzle.

JET NOZZLE AREA
The area of the final opening of an Exhaust Duct through which engine exhaust gases pass out to the atmosphere.

JET PIPE
Another name for Exhaust Duct or Tailpipe.

JET PIPE TEMPERATURE (JPT)
British term for Exhaust Gas Temperature (EGT).

JET PUMP
A small accessory for aircraft or engine. A pump that operates by directing a fluid as a motive force through a venturi for the purpose of carrying surrounding fluid or gas along with it. Jet Pumps are common to scavenge oil systems, being used as oil pumps, and to fuel systems, being used as boost pumps and vapor eliminator pumps. *See also: Ejector Pump (fuel boost).*

JET SILENCER
Another name for Noise Suppressor.

JET WAKE (exhaust)
Another name for Jet Efflux.

JETCAL ANALYZER™
A unit of test equipment. A Bell and Howell Company trade-name for an electronic test apparatus used to check the calibration of engine indicating systems such

as Exhaust Gas Temperature and Percent RPM.

JP FUEL
Military designation for jet fuel, short for Jet Propellent. *See also: Fuel Types (military).*

JOULE RATING (ignition system)
The power output rating for Gas Turbine Engine ignition systems. The range of turbine ignition systems is from approximately 4-Joules to 20-Joules. A joule is a unit of electrical energy expended in maintaining a current flow of 1 Ampere against a resistance of 1 Ohm for 1 Second. One joule also equals 0.73732 ft.lbs. of work. In turbine engine ignition circuits.

$$Joules = W \times T$$

Where W = Watts expended, T = Time for the spark to jump the plug gap. *See also: Ignition Systems.*

JUMBO JETS
Refers to wide-bodied jet aircraft. Locate under Wide-Bodied Aircraft.

JUNKERS JUMO (model 109-004)
The axial flow Turbojet engine, commonly called the Jumo-004, which powered the ME-262 "Swallow". The Swallow was Germany's first production jet-fighter aircraft. *See also: 1) Franz, Anselm, 2) Heinkel 178.*

K

KAWASAKI COMPANY (Japan)
Manufacturer of Gas Turbine Engines. Locate in section entitled "Gas Turbine Engines for Aircraft".

KELVIN TEMPERATURE SCALE
Absolute temperature scale with minus 273° Celsius as absolute zero. Used in many calculations of turbine engine performance.

KEROSENE FUEL
Locate under Civil Aviation Kerosene.

KEVLAR™
A Dupont Company trademark. A man-made plastic resin and synthetic fiber material. Ounce for ounce Kevlar is stated to be five times stronger than steel. This material is widely used as a bonded, light weight, outer layer on engine casings. *See also: 1) Composite Materials, 2) Nomex™.*

KEY WASHER
A small expendable part. A type of lockwasher that takes the place of a conventional washer. It keys to a slot or hole in a mounting base and has small ears that crimp over to lock in place an individual nut or bolt. *See also: Tab Washer.*

KIDDE™ FIRE DETECTOR
A Kidde Company produced continuous loop type detector which completes an electrical cockpit warning circuit in the presence of heat generated by a hot air leak or a fire. The detector loops are located in the engine nacelle surrounding the exterior of an engine. *See also: Fire Detector.*

KIMPACK™
A type of packaging material used to cushion turbine engine parts in shipment.

KINEMATIC VISCOSIMETER
A unit of laboratory test equipment. One such device measures the falling rate of a small sphere in a fluid. A viscosity measuring device for use in lower viscosity ranges, where the traditional Saybolt Seconds Universal Viscosimeter loses accuracy. The viscosity is given in terms of Kinematic Centistokes at 100°F., 180°F., and 210° F. Kinematics in this case is the study of motion of fluids. *See also: 1) Saybolt Universal Seconds Viscosimeter, 2) Viscosity Index, 3) Centistoke.*

KINEMATIC VISCOSITY
A ratio of absolute viscosity and density, expressed in units of centistokes. It can be stated that centistoke rating is to synthetic oils as SAE numbers are to mineral (petroleum) oils. Turbine oils are typically between 3 and 7 centistoke in viscosity at 210°F. By comparison, water at 68° F. has a viscosity of 1 centistoke. *See also: Synthetic Lubricating Oils.*

KINETIC ENERGY
Energy of motion as in the thrust produced by the gas flow from an engine tailpipe. Mathematically expressed as:

$$Ke = 1/2\ MV^2$$

Where M is mass and V is Velocity. U.S. Units expressed in ft.lbs. *See also: Energy.*

KINETIC HEATING
The surface temperature rise on an airfoil due to drag. The formula for approximate skin induced Temperature Rise (ΔT) in °C. is as follows:
$T = (\text{speed in Mph} \div 100)^2$. For Example, at 1,350 MPH the approximate skin temperature rise of Concorde is $(1,350 \div 100)^2 = 180°$ Celsius. *See also: Drag.*

KINETIC VALVE
A component part of a Fuel Control. A servo mechanism which operates by the force of a moving stream of fluid.

KIRILIAN PHOTOGRAPHY
A photographic process used to detect flaws in metal alloys and composites. A high frequency, high voltage electrical field is directed onto an object which causes the object to radiate a light pattern. An Electro-Magnetic Discharge (EMD) system is then employed to produce a photograph which can be used for detecting surface and subsurface flaws.

KLOECKNER COMPANY (West Germany)
Manufacturer of Gas Turbine Engines. Locate in section entitled "Gas Turbine Engines for Aircraft".

KNIFE EDGE FILTER (oil system)
Locate under Oil Filter (knife edge).

KNIFE EDGE SEAL
Locate under Air Seals.

KNIFE EDGE TIP
An air seal tip. One or more thin metal rims on a shrouded turbine blade. These sealing tips establish a clearance between turbine blades and the stationary turbine shroud ring. The shroud ring is often constructed of a porous metal. When high loading on the turbine blades causes contact, the knife-edge will cut a running groove in the shroud. *See also: Turbine Blade.*

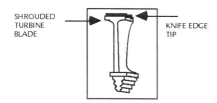

KNOB ALIGNMENT (throttles)
Another name for Throttle Stagger.

L

LAB SEAL
Short for Labyrinth Seal.

LABYRINTH AIR SEAL
A small component part of an engine. This seal contains thin metal rims that control air leakage between rotating and stationary sections of the engine. In some locations, labyrinth seals are positioned to exclude airflow, but in others the seals act as an orifice to meter airflow between the gaspath and the inner portions of the engine. Air metered for this purpose is called Engine Bleed Air. *See also: Engine Bleed Air.*

LABYRINTH OIL SEAL
A small component part of an engine. A soft metal main bearing oil seal, consisting of a rotor and a stator. Three main categories of labyrinth oil seals are used: 1) Continuous Groove Type, 2) Thread Type, 3) Brush Type. The Continuous Groove and Thread Types maintain a clearance to a rotating shaft-runner, but the Brush Type is a static ring of fine wire bristles in continuous contact with a rotating shaft-runner. Labyrinth Oil Seals allow Engine Bleed Air to leak-by into the bearing sumps in order to keep the air-oil mist within the sumps from escaping. *See also: Carbon Oil Seal.*

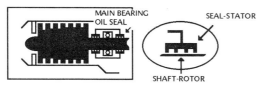

LABYRINTH RUNNER
One of the two main components of a labyrinth seal. The Runner (or rotor) has a smooth surface while the labyrinth stator has annular grooves or fine bristles. *See also: Labyrinth Seal.*

LABYRINTH SEAL DIRT SEPARATOR
A small engine accessory. An in-line filter between an Engine Bleed Air source and a labyrinth oil seal. The Dirt Separator is positioned to prevent airborne contaminants in the gaspath from reaching the seal and possibly contaminating the oil system.

LACQUER (oil)
A hard amber colored film (build-up) on the inner walls of the oil wetted sections of the engine. Lacquer formation is more prone to occur where oil laden air is subjected to high heat. *See also: Coking (oil).*

LAG (engine)
A transient change in RPM of a Turboshaft or Turboprop engine that results when the Collective Lever is moved. RPM remains off-schedule until the fuel governor stabilizes the engine back at the original speed.

LAMILLOY™
A General Motors proprietary material. A type of high temperature strength, laminated sheet metal. A bonded, multi-layer type material with "transpiration cooling" properties resulting from its porous structure and photo-etched channels. Future uses are predicted to include construction of: 1) combustion liners, 2) afterburners, 3) other parts requiring sheet metal construction.

LAMINAR AIRFLOW
Non-turbulent flow. Airflow over an airfoil in thin layers with no air transfer between layers. In other words, streamlined airflow where viscous forces are more significant than inertial forces, and free-stream flow over an airfoil surface is smooth and without turbulence. *See also: 1) Boundary Layer Air, 2) Reynolds Numbers.*

LAND (turbine)
British term for Shroud Ring.

LAPSE RATE (adiabatic)
The rate at which pressure and temperature changes occur with increasing altitude. The rate is 3.37°F. and 0.934 in. hg. per 1,000 feet.

LASER ANEMOMETRY
The process of studying airflow patterns over airfoils using laser light, photoelectric cells, and laser holographic photography. In the case of the Turbine Engine, airflow patterns through compressors and turbines.

LASER BEAM DRILLING
A manufacturing procedure that employs laser heat to drill small holes down to 5 mils in diameter in very hard metals such as turbine blades and vanes. Laser drilled cooling air holes can be other than round in shape, as some are distinguished by their elongated appearance. This occurs when the drilling angle to the piece in other than 90 degrees. *See also: Gill Holes.*

LASER BEAM WELDING
A fusion welding process accomplished in normal atmosphere. Laser Beam Welding is especially useful on highly stressed parts such as fan blades, compressor blades, and turbine blades. It is also useful in assembling parts not conducive to conventional welding methods. *See also: 1) Electron Beam Welding, 2) Inertia Welding, 3) Heli-arc Welding.*

LASER DOPPLER VELOCITY (LDV)
A technique of measuring flow fields within a turbine engine gaspath with laser light. The LDV method eliminates the use of conventional probes which are said to disturb the normal airflow patterns.

LASER-GLASS™
A United Technologies Corporation super-alloy forming process. The basic material, in the form of wire or powder, is dropped onto a rotating disc and fused into a solid piece by heat from a carbon dioxide electrical discharge laser beam. This action creates a material of very high strength, with a radial grain pattern.

LASER GLAZING
A process of melting a ceramic coating over a Plasma Sprayed metallic coating. Laser Glazing results in a hard erosion resistant surface. *See also: 1) Plasma Spray Coating, 2) Thermal Barrier Coating.*

LASER IGNITION
Locate under Ignition System (laser).

LASER VELOCIMETRY (airflow)
Another name for Laser Doppler Velocity.

LAST CHANCE FILTER
Locate under 1) Oil Filter (last chance), 2) Fuel Filter (last chance).

LATENT HEAT OF EVAPORATION
A physics term. A principle by which thrust augmentation by water injection can be explained. As compression tends to increase the temperature of air, water injected into the gaspath changes state to a vapor and absorbs heat from the air. Heat is absorbed at the rate of approximately 1000 BTU per pound of water. The transfer of energy that takes place increases air density and enhances the compression process. *See also: Water Injection.*

LBT (lbt)
Abbreviation for pound of thrust. See section entitled "Gas Turbine Engines for Aircraft" for specific engine applications. For conversion to kN multiply by 0.004448.

LEAKAGE LIMITS (seal drains)
Fluid leakage limits established for each accessory drain point on the Accessory Drive Gearbox. The fluid may be oil that is leaking past a drive pad seal on the gearbox, or the fluid could be fuel, oil or hydraulic fluid leaking from a shaft seal on an accessory. Minor leakage is allowable up to the established limit. Leakage is typically given in drops or cubic centimeters per minute. *See also: Fuel Leakage Check.*

LEAKY TURBOJET
Military jargon for a Low Bypass Turbofan Engine.

LEAN DIE-OUT (engine)
An engine flameout condition. Lean Die-Out is caused by too little fuel to the combustor and a resultant weak fuel-air mixture. This condition weakens the flame and allows the force of the airstream to separate the flame from the head of the fuel nozzle. A normal mixture is considered to be 15 : 1 by weight with a Lean Die-Out mixture point at approximately 25 : 1. *See also: 1) Flame-Out, 2) Stoichiometric Mixture.*

LED DISPLAY (cockpit)
Locate under Cockpit Displays.

LEFKOWELD™
An epoxy resin used to tighten loose compressor stator vanes in their inner and outer supporting shrouds and for filling small areas of surface damage.

LENTICULAR PROFILE (fan blade)
Refers to fan blades that have both sharp leading and trailing edges. Also called a biconvex airfoil shape.

LETTER CHECKS
Generic name for inspections performed on larger aircraft in airline service, as follows:
A-Check - An operational check of major aircraft systems (including engines) along with visual inspections.
B-Check - Incorporates an A-Check along with detailed testing and inspection of selected systems.
C-Check - Incorporates A and B Checks along with detailed testing and inspection beyond lower level checks. May include flight tests.
D-Check - A major inspection to include removal for inspection of aircraft components such as landing gear and engines. D-Checks are the highest level of maintenance an airliner will undergo. It is scheduled to be performed at a time interval of approximately 10 years. *See also: Progressive Inspection.*

LFEC
Abbreviation for Low Frequency Eddy Current test apparatus.

LNG
Abbreviation for Liquefied Natural Gas.

LIFE CYCLE COST (engine)
Abbreviation (LCC). Refers to the average cost of operating and maintaining a selected turbine engine throughout its service life.

LIFE-LIMITED PART
An engine Accessory, an engine Module, or a Component or Component Part that has a specified number of operating hours or operating cycles at which it must be removed for overhaul. Also called Time Restricted Part. *See also: 1) Low-Cycle Fatigue Time, 2) On-Condition Maintenance.*

LIFT
The aerodynamic force on airfoil surfaces that opposes the pull of gravity on that surface.

LIFT/DRAG RATIO
A measure of the effectiveness of an airfoil for a given angle-of-attack. For example, a ten to one L/D indicates (10) units of lift for 1 unit of drag. *See also: 1) Aerodynamic Efficiency, 2) Drag.*

LIFT-FAN ENGINE
A Turbofan engine configured with exhaust nozzles that can be swiveled downward to provide upward thrust from a propelling jet of gas. This High Bypass Ratio engine is used in military, fixed-wing vertical takeoff aircraft such as the Harrier Jump-Jet and its derivative, the McDonnell AV-8. The lift-fan is also seen in other transportation devices which ride on a cushion of air produced by the turbine engine. Also referred to as a Lift-Jet. *See also: 1) Plenum Chamber Burning, 2) Remote Augmented Lift System.*

LIFT-JET
Locate under Lift-Fan.

LIFT-THRUST RATIO (L/T)
A physics term. The ratio of lift to thrust of a vectored thrust Lift-Fan engine.

LIGHT-OFF (engine)
The point at which combustion occurs during the engine starting cycle as indicated by a rise in Exhaust Gas Temperature. Also called Light-Up.

LIGHT-OFF SPEED
The compressor speed at which Light-Off will typically occur for a given engine.

LIGHT-OFF TIME (engine)
The time between introduction of fuel to the engine and initial combustion. There is a prescribed limit for each engine type, approximately 20 seconds, after which an unsafe condition exists and the engine start cycle must be aborted.

LIGHT-UP SPEED
Another name for Light-Off Speed.

LIMIT LOAD (engine)
The maximum cyclic and heat loads expected to occur during the life of an engine and a load point where structures will not deform. *See also: Ultimate Load.*

LIMIT LOAD (of materials)
The load stress which the primary structure of a part can withstand without permanent deformation. See also: Material Strength.

LINE MAINTENANCE (engine)
Refers to inspections and repairs to engines mounted on the aircraft.

LINE REPLACEABLE UNIT (LRU)
Any separate unit on an engine that can be removed and replaced while the engine is installed in the aircraft. LRU's generally refer to engine accessories, component parts and modules. *See also: Shop Replaceable Unit.*

LIP SEAL (pump)
A type of rubber seal used on the drive shaft of fluid carrying pumps. This seal fits around the shaft with a squeezing action to prevent fluid leakage. Variations of this seal are called Garlock™ seals and garter-spring seals. *See also: Seal (accessory drive gearbox).*

LIQUID HYDROGEN FUEL
Locate under Hydrogen Fuel.

LIQUID TO LIQUID COOLER (oil system)
Another name for Fuel-Oil Cooler.

LIQUEFIED NATURAL GAS (LNG)
A largely cryogenic methane fuel proposed for use in high speed jet aircraft of the future. LNG offers higher heat release and cleaner burning characteristics than present hydrocarbon fuels. *See also: Hydrogen Fuel.*

LOAD COMPRESSOR (GTCP)
A compressor in a Gas Turbine Compressor-Power Unit which supplies air to aircraft pneumatic systems. A separate unit from the engine compressor which supplies air to the combustor. The Load Compressor is connected directly to the engine compressor, or is driven by a separate turbine as seen in the illustration. *See also: Gas Turbine Compressor-Power Unit.*

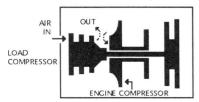

LOAD CYCLE
British term for Engine Cycle.

LOAD LIMIT (engine)
Locate under Limit Load.

LOADS (engine)
The compression and tension stresses placed on turbine parts by centrifugal loading, gyroscopic motion and heat loading. The two categories of loads are Limit Loads and Ultimate loads. *See also: 1) Limit Loads, 2) Ultimate Loads.*

LOADCELL (engine test cell)
A Test Cell Component. An electronic strain gauge type device in a bridge electrical circuit used to measure engine thrust. The engine is positioned to exert a mechanical pushing force on the load cell and the strain gauge electrical circuit measures the resistance change imposed by the load. The load cell then transmits a signal to a thrust indicator on the test cell control panel.

LOADCELL MASTER (engine test cell)
A precision test device used to calibrate operating Loadcells in engine test facilities.

LOADING
Locate under, 1) Blade Loading, 2) Disc Loading, 3) Stage Loading.

LOBE-TYPE NOISE SUPPRESSOR
An old style noise suppressor unit used on early Turbojet powered airliners. The jet exhaust was configured with several separate nozzle openings that promoted mixing of the free airstream with the engine exhaust. The jet efflux, when broken up into several smaller airstreams, reduces the large eddies that produce the loudest, low frequency noises. The mixing of gases in this manner reduces the shearing noise effect which takes place when hot exhaust air meets cold ambient air. *See also: Noise Suppressor.*

LOCAL SPEED OF SOUND
Locate under Cs Symbol.

LOCKTAB
Locate under Tablock.

LOCKWASHER
A small expendable part. Any of the various devices used to prevent loosening of fasteners such as nuts, bolts and screws. Typical of these are split lockwashers, external tooth lockwashers, and Internal tooth lockwashers.

LOCKWIRING
A maintenance procedure to secure fasteners of various types, with specially annealed, fatigue resistant metal wire. Lockwiring prevents loosening of fasteners such as nuts, bolts, etc., which results from vibration induced by engine operation. Also called Safety-Wiring.

LOGIC TREE
Another name for Fault Tree.

LONG DUCTED TURBOFAN
Another name for Fully-Ducted Turbofan.

LONG RANGE CRUISE SPEED (LRC)
For turbine powered aircraft, an industry agreed upon value described as "the aircraft speed that will provide a minimum of 99% of an aircraft's maximum fuel mileage". LRC is slightly faster than Maximum Range Cruise (MRC). *See also: 1) Maximum Range Cruise speed, 2) Flight Management System.*

LOOP CUSHION CLAMP
A small fastener on an engine. A one piece circular metal clamp retained by one bolt and with a rubber or Teflon™ liner to prevent chafing. Loop clamps secure air lines, fuel lines, oil lines, and electrical wiring to exterior locations on engines. Loop Clamps also have various aircraft applications. When two clamps are used together back to back or end to end, the configuration is often called a butterfly clamp. Often referred to as Adel Clamps™. *See also: Split-Clip.*

LORIN ENGINE
A Jet Engine design named after its French inventor in the early 1900's. Today it is called a Ramjet or Aero-Thermodynamic Duct (athodyd). This engine is not considered to be a gas turbine because it has no turbine wheel driven compressor.

LOW ASPECT RATIO BLADING
An advanced technology design of fan, compressor, and turbine blades that feature shorter lengths and wider chords than traditional blades. These blades also have higher Reynold's Number characteristics and are designed for use with rotors of high rotational speeds. Low Aspect Ratio Blades are stated to have improved weight distribution and resultant fuel savings. *See also: Wide-Chord fan blades.*

LOW BYPASS TURBOFAN
A general classification of engines having less than a 2:1 bypass ratio. For example, a 1:1 bypass ratio refers to the fact that the same amount of mass airflow passes out of the fan exhaust as passes out of the core engine. *See also: Turbofan.*

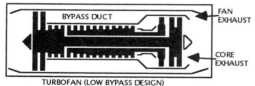
TURBOFAN (LOW BYPASS DESIGN)

LOW COMPRESSOR TURBINE
Another name for Low Pressure Turbine (LPT).

LOW CYCLE COUNTER
Locate under Low Cycle Fatigue.

LOW CYCLE FATIGUE (LCF)
Fatigue (cracks, fractures etc.) occurring in rotating airfoils from direct centrifugal and thermal stresses. Stresses result with each load change, as the engine is accelerated from a low RPM to a high RPM. The time frame is generally in terms of several seconds per cycle. Many aircraft employ automatic counters which record the number of power changes. The data is then used to plot a trend analysis of part expected service life. *See also: 1) Event Counter, 2) Fatigue Failure, 3) Hobbs Meter, 4) Material Stress.*

LOW CYCLE FATIGUE-TIME (or life)
LCF Time refers to the manufacturer's authorized total operating hour limit or cycle limit on selected turbine engine accessories, components, component parts, or modules. When the limit is reached, the unit is considered unsafe to operate and must be removed from service. Similar to the Life-Limited Part concept.

LOW FREQUENCY EDDY CURRENT TESTER (LFEC)
Locate under Eddy Current Tester.

LOW OIL LEVEL WARNING LIGHT
A cockpit warning light. This light warns the flight crew that an unsafe condition exists when oil level drops to a predetermined low quantity, generally 20 to 30 percent of tank capacity.

LOW OIL PRESSURE SWITCH (APU)
A switch that will shut down the engine when oil pressure drops to an unsafe level. Also called LOP switch *See also: Oil Pressure Switch.*

LOW PRESSURE AIR START VALVE
A small accessory for engines. A valve that directs a low pressure air source from the engine compressor to the starting cross-bleed manifold in the aircraft. *See also: High Pressure Air Start Valve.*

LOW PRESSURE COMPRESSOR (LPC)
A major engine component. The front compression section of a dual or triple spool axial-flow compressor. On many Turbofan engines, the LPC consists of only a fan rotor, on others the fan may be followed by one or more compression stages. The LPC may also be the front compression section of a dual, combination axial-centrifugal compressor. Other references to the low pressure compressor include: 1) LP Compressor, 2) N1 Compressor, 3) Low Speed Compressor. *See also: High Pressure Compressor.*

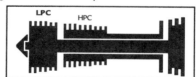
TURBOJET (DUAL SPOOL DESIGN)

LOW PRESSURE ROTOR SYSTEM
An assembly consisting of a Low Pressure Compressor connected to its Low Pressure Turbine.

LOW PRESSURE TURBINE (LPT)
A major engine component. The rear turbine section of a dual or triple spool axial-flow turbine engine. The LPT mechanically attaches to and drives the low pressure compressor. It is named as such because it receives the lowest gas pressure of all the turbines. LPT functions include: 1) drive the fan rotor on the triple-spool Turbofan engine, 2) drive the fan rotor on a dual-spool Turbofan engine, 3) drive the front compressor on a dual-spool Turbojet, Turboprop or Turboshaft engine. Also called N1 Turbine or Low Speed Turbine. *See also: High Pressure Turbine.*

TURBOJET (DUAL SPOOL DESIGN)

LOW SPEED EXTERNAL GEARBOX
An auxiliary gearbox driven by the Low Pressure (low speed) Compressor. On many engines, this gearbox is located on the front of the engine inlet case.

LOW SPEED ROTOR
Refers to: 1) The Low Pressure rotor of a Turbofan or Turbojet engine because its wider diameter relegates it to a lower speed than the High Pressure rotor system. 2) The Free Power Turbine of a Turboprop or Turboshaft engine because its wider diameter than the Gas Producer Turbine relegates it to a slower speed. *See also: 1) High Speed Rotor, 2) Tip Speed.*

LOW SPOOL SPEED
Refers to speed of the Low Pressure Compressor.

LOW VOLTAGE IGNITER PLUG
Locate under Igniter Plug (low voltage).

LOW VOLTAGE IGNITION SYSTEM
Locate under Ignition System (low voltage).

LP COMPRESSOR (LPC)
Locate under Low Pressure Compressor.

LP Fuel
Symbol for liquid propellant. *See also: 1) JP (fuel). 2) RP (fuel).*

LP SPEED SENSOR
Locate under Fan Speed Sensor.

LP THRUST BEARING
Refers to a main bearing of the ball-type which prevents axial movement of the Low Compressor. Main ball bearings absorb thrust (axial) loads when high pressure forces at the rear exert a push forward on the rotor. *See also: Bearings.*

LP TURBINE (LPT)
Locate under Low Pressure Turbine.

LRU
Abbreviation for Line Replaceable Unit.

LUBRICATION
A means of reducing rolling and sliding friction. In a turbine engine, lubricants provide a film of synthetic oil molecules between moving parts to fill surface irregularities and to prevent metal to metal contact. Lubricants also cool, cushion and remove contaminants. *See also: Synthetic Lubricanting Oil.*

LUBRICATION SYSTEM
The Gas Turbine Engine typically utilizes a low viscosity, low volume, self-contained (closed loop), pressurized system to lubricate its bearings, seals shafts and gears. The oil reservoir can be aircraft mounted, but is generally mounted on or within the engine. Also called Oil System. *See also: 1) Dry Sump (lubrication system), 2) Several listings under Oil, 3) Wet Sump (lubrication system).*

LUBRICATION SYSTEM (categories)
The two categories of turbine engine lubrication systems are as follows:
 1) Pressure Relief Valve Oil System - A system which contains a bypass relief valve in its Pressure Subsystem to establish a constant supply pressure to the oil jets over varying engine speeds. In this design, when the Vent Subsystem pressure increases at high RPM, the oil flow reduces. If this creates a lubrication problem, the relief valve can be arranged with vent pressure to assist the bypass spring so that less oil bypass and higher pumping pressure occurs at high engine speeds.
 2) Full Flow Oil System - A system which contains no regulating relief valve. Used in engines with normally high Vent Subsystem pressures. When the engine speeds up, more oil is delivered to the oil jets to combat the increasing vent pressure and thus maintain correct flow. This system does, however, contain a Cold Start Relief Valve to prevent excessive pressure buildup during starting when the oil viscosity is high. See also sub-categories such as: 1) Cold Tank Lubrication System, 2) Dry Sump Lubrication System, 3) Hot Tank Lubrication System, 4) Total Loss Lubrication System, 5) Wet Sump Lubrication System.

LUBRICATION SYSTEM (sub-systems)
The engine lubrication system contains three subsystems as follows:
1) Oil Pressure Subsystem - supplies oil to meet the needs of the engine.
2) Oil Scavenge Subsystem - returns oil to the storage tank.
3) Oil Vent Subsystem - removes excess air back to the atmosphere.
See also: above subsystem titles for more detailed explanations.

LUBRICATION SYSTEM (wet sump)
Locate under Wet Sump (lubrication system).

LUBRICATION SYSTEM FILTERS
Locate under Oil Filters.

LUBRICATION SYSTEM RELIEF VALVE
Locate under Oil System Relief Valve.

LUG
A small metal protrusion used for alignment of parts, attachment of parts, or as a point to attach lockwire.

M

MACH NUMBER
The mathematical ratio of "actual air velocity" to "acoustic velocity" (local speed of sound). Mathematically expressed as:

$$Mn = V \div Cs$$

Where: V = Air Velocity, Cs = Acoustic Velocity. At International Standard Day (ISD) condition, "Cs" is 1,120 ft./sec. For example, when an aircraft is traveling at 2,240 ft./sec. and ambient temperature is at an ISD condition, the aircraft is said to be flying at twice the speed of sound (Mach-2). *See also: 1) Flight Regimes, 2) Speed of Sound.*

MACHINE KEY
A small metal piece that fits in a slot formed between a shaft and a gear. In this manner the key prevents the gear from rotating on the shaft. Another name for Woodruf Key.

MACHINED RING COMBUSTOR
Locate under Combustion Liner (machined ring).

MACRO-ETCH INSPECTION
An etching inspection used to detect surface cracks in turbine engine parts, especially titanium.

MAGNAFLUX™ INSPECTION
A crack detection process employing a ferrous grit, either dry or wet, on a magnetized test piece. The magnetic field in the piece causes the grit to align with a discontinuity to show visible evidence of a fault. *See also: Non-Destructive Inspection.*

MAGNESIUM ZIRCONATE
A thermal barrier ceramic coating, used primarily on combustor section parts. Similar to Brunsbond Coating. *See also: Thermal Barrier Coatings.*

MAGNESYN TRANSMITTER
An A. C. electrical transmitter which powers a cockpit gauge. Its mechanism functions by use of a rotating permanent magnet in a magnetic field. When energized, the field coils transmit to the cockpit gauge. A fuel flowmeter indicator is one such gauge. *See also: 1) Autosyn, 2) Selsyn.*

MAGNETIC CHIP DETECTOR
Locate under Chip Detector.

MAGNETIC DRAIN PLUG (lubrication system)
A magnetic plug placed in a lower section of a lubrication system, such as the sump of an Accessory Drive Gearbox. This plug serves as a drain point and also as a metal Chip Detector. *See also: 1) Chip Detector, 2) Swarf.*

MAGNETIC FIELD DETECTOR
Another name for Bearing Field Detector.

MAGNETIC SEAL (Accessory Drive Gearbox)
A small shaft seal arrangement with a magnetic carbon seal housing to hold a seal onto a metal rub ring. Similar to a Carbon Seal (face type) in a main bearing location. Designed to prevent oil from leaking past the shaft and out of the gearbox.

MAIN BEARING
The Bearings located along the compressor and turbine shafts which absorb the axial and radial loading. *See also: Bearings.*

MAIN ENGINE CONTROL (MEC)
Another name for Main Fuel Control.

MAIN ENGINE DIFFUSER
Refers to the Diffuser located at the engine mid-section. See also: Diffuser.

MAIN FRAME
Another name for Diffuser Case.

MAIN GEARBOX
Another name for Accessory Drive Gearbox.

MAIN FUEL (duplex fuel manifold)
Another name for Secondary Fuel.

MAIN FUEL CONTROL (MFC)
An engine accessory. The central fuel controlling device of a turbine engine. Locate under Fuel Control Unit (main).

MAIN FUEL SYSTEM (engine)
The fuel distribution system to the combustor that is used for all normal engine operating conditions, from start-up, to idle, to takeoff power. *See also: 1) Emergency Fuel System, 2) Primer Fuel System.*

MAIN GEARBOX
Locate under Accessory Drive Gearbox.

MAIN METERING VALVE (fuel control)
A valve with a variable orifice which schedules fuel by weight to the engine combustor. This valve functions in conjunction with a Differential Pressure Regulating Valve to schedule fuel in response to movement of the cockpit power lever. Also called Contour Valve. *See also: 1) Differential Pressure Regulating Valve, 2) Weight of Fuel.*

MAINTENANCE OPERATION CENTER (MOC)
The control center for aircraft and engine maintenance in both major repair facilities and airline maintenance facilities.

MAINTENANCE SIGNIFICANT ITEM (MSI)
A controlled item requiring special supply procedures. In reference to engine parts, an item classified on the basis of safety, initial cost, removal rate, and maintenance cost.

MAJOR ENGINE SECTIONS
Locate under Engine Major Sections.

MAJOR ALTERATION (engine)
In FAA terms, an Alteration not listed in the aircraft engine specifications, and one that might appreciably affect balance, structural strength, performance, weight, or other qualities that ensure engine airworthiness.

MAJOR OVERHAUL (engine)
A complete disassembly, cleaning, inspection, repair as necessary, and re-assembly of an engine. This type of overhaul also includes replacing all Life-Limited parts, Life-Cycle Time parts, Modules which are due for time change, and the compliance with all FAA Airworthiness Directives (commonly called AD Notes).

MAJOR REPAIR (engine)
In FAA terms, a repair which, if improperly accomplished, might appreciably affect balance, structural strength, performance, weight, or other qualities which ensure engine airworthiness. *See also: Heavy Maintenance.*

MALFUNCTION
A general term used to denote the occurrence of failure of a product to give satisfactory performance. It need not constitute a failure if readjustment of operator's controls can restore an acceptable operating condition.

MANIFOLD (fuel)
Locate under Fuel Manifold.

MANIFOLD BLEED PRESSURE
The pressure in an aircraft bleed air system. Air that is supplied by an Auxiliary Power Unit or by Cross Bleed air valves on operating engines. This air has many uses, most important of which is to provide pressurized air to Pneumatic Starters and Air Conditioning Units. Pressure in this manifold is received at

APU compressor discharge pressure of approximately 50 psig. Also Called Customer Bleed Air. *See also: Pneumatic Starter.*

MANUFACTURER'S SERVICE BULLETIN
Locate under Service Bulletin.

MARGIN (instruments)
The difference between the observed indication and the instrument Red Line Value.

MARKING PENCIL (hot section)
A special marking device that does not contain carbon or zinc, such as a felt tip applicator or Colorbrite™ Pencil. Common lead pencils or grease pencils are not used because they may contain carbon or zinc, which are readily absorbed into heated metals causing brittleness and a serious stress cracking hazard.

MARMAN CLAMP™
A type of V-Band Coupling.

MASS
The quantity of matter in a body. Near the earth the force of gravity creates a condition where mass is equal to weight. In outer space a body can be weightless, but it will still retain its mass. The standard units of Mass are as follows: U.S. System (lbs.), Metric System (grams), and for some Aeronautical Computations (Slugs).

MASS AIRFLOW (Ms)
The amount of air flowing through an engine at a particular power setting. Locate specific values in section entitled "Gas Turbine Engines for Aircraft.

MASTER LOADCELL
Locate under Loadcell Master.

MATCHMARK (engine)
A mark imprinted into two or more adjoining parts for purposes of realignment when the parts are reassembled. Another name for zero ("0") balance mark on rotating parts.

MATERIAL STRENGTH (terms)
Locate under 1) Endurance Limit, 2) Limit Load, 3) Ultimate Strength, 4) Yield Strength. *See also Fatigue Failure.*

MAXIMUM CLIMB POWER
An Engine Power Rating as set by FAA Standards. A power level higher than Cruise Power Rating to be used for normal climb to cruise altitude. No time limit is imposed on this rating. *See also: Engine Power Ratings.*

MAXIMUM CONTINGENCY POWER (MCP)
A Military Power Rating above Maximum Continuous Power. Operation is restricted to a limited time period and is mainly used in a one-engine-out condition. *See also: Engine Power Ratings.*

MAXIMUM CONTINUOUS POWER
An Engine Power Rating as set by FAA Standards. A power level higher than Cruise Power Rating to be used during emergency situations at the discretion of the pilot. This rating is often the same as Maximum Cruise Power Rating. No time limit is imposed on this rating. *See also: Engine Power Ratings.*

MAXIMUM CRUISE POWER
Locate under Cruise Power Setting.

MAXIMUM POWER (engine)
A prescribed engine thrust rating at which a military engine must perform during take-off in afterburner. In civil aviation it is called "take-off wet thrust". *See also: Military Power.*

MAXIMUM RANGE CRUISE (MRC) SPEED
The speed at which an aircraft must fly in order to obtain its maximum range. *See also: 1) Flight Management System, 2) Long Range Cruise Speed.*

MCD (engine)
Abbreviation for Magnetic Chip Detector. Locate under Chip Detector.

MEAN GAS TEMPERATURE (MGT)
A cockpit indication similar to Turbine Outlet Temperature (TOT).

MEAN TIME BETWEEN FAILURE (MTBF)
The actual average time between component failures, a factor in Life-Cycle Cost analysis.

MEC (engine)
Abbreviation for Main Engine Control. Locate under Fuel Control Unit.

MECCANICHE RINALDO COMPANY (Italy)
Manufacturer of Gas Turbine Engines. Locate in section entitled "Gas Turbine Engines for Aircraft".

MECHANICAL BLOCKAGE (thrust reverser)
Often an aircraft manufacturer's supplied component. The two most common types are the Clamshell design and the Target-Bucket design. A Clamshell reverser receives its name because of its two rounded and ribbed shape sections. The Target-Bucket design uses two or more rectangular reverser panels. Both types

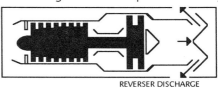

REVERSER DISCHARGE

of reversers are seen in core (hot) exhausts and mixed exhausts of Turbofan engines. *See also: Thrust Reverser.*

MEDIUM BYPASS TURBOFAN (engine)
A general category of Turbofan engines with fan bypass ratios of 2:1 to 3:1. The fan bypass ratio refers to the quantity of mass airflow through the fan section as compared to the mass airflow passing through the core engine. Also called Moderate Bypass Turbofan. *See also: 1) High Bypass Turbofan, 2) Low Bypass Turbofan.*

MERL 80™
A recent development in turbine engine super-alloys produced by a powder metallurgy process. This material is used almost exclusively in the hot section. Another metal in this series is Merl 200. *See also: Powder Metallurgy.*

MESH RATING (filter screens)
A U.S. Sieve Number system which relates to engine fuel filters ratings. For example, a filter with a U.S. Sieve No. 200 is stated to have 200 meshes per linear inch. The mesh opening is 0.0029 inches square and the filter will trap all particles larger than 0.0029 inches in diameter.

MESSERSCHMITT ME-262
Germany's first production fighter aircraft, in service in early 1944. Some fighter-bomber versions were also produced. The aircraft was given the name "Swallow". *See also: Junkers Jumo.*

METAL BRIDGING DRAIN PLUG
Another name for Chip Detector, it could be of the Continuity Type or of the Indicating Type.

METAL FATIGUE
Locate under Fatigue.

METAL FORMING
Locate under 1) Casting, 2) Directional Solidification, 3) Forging, 4) Hot Isostatic Pressing, 5) Powder Metallurgy, 6) Rapid Solidification Rate.

METAL INERT GAS WELDING (MIG)
A common welding process used for Gas Turbine Engine construction. A welding process which uses an inert gas, such as argon, to prevent oxidation, and uses an electrode which also serves as the filler rod. A production welding method faster than hand TIG welding.

METAL-MATRIX COMPOSITE (MMC)
A family of metal materials under development for the Gas Turbine Engine. MMC is a high strength and low weight metal matrix, combined with fibers arranged in the direction of the anticipated stress. This process is intended for use with such materials as titanium, aluminum and nickel. One such material is constructed of layers of fabric containing silicon carbide filaments and titanium aluminide foil and is said to have only half the weight of present day titanium alloy.

METALIZING
Locate under Engine Inspection and Distress Terms in Appendix.

METCO 450™
A tungsten carbide, corrosion resistant coating applied to high pressure compressor blades and stator vanes.

METERED FUEL
Fuel that leaves the fuel control in a precisely metered condition on its way to the engine combustor. Fuel is metered in relation to mass airflow through the engine, which under ideal conditions represents a 15 : 1 air/fuel mixture.

METERING HEAD REGULATOR VALVE (fuel control)
Another name for a Differential Pressure Regulating Valve in a fuel control unit.

MICE
Jargon for Tailpipe Insert. Called mice because of their black sooty appearance.

MICROINCH
One millionth of an inch. A common measurement in surface flatness checks for such items as carbon seal contact surfaces and bearings.

MICRON
A Metric Unit of measurement equal to one millionth of a meter and in the U.S. System of measurement equal to 1/25,000 inch.

MICRONIC RATING (metal filter screens)
A rating system that defines the mesh openings in microns. For example, a 20 micron filter will have an open flow area of 20 microns in diameter in its mesh and will trap all particles larger than 20 microns. The micron rating is common to most turbine engine fuel and lubrication system filters. *See also: 1) Filter Ratings 2) Micron, 3) Mesh Rating.*

MICRONIC RATING (paper filter)
A rating system that defines the nominal filtration capability in microns. For example, a 20 micron filter will trap particles over 20 microns in diameter. The micron rating applies to most fuel and lubrication system filters. *See also: 1) Filter Ratings, 2) Micron, 3) Mesh Rating.*

MICROTURBO COMPANY (France)
Manufacturer of Gas Turbine Engines. Locate in Section entitled "Gas Turbines for Aircraft".

MID-FRAME (engine)
Another name for Diffuser Case.

MID-SPAN DAMPER (fan blade)
Another name for Mid-Span Shroud.

MID-SPAN SHROUD (fan blade)
A supporting lug located at mid-point along a fan blade. The span shrouds contact adjacent shrouds,

forming a circular support ring which reduces the stress concentrations at the base of the blades. See also: Span-Shroud.

MIDDLE DISTILLATE (fuel)
A faction of petroleum base fuels in the middle range of refinery products, neither light not heavy. For example, Jet-A, Diesel Fuel, Kerosene, and Stove Oil are middle distillates. *See also: Fuel Types-Jet.*

MIG
Abbreviation for Metal Inert Gas welding.

MIL SPEC
Refers to Military Specification. A specification to describe a standard of acceptance for a product. The aviation industry widely uses products manufactured under Military Specifications. For example, Mil-L-23699 Synthetic Turbine Lubricating Oil is the most commonly used oil in both civil and military aviation.

MILITARY POWER RATINGS (comparison to civil engines ratings)
Maximum Rated Power - FAA Take-Off Wet Power Rating.
Military Rated Power - FAA Take-Off Dry Rating.
Military Normal Rated Power - FAA Maximum Continuous Power Rating.
Military Cruise Rated Power - FAA Cruise Power Rating.

MILITARY RATED THRUST
Locate under Military Power Ratings.

MILITARY SPECIFICATION   Locate under Mil Spec.

MILLIPORE FILTER
A type of fuel filter which indicates the level of solid contaminants and water present in fuel when inspected under a special light.

MINERAL BASE OIL
Another name for Petroleum Base Oil.

MINIMUM EQUIPMENT LIST (MEL)
An FAA approved listing of items in aircraft which are not considered critical to flight safety. Some of these items include engine indicating and warning devices. Should items on this list become inoperative, the aircraft is allowed to fly in normal service.

MINIMUM FLOW STOP (fuel system)
Component part of a fuel control. Refers to a fuel control design feature that prevents the power lever from shutting off fuel to the combustor. When the throttle is at its full rearward position in the cockpit, the fuel control main metering valve is against its Idle Flow Stop and the engine is operating at idle speed. In the event the Idle Stop is maladjusted, the Minimum Flow Stop will prevent engine shutdown. A separate shut-off valve is present to terminate fuel flow.

MINIMUM PRESSURIZING AND SHUT-OFF VALVE
A small engine accessory. A check valve type device located downstream of the fuel control unit which prevents fuel flow past that point until a predetermined fuel pressure build-up in the system is achieved. The presence of this valve accomplishes the following: 1) It tends to keep working pressure at correct values for operation of fuel control servo mechanisms, 2) It prevents flow to the fuel nozzles until sufficient pressure is available to ensure a good spray pattern, 3) It closes at low flow condition during engine shut-down to lock in the system working pressure and prevent formation of air pockets in the system. See also: Pressurizing Valve.

MINOR ALTERATION (engine)
In FAA terms, any alteration to an engine that is not considered a major alteration.

MINOR REPAIR (engine)
In FAA terms, any repair to an engine that is not considered a major repair.

MIST OIL JET
Another name for Fogger Oil Jet.

MITSUBISHI COMPANY (Japan)
Manufacturer of Gas Turbine Engines. Locate in section entitled "Gas Turbine Engines for Aircraft".

MIXED AFTERBURNER
An afterburner in which fan bypass air and core engine hot gases mix prior to combustion in the afterburner.

MIXED COMPRESSION INLET
Another name for Convergent-Divergent Inlet. A flight inlet utilized on supersonic aircraft in which part of the inlet compression is a result of shock wave formation and part is from its convergent-divergent shape. See also: Convergent-Divergent Inlet Duct.

MIXED EXHAUST DUCT
An engine component. An exhaust duct on a fully ducted Turbofan engine. A design that allows the primary (hot-core exhaust) and the secondary (cold-fan exhaust) to combine in a common exhaust duct (tailpipe) prior to leaving the engine. Also called Integrated Exhaust Duct. See also: Exhaust Duct.

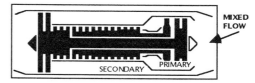

MIXED-FLOW COMPRESSOR
A major engine component. An early design not in current use. This compressor discharges its air at an angle rearward, somewhere between an axial and a radial direction. Mixed Flow is again undergoing research and development, in that new manufacturing techniques now make it a feasible design. The Fairchild J-44 Turbojet of the late 1940's contained a compressor of the mixed flow type. See also: Integrated Blade and Rotor.

MIXER (Exhaust Duct)
Locate under Exhaust Duct (configurations).

MIXER FRAME
A main engine frame located between the Engine Core and its rear mounted Propulsor Unit on a UDF™ Engine.

MMC
Abbreviation for Metal Matrix Composites.

Mn
Symbol for Mach Number. The letter "M" alone is also used. Locate under Mach Number.

MOC
Abbreviation for Maintenance Operation Center.

MODERATE BYPASS TURBOFAN
Locate under Medium Bypass Turbofan.

MODERN TECHNOLOGY DEMONSTRATOR ENGINE (MTDE)
A preproduction turbine engine. An industry concept of constructing a demonstration engine upon which a sales promotion program can be built.

MODULAR ENGINE
Those engines consisting of several independent assemblies called modules, which by design can be removed/replaced without major disassembly of the engine or other modules. For example, the compressor, combustion, turbine, afterburner, gearbox, torquemeter, or combination thereof.

MODULAR MAINTENANCE
A maintenance concept in which large sections of an engine, called modules, can be pre-assembled and stocked as spare units. Modules can often be replaced on-the-wing to preclude engine removal. Most modern engines are of modular construction because of the savings in flow-time and flexibility of maintenance it provides. *See also: Module (engine).*

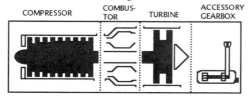

MODULATED BLEED VALVE
A component part of an Active Clearance Control System. *See also: Active Clearance Control.*

MODULATED CLEARANCE CONTROL
Another name for Active Clearance Control.

MODULE (engine)
A large engine section which can be removed and/or replaced as a unit. For example, a Power Turbine Module on a Turboshaft engine can be replaced with a similar module with only limited testing required. Before modular construction design was developed, the Power Turbine would have been rebuilt and reinstalled, followed by extensive engine operational checks. The four typical modules of a Turbofan engine are as follows: Fan Module, Core Engine Module, Hot Section Module, Accessory Section Module. Modules are generally tracked independently of engine operating time using module operating hours and/or cycles. Also called Engine Maintenance Unit. *See also: 1) Modular Maintenance, 2) Sub-Module.*

MOMENT-WEIGHT BALANCING (compressor & turbine rotors)
A method of static balancing larger disk and blade assemblies by installing blades in pairs. In this way, two blades of the same moment-weight are always opposite each other. When the assembly is completed, the unit will require little or no additional static balancing. *See also: Balance Checks.*

MOMENT-WEIGHT CODE (compressor & turbine blades)
An identification number or letter system indicating a moment arm measurement that includes both the blade's mass weight and its center of gravity. Moment-Weight is the product of the blade's mass weight and the distance between the center of gravity and its base. In the U.S. System, Moment Weight is given in ounce-inches, in the Metric System, gram-millimeters.

MOMENTUM THRUST
Another name for Reactive Thrust.

MONITORING SYSTEM (engine)
Locate under Engine Monitoring System.

MONO-CRYSTAL TURBINE BLADE
Another name for Single Crystal Turbine Blade.

MONTI CARLO RISK ANALYSIS
Locate under Risk Analysis.

MOTIVE FLOW PUMP (fuel)
Another name for Jet Pump.

MOTORING (engine)
Another name for Motor-over.

MOTORLESS MASSFLOW TRANSMITTER (fuel)
Locate under Fuelflow Transmitter.

MOTOR-OVER (dry)
A process of rotating a turbine engine at low speed with the starter but without introducing fuel or ignition. This procedure is typically used during the following maintenance procedures: 1) Clearing the engine of vapors, 2) Field Cleaning, 3) Preoiling, 4) Troubleshooting.

MOTOR-OVER (wet)
The process of rotating an engine at low speed with the starter and with the fuel lever open but no ignition. This procedure is used primarily for engine troubleshooting and for bleeding air from the fuel system.

MOTOR-OVER WASH (engine)
Another name for Field Cleaning.

MOTOREN-UND TURBINEN-UNION (West Germany)
Manufacturer of Gas Turbine Engines. Locate in section entitled "Gas Turbine Engines for Aircraft".

MOUSE MILK
A type of high temperature anti-seize compound used on threaded fasteners in hot sections of turbine engines.

MOVABLE WEDGE (aircraft inlet)
Another name for Vari-Ramp. *See also: Variable Geometry Inlet.*

MOVABLE SPIKE (aircraft inlet)
Another name for Inlet Spike.

MOVABLE WEDGE (inlet)
Another name for Vari-Ramp.

MRT
Abbreviation for Military Rated Thrust.

MS
Symbol for Military Standard. Locate under Aircraft Standards.

Ms
Symbol for Mass Airflow.

MSI
Abbreviation for Maintenance Significant Item.

MTBF
Abbreviation for Mean Time Between Failure.

MTBO
Abbreviation for Mean Time Between Overhaul.

MTBR
Abbreviation for Mean Time Between Removal.

MTDE
Abbreviation for Modern Technology Demonstration

Engine.

MULTI-NATIONAL GAS TURBINE ENGINE
A Gas Turbine Engine that is produced by multiple companies located in different countries. This is currently a method of sharing technologies and development costs and with a predictable sales market in the countries concerned. Locate specific engines under Multi-National listing in section entitled "Gas Turbine Engines for Aircraft".

MULTI-TUBE NOISE SUPPRESSOR
Locate under Multiple-Tube Noise Suppressor.

MULTI-VISCOSITY OIL
An oil that is designed to have good pour-ability at low temperatures and good viscosity retention at operating temperatures. For example: An SAE 10W-50 oil has a viscosity similar to SAE-10 when at 0° F. and a viscosity similar to SAE-50 when at 210° F. "W" indicates winterized (includes additives) for good low temperature pour point. Turbine lubricating oils are not identified by SAE numbers. They are however, synthesized to contain multi-viscosity qualities of approximately SAE 5W-20.

MULTIPLEXER (engine)
Locate under Propulsion Multiplexer.

MULTIPLE-CAN COMBUSTOR
A combustor design with multiple individual combustion chambers, each with a liner and outer case. Its Ring and Tube assembly is fitted coaxially around the turbine shaft housing. The multiple-combustor design is rarely used today. It is more common today to see single combustors of the can-type on small flight engines and auxiliary power units. Also called 1) Can-Type Combustor, 2) Tubular Combustor. *See also: Ring and Tube Assembly.*

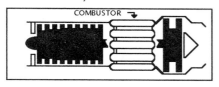

MULTIPLE-PASS COOLING (Turbine)
Turbine blade and vane cooling in which cooling air simultaneously passes through multiple separate passages. Typically one passage supplies air to the leading edge slots, one for the center section, and one for the trailing edge slots. *See also: Turbine Cooling.*

MULTIPLE-TUBE NOISE SUPPRESSOR
An older type noise suppressor that takes the place of a conventional tailpipe. It is a series of exhaust tubes manifolded together to break up the normally large eddies in the jet efflux that produce the loudest low frequency noises. *See also: Noise Suppressor Unit.*

MULTIPLE WAFER OIL FILTER
An engine accessory. A filter unit containing a filtering element made up of several pancake shaped wafers of stainless steel wire mesh. *See also: Oil Filter (cleanable)*

MURPHY PROOF INSTALLATION
Refers to a one-way installation, not easily misaligned, maladjusted, or otherwise improperly installed.

MUSCLE PRESSURE (air)
Air pressure employed as a motive force to actuate a mechanism.

N

N (symbols)
Symbol for rotational speed of the Compressor and Turbine Rotor system, in percent (%) RPM. The "N" symbol may also at times represent actual RPM. Cockpit gauges, engine operational charts, and graphs all use the "N" symbol to depict speeds of a specified rotor system. *See also: Appendix for complete listing of symbols.*

N1 COCKPIT INDICATOR
A gauge which indicates speed of the Low Pressure Rotor. This gauge is often used to indicate engine thrust output. When this is the case, thrust is stated to be linear with N1 Speed within ± 1 percent. The N1 Indicator itself is accurate within approximately ± 0.2 percent.

N1 COMPRESSOR
A major engine component. Depending on the type of engine, the following N1 Compressor designations apply: 1) The low pressure compressor of a dual-spool axial flow engine, 2) The compressor of a single-spool axial flow engine, 3) The compressor of a centrifugal flow engine.

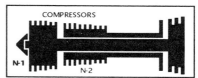
TURBOJET (DUAL COMPRESSOR DESIGN)

N1 POWER EXTRACTION
Refers to power taken from the Low Pressure Compressor front shaft to drive a fan through a reduction gearbox.

N1 Speed (%)
1) Rotational speed of as a single compressor engine. 2) Speed of a Low Pressure Rotor System (N1) in a dual-spool Turbojet or Turbofan engine. Sometimes the letter "L" is used for (low) instead of the number 1. For example, NL. *See also: Fan Speed.*

N1 TURBINE
A major engine component. Depending on the type of engine the N1 Turbine may be: 1) The low pressure turbine(s) of dual compressor Turbojet and Turbofan engines, 2) The gas producer turbine(s) of Turboprop or Turboshaft engines.

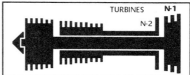
TURBOJET (DUAL SPOOL DESIGN)

N2 COMPRESSOR
A major engine component. Depending on the type of engine, the following N2 Compressor designations apply: 1) The high pressure compressor of a dual-spool

axial flow engine, 2) The intermediate compressor of a triple-spool axial flow engine, 3) The second impeller of a dual-Impeller centrifugal flow engine.

N2 POWER EXTRACTION
Power taken from the High Pressure Compressor drive train by a radial shaft which powers the Accessory Drive Gearbox.

N2 SENSOR
Another name for Tachometer Generator for the N2 rotor system.

N2 Speed (%)
1) Speed of a High Pressure Rotor System (N2) in a dual-spool Turbojet or Turbofan engine. Sometimes referred to as NH, with "H" indicating high.
2) Speed of an Intermediate Pressure Rotor in a triple-spool Turbofan engine.
3) Speed of a Free Power Turbine in a Turboprop or Turboshaft engine, although (NF) is more common.

N2 TURBINE
A major engine component. Depending on the type of engine, the following designations apply: 1) The High pressure turbine(s) of dual compressor engines, 2) The intermediate pressure turbine(s) of a triple-spool axial flow engine, 3) The Free Power turbine of a Turboprop or Turboshaft engine.

N3 COMPRESSOR
A major engine component. The high pressure compressor of a triple-spool axial flow Turbofan engine.

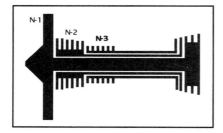

N3 Speed (%)
Speed of a High Pressure Rotor (N3) in a triple-spool engine.

N3 TURBINE
A major engine component. A low pressure turbine on a triple-spool axial flow Turbine engine.

Nc Speed (%)
Speed of any single (compressor) engine. N1 is also used at times to indicate this speed. Nc is sometimes used to mean speed of the Core portion of an engine.

NF Speed (%)
Speed of a (Free) Power Turbine rotor in a Turboprop or Turboshaft engine. Also Speed of a fan in a Turbofan Engine

NG Speed (%)
Speed of a (Gas) Producer Rotor in a Turboprop or Turboshaft engine.

NH Speed (%)
(High) Pressure Rotor speed of a recently developed, dual centrifugal compressor, three shaft Turboprop or Turboshaft engine. This engine most likely will have three cockpit gauges indicating: NH Speed, NL Speed, and NF Speed.

NL Speed (%)
(Low) pressure rotor speed of a recently developed, dual centrifugal compressor, three shaft Turboprop or Turboshaft engine.

NP Speed (%)
Free (Power) Turbine Speed of a Turboprop or Turboshaft engine. Another name for NF Speed. Also used to mean propeller speed in either actual RPM or % RPM.

NR Speed (RPM)
The actual RPM of the (Rotor) in a rotorcraft.

NACELLE (engine)
A streamlined structure of detachable cowlings that fit around an engine. Generally an aircraft manufacturer's supplied component. Also called a pod when the engine is mounted on a pylon in an under-wing or side-fuselage location.

NACELLE CHINE
A large vortex generator placed on the outer structure of an engine nacelle to direct airflow and reduce aircraft surface drag. A nacelle chine has more to do with aircraft aerodynamics than with engine function. *See also: Vortex Generator.*

NACELLE VENTILATING & COOLING
A means of cooling and purging of fumes from the area between the exterior of the engine and the interior of the nacelle. Ram air scoops at the front of the nacelle and discharge vents at the rear allow a constant airflow in flight to cool and prevent dangerous vapor from accumulating. On a typical large engine, this system provides approximately 1 lb./sec. of ventilating and cooling airflow.

NAPHTHA BASE FUELS
A light volatile faction of petroleum oil. In jet fuels, naphtha content indicates a blended gasoline fuel containing some kerosene. *See also: Fuel Types-Jet.*

NAS
Locate under Aircraft Standards.

NASA
Abbreviation for the National Aeronautics and Space Administration. This federal agency deals with aerospace research and development, including projects concerning the Gas Turbine Engine. The NASA Turbine Facility is located at Langley Research Center, Virginia.

NATIONAL AEROSPACE LABORATORY (Japan)
Manufacturer of Gas Turbine Engines. Locate in section entitled "Gas Turbines for Aircraft".

NATIONAL TRANSPORTATION SAFETY BOARD (NTSB)
The Federal authority, under the Department of Transportation, that investigates aircraft transportation incidents and accidents. The NTSB also recommends preventive and corrective measures to the Federal Aviation Administration (FAA).

NATURAL FREQUENCY
Term that means the rotational speed of a part is also its natural frequency, if acted upon by no other force. See also: Frequency Tuning.

NATURAL FREQUENCY OF VIBRATION
In relation to rotating parts, the rotational speed at which vibration starts to occur. In other words, the lowest speed at which an engine part will vibrate. Engine parts are designed to rotate at speeds below the point of Natural Frequency of Vibration. *See also: 1) Harmonics, 2) Resonance of Blades.*

Nc SPEED
Locate under N (symbols).

NDI PROCEDURES
Abbreviation for Non-Destructive Inspection procedures. Also known as Non-Destructive Testing (NDT) procedures.

NEGATIVE GRAVITY VALVE (oil tank)
A component part of an oil system. A valve in the oil tank vent line that closes off to prevent oil loss during inverted flight. Primarily used in military aircraft.

NEGATIVE TORQUE SYSTEM (NTS)
A Turboprop system. A safety system which prevents propeller overspeeding. Negative torque occurs when the oncoming airstream begins to drive the propeller. The propeller can easily over-speed in this condition, so the NTS system signals the propeller governor to increase blade angle in order to regain positive torque. *See also: Safety Coupling.*

NET THRUST
Locate under Thrust (net).

NEUTRALIZER (EGT circuit)
An electrical resistor unit made of a material that loses some of its resistance when the material temperature increases. In an EGT circuit, the neutralizer looses as much resistance as the chromel-alumel portion of the circuit gains due to heat applied. By this means, the circuit remains calibrated to a specific resistance level.

NEUTRON RADIOGRAPHY (NR)
A newly emerging type of non-destructive inspection which employs a transportable cyclotron to create a flow of neutrons through a test piece. By this penetrating neutron radiation technique, video impulses in real time can show clearances and growth factors in engines and also show fuel and oil movement within lines and passageways. Neutron Radiology is said to produce less scatter and better imaging than X-ray techniques. *See also: Non-Destructive Inspection.*

NEWTON'S LAWS
Physical Laws published by the seventeenth century inventor, scientist, and mathematician, Sir Isaac Newton. He published three laws which today help to explain the physics of jet propulsion. Briefly stated they are: 1) Objects in motion tend to stay in motion, objects at rest tend to stay at rest, 2) Force equals mass times acceleration, 3) For every acting force there is an equal and opposite reacting force.

Ng TOPPING CHECK (ground)
An engine performance check on a helicopter engine. This check assures that for prevailing ambient conditions, the prescribed Gas Generator (Ng) Speed is attainable with the engine operating at the part power stop on the Fuel Control. This check is preliminary to an Ng Topping Check in flight.

Ng TOPPING CHECK (flight)
An in-flight engine power check to ensure maximum Gas Generator (Ng) Speed is attainable without exceeding limits of torque or internal temperatures. This adjustment is accomplished at the Fuel Control full-open stop.

NF, NG, NH SPEED
Locate under N (symbols).

NICK
Locate under Bearing Inspection and Distress Terms and Engine Inspection and Distress Terms in Appendix.

NICKEL ALUMINIDE (Ni3AL)
An improved nickel alloy in development, known as an "inter-metallic" material because it exhibits the characteristics of both metal and ceramic. Ni3AL does not depend on chromium or cobalt for its temperature strength as do nickel alloys. This material is six times stronger than stainless steel and proposed for future use in construction of turbine discs. It is approximately 15% lighter in weight than currently used nickel-base super-alloys. *See also: Super-Alloys.*

NICKEL-BASE ALLOY
Nickel-base metal is in the range of 70 to 80 percent nickel alloy with varying percentages of aluminum, chromium, cobalt, tanilum, and titanium. Nickel is either cast or forged, when in the family of Gas Turbine Engine super-metals. Nickel is primarily used in engine hot sections for its high fatigue strength under heat loads and its ability to retain its strength when nicked, dented, or eroded. It is also used in the highly stressed rear compression stages of many large compressors. *See also: 1) ODS Super-alloy, 2) Super-Alloys.*

NITRIC OXIDE (NO)
An atmospheric gas which originates mainly from soil bacteria and fertilizers. It is also present in turbine engine exhaust emissions, in small amounts that adversely affect (reduce) atmospheric ozone. The change from ozone to NO results when hydrocarbon fuel is combusted in atmospheric air. Gas turbine Engine manufacturers are required to meet strict governmental standards pertaining to NO emission. *See also: Emissions (engines).*

NL SPEED
Locate under N (symbols).

NO-TAIL-ROTOR
A Gas Turbine Powered Helicopter design which uses tail-boom thrusters instead of a tail rotor. A fan driven by the engine takes in ambient air and discharges it rearward to vectoring exhaust nozzles in the tail-boom. The thrusters provide side force thrust for anti-torque control.

NOEL PENNY TURBINES LTD.
Manufacturer of Gas Turbine Engines. Locate in section entitled "Gas Turbine Engines for Aircraft".

NOISE (engine)
In regard to design of Gas Turbine Engines, the two categories of aircraft engine noise are as follows:
1) *Discrete Tones* - a single frequency noise produced by movement of blades in air, fan wake, and hot exhaust wake.
2) *Broadband Tones* - a wide range of frequencies produced by movement of air over blades and from combustion.

NOISE (far-field)
With respect to FAA aircraft engine noise standards, the noise (unwanted sounds) experienced by the community in and around an airport.

NOISE (near-field)
With respect to FAA aircraft engine noise standards, the noise (unwanted sounds) experienced in the cabin of an air aircraft.

NOISE ABATEMENT TAKE-OFF
A take-off at less than full thrust. This procedure is often required to meet governmental regulations concerning the impact of atmospheric noise pollution on local communities.

NOISE SUPPRESSION (FAA)
FAR Part 36 requirements concerning aircraft noise. The FAA places controls and restrictions on aircraft and engine noise emissions. Many types of acoustic materials and airflow chamber designs are employed to restrict noise generation in order to meet FAA requirements. *See also:* 1) *Ground Runup Enclosure,* 2) *Hush Kit,* 3) *Noise Suppressor Unit.*

NOISE SUPPRESSOR UNIT (aircraft)
A sound attenuating device fitted to the exhaust duct. Often an aircraft manufacturer's supplied component. In the early days of commercial jet aviation, noise suppressor units were quite common. At that time, Turbojet Engine hot exhaust velocities were much higher and noise was greater than what is experienced today with the use of Turbofan Engines. Current practice with Turbofan Engines is to attenuate noise using light-weight acoustic liner materials at the inlet and exhaust and to employ core to fan air mixers. There are, however, some add-on noise suppressor units being developed to meet newer, more stringent requirements being introduced. Also called 1) Exhaust Silencer, 2) Jet Silencer, 3) Sound Suppressor. *See also:* 1) *Corrugated Perimeter Noise Suppressor,* 2) *Ejector Exhaust Suppressor,* 3) *Exhaust Mixer,* 4) *Lobe-Type Noise Suppressor,* 5) *Multiple-Tube Noise Suppressor.*

NOISE SUPPRESSOR UNIT (ground)
Locate under Sound Suppressor (ground mounted).

NOMEX™
A Dupont Company trademark. A composite honeycomb material, often covered with Kevlar™, used in the construction of engine nacelles. *See also:* 1) *Composite Materials,* 2) *Kevlar™.*

NOMOGRAPH (viscosity index)
A chart produced by the American Society of Testing Materials (ASTM) which is used to plot the viscosity change of turbine oils when the oil is heated to its normal operating temperature. The typical temperatures at which plots are taken are 100° and 280° Fahrenheit. The reduction in viscosity which results with heat added to oil is converted to a Viscosity Index (VI) Number. *See also: Viscosity Index.*

NON-AXISYMMETRIC EXHAUST NOZZLE
Locate under Exhaust Nozzle 2D.

NON-AXISYMMETRIC THRUST
Thrust generated off-center with reference to the center line of the aircraft, as would be experienced in a one-engine-out condition during flight.

NON-DESTRUCTIVE INSPECTION (NDI)
Inspection or testing procedures using techniques that detect faults in the surface and subsurface but do not damage or change the structure of the part. The most common NDI procedures used on turbine engines are: 1) Dye Penetrant (red), 2) Dye Penetrant (fluorescent) with Black Light, 3) Eddy Current, 4) Magnaflux™, 5) Ultra-Sonic, 6) X-Ray. *See also:* 1) *Computed Tomography,* 2) *Neutron Radiography,* 3) *Positron Emission Tomography.*

NON-DESTRUCTIVE TESTING PROCEDURES (NDT)
Tests performed as part of the Non-Destructive Inspection procedure.

NORMAL RATED THRUST (NRT)
Locate under Military Power Ratings.

NORMAL SHOCK WAVE
A shock wave perpendicular to an objects leading edge, which occurs at Mach -1 airspeed. Inlets designed for supersonic aircraft utilize shock waves to slow airflow and increase static air pressure. *See also:* 1) *Oblique Shock Wave,* 2) *Shock Wave.*

NOSE BULLET
British term for nose dome.

NOSE CASE
Another name for Inlet Gearbox.

NOSE CONE
Another name for Nose Dome or Spinner.

NOSE COWL
The Flight Inlet portion of an engine nacelle.

NOSE DOME
A stationary inlet fairing which attaches to and covers the engine inlet case. Its function is to reduce turbulence in air entering the engine compressor. Also called Inlet Centerbody.

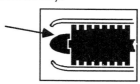

NOSE GEARBOX
Another name for Auxiliary Accessory Gearbox.

NOTAR
Abbreviation for No-Tail-Rotor.

NOZZLE-2D
Locate under Exhaust Nozzle (vectored Thrust).

NOZZLE (gas flow)
A chamber in which flowing gases experience a decreases in pressure and an increase in velocity. *See also: 1) Exhaust Nozzle, 2) Subsonic Exhaust Nozzle, 3) Supersonic Nozzle, 4) Variable Area Exhaust Nozzle.*

NOZZLE AREA CONTROL VALVE
An engine accessory. A mechanism designed to control the position of a Variable Area Exhaust Unit.

NOZZLE DIAPHRAGM
A nomenclature used in early turbine engines, rarely used today. Another name for Turbine Nozzle Guide Vane Assembly.

NOZZLE EXIT
Locate under Exhaust Nozzle Exit.

NOZZLE FLAPS
A component part of an afterburner. Movable segments that form a variable flow area at the nozzle exit.

NOZZLE GUIDE VANE
Another name for Turbine Nozzle Guide Vane.

NOZZLE PLUG
A component part of an exhaust duct. The center section of a plug type exhaust nozzle. Also called Exhaust Centerbody. *See also: Plug Nozzle.*

NOZZLE PRESSURE RATIO (exhaust)
The ratio of nozzle inlet pressure to nozzle discharge pressure. In reference to a Turbine Engine, the pressure change across the Exhaust Duct.

 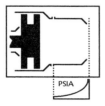

NOZZLE RING (turbine)
An engine component part. The turbine stator nozzle of a radial inflow turbine.

NOZZLE STREAKING
Another name for Hot Streaking.

NOZZLE THROAT
Locate under Exhaust Nozzle Throat.

NOZZLE TRIMMER TAB (tailpipe)
Another name for Tailpipe Insert.

NP SPEED
Locate under N (symbols).

NR SPEED DROOP
In a rotorcraft, the tendency for rotor speed to reduce when rotor blade pitch is increased.

NR SPEED
Locate under N (symbols).

NRT
Abbreviation for Normal Rated Thrust.

NULL POSITION
In reference to having a neutral or zero force. A mechanism in a nullified or locked position where it neither receives nor sends a signal and movement is temporarily interrupted.

NULL POSITION (flyweight governor)
In a Fuel Control Unit or Propeller flyweight governing system, null refers to having a zero force. The system is said to be in an on-speed condition when flyweight force is in equilibrium with its speeder spring force.

NUT PLATE
Another name for Channel Nut Plate.

NTSB
Abbreviation for National Transportation Safety Board.

O

O-RING
A circular rubber or synthetic packing which fits around a shaft or against a flat surface to prevent leakage of air or fluids. The O-Ring becomes compressed when mating parts are fitted together providing a sealed surface.

OBLIQUE SHOCK WAVE
A shock wave in air angled back, in the shape of a bow-wave on a ship. The oblique waves occur when normal shock waves are bent back by the oncoming airstream, as the airfoil continues to increase in airspeed. *See also: Normal Shock Wave.*

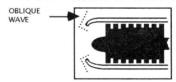

OBLIQUE SHOCK WAVE (flight inlet)
A controlled shock wave condition designed to occur in a supersonic aircraft flight inlet. As the airstream penetrates the shockwave, its velocity decreases and static air pressure increases. *See also: 1) Normal Shock Wave.*

ODS SUPER-ALLOY (inconel™)
A recent development in materials for turbine blade construction called Oxide-Dispersion-Strengthened Inconel (ODS). The blade is forged in a near net shape in an unrecrystallized state and then undergoes a Directional Solidification (recrystallizing) heat treatment. Inconel blades formed in this way are said to have heat strength up to 1000°C. This value exceeds the heat strength of other high-temperature-strength materials, such as Single Crystal materials. *See also: 1) Inconel, 2) Super-Alloys.*

OEI
Abbreviation for One-Engine-Inoperative.

OFF-DESIGN SPEED
Off-design speed is any engine speed lower than cruise speed. Some engines experience airflow problems in this range and anti-stall systems such as variable angle compressor vanes and compressor bleed valves are installed. These anti-stall systems maintain the correct

relationship between engine speed and engine mass airflow.

OFF-IDLE SPEED
Typically any speed between Idle RPM and Cruise RPM.

OFF-IDLE STALL
A mild compressor stall which occurs when the power lever in the cockpit is advanced from the ground idle stop. Often times this stall is not considered an airworthiness concern because engine instruments all remain in the normal range, and the stall condition disappears on spool up. *See also: Compressor Stall.*

OGV
Abbreviation for Outlet Guide Vanes.

O/H
Abbreviation for Overhaul.

OIL ADDITIVES
Materials added to oil during manufacture, such as anti-oxidants, detergents, corrosion inhibitors, anti-foaming agents, pour point depressants, viscosity index improvers, and lubricity improvers.

OIL (engine)
Locate under Synthetic Lubricants.

OIL BREATHER VALVE
Another name for Breather Pressurizing and Vent Valve.

OIL BOOSTER PUMP
Locate under Oil Pump (booster).

OIL BOWSER
A unit of ground support equipment. An engine servicing cart used to filter and pump lubricating oils into the oil tank via a pressure fill port as opposed to hand filling through the gravity filler cap. This fill method will prevent oil can fragments and atmospheric contaminants from entering the oil system during servicing. *See also: 1) Gravity Fill Point, 2) Pressure Fill Port.*

OIL COKING
Locate under Coking (oil).

OIL CONSUMPTION (high)
Engine oil consumption rate which exceeds the maximum allowable limit. Oil is most often lost by one of the following: 1) external leaks at lines and fittings, 2) overboard via the oil vent system when vent pressures are high, 3) through the gas path when leaking past main bearing seals.

OIL COOLER (lubrication system)
An engine accessory. A heat exchanger unit used to cool the oil by transferring heat into atmospheric air or into engine fuel. *See also: 1) Air-Oil Cooler, 2) Fuel-Oil Cooler.*

OIL DAMPED BEARING
Locate under Bearing (oil damped).

OIL DEFLECTOR
Another name for Oil Seal.

OIL FILTER
An engine accessory. A filter housing or bowl containing a filtering element for straining out contaminant particles in oil. Contaminants common to engine oil systems are: 1) carbon which results when oil breaks down chemically, 2) wear-metals generated as oil wetted engine parts gain service time, 3) foreign particles that enter from the atmosphere through bearing seals or during servicing.

OIL FILTER (barrier type)
Refers to the conventional cleanable or disposable oil filters which provide a barrier for trapping contaminants.

OIL FILTER (centrifugal)
A filter configured with a rotary filtering element that throws contaminant particles outward into a cleanable sediment trap. Not a common type filter.

OIL FILTER (cleanable screen)
A filter containing a steel mesh filtering element. Oil filters of this type normally have micronic ratings. For example, a 20 micron filter in which each mesh opening has a flow area of 20 microns diameter, will filter out particles larger than 20 Microns. *See also: 1) Multiple Wafer Filter, 2) Pleated Cylindrical Filter, 3) Stacked Disk Filter.*

OIL FILTER (disposable)
A filter containing a disposable filtering element made of pressed paper or cellulose. This element is generally pleated to provide a maximum surface area. Filters of this type normally have a nominal micronic rating, even though they have no actual mesh diameter. The overlapping fibers provide an average filtration value.

OIL FILTER (knife edge)
A filtering element containing a solid metal cylinder, shaped such that its surface is a series of incoming and outgoing longitudinal, sharp-cut grooves. When inserted into a close tolerance oil passageway, contaminants in oil are trapped in the incoming grooves and clean oil squeezes between the knife-edges of the filter and then through the outgoing grooves to the oil jet. Most often used as a Last Chance Filter. Not a common type filter. *See also: Oil Filter (last chance).*

OIL FILTER (last chance)
A small engine accessory. A filter, normally a metal screen type, located very close to an oil jet and generally only cleaned at engine overhaul because of its remote location within the engine.

OIL FILTER (main)
A cleanable screen or disposable filtration element through which all of the lubrication system oil is pumped on its way to the engine lubrication points. The entire oil supply, on average, will pass through this filter 3 to 5 times per minute to maintain the required cleanliness of the oil. Most filters of this type have a bypass indicator. Locate under Filter Bypass Indicator. *See also: 1) Oil Filter (cleanable screen), 2) Oil Filter (disposable), 3) Oil Filter Debris Analysis.*

OIL FILTER (scavenge)
A filter in the return side of the lubrication system that assists the main oil filter. A scavenge filter is not considered to be necessary in most lubrication systems.

OIL FILTER BYPASS INDICATOR
Locate under Filter Bypass Indicator.

OIL FILTER DEBRIS ANALYSIS
Refers to the basic procedure of visual inspection of oil filters by technicians. It may also refer to a detailed laboratory inspection and analysis of contaminant types and levels found in lubrication system filters. *See also: 1) Qualitative Debris Monitoring, 2) Spectrometric Oil Analysis.*

OIL FILTER IMPENDING BYPASS SWITCH
An oil pressure operated switch that actuates a cockpit warning light should the main filter screen clog sufficiently to cause a pressure drop equal to the alarm value of the switch. Often called an FIB Switch. In the event of further pressure drop, the filter will bypass. This is often signaled by an oil Filter Actual Bypass Switch (FAB Switch). *See also: Filters.*

OIL FILTER SHUT-OFF VALVE
Locate under Service Shut-Off Valve.

OIL GALLERY
A oil flow groove in an oil lubricated part such as a bearing, gear, or shaft.

OIL JET (air-mist)
A small engine component part. A pressurized air and oil spray type of calibrated oil jet used to create an air-oil mist. The mist lubricates main bearings, as opposed to the more traditional oil stream. The air-mist jet is used where a wide area of minimal oil film is sufficient. Also called a Fogger oil jet.

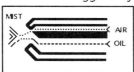

OIL JET (stream-flow)
A small engine component part. A small nozzle (flow orifice) that directs a calibrated stream of oil onto a point to be lubricated such as bearings, carbon seals, gears, etc. Stream-flow type oil jets are the most common lubricating method, and the most effective way of lubricating when loads are heavy and speeds and temperatures are high.

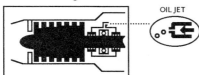

OIL LACQUER BUILD-UP
Locate under Lacquer (oil).

OIL MIXING
Gas Turbine Oils are engine manufacturer approved synthetic lubricants that have specific restrictions concerning mixing. Typical guidelines are as follows: 1) Synthetic lubricants can never be mixed with petroleum oils. 2) Synthetic Type-1 lubricants can be mixed with Synthetic Type-2 lubricants only with engine manufacturer's approval. 3) Different brands of oils can be mixed only with engine manufacturer's approval. Inadvertent mixing generally requires an oil system drain and flush. *See also: Synthetic Lubricanting Oils.*

OIL NOZZLE
Another name for Oil Jet.

OIL POUR POINT
Locate under Pour Point.

OIL PRESSURE GAUGE (cockpit)
A cockpit engine condition instrument that indicates, in PSI, the oil pressure in the engine lubrication system. When indicating that oil pressure is within limits, the operator is assured that oil flow to all parts of the engine is correct. Sometimes more than one gauge is necessary to accomplish this, for instance a gauge upstream of the oil filter and another downstream might be required. Generally this gauge is in an electrical type, receiving its signals to operate from an engine mounted transmitter. *See also: Transmitter (oil).*

OIL PRESSURE INDICATOR
Locate under Oil Pressure Gauge.

OIL PRESSURE RELIEF VALVE (cold start type)
A small engine accessory. A relief valve in the Pressure Subsystem of the engine lubrication system which prevents excessive pressure build-up at low ambient temperatures. This valve is set to crack (offseat) at the very high pressures associated with high viscosity conditions that result during a cold weather engine start. Generally when the cold start relief valve is utilized, the system will not contain a pressure regulating relief valve and engine oil pressure will have a wide range of readings.

OIL PRESSURE RELIEF VALVE (pressure regulating type)
A small engine accessory. An adjustable relief valve, generally of the spring and cone type, located in the Pressure Subsystem of the engine lubrication system. This valve is set to crack open at normal system pressure and bypass oil back to the oil pump inlet, in order to maintain a fairly constant engine operating oil pressure. For example, a typical allowable oil pressure might be 50 ± 5 psig maximum. Maintenance personnel will set the oil pressure anywhere between 45 to 55 psig, and it will not need to be reset as long as system pressure remains in this range. Also called Oil Pressure Regulator. *See also: DeOiler Valve.*

OIL PRESSURE SUBSYSTEM
The oil supply side of the engine lubrication system. It is one of three subsystems in the lubrication system. The Pressure Subsystem carries oil under pressure to the oil jets. *See also: Lubrication System (subsystems).*

OIL PRESSURE SWITCH
A small engine accessory. Generally a diaphragm type device with oil pressure on one side and an electric microswitch mechanism on the other. This oil pressure operated switch activates a warning light in the cockpit when a system malfunction is sensed. *See also: 1) Low Oil Pressure Switch, 2) Filter*

Bypass Indicator.

OIL PRESSURE SWITCH (low oil pressure)
A small engine accessory. An oil pressure operated microswitch that is set to power a warning light in the cockpit to indicate that a hazardous low engine oil pressure condition exists.

OIL PRESSURE SWITCH (filter bypass)
Locate under Filter Bypass Indicator.

OIL PRESSURE TRANSMITTER
Locate under Transmitter (oil).

OIL PUMP (boost type)
A small engine accessory. A boost pump, typically a centrifugal type pump, located upstream of the main pump to provide a head of pressure and prevent main pump cavitation. Not commonly used today. *See also: Pump Cavitation.*

OIL PUMP (gear type)
A rotating, positive displacement pump, that utilizes traditional gear-shaped pumping elements. An element consists of a drive gear and an idler gear. Oil enters the pump inlet and is carried around in the spaces between the gear teeth and the outer casing. Resistance to flow downstream of the pump creates system pressure.

OIL PUMP (gerotor type)
A rotating oil pump design utilizing a driven, lobe-type male gear, within a female lobe-type idler gear. The outer idler gear has one extra lobe which fills with oil at the pump inlet, and is then driven around to a zero clearance, forcing the oil through the outlet under pressure. Also called a Trochoidal Pump due to the plane of rotation of its pumping elements.

OIL PUMP (lube)
An engine accessory. A lube (supply) pump, typically a positive displacement pump mounted on the accessory drive gearbox. Most engines utilize only one such pump to provide lubricating oil to all the lube points in the engine. This pump may be a separate unit or part of a combination lube and scavenge pump. Also called 1) main oil pump, 2) supply oil pump. Refer also to Oil Pressure Subsystem. *See also: The most common Oil Pump types as follows: 1) Gear Type, 2) Gerotor Type, 3) Vane Type.*

OIL PUMP (lube and scavenge)
An engine accessory. A pump which contains both pressure (supply) and scavenge (return) oil pumps in a common housing. This pump can be of the gear, gerotor, or vane type.

OIL PUMP (scavenge)
A small engine accessory. Generally a positive displacement pump. Most engines contain a main scavenge pump mounted on the accessory drive gearbox and several small scavenge pumps, either located within the main pump housing or within the engine near the point to be scavenged. The function of these pumps is to return oil back to the oil tank from bearing and gearbox locations. Also called 1) scavenger pump, 2) suction pump. See also: Oil Scavenge Subsystem.

OIL PUMP (vane type)
A positive displacement oil pump design that utilizes a sliding vane type pumping element. The spaces between vanes fill with oil at the pump inlet and, as the vanes rotate moving to a zero clearance, the oil is forced through the outlet port under pressure.

OIL PUMP AND ACCESSORY DRIVE HOUSING (OPAH)
Locate under Accessory Drive Gearbox.

OIL QUANTITY GAUGE
A cockpit gauge which indicates the amount of oil remaining in the engine oil tank. Its operating mechanism may be a float-switch, a capacitance probe, or an electrical densiometer. *See also: Low Oil Level Warning Light.*

OIL SAMPLE
A small quantity of oil removed at periodic intervals from an oil supply tank for analysis. Normally the sample is taken just after engine shutdown, when the oil is still warm and the contaminants remain suspended. The sample is then analyzed for the types and levels of wear-metals and other contaminants it might contain. In some instances, the sample base data is such that sampling can be accomplished cold, even after long periods of inactivity of the system. *See also: Spectrometric Oil Analysis.*

OIL SAMPLE PLUG (oil tank)
A plug that can be removed when taking an oil sample from the oil supply tank. The plug is located at the approximate mid-point on the tank to ensure an accurate sample.

OIL SCAVENGE SUBSYSTEM
The return side of the engine lubrication system. The scavenge subsystem carries oil back to the oil tank under pressure from the bearing sumps and gearboxes. Because this oil becomes greatly aerated, the scavenge subsystem, by FAA Regulation, must be at least twice the capacity of the pressure subsystem. *See also: Lubrication System (subsystems).*

OIL SCREEN
Another name for Oil Filter or may mean a metal screen-type filtering element in a filter assembly.

OIL SEAL
Locate under Carbon Oil Seal or Labyrinth Oil Seal.

OIL SEAL RACE
Another name for Carbon Seal Rub Ring. See also: Carbon Oil Seal.

OIL SEPARATOR
Locate under Air-Oil Separator.

OIL SERVICING CART
Another name for Oil Bowser.

OIL SLINGER (lubrication system)
A small component part. A rotating metal part used to centrifugally load and direct oil flow as follows: 1) Onto bearings and carbon seals for lubrication, 2) Away from bearing sumps and towards a scavenge point.

OIL SUCTION SYSTEM (engine)
Another name for Oil Scavenge Subsystem.

OIL SUCTION PUMP
Another name for Oil Scavenge Pump.

OIL SUPPLY SYSTEM (engine)
Another name for Oil Pressure Subsystem.

OIL SYSTEM
Locate under Lubrication System.

OIL SYSTEM (emergency)
Locate under Emergency Oil System.

OIL TANK (lubrication system)
An engine component. The main oil storage reservoir on an engine. Oil tanks may be externally or internally located. Generally oil tanks are pressurized 3 to 5 psi above ambient pressure to reduce foaming and to place a pressure head on the surface of the oil. *See also: 1) Dry Sump (lubrication system), 2) Integral Dry Sump (lubrication system), 3) Wet Sump (lubrication system), 4) Saddle Mounted Tank.*

OIL TANK BAFFLE
An internal component part of an oil tank. A divider that reduces oil sloshing and prevents oil re-aeration within the tank.

OIL TANK CHECK VALVE
Another name for Oil Tank Pressurization Valve.

OIL TANK DEAERATOR
Locate under Deaerator.

OIL TANK EXPANSION SPACE
The space above the maximum oil level in an oil tank. By FAA Regulation it must equal to at least 10% of tank volume or one-half gallon, whichever is greater.

OIL TANK FILL POINT
Locate under 1) Gravity Fill Point, 2) Pressure Fill Point.

OIL TANK PRESSURIZING VALVE
A component part of an oil tank. A check valve located in the tank expansion space used to trap some of the air that returns to the tank with scavenged oil. The check valve releases the majority of air to the atmosphere, but retains enough to pressurize the tank to approximately 3 to 5 psig. A bleed orifice is generally provided to allow the tank to depressurize after engine shut-down.

OIL TEMPERATURE BULB (lubrication system)
A small engine accessory. A probe-shaped transducer or sensor that is fitted directly into the engine oil flow to send a temperature sense to a cockpit instrument system. *See also: 1) Resistance Temperature Detector, 2) Thermocouple (oil temperature).*

OIL TEMPERATURE GAUGE
A cockpit indicator displaying the temperature of engine oil as an indication of engine condition. The gauge receives an electrical signal from a transducer (oil temperature bulb) located in the oil stream.

OIL TEMPILABEL
An oil tank stick-on label containing a heat sensitive material which changes color in the presence of an oil over-temperature condition. In some instances, this method may be used in place of an oil temperature indicator in the cockpit.

OIL THERMOSTATIC VALVE
Locate under Thermostatic Valve.

OIL WATCH
Refers to increased surveillance by flight and ground crews of an engine experiencing a high oil consumption rate. *See also: Oil Consumption.*

OIL VENT SUBSYSTEM
A system of interconnecting air passageways between oil sumps which directs unwanted air back to the atmosphere. This subsystem removes air that enters the lubrication system via main bearing seals. An excess of vent air pressure, if allowed to build to unacceptable levels, interferes with normal fluid flow-rate from the oil jets. *See also: Lubrication System (subsystems).*

OIL WETTED SECTIONS (engine)
The portions of the engine where lubricating oil flows for the purpose of reducing friction, cleaning and cooling.

OMNIPOL FOREIGN TRADE GROUP (Czechoslovakia)
Manufacturer of Gas Turbine Engines. Locate in section entitled "Gas Turbine Engines for Aircraft".

ON-CONDITION MAINTENANCE
A concept under which many engine accessories and components are maintained. Refers to engine parts with no life limiting requirements. If during inspection the part shows no defects, it is said to be in satisfactory condition to remain in service. *See also: 1) Engine Cycle, 2) Engine Monitoring System, 3) Life-Limited Part.*

ON-CONDITION OVERHAUL
Refers to Engines, Engine Modules, and Component Parts of Engines where no TBO is suggested by the manufacturer. In this case, inspection schedules and trend analyses of performance are established, and as long as the engine and its components meet the airworthiness criteria, they remain in service. *See also: 1) Soft-Time Overhaul, 2) Time Between Overhaul.*

ONE-ENGINE-INOPERATIVE (OEI)
Refers to flight operation with one of the aircraft's engines shut down.

OPAH
Abbreviation for Oil Pump and Accessory Drive Housing.

OPEN CYCLE ENGINE (thermodynamic)
An engine in which atmospheric air, referred to as the working fluid, is processed and then returned to the atmosphere. In a Gas Turbine engine, air enters the engine inlet, is compressed, expanded by combustion to perform work (drive the turbine and compressor), and is then discharged back to the atmosphere not to be used again. This process is completed in essentially what could be called one chamber. In a closed cycle engine, the working fluid is compressed, expanded to perform work, and is then cooled to be compressed again. Combustion and Exhaust in a closed cycle process are also completed at separate locations, not necessarily within the engine. *See also: 1) Brayton Cycle, 2) Closed Thermodynamic Cycle.*

OPEN PERIMETER TURBINE BLADE
Another name for open tip blades.

OPEN TIP BLADES (compressor)
Locate under Compressor Blade (shapes).

OPEN-TIP BLADE (turbine)
Turbine blades with no interlocking shrouds attached to their tip areas. The blades can be either of the following types: 1) Flat-machine cut, 2) Squeeler-tip cut. *See also: Turbine Blades.*

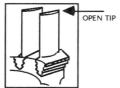

OPPOSITE POLARITY IGNITION SYSTEM
An older type turbine engine ignition system utilizing a coil discharge as opposed to the higher wattage, capacitor discharge system in common use today.

ORTHOGONAL COMPRESSOR DESIGN
A compressor with blading set perpendicular to the flow path. The compressor blading forms a rounded shape when seen from the side view, as opposed to a straight taper seen in most conventional compressors. The blades are set at right angles to the compressor disc rather than at right angles to the axis of the engine. Orthogonal design is a recent development to enhance aerodynamic performance. It is said to produce more compression for a given compressor size when compared to compressors with traditional, radially placed blades. *See also: Compressor Blade (shapes).*

OTW
Abbreviation for Over-The-Wing location of an engine nacelle. *See also: UTW.*

OUTLET GUIDE VANES (OGV)
Another name for Exit Guide Vanes.

OUTER LINER (combustor)
An engine component part. Refers to the outer portion of a dual-wall type liner. Some older can-annular liners were of this configuration. *See also: Combustor Types.*

OUTPUT REDUCTION GEARBOX
An engine component. A reduction gearbox on a Turboshaft engine that converts a high speed, low torque input from a turbine shaft into low speed, high torque via the reduction gearing mechanism. The gearbox in turn powers the output drive shaft. *See also: Gearbox.*

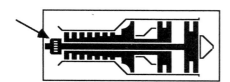

OVERALL EFFICIENCY (engine)
A physics term. The product of multiplying Propulsive Efficiency and Thermal Efficiency and expressed as a percentage. Supersonic aircraft engines show the best Overall Efficiency because of the positive influence of aircraft speed on both Propulsive and Thermal Efficiencies. *See also: Efficiencies.*

$$\text{Overall} = \text{Thermal} \times \text{Propulsive}$$

OVER-BOOST (engine)
Refers to an engine being operated at a power level over the established limit. When an over-boost is noted on a cockpit instrument, it indicates to the operator that the engine has been over stressed and implies a flight safety hazard may exist. *See also: Over-Shoot.*

OVER-EXPANDED NOZZLE
A nozzle in which the engine exhaust leaves the jet nozzle at less than ambient pressure.

OVER-FLOW LINE (oil tank)
A component part of an oil tank. A tank that is configured with a pressure filling capability. The overflow line gives a visual indication by spilling oil when the tank is full. *See also: 1) Oil Bowser, 2) Pressure Fill Port.*

OVERHAUL (engine)
Locate under Time Between Overhaul (TBO).

OVERHEAT DETECTOR
Locate under Fire Detector.

OVER-HEATING
Locate under Engine Inspection and Distress Terms in Appendix.

OVER-RUNNING CLUTCH (starter)
A component part of a starter. Generally a pawl and ratchet mechanism. The over-running clutch is designed to permit the starter to drive the engine but will not allow the engine to drive the starter. The pawls will ride freely over the ratchet surface (overrun) to prevent the engine from driving the starter if normal disengagement does not occur. Typically a retract spring or centrifugal loading on the pawls will lift the pawls off the ratchet to completely disengage the starter from the engine as the engine spools up to idle speed. Also called 1) Pawl and Ratchet Clutch, 2) Sprag Clutch.

OVER-SHOOT (engine instruments)
Slight gauge peaks that come back quickly to the normal range. Over-shoots can be a normal occurrence or a malfunction depending on the severity of the overshoot and the particular system involved. *See also: Bloom (engine instrument).*

OVER-SPEED CONDITION
A condition wherein the engine operates over the manufacturer's prescribed rotational limit. Depending on the speed attained, this condition will warrant, at a minimum, a special visual compressor and turbine inspection (in most cases accomplished on the flightline). In an extreme case, an engine removal for complete overhaul may be required. *See also: Overspeed Loading.*

OVER-SPEED GOVERNOR (engine)
A small engine accessory. A flyweight operated fuel metering device that: 1) limits free power turbine

speeds on torque producing engines by reducing fuel flow to the gas generator portion of the engine, 2) limits N2 Speed on a dual-spool Turbofan engine by reducing fuel flow to the HP Rotor System.

OVER-SPEED LOADING
The centrifugal loading on a rotating part equal to the square of its speed. For example, an over-speed to 106% when maximum allowable speed is 100% results in a 12.4% overload ($1.06^2 = 1.124$). An over-speed to 120% results in a 44% overload ($1.20^2 = 1.44$).

OVERSPEED TRIP SYSTEM (turbofan)
An overspeed protection system utilized on some Geared-Fan engines. Should a loss of load occur and the Fan and Low Pressure Turbine attempt to overspeed, the Trip System actuates a fuel solenoid which shuts down the engine. This system prevents catastrophic fan disintegration which could rupture the engine casings and damage the aircraft.

OVER-SPILL DRAIN (oil tank)
Another name for Scupper Drain.

OVER-TEMPERATURE CONDITION
A condition wherein an engine is operated over the manufacturer's prescribed internal temperature limits (generally TIT, ITT or EGT). This condition will warrant, at a minimum, a special visual inspection of the hot section (with the engine remaining in the aircraft). In an extreme case, an engine removal for hot section overhaul may be required. An over-temperature condition is most common during starting of the engine or during spool-up to high power settings. *See also: Exhaust Gas Temperature Limits.*

OVER-TEMPERATURE LIMITS
Locate under Exhaust Gas Temperature Limits.

OVER-THE-WING (OTW)
Refers to an engine mounted higher than the wing leading edge. Typically with a Turbofan engine, this would indicate that the fan will exhaust is gases over the wing. *See also: Over-Wing Blowing.*

OVER-WING BLOWING
Fan exhaust air from a Turbofan Engine which is directed over the top of a wing surface to enhance an aircraft's Short Take-Off and Landing (STOL) capability.

OXIDATION
Locate under Intergranular Oxidation listed in Engine Inspection and Distress Terms in Appendix.

OZONE (O_3)
Ozone is a tri-atomic form of oxygen symbolized as O_3, occurring naturally in the atmosphere. Ozone is formed when oxygen (O_2) is broken down into separate oxygen atoms, which then combine with oxygen to form O_3. Ozone provides an important filtering-out effect of ultra-violet light which would otherwise strike the earth. Ozone, in small amounts, is destroyed when it passes through a turbine engine and mixes with nitric oxides in the combustion process. It is for this reason that the FAA and EPA constantly monitor aviation equipment, industrial equipment, and other processes that have an ozone depletion effect.

P

P (symbols)
Upper case "P" generally means Pressure of gases within a turbine engine. When subscripts are added they identify particular pressures and engine station locations. *See also: Appendix for complete symbol listing.*

Pb
Symbol for Pressure-burner, meaning the static pressure within the combustion liner. This pressure signal is generally sent to the Fuel Control Unit as a measure of mass airflow, against which a fuel schedule is established.

Ps
Symbol for Static Pressure.

Pt
Symbol for Pressure-total, meaning the total pressure of gas in a duct, to include both ram pressure and static pressure. *See also: Total Air Pressure.*

Pt_2
Symbol for Pressure-total, engine station two. On many engines this symbol represents Compressor Inlet Pressure, an important engine parameter sent to the fuel control for fuel scheduling purposes. This parameter is also sent to the Engine Pressure Ratio Indicating system in the cockpit. Also referred to as Compressor Inlet Pressure (CIP).

Pt_7
Symbol for Pressure-total, station seven. On many engines this symbol represents Turbine Discharge Pressure, an important engine parameter sent to the Engine Pressure Ratio indicating system in the cockpit. *See also: Total Air Pressure.*

Pt_7 GAUGE
A device to measure turbine discharge pressure. It is used by maintenance personnel when performing engine trimming procedures and in some cockpits as an engine performance gauge. Also called P7 gauge.

PACKING
A name given to the common O-Ring seal.

PADDLE PUMP
A small engine component. A centrifugal-type oil scavenging pump, located mainly at turbine bearing sumps to quickly remove oil back to the Accessory Drive Gearbox.

PARACYMENE
A liquid utilized in capillary sensor tubes due to its ability to change volume readily with temperature change. Paracymene, as an example, is used in Compressor Inlet Temperature (CIT) sensor units to send a temperature signal to a main fuel control servo mechanism. *See also: T2 Sensor.*

PARAMETERS (fuel control)
Locate under Signals (fuel control).

PARASITE DRAG
Locate under Drag (parasite).

PARASITIC AIR (engine)
Air that is stagnant on a surface. Within an engine it is often another name for Engine Bleed Air because this air is drawn away from the surface of a compressor drum or disk where boundary layer air is rather stagnant.

PART CONDITION CODE
Generally a numerical indicator which engine manufacturers use to show the condition of inspected parts. These indicators are a quick means of identification on parts tags, work orders, etc. For example, a manufacturer might use a three digit coding system such as (056) meaning chipped and (065) meaning burred.

PART POWER ENGINE
Refers to a turbine engine that is capable of undergoing a Part Power Trim Check. *See also: Full Throttle Engine.*

PART POWER TRIM CHECK (engine)
A performance check with the power lever linkage forward and against the part power trim stop. The engine performance is compared to standard charts showing acceptable EPR or N1 Fan Speed at part power operating condition. In years past, the trim check was accomplished at full engine power. Later, Part Power Trim Checking procedures were developed to reduce fuel consumption, high temperature loading on engine parts, and to reduce noise. Part Power in terms of this check is approximately cruise power lever angle. *See also: Part Power Trimming.*

PART POWER TRIMMING (engine)
A maintenance procedure in which the fuel control is adjusted to meet the required trim values for prevailing ambient conditions, during the part power trim check.

PART POWER TRIM STOP (fuel control)
An obstruction that is placed in the path of the power lever linkage. A small cam or a rig pin that the linkage rests against during an engine trim check. Also called Partial Power Trim Stop. *See also: 1) Part Power Trim Check, 2) Trimming (full power).*

PART-SPAN SHROUD (fan blade)
A support lug on a fan blade. A span shroud with a function similar to that of a Mid-Span Shroud. *See also: 1) Mid-Span Shroud, 2) Span Shroud.*

PART THROTTLE ENGINE
Refers to a turbine engine that is Flat Rated. *See also: 1) Flat Rating (engine), 2) Full Throttle Engine.*

PART THROTTLE REHEAT
Refers to use of an afterburner at less than its full capability.

PART THROTTLE TRIMMING
Another name for Part Power Trimming.

PARTIAL POWER TRIM CHECK
Another name for Part Power Trim Check.

PARTICLE SENSOR (oil)
Another name for Chip Detector.

PARTICLE SEPARATOR (flight inlet)
Locate under Inlet ice and Sand Separator.

PASSIVE CLEARANCE CONTROL (compressor)
A cold section blade tip clearance control system. The two systems in common use function as follows: 1) air extracted form the compressor is channeled over the exterior surfaces of major cases to minimize heat expansion, 2) air is routed into the interior of a compressor drum or disc to expand the Rotor. In this manner, compressor blade to case clearances are maintained and compressor efficiency is increased. *See also: 1) Active Clearance Control, 2) Thematic Compressor Rotor.*

PASSIVE CLEARANCE CONTROL (turbine)
A hot section cooling system utilizing air extracted form the compressor to flow over the outer surfaces of turbine cases and minimize heat expansion. In this way, turbine blade to case clearances can be maintained and turbine efficiency increased. *See also: Active Clearance Control.*

PAVLECKA, VLADIMIR
A design engineer for the Douglas Aircraft Company, who in 1933 proposed to build a Turboprop engine to replace the conventional piston engine in a Douglas Aircraft for military use. It was not until WWII, however, that he was able to interest the military in his project. At that time he designed the T-40 Turboprop Engine, built by the General Motors Allison Division.

PAWL & RATCHET CLUTCH
A clutch using a small curved metal part called a Pawl which engages a ratchet-shaped wheel, creating a driving force in only one direction. Many Gas Turbine Engine starters utilize a drive mechanism of the pawl and ratchet clutch type that allows the starter to drive the engine but does not allow the engine to drive the starter. Also called 1) Over-Running Clutch, 2) Sprag Clutch.

Pb
Locate under P (symbols).

PCB
Abbreviation for Plenum Chamber Burning.

PCU
Abbreviation for Propeller Control Unit.

PCD AIR
Abbreviation for Pressure Compressor Discharge. Another name for CDP (compressor discharge pressure).

PEDAL DOOR THRUST REVERSER
A blocker door and deflector, used in place of turning vanes to reverse engine exhaust gases. *See also: Thrust Reverser.*

PEENING
1) A maintenance procedure of displacing metal with a center punch to lock an object in place or to identify a location with a peen-mark. 2) A distressed surface,

locate under Bearing Distress Terms and Engine Inspection Terms in Appendix.

PELLET SEMI-CONDUCTOR
Locate under Igniter Plug.

PERCENT RPM
A reading on a cockpit tachometer indicator to show rotor speed (compressor and turbine) in percent RPM. The pilot (or operator) will not see the actual RPM displayed, but from standard data he can compute actual RPM. For example, if 100% N2 Speed were to equal a 9,000 RPM actual compressor speed, then a 50% N2 speed indication would equal 4,500 RPM.

PERFORMANCE INDICATOR (engine)
A cockpit gauge to show the pilot (or operator) the ongoing performance level of an engine. The two most commonly seen performance instruments are: 1) EPR Gauge, 2) N1 Fan Speed Gauge. Both inform the pilot of ongoing engine thrust when compared to standard charts. *See also: Condition Indicator (engine).*

PERFORMANCE MANAGEMENT SYSTEM
Another name for Flight Management System. *See also: Electronic Flight Instrument System (EFIS).*

PERFORMANCE MULTIPLEXER
An electronic device that receives signals from engine mounted sensors and presents the information on a cockpit display.

PERFORMANCE RECOVERY WASH (compressor)
A maintenance procedure of running an engine at idle speed or slightly above while injecting a wash solution into the engine inlet. This procedure is normally followed by a pure water rinse. Performance recovery wash is accomplished to remove contaminants from compressor airfoils as follows: 1) As a scheduled preventive maintenance procedure, 2) Whenever engine power is lagging. *See also: 1)Desalination Wash, 2) Field Cleaning.*

PERMANENT MAGNET GENERATOR (PMG)
A small engine accessory. An electrical generator of very small output, typically used to power an electronic circuit for a fuel control. Also called Control Alternator.

PERTURBATION (gas path)
An instability within the engine gaspath that disturbs normal airflow at a specific location. Pertubations can lead to more serious compression problems, but most often correct themselves and do not result in engine stall.

PETROLEUM OILS
Depending on the crude oil source, the two general classifications of petroleum oils (mineral oils) are as follows: 1) naphthalenic oils, 2) paraffinic oils. Both are considered general purpose oils with a useful heat range of -40°F. to 250°F. A common commercial light weight petroleum oil used in early turbine engines was SAE Grade 1010, Mil-L-6081. *See also: Synthetic Lubricants.*

PHASED-SHIFT TORQUEMETER SYSTEM
Locate under Torquemeter Phased-Shift System (turboprop).

PHONIC SPEED SENSOR
A Speed Sensor component. Part of an electronic Compressor Speed Indicating System, or a Turbine Overspeed Protection System. The Phonic Sensor's function is to receive a sound signal created by a rotating engine part and convert that signal to a percent RPM indication. The sensor, located on the engine, transmits its data to a cockpit gauge or an engine control system. *See also: 1) Tachometer System (phonic), 2) Turbine Overspeed Sensor System.*

PHYSICAL VAPOR DEPOSITION PROCESS
A plating process designed to increase a metal's resistance to corrosion and wear.

PIAGGIO SPA COMPANY (Italy)
Manufacturer of Gas Turbine Engines. Locate in section entitled "Gas Turbine Engines for Aircraft".

PICKLING (engine)
Generic term for preservation of metals. Colloquial for an engine preservation process prior to engine extended storage. *See also: Engine Preservation.*

PICKUP
In reference to metal distress, locate under Engine Inspection and Distress Terms in Appendix.

PICKUP (sensor)
Another name for: 1) Probe (total pressure), 2) Probe (total temperature). *See also Vibration Pickup.*

PIG TAIL (fuel system)
A small engine component part. A small curved fuel line which allows for thermal expansion. It is generally used as a fuel supply line to an afterburner spray bar.

PILOT FUEL (duplex fuel manifold)
Another name for Primary Fuel.

PIN ROOT (compressor blade)
A type of compressor blade root formed with a retaining pin hole for attachment to the compressor disc. *See also: Blade Roots.*

PING TEST
A maintenance test conducted by lightly tapping an object with a small hammer. For example, when tapping on the vanes in an inlet guide vane assembly, a change of sound from one vane to the next indicates a crack or looseness of fit.

PING TEST (rotor blades)
A test conducted by lightly tapping on a blade with a small hammer and measuring the noise distribution in order to determine the blade's natural frequency. *See also: Frequency Tuning.*

PINION GEAR
The smaller gear of two gears in contact. *See also: 1) Bevel Gear Terms, 2) Bevel Pinion Gear, 3) Planet Pinion Gear.*

PIPE DIFFUSER (engine)
Part of a main engine diffuser section. A series of external pipes that carry compressor discharge air from the Low Pressure Impeller to the High Pressure Impeller of a two-stage Centrifugal Flow Compressor. *See also: 1) Diffuser, 2) Vane Diffuser*

PISTON FUEL PUMP
Locate under Fuel Pump (piston).

PITCH DISTRIBUTION (blade)
Similar to the Blade Twist concept of blade design in engine compressors. Pitch Distribution is thought of as giving the ideal angle of attack at each blade station, whereas Blade Twist is thought of as giving a constant axial velocity off the trailing edge at each blade station. *See also: Blade Twist.*

PITTING (bearings)
Locate under Bearing Inspection and Distress Terms in Appendix.

PLA
Abbreviation of Power Lever Angle.

PLAIN BEARING
Locate under Bearings.

PLAIN COMPOUND EXHAUST DUCT
Locate under Exhaust Duct (configurations).

PLANET PINION GEAR
A pinion gear used in gear reduction systems that takes power from a planetary gear system and drives another unit. *See also: 1) Pinion Gear, 2) Planetary Gear System.*

PLANETARY GEAR SYSTEM (single stage)
A gear reduction system commonly found in Output Reduction Gearboxes of torque producing turbine engines. Planetary Gear Systems receive a high speed input from a turbine rotor and output at a much lower drive speed. The planetary gear system consists basically of three components, 1) turbine shaft driven sun gear, 2) planet gears surrounding the sun gear, 3) ring gear, driven by the planet gears, connected to the output drive shaft. Also called Star-Type Gearing.

PLANETARY GEAR SYSTEM (two stage)
A gear reduction system commonly found in Propeller Reduction Gearboxes. The two stage system is similar to a single stage planetary gear system except that it contains two sets of planet gears in series to provide additional speed reduction. This system is more common to Turboprop engines because all of the gear reduction from turbine speed to propeller speed must be accomplished within the engine. A rotorcraft powerplant provides only a part of the gear reduction and the aircraft transmission provides the remainder.

PLASMA SPRAY COATING
A process of applying a coating of highly wear resistant or heat resistant material onto a turbine engine part to prolong its service life or to rebuild its surfaces to original dimensions. The process is accomplished by a spraying technique using extremely high temperature (approximately 50,000 °F) to melt together the base and coating materials. This process is often referred to as an unglazed coating. *See also: 1) Laser Glazing, 2) Thermal Barrier Coating, 3) Vacuum Plasma Coating.*

PLASTICIZING
A plastic anti-corrosion coating process used on inner surfaces of compressor cases and on outer surfaces of compressor vanes.

PLEATED CYLINDRICAL FILTER (oil)
An engine accessory. A main oil filter, generally cleanable, with a stainless steel wire mesh filtering element. It is cylindrical in shape and pleated to provide maximum surface area for filtration. *See also: Oil Filter (cleanable).*

PLENUM CHAMBER (inlet air)
A component part of an inlet duct, either aircraft or engine. A special purpose type Flight Inlet acting as an air collection area. Its function is to intake ambient air and then distribute it to the engine compressor and also to other points requiring ram air.

PLENUM CHAMBER BURNER (PCB)
British for a type of thrust augmentation. An afterburner in a Lift Fan bypass duct located at the Fan Exhaust. Often called Lift Burning. *See also: Remote Augmented Lift System.*

PLUG (igniter)
Locate under Igniter Plug.

PLUG NOZZLE (engine)
An engine component. A tailpipe with a protruding tailcone (plug). On some engines axial gas velocity is increased by using this design instead of the more conventional Internal-Flow Nozzle. A shorter tailpipe length also results with a plug nozzle and a corresponding weight saving. *See also: 1) Nozzle Plug, 2) Internal-Flow Nozzle.*

PLUNGER-TYPE PUMP
Locate under Fuel Pump (piston).

PMUX
Abbreviation for Propulsion Multiplexer.

PNdB
Abbreviation for Perceived Noise Decibels.

PNEUMATIC-MECHANICAL FUEL CONTROL
An engine accessory. A type of fuel control utilizing pneumatic and mechanical forces to operate its internal mechanisms. Air tapped from the engine compressor is routed into the fuel control to provide a motive force for many of its functions. The traditional power lever and flyweight governor are also present in the mechanical portion of the unit. *See also: Fuel Control Unit.*

PNEUMATIC FIRE DETECTION SYSTEM (engine)
An engine nacelle fire and overheat detection system. The system includes a long hollow gas filled tube installed on the inner walls of an engine nacelle. The gas within the tube expands in the presence of heat and the gas pressure closes an electrical switch to a warning light in the cockpit. One of the more common types is the Systron-Donner System. *See also: Fire Detector.*

PNEUMATIC STARTER
An engine accessory. A unit typically mounted on the Accessory Drive Gearbox. An air motor which provides starting torque and initial rotation of a turbine engine. This starter operates by receiving a large volume, low pressure air supply directed through an internal, high speed turbine that is gear reduced to its output shaft. The Pneumatic Starter is the unit of choice for most medium to large Gas Turbine Engines because of its

high power to weight ratio. Also called 1) Air Turbine Starter, 2) Compressed Air Starter Motor (British). *See also: 1) Customer Bleed Air, 2) Manifold Bleed Pressure.*

PNEUMATIC STARTER (High-Low Pressure)
A derivative of the basic Pneumatic Starter. This unit can utilize the traditional low pressure, high volume air supply from the aircraft manifold or a high pressure air supply from aircraft storage bottles. The bottle-start provides an engine self-starting capability in the event of failure of the low pressure air supply.

POINT CAPACITOR (ignition system)
A small component part of the ignition exciter unit. The capacitor is wired in parallel across the points in the primary transformer circuit. Its function is to prevent arcing by absorbing the charge when the points begin to open.

POLISH MANUFACTURED ENGINES
Locate under Polskie Zaklady Lotnice in section entitled "Gas Turbine Engines for Aircraft".

POLLUTANTS
Locate under Emission (engines).

POLSKIE ZAKLADY LOTNICE (PZL Poland)
Manufacturer of Gas Turbine Engines. Locate in section entitled "Gas Turbine Engines for Aircraft".

POLYTROPIC EFFICIENCY
A physics term. Refers to the isentropic efficiency of a compressor stage. The modern compressor is designed with a fairly constant isentropic efficiency over each stage. Older compressors were designed with a fairly constant compression ratio over each stage. *See also: 1) Compressor Efficiency, 2) Isentropic Efficiency.*

POP-OPEN NOZZLE
A type of afterburner exhaust nozzle that cycles (pops) to the full open position at idle engine speed. The pop-open nozzle provides a low exhaust velocity and low thrust operation for flightline safety during ground run-up and during aircraft movement on the ground.

POP-OUT INDICATOR
Visual indicators which show: 1) a partially clogged fuel or oil filter condition exists, 2) overloads of electrical equipment have taken place. *See also: Filter Bypass Indicator (pop-out).*

POSITIVE DISPLACEMENT PUMP
Locate under Pump (positive displacement).

POSITRON EMISSION TOMOGRAPHY
An inspection technique using a Positron-Emitting radio-isotope generator. When positrons encounter fluid such as fuel or oil in an engine, photons are produced which generate video images. The images can be viewed in real time or as static images.

POST-EXIT THRUST REVERSER
Locate under Thrust Reverser.

POTENTIAL ENERGY
A physics term. It is stored energy that has the capacity to perform work. Energy of position, such as the energy stored in an airplane at altitude, is an example. Potential Energy (Pe) is expressed as Force (weight) x Distance. Pe is expressed in foot pounds in U.S. Units and Newton Meters in the Metric System. For example, Pe = 60 ft.lbs. (81.4 Nm). *See also: Energy.*

POUR POINT (oil)
A quality of lubricating oil referring to the lowest oil temperature (point) at which oil will gravity flow. The pour point affects the time it takes to prime the various oil wetted engine components when an engine is first started at low ambient temperatures.

POWDER METAL DISC (turbine)
A turbine disc manufactured by a process which produces high density to weight ratio materials. This disc is also stated to be relatively free of defects because no defect can be greater than the largest particle of the fine grain metallic powder. The disc can therefore be made lighter, permitting higher wheel speeds. *See also: Powder Metallurgy.*

POWDER METALLURGY (PM)
A recent development in alloy production utilizing metal in powder form, rather than in the form of a metal ingot. A process of extremely high pressing force which produces sufficient heat to melt the fine gain powder to form a billet of very high density. Currently used in the production of many engine parts constructed of such alloys as aluminum, titanium, nickel, and cobalt. *See also: 1) Hot Isostatic Pressing, 2) Rapid Solidification Rate, 3) Super Alloy.*

POWER
Physics term. Power is expressed as Force x Distance ÷ Time in the U.S. System and Joules per second in the Metric System. For example, 50 ft.lbs./min. (1.13 Joules/sec.). The power formula is the basis for calculating horsepower and thrust horsepower of a turbine engine.

$$\text{Power} = \text{Force} \times \text{Distance} \div \text{Time}$$

POWER ANALYZER RECORDER™ (PAR)
A Teledyne Company trademark. A continuous engine monitoring and diagnostic system. This system records such data as: start times, cool down times, number and severity of hot starts, and various engine parameters. The data can be displayed on cockpit instruments for use by the operator, or recorded on tape for later use in planning maintenance.

POWER ASSURANCE CHECK (rotorcraft)
An operational check to determine whether Rated Torque and Rotor Speed are being produced by a Turboshaft engine. *See also: Engine Power Ratings.*

POWER-BACK PROCEDURE (aircraft)
A procedure in which main engine reverse thrust is used to move an aircraft backwards on the ground. Power Back is used in backing an aircraft from a loading ramp rather than using a Push-Back Procedure using a tow vehicle. Also called Reverse Taxi Procedure. *See also: Push-Back Procedure.*

POWER LEVER
The cockpit lever used by the operator to transmit a manual signal to the fuel control unit on the engine. A manual means of scheduling fuel flow to the combustor. Also called the Throttle Lever, a hold-over from year's past when power was managed by throttling airflow into intake manifolds of reciprocating engines.

POWER LEVER ANGLE (PLA)
The angle formed when the power lever linkage on the fuel control unit is moved from zero angle at idle speed to high angle at full power. A protractor on the fuel control indicates movement in degrees. The degrees of movement have a relationship to engine speed.

POWER LEVER ALIGNMENT
Locate under Throttle Alignment.

POWER MANAGEMENT CONTROL
An electronic over-ride system on a hydro-mechanical fuel control unit which protects against engine overshoots of EGT and RPM. This system also provides a consistent power lever angle (PLA) for take-off regardless of the outside air temperature and inlet conditions.

POWER OUTPUT GEARBOX
A gearbox which reduces engine turbine wheel speed to correct propeller drive speed in a Turboprop engine, or to correct transmission drive speed in a Turboshaft engine for rotary-wing or tilt-wing aircraft. *See also: Gearbox.*

POWER-PACK
Name given to a Gas Turbine Engine in its most complete form, ready for installation in an aircraft. For example, a wing pod Power Pack would typically include the following: 1) an engine with its quick engine change kit, 2) inlet cowl, 2) side cowls, 3) exhaust duct, 4) thrust reverser, 5) exhaust plug, and 6) strut assembly which holds the nacelle to the pylon. *See also: 1) Basic Engine, 2) Quick Engine Change Kit.*

POWERPLANT
In FAA terms, a prime mover engine used to propel an aircraft. *See also: Gas Turbine Engine.*

POWERPLANT RATING
A rating granted to individuals under FAR 65, who meet the experience and testing requirements set forth by the Federal Aviation Administration (FAA). These individuals are authorized to perform "return to service" maintenance on all powerplants in aircraft with U. S. Registry. By reciprocal agreement, these individuals are also able to perform maintenance on aircraft of foreign registry when they are under FAA jurisdiction.

POWER RATINGS (engine)
Locate under Engine Power Ratings.

POWER TAKE-OFF GEARBOX (PTO)
An engine component. A small bevel gear system driven by the compressor. The PTO in turn drives the Accessory Drive Gearbox or an Auxiliary Drive Gearbox. In some older engines, the PTO was externally mounted on the front of the compressor. Today it is more common to locate the PTO within the engine.

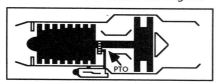

POWER TRANSMISSION SYSTEM (engine)
Another name for Accessory Drive Gearbox.

POWER TRANSFER GEARBOX
Another name for Auxiliary Drive Gearbox.

POWER TURBINE (PT)
Another name for Free Power Turbine.

POWER TURBINE MODULE
Locate under Module.

POWER UNIT (gas turbine)
Short for Auxiliary Power Unit or Ground Power Unit.

PRATT & WHITNEY (P&W)
A division of United Technologies Corporation. The manufacturer of many commercial and military turbine engines. The commercial division is located in East Hartford, Connecticut and the military division is located in West Palm Beach, Florida. Locate in section entitled "Gas Turbine Engines for Aircraft".

PRATT & WHITNEY OF CANADA LTD. (P&WC)
A division of United Technologies Corporation, located in Montreal, Canada. Manufacturer of many small Gas Turbine Engines. Locate in section entitled "Gas Turbine Engines for Aircraft".

PRE-CHAMBER TYPE COMBUSTOR
An engine component. A recently developed annular combustor with two separate combustion chambers. The forward or head end of the combustion liner is configured with a small separate fuel and air mixing chamber that contains its own fuel nozzle. After initial combustion, the gases flow into a larger section of the liner where the main (secondary) fuel nozzle is located. This arrangement is said to provide the following advantages over traditional combustor designs: 1) low emission of pollutants, 2) ease of starting in cold weather, 3) flameout protection on deceleration.

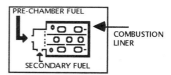

PRE-EXIT THRUST REVERSER
Locate under Thrust Reverser.

PRE-TURBINE INJECTION (PTI)
An older method of thrust augmentation. During flight, fuel was introduced forward of the turbine nozzles to create additional expansion of the gases. However, the severe impact on service life of turbine components limited its usefulness.

PRESERVATION (engine)
A method of preparing an engine for storage. Preservation generally includes flooding the fuel system and sometimes the oil system with special preservative fluids, placing desiccants such as Silica-Jel™ in the inlet and tailpipe, and wrapping the engine in moisture proof paper. For very extended storage, the engine is placed in a pressurized metal container after preservation.

PRESSURE
A physics term. A measurement of force per unit of area. For example, within the mechanisms of an air pressure gauge, it is the force exerted by the number

of molecules and the activity of those molecules. In this case, molecules will hit and rebound with such rapidity that a continuous pressure can be measured. The typical measurements are 1) lb./in.^{2} in the U.S. System and kg/m^2) in the Metric System.

PRESSURE (absolute)
Locate under Absolute Pressure.

PRESSURE (atmospheric)
Locate under Barometric Pressure.

PRESSURE (gauge)
Locate under Gauge Pressure (psig).

PRESSURE (ram, static, total)
Locate under 1) Ram Pressure Rise, 2) Static Air Pressure, 2) Total Air Pressure.

PRESSURE (Po)
At International Standard Day conditions, 29.92 in. hg., 14.696 lb./in.^{2}, 2116.22 lb./ft.^{2}.
See also: 1) Temperature (To), 2) Density (of air).

PRESSURE ALTITUDE
The altitude based on pressure of air only, as sensed by a sensitive barometer type altimeter. The altitude in standard air that corresponds to an existing air pressure. For instance, at 24.90 in.hg. absolute (12.23 psia) the pressure altitude is 5,000 feet. This represents a convenient means of expressing the ambient pressure as a representative value for the actual altitude. See also: Density Altitude.

PRESSURE DIFFERENTIAL VALVE
Another name for Bypass Relief Valve.

PRESSURE FILL PORT (oil tank)
An oil tank fill point other than at the oil filler cap (the gravity fill port). The pressure fill port is generally a quick disconnect type fitting to which a hand-pump type oil servicing cart can be attached. The cart is often called an Oil Bowser. Servicing in this manner provides a means of: 1) preventing atmospheric contaminants from entering the engine during servicing, 2) filtering oil before service to ensure no oil can materials enter the engine. See also: 1) Oil Bowser, 2) Oil Tank, 3) Overflow Line (oil tank).

PRESSURE RAKE
Locate under Rake (pressure).

PRESSURE RATIO
A physics term. One pressure divided by another to form a mathematical ratio of the two values. Pressure ratios are used to describe numerous engine functions. See also: 1) Compressor Pressure Ratio, 2) Engine Pressure Ratio, 3) Inlet Pressure Ratio.

PRESSURE REGULATING RELIEF VALVE
Locate under Relief Valve.

PRESSURE SIDE COOLING (turbine blades & vanes)
Refers to film cooling on the under (convex) side of the blade or vane. See also: Film Cooling.

PRESSURE SUBSYSTEM (lubrication system)
Locate under Oil Pressure Subsystem.

PRESSURE THRUST
Thrust produced when an exhaust nozzle is operating in a choked condition. Pressure thrust adds to the ongoing Reactive Thrust. Pressure Thrust is calculated by multiplying the area of the nozzle by the pressure above ambient pressure across the nozzle. The hot tailpipes of most turbine powered aircraft flow at a choked condition while in the cruise to take-off power range. The formula for calculating Pressure Thrust is as follows: Aj(Pj-Pam); where Aj = area of jet nozzle, Pj = pressure absolute across jet nozzle, Pam= pressure ambient. See also: Thrust.

PRESSURE TOTAL (Pt)
Locate under Total Air Pressure.

PRESSURIZING AND DUMP VALVE (fuel system)
An engine accessory. A fuel distributor valve supplying fuel to the primary and secondary manifolds of a duplex fuel nozzle system. The cycle of operation is as follows: 1) The dump valve closes to allow fuel to enter the primary fuel manifold for engine starting and ground idle, 2) The pressurizing valve opens, flowing fuel to the secondary fuel manifold at approximately flight idle speed, 3) On shutdown the dump valve reopens draining fuel from both manifolds to keep fuel from forming carbon in the fuel nozzles. Also called 1) P & D Valve, 2) Pressurizing and Drain Valve. See also: Dual-Line Duplex Fuel Manifold.

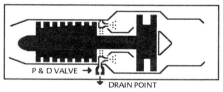

PRESSURIZING VALVE (fuel system)
A small engine accessory. A check valve type device located downstream of the fuel control unit which prevents fuel flow past that point until a predetermined fuel pressure build-up in the system is achieved. The presence of this valve accomplishes the following: 1) It tends to keep working pressure at correct values for operation of fuel control servo mechanisms, 2) It prevents flow to the fuel nozzles until sufficient pressure is available to ensure a good spray pattern. Also called Minimum Pressurizing Valve. See also: Minimum Pressurizing and Shut-Off Valve.

PREVENTIVE MAINTENANCE
FAA designated routine maintenance tasks, performed at defined intervals to ensure that aircraft and engines remain in airworthy condition. The FAA allows flight personnel, other than certificated maintenance technicians, to perform this maintenance.

PRIMARY AIR (combustor)
Approximately 25% to 40% of the total compressor discharge air. Primary air is routed to the head (inlet) of the combustion liner to support combustion of fuel. See also: 1) Combustion Zone, 2) Secondary Air.

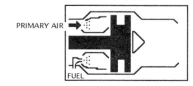

PRIMARY AIRSTREAM (turbofan engine)
The portion of total airflow from the flight inlet that is routed into the core engine. *See also: Secondary Airstream (turbofan engine).*

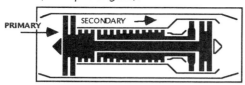

PRIMARY COMBUSTION ZONE
Locate under Combustion Zone.

PRIMARY EXHAUST DUCT AFTERBURNER
The forward (inner) ring of flaps in a Flap Exhaust Nozzle. When opened, the flaps change the geometry of the exhaust to match the flow condition of the engine when operating in afterburner. Also called Primary Exhaust Nozzle. *See also: Flap Exhaust Nozzle (afterburner).*

PRIMARY EXHAUST DUCT TURBOFAN
The point at which the Core of a Turbofan engine discharges its gases. Depending on whether the engine has an unmixed or mixed exhaust duct, the primary nozzle will flow directly back to the atmosphere or it will mix with the fan discharge air in a common tailpipe. *See also: 1) Secondary Exhaust Duct, 2) Mixed Exhaust Duct, 3) Unmixed Exhaust Duct.*

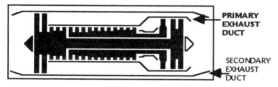

PRIMARY EXHAUST NOZZLE EXIT
Locate under Exhaust Nozzle Exit.

PRIMARY EXHAUST NOZZLE THROAT
Locate under Exhaust Nozzle Throat.

PRIMARY FUEL FLOW (combustor)
The initial fuel flow to the combustor during engine start. Primary Fuel Flow will be supplemented by Secondary Fuel Flow to maintain all engine speeds higher than idle speed. Primary Fuel generally flows from the center orifice of a duplex fuel nozzle. In some installations, It may also flow from a separate fuel nozzle. Also called Pilot Fuel. *See also: 1) Pressurizing and Dump Valve, 2) Secondary Fuel Flow (combustor).*

PRIMARY GEAR ELEMENT (fuel pump)
A component part of a parallel-stage fuel pump. A pair of pumping gears placed in parallel with a second pair of pumping gears called the Secondary Gear Element. *See also: Fuel Pump (parallel-stage).*

PRIMARY JET NOZZLE
Another name for Primary Exhaust Nozzle.

PRIMARY LEAD (ignition system)
Another name for High Tension Lead.

PRIMARY ZONE (combustor)
Refers to the area within the head end of a combustion liner where the Primary Air enters and mixes with fuel. *See also: Primary Air.*

PRIMER FUEL SYSTEM
A secondary fuel system used in conjunction with a main fuel system. The primer system contains an on-off valve, a manifold, and a set of fuel nozzles, generally of the atomizing type. *See also: 1) Primer Fuel System (starting), 2) Primer Fuel System (continuous).*

PRIMER FUEL SYSTEM (starting)
A low volume fuel delivery system used mainly for ground starting, but can also be used for an in-flight Windmilling Start. Primer Nozzles are generally atomizing types and are used where the main distribution system utilizes a vaporizing type nozzle. Also called a Starting Fuel System. *See also: 1) Fuel Nozzles, 2) Fuel Vaporizing Tube.*

PRIMER FUEL SYSTEM (continuous)
An extension of the Primer Fuel Starting system that can be activated to prevent engine stalls during undershoots of engine power after a rapid throttle chop to the idle stop.

PRIST™
A fuel additive. An anti-icing and anti-biocidal fluid additive in Jet Fuel to prevent formation of ice or biological growth in fuel tanks and other fuel system components of aircraft or engine. Prist is normally sprayed directly from an aerosol container into the fuel tank during aircraft refueling. *See also: Fuel System Icing Inhibitor.*

PROBE (total pressure)
A small engine accessory. A metal tube-type sensor that extends into the engine gaspath to be acted upon by the gas stream. The probe senses both static pressure and ram pressure and when connected to a gauge will indicate total pressure at the point of insertion. For example, a Turbine Discharge Pressure probe in an Engine Pressure Ratio (EPR) system. Also called Total Pressure Probe.

PROBE (total temperature)
A small engine accessory. A metal tube-shaped sensor that extends into the engine gaspath to be acted upon by the gas stream. The probe senses both static temperature and ram (friction) temperature and when connected to a gauge will indicate total temperature at the point of insertion. An example of this type probe is a Tt_7 probe in an Exhaust Gas Temperature (EGT) Indicating system. Also called 1) Total Temperature Probe, 2) Thermocouple.

PROGRESSIVE INSPECTION
An inspection program that eliminates D-Checks and combines the individual tasks required therein into each of the C-Checks that normally occur within the D-Check cycle. Also known as 1) Equalized Maintenance, 2) Phased Maintenance, 3) Continuous Maintenance. *See also: Letter Checks.*

PROFILE-TIP BLADES (compressor & turbine)
Another name for Squeeler-Tip Blades.

PROOF-OF-CONCEPT TEST (engine)
An FAA testing procedure to determine that an engine undergoing research and development meets certain

minimum performance standards.

PROP
Shortened term for Propeller.

PROP-FAN BLADE
A rotating airfoil (blade) mounted on a Prop-Fan Engine. It function is similar to that of a propeller blade, but it is of newer aerodynamic design. Because of it size, newer materials are being developed for this type blade, one of which is a foam material over a central aluminum core, covered with fiberglass.

PROP-FAN™ ENGINE
A Hamilton-Standard Company trademark. Commonly used today to describe a class of UHB engines. Locate under Ultra-High-Bypass Engine (UHB).

PROP ROTOR
A derivative of the aircraft propeller and helicopter rotor that is now utilized on the newer Tilt-Rotor aircraft. This rotor can operate in forward flight as a puller propeller or in vertical flight similar in function to a helicopter rotor. *See also: Cross Shaft Transmission.*

PROPELLER
A component attached to an aircraft engine. A rotating set of airfoils designed to move air rearward and create forward thrust perpendicular to the propeller's plane of rotation. Also called an Airscrew (British). *See also: 1) Prop Rotor 2) Propulsor.*

PROPELLER (CRP)
Locate under Counter-Rotating Propeller.

PROPELLER (puller)
The conventional propeller arrangement with the propeller mounted in front of a Turboprop engine. The propeller creates a reactive thrust as it moves air rearward. The propeller in turn pulls the aircraft through the air. In these terms, the new front mounted propfans (propellers) on Ultra-High-Bypass engines fall into this category. Also called a Tractor-Mounted Propeller.

PROPELLER (pusher)
A propeller located at the rear of a Turboprop engine. The propeller creates a reactive thrust as it moves air rearward. The propeller, in turn, pushes the aircraft through the air. In these terms, the new rear mounted propellers on Ultra-High-Bypass Fan engines fall into this category. *See also: Hot Prop.*

PROPELLER (SRP)
Abbreviation for Single-Rotating Propeller.

PROPELLER BRAKE
An engine component. 1) A safety device that locks the propeller reduction gearbox. The brake prevents a feathered propeller from windmilling and creating excessive drag when an engine is shut down in flight. In some cases this brake can also be employed to decrease rotational time during normal shutdown. 2) A braking device (usually hydraulically operated) used to stop rotation of the Free Turbine and Propeller while the Gas Generator portion of the engine remains operating at idle speed. This arrangement allows the engine to act as an Auxiliary Power Unit during quick turn-around operations. Also called Arrested Propeller System. *See also: Brake (propeller or rotor).*

PROPELLER CONTROL UNIT (PCU)
An engine accessory. A device used to control the propeller condition on a Turboprop Engine.

PROPELLER EFFICIENCY
A physics term. The ratio of Thrust-Horsepower, to Shaft Horsepower delivered to the propeller. The average thrust horsepower developed is approximately (0.8) of the shaft horsepower input, which results in a propeller efficiency of 80%. This figure is true of propellers on piston and Turboprop engines up to approximately Mach 0.6 flight speed, after which efficiency begins to decrease. *See also: Prop-Fan Engine.*

$$\mathrm{Prop}_{(eff)} = \mathrm{THP} \div \mathrm{SHP}$$

PROPELLER GEARCASE (PGC)
Another name for Propeller Reduction Gearbox.

PROPELLER GOVERNING MODE (engine)
A Turboprop Engine governing mode of operation wherein the propeller governor selects blade pitch to control engine RPM.

PROPELLER PITCH CHANGE UNIT
Locate under Autonomous Electro-Mechanical Pitch Change Unit.

PROPELLER REDUCTION GEARBOX (turboprop)
An engine component. A reduction gearbox on a Turboprop engine that converts a high speed, low torque input from a turbine wheel into low speed, high torque to a propeller shaft. This gearbox is generally a two stage planetary gear system and described as a very heavily loaded engine component. In some cases the reduction capability is as high as 30:1. *See also: 1) Cantilevered Propeller Gearbox, 2) Output Reduction Gearbox (turboshaft).*

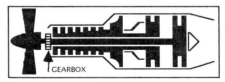

PROPELLING NOZZLE (British)
Another name for Exhaust Nozzle.

PROPJET
Common name for Turboprop powered aircraft.

PROPULSION MULTIPLEXER (PMUX)
Often an engine mounted unit which converts signals from engine sensors to digital data to be used in an engine monitoring system. *See also: Ground-Based Engine Monitoring (GEM) system.*

PROPULSIVE DUCT
Another name for Ram-Jet.

PROPULSIVE EFFICIENCY
A physics term. A measurement of how efficiently the available energy at the exhaust nozzle(s) of a turbine engine is used to create propulsive power. It can also be described as a measure of energy losses in the exhaust to the atmosphere. In general terms,

Propulsive Efficiency is a comparison of an aircraft's forward speed and its engine's exhaust velocity. Mathematically expressed as a percentage by the following formula:

$$P_{eff} = 2 \div [1 + (V_2 \div V_1)]$$

Where V_2 = Exhaust Velocity, V_1 = Aircraft Speed.
See also: 1) Efficiencies (engine), 2) Overall Efficiency, 3) Thermal Efficiency.

PROPULSOR UNIT (UDF)™
An engine component. A recently developed pusher-type fan module which mounts externally to an Unducted Fan Engine (UDF). The Propulsor contains a tandem mounted, contra-rotating fan module which is driven by turbine wheels within the Propulsor Unit. The Propulsor contains no stationary turbine nozzle vanes, but rather bladed-wheels acting as nozzles placed in front of the traditional turbines wheels. Six bladed wheels rotate in one direction while an additional 6 bladed wheels rotate in the opposite direction. This design can be thought of as employing two, six-stage, independent, counter-rotating turbines. Each six-stage turbine unit drives a Unison Ring assembly to which a series of fan blades are mounted. The rotating nozzle vane concept is said to double the energy onto the adjacent wheels, providing enough energy to directly drive the fan without benefit of reduction gearing. *See also: 1) Unducted Fan Engine, 2) Unison Ring.*

Ps SYMBOLS
Locate under P (symbols).

PSEUDO FLANGE
An engine component part. A flange on a hot section outer case containing an air cooling passageway designed to minimize heat expansion and maintain established clearances. *See also: Passive Clearance Control.*

PSIA, PSID, PSIG
Locate under Absolute Pressure, Differential Pressure, Gauge Pressure.

Pt SYMBOLS
Locate under P (Symbols).

PTIT
Abbreviation for Power Turbine Inlet Temperature. A temperature indication taken by a thermocouple probe located between the Gas Producer Turbine and the Power Turbine.

PTO
Abbreviation for Power Takeoff. Locate under Power Takeoff Gearbox.

PULLER PROPELLER
Locate under Propeller.

PULSE-JET
A type of ram-jet engine configured with inlet shutters. The German military, during WWII, developed a flying bomb called a V-1 "buzz-bomb" that was powered by this type engine. No further development was done with the pulse-jet because of its poor power to weight ratio as compared to the traditional ram-jet. The Pulse-Jet was also known as an Impulse Duct.

PULSED CHIP DETECTOR
Another name for Chip Detector (capacitive discharge)

PULSE PICK-UP (engine speed)
A unit similar in function to a Fan Speed Sensor. The Pulse Pick-up is employed to sense compressor speeds and transmit a signal to a cockpit indicator by way of an electronic signal processor. Another name for Speed Sensor.

PUMP (centrifugal)
A pump utilizing a rotating impeller to pressurize and move a fluid. This type pump is most often used as a boost pump, delivering fuel to a second, higher pressure type pump. It can also be used as a primary delivery pump. *See also: Fuel Pump.*

PUMP (positive displacement)
A pump that utilizes a pair of rotating pumping elements, such as spur gear, gerotor, or sliding vane, to pressurize and move a fluid in a fixed volume per revolution. The output of a Positive Displacement Pump is directly proportional to its speed. Also called: 1) Constant Delivery Pump, 2) Constant Displacement Pump, 3) Fixed Displacement Pump. *See also: 1) Fuel Pump, 2) Oil Pump.*

PUMP (variable displacement)
Locate under Fuel Pump (piston).

PUMP CAVITATION
Introduction of air in quantity sufficient to reduce pumping action. Cavitation occurs when a fluid pump is pulling in air along with fluid and losing efficiency. Cavitation generally occurs when fluid is low in the reservoir feeding the pump.

PURGE CYCLE
Another name for Clearing the Engine.

PURGE TANK (fuel system)
A small engine accessory. A fuel drain tank into which the fuel manifolds deposit fuel during engine shutdown. The tank is pressurized with air and purges the fuel back into the combustor to evaporate. The Purge Tank system prevents boiling of fuel and carbonizing of fuel nozzles after engine shutdown.

PURGE VALVE (fuel control)
A component part of a fuel control. A valve which opens during the engine starting cycle to clear the fuel control of vapor. This entrained air is generally returned back to the fuel supply tank where it is vented out of the system.

PUSH-BACK PROCEDURE
Moving an aircraft back out of the loading area by means of a tow-bar and tractor as opposed to a Power-Back Procedure using the main engines in reverse thrust. *See also: Power-Back Procedure.*

PW (P&W)
Abbreviation for Pratt & Whitney Company.

PYLON
An aircraft component. A support used in attaching an engine to an aircraft. The pylon fits between an aircraft fuselage or an aircraft wing and the engine nacelle (pod).

PYROMETER (turbine temperature)
A small engine accessory. A non-intrusive sensor which monitors exhaust gas temperature. An optical device is used measure infrared emissions via a small opening in the exhaust casing. The pyrometer sends a cut-back signal to the Fuel Control Unit when turbine temperature reaches its maximum limit.

Q

Q.A.D. CLAMP
Abbreviation for Quick-Attach-Detach Clamp.

Q.E.C. ACCESSORY
An aircraft accessory that is installed on a turbine engine along with basic engine accessories. Q.E.C. accessories are considered non-essential to basic engine operation. For example, the hydraulic pump, electrical generator, fluid pressure transmitters, etc. *See also: Quick Engine Change Kit.*

Q.E.C. KIT
Abbreviation for Quick Engine Change Kit.

QUANTITATIVE DEBRIS MONITORING (QDM)
An electronic means of detecting mechanical failures in oil-wetted parts of a Gas Turbine Engine. Ferrous debris passing through an engine mounted QDM Sensor along with the normal engine oil flow produces electrical pulses. The mass of the debris scales the size of the electrical pulses. The pulses, when of sufficient magnitude, alert a fault indicator on a cockpit display. *See also: 1) Chip Detector, 2) Oil Filter Debris Analysis.*

QUICK ATTACH-DETACH CLAMP
A circular two-piece clamp utilized to fasten accessories to their mounting points on the Accessory Drive Gearbox. Both the accessory and the gearbox mounting point are configured with interlocking flange rings. When installing an accessory, the Q.A.D. clamp can be secured by a single fastener which locks the rings together.

QUICK ENGINE CHANGE KIT (QEC Kit)
A set of fluid lines, electrical harnesses, and engine accessories that attach to a Basic Engine to configure it for installation in a particular aircraft. Sometimes called an Engine Build-Up Kit or Airframe Company Furnished Accessories Kit. *See also: 1) Power Pack, 2) QEC Accessory.*

QUILL SHAFT
Name given to a shaft with a series of parallel grooves (splines) cut into its outer surface at the drive end.

R

RABBETED FLANGE
A type of overlapping, circular flange that provides a sealing and strengthening feature to two adjoining parts.

RADIAL AIR INLET (intake)
A screened air inlet circling the outer surface of a turbine engine. Radial intakes are seen mainly on Turboprop and Turboshaft engines. Inlet air is directed into the screen from a Plenum within the aircraft.

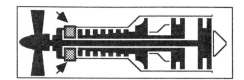

RADIAL CARBON OIL SEAL
Locate under Carbon Oil Seal.

RADIAL CLEARANCE CHECK
A clearance measurement taken in a direction 90 degrees to the axis of the engine center-line. Compressor and Turbine blade tip clearances are examples of radial clearances.

RADIAL DRIVE SHAFT
Another name for Accessory Gearbox Drive Shaft.

RADIAL LOADING
Centrifugal loading on bearings and other rotating components which emanates outward from the engine axis. *See also: Bearing (roller).*

RADIAL INFLOW TURBINE
A turbine wheel with the appearance of an impeller. Unlike an impeller, however, this turbine receives gases at its blade tips from a turbine stator nozzle and then turns the gases inward and outward into the exhaust duct. This turbine system is common to Auxiliary Power Units. *See also: Nozzle Ring.*

RADIAL OUTFLOW COMPRESSOR
Another name for Centrifugal Flow Compressor.

RADIAL RUNOUT CHECK
A dial micrometer measurement performed to check for bend, distortion, or excessive radial movement of rotating components such as compressor and turbine shafts and discs. Also called Diametral Runout. *See also: 1) Axial Runout Check, 2) Runout Check.*

RADIOLOGY
Locate under Non-Destructive Inspection.

RAG AND GLUE MATERIAL
Jargon for epoxy materials with a fabric material base.

RAIN STEP
A baffle in a flight inlet of a rotorcraft that helps prevent precipitation in the atmosphere from entering the compressor. Often the rain step is part of an Inertial Sand and Ice Separator.

RAKE (pressure)
A pressure sensing point formed by small forward facing holes in a stationary engine vane. The rake serves as a total pressure probe to sense both ram and static pressure. Often this signal is sent to the fuel control for fuel scheduling purposes.

RAM ACCELERATION
Another name for Throttle Burst.

RAM AIR PRESSURE RISE (flight inlet)
The pressure buildup above ambient pressure in the Flight Inlet due to the forward motion of the aircraft. The result of speed (kinetic energy) of a unit volume of air diffusing to cause a pressure rise. For example,

the pressure rise at Mn = 0.85 flight speed is approximately 1.6 times ambient. Or it could be stated that the inlet ram compression is 1.6 : 1. Also called Dynamic Pressure rise.

RAM AIR TEMPERATURE RISE (flight inlet)
Compressor Inlet Temperature (CIT) rise due to the increased friction in air which results as a function of airspeed and inlet pressure rise. *See also: Ram Pressure Rise.*

RAM COMPRESSION
Locate under Ram Pressure Rise.

RAM COMPRESSION RATIO
Another name for Ram Ratio.

RAM DRAG
A physics term. Mathematically expressed as: 1) Gross Thrust minus Net Thrust, or 2) Engine Mass airflow times Aircraft Speed. Ram Drag increases as aircraft speed increases due to momentum losses to Mass Airflow. Ram drag is zero with maximum velocity change of engine mass flow, this occurring when a turbine engine is operating at full power but not moving forward. Also called Inlet Momentum Drag.

RAM HORSEPOWER RATING (turboprop)
A power rating slightly above an engine's static rated, Take-Off Shaft Horsepower (SHP). The horsepower available during take-off after ram inlet pressure rise boosts the Static Rated SHP. For example, one model of the Allison T-56 Turboprop Engine gains 90 SHP due to the ram inlet pressure rise (4,910 to 5,000).

RAM INDUCTION COMBUSTOR
A recent development in annular combustor design utilizing a liner that incorporates a diffuser shaped fairing around its dome in the area of the fuel nozzle. The fairing affects a slight air pressure rise prior to combustion to promote an increased heat release.

RAM-JET
An open cycle engine in the family of Jet Engines. The ram-jet is an engine with few if any moving parts in its gaspath. A Ram-Jet creates thrust by ram air compression in its inlet, followed by combustion and expansion of gases from a jet nozzle.
See also: 1) Athodyd, 2) Scramjet.

RAM PRESSURE RISE (flight inlet)
Locate under Ram Air Pressure Rise.

RAM RATIO
The mathematical ratio of Compressor Inlet Pressure (CIP) to Ambient Pressure. Also called Ram Compression Ratio.

RAM RECOVERY
The ability of a flight inlet to take advantage of Ram Pressure.

RAM RECOVERY POINT (flight inlet)
Recovery from a negative air pressure to a positive air pressure. Ram Recovery Point is the point at which the suction (negative pressure) within the inlet returns to a value equal to ambient pressure as the aircraft moves forward on the ground. This will typically occur at or near take-off airspeed. Also called Duct Recovery Point.

RAM TEMPERATURE RISE
Locate under Ram Air Temperature Rise.

RANKINE TEMPERATURE SCALE
The absolute temperature scale in the U.S. System of measurements where the absolute zero temperature is zero degrees Rankine. Conversion can be made to degrees Fahrenheit such that absolute zero is minus 460 °F. For example, conversion of 519 °R is (519-460) = 59° Fahrenheit.

RAPID SOLIDIFICATION RATE (RSR)
A powder alloy process which produces very uniform small grain size and very few internal flaws when a billet is formed. The process involves cooling molten alloy at a rate of millions of degrees Celsius per second. Afterwards, the powder is compacted into a billet by Hot Isostatic Pressing or by Forging. The final piece is machined from the billet. This process is used to upgrade the strength of alloys such that magnesium, for example, can be used in place of aluminum. *See also: 1) Hot Isostatic Pressing, 2) Powder Metallurgy, 3) Super Alloys.*

RATED THRUST (Turbojet or Turboshaft)
The manufacturer's guaranteed thrust rating. This thrust is present up to a specified ambient temperature as stated in the engine Type Certificate Data Sheet. For example, an engine rated at 3,500 pounds of thrust, guaranteed up to 84° F. See also: 1) Flat-Rating, 2) Thrust.

RATIOMETER SYSTEM (oil temperature)
Another name for Resistance Bulb.

RATO
Abbreviation for Rocket Assisted Take-off. Another name for JATO (Jet Assisted Take-off).

REACTION ENGINE
Refers to all types of jet engines that produce thrust in reaction to the rearward expulsion of a gaseous mixture.

REACTION TURBINE
A type of turbine assembly not common to flight engines. The stator vanes do not converge; they angle but do not accelerate the gases. The blading in this design forms convergent passageways and the turbine rotates in reaction to the squirting action of the gases leaving the trailing edge of the blades. This system is commonly seen in turbine driven accessories such as Fuel-Air Combustion and Pneumatic Starters. *For comparison see 1) Impulse Turbine System, 2) Impulse-Reaction Turbine.*

REACTIVE THRUST (reaction thrust)
Thrust in one direction obtained in reaction to the expulsion of a gas in the opposite direction. Reactive thrust occurs in accordance with Newton's Second Law, "F = MA" and his third law, "For every Acting Force there is an Equal and Opposite Reacting Force". See also: Thrust.

REAR FAN (turbofan engine)
Another name for Aft-Fan Engine.

REAR FRAME
Locate under Compressor Rear frame.

RECOUP AIR
Air which leaks from the gas path past compressor rotor seals that is captured and put to use in cooling and pressurizing other parts of the engine. For example: Turbine Cooling Air and Bearing Seal Pressurization Air.

RECUPERATED BRAYTON CYCLE
Refers to the use of a Recuperator as seen on some Turboshaft Engines to improve Thermal Efficiency. *See also: Brayton Cycle.*

RECUPERATOR
A type of heat recover device used on some Turboshaft Engines. A Recuperator contains stationary baffles which radiate exhaust heat back into gaspath air prior to its entering the combustion process. This heat recovery process reduces fuel needed to attain the required internal temperatures for a specific power output. *See also: Regenerator.*

RED DYE PENETRANT
Locate under Dye Penetrant.

REDUCTION GEARBOX (engine)
A gearbox that receives a high drive speed input and produces a reduced speed to its output drive(s). For most common types see: 1) Accessory Drive Gearbox, 2) Auxiliary Gearbox, 3) Output Reduction Gearbox, 4) Propeller Reduction Gearbox.

REDUCTION GEARING
Locate under Planetary Gearing.

REGENERATOR
A ducting system that directs both exhaust gases and compressor discharge gases through adjacent compartments in a rotating drum-type heat exchanger. By heat of radiation, temperature of compressor discharge air is raised prior to the gases entering the combustor, thus reducing fuel requirements for a specific power. The regenerator system is presently considered to be a prohibitive weight penalty for flight engines but it is common to stationary and surface turbine engines. *See also: Recuperator.*

REHEATER
Another name for Afterburner (British).

RELATIVE PRESSURE GAUGE
A cockpit instrument or a test gauge. An electric or direct pressure (Bourdon tube or bellows) gauge that does not include ambient pressure in its readout and therefore reads zero when no pressure is applied. Oil and fuel pressure gauges are examples of relative pressure gauges. The readout is called "gauge pressure". Cockpit gauges are AC or DC powered. Gauges used for maintenance and troubleshooting checks are generally direct pressure types. *See also: 1) Absolute Pressure Gauge, 2) Direct Pressure Gauge.*

RELIEF VALVE (oil system regulating)
A small engine accessory, often a component part of an oil pressure pump. A valve providing a means of regulating system oil pressure. It is adjusted to a minimum "cracking pressure" and will bypass oil in reaction to the pressure applied. The valve mechanism includes an adjusting screw, a spring and a ball, and a cone or poppet shaped bypass valve. *See also:*

Cold Starting Relief Valve.

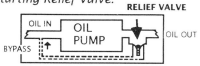

RELIEF VALVE (fuel system)
Typically a component part of a fuel pump. This valve is not generally a pressure regulating valve but rather a type of emergency blow-off valve which cracks open only when normal system pressure is exceeded.

REMOTE AUGMENTED LIFT SYSTEM (RALS)
An afterburner in a Lift-Fan Engine, located in a separate section quite remote from the core engine exhaust or the fan exhaust. *See also: Plenum Chamber Burning.*

REMOTE FILL POINT (oil tank)
Another name for Pressure Fill Point (oil tank).

REMOTE TRIM
A maintenance procedure using a Remote Trim Device. One end of a long electrical hook-up is attached to the Fuel Control and the other end is attached to a control box placed in the cockpit. During engine run-up for trimming, a rheostat on the control box is used to turn the trim adjuster at the Fuel Control. This Trimming method, as opposed to a manual trim screw adjustment, is widely used on Turbofan engines where the fan wake makes trimming a personnel safety hazard. *See also: 1) Trim Kit, 2) Trimming.*

RENE METALS
A nickel-chromium super-alloy produced for turbine engines that are manufactured by the General Electric Corporation. Examples: Rene 80 and Rene 100, both used in turbine blade manufacture.

RERATED ENGINE
Rerating is a manufacturer's method of reducing an engine's power rating when the aircraft in which the engine is installed does not require the maximum power attainable. Rerating generally means only a change in the engine's flight and technical manuals and few if any physical changes to the engine. *See also: 1) Derated Engine, 2) Flat-Rating (engine).*

RESERVE THRUST RATING (RTR)
The maximum allowable thrust available to a Turbofan or Turbojet engine that can be used in emergency situations during take-off or in flight. For example, RTR can be applied to the remaining engine(s) in the event that one engine loses power on take-off. The operating time in RTR is restricted in accordance with the flight manual. This rating is approximately 5% above the normal thrust rating. Also called 1) Reserve Take-Off Rating, 2) Alternate Thrust Rating. *See also: Engine Power Ratings.*

RESERVE TAKEOFF THRUST SYSTEM (RTT)
An aircraft system that automatically boosts the thrust of the remaining engine(s) when power is lost on one engine during take-off. For example, Rolls-Royce Tay Turbofan engines in the Gulfstream IV aircraft, will automatically boost thrust by 6%, from 13,300 lbt to 14,100 lbt, during a one engine out condition. Also called Take-off Thrust Reset (TTR) system.

RESERVOIR CAPACITOR
Another name for Storage Capacitor in an ignition exciter unit.

RESIDUAL FUEL (engine)
Refers to fuel remaining in the fuel system components and fuel lines when the engine is not operating.

RESIDUAL OIL (engine)
Oil that remains in the oil system components, oil lines, and accessory drive gearbox when the engine is not operating.

RESIDUAL PROPULSIVE FORCE
British term for Thrust (net).

RESIN COMPOSITE
Locate under Composite Materials.

RESISTANCE BULB
Locate under Resistance Temperature Detector.

RESISTANCE SPOOL (thermocouple circuit)
A fixed resistor used to calibrate the circuit to a predetermined resistance value. Typically in the Alumel side of Exhaust Gas Temperature circuit or the Constantan side of an oil temperature circuit.

RESISTANCE TEMPERATURE DETECTOR (RTD)
A small engine accessory. A detector (bulb) that operates on the principle that metals produce a positive change in electrical resistance for a positive change in temperature applied. The detector is actually a variable resistor in a bridge circuit. *See also: Resistance Temperature Detector (oil temperature).*

RESISTANCE TEMPERATURE DETECTOR (oil temperature)
A resistance bulb (RTD) positioned so that it protrudes into the engine oil flow. As temperature of oil increases, resistance of the bulb increases and causes an electrical current flow change in the cockpit gauge circuit. The gauge reads out in degrees Celsius in response to the current flow rate.

RESISTANCE TEMPERATURE DETECTOR (inlet temperature)
A device which sends a compressor inlet temperature signal to the engine Fuel Control Unit for fuel scheduling purposes. Commonly called a T2 Sensor, Tt_2 Sensor, or CIT Sensor. *See also: T2 Sensor.*

RESISTOR (ignition)
Locate under Bleeder Resistor.

RESLAM ACCELERATION PROCEDURE
An engine deceleration from full power to idle speed followed by a rapid slam-acceleration back to full power. The Reslam check is performed primarily to check engine stall tendencies. It must be accomplished within strict guidelines due to low blade tip clearances which may be present in the compressor and turbine.

RESONANCE METHOD (ultra-sonic inspection)
Locate under Ultra-Sonic Inspection.

RESONANCE of BLADES (critical)
The point at which vibration stresses, once started and if allowed to continue, may fatigue rotating blades to a point of failure. Proper tuning of new blades pushes Critical Resonance out of the operating range of the engine. Wear of the blades, however, can change the blade tuning and in this way, Critical Resonance can appear within the normal operating range. *See also: Frequency Tuning.*

RESONANT FREQUENCY
The frequency of a source of vibration that is approximately the same as the Natural Frequency of Vibration. Resonant Vibration (resonance) induced into engine parts, if uncorrected, can build to a point of component failure. *See also: 1) Natural Frequency of Vibration, 2) Vibration.*

RETROFIT (engine)
A term to indicate a major engine modification. Retrofit, in effect, means going back and refitting an engine with a newly designed part.

RETURN FLOW ANNULAR COMBUSTOR
Another name for Annular Combustor (reverse-flow).

REVERSE FLOW ANNULAR COMBUSTOR
Locate under Annular Combustor (reverse-flow).

REVERSE FLOW ENGINE
A Turbine Engine with an axial flow compressor which takes in its air at the rear and moves air forward, after which air is turned rearward into the combustor. Not a design seen in flight engines.

REVERSE FLOW COMBUSTOR
Presently there is only one reverse flow combustor in use, the Annular, Reverse-Flow. Locate under Annular Combustor (reverse flow).

REVERSE IDLE
The lowest power lever angle (PLA) attainable during reverse thrust operation. At Reverse Idle power lever position, the thrust reverser is deployed with the engine operating at idle speed.

REVERSE LEVER (cockpit)
A lever typically attached to the cockpit power lever and used to initiate thrust reverse. Actuation of this lever deploys the thrust reverser at idle engine speed and is then used to accelerate the engine to full power if desired.

REVERSE PITCH (propeller)
A ground only propeller blade setting which reverses the direction of airflow to create reverse thrust without changing direction of rotation. Refers also to variable pitch Turbofan Engines.

REVERSE TAXI PROCEDURE
Another name for Power-Back Procedure.

REVERSE THRUST
Locate under Thrust Reverser.

REVERSER
Locate under Thrust Reverser.

REVERSER BUCKET
Another name for Clamshell, referring to the blocker-door portion of a Mechanical Blockage Thrust Reverser.

REYNOLDS NUMBER EFFECT (compressor blade)
A measure of laminar flow over airfoils. A high Reynolds

Number indicates an efficient compressor blade design and excellent laminar flow characteristics. High altitude has a negative effect on laminar flow resulting in a low Reynolds Number. Also called Reynolds Index Number. *See also: Laminar Flow.*

RICE HULL CLEANING
Locate under Compressor Rice Hull Cleaning.

RICH BLOW-OUT (engine)
An engine flameout caused by an over rich air-fuel mixture in the combustor. This condition generally occurs during a rapid advance of the cockpit power lever during a time when the flight regime is providing a weak airflow into the engine inlet. The Rich Blow-Out occurs when the mixture enriches from the normal 15 : 1 to approximately 5 : 1. *See also: 1) Flameout, 2) Stoichiometric Mixture.*

RIG TEST
A generic term to indicate any number of test procedures performed on engine components in a special test apparatus.

RIG PIN (fuel control)
A small tool. A small metal pin that is inserted into a linkage system to attain a predetermined position during rigging checks. After the adjustment is completed the rig pin is removed. *See also: Rigging.*

RIGGING (fuel control)
Refers to adjusting the mechanical linkages between the cockpit power lever and the engine mounted fuel control unit. For example, when the cockpit power lever is set at idle speed position, the fuel control linkage must also be at the angle that provides idle speed. *See also: Rig Pin.*

RIM FLOW DISCOURAGER SEAL
Air seals in the turbine area, usually constructed as rotating metal rims designed to prevent excessive inward leakage of gas path air.

RIM SPEED
Speed of a turbine wheel configured with shrouded blade tips. Also called Tip Speed. Generally expressed in units of ft./sec., m/sec. or Mach number.

RIME ICE
Locate under Icing (rime).

RING AND TUBE ASSEMBLY
A major engine component of an older, multiple-can combustor. This unit consists of both front and rear support rings and all of the combustion chambers.

RING GEAR
Locate under Planetary Gear System.

RING-TYPE COMBUSTOR
Another name for Annular Combustor (machined ring).

RISK ANALYSIS (turbine engine)
A safety-related engineering study used to improve designs of engine parts. It is a method of calculating the number of expected part failures based on the in-service performance of similar parts. Two common methods are the Weibull Risk Analysis and the Monte Carlo Risk Analysis.

ROLL PIN
A small engine component part. A type of locking pin made of rolled steel. Some engines use roll pins to retain turbine blades in the disc, preventing axial movement.

ROLL-RING COMBUSTOR
Another name for Annular Combustor (machined ring).

ROLL-BACK (RPM)
Locate under Engine Rollback.

ROLL OVER STAND
A maintenance stand in which the engine can be rolled about its axis to facilitate maintenance.

ROLLER BEARING
Locate under Bearings.

ROLLER END-WEAR (bearings)
Locate under Bearing Inspection and Distress Terms in Appendix.

ROLLS-ROYCE LTD (Great Britain)
Manufacturer of Gas Turbine Engines. Locate in section entitled "Gas Turbine Engines for Aircraft".

ROTARY AIR-OIL SEPARATOR
Locate under Air-Oil Separator (rotary).

ROTARY AUGMENTOR
Locate under Augmentor.

ROTARY BREATHER
Another name for Air-Oil Separator (rotary).

ROTARY DEAERATOR
Another name for Air-Oil Separator (rotary).

ROTARY WING AIRCRAFT
Helicopters and other aircraft without fixed wings. Most rotary wing aircraft are powered by Gas Turbine Engines.

ROTATING GUIDE VANE
Another name for Inducer (impeller compressor).

ROTATING RIM
A type of air seal constructed as sheet metal rims attached to a solid metal ring. The rims are positioned such that they have a small clearance to an adjacent stationary engine part. When rotating, the rims form a metering orifice, allowing only a predetermined amount of air to pass.

ROTOR
Refers to a rotating engine component such as a Compressor Rotor or Turbine Rotor.

ROTOR ACTIVE CLEARANCE CONTROL (compressor)
Locate under Active Clearance Control.

ROTOR BALANCING
Locate under Balance Checks.

ROTOR-BRAKE (rotorcraft)
A component part of a helicopter transmission. A braking device used to reduce personnel and equipment hazards. When applied, the brake quickly stops

the flight-rotor shaft (in 3 or 4 revolutions) after engine shutdown. The brake is also used during engine starting to lock the free power turbine. At approximately 25% compressor speed the brake is released. This procedure ensures sufficient energy is available to the free power turbine to rotate the flight-rotor blades with the required force. Otherwise, blade droop in heavy winds can result in rotor contact with the tailboom. *See also: Brake (propeller or rotor).*

ROTORCRAFT
Another name for Rotary Wing Aircraft.

ROTOR PATH
British for Gas Path.

ROYAL AIRCRAFT ESTABLISHMENT
Located in Great Britain. A government controlled, primary research and development testing facility for Gas Turbine Engines.

RPM INDICATOR GENERATOR
British term for Tachometer Generator.

RPM MONOPOLE PICKUP
A small engine accessory. An electronic RPM transmitter that is fitted into the engine. It is placed adjacent to a magnet imbedded in a main rotor-shaft. Rotation of the shaft is sensed by the Pickup, which in turn transmits a signal to a cockpit percent RPM indicator. Similar to a Fan Speed Sensor System.

RPM SIGNAL TRANSMITTER
British term for signal sensor (tachometer).

RP (fuel)
Symbol for Rocket Propellent.

R-R
Abbreviation for Rolls-Royce LTD.

RTV SEALANT™
Abbreviation for Room Temperature Vulcanizing. A family of silicon rubber sealer materials used for many sealing purposes in Gas Turbine Engines. RTV is also available in several heat ranges.

RUB RING (shaft)
A small part of an accessory. One of two parts of a fuel or oil pump drive shaft seal. A rub ring is constructed of hardened steel with a highly polished surface upon which a carbon or bronze seal rides to form a fluid seal. Also called Seal Runner.

RUBBEROID FILLER
A rubber substance with a vibration damping quality. Often used to fill hollow compressor vanes. This filler requires special handling because a danger of phosgene gas emission exists if overheated by welding during assembly.

RUMBLE (combustor)
A combustor noise. Term to indicate a low frequency vibration associated with poor fuel scheduling and a subsequent fuel and airflow mismatch condition. Rumble results from choking and unchoking of the gas flow at the turbine nozzle vanes. Rumble is usually a transitory condition at low to mid power range. Rumbling is heard as a low volume noise which dissipates during spool-up and is not usually an airworthiness concern. *See also: Combustor Instability.*

RUN-IN TEST (engine)
A run procedure after maintenance or overhaul to operationally check the engine systems and to seat gas path seals.

RUN-ON TORQUE
Locate under Running Torque.

RUNOUT CHECK
A dial micrometer check to measure the trueness to the plane of rotation of a rotating part.

RUNOUT CHECK (axial)
Locate under Axial Runout Check.

RUNOUT CHECK (radial)
Locate under Radial Runout Check.

RUNNING TORQUE
The torque required to overcome run-on friction of a self-locking nut being placed on a threaded fastener before it makes contact with its mating surface. Also called Run-On Torque.

S

SADDLE MOUNTED TANK (oil)
An engine component. A sheet metal, externally mounted oil tank constructed with the same contour as the compressor outer casing to facilitate a flush mounting. Also called Quarter saddle tank.

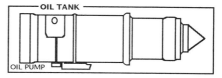

S.A.E.
Abbreviation for Society of Automotive Engineers.

SAFETY COUPLING (turboprop)
Automatic safety device to uncouple the propeller reduction gearbox from the engine in a dangerous, negative torque condition. This system is a backup for the Negative Torque System and will disconnect and reduce drag of the propeller in the event of an engine failure on take-off. *See also: Negative Torque System.*

SAFETY RESISTOR (ignition system)
Another name for Bleeder Resistor.

SAFETY WIRING
Another name for Lockwiring.

SAND AND DUST SEPARATOR
Another name for Inertial Sand and Dust Separator.

SAYBOLT SECONDS UNIVERSAL VISCOSIMETER
A device that measures the viscosity of lubricating oils by indicating the time (in seconds) required for a measured amount of oil to flow through a calibrated orifice. The time is then converted to a viscosity number. For example, SAE Grade 30 petroleum lubricating oil. Also called 1) SSU Viscosimeter, 2) SUS Viscosimeter. *See also: Viscosity Index Number.*

SCALLOPING (blades)
Locate under Blending.

SCAVENGE OIL FILTER
Locate under Oil Filter (scavenge).

SCAVENGE OIL PUMP (lubricating system)
Locate under Oil Pump (scavenge).

SCAVENGE OIL SUBSYSTEM
Locate under Oil Scavenge Subsystem.

SCHEDULED MAINTENANCE
Inspections, modifications and repairs planned in advance. Scheduled Maintenance is performed at intervals based on operating cycles, operating hours, or calendar times. *See also: Unscheduled Maintenance.*

SCHENECK BALANCE MACHINE™
A single plane balancing apparatus used to dynamically balance compressor and turbine rotor assemblies. *See also: Spin Pit Balancing.*

SCOOP-PROOF CONNECTION (electrical system)
An electrical connector plug requiring more torque to remove than to install. A connector that eliminates the need for lockwiring.

SCORING
Locate under Bearing Inspection Terms and Engine Inspection Terms in Appendix.

SCP
Abbreviation for Super Contingency Power.

SCRAMJET
Acronym for Supersonic Combustion Ramjet.

SCRATCHING
Locate under Bearing Inspection Terms and Engine Inspection Terms in Appendix.

SCREECH (afterburner)
A noise created by combustion of fuel in an afterburner. An acoustic phenomenon caused by high frequency combustion oscillations due to the high fuel-air ratios that occur near or at maximum afterburner operation. *See also: Screech Liner.*

SCREECH LINER
A component part of an afterburner. This liner forms the inner surface of an afterburner. It is constructed of perforated sheet metal and is designed to reduce high pitch noise and resultant metal fatigue caused by vibrations in air. This vibration is the result of combustion instability in the afterburner gases. *See also: Screech.*

SCRIVIT
A recent development in fastener design. A screw-like fastener in the form of a combination screw and rivet. It is most useful in construction of sheet metal parts.

SCROLL CAN (inlet separator)
A component part of an inertial ice and sand separator. The Scroll Can directs contaminant particles away from the inlet and back to the atmosphere through a discharge duct.

SCROLL CARBURETION (combustor)
A recent development in the mixing process of hydrocarbon fuel and air. Scroll carburetion refers to a circumferential rotation of fuel and air in a scroll-shaped pre-combustor chamber, prior to entering the main combustor. The fuel and air counter-rotate, allowing atomization to take place without the high pressure required of traditional fuel nozzles fitted with spin chambers.

SCRUBBING DRAG
Locate under Drag (parasite).

SCROLL TYPE COMBUSTOR
An engine component. A type of circular combustor widely used in auxiliary gas turbine units. This combustor completely encircles the turbine nozzle assembly, feeding combustion gases radially into the nozzle vanes and ultimately to a radial inflow turbine system. Also called Volute Combustor. *See also: Auxiliary Power Units.*

SCUFFING
Locate under Engine Inspection and Distress Terms in Appendix.

SCUPPER DRAIN (oil tank)
A component part of an oil tank. A spill cup encircling the oil filler opening. The scupper is designed to catch spilled oil during servicing and direct it to a central fluid drain system at the bottom of the engine, called a Gang Drain.

SD
Abbreviation for System Display.

SEAL (rotating)
Locate under 1) Air Seals, 2) Oil Seals.

SEAL (case flange)
Locate under Air Seals.

SEAL (accessory drive shaft)
Seals that fit on a drive shaft of an accessory or on its drive shaft receptacle in an Accessory Drive Gearbox, to prevent fluid leakage. These seals are generally one of the following types: 1) Spring Seal, one that uses a spring to pre-load a carbon ring onto a metal rub ring. Sometimes called a Sealol Seal™, 2) Magnetic Seal, one similar to a Spring Seal but which uses a magnetic attraction to pre-load its carbon ring onto a rub ring, 3) Lip Seal, a rubber seal that fits around a shaft, often with a garter spring insert to maintain its shape. Sometimes called a Garlock Seal™.

SEAL (turbine blade tips)
Locate under Shrouded-Tip Turbine Blades.

SEAL LEAKAGE (fuel and oil)
Locate under Leakage Limits.

SEAL RUNNER
Another name for Rub Ring.

SEAL WITNESS DRAIN
Another name for Gang Drain.

SEALOL SEAL
Locate under Seal (accessory drive shaft).

SECONDARY AIR (combustor)
Refers to the approximately 60 to 75 % of total compressor discharge air which is utilized to cool the combustion liner and dilute the combustion temperatures prior to the gases reaching the turbine.

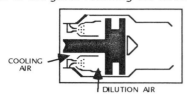

SECONDARY AIRSTREAM (turbofan engine)
The portion of air in the flight inlet of a Turbofan engine that passes through the fan. The remainder of inlet air flows through the core-exhaust.
See also: 1) Core-Engine, 2) Primary Airstream.

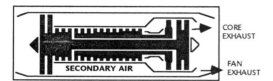

SECONDARY EXHAUST DUCT (afterburner)
A component part of an afterburner. The aft (outer) exhaust duct of a Flap-Type Exhaust Nozzle. The nozzle is formed by a ring of movable segments that change the geometry of the nozzle to match the flow conditions of afterburner or non-afterburner operation. Also called Secondary Exhaust Nozzle. See also: Flap Exhaust Nozzle (afterburner).

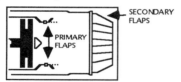

SECONDARY EXHAUST DUCT (turbofan)
A component part of an engine nacelle. The point at which the fan mass airflow of a Turbofan engine is discharged from the engine. This nozzle will pass fan air directly back to the atmosphere or it will flow air into a mixed exhaust duct. Also called, 1) Cold Nozzle, 2) Secondary Exhaust Nozzle. See also: 1) Primary Exhaust Duct, 2) Mixed Exhaust Duct.

SECONDARY FUEL FLOW (combustor)
Refers to the fuel that flows along with the Primary Fuel Flow when the engine is in operation above idle speed. Secondary Fuel Flow discharges from either the outer orifice of a duplex fuel nozzle or from a separate fuel nozzle. Also called Main Fuel. See also: Primary Fuel Flow (combustor).

SECONDARY POWER SYSTEM (aircraft)
Refers to an Auxiliary Power Unit.

SECONDARY POWER SYSTEM (engine)
The name given to the process of: 1) Afterburning, 2) Duct Burning, 3) Plenum Chamber Burning.

SECTIONS (engine)
Locate under Engine Sections.

SEEPAGE CHECKS (fuel)
Seepage implies a slight fluid leak. Locate under Fuel Seepage Checks. See also: Leakage Limits.

SEIZURE
Locate under Engine Inspection and Distress Terms in Appendix.

SELF-SUSTAINING SPEED (engine)
Speed at which the engine can operate without depending on power from the starter unit.

SELSYN TRANSMITTER™
A General Electric trademark. A transmitter powered by D.C. voltage which in turn powers a cockpit gauge. Its mechanism functions by using a permanent magnet and resistor arrangement. See also: 1) Autosyn Transmitter, 2) Magnesyn Transmitter.

SEMI-CONDUCTOR IGNITER PLUG
Another name for Igniter Plug (low voltage).

SERMALOY COATING™
A coating on metal parts in hot sections which resist hot corrosion and hot oxides. See also: 1) Sermetal™, 2) Thermal Barrier Coating.

SERAMEL COATING™
Similar coating to Sermetal-W™.

SERMETAL-W COATING™
An anti-corrosion and anti-erosion coating used on compressor blades and vanes to provide a very smooth and durable surface for improved aerodynamic performance. See also: Sermaloy™.

SERPENTINE COOLING (turbine blades and vanes)
Blades and Vanes configured with air flow passageways and twisted metal inserts called turbulator fins. The turbulators create slight turbulence to slow airflow and provide a better heat transfer from the metal airfoil surfaces to the cooling air. See also: 1) Turbine Cooling 2) Turbulators.

SERVICE BLEEDS (compressor)
Points on an engine compressor or diffuser outer case where air is extracted and then routed into aircraft systems. Also called Extraction Pads. See also: Customer Service Air.

SERVICE BULLETIN
A printed information sheet. The means by which most manufacturers communicate with customers to inform, to specify required maintenance, to limit use of equipment, etc. If the situation warrants, the FAA will issue an Airworthiness Directive on the same subject. Sometimes Service Bulletins are known as Service Letters or Service Instructions, but more often the word "Bulletin" is used to indicate the most urgent information, and the words "Letter" or "Instruction" indicate information that is less urgent.

SERVICE SHUT-OFF VALVE (oil filter)
A spring loaded oil shut-off valve which actuates whenever the oil filter is removed, preventing drain down of oil from the oil tank.

SERVICING OVERFLOW LINE
Locate under Overflow Line (oil tank).

SERVO FUEL FILTER
Locate under Fuel Filter (servo).

SERVO-FUEL HEATER
A small engine accessory. A small air-fuel heat exchanger that heats only the servo-fuel (working fluid) going to the inner mechanisms of the fuel control. This unit provides icing protection by keeping entrained water above its freezing point. Generally used only on engines that are seriously affected by even the slightest fuel icing conditions. This system augments the main Fuel-Oil Cooler. *See also: Fuel-Oil Cooler.*

SFC
Abbreviation for Specific Fuel Consumption.

SHAFT HORSEPOWER (corrected)
Refers to correcting an observed SHP reading to ambient conditions in order to find it actual value. The formula is as follows:

$$\text{Corrected SHP} = [\text{SHP (observed)} \times 29.92 \div \text{Pam}] \times [\sqrt{288°K \div \text{Tam}°K}]$$

Where: 29.92 in. hg. = International Standard Day, Pam = barometric pressure (in.hg.), Tam = temperature ambient in absolute units (°C. + 273).

SHAFT HORSEPOWER (turboprop, turboshaft engine)
The horsepower delivered to the output drive of a Turboprop Engine or a Turboshaft Engine. Abbreviated as SHP. *See also: 1) Brake Horsepower, 2) Friction Horsepower.*

SHAFT BUSHING
Locate under Bushing (pump shaft).

SHAFT SEALS
Locate under Seal (accessory drive shaft).

SHAKEPROOF WASHER
Small expendable part. A type of lockwasher with serrations on either its inner or outer perimeter and used to prevent loosening of a nut or bolt. The two types commonly seen are: 1) Internal Tooth Lockwasher, 2) External Tooth Lockwasher. Also called Star Washers.

SHAKER TABLE (vibration pickup)
A test apparatus used to bench check Vibration Pickups under laboratory controlled vibratory loads.

SHEAR FAILURE
Locate under Engine Inspection and Distress Terms in Appendix.

SHEAR NUT
A thin nut used where loads on a bolt are entirely in shear. The nut in this case merely retains the bolt.

SHEAR STRESS
A stress that occurs as parts tend to slide past each other. *See also: Tensile Stress.*

SHEAR NECK
Another name for Shear Point Shaft.

SHEAR POINT SHAFT
Refers to an intentionally weakened point on a drive shaft which will fracture (shear) to protect the remainder of the system from excessive loads. For example, a shear point on a starter shaft protects the starter drive spline in the Accessory Drive Gearbox. Similarly, a shear point found on a dual-element fuel pump ensures continued fuel supply to the engine by shearing in the event that one element seizes. Shearing of one shaft leaves the other element still functioning. Also called 1) Shear Neck, 2) Shear Spline.

SHELL MOULDING
The metal forming process most used for large turbine engine parts. *See also: Investment Casting Process.*

SHINGLE LINER (combustor)
Refers to the inner, movable, flap-type shingles of a Float-Wall Combustor. *See also: Combustion Liner (floatwall).*

SHINGLING
Locate under Fan Blade Shingling. Also locate under Engine Inspection and Distress Terms in Appendix.

SHOCK COMPRESSION (inlet)
Compression which occurs as the airstream passes through a shock wave. For a normal shock wave, the ratio of pressures formula is as follows:

$$P_1/P_2 = (7 \times Mn^2 \div 6.167)$$

Where: P = Pressure and Mn = Mach Number.

SHOCK DIAMONDS (exhaust)
Diamond shaped shock waves being reflected between the boundaries of the exhaust plume and the atmosphere while a Turbine Engine is operating in afterburner mode.

SHOCK DRAG
Locate under Drag (shock).

SHOCK EXPULSION (inlet)
Another name for Inlet Buzz.

SHOCK FAILURE
Locate as Stress Failure in Engine Inspection and Distress Terms in Appendix.

SHOCK SPIKE
Locate under Inlet Spike.

SHOCK STALL SPEED (airfoils)
In reference to the Gas Turbine Engine, the rotational speed of a compressor or turbine blade at which a shock wave occurs at the blade tips. The airflow disturbance accompanying the shock wave degrades the efficiency of the airfoil. If not corrected, this condition can to lead to engine stall or even engine structural failure. *See also: Compressor Stall.*

SHOCK WAVE
A "compression wave" formed in air when an airfoil moves at speeds greater than the speed of sound. At this speed, air does not have time to separate as the airfoil approaches, and a thin wall of air pressure forms (approximately 1/25,000 inch in diameter). *See also: 1) Local Speed of Sound, 2) Normal Shock Wave, 3) Oblique Shock Wave.*

SHOP REPLACEABLE UNIT
An item which, upon failure, is designated to be removed or replaced by a higher level maintenance facility (intermediate or depot maintenance activity), and is to be tested as a separate entity. *See also: Line Replaceable Unit.*

SHOP VISIT (SV)
Refers to engines removed from aircraft and returned to a maintenance facility, during which a lower echelon of maintenance than a Time Since Overhaul procedure (Time Since Restoration) is performed.

SHOP VISIT MANUAL
A written set of instructions used to perform a

SHOP VISIT RATE (SVR)
Refers to the frequency (cycles or hours) at which a particular engine model requires removal from the aircraft for shop maintenance.

SHOT-GUNNING
Jargon for a scattered approach or lack of logical sequence when troubleshooting.

SHOT PEENING
A process using an air propelled shot material aimed at a surface. Titanium is often shot peened to harden its surface and make it more crack resistant.

SHORT DUCTED TURBOFAN
An engine design wherein the fan exhaust duct does not extend to the rear of the engine. In this design, the fan discharges its air at a forward location in relation to the core-engine exhaust. *See also: 1) Turbofan Engine, 2) Unmixed Exhaust Duct.*

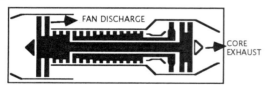

SHORT TAKE-OFF AND LANDING AIRCRAFT (STOL)
Aircraft designed for short rolling distances before lift-off.

SHORT TAKE-OFF AND VERTICAL LANDING AIRCRAFT (STOVL)
Gas Turbine Engine powered fixed-wing military aircraft with vertical take-off and hover capabilities, such as the Harrier Jet Fighter. *See also: Lift-Fan Engine.*

SHOWER HEAD COOLING (turbine blades and vanes)
Refers to air leaving hollow leading edges of airfoils in a forward direction and spilling back over both top and bottom sides, forming a heat barrier. *See also: Air Cooled Blades and Vanes.*

SHP
Abbreviation for Shaft Horsepower.

SHROUD LOCK (turbine blade)
The notched or interlocking area of a shrouded tip turbine blade. The shrouds fit together one blade to the next in what is called a Z-Form, to provide a circular support ring and reduce blade bending and untwisting stresses. Also called Z-Cut. *See also: 1) Shrouded Tip Blade (turbine), 2) Stellite Alloy.*

SHROUD NOTCH
Another name for Shroud Lock.

SHROUD RING
An engine component. A stationary air sealing ring positioned just outside the rotational plane of a set of rotor blades. In compressors the outer casing generally forms the shroud ring. In the turbine area the shroud ring is often a removable part. *See also: 1) Feltmetal Seal, 2) Honeycomb Shroud-Ring, 3) Abradable Shroud 4) Tip Clearance.*

SHROUD RING TRENCHING
A technique of machining a recess in a turbine shroud ring within which turbine blade tips rotate. The trench, in turn, controls tip leakage.

SHROUDED BLADE
Locate under Shrouded Fan or Shrouded Tip Blade (turbine blade).

SHROUDED FAN
A fan in a Turbofan Engine in which the fan blades are configured with blade span shrouds. *See also: Span Shroud (fan blade).*

SHROUDED-TIP BLADE (turbine)
Turbine blades configured with tip platforms (shrouds) attached such that they contact one another to form a circular support ring, providing added tip stability. It is common practice today to cut the shrouds in what is called a Z-Form Interlock. *See also: Shroud Lock.*

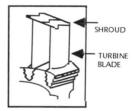

SHROUDED-TIP & KNIFE EDGE SEAL (turbine blade)
A turbine blade shroud platform configured with thin, sharp-edged metal seals. When all of the blades are

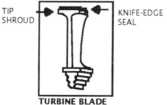

in place in the disc, the knife-edge seals form a continuous sealing ring. The seals rotate within the plane of the Turbine Shroud Ring to establish a clearance and to control tip leakage. *See also: Shroud Ring.*

SHUNTED GAP IGNITER PLUG
Another name for Igniter Plug (low voltage).

SHUTDOWN
Cessation of engine operation for any reason other than training or normal operating procedures.

SIAMESE THERMOCOUPLE
Another name for Dual Thermocouple.

SIGNAL CONDITIONER
Locate under 1) Speed Sensor, 2) Torque Sensor.

SIGNALS (fuel control)
The engine parameters that are used as inputs to the fuel control to assist in correctly scheduling fuel to the combustor. Common signals are: 1) Pressure and Temperature in the flight inlet, 2) Compressor Discharge Pressure, 3) Burner Can Pressure, 4) LP and HP compressor speeds.

SIGHT GAUGE (oil tank)
A visual means of checking the oil tank level. An FAA requirement.

SILENCER (exhaust)
Another name for Noise Suppressor.

SILICA-GEL
A commercial calcium oxide type desiccant material which has high affinity for water. It is placed in engine and engine parts containers to absorb moisture and prevent corrosion.

SILICON
A solid non-metallic element used in turbine alloys and, in a pure form, used as a semi-conductor material. When found in Spectrometric Oil Analysis, it is considered to be dirt.

SILICON CARBIDE
A family of very hard ceramic materials with high heat strength, used in many turbine materials. Silicon Nitride also falls into this category.

SILICONE
A semi-inorganic polymer. A pliable substance used as adhesives, lubricants, and protective coatings.

SIMPLEX FUEL CONTROL
Locate under Fuel Nozzle (simplex).

SING (blade)
Locate under Blade Sing.

SINGLE-CRYSTAL (turbine parts)
Engine component parts constructed of nickel-cobalt alloy by an improved casting process which produces higher temperature strengths than conventional casting or the Directional Solidification process. The entire mono-crystal part, being one solid crystal or grain of metal, contains no grain boundaries. Parts manufactured by this process are of near-net-shape, requiring little machining after casting. This process also produces alloys of very high creep strength, which increases durability without increasing the need for internal cooling. Single-Crystal is used primarily for turbine blades. Single-crystal technology also applies to Ceramic parts. Also known as 1) Grown Turbine Blades, 2) Mono-Crystal Turbine Blades. *See also: Super Single-Crystal Alloy.*

SINGLE-ENTRY COMPRESSOR
Refers to a single-stage, single-sided impeller type compressor. *See also: Dual-Entry Centrifugal Flow Compressor.*

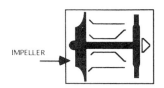

SINGLE PASS COOLING (turbine)
Locate under Convection Cooling.

SINGLE-PLANE BALANCING (compressor & turbine rotors)
Compressor and Turbine rotor balancing (generally smaller units) in which each disk is statically balanced before assembly. Then blade pairs of similar mass weight are mounted 180° apart, until all of the blades are installed. Also called Static Balancing. *See also: Balance Checks.*

SINGLE REDLINE LIMIT
A cockpit gauge with only red-line values for all flight operating regimes. *See also: Variable Redline.*

SINGLE-ROTATING PROPELLER (engine)
A Propfan or Turboprop Engine configured with one propeller as opposed to contra-rotating propellers.

SINGLE-SHAFT ENGINE
A Turbine Engine in which all of the compressor stages are connected to a common shaft, which is attached to its turbine wheel(s). Another name for Single-Spool Compressor in reference to an axial flow compressor.

SINGLE-SPOOL COMPRESSOR
Refers to a single axial-flow compressor, driven by a turbine rotor. Single-spool compressors typically contain 8 to 16 stages of compression. The single-spool compressor is not a popular design in the modern engine. *See also: Compressor (engine).*

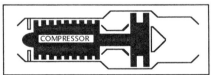

TURBOJET (9-STAGE AXIAL FLOW)

SINGLE STAGE COMPRESSOR
Refers to an impeller compressor, of either the Single-Entry or the Dual-Entry type.

SINGLE STAGE TURBINE
A turbine system with only one set of Turbine Nozzle Guide Vanes and one Turbine Wheel.

SINTERING PROCESS
A Powder Metallurgy process in which metal particles contacting each other bond together from compression forces. Bonding occurs at approximately one-half the melting-point temperature and without the metal becoming molten. *See also: Powder Metallurgy.*

SKIDDING (bearings)
Locate under Bearing Inspection and Distress Terms in Appendix.

SKIN FRICTION DRAG
Locate under Drag (parasite).

SLAM-ACCELERATION
Another name for Throttle Burst.

SLAVE NUT
Another name for Captive Nut.

SLEEVE BEARING
Locate under Bearing (plain).

SLIDING VANE PUMP
Locate under Oil Pump (vane).

SLIP-CLUTCH MECHANISM (electrical starter)
A component part of an engine starter unit. A clutch mechanism installed at the drive end of an electrical starter. This clutch has the appearance of a miniature Rotating Disk aircraft brake. Its function is to prevent sudden shock loads into the engine gearbox by slipping during initial engagement. The clutch plates (disks) slip until torque on the gearbox side reduces to the torque on the clutch side. Also called Friction Clutch.

SLUG (air)
A physics term. A Mass unit of air equal to the mass weight divided by the gravity constant. A measurement of airflow in turbine engines equal to the weight of mass airflow in lbs. divided by 32.2. Defined as that mass which, when acted upon by a force of 1 Lb., will be given an acceleration of 1 ft./sec.^{2}.

SLUG (fuel)
Refers to water that visibly settles out of fuel as opposed to entrained water suspended in fuel.

SLUGGING (fuel pump)
Fuel pump induced pulsations in the fuel that occur at altitude, when less back-pressure is acting on the system. See also: Slugger Control.

SLUGGER CONTROL (fuel system)
A small engine accessory. A scheduling device employed in slowing fuel flow at altitude flight condition, to facilitate responsive and stall-free engine RPM changes. See also: Slugging (fuel pump).

SMALL GAS TURBINE ENGINE
A general category of engines for installation in small, private or corporate aircraft as opposed to larger commercial aircraft. Also used to denote engines for auxiliary power units.

SMOKE CHECK
Jargon for a maintenance procedure of: 1) directing smoke or shop air into a fluid or air line as a visual check for leaks or restrictions to flow, 2) applying power to an electrical system during troubleshooting, to watch for presence of short circuits and smoke, 3) application of power for the initial operation of any electrical system.

SNAP DIAMETER
Refers to a Snap-Fit Flange.

SNAP-FIT FLANGE
An overlapping flange, such as found between major engine cases to: 1) prevent leakage of the gases, 2) maximize alignment of the mating sections.

SNAP RING
A small expendable part. A circular spring steel ring in a horseshoe-shape which pushes outward into a retaining channel or squeezes inward into a retaining channel to act as a retainer ring.

SNAP-RING PLIERS
A special type of pliers used to remove Snap Rings.

SNECMA (France)
Manufacturer of gas turbine engines. Locate in section entitled "Gas Turbine Engines for Aircraft".

SNUBBER (fan blades)
Another name for Span-Shroud or Clapper.

SOCIETY OF AUTOMOTIVE ENGINEERS (SAE)
A society of professional engineers who have established standards of design and quality for many products in the automotive, aviation, and other industries.

SODA STRAW (fuel-oil cooler)
Refers to the shape of the numerous thin wall fuel passageways, which are surrounded by oil, providing heat transfer into the cooler flow medium (fuel).

SODIUM SULFATES
Locate under Hot Corrosion.

SOFT CASE CONTAINMENT
Refers to outer engine cases constructed of materials other than metals, that act as a containment ring. For example, a Kevlar™ outer fan case. See also: Containment Ring (fan).

SOFT-LIFE TIME OVERHAUL
Refers to certain engine modules, components, component parts, and accessories with no specified time change interval. They are sent to overhaul only when their condition warrants replacement. See also: Hard-Life Time. Also Called On-Condition Overhaul.

SONIC SPEED
Speed in air at Mach-One.

SONIC VENTURI
A venturi in which air reaches sonic speed and a pressure build up occurs. Seen in many pneumatic systems as a pressurizing device.

SOUND SUPPRESSOR (aircraft mounted)
Another name for Noise Suppressor Unit.

SOUND SUPPRESSOR (ground mounted)
A sound attenuating device in an Engine Test Cell or on an Aircraft Run-Up Pad. The engine exhaust is directed into the suppressor unit, containing sound absorbing baffles, and often times a water spray that reduces gas heat and velocity. This arrangement reduces the noise level of the gases leaving the unit and entering the atmosphere. Also called Noise Suppressor (ground mounted).

SPALLING
Locate under Bearing Inspection and Distress terms and Engine Inspection and Distress Terms in Appendix.

SPAN-SHROUD (fan blade)
A component part of a fan blade. Span-shrouds are lugs located on the blade face and back that contact one blade to the next to form a circular support ring. Unlike turbine blade shrouds which are located at the

blade tips, fan blade shrouds are located part way along the blade length. The shrouds enhance blade strength and provide a vibration damping effect to the fan assembly. A disadvantage of Span-Shrouds is that they block airflow. Also called 1) Clappers, 2) Mid-Span Shrouds, 3) Snubbers, 4) Span-Supports. *See also: 1) Fan Blade Shingling, 2) Wide-Chord Blades.*

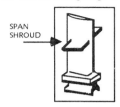

SPAN SUPPORT (fan blade)
Another name for Span-Shroud.

SPANNER NUT
A retainer with either external or internal threads that can be tightened and loosened by engaging a spanner-type wrench to the notches on the face of the nut. Spanner nuts are commonly used on turbine engine shafts to retain bearings and bearing seals.

SPAR-SHELL BLADE
An airfoil constructed with a metal center spar encased in a composite outer shell, filled with a foam material.

SPARK EROSION
The most common damage seen at the firing end of a turbine igniter plug. Outer shell, insulation, and center electrode materials all erode away as a result of the blasting effect of the high voltage discharge (flashover).

SPARK IGNITER
Another name for Igniter Plug.

SPARK SPLITTER (ignition system)
Generally an internal component part of a main ignition system exciter unit. This unit receives a high voltage output from the exciter storage capacitor which it divides it into two separate high voltage paths, one to the right-hand igniter plug and one to the left-hand igniter plug. Sometimes this unit is an external splitter when used with a four igniter plug system.

SPECIFIC FUEL CONSUMPTION (SFC)
A physics term. An important vital statistic of a Turboshaft Engine. SFC is the ratio of fuel being consumed in pounds per hour to engine power output in shaft horsepower units. Formula: $SFC = W_f \div SHP$
For example, a Turboshaft Engine might be stated to have a SFC of 0.49 lb./hr./shp, (pounds per hour of fuel per shaft horsepower). By comparison, Turbojet and Turbofan Engine fuel consumption ratios are generally given as Thrust Specific Fuel Consumption (TSFC). *See also: 1) Equivalent Specific Fuel Consumption, 2) Thrust Specific Fuel Consumption.*

SPECIFIC GRAVITY (fluids)
A ratio of the weight of a volume of fluid to the same volume of pure water at a specified temperature.

SPECIFIC GRAVITY (jet fuels)
The Specific Gravity of jet fuels at 15°C. is as follows: Jet-A = 0.816, Jet-A1 = 0.806, Jet-B = 0.764.

SPECIFIC GRAVITY ADJUSTMENT (fuel control)
An adjustment to a fuel control unit that may be required prior to Trimming or Full Power Testing of a turbine engine. The Specific Gravity adjustment must match the jet fuel being used to obtain an accurate operational check. On many fuel control units this adjustment changes the spring tension applied to the differential pressure regulating valve. When the tension matches the specific gravity of the fuel being used, the fuel control will establish the correct bypass condition and accurately schedule fuel to the combustor.

SPECIFIC HEAT RATIO (Cp)
A physics term. A ratio to mean Specific Heat at Constant Pressure.

$$C_p \text{ is defined as } 0.24 \text{ BTU/lb.°F}$$

This ratio indicates that it requires 0.24 BTU to raise the temperature of one pound of air 1°F. By comparison, at constant pressure (Cp) it takes 1.0 BTU to raise 1 lb. of water 1°F. Cp is used in the formula: Horsepower (to Drive The Compressor) of a gas turbine engine.

SPECIFIC IMPULSE
A physics term. A ratio defined as the number of units of thrust produced per unit of fuel (weight) consumed.

SPECIFIC POWER (PS)
A physics term. A ratio of horsepower produced to engine Mass Airflow (lb./sec.). A performance calculation for torque producing engines, namely the Turboprop (ESHP) and Turboshaft (SHP).

SPECIFIC POWER (ESHP)
Mathematically defined as: $P_s = ESHP \div M_s$

Where Ps = Specific ESHP, ESHP = total power output, Ms = Mass airflow (lb./sec.). Specific ESHP is one means of comparison between engines. For example, a Turboprop Engine with a rating of 2,000 ESHP, which has a Mass airflow of 20 lb./sec., would have a Specific ESHP of 100 to 1.

SPECIFIC POWER (SHP)
Mathematically defined as: $P_s = SHP \div M_s$

Where Ps = Specific SHP, SHP = total power output, Ms = Mass airflow (lb./sec.). Specific SHP is one means of comparison between engines. For example, a Turboshaft Engine with a rating of 2,000 SHP, which has a Mass airflow of 20 lb./sec., would have a Specific SHP of 100 to 1.

SPECIFIC STRENGTH (materials)
Material strength vs. material density. For example, materials such as Carbon-Carbon Composite, Ceramic Matrix, Fiber-Reinforced Metals, and Nickel-Base alloys, are said to have high Specific Strength, because they are light in weight for the strength characteristics they exhibit.

SPECIFIC THRUST (Fs)
A physics term. A ratio of thrust output per unit (lb.) of engine mass airflow. A performance calculation for thrust producing engines, namely the Turbojet and Turbofan.
Mathematically defined as: $F_s = F_n \div M_s$

Where Fs = Specific Thrust, Fn = Thrust, Ms = Mass

airflow (lb./sec.). Specific Thrust is one means of comparison between engines. For example, an engine with a thrust rating of 60,000 lbs., which has a Mass airflow of 1,500 lb./sec., would have a Specific Thrust of 40 to 1.

SPECIFIC WEIGHT (air)
A physics term (ratio). 1) Commonly described as air density in units of lb. per cubic foot and mathematically defined as 0.076474 lb./ft^3 (1.2250 kg/m^3), 2) For other computations, defined as Density ÷ Gravity (0.0023769 lb-sec^2/ft.^{4}) *See also: Density (air).*

SPECTROMETER (oil analysis)
A device employed to analyze wear-metal contaminant levels and airborne contaminant levels in oil samples taken from turbine engines. The spectrometer measures illumination intensity (indicating the amount of metal) and illumination color (indicating the type of metal) of an oil sample being burned in a light spectrum measurement device. The spectrometer reads out in parts per million (PPM). *See also: Spectrometric Oil Analysis Program.*

SPECTROMETRIC OIL ANALYSIS PROGRAM (SOAP)
A maintenance program employing a Spectrometer to analyze engine oils for entrained wear-metals generated by oil wetted engine parts and also for airborne contaminants which find their way into engine oil wetted areas. This procedure is generally thought of as measuring particles of 8 microns or less that are in suspension in engine oil. The SOAP readings are used to plot a graph which becomes a trend analysis of engine condition. Particles larger than 8 microns are dealt with by Filter Debris Analysis. *See also: Filter Debris Analysis.*

SPEED FORMULA
A physics term.
Mathematically defined as: $S = d \div t$

Where S = speed, d = distance, t = time. U.S. Units of ft./sec. and mi./hr. in U.S. Units. Meters/sec. and Kilometers/hr. are common Metric Units.

SPEED OF SOUND
A physics term. The speed of sound occurs where the actual air velocity and the acoustic velocity are equal to a ratio of one. Expressed as Mach 1 or Mn = 1. Mach number = velocity of a body ÷ velocity of sound.

$Mn = V \div Cs$

For example, if an object is moving at 2240 fps at a Standard Day condition (when speed of sound is 1,120 fps), the object is said to be moving at Mach Two. *See also: 1) Cs Symbol, 2) Mach Number.*

SPEED RATED ENGINE
A Turbofan engine in which "Engine Ratings" are guaranteed to occur at certain fan speeds. *See also: 1) Engine Ratings, 2) EPR Rated Engine.*

SPEED SELECTOR LEVER (rotorcraft)
A cockpit control lever that connects to the fuel control. The Speed Selector Lever provides a means of manually adjusting engine speeds from Ground Idle setting up to Take-Off Power setting. *See also: Collective Pitch Lever.*

SPEED SENSITIVE SWITCH
A flyweight operated sequencing switch, driven by the Accessory Drive Gearbox. This switch is used to automatically open and close electrical circuits to the engine starter, ignition, and fuel flow valves during the engine starting cycle.

SPEED SENSOR (tachometer)
A sensing unit employed in an electronic tachometer system to send a speed signal to either a Fan Speed or Core Speed indicator. Locate under 1) Fan Speed Sensor, 2) HP Speed Sensor.

SPEEDER SPRING
A spring which provides an opposing force to a set of flyweights in a mechanical flyweight governor.

(SPF/DB)
Abbreviation for Superplastically Formed, Diffusion Bonding.

SPIGOT
An overlapping rim that provides a self-alignment feature between mating parts.

SPIKE INLET
Locate under Movable Spike Inlet.

SPIN CHAMBER (fuel nozzle)
A fuel passageway within a fuel nozzle, cut in the shape of small slots or flutes. The spin chamber imparts a turning moment in the exiting fuel to promote atomization. Also called Conical Vortex Chamber.

SPIN-PIT BALANCING
Another name for Dynamic Balancing.

SPIN TESTING
Locate under Balance Checks.

SPINNER (turbofan engine)
A rotating fairing mounted at the front center of a fan. This aerodynamic fairing reduces turbulence in air entering the rotating airfoils and also prevents heavy ice build-up. Also called Spinner Cone.

SPINNER (turboprop engine)
A rotating fairing mounted at the front center of a propeller. The spinner provides an aerodynamic function of reducing turbulence in air entering the propeller blading near the hub.

SPIRAL BEVEL GEAR
Locate under Bevel Gear.

SPIRAL GEAR
A type of curved gear which meshes at right angles to another gear. This gear is typically used in propeller blade pitch changing mechanisms. *See also: Gear Designs.*

SPLASH LUBRICATION (accessory drive gearbox)
Refers to the gears in contact with residual oil in the lower portions of the gearbox. These gears act as slingers to distribute oil throughout the unit to facilitate

lubrication. A design applicable to both Dry Sump and Wet Sump Lubrication systems.

SPLASH LUBRICATION SYSTEM
A Wet Sump lubricating system wherein the lower rotating parts contact oil in the reservoir to create a splashing action that distributes lubricating oil over a wide area. *See also: Wet Sump (lubrication system).*

SPLIT-CLIP
A two-piece metal clamp fitted with a rubber or Teflon™ chafing pad. This clamp is used to secure in place fluid lines, air lines, and electrical leads. *See also: Loop Cushion Clamp.*

SPLIT-RING CARBON SEAL
A Carbon Oil Seal of the ring type, with dual side by side segments. *See also: Carbon OIL Seal.*

SPLIT-SPOOL COMPRESSOR
Another name for Dual-Spool Compressor.

SPLIT TURBINE
Refers to the High Pressure and Low Pressure turbine arrangement in an engine with a dual-spool compressor.

SPLITTER (ignition)
Locate under Spark Splitter.

SPLITTER FAIRING (turbofan)
An airflow divider located aft of the fan which directs a portion of the total airflow into the core (low pressure compressor) and the remainder down the bypass duct. The splitter has a secondary function of directing airborne contaminants away from the core inlet and along the fan air flow path. Airflow into the core portion of the engine is called primary air, airflow through the fan is referred to a secondary air. *See also: 1) Primary Airstream (turbofan), 2) Secondary Airstream (turbofan).*

SPLITTER NOSE
The inlet cowl of a Splitter Fairing.

SPLITTER VANE (centrifugal compressor)
A half-vane located between the full length vanes on an impeller compressor disc.

SPOILER (thrust)
Locate under Thrust Spoiler.

SPOOL
Another name for Axial Flow Compressor. For example, Single-Spool Compressor, Dual-Spool Compressor.

SPOOL-DOWN
Refers to a decay in compressor (spool) speed, which in fact amounts to a decrease in engine speed.

SPOOL-DOWN TIME
Another name for Coast Down Time. See also: Coast Down Check.

SPOOL-UP
Refers to increasing compressor (spool) speed, which in fact amounts to an increase in engine speed.

SPOOL-UP CHECK
Another name for Acceleration Check.

SPOT FACE (turbine blade)
A smooth machined spot on a turbine blade that facilitates a blade stretching check.

SPRAG CLUTCH (engine starter)
A type of over-running clutch, often of a pawl and ratchet type. See also: Over-running Clutch.

SPRAY BARS (afterburner)
Tube shaped fuel injectors that protrude radially into the airstream. Each bar contains numerous small flow orifices which direct atomized fuel into the gaspath.

SPRAY NOZZLE (fuel)
Another name for Fuel Nozzle.

SPRING WASHER
A spring steel washer shaped such that it applies a spring type pressure parallel to the shaft on which it is located. Spring washers are commonly used to preload bushings and shaft seals in many accessories. Spring Washers provide strong compression force with limited movement. Three common names for Spring Washers are: 1) Bellville Washer 2) Finger Washer, 3) Wave Washer.

SPUR GEAR
A type of mechanical gear with straight-cut gear teeth, as opposed to a helical-cut.

SPUR GEAR PUMP
A fluid pump with straight cut gears. Most often refers to a conventional driven gear and idler gear arrangement with the gears meshing at their outer perimeters. By definition, also a gerotor pump which also contains straight cut gears, but with its drive gear located within the circumference of the idler gear.

SQUEALER-TIP (compressor or turbine blade)
A reduced thickness along the blade chord at the tip of the blade. The tip is designed to wear away with minimal damage to its adjacent casing should contact occur between the two. Also called a Profile-Tip. *See also: Vortex-Tip (compressor blade).*

SQUEEZE FILM BEARING
Another name for Bearing (oil damped).

SQUEEZE FILM DAMPING
Refers to Squeeze Film Bearings which dissipate energy caused by unbalanced motion in rotating parts.

SQUEEZE FILM DAMPER
A type of main bearing mount which contains oil under pressure in a squeeze gap (approximately 0.010 in.) between the bearing outer race and its housing.

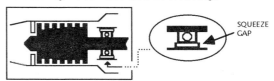

SQUAT SWITCH (thrust reverser)
A landing gear operated safety switch that actuates when the aircraft weight is on the landing gear struts. This switch allows the reverser to be used only during ground engine operation. Note: Some thrust reversers are designed to be used in flight, in which case no squat switch is needed.

SRL
Abbreviation for Single Redline Limit.

SRP
Abbreviation for Single-Rotating-Propeller or Single-Rotation-Propfan. Locate under Propellers.

SST
Abbreviation for Supersonic Transport.

SSU
Abbreviation for Saybolt Seconds Universal viscosimeter.

STACKED DISK OIL FILTER
An engine accessory. A filter unit containing an edge-type filtering element made up of numerous flat screens separated by spacers. Typical of Pratt and Whitney Engines. *See also: Oil Filter (cleanable).*

STAGE LOADING (engine airfoils)
Refers to fan, compressor, and turbine rotor stages and the total amount of Horsepower absorbed by the unit, divided by the number of stages it contains. Higher metal strengths available in the modern turbine engine means higher stage loading can be used in the form of fewer blades per stage. *See also: Blade Loading.*

STAGGER ANGLE (rotor blade)
Refers to the twist in fan, compressor and turbine blades. The twist can be seen when looking from tip to base of the blade. The stagger angle provides the blades with a uniform trailing edge velocity and thus a uniform work distribution along its length. *See also: Impulse-Reaction Turbine System.*

STAGGER (throttle)
Locate under Throttle Stagger.

STAGNATED START
Another name for Hung Start.

STAGNATION (compressor)
Another name for Compressor Stall.

STAGNATION PRESSURE
Another name for Total Air Pressure (Pt).

STAGNATION STALL (engine)
Another name for Compressor Surge.

STAGNATION STALL (flight inlet)
A pressure condition occurring in flight inlets when the back pressure equals the total pressure of flow. This condition essentially stops air movement within the inlet.

STAGNATION UNSTART
Locate under Unstart (inlet).

STAGNATION TEMPERATURE
Another name for Total Air Temperature (Tt).

STAINING (bearings)
Locate under Bearing Inspection and Distress Terms in Appendix.

STAKING
A hand impact method of flattening or spreading out of metal to lock a small part in place. Sometimes another name for Peening.

STALL (airfoil)
A loss of lift. See also: Compressor Stall.

STALL (cold)
Locate under Cold Stall. *See also: Compressor Stall.*

STALL (hot)
Locate under Hot Stall. *See also: Compressor Stall.*

STALL (starting)
A compressor stall condition, generally induced by excessive back-pressure in the combustor and the result of high fuel flow. This stall may also be accompanied by a hot-start.

STALL (sub-idle)
Locate under Sub-Idle Stall.

STALL BUCKET
Refers to a curved shape in the stall line on a fuel control (Wf/Pb vs. % RPM) operating graph. The bucket is located at the approximate mid-point of the stall line between Idle Speed and Take-Off Speeds. Even slight mismanagement of fuelflow at this point can cause compressor stall.

STALL MARGIN CURVE OR MAP
Another name for Compressor Stall Margin Graph.

STALL MAP (compressor)
Another name for Compressor Stall Margin Graph.

STANDARD DAY INTERNATIONAL
Locate under International Standard Day.

STAND-BY ROCKET POWER (SRP)
Small rocket engines that augment aircraft thrust. They are certificated by the FAA for use on many types of aircraft as an emergency source of thrust.

STAND-BY IGNITION
An emergency ignition system powered by a small aircraft inverter. Stand-by ignition generally powers only one igniter plug.

STAR WASHER
Locate under Shakeproof Washer.

START
Refers to normal engine starting by use of a starter unit to initiate compressor rotation and to provide sufficient air to support combustion. *See also: 1) Aborted Start, 2) Hot Start, 3) Hung Start.*

START LOCK (propeller)
A mechanical latching device on a Turboprop propeller; used to position the propeller at minimum pitch (low drag) position during engine starting.

START PLUG
Point where the air source of a Ground Power Unit connects to the aircraft starting manifold.

STARTER (types)
Locate under 1) Cartridge-Pneumatic Starter, 2) Electrical Starter, 3) Fuel-Air Combustion Starter, 4) Gas Turbine Starter, 5) Hydraulic Starter, 6) Jet Fuel Starter,

7) Impingement Starting, 8) Iso-Propyl-Nitrate Starter, 9) Pneumatic Starter, 10) Starter-Generator, 11) Super-Integrated Power Unit.

STARTER AIR VALVE
Another name for Starter Pressure-Regulating and Shut-Off Valve.

STARTER ASSISTED AIR START
Use of the Starter to assist in spool-up during an in-flight engine starting procedure. The starter aids engine rotation when windmilling speed is too low to effect a successful in-flight restart. *See also: Air Start.*

STARTER CONTROL VALVE (pneumatic starter)
Another name for Starter Pressure-Regulating and Shut-Off Valve.

STARTER CLASH RE-ENGAGEMENT (pneumatic starter)
A procedural error by the engine operator. The most common conditions under which clash engagement may occur are as follows: 1) The operator selects starter engagement into a rotating engine at engine spool-down. 2) The operator, during an aborted start, when engine speed is above the clutch pawl lift-off speed, does not manually shut off the air supply when he cuts off fuel and a high free-running starter speed takes place. When the engine slows to pawl drop-in speed, a violent (clash) re-engagement occurs and damage to the starter and the accessory drive gearbox can result.

STARTER CROSS-BLEED AIR
Locate under Cross-Bleed Air.

STARTER-GENERATOR
A combination electric starter and generator. After engine start-up, the starter undercurrent relay actuates to convert the unit into an electrical generator. A commonly used unit on smaller engines, but prohibitively heavy if sized to accommodate the starting power requirements of a large engine.

STARTER PRESSURE-REGULATING AND SHUT-OFF VALVE (R & S Valve)
An electrically controlled and air pressure operated valve located in a Pneumatic Starter systems. The R & S Valve opens electrically via the cockpit starter switch to allow pressurized air to reach the starter. This valve also regulates air pressure to protect the system from over pressurization.

STARTER SYSTEMS
Locate under Starter (types).

STARTING AIR MANIFOLD
An aircraft mounted air manifold connecting the APU to each engine Pneumatic Starter Unit.

STARTING BLEED VALVE
A small engine accessory. A compressor case mounted valve that dumps some compressor discharge air overboard to unload the compressor. The work required by the starter is thereby reduced in turning the compressor. This valve is scheduled closed after engine start-up.

STARTING FUEL CONTROL
Locate under Fuel Control (starting).

STARTING FUEL SYSTEM
Another name for Primer Fuel System.

STARTING STALL
Locate under Stall (starting).

STATIC AIR-OIL SEPARATOR (oil tank)
Another name for Deaerator (oil tank).

STATIC AIR PRESSURE (Ps)
Pressure of air not in motion. Still air pressure resulting from kinetic energy of air (molecular motion) when a unit volume of air is confined in a container. The greater the molecular motion or the greater the number of molecules compressed into the confined space, the greater the resulting air pressure. *See also: 1) Ram Air Pressure, 2) Total Air Pressure.*

STATIC AIR TEMPERATURE (Ts)
Still air Temperature. Temperature of air not in motion or temperature taken with a thermometer moving along with an airstream taking the true temperature due to random motion of air molecules. *See also: Total Air Temperature.*

STATIC ANTI-LEAK CHECK VALVE (lubrication system)
A small engine accessory. An "in direction of flow" check valve with a cracking pressure just above fluid gravity pressure. This valve is located in the pressure (supply) subsystem, downstream of the oil pressure pump. Its function is to prevent oil tank seepage into lower portions of the lubrication system during periods of engine inactivity.

STATIC BALANCING
The process of obtaining equilibrium of a body that has its center of mass on its rotational axis. Components such as compressor rotors, turbine rotors, and other rotating parts are statically balanced by adding or removing small amounts of weight prior to dynamic balancing. *See also: 1) Balance Checks, 2) Balance Weights, 3) Dynamic Balancing.*

STATIC TAKE-OFF THRUST
Another name for Thrust (gross).

STATIC FUEL FILTER
A common fuel filter with a stationary filtering element, as opposed to a filtering element designed to rotate in order to centrifuge out entrained contaminants.

STATIONS (engine)
Locate under Engine Stations.

STATOR CASE
An outer engine case which houses either a compressor or turbine stator vane assembly.

STATOR VANE
Any of the stationary vanes in the gaspath of a turbine engine that control the direction of airflow. Stator vanes include: 1) Compressor Inlet Guide Vanes, 2) Compressor Stator Vanes, 3) Compressor Exit Vanes, 4) Turbine Nozzle Vanes, 5) Turbine Exit Guide Vanes.

STATION DESIGNATIONS
Locate under Engine Stations.

STEEL MESH FILTER
Located under Pleated Cylindrical Filter.

STELLITE ALLOY
A very hard, wear resistant material. In the turbine engine, it is bonded to the notch surfaces of shrouded-tip turbine blades. *See also: Shroud Locks.*

STIFF CONSTRUCTION (engine)
Refers to an engine design technique in which major parts are formed in one-piece units to provide maximum mechanical stiffness. *See also: Stiffness.*

STIFFENER (compressor blade)
A thickened chordwise section shaped like a rib and located in the outer half of the blade. A means of adding structural strength to a compressor blade. Unlike span shrouds, stiffeners do not make contact with adjacent blades. *See also: Span Shrouds.*

STIFFNESS
The ability of a unit to resist changes in its physical dimensions. Turbine Engine outer casings have predictable stiffness, to provide structural strength and prevent rotor to stator contact. *See also: Stiff Construction.*

STINGER MOUNT
A nacelle pod which projects forward from the wing leading edge, as opposed to an under-the-wing pod location.

STOICHIOMETRIC COMBUSTION (jet fuels)
Combustion occurring at the full heat potential of jet fuel, between 3500° and 4000°F.

STOICHIOMETRIC MIXTURE (turbine engine)
The chemically correct air-fuel mixture, where oxygen molecules and hydrocarbons are capable of complete combustion. For Gas Turbine Engines, generally given as an air to fuel mixture of 15 : 1 by weight, in the combustion zone. *See also: Flame-Out.*

STOICHIOMETRIC TEMPERATURE (turbine engine)
The temperature that results from complete combustion of a stoichiometric mixture. In Jet Fuels this temperature is between 3500° and 4000° Fahrenheit.

STOL
Abbreviation for Short Takeoff and Landing (aircraft). STOL aircraft are generally powered by Gas Turbine Engines. *See also: STOVL.*

STOP COCK (procedure)
To Stop-Cock an engine is to move the cockpit fuel shut-off lever to the off position to terminate an engine run.

STOP COCK VALVE
A component part of a fuel control. A small valve actuated by a cockpit Fuel Shut-Off lever to block the fuel being supplied to the engine at engine shut-down.

STOP COCK POSITION
The off position of a cockpit Fuel Shut-Off Lever.

STOP DRILL REPAIR
A method of reducing stress concentrations in solids to retard crack growth. The repair includes drilling a small hole at the extreme end of a crack to produce an increased radius. This procedure is accomplished mainly on sheet metal parts common to combustion and exhaust sections, and is generally considered a temporary repair. *See also: Enhanced Stop Drill Repair.*

STOP NUT
Refers to a fiber locknut.

STORAGE CAPACITOR (ignition)
Locate under Capacitor.

STOVL
Abbreviation for Short Take-Off and Vertical Landing (aircraft). Almost all STOVL aircraft are Turbine Powered.

STRAIN GAUGE LOADCELL
Locate under Loadcell.

STRAINER (bearing oil)
Another name for Last Chance Filter.

STRAINER (fuel or oil)
Another name for Filter.

STRATOSPHERE
The altitude above 36,084 feet where many jet aircraft operate. *See also: Troposphere.*

STRESS (engine)
Stresses acting on parts of an operating gas turbine engine are generally thought of as one of the following: 1) Bending Stress, 2) Centrifugal Stress, 3) Thermal Stress. *See also: Fatigue Failure.*

STRESS (materials)
Locate under Material Stress.

STRESS CORROSION
Metal cracking from residual stress due to operational loads.

STRESS FAILURE
Locate under Engine Inspection and Distress Terms in Appendix.

STRESS RUPTURE
Locate under Engine Inspection and Distress Terms in Appendix.

STRESS RUPTURE CRACKS (turbine blades and vanes)
A condition seen as small chordwise cracks along the leading and trailing edge of blades and vanes. Stress Rupture results from either an over temperature operating condition, overstress from centrifugal loading, or a combination of both. Also locate under Engine Inspection and Distress Terms in Appendix.

STRESS RUPTURE LIFE
Refers to the service life expected of turbine parts.

STRIPPED THREAD
Locate under Engine Inspection and Distress Terms in Appendix.

STRUT (main case)
A component part of an engine. Struts are located in major engine casings and have the appearance of spokes on a wheel to: 1) provide support between an inner and outer stationary section, 2) provide a routing for air and oil lines from the exterior to the interior of the engine, 3) provide support for bearing hubs that, in turn, support the rotor system.

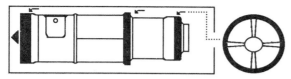

STRUT (exhaust)
Locate under Exhaust Strut.

STUB SHAFT (compressor)
The fore and aft supporting shafts, generally with main bearings attached. The front compressor stub shaft typically attaches to the fan of a Turbofan engine or a gearbox in Turboprop and Turboshaft engines. The rear compressor stub shaft attaches to the turbine shaft.

SUB-IDLE STALL (compressor)
An in-flight engine stall condition occurring with rapid power lever movement, when the engine is operating below flight idle speed. *See also: Compressor Stall.*

SUB-MODULE
A major section of a module. For example, a High Pressure Compressor sub-module and a High Pressure Turbine sub-module which, together, make up a Core Engine Module. *See also: Module.*

SUBSONIC DIFFUSER
A divergent shaped air flow chamber through which air under subsonic flow passes, causing a velocity decrease and a static pressure increase. *See also: Diffuser.*

SUBSONIC FLIGHT INLET (engine)
A divergent air flow duct that acts as a Subsonic Diffuser. The flight inlet facilitates a decrease in air velocity and a rise in static pressure. This resulting ram compression supercharges the engine compressor for an overall increased compressor pressure ratio. *See also: Cycle Pressure Ratio.*

SUBSONIC EXHAUST NOZZLE
In reference to a tailpipe on a subsonic airplane, a final opening formed as a convergent flow chamber through which air moves with a resultant static pressure decrease and velocity increase. In reference to a Rotorcraft Exhaust Nozzle, see Exhaust Diffuser. *See also: Exhaust Nozzle.*

SUCTION SIDE COOLING (airfoil)
Refers to the cambered (top) side of turbine blades or vanes configured with cooling air holes. *See also: Pressure Side Cooling.*

SUGAR SCOOP (APU)
Jargon for an inlet air scoop which provides atmospheric air to the compressor of an auxiliary power unit.

SULFIDATION (turbine)
An accelerated oxidation process in the presence of sulfur ions. Corrosion damage to the surface of turbine parts from contact with salt in air, chlorides in air from industrial pollution, and sulfur in jet fuels. Both substances create a chemical reaction to create sodium sulfates when mixed with impurities in jet fuel such as lead and sulfur. The resulting Sulfidation corrosion has the appearance of scaling, blistering, or delaminating of surfaces. Also called Hot Corrosion.

SUMP
A point where oil collects. The lowest point of an oil wetted cavity such as a gearbox or bearing sump.

SUMP PRESSURIZATION AIR
Air bled from the compressor and directed to the area of main bearing oil seals to prevent oil loss. Also called Engine Bleed Air.

SUMP VACUUM RELIEF VALVE
A small engine accessory. A relief valve found in lubrication vent subsystems to limit negative pressure (suction). Suction is present in some lubrication systems from tight air sealing and vigorous action of the scavenging pumps at high RPM. A typical suction under these conditions is approximately 2 to 4 psi below ambient.

SUMP VENT SYSTEM
Another name for Vent or Breather Subsystem of the engine lubrication system.

SUN GEAR
A gear that drives a set of planet gears in a reduction gearbox. Typical of these gearboxes are: 1) Output Reduction Gearbox of a Turboshaft Engine, 2) Propeller Reduction Gearbox of a Turboprop Engine. *See also: Planetary Gear System.*

SUPER ALLOYS
A family of nickel and cobalt materials used in the manufacture of many turbine parts, primarily turbine blades and vanes. Super Alloys have high strength to density ratio. These alloys also have very high thermal strength and resistance to corrosion. Super alloys are formed by such processes as: 1) Directionally Solidified Eutectics, 2) Hot Isostatic Pressing, 3) Rapid Solidification Rate. *See also: 1) Cobalt Base Alloys, 2) Hastelloy, 3) Inconel, 4) Nickel Aluminide, 5) Nickel Base Alloy, 6) Rene Metal, 7) Waspaloy.*

SUPER-BALANCED ROTORS
Compressor and Turbine Rotors balanced on highly sensitive balancing machines. One such balancing

device employs laser cutting to automatically remove material and produce the most precise balancing. *See also: Geometral Multiplane Balancing Procedure.*

SUPER-CHARGED CORE ENGINE
A Turbofan Engine in which the low pressure compressor is thought of as a super-charger. *See also: Compressor Super-Charger.*

SUPER-CHARGER COMPRESSOR
Locate under Compressor Super-Charger.

SUPER-CONTINGENCY POWER (SCP)
Refers to use of all the engine power available, even to the point of over-temperature, whereby an engine change may later be required. SCP is a military combat power setting. *See also: Engine Ratings.*

SUPER-CRITICAL AIRFOIL
An airfoil with a high critical Mach Number, meaning a delayed airflow separation as air flow approaches Mach One. This airfoil will operate at a higher speed, approaching Mach One at its leading edge. It exhibits good laminar airflow characteristics over its suction (top) surface with minimal shock induced air separation. *See also: Critical Mach Number.*

SUPER-CRUISE
Refers to cruise altitude and level flight at supersonic airspeed without the use of afterburner.

SUPER-INTEGRATED POWER UNIT
A recent development in small Auxiliary Power Units primarily for use in jet fighters. An Auxiliary Power Unit which operates on Liquid Oxygen and Jet Fuel. This unit is a non-air breathing APU for engine in-flight restarting independent of altitude. *See also: Starters.*

SUPERPLASTICALLY FORMED DIFFUSION BONDING
Locate under Diffusion Bonding Process.

SUPER SINGLE-CRYSTAL ALLOY (SSC)
An improved metallic material used for turbine vane manufacture. This material is said to exceed the temperature strength of other Single Crystal Alloys by approximately 100° F. *See also: Single-Crystal Alloy (turbine blades).*

SUPERSONIC COMBUSTION RAMJET (Scramjet)
A type of Jet Engine under development. A Ram-Jet in which the gas path airflow is allowed to maintain supersonic speeds while a portion of its boundary layer air is extracted and slowed for combustion. Scramjets will more likely be powered by a cryogenic fuel, such as liquid hydrogen, mixed with oxygen drawn from the atmosphere. Engine fuel will also be employed as a heat sink to cool engine and aircraft surfaces. *See also: 1) Ram-Jet, 2) Hypersonic Engine.*

SUPERSONIC COMPRESSOR
Another name for Transonic Compressor.

SUPERSONIC DIFFUSER
Locate under Convergent-Divergent Inlet Duct.

SUPERSONIC INLET DUCT
Locate under Convergent-Divergent Inlet Duct.

SUPERSONIC NOZZLE (tailpipe)
The divergent section of a convergent-divergent tailpipe. A duct into which pressurized air at a constant sonic flowrate enters and then expands, with resultant gas pressure decrease and velocity increase to supersonic flowrate. *See also: Convergent-Divergent Exhaust Duct.*

SUPERSONIC SPEED
Classified as speeds between Mach One and Mach Five. *See also: Hypersonic Speed.*

SUPERSONIC TIP SPEED
Blade tip speeds that exceed Mach One. For example, many fan blades are allowed to rotate faster than the speed of sound. When properly controlled, the slight shock wave produced is not detrimental to mass air flow.

SUPERSONIC TRANSPORT (SST)
The name given to transport aircraft with the capability of supersonic flight speeds. The British-French Concorde and the Russian TU-144 are the only two such aircraft presently in service. The Boeing 2707 was to be the American counterpart, but its development ceased in the early 1970's. Present SST's fly at approximately 60,000 foot altitudes and at Mach 2.0 airspeeds.

SUPPLEMENTAL TYPE CERTIFICATE
A document issued by the FAA authorizing alteration of aircraft, engine, or propeller, operating under an FAA Type Certificate.

SURGE (compressor)
Locate under Compressor Surge.

SURGE BLEED SYSTEM
Another name for Surge Control System.

SURGE BLEED VALVE (SBV)
Another name for Compressor Bleed Valve.

SURGE CONTROL SYSTEMS
Locate under: 1) Compressor Bleed Valve, 2) Variable Stator Vanes.

SURGE CONTROL VALVE (auxiliary engine)
An accessory of an Auxiliary Power Unit or Ground Power Unit, located in its pneumatic bleed system. When excessive air is being bled off the APU's compressor, this valve reduces airflow to prevent compressor surge and engine overheating.

SURGE MARGIN
Refers to the spread between the compressor operating line and the surge line on a Compressor Surge Margin Map. The three parameters that must stay in match to prevent the operating line from shifting up to the surge line are Compression Ratio, Mass Airflow, and RPM. Also called Stall Margin. *See also: Surge Point (compressor).*

SURGE POINT (compressor)
The intersecting points of Mass Airflow (lb./sec.) and Compressor Pressure Ratio on a Compressor Surge Margin Map, at which compressor surge is likely to

result. This condition will usually only occur when Mass Airflow is too high for the Compression ratio or vise versa. In flight, the surge point can be reached when inlet conditions cause a mismatch in the two parameters. *See also: 1) Compressor Stall, 2) Compressor Stall Margin Graph, 3) Compressor Surge.*

SURGE VALVE (lubrication system)
An engine accessory. A small back-up valve for the relief valve in the Breather Pressurizing and Vent Valve assembly. When high volume of vent air flow is present, the Surge Valve opens providing a second relief point to quickly evacuate the vent system to the atmosphere.

SUSTAINING SPEED (engine)
Locate under Self-Sustaining Speed.

SVR Abbreviation for Shop Visit Rate.

SWARF
Name given to the machining debris left in an engine during assembly. This material may be found after initial runup on magnetic chip detectors and magnetic drain plugs as thin curled bits of metal. Swarf particles are generally distinguishable from a more serious condition of metallic flakes or chips which indicate possible bearing failure, bearing journal failure, or gear failure. *See also: Fines.*

SWEDISH MANUFACTURED ENGINES
Locate under Volvo Engine Corp. in section entitled "Gas Turbine Engines For Aircraft".

SWEENEY POWER TOOLS
Special tools made for the aircraft industry by the Sweeney Company. Many Sweeney Tools are types of torque wrenches that operate on a gear reduction principle and with very high torque tightening capabilities.

SWEPT-BACK COMPRESSOR BLADE
A recent development in compressor blade design wherein the blades sweep back, similar in appearance to newer propeller blade designs. Swept blades are allowed to operate at higher speeds without shock stall and as a result, produce the highest per stage compression in axial flow compressor designs. *See also: 1) Compressor Blade Shapes, 2) Swept-Back Propeller (UHB).*

SWEPT-BACK PROPELLER (UHB)
A recent development in propeller design, with a scimitar shape or sweep-back. Swept Propellers were developed for Prop-Fan engines, commonly called Ultra High Bypass (UHB) engines. Sweep Back delays shock wave formation, thereby reducing compressibility problems when rotating at supersonic tip speeds. Pressure rise across this propeller is stated to be much higher than possible in traditional propellers.

SWEPT-TIP PROP-FAN
A Prop-Fan engine with a Swept-Back Propeller.

SWIRL (air)
The turning moment in air as it leaves the trailing edge of a rotating airfoil such as a compressor blade, or propeller. *See also: Counter-Rotating Propeller.*

SWIRL CHAMBER (fuel)
Another name for Spin Chamber (fuel nozzle).

SWIRL CHAMBER (oil)
Another name for Deaerator (oil tank).

SWIRL CUP (combustor)
Another name for Swirl Vanes (combustor).

SWIRL DOME (combustor)
The location at the head end of a combustion liner. The dome contains the Swirl Vanes which provide physical support to the fuel nozzles. *See also: Swirl Vanes.*

SWIRL FRAME
A component part of an Inertial Sand and Ice Separator on the inlet case of a Turboshaft or Turboprop engine. This unit angles incoming air to separate heavy airborne particles, allowing only clean air to enter the engine. *See also: Inertial Sand and Ice Separator.*

SWIRL NOZZLE (combustor)
Another name for Swirl Vanes (combustor).

SWIRL VANES (combustor)
A series of small air flow passageways encircling the discharge end of the fuel nozzle located at the head of a combustion liner. Swirl Vanes are shaped to angle air as it mixes with fuel being sprayed into the combustor. This action creates a vortex, thereby increasing mixing time and promoting more complete combustion. *See also: Variable Geometry Combustor.*

SWIRL VANE (fuel nozzle)
Another name for Spin Chamber (fuel nozzle).

SWIRL VANES (impeller)
Vanes which serve as the outlet guide vanes of an impeller type compressor to change the induced radial flow of air back to axial flow. *See also: Diffuser Vanes.*

SWIRLER (oil tank)
A component part of an engine oil reservoir. Located at the scavenge oil return port. An air-oil separator shaped as a thread-like deflector that accelerates scavenged oil in a helical direction to provide a de-aerating effect. *See also: De-aerator (oil tank).*

SWIRLER VENTURI (combustor)
A type of Swirl Vane. Locate under Swirl Vane (combustor).

SYMBOLS OF THE GAS TURBINE ENGINE
Letters, with number and/or letter subscripts, used to represent engine pressures, speeds, temperatures, locations, etc. For example Pt_2 (Pressure total at station two) is generally used to mean compressor inlet pressure of a Pratt and Whitney Turbine Engine. *See also: 1) P (symbols), 2) T (symbols), 3) Appendix for complete listing.*

SYMMETRICAL COMPRESSOR
A name not commonly used today. All modern axial flow compressors are of this type. A design in which approximately one-half of the diffusion and static pressure rise takes place in the rotor blades and the other half takes place in the stator vanes located downstream of the blading.

SYNCHRONIZER (engine)
A small engine accessory. An electrical power lever

positioning device which provides a means of synchronizing the low or high compressor speed of a slave engine to that of a master engine. Engine Synchronizers are used on Turbofan engines and are similar to a propeller synchronizer of a Turboprop engine. Also called Trim Synchronizer.

SYNCHRONIZER (fan)
An electronic means of scheduling the same operating speed on two or more fans in Turbofan Engines. Fan synchronizing reduces engine vibration and out-of-sync noises that would find their way into the aircraft.

SYNCHRO-PHASING SYSTEM
A system by which two propellers or two stages of fan blades in a Propulsor Unit pass each other at the same point in space at each revolution. Synchro-phasing reduces noise and certain vibratory stresses in the rotating turbo-machinery.

SYNCHRONOUS MOTOR
An electronic motor that operates on alternating current, in which the rotor is synchronized with the rotating field produced by the stator. The speed of rotation is always in time with the frequency of the applied alternating current. For example, a Tachometer Indicator in an engine speed indicating system contains a synchronous motor, which remains in synchronization with its engine driven tachometer generator. *See also: Tachometer System.*

SYNCHRONOUS VIBRATION (engine)
A vibration that occurs when a vibratory force is imparted to an engine component having the same natural frequency. If undetected, the magnitude of synchronous vibration can increase until part failure results. *See also: Vibration.*

SYNJET-A
A synthetic jet fuel similar to Jet-A petroleum base fuel. Synjet-A is produced from coal, oil shale or tar sand, or a combination. *See also: Fuel Types-Jet.*

SYNTHETIC FUEL (jet)
See Synjet-A.

SYNTHETIC LUBRICATING OIL (classifications)
Two classifications of synthetic lubricating oils are used in turbine engine lubrication systems: 1) Diesters, which have high operating temperature capabilities and are the most commonly used. 2) Fluorocarbons, which also have high operating temperature capabilities but which do not have the good corrosion protection properties of diesters, and are therefore less often used.

SYNTHETIC LUBRICATING OIL (composition)
Synthesized oils are a mixture of low viscosity base materials such as mineral, animal and vegetable oils blended with certain chemical additives to give very high quality lubricating qualities such as: low foaming, low pour point, low volatility, low lacquer and coke formation, high viscosity index, high flash point, high thermal oxidation limits, and wide temperature range. A main difference between synthetic and petroleum oils is that synthetic oils are produced from a manufactured base stock and the petroleum oils are produced from natural crude stock. Viscosity of synthetic lubricants is measured in Centistokes; the range is normally between 3 to 7 centistokes. A mid-range oil at 5 centistokes is equivalent in viscosity to SAE 5W-10 petroleum base oil. Weight of this product is approximately 7.75 lb./gal. *See also: 1) Centistoke, 2) Kinematic Viscosity, 3) Oil Mixing.*

SYNTHETIC LUBRICATING OIL (types)
Commercial operators generally use one of the following:
1) Type-1- lubricant conforming to Military Specification Mil-L-7808 with commercial use additives.
2) Type-2- lubricant conforming to Mil-L-23699 with commercial use additives. Type-2 oils are the most recent to be developed and are found in most modern turbine engines.

SYSTEM DISPLAY (GLASS COCKPIT)
CRT displays in an electronic cockpit that allow the flight crew access to specific information regarding the nature of an engine malfunction that has been seen as an alert message on other CRT displays.

SYSTEMS
Locate under Engine Systems.

SYSTRON-DONNER FIRE DETECTOR™
A company which markets a type of gas-filled continuous loop fire detector system. Gas pressure within the loop is employed to actuate an electrical microswitch. The detector loop completes a cockpit warning circuit in the presence of an overheat (pressure rise) condition from engine hot air leaks or fire acting on its outer surface. The loop is located in the nacelle area surrounding the exterior of an engine. *See also: Fire Detector.*

T

T
Abbreviation for Turbine Engine as the first letter of many engine designations. For example the General Electric T-58 and the Textron-Lycoming T-55, Turboshaft Engines.

T2 CUTBACK
A fuel control mechanism that limits maximum engine RPM as per Compressor Inlet Temperature (at station two) to prevent compressor stall and possible flameout during take-off.

T2 SENSOR
A compressor inlet temperature sensing device which sends a signal to the engine Fuel Control Unit as a measure of inlet density. A Helium filled tube is one method used for this purpose, while another is a tube with a liquid called Paracymene. Both change pressure quickly in the presence of air temperature changes. Also called Tt_2 Sensor. *See also: Resistance Temperature Detector (inlet temperature).*

T7 PROBE
A thermocouple in an Exhaust Gas Temperature Measuring system. Also called Tt_7 Probe (Temperature total, station 7)

T (symbols)
Upper case "T" generally means Temperature of gases within an engine. When subscripts are added they identify particular temperatures and engine station locations. *See also: Appendix for complete symbols listing.*

Ts
Symbol for Temperature-static.

Tt
Symbol for Temperature-total, meaning the total temperature of gas in a duct to include both ram (friction) influence on temperature and static temperature. *See also: Temperature Total.*

Tt_2
Symbol for Temperature-total, engine station two. On many engines this symbol represents Compressor Inlet Temperature, an important engine parameter sent to the fuel control for fuel scheduling purposes and to schedule operation of Variable Compressor Vane systems.

Tt_7
Symbol for Temperature-total, station seven. On many engines this symbol represents Turbine Discharge Temperature, an important engine parameter sent to the cockpit EPR indicating system. Also called Exhaust Gas Temperature (EGT) on many engines. *See also: Total Air Temperature.*

TABS (tailpipe)
Another name for Tailpipe Inserts.

TABLOCK
A small expendable part. Any of the thin, washer-like metal locking devices which are bent into position to hold in place such items an nuts, bolts, compressor blades, turbine blades, and many other items. Also called Locktab.

TAB WASHER
A small expendable part. A metal locking plate of various shapes that take the place of a conventional washer. Tab washers are crimped to lock fasteners such as hex nuts, spanner nuts, etc. in place. *See also: Key Washer.*

TACHOMETER GENERATOR
A small engine accessory mounted on the Accessory Drive Gearbox. This small generator produces a three-phase alternating current output of approximately 20 volts and is used to power a cockpit indicator containing a synchronous motor-driven mechanism.

TACHOMETER SYSTEM (electric)
A traditional rotor-speed indicating system which utilizes an engine driven tachometer generator and a synchronous motor driven indicator. This system provides both Low and High Pressure Rotor percent RPM indications in the cockpit. However, newer engines are turning to electronic systems for improved accuracy. *See also: 1) Synchronous Motor, 2) Tachometer System (electronic).*

TACHOMETER SYSTEM (electronic)
An improved percent RPM indicating system which utilizes a computer and Speed Sensor. One such system counts fan blades past a pick-up point on the fan casing. Another places a sensor adjacent to a master tooth on a main shaft gear. Both produce signals which a computer converts to a percent RPM indication on a cockpit gauge. *See also: Speed Sensor.*

TACHOMETER SYSTEM (phonic)
The most recently developed percent RPM indicating system which utilizes a computer and a Phonic Probe. The probe penetrates radially into the engine and down to a check point near a rotating shaft. Sound transmission by the moving piece is picked up by the stationary probe and is converted to a percent RPM indication at the cockpit gauge.

TACHOMETER SENSOR
Another name for Fan Speed Sensor.

TAILCONE
An engine component. A cone shaped fairing centered in the exhaust stream in back of the last stage turbine wheel. The Tailcone forms the inner section of the Exhaust Collector. Its function is to act as a fairing and prevent gases from circulating over the rear face of the turbine wheel and creating turbulence. Also called Exhaust Centerbody.

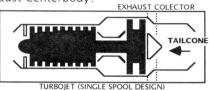

TURBOJET (SINGLE SPOOL DESIGN)

TAILPIPE
Common name for Exhaust Duct.

TAILPIPE INSERT
A small component part. Small sheet metal tabs that are attached to the inner surface of the exhaust nozzle of some older engines to reduce flow area. The smaller area for a given weight of gas flow increases exhaust velocity and thrust. Also called 1) Tabs, 2) Mice, 3) Nozzle Trimmer Tabs.

TAILPIPE TEMPERATURE
Another name for Exhaust Gas Temperature. Not in common use today.

TAKE-OFF (dry)
Symbol TOD. An Engine Power Rating as set by FAA standards. A maximum power setting during which no augmentation such as afterburning or water injection is used. Generally a 5 minute time limit is placed on this power setting. *See also: Engine Power Ratings.*

TAKE-OFF (wet)
Symbol TOW. An Engine Power Rating as set by FAA standards. A maximum power setting during which augmentation such as afterburning or water injection is used. Water injected take-off is often thought of as a hot-day thrust recovery system. Afterburning is used for many power assist purposes such as additional aircraft speed and maneuverability or enhanced ability to take off under heavy load conditions. *See also: 1) Engine Power Ratings, 2) Water Injection, 3) Afterburning.*

TAKE-OFF AND GO-AROUND SWITCH
A cockpit switch on the engine Power Lever. In an emergency, actuating this switch automatically resets engine power to the take-off rating previously set, without the pilot having to move the power lever or monitor a Power indicating instrument.

TAKE-OFF POWER RATING
An FAA Engine Power Rating on a Type Certificate Data Sheet. A manufacturer's guaranteed engine power

available upon command of the operator. *See also: Engine Power Ratings.*

TAKE-OFF RATED THRUST (TRT)
Locate under Military Power Ratings.

TAKE-OFF RESERVE POWER
Another name for Reserve Take-Off Power.

TAKE-OFF THRUST RESET SYSTEM
Another name for Automatic Reserve Power System.

TANTALUM
An element added to Inconel Alloy to improve creep strength and oxidation resistance.

TARGET THRUST REVERSER
Locate under Mechanical Blockage (thrust reverser).

TBC
Abbreviation for Thermal Barrier Coating.

TBO (engine)
Abbreviation for Time Between Overhaul.

TCDS
Abbreviation for Type Certificate Data Sheets and Specifications.

TEAR
Locate under Engine Inspection and Distress Terms in Appendix.

TEARDOWN (engine)
Refers to engine disassembly.

TECHNICAL R & D INSTITUTE (Japan)
Manufacturer of gas turbine engines. Locate in section entitled "Gas Turbine Engines for Aircraft".

TEHP
Abbreviation for Total Equivalent Horsepower.

TELEFLEX CABLE
A stiffened cable with a thread like groove, used to connect engine accessories to each other or to cockpit controls. For example, a Feedback Cable is often a type of teleflex cable.

TELEDYNE CAE COMPANY
Manufacturer of gas turbine engines. Locate in section entitled "Gas Turbine Engines for Aircraft".

TEMPER
Abbreviation for Turbine Engine Module Performance Estimation Routine.

TEMPERATURE
Physics term. The intensity of heat contained in a substance. A measure of the internal energy or molecular activity of a substance as measured by a thermometer device. In the U.S. System, temperature units are degrees Fahrenheit and degrees Rankine. In the Metric System, degrees Celsius and degrees Kelvin are used.

TEMPERATURE (exhaust)
Locate under 1) Exhaust Gas Temperature, 2) Intermediate Turbine Temperature, 3) Turbine Gas Temperature, 4) Turbine Inlet Temperature, 5) Turbine Outlet Temperature.

TEMPERATURE (To)
At International Standard Day conditions: 59° Fahrenheit, 519° Rankine, 15° Celsius, 288 Kelvin.

TEMPERATURE AMPLIFIER (fuel control)
A small engine accessory. An electronic device which amplifies an exhaust gas temperature signal. The signal is then used to power a fuel control scheduling mechanism. Also called Temperature Control Amplifier. *See also: Temperature Datum System.*

TEMPERATURE DATUM SYSTEM (fuel control)
An electronic fuel scheduling circuit powered by a Temperature Amplifier in an electronic or electro-hydromechanical fuel controlling system.

TEMPERATURE LAPSE RATE
The rate of change of temperature in atmospheric air from sea level to isothermal altitude. Given as minus 3.57° F. per 1,000 feet. Also called Adiabatic Lapse Rate due to the fact that the temperature of the air changes, but the air has neither gained nor lost heat energy. The temperature change occurs as a result of atmospheric pressure change.

TEMPERATURE LIMITED
Refers to the idea that all turbine engines are Turbine Inlet Temperature (TIT) limited and that strict adherence must be given to TIT limits during engine operation by way of the EGT, ITT or TIT Gauge.

TEMPERATURE OVERSHOOT
Combustor temperature elevations above maximum values which quickly reduce to normal. The overshoot, depending on its severity, may or may not register on a cockpit gauge. Numerous sensors are employed to reduce temperature overshoot because even small temperature peaks can adversely affect engine service life.

TEMPERATURE RISE (friction)
Locate under Kinetic Heating.

TEMPERATURE RISE (Tr)
A physics term. Temperature rise above ambient temperature due to the effect of the compression process on air within an engine. Term used in the formula to compute "Horsepower to Drive the Compressor".

TEMPERATURE PICKUP
A small engine accessory. A temperature sensor, often a small bellows unit in an engine inlet, that sends a signal to a fuel controlling device as an indication of inlet air density.

TEMPERATURE TOTAL (hot exhaust)
Symbol (Tt). Exhaust temperature is measured by use of a thermocouple which includes in its readout both static temperature and temperature effect due to ram. The temperature probe in the gas stream will register the presence of static heat and also velocity energy as heat energy. If the probe is located at Engine Station 7, the temperature will be described in symbols as Tt_7. *See also: Total Air Temperature.*

TEMPERATURE TOTAL (flight inlet)
Symbol (Tt). The temperature as measured by a compressor inlet temperature probe which includes in its

readout both static temperature and temperature due to ram effect. The temperature probe in the gas stream will register the presence of static heat and also velocity energy as heat energy. If the probe is located at Engine Station 2, the temperature will be described in symbols as Tt_2. *See also: Total Air Temperature.*

TEMPERATURE TRANSIENT (engine)
Another name for Temperature Overshoot.

TEMPERATURE TRANSDUCER (exhaust)
Any of the types of heat sensors that transmit engine exhaust temperatures to a cockpit gauge. The most common transducer is the chromel-alumel type thermocouple.

TEMS
Abbreviation for Thrust Engine Monitoring System.

TENSILE STRESS
Stress produced by two external forces acting in opposite directions on a part. *See also: Shear Stress.*

TENSION FAILURE
Locate under Engine Inspection and Distress Terms in Appendix under Stress Failure.

TEO SENSOR
Abbreviation for Temperature, Engine Oil. Locate under Oil Temperature Bulb.

TERTIARY COMBUSTION ZONE
Another name for Dilution Zone (combustion).

TERTIARY DOORS (nacelle)
Small doors in engine nacelles that have a function of allowing air in or out, for ventilating or cooling purposes.

TEST BED
British term for Test Cell.

TEST BENCH (engine)
British for Test Cell.

TEST CELL (engine)
A facility in which engines are test run to check performance or for troubleshooting that cannot be accomplished in the aircraft. Larger test cells are completely enclosed and provide sound suppression. Open air Test Cells are commonly called Test Stands. *See also: Ground Runup Enclosure.*

TEST CHAMBER
An enclosed, environmentally controlled test cell in which temperature and altitude operating conditions can be achieved during engine operation.

TEST STAND (engine)
Locate under Test Cell.

TET
Abbreviation for Turbine Entry Temperature.

TEXTRON-LYCOMING CORP.
Manufacturer of Gas Turbine Engines. Formerly Avco-Lycoming. Locate in section entitled "Gas Turbine Engines for Aircraft".

TGT
Abbreviation for Turbine Gas Temperature.

THERMAL BARRIER COATING (TBC)
Any of the insulating and oxidation resistant coatings placed on hot section parts. TBC slows the oxidation process, reduces cooling air requirements, and increases component life. Coating thickness is approximately 1 to 8 thousandths inch, with heat-transfer reduction to coated surfaces of up to 300°F. The ceramic material Zirconia with Zttria added (stabilizing agent) is a common coating that is often deposited by a plasma spraying process. The Zirconia coating can also be "laser glazed", an additional outer coating melted on the surface by a laser process. Other common substances used as barrier materials are Brunsbond™, Calcium Silicate, Calcium Titanate, Magnesium Zirconate, and Sermaloy™. *See also: 1) Aluminide Coating, 2) Gator Guard™, 3) Plasma Spray Coating, 4) Sulfidation, 5) Wear Resistant Coatings.*

THERMAL EFFICIENCY
A physics term. Internal engine efficiency in terms of fuel energy available versus work produced. Mathematically expressed as a percentage by the following formula:

$$Teff = Fn \times V_1 + Ms/2g\ (V_2-V_1)^2 \div Wf \times 778.26\ ft.lb./BTU \times Fuel\ BTU\ Value$$

Where: Fn is net thrust, g is the gravity constant, V_1 is aircraft speed, V_2 is exhaust velocity, Ms is mass airflow, Wf is fuelflow. *See also: 1) Efficiencies (engine), 2) Overall Efficiency, 3) Propulsive Efficiency.*

THERMAL FATIGUE (turbine metals)
A compression and tension load condition in turbine engine metals caused by heating and cooling each time an engine power change occurs. *See also: Fatigue Failure.*

THERMAL LOCKOUT SWITCH (oil filter)
A small component part of an engine oil filter. A thermal lockout prevents the pop-out bypass indicator button from actuating during a cold weather engine start. Oil temperature must be above a preset value of approximately 40°C. for pop-out of the filter warning button to occur. The pop-out button is often called a Delta-P (ΔP) Button. *See also: Filter Bypass Indicator (pop-out).*

THERMAL POWER
A term referring to the shaft horsepower or kilowatt rating of a Turboshaft Engine.

THERMAL SHOCK (turbine metals)
Stress exerted on a metal's structure. The result of rapid temperature change which can result in cracking or even complete metal failure.

THERMATIC CLEARANCE CONTROL SYSTEM
Locate under Thematic Compressor Rotor.

THERMATIC COMPRESSOR ROTOR
An Engine Component. A recent development in drum or disc type compressor design providing rotor expansion via an internal, pressurized air system. Expansion, in turn, tightens compressor blade tip clearances during engine operation. This system is considered a Passive Clearance Control and is less expensive than the Active Clearance Control used to control turbine

tip clearances. With Thermatic Clearance Control, the resulting increase in compressor efficiency decreases fuel consumption. The conventional method in the past has been to cool and contract the compressor outer case rather than to expand the compressor drum or disc. *See also: Tip Clearance.*

THERMISTOR (fire detection)
A semi-conductor type temperature sensing device. Unlike a thermocouple fire detector, a Thermistor requires an external power source. Thermistors exhibit large parameter changes with changes in temperature and are utilized as a type of switching mechanism in fire detection circuits.

THERMOCOUPLE (bi-metallic)
Any of the self-generating temperature sensors that operate on a dissimilar metal principle of creating a thermo-electric current flow when under a heat load. The junction of the two metals placed in the engine is called the hot junction; the junction at the gauge end is called the cold junction. The cockpit gauge serves as the reference junction and indicates the temperature by way of the current flow through its mechanisms. The hot junction placed in the engine exhaust is also called the measuring junction. *See also: Trim Thermocouple.*

THERMOCOUPLE (exhaust temperature)
A small component part of an Exhaust Gas Temperature Indicating System. Typically a chromel-alumel probe which fits radially into the exhaust gas stream. Chromel is a nickel-chromium alloy, is a non-magnetic material, has a positive electrical potential, and is color coded "white". Alumel is a nickel-aluminum alloy, is magnetic material, has a negative electrical potential, and is color coded "green". The exact position of the probe in the engine often determines the name given the readout on the cockpit gauge. *See also: Temperature (exhaust).*

THERMOCOUPLE (dual-terminal)
A thermocouple probe fitted with two sets of output terminals. One set connects to the cockpit gauge circuit, the other set connects to either: 1) A test circuit junction box, 2) An electronic fuel control circuit.

THERMOCOUPLE (multiple-element)
A thermocouple configured with two or more hot junction loops within the probe. The hot junctions, when located at different depths in the gas flow, provide an improved average gas temperature indication.

THERMOCOUPLE (immersion type)
Another name for Thermocouple (exhaust temperature), because the probe penetrates into the exhaust stream.

THERMOCOUPLE (oil temperature)
A small component part of an Oil Temperature Indicating System. A thermocouple positioned in the oil flow to provide a cockpit indicator with a self-generating oil temperature signal. The bimetallic materials in this thermocouple are generally of Iron and Constantan alloys, or Chromel and Alumel alloys.

THERMOCOUPLE COLD/HOT JUNCTION
Locate Under Thermocouple (bi-metallic).

THERMOCOUPLE FIRE DETECTOR
A small component part of a Fire and Overheat Detection System. A thermo-electric generating device that powers a cockpit warning light circuit when a rapid rate-of-heat rise occurs in the area of the thermocouple. *See also: Fire Detector.*

THERMODYNAMIC CYCLE
Locate under 1) Brayton cycle, 2) Open Cycle Engine.

THERMODYNAMIC POWER RATING (engine)
The Thrust or SHP capability of an engine operated at maximum permitted Turbine Inlet Temperature, and International Standard Atmosphere (ISA) conditions. This rating represents the theoretical power available to an engine. For example, an engine's Thermodynamic Thrust might be 5,300 lbt, but when installed to meet a particular aircraft requirement, it is Flat-Rated to only 5,000 lbt. *See also: Flat-Rating (engine).*

THERMODYNAMIC SHP
A Dynamometer measurement of engine power available to its output shaft. The shaft horsepower delivered to a Turboprop Engine from its power turbine while the engine is operating at maximum permitted Turbine Inlet Temperature, and International Standard Day (ISD) conditions. The Thermodynamic SHP rating indicates the theoretical power available to an engine. *See also: Thermodynamic Power Rating.*

THERMODYNAMIC THRUST
Engine thrust available while operating at maximum permitted Turbine Inlet Temperature, at International Standard Day (ISD) conditions. The Thermodynamic Thrust rating indicates theoretical power available to an engine. *See also: Thermodynamic Power Rating.*

THERMOPLASTIC COMPOSITES (TP)
A family of glass-fiber and carbon-fiber reinforced high temperature thermoplastic composites. These materials demonstrate good mechanical properties in the 500 to 600° F. range. *See also: Composite Materials.*

THERMOSTATIC VALVE (oil cooler)
A small component part of an engine oil cooler. A bimetallic-spring operated valve set to the open position when cold, allowing unheated oil to bypass the cooling chamber. Then as the oil approaches normal operating temperature, this valve closes completely to force all of the lubrication system oil through the cooling chamber portion of the oil cooler. *See also: 1) Air-Oil Cooler, 2) Fuel-Oil Cooler.*

THERMOSWITCH (fire detector)
A small component part of a Fire and Overheat Detection System. A device in which thermal expansion of its metal outer casing mechanically closes an electrical contact within the unit. Heat generated by hot air leaks, fuel, oil, or electrical fires will close the contacts to cause a fire warning light to illuminate in the cockpit. *See also: Fire Detector.*

THETA TABLES
Tables that compare ambient temperature (T) to International Standard Day (ISD) temperature (To). The symbol "Ø" (Greek letter Theta) indicates the ratio of ambient temperature to ISD temperature (T/To).

THICKENED FUEL
Refers to the jel-type fuel. Locate under Anti-Misting

Fuel Additive.

THP
Abbreviation for Thrust Horsepower.

THROAT AREA (C-D nozzle)
The narrowest flow area, the point where the convergent and divergent sections meet in a supersonic inlet or exhaust duct. Can also refer to the discharge point at the trailing edge of a Turbine Nozzle Vane assembly. Sometimes called a "waist area".

THREE-D CAM (fuel control)
A small component part of a fuel control. A multi-lobed cam that can simultaneously rotate about it axis and move up and down. This movement allows a cam follower riding on its surface to seek an infinite number of positions. The Three-D Cam is used to minimize fuel controlling linkages within the fuel control.

THREE-D COMPUTATIONAL FLUID DYNAMICS (CFD)
A recent development in aero-thermo-dynamic design of turbo-machinery. CFD is a computer based process which can predict performance factors of computer designed engine models before actual engine construction begins.

THREE-D TURBINE VANE
A vane designed by a computer based technique called Three-D Computational Fluid Dynamics.

THROTTLE BURST
Rapid cockpit Power Lever (throttle) movement from idle position to take-off position. An emergency procedure for quick engine power response, often at the expense of component life. Also called Jam or Slam Acceleration.

THROTTLE CHOP
Rapid cockpit power lever movement from normal engine operating speeds to idle. An emergency procedure for quick engine power reduction, often at the expense of component life.

THROTTLE LEVER
Another name for Power Lever.

THROTTLE LEVER ANGLE (TLA)
Another name for Power Lever Angle.

THROTTLE KNOB ALIGNMENT
Another name for Throttle Stagger.

THROTTLE PIGGYBACK HANDLE
A thrust reverser lever that is mounted on the cockpit throttle lever.

THROTTLE PUSHING
Jargon for increasing the power in a follow-on model of an engine currently in production.

THROTTLE QUADRANT
The Throttle Lever and its associated mechanisms in the cockpit.

THROTTLE RESOLVER
A unit which receives mechanical signals from the aircraft throttles and transmits electronic signals to a computerized fuel controlling system.

THROTTLE STAGGER
Misalignment of throttle-knobs in the cockpit when more than one engine is operating at the same engine power setting. Some minor deviation is allowed before re-rigging of the linkage system is required. Also called Throttle Knob Alignment.

THROTTLE STOPCOCK POSITION
The rearward most position of a cockpit throttle, in a throttle type having the capability of shutting down an engine. However, the rearmost position of most cockpit throttle levers is the idle engine speed position, in which case engine shut-down is accomplished by use of a separate Fuel Shut-Off lever or switch.

THROUGH-FLOW COMBUSTOR
A conventional straight through combustor in which air from the diffuser enters the combustor, fuel is added, and the gases expand axially rearward into the turbine. Also called an In-Line Combustor. See also: Annular Combustor (reverse-flow).

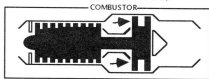
TURBOJET (THROUGH-FLOW COMBUSTOR)

THRUST
A physics term. A forward force on an object in reaction to a rearward movement of a mass from or through the object. Thrust is often defined in accordance with Newton's Third Law, that for every action there is an equal and opposite reaction. Thrust is measured in accordance with Newton's Second Law as follows:

$$F = MA$$

Where F = thrust force, M = mass, A = acceleration. *See also: 1) Pressure Thrust, 2) Rated Thrust, 3) Reactive Thrust.*

THRUST (ambient corrected)
A means of computing actual thrust from observed thrust. The formula is as follows:

$$F_{(cor)} = \text{Thrust (observed)} \times 29.92 \div Po$$

Where 29.92 in. hg. = International Standard Day pressure and Po = actual barometric pressure (in. hg.).

THRUST (dry or wet)
Locate under Take-off (dry), Take-off (wet).

THRUST (gross)
Symbol (F_g). Thrust when the engine is operating but not moving forward. Mathematically expressed as:

$$F_g = M_s (V_2 - V_1) \div g$$

Where: M_s = mass airflow (lb./sec.), V_1 is aircraft speed in ft./sec., V_2 is exhaust velocity (ft./sec.), g is the gravity constant 32.17 ft./sec.^{2}. Gross thrust occurs when V_1 is zero. Gross Thrust is also called Static Thrust. *See also: Thrust (net).*

THRUST (net)
Symbol (F_n). Thrust when the engine is operating and moving forward on the ground or in flight. Math-

ematically expressed as shown in Thrust (gross). Net thrust occurs when V_1 has a value. *See also: Thrust (gross) for explanation of symbols.*

$$Fn = Ms (V_2 - V_1) \div g$$

THRUST (nozzle choked)
Gross Thrust or Net Thrust when the exhaust (jet) nozzle is flowing in a choked condition. The formula for computing is:

$$Fg = [Ms (V_2 - V_1) \div g] + [Aj (Pj - Pam)]$$

Where Ms = mass air flow in lbs./sec, V_2 = exhaust velocity in feet/sec., V_1 = aircraft speed in feet/sec., g = the gravity constant of 32.17 ft./sec.^{2}, Aj = area of the jet nozzle, Pj = pressure at the jet nozzle, Pam = pressure ambient. *See also: Thrust (gross).*

THRUST (Wf added)
Fuelflow (Wf) if often omitted from Thrust calculations because its weight of approximately 2% of the engine's mass air flow rate is also approximately the weight of air extracted from the engine as Bleed Air. When Wf is included the formula is:

$$Fg = [Ms (V_2 - V_1) \div g] + [Aj (Pj - Pam)] + [(Wf \div g) \times V_2]$$

See also: Thrust (gross) for symbol identification.

THRUST (rated)
Locate under Rated Thrust.

THRUST AUGMENTATION (engine)
A means of increasing engine thrust. The two most commonly used methods are: 1) Afterburner, 2) Water Injection.

THRUST BALANCE CHAMBER
Another name for Balance Chamber (thrust). *See also: Hydraulic Thrust Rotor Control.*

THRUST BEARING (main bearing)
Name given to ball-type main bearings in Gas Turbine Engines. The ball bearing is caged and absorbs thrust loads on compressors and turbines. Locate under 1) HP Thrust Bearing, 2) IP Thrust Bearing, 3) LP Thrust Bearing, 4) Turbine Thrust Bearing.

THRUST BEARING (prop)
A ball bearing in a Turboprop reduction gearbox designed to absorb axial (thrust) loads from the propeller. This bearing is not generally considered one of the Main Engine Bearings.

THRUST BRAKE
Another name for Thrust Reverser.

THRUST DIRECTING EXHAUST NOZZLE
Locate under Exhaust Nozzle (vectored thrust).

THRUST EQUIVALENT HORSEPOWER (TEHP)
Another name for Equivalent Shaft Horsepower (ESHP).

THRUST HORSEPOWER (THP)
A physics term. A means of computing horsepower from Net Thrust (Fn in flight). Mathematically expressed as:

$$THP = Fn \times \text{Aircraft Speed (MPH)} \div 375 \text{ mi.lb./hr}$$
$$THP = Fn \times \text{Aircraft Speed (ft/min)} \div 33{,}000 \text{ ft.lb./min}$$
$$THP = Fn \times \text{Aircraft Speed (ft/sec)} \div 550 \text{ ft.lb./second}$$

These formulas provide a means of comparing Thrust Horsepower factors between engines of the same thrust at varying aircraft speeds.

THRUST INDEX
A numeric scale on a cockpit gauge which indicates the thrust being produced by the engine. This display is not very common in present day aircraft, with engine pressure ratio and fan speed being the preferred methods of indicating engine thrust. *See also thrustmeter.*

THRUST LEVER
Another name for Power Lever.

THRUST LEVER ANGLE (TLA)
Another name for Power Lever Angle (PLA).

THRUSTMETER
An instrument to measure an engine's thrust. It is used primarily in test cells, test stands, and other ground engine test facilities. It was used in some early turbine engine powered aircraft as an engine performance gauge. *See also thrust index.*

THRUST MOUNT
An engine component. A main engine mount by which engine thrust is transferred to the aircraft. Also called Thrust Link.

THRUST POWER
A physics term. Defined as Thrust x Aircraft Speed. Mathematically expressed as: 1) $Fn \times V_1$, 2) $[Ms (V_2 - V_1) \div g] V_1$. *See also: Thrust (gross) for symbol identification.*

$$F_{(power)} = [Ms (V_2 - V_1) \div g] V_1$$

THRUST PRODUCING ENGINE
A Gas Turbine Engine that produces reactive thrust for propulsion, namely Turbojets and Turbofans. See also: Torque Producing Engine.

THRUST RESERVE POWER
Locate under Reserve Thrust.

THRUST REVERSE (capability)
The stopping force available from an engine Thrust Reverser. Thrust Reverse is most effective immediately after touchdown when aircraft momentum is high and brakes are less effective. At that time Engine Thrust Reverse and Aircraft Drag account for approximately 55% of the stopping force on a dry runway and up to 80% on a wet and icy runway. Thrust reverse is selected at Ground Idle Speed, after which the pilot raises power as required by runway condition, runway length, and aircraft weight.

THRUST REVERSE LIMITER
A cockpit control which limits the amount of reverse thrust. The common methods used are to mechanically or electrically limit the travel of the thrust lever.

THRUST REVERSER
A component part of an engine Q.E.C Kit. A mechanical obstruction to normal rearward flow of fan and/or

core engine exhausts. The reverser blocks the rearward flow and turns the gases forward to provide thrust reverse assistance to the aircraft brakes and to provide directional control of the aircraft upon landing or during a rejected take-off. Reversers are especially helpful on wet and icy runways. Some aircraft also allow thrust reverse in flight for rapid rate of descent. Fully ducted turbofan engines with a common hot/cold tailpipe have only one reverser. Turbofan engines with unmixed exhausts may have two reversers. If only one reverser is installed, it will more likely be located on the fan exhaust duct. *See also: 1) Aerodynamic Blockage (thrust reverser), 2) Mechanical Blockage (thrust reverser), 3) Thrust Spoiler (turbofan).*

THRUST REVERSER (directed flow)
The name given to a Thrust Reverser designed to direct exhaust gases forward, only from the upper two quadrants of the exhaust duct. This arrangement reduces the tendency of ground debris to recirculate into the engine.

THRUST REVERSER BUCKET
The clamshell (blocker door) portion of a Mechanical Blockage Thrust Reverser.

THRUST SENSITIVE SYSTEM (TSS)
A Turboprop auto-feather system. A system which provides a means of detecting a loss of engine torque on take-off from a failed engine. When alerted, the TSS auto-feathers the failing engine to reduce drag..

THRUST SPECIFIC FUEL CONSUMPTION
A physics term. An important statistic of a Turbojet or Turbofan Engine because it indicates the fuel consumption rate. TSFC provides a means of comparing efficiency values between engines of the same power. It is a ratio of the weight of fuel being consumed per hour to engine thrust. Mathematically expressed as:

$$TSFC = Wf \div F$$

Where Wf is fuel flow in lb./hr. (PPH), and (F) is either Gross Thrust (Fg) or Net Thrust (Fn) in pounds. For example, the General Electric Unducted Fan Engine (UDF) is stated to have a very low fuel consumption rate with a TSFC of 0.24 lb./hr./lbt. (pound of fuel/hr. per pound of thrust. Additional information is located under Specific Fuel Consumption (SFC). See also section entitled "Gas Turbine Engines for Aircraft" for specific engine applications.

THRUST SPOILER (Turbofan)
A device attached to some Turbofan Engine hot exhausts to redirect the gases in the manner of a thrust reverser. The spoiler, however, does not reverse the thrust, but rather directs the gases radially to nullify thrust. *See also: Thrust Reverser.*

THRUST TO WEIGHT RATIO
The ratio of maximum thrust output of an engine to its physical weight. In the modern fighter engine operating in full afterburner, the ratio is approximately 10 to 1.

THRUST VECTORING EXHAUST NOZZLE
Locate under Exhaust Nozzle (vectored thrust).

TIG
Abbreviation for Tungsten Inert Gas (welding process).

TILT-NOZZLE AIRCRAFT
Another name for Vectored Thrust Engine.

TILT-PROPELLER AIRCRAFT
A type of Vertical Take-Off and Landing (VTOL) aircraft design in which the propellers of its Turboprop Engines are capable of angling to a vertical position. The function of this aircraft is similar to that of a Tilt-Rotor Aircraft.

TILT-ROTOR AIRCRAFT
A type of Vertical Take-Off and Landing (VTOL) aircraft. It has fixed wings and turbine engines fitted with wide diameter rotors located at the wing tips. The engines can tilt upwards for vertical take-off, hover, and landing, or move to the horizontal forward thrust position for cruise flight in the manner of a turboprop. The Boeing V-22 Osprey is an example of a Tilt-Rotor aircraft. *See also: 1) Prop Rotor, 2) Short Take-Off and Vertical Landing Aircraft.*

TILT-WING AIR
A type of Vertical Take-Off and Landing (VTOL) aircraft. Its wings are capable of tilting until their leading edges are in the vertical position. The Gas Turbine Engines rotate along with the wings to vertical. The function of this aircraft is similar to that of a Tilt-Rotor Aircraft. The Ishida TW-68 is an example of a Tilt-Wing aircraft.

TIME BETWEEN OVERHAUL (TBO)
The time in engine operating hours or engine cycles that the manufacturer recommends as the total service life from new to overhaul or from one overhaul until the next. TBO requirements apply quite differently from one engine to another. For example, one engine might have an established "Engine TBO", while another might have a TBO requirement only on its individual Modules. In addition, TBO is used for Components and Accessories. TBO is not an FAA requirement for private aircraft, but TBO procedures are required by FAA Parts 121 and 135 on air carrier and commercial operator aircraft. Should Cycles be used as a time change criteria, the service life is often known as Cycles Between Overhaul (CBO). *See also: 1) Engine Cycle, 2) On-Condition Overhaul, 3) Time Since Restoration.*

TIME RESTRICTED PART
Another name for Life-Limited Part.

TIME SINCE RESTORATION (TSR)
Newer term which is replacing Time Between Overhaul. Under this concept, when cycles are used as time change critetia, the service life is known as Cycles Since Restoration (CSR).

TIME SINCE INSTALLATION (TSI)
The operating hours accumulated since an engine part was first installed on an engine. TSI is normally recorded in conjunction with the Time Between Overhaul data.

TIME SINCE NEW (TSN)
The operating hours accumulated since an engine part was first placed in service. TSN is normally recorded in conjunction with the Time Between Overhaul data. Could also be used to mean Engine Time Since New, but Total Engine Time (TET) is more often used for that purpose. *See also: Total Engine Time (TET).*

TIP CLEARANCE
The spacing between compressor or turbine blade tips and their stationary shroud rings. Should clearances be too tight when cold, contact may occur when operational loads are applied. When running clearances are too large, blade efficiency is degraded. Tip clearances during operation are often very close, in the range of 0.003 to 0.010 inch. These clearances are generally given in the table of limits of the engine service manual. *See also: 1) Blade Tip Vortices, 2) Compressor Tip Clearance, 3) Shroud Ring, 4) Thermatic Compressor Rotor.*

TIP/HUB RADIUS RATIO
Locate under Hub to Tip Radius Ratio.

TIP SHAKE
The amount a compressor or turbine blade will move circumferential at its tip from looseness in its attachment slot at the base. Tip Shake limits are generally given in thousandth of inches and are located in the table of limits of engine service manuals.

TIP SHOCK STALL
Locate under 1) Shock Stall, 2) Tip Speed.

TIP SPEED
The speed of a rotating airfoil at its tip area. For example, Centrifugal Compressor Impeller blade tips and Fan Blade tips generally rotate at supersonic speeds of Mach 1.1 to 1.5 at full engine RPM. Tip speeds are given in feet per second, meters per second, or Mach Number. Tip Speeds are limited in the Gas Turbine Engine due to shock stalling and the loss of airfoil efficiency which results. *See also: Shock Stall Speed.*

TIR
Abbreviation for Total Indicator Reading.

TITANIUM ALLOY
An important Gas Turbine Engine alloy used mainly in the cold section, but recent advances in heat strength have also created limited uses in the hot section. Titanium has a high rigidity to density ratio and only one-half the weight of stainless steel. It also has about 90 percent of the strength of stainless steel and equal corrosion resistant qualities. Titanium has only minimal application in the hot section because it lacks the high heat strength of Nickel or Cobalt alloys. *See also: Titanium Aluminide.*

TITANIUM ALUMINIDE (Ti3AL)
An intermetallic metal still undergoing development. It is a titanium-aluminum alloy of much higher heat strength than other titanium alloys, and presently used in the manufacture of high pressure compressor cases and some rear stage turbine discs. This material has 30 to 40 percent less weight than nickel-base alloys and is expected to be a successor to many such metals. Present development problems include brittleness at room temperature and difficulty of manufacture.

TITANIUM SPONGE
The initial form of titanium that is produced from the basic mined material. The sponge is later turned into ingots from which engine parts are manufactured.

TLA
Abbreviation for Throttle Lever Angle.

TOBI-DUCT
An engine component. Acronym for Tangential On-Board Injector Duct. This duct is an air manifold that directs cooling air onto a turbine wheel in an APU installation.

TOD
Abbreviation for Take-Off Dry.

TOGA SWITCH
Locate under Take-off and Go-Around Switch.

T.O.P.
Abbreviation for Torque Oil Pressure.

TOPPING CHECK (rotorcraft)
Locate under Ng Topping Check.

TORCH IGNITER
An engine component. An open flame fuel fed device similar to a blow-torch, used to initiate combustion in an afterburner. An electrical spark and a separate fuel source are used to ignite the torch, which in turn lights off the afterburner main fuel being injected by its spray bars. *See also: Hot Streak Ignition.*

TORCHING
A visible flame from the tailpipe. A condition that may occur during a Fuel Enrichment engine start or from puddled fuel in the combustor or tailpipe. Generally torching occurs during a late start, after fuel fogging is noticeable at the tailpipe. Not a serious condition if Exhaust Gas Temperature on the cockpit gauge stays within acceptable limits. Also called Enrichment Burst. *See also: Fuel Enrichment Valve.*

TORQUE (Tq)
Refer to Torque-Meter.

TORQUE CONE (compressor)
A component part of a compressor rotor. The rear portion of a drum type compressor. It is a hollow sheet metal cone fitted to the rear of a compressor drum and positioned such that it supports the rear stub shaft. The Torque Cone is designed to absorb torque loads induced by the turbine shaft that connects to the rear stub shaft. *See also: Compressor Drum.*

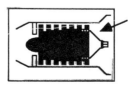

TORQUE CONE (turbine)
Locate under Turbine Drum.

TORQUE COUPLING (turbine)
A component part of a turbine rotor. A spacer located between fixed turbine stages. The Torque Coupling is designed to transmit torque loads forward from one turbine disc to the next. Also called Torque Ring.

TORQUE-METER
Refers to Torquemeter Indicating Systems which produce a cockpit display of engine power output. A required system for Turboprop and Turboshaft powered aircraft. *See also: 1) Torquemeter, 2) Torquemeter System (phased-shift).*

TORQUE-METER INDICATOR
A Turboprop or Turboshaft cockpit instrument used to indicate engine power output. The propeller or rotor inputs a twisting force to an electronic phased-shift or oil operated torquemeter transmitter, which in turn sends a signal to the cockpit indicating system. *See also: Torquemeter Pickup.*

TORQUE-METER PICKUP
A small engine accessory. A transmitting device employed in a Torquemeter Phase-Shift System. Also called Torque Sensor.

TORQUE-METER SYSTEM (balanced-piston)
A system which utilizes a traditional hydromechanical, oil operated mechanism. Torque forces at the output drive cause oil pressure changes in a piston and cylinder arrangement. The oil pressure changes are then directed to a Torquemeter transmitter which powers a cockpit gauge.

TORQUE-METER SYSTEM (phased-shift)
A system which utilizes an electronic pickup (sensor) and a transmitter. The system contains a magnet and coil mechanism that measures the angular deflection or twist of the torque shaft connecting the propeller to the engine. When a load is applied, the sensing system sends a signal to a Torquemeter indicator in the cockpit.

TORQUE MOTOR (APU fuel control)
A component part of an APU Fuel Control. A mechanism used in place of a traditional flyweight governor to schedule fuel to the combustor during spool-up to full power. The torque motor is powered by an electronic control circuit that monitors selected parameters such as EGT, RPM, Ambient Pressure and Ambient Temperature.

TORQUE OIL PRESSURE
A cockpit indication of engine power from a Torquemeter located within a Turboprop or Turboshaft engine. The cockpit gauge reads out in Torque Oil Pressure, ft.lbs., PSI, or Percent of torque. The cockpit readings can be, if required, converted to Horsepower using a conversion factor supplied by the manufacturer. *See also: Torquemeter.*

TORQUE PRODUCING ENGINE
A Gas Turbine Engine that produces torque to an output power shaft, namely a Turboshaft Engine or Turboprop Engine. *See also: Thrust Producing Engine.*

TORQUE RING (turbine)
Another name for Torque Coupling.

TORQUE SENSOR
Another name for Torquemeter Pickup.

TORQUE SIGNAL CONDITIONER (TSC)
A small engine accessory, used especially on Twin-Pac™ Engines. It has one or more of the following functions: 1) Processes inputs from a torquemeter pickup and creates a signal to a cockpit indicator, 2) Acts as a low torque warning system that signals the cockpit of a failed engine and also alerts a propeller auto-feather system, 3) Alerts mechanisms which can up-trim the remaining engine(s) to a maximum safe power level.

TORQUE & TEMPERATURE LIMITING (TTL)
A Turboprop and Turboshaft fuel limiting system controlled by a cockpit switch. When armed, this system monitors both engine torque and exhaust gas temperature (EGT). Should either parameter attempt to overshoot, the TTL system automatically reduces fuel to the combustor to ensure torque and temperature remain within operational limits.

TORQUE WRENCH
A wrench with a means of indicating the twisting force in pounds-feet (lb. ft.) or pounds-inches (lb. ins.) being applied to a threaded fastener such as a nut or bolt.

TOT
Abbreviation for Turbine Outlet Temperature.

TOT GAUGE
A cockpit indicator. A gauge in a system identical to an Exhaust Gas Temperature System. Both Gauges provide an indication of turbine discharge temperature. TOT displays are more often used in rotorcraft and small turboprop aircraft.

TOTAL AIR PRESSURE (Pt)
A pressure measurement of air in motion. In a unit volume of air moving through a duct, the sum of static pressure and ram pressure. The total energy it would take to stop the unit volume of air from moving. Also called Stagnation Pressure.

TOTAL AIR TEMPERATURE (Tt)
A temperature measurement of air in motion. The sum of static temperature and temperature rise due to friction in moving air. Also called Stagnation Temperature. *See also: 1) Temperature Total (hot exhaust), 2) Temperature Total (flight inlet).*

TOTAL ENGINE CYCLES (TEC)
Engine operating cycles accumulated since the engine was first placed in service. TEC is normally recorded in conjunction with the Time Between Overhaul data. Turbine Engines are rarely zero timed after overhaul and TE Cycles continue throughout the life of the engine. *See also: Time Since New.*

TOTAL ENGINE TIME (TET)
Engine operating hours accumulated since the engine was first placed in service. TET is normally recorded in conjunction with the Time Between Overhaul data. Turbine Engines are rarely zero timed after overhaul and TE hours continue throughout the life of the engine. *See also: Time Since New.*

TOTAL EQUIVALENT HORSEPOWER (TEHP)
Another name for Equivalent Shaft Horsepower (ESHP).

TOTAL INDICATOR READING (TIR)
The full pointer movement on a dial micrometer when measuring the out-of-round or runout condition of engine rotating parts. Also called Full Indicator Reading (FIR).

TOTAL LOSS LUBRICATION SYSTEM
An auxiliary engine oil system in which oil is not returned to the oil tank but rather sent overboard. Used primarily in engines with short duration operating times such as an auxiliary boost engine in a vertical take-off aircraft. Not a main engine system.

TOTAL PRESSURE
Locate under Total Air Pressure.

TOTAL PRESSURE PROBE
Locate under Probe (gas pressure).

TOTAL TEMPERATURE PROBE
Locate under Probe (gas temperature).

TOTAL TEMPERATURE
Locate under Total Air Temperature.

TOW
Abbreviation for Take-Off Wet.

TOWER SHAFT (accessory drive gearbox)
A radially placed shaft that drives the Accessory Drive Gearbox from a bevel gear system rotated by the engine compressor. Also called 1) Accessory Drive Shaft, 2) Radial Drive Shaft.

Tq
Symbol for torque produced by a Turboshaft or Turboprop Engine. This symbol is used on the face of Cockpit instruments and in technical data.

TRACKING (igniter plug)
Locate under Igniter Plug Tracking.

TRACTOR-MOUNTED PROPELLER
Another name for Puller Propeller.

TRAILING EDGE SLOTS (turbine blades & vanes)
Refers to the shape of the cooling air discharge ports located below the trailing edges of turbine rotor blades and turbine nozzle vanes.

TRAINING POWER RATING (TPR)
Flight training procedure to manage engine power with one engine inoperative. *See also: Engine Power Ratings.*

TRANS-ATMOSPHERIC VEHICLE (TAV)
A combination Gas Turbine and Rocket powered aircraft undergoing study. A high speed aircraft (perhaps Mach 20) that is expected to operate at up to 500,000 feet altitude. See: 1) Turbo-Ramjet Engine, 2) Turbo-Rocket Engine, 3) Variable Cycle Engine, 4) WaveRider Vehicle.

TRANS COWL
Another name for Translating Cowl.

TRANSDUCER (fuel & oil pressure)
Locate under Transmitter.

TRANSDUCER (vibration)
Locate under Vibration Pickup.

TRANSFER GEARBOX
Another name for Auxiliary Gearbox. Also called Transfer Auxiliary Gearbox.

TRANSFER TUBE (fluids)
A small tube-type connector, generally with o-rings at either end, that carries air, fuel, or oil between two mating parts of an engine.

TRANSFER TUBE (compressor discharge air)
A tube that acts as an engine diffuser, carrying compressor discharge air to the combustor.

TRANSFORMER UNIT OR BOX (ignition system)
Locate under Ignition Exciter.

TRANSITION CASE OR DUCT
Locate under Combustor Transition Duct.

TRANSLATING CENTERBODY (inlet)
A movable device in a supersonic inlet. A centerbody which can move laterally to control airflow. Also called Inlet Spike.

TRANSLATING COWL (thrust reverser)
A component part of an engine nacelle. A movable portion of the engine cowling. When actuated to its rearward position, this cowl forms a fan exhaust nozzle that redirects fan discharge gases forward creating reverse thrust. *See also: Thrust Reverser.*

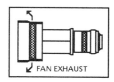

TRANSLATING NOZZLE (thrust reverser)
A component part of an engine nacelle. A movable portion of the engine cowling. When actuated to its rearward position this cowl forms a core exhaust nozzle that redirects hot discharge gases forward creating reverse thrust. *See also: Thrust Reverser.*

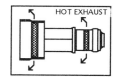

TRANSLATING COWL THRUST REVERSER
Another name for Aerodynamic Blockage Thrust Reverser.

TRANSMISSION (rotary-wing aircraft)
An aircraft mounted reduction gearbox driven by a Turboshaft Engine. The Transmission provides the necessary reduction gearing between the engine and the rotor head. In cases where the engine contains no reduction gearing, all of the speed reduction takes place in the aircraft mounted transmission. Some transmissions are capable of upwards of 100 to 1 gear reduction. In dual-rotor helicopters, the transmission provides shaft power to both rotor heads. *See also: Cross Shaft Transmission.*

TRANSMITTER (fuel pressure)
A small engine accessory. An electrical device in the fuel pressure indicating system. The transmitter receives a fuel pressure sense from the engine and in turn transmits an electrical signal to a cockpit gauge. The cockpit gauge reads out in Fuel Pressure (PSI). See also: Fuel Pressure Gauge.

TRANSMITTER (oil pressure)
A small engine accessory. An electrical device in the oil pressure subsystem of the engine lubrication system. The transmitter receives an oil pressure sense from the engine and in turn transmits an electrical signal to a cockpit gauge. The gauge reads out in Oil Pressure (PSI). *See also: Oil Pressure Gauge.*

TRANSPLY COMBUSTOR LINER
Locate under Combustion Liner (transply).

TRANSONIC AIRFOIL
An airfoil in motion in which airflow is generally thought of as being between Mach 0.8 and 1.3 at different locations over its surfaces.

TRANSONIC AXIAL FLOW COMPRESSOR
A compressor in which airflow over the blade tips travels at supersonic speeds while airflow elsewhere over the blading remains at sonic to subsonic velocities.

TRANSONIC BLADE (compressor)
A blade designed for use in a Transonic Axial Flow Compressor. Increased airflow created by transonic blades designs generally means higher per stage compression. *See also: Compressor Stage.*

TRANSONIC BLADE (fan)
A fan blade design in which air over its surfaces is subsonic in the area nearer to the base, while at the same time airflow is slightly supersonic in the area nearer the tip. *See also: Transonic Speed Range.*

TRANSONIC BLADE (turbine)
A turbine blade design in which airflow varies from blade base to blade tip, reaching slightly supersonic speeds over the blade tips. A transonic blade has increased loading exerted on its surfaces but extracts more energy from the flowing gases for a given flow area as compared to subsonic turbine blade designs. Transonic blading results in a narrow diameter, high speed turbine rotor. The enhanced speed is necessary to drive high pressure compressors designed for very high compressor pressure ratios.

TRANSONIC DRAG
The drag on an airfoil or aircraft caused by the presence of supersonic airflow over portions of its surfaces, even though the object itself is traveling at less than Mach-one. *See also: Drag.*

TRANSONIC SPEED RANGE (airflow)
Generally stated as the speed range of Mach 0.8 to Mach 1.3. When an airfoil is moving at a speed of Mach 0.8, some portions of that airfoil may have supersonic air flow while other areas have subsonic flow. This is an unstable region of speed where the laws of subsonic flow and supersonic flow both apply simultaneously. After approximately Mach 1.3 airspeed, all airflow over the airfoil or aircraft is supersonic and stability returns. *See also: 1) Transonic Drag, 2) Transonic Blade.*

TRANSPIRATION COOLING (turbine blades & vanes)
An internal cooling process wherein air is allowed to exit through porous walls of turbine blades and vanes. Not a common manufacturing method today. *See also: Air Cooled Blades and Vanes.*

TREMBLER UNIT
A point system in the primary circuit of a D. C. input type Ignition Exciter Unit. *See also: 1) Ignition Exciter Box, 2) Ignition Point Capacitor.*

TRI-CRESYL PHOSPHATE
An additive to synthetic turbine engine oils which promotes oiliness, film strength, extreme pressure strength, and anti-wear characteristics. By chemical reaction, tri-cresyl phosphate forms a protective film on metal surfaces which prevents metal to metal contact in the event of normal oil film rupture. This substance is toxic and requires special handling.

TRIM
Locate under Trimming.

TRIM BALANCE KIT
A unit of maintenance support equipment used to balance the Fan of a Turbofan engine or the Free Turbine of a Turboprop or Turboshaft engine. The balancing process consists of an oscilloscope type vibration check and the adding or removing of balance weights. In some cases this procedure can be accomplished on engines installed in the aircraft, otherwise it is done in an engine test cell.

TRIM KIT (engine)
A maintenance and troubleshooting kit containing gauges, air lines, and electrical leads needed during engine trimming. Such a kit might include a precision thermometer, an absolute pressure gauge for trimming EPR rated engines, a precision tachometer for trimming Fan Speed Rated engines, and often times a cable and control box for remotely adjusting the fuel control from the cockpit. *See also: 1) Remote Trim, 2) Trimming.*

TRIM SYNCHRONIZING SYSTEM
Locate under Synchronizer (engine).

TRIM THERMOCOUPLE
A thermocouple located outside of the engine gas path to maintain a constant relationship between the true, average exhaust temperature and the indicated cockpit readout. In this arrangement, ambient temperature changes are compensated for in the readout.

TRIMMING (full power)
A term to describe an engine performance check and the necessary fuel control adjustments to obtain full power output. In year's past part of the tailpipe could be trimmed away with tin snips to affect flow area and engine performance. Trimming is done today only with changes to fuel scheduling during engine operation. Operational Trim adjustments were formerly carried out only at full engine power, but today, in the interest of fuel conservation, this procedure is accomplished at less than full power. *See also: Part Power Trimming.*

TRIMMING (part power)
Locate under Part Power Trimming.

TRIPLE-SPOOL COMPRESSOR
A Compressor design with three independently rotating axial flow compressors, each driven by its own turbine wheel(s). The front compressor is referred to as the Low Pressure Compressor (LPC), the middle compressor is referred to as the Intermediate Pressure Compressor

(IPC), and the rear compressor is referred to as the High Pressure Compressor (HPC). See also: Compressor (engine).

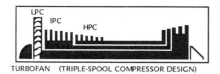

TURBOFAN (TRIPLE-SPOOL COMPRESSOR DESIGN)

TRIT
Abbreviation for Turbine Rotor Inlet Temperature. The more common abbreviation, however, is TIT.

TROPOSPHERE
The portion of the altitude below the Stratosphere (36,084 ft.), where temperature decreases with increasing altitude. Engine thrust in this range is not as adversely affected by altitude change as in the Stratosphere because cooler air helps to maintain thrust output. With increased altitude in the Stratosphere, temperature remains stable and ambient pressure decreases, more seriously affecting inlet density and thrust.

TROUBLESHOOTING (Gas Turbine Engine)
A systematic investigation, in logical sequence by trained technicians, to determine the cause of malfunctions, defects, component failures, and the like. See also: 1) BITE Panel, 2) Fault Tree, 3) Flight Environment Fault Isolation, 4) Shotgunning.

TROCHOIDAL PUMP
Another name for an Oil Pump of the Gerotor design.

TRT
Abbreviation for Take-Off Rated Thrust.

TRUE AIR TEMPERATURE (TAT)
The total air temperature. A temperature value equal to ambient air temperature plus ram air temperature (ram air temperature occurring from friction on a surface as it moves through air).

TRUNNION MOUNT
A component part of an engine. A type of main engine mount with a shape similar to a tapered steel pin. The pin fits into a mounting fixture in the aircraft. Usually two main trunnions are mounted at the engine center of mass in the three and nine o'clock positions. Sometimes called a Cone Bolt Mount. See also: Engine Mount.

TSCP UNIT
Abbreviation for Twin-Spool Compressor Power Unit. See also: Ground Power Unit.

TSFC
Abbreviation for Thrust Specific Fuel Consumption. U.S. Units are given as (lb./hr./lbt). Metric units are given in a variety of ways, the most common of which are (mg/Ns) and (kg/hr./kNt). The (t) refers to thrust in both examples. For conversion to mg/Ns multiply TSFC by 28.325. Locate examples in section entitled "Gas Turbine Engines for Aircraft".

TSI
Abbreviation for Time Since Installation.

TSN
Abbreviation for Time Since New.

TSR
Abbreviation for Time Since Restoration.

Tt_2, Tt_7
Locate under T symbols.

Tt_2 SENSOR
Locate under T2 Sensor.

Tt_7 PROBE
Locate under Thermocouple (exhaust gas).

TTL
Abbreviation for Torque & Temperature Limiting.

TUBULAR COMBUSTOR
Another name for Can-Type Combustor.

TUNGSTEN INERT GAS WELDING (TIG)
A welding process used for many Gas Turbine Engine repairs. A welding process using a hand held probe containing a tungsten electrode, an inert gas such as helium or argon to prevent oxidation, and a hand fed filler rod. Also called 1) Gas Tungsten Arc Welding (GTAW), 2) Heli-Arc Welding. See also: Welding.

TUNING (blades)
Locate under Frequency Tuning.

TURBINE
Shortened form of Turbine Engine or Turbine Wheel.

TURBINE (axial flow)
A turbine rotor system which receives combustor discharge gases in an axial direction. Turbine systems in aircraft engines are of this type. See also: Turbine Rotor Assembly.

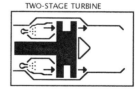

TWO-STAGE TURBINE

TURBINE (radial-inflow)
Locate under Radial-Inflow Turbine.

TURBINE ACTIVE CLEARANCE CONTROL
Locate under Active Clearance Control.

TURBINE ADIABATIC EFFICIENCY
The ratio of the actual temperature drop across a turbine assembly to the isentropic temperature drop (Isentropic meaning without entropy/turbulence). See also: Adiabatic Efficiency (turbine).

TURBINE AREA HAZARD MARKER
A red line or mark on an engine nacelle to indicate a hazard. The indicating mark is placed near the plane of rotation of the turbine wheel to alert ground personnel to the danger in the event of turbine failure. See also: Inlet Area Hazard Marker.

TURBINE BALANCE PLATE
A plate fitted near a turbine disc center of mass to which balance weights are attached. See also: Turbine

Wheel Balancing.

TURBINE BLADE
An airfoil-shaped rotating blade attached to a turbine disc. Gases leaving the combustor pass across the turbine blades and Kinetic Energy of flow changes to Mechanical Work. The turbine assembly, via a connecting shaft, drives the compressor and accessory drive gearbox. Many types of turbine blade designs are available. *See also: 1) Abrasive-Tip, 2) Dovetail Serration, 3) Extended-Root, 4) Fir-Tree Root, 5) Impulse-Reaction Turbine Sysyem, 6) Open-Tip, 7) Shrouded-Tip, 8) Single Crystal, 9) Squeeler-Tip, 10) Winglet Tip.*

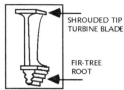

TURBINE BLADE COOLING
Locate under Turbine Cooling.

TURBINE BLISK
Locate under Blisk.

TURBINE BRAZE REPAIR (TBR)
A process of turbine vane replacement by specialized brazing techniques.

TURBINE CASE
The outer casing in which turbine wheels are housed. Also called Turbine Stator Case.

TURBINE CLEARANCE CONTROL SYSTEM
Another name for Active Clearance Control System.

TURBINE CLEARANCE CONTROL VALVE
Part of the Active Clearance Control System. A valve that regulates the compressor bleed air delivered to the Turbine Clearance Control system.

TURBINE CONTAINMENT RING
Locate under Containment Ring (turbine).

TURBINE COOLING
Refers to cooling of turbine blades, discs, and vanes using air bled away from the engine compressor. *See also: 1) Convection Cooling, 2) Eloxed Holes, 3) Engine Bleed Air, 4) Film Air Cooling, 5) Multiple-Pass Cooling, 6) Serpentine Cooling, 7) Shower Head Cooling, 8) Transpiration Cooling.*

TURBINE DISC (hollow)
Locate under Turbine Drum.

TURBINE DISC
The solid center portion of a turbine wheel to which the turbine blades and shaft are attached to form a Turbine Disc and Blade Assembly.

TURBINE DISC & BLADE ASSEMBLY
An engine component part. A heavy metal disc configured with blades around its perimeter. The blades receive combustor discharge gases in an axial direction and vector the gases at release. This process extracts the energy necessary to create wheel rotation. More commonly referred to as a Turbine Wheel. *See also:* *1) Turbine Blisk, 2) Turbine Drum.*

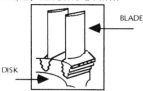

TURBINE DISTRESS
Locate under 1) Hot Section Distress, 2) Locate in the Appendix.

TURBINE DISCHARGE PRESSURE
Generally means total pressure in the exhaust gases, aft of the last turbine stage. On many engines, this engine parameter is one of the inputs to the Engine Pressure Ratio (EPR) System along with Compressor Inlet Pressure. On many dual-spool Pratt & Whitney engines, this location is identified as Pt_7 (pressure total, station 7). Turbine Discharge Pressure can also mean gas pressure discharging from an intermediate turbine stage. Again this pressure is also an input to the EPR system.

TURBINE DRUM
A cone-shaped hollow drum of sheet metal construction, with turbine blades attached to form a turbine rotor assembly. The narrow section of the cone forms the turbine shaft (often called a Torque Cone).

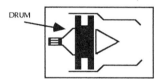

TURBINE EFFICIENCY
A physics term. A ratio of measured work performed by the turbine wheel in units of ft.lbs./BTU, as compared to the laboratory standard of 778 ft.lbs./BTU, expressed as a percentage. The modern Turbine Assembly is in the range of 90 to 95 percent efficient. The 5 to 10 percent loss in efficiency is due primarily to blade tip air energy losses and aerodynamic losses (turbulence) over vanes and blades. Also called Turbine Thermal Efficiency. *See also: 1) Adiabatic Efficiency (turbine), 2) Efficiencies.*

TURBINE ENGINE
Shortened form of Gas Turbine Engine.

TURBINE ENGINE MODULE PERFORMANCE ESTIMATION ROUTINE (TEMPER)
Part of a GEM computer program system used for both on-the-wing and engine test cell engine monitoring. The TEMPER program's function is to collect data on module performance, which it in turn translates into engine performance. *See also: Ground-Based Engine Monitoring (GEM) system.*

TURBINE ENGINE TRANSIENT ROTOR ANALYSIS (TETRA)
A computer system designed to predict stresses in rotating components in the event of sudden rotor imbalance resulting from turbine blade loss during engine operation.

TURBINE ENTRY TEMPERATURE (TET)
British term, similar to Turbine Inlet Temperature.

TURBINE EXHAUST CASE (turbofan)
Another name for Exhaust Cone (outer). On a mixed flow Turbofan engine, the Turbine Exhaust Case forms the primary exhaust nozzle for the core engine.

TURBINE EXHAUST DUCT
Another name for Exhaust Cone (outer).

TURBINE EXHAUST FAIRING
An engine component part. An airfoil-shaped fairing which fits over an exhaust strut at the rear of the turbine section. Its function is to promote good axial flow in the exhaust duct.

TURBINE EXIT GUIDE VANES
Locate under Exit Guide Vanes.

TURBINE FUELS
Locate under Fuel Types.

TURBINE GAS TEMPERATURE (TGT)
Total temperature taken at an intermediate location between sets of turbine wheels. Another name for Intermediate Turbine Temperature (ITT). *See also: Temperature (exhaust)*.

TURBINE GUIDE VANES
Another name for Turbine Nozzle Guide Vanes.

TURBINE HUB to TIP RADIUS RATIO
Locate under Hub/Tip Radius Ratio.

TURBINE INLET TEMPERATURE
Total Temperature of the gases directly in front of the first stage turbine nozzle guide vanes. One of the most critically controlled engine parameters. The ideal point to measure internal engine temperatures, but generally too hot for locating measuring devices such as thermocouples. Early engines were designed for maximum TIT values of approximately 1,300°F, but today, through advances in metallurgy and cooling techniques, allowable TIT values are approaching 3,000° Fahrenheit. See also: Temperature (exhaust).

TURBINE MECHANICAL EFFICIENCY
Another name for Turbine Efficiency.

TURBINE MODULE
Locate under Module.

TURBINE NOZZLE BLADE
Another name for Turbine Nozzle Vane.

TURBINE NOZZLE CLASS
A method by which nozzle vane assemblies are identified to indicate the actual gas flow area (in square inches) through the assembly. When the turbine nozzle is assembled with the correct flow area, it will not only supply the turbine wheel with the correct amount of airflow but will also maintain the established back pressure on the flow system.

TURBINE NOZZLE GUIDE VANE ASSEMBLY
A ring of stationary airfoils located directly in front of each turbine wheel. The function of this unit is to increase gas velocity and also to direct the gases into the turbine blades at the optimum angle. Also called Turbine Inlet Guide Vanes.

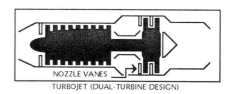
TURBOJET (DUAL-TURBINE DESIGN)

TURBINE OUTLET TEMPERATURE
Another name for Exhaust Gas Temperature. See also: Temperature (exhaust)

TURBINE OVERSPEED SENSOR SYSTEM
A speed sensor system similar to an electronic tachometer system. Rather than provide a cockpit indication, this system produces a signal to the Fuel Control Unit which in turn reduces fuel flow and prevents overspeed.

TURBINE POLYTROPIC EFFICIENCY
A physics term. Refers to the isentropic efficiency of each turbine stage. In a modern turbine assembly, the polytropic efficiency is designed to be fairly constant per stage.

TURBINE RADIAL-INFLOW
Locate under Radial Inflow Turbine.

TURBINE REAR FRAME
Another name for Turbine Case. Also called Turbine Exhaust Frame.

TURBINE REVERSER
An engine component. The primary airflow (core engine) thrust reverser of a turbofan engine.

TURBINE ROTOR ASSEMBLY
A major engine component. A complete rotating assembly of a turbine system, including the turbine wheel and its drive shaft, but excluding the stators. For example, a three-stage High Pressure Turbine assembly consists of three turbine wheels fitted to a common shaft. The Turbine Wheels transform kinetic energy of the flowing gases into mechanical work to drive compressors, and also output reduction gearboxes to which propellers or rotorcraft transmissions are attached. *See also: 1) Free Power Turbine, 2) Fixed Turbine, 3) Gas Generator Turbine, 4) High Pressure Turbine, 5) Impulse-Reaction Turbine, 6) Intermediate Pressure Turbine, 7) Low Pressure Turbine.*

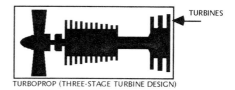
TURBOPROP (THREE-STAGE TURBINE DESIGN)

TURBINE ROTOR INLET TEMPERATURE (TRIT)
Another name for Turbine Inlet Temperature.

TURBINE SECTION
The section within a turbine engine in which the kinetic energy of gas flow is converted to mechanical work to rotate the turbine wheels. Gases leaving the turbine section enter the Exhaust Section. *See also: Engine Sections.*

TURBINE SHROUD RING
Locate under Shroud Ring.

TURBINE SHROUD COOLING
Air cooling of the turbine shroud with air that passes radially through the rotor blades and onto the inner surface of the turbine shroud. *See also: Active Clearance Control System.*

TURBINE SPACER
An engine component part. A circular metal ring that provides the correct spacing between the turbine discs of a multiple turbine wheel.

TURBINE STAGE
A stage of turbine consists of a turbine nozzle vane assembly positioned in front of a turbine wheel assembly. Gases directed through the turbine nozzle guide vanes experience a change in direction and an increase in velocity in order to impinge on the turbine rotor at the optimum angle and with maximum energy.

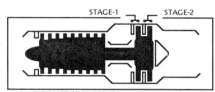
TURBOJET (TWO-STAGE TURBINE DESIGN)

TURBINE STATOR VANES
Another name for Turbine Nozzle Vanes.

TURBINE TEMPERATURE PYROMETER
Locate under Pyrometer (turbine temperature).

TURBINE THRUST BEARING
Refers to a ball-type main bearing which prevents axial movement of a Turbine Rotor. Main ball bearings absorb thrust loads created by high gas pressure forces at the front of the turbine which exert a push rearward on the rotor. *See also: Bearing.*

TURBINE WHEEL
Another name for Turbine Disc and Blade Assembly.

TURBINE WHEEL BALANCING
Turbine Wheels are balanced in several ways. The disc is balanced both statically and dynamically with no blades installed, then the blades are installed in sets at opposing angles to maintain static balance. The wheel is rotated and dynamically balanced by removing material or by adding or removing small weights.

TURBO-ANNULAR COMBUSTOR
Another name for Can-Annular Combustor.

TURBO-COMPOUND-DIESEL ENGINE
A combined gas turbine and diesel engine design presently undergoing research for future rotorcraft applications. The output shaft of the gas turbine will be gear reduced and provide an input to the diesel portion of the engine. The diesel output will be gear reduced and act as the rotorcraft transmission.

TURBO-COMPRESSOR (APU)
Name given to a Gas Turbine Auxiliary Power Unit configured with a bleed air capability as well as an electrical generating capability. Turbo-Compressor is also a name given to an air producing unit in an aircraft air conditioning system.

TURBO-DIESEL ENGINE
Another name for Turbo-Compound-Diesel Engine.

TURBOFAN ENGINE
A type of Gas Turbine Engine. A simple Turbojet core with the addition of a Fan. Generally the Fan is located in front of the engine compressor. The fan serves as a ducted fixed pitch propeller to create a pressurized bypass airflow and to aid in core compression. *The four basic types of Turbofans are*: 1) Low Bypass Turbofan, 2) Medium Bypass Turbofan, 3) High Bypass Turbofan, 4) Ultra-High Bypass Turbofan. *Variations of types are:* 1) Aft-Fan Turbofan, 2) Fully Ducted Turbofan, 3) Geared Turbofan, 4) Short-Ducted Turbofan, 5) Unducted Fan, 6) Variable Pitch Turbofan.

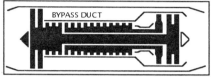
TURBOFAN (TWO-STAGE FRONT FAN DESIGN)

TURBOFAN AIRFLOW
Air passing through a Turbofan Engine falls into two categories: 1) Air bypassing the core is called Bypass Air or Secondary Air, 2) Air passing through the core is called Core Air or Primary Air.

TURBOJET ENGINE
A basic Gas Turbine Engine design with Inlet, compressor, combustor, turbine and exhaust sections. The Turbojet was the original Gas Turbine Engine. The two persons to make the most significant developments in this field were Sir Frank Whittle in England, and Dr. Hans Von Ohain in Germany, between 1935 and 1940.

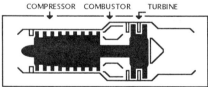
TURBOJET (SINGLE-SPOOL DESIGN)

TURBOMACH-SUNSTRAND COMPANY (USA)
Manufacturer of Gas Turbine Engines. Locate in section entitled "Gas Turbine Engines for Aircraft".

TURBOMACHINE
Sometimes used to refer to the "Core" portion of a Turboprop or Turboshaft Engine, excluding the propeller and output reduction gearbox.

TURBO-MACHINERY
Refers to the major working components of a Gas Turbine Engine such as a Fan, Compressor, and Turbine Rotors.

TURBOMECA (France)
Manufacturer of Gas Turbine Engines. Locate in section entitled "Gas Turbine Engines for Aircraft".

TURBOPROP ENGINE
A type of Gas Turbine Engine. A simple Turbojet core with the addition of a propeller output reduction gear-

box and a propeller shaft. The two most common engines of this type are: 1) Free-Turbine Turboprop, which has its propelling gearbox driven by an independent turbine rotor, 2) A Fixed-Turbine Turboprop, which drives its propelling gearbox by direct attachment to the front of the compressor. The newer Propfan Engines and Unducted Fan Engines are adaptations of the free-turbine Turboprop Engine.

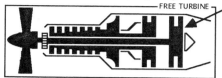

TURBOPROP (FREE TURBINE DESIGN)

TURBO-RAMJET ENGINE
A combination Turbojet and Ramjet design presently undergoing development, configured with a Turbojet core within a ram bypass duct. An engine for high speed aircraft of the future, perhaps Mach 5 to Mach 6. The aircraft would take off and climb on the Turbojet portion of the engine and then convert over to a bypass duct-burning Ramjet portion for maximum speed. Also called Air-Turbo-Ramjet (ATR). See also: 1) Hypersonic Engine, 2) Variable Cycle Engine.

TURBO-RAMJET ENGINE (SR-71 aircraft)
The Pratt & Whitney J-58 is the only operational engine of this type. It is also referred to as the JT11D-20 engine. This is an example of a unique Turbojet with special features, and not a true Turbo-Ram Engine. It diverts some compressor inlet air and also some of the compressor discharge air directly into the afterburner at high Mach Speeds to increase exhaust thrust. The Afterburner in this engine acts as a Ramjet because it is operating at all times, unlike conventional afterburners which operate on an intermittent duty cycle only.

TURBO-ROCKET ENGINE
A combination Turbojet and Rocket-Jet design presently undergoing development, configured with a Turbojet core and rocket power available to its exhaust nozzle at high Mach speeds. A design being considered for future hypersonic aircraft such as Trans-Atmospheric Vehicles proposed for Mach 5 to Mach 20 speeds.

TURBOSHAFT ENGINE
A Gas Turbine Engine designed for Rotary Wing Aircraft. The Turboshaft engine has a simple Turbojet as its core, as do all current types of Gas Turbines. Turboshaft engines are similar to Turboprop engines except they provide power to a transmission unit connected to a rotor, rather than to a propeller. Similar engines are

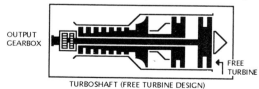

TURBOSHAFT (FREE TURBINE DESIGN)

used as aircraft installed Auxiliary Power Units, as Ground Power Units, and for marine and industrial uses. Turboshaft Engines are of either Free Turbine or Fixed Turbine designs. See also: Turboprop Engine.

TURBOSHAFT STARTER
Another name for Jet Fuel Starter.

TURBO-SUPERCHARGER
A reciprocating engine Accessory. The simplest form of a Gas Turbine device. It attaches to the exhaust of a reciprocating engine to increase compression of intake airflow. The turbo-supercharger was the earliest application of the gas turbine principle in a flight engine. Dr. Sanford Moss, a General Electric Engineer of the early 1900's, contributed most significantly to its development.

TURBULATORS
Twisting passageways within engine parts, such as oil cooler cores, turbine blades, and turbine vanes. As a fluid or gas passes over the Turbulators, movement is slowed down enhancing the effectiveness of heat transfer. Sometimes called Turbulator Fins. See also: Serpentine Cooling.

TURBULENCE UPSET
An FAA term. Loss of normal flight attitude due to flight into turbulent air conditions. Engine flameout may result if this condition is not corrected.

TURKEY FEATHERS (engine cowling)
Jargon for an aerodynamic exhaust fairing used around the nacelle opening of an afterburner. A circular sheet metal or graphite polyimide cowl section with split edges surrounding the engine tailpipe to act as an air seal.

TURKEY FEATHERS (afterburner)
Jargon. Refers to some flap type afterburner exhaust nozzles that have the appearance of being feather-shaped.

TWIN-PAC™ (engine)
A pair of engines which mount side by side in an aircraft. Both engines are also connected to the same output drive unit. Turboprop and Turboshaft Engines are both utilized in this configuration. Should one engine fail, the other will remain on-line to provide partial power for sustained flight. See also: Combining Gearbox.

TWIN-SPOOL COMPRESSOR
Another name for Dual-Spool Compressor.

TWIN-SPOOL COMPRESSOR-POWER UNIT (TSCP)
A Gas Turbine Compressor Power Unit (GTPC) with two independently rotating compressors, referred to as an Engine Compressor and a Load Compressor. The Engine Compressor provides compressed air to the combustor and the Load Compressor provides air to aircraft systems. See also: Load Compressor.

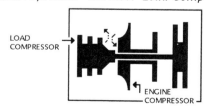

TWINNED TURBOSHAFT ENGINE
Another name for Twin-Pac Engine.

TWO-DIMENSIONAL EXHAUST NOZZLE (2-D nozzle)
Locate under Exhaust Nozzle (vectored thrust).

TWO-POSITION EXHAUST NOZZLE
British term for an afterburner final nozzle having only two operating modes, either fully open or fully closed.

TWO ZONE COMBUSTOR
Locate under Annular Combustor (dual-zone).

TYPE-1 (and Type-2) LUBRICANTS
Synthetic engine oils produced from a manufactured base stock. Both Type-1 and Type-2 were originally designed for military use and later became the two most commonly used commercial engine oils. Type-1 conforms to MIL-L-7808, Type-2 conforms to MIL-L-23699.

TYPE CERTIFICATE DATA SHEETS & SPECIFICATIONS (TCDS)
A document issued by the Federal Aviation Administration (FAA) to indicate that a manufacturer has met all FAA requirements and is eligible to obtain a U.S. Airworthiness Certificate for an aircraft or engine.

TYPE CERTIFICATION TESTING
The testing requirements of the FAA that are met before a Type Certificate is issued.

U

UBE
Abbreviation for Ultra-Bypass Engine. The more common abbreviation is UHB.

UBF ENGINE™
An alternate abbreviation for Ultra-High Bypass Turbofan Engine. The more generally used abbreviation is UHB for Ultra-High Bypass Turbofan Engine.

UDF ENGINE (trademark)
General Electric Company abbreviation for Unducted Fan engine.

UHB ENGINE
Generic abbreviation for Ultra-High Bypass Fan type engines.

ULTIMATE LOAD (engine)
Generally thought of as 150 percent of the engine Limit Load. The point where a part will permanently deform, seize, or loose a portion of its mechanism, but a point where the engine will not separate from the aircraft. See also: Material Strength.

ULTIMATE STRENGTH (of materials)
Generally thought of as 150 percent of the Limit Load of that material. The point up to which the part may deform but not fail due to fatigue. Also called Burst Margin. See also: Material Strength.

ULTRA BYPASS ENGINE (UBE)
Another name for Ultra-High-Bypass Engine. Also a type of Unducted Prop-Fan engine under study which has long, open propeller blades and a bypass ratio of 100 : 1 or higher.

ULTRA-HIGH-BYPASS (UHB) ENGINE
A recent development in design of Gas Turbine Engines. A combination of the Turbofan and Turboprop engine designs. The propeller is of radically different design to a conventional propeller and is called a propulsor. The Propulsor section can be either an unducted or a ducted blading configuration. Either type has two thrust producing gas streams, a cold stream and a hot stream. The hot stream contributes only 5% to 10% of total engine thrust. The unducted propulsor takes air directly from the atmosphere rather than from the aircraft inlet. The ducted type takes air from a conventional aircraft type inlet. Two types of UHB's are under development, one with the propulsor section in front of the compressor and one at the rear of the turbine section. The front propulsor is mounted in the manner of a propeller, whereas the rear mounted propulsor can be thought of as an extension of the turbine wheel(s). Both types have very high bypass ratios in the order of 30 : 1 or higher. Current designs have 8 to 10 variable pitch blades with noticeable sweep-back. The ultra-high-bypass ratio is said to result in very low fuel consumption as compared to traditional high bypass turbofan engines. Also called 1) ADP, 2) CRISP, 3) Free Fan Turbofan, 4) Prop-Fan. See also: Very High Bypass Engine.

UHB (front fan design)
A prop-fan engine with its fan rotor located in front of the compressor and driven at a gearbox reduced speed from free power turbine speed. The front fan design is suitable for an under wing location. UHB blading is curved to reduce shock stall and allow Mach 0.8 flight speeds before loss of blade efficiency starts to occur. Note: A traditionally shaped propeller starts to loose efficiency at approximately Mach 0.6 airspeed.

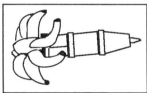

PROP-FAN ENGINE DESIGN

UHB (rear fan design)
A prop-fan engine with its fan rotor located at the rear of the engine as an extension of the blades on the turbine wheel(s). When this system is not gear reduced, the fan rotates at turbine wheel speed. As with the front-fan, the blades have swept tips which delay shock stall and allow this engine to maintain efficiency up Mach 0.8 flight speeds. One disadvantage of this design is that it is not a likely candidate for large engine development because it is not suitable for an under wing location presently required for wide-bodied aircraft. See also: Unducted Fan (UDF).

ULTRA-SONIC CLEANER
A type of cleaning apparatus that transmits sound (pressure) waves through a cleaning fluid in which a part to be cleaned is immersed. It is used for cleaning of many small parts such as filters and bearings.

ULTRA-SONIC INSPECTION (resonance method)
A maintenance check. A non-destructive inspection procedure to check for internal damage of solid materials by introducing sound energy. When a subsurface defect is present, the return sound level is diminished and a trace on a CRT display indicates the location and extent of the defect.

UNBALANCE CONDITION
Locate under Engine Inspection and Distress Terms in Appendix.

UNCONTAINED FAN FAILURE
Locate under Fan Failure (uncontained).

UNDERSPEED GOVERNOR (turboprop)
A small accessory or part of an accessory. A flyweight operated fuel metering device that establishes engine RPM during Beta Mode of operation. See also: Overspeed Governor.

UNDUCTED FAN ENGINE (UDF™)
A General Electric Trade Mark. A type of Ultra-High Bypass Turbofan Engine with its fan located at the rear of the engine. The GE36 Unducted Fan Engine is the first of such engines to be successfully test flown. The UDF has a two-stage counter-rotating fan module called a Propulsor Unit which acts as a pusher propeller. Counter-rotation produces high propulsive efficiency and reduces gyroscopic effect, resulting in low engine vibration. The Propulsor is not gear reduced, so the fan rotates at turbine speeds. This engine is in the 25,000 lb. thrust range, has a fan bypass ratio of 30:1, and is capable of powering a narrow-bodied airliner at Mach 0.8 airspeeds. See also: 1) Propulsor Unit, 2) Ultra-High Bypass Engine.

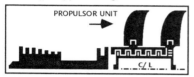

UNDER-THE-WING ENGINE
An engine mounted below the level of the wing leading edge, with the engine nacelle mounted on a pylon. See also: Over-The-Wing.

UNISON RING (compressor)
An engine component part. A metal ring that interconnects each of the Variable Angle Stator Vanes in a particular stage. See also: Variable Angle Stator Vanes.

UNISON RING (propulsor)
A pitch change mechanism connected to rotor blades. Used in place of a traditional rack and pinion gearing system. This system is used in the Propulsor of a Unducted Fan Engine to permit synchro-phasing of blades. See also: Synchro-Phasing System.

UNIT TURBINE
Another name for Turbine Module. A turbine rotor asssembly built and tested as a separate unit. Modules are generally tracked independently of engine operating time using module operating hours and/or cycles.

UNITED STATES MANUFACTURED ENGINES
Locate by manufacturer in section entitled "Gas Turbine Engines for Aircraft".

UNITED TECHNOLOGIES CORPORATION
Parent corporation of the Pratt & Whitney Company of Hartford, Conn., Commercial Engine Division, and its Military Engine Division in Palm Beach, Florida. Locate also in section entitled "Gas Turbine Engines for Aircraft".

UNMETERED FUEL
Fuel that enters the fuel control from the fuel pump before the metering process begins.

UNMIXED EXHAUST DUCT (turbofan engine)
A type of tailpipe on a turbofan engine in which the primary discharge from the core and the secondary discharge from the fan are completely separated. See also: Exhaust Duct.

UNSCHEDULED MAINTENANCE
Maintenance performed as a result of discrepancies found during inspections by maintenance personnel and during operation by flight personnel. See also: Scheduled Maintenance.

UNSHROUDED FAN
A fan of a Turbofan Engine in which the fan blades contain no supporting span shrouds. See also: Span Shrouds (fan blade).

UNSTART (inlet)
A phenomenon that occurs in some high speed flight inlets. Inlets of this type are designed to provide shock wave formations to reduce inlet airflow velocity from supersonic to subsonic speeds. The inlet Unstart occurs when shock waves move too far out of the inlet causing a large boundary layer separation and drag on the air entering the flight inlet. This condition can completely stall and flame-out an engine if not corrected. Unstart can usually be controlled by changing the aircraft attitude or airspeed. See also: Inlet Buzz.

UNSYMMETRIC THRUST
Thrust of an aircraft with one engine away from the centerline out of commission. Also called Asymmetric Thrust.

U.S. SIEVE RATING
A rating system for fluid filters. A numeric rating based on the number of openings per linear inch.

UTW
Abbreviation for Under-The-Wing.

V

V-1 FLYING BOMB
Locate under Pulse-Jet.

VACUUM PIT BALANCING
Another name for Spin Pit Testing.

VACUUM PLASMA COATING
Metallic spray coating process used on new parts and also used to restore worn parts to original dimensions. The process results in a surface resistant to abrasion, oxidation, and sulfidation. Unlike sprays applied in free air, this method provides a dense and void free layer, typically of nickel, cobalt, and chromium. It is applied to oxidation sensitive substrates such as titanium and zirconium. See also: Plasma Spray.

VACUUM RELIEF VALVE (oil tank)
A component part of an oil tank. A valve which cracks open in the presence of negative air pressure, approximately 0.5 PSI. This occurs during certain flight maneuvers, when the pressure oil pump is drawing down the normally positive air pressure in the oil

tank. The relief valve allows the tank expansion space to return to ambient pressure until oil returning via the scavenge system again re-pressurizes the oil tank.

VANE
Term generally used to describe many non-rotating airfoils within a Gas Turbine Engine. *See also: Stator Vane.*

VANE COOLING
Locate under Turbine Cooling.

VANE DIFFUSER
Refers to the traditional diffuser section at the outlet of a centrifugal compressor. This diffuser utilizes radially placed vanes to guide airflow into the combustor. *See also: Pipe Diffuser.*

VANE OIL PUMP
Locate under Oil Pump (vane).

VAPOR ELIMINATOR
A small relief valve placed at various locations in fuel systems to remove air from fuel carrying lines and passageways. Normally this vapor is returned to the expansion space at the top of the fuel tank. *See also: Vapor Lock.*

VAPOR LOCK
A dangerous operating condition in which vapor is trapped within fuel system lines or components. Vapor is harmful to fuel system function because it takes up fuel space and is compressible. When a pocket of vapor occurs, it hampers normal operation of hydro-mechanical mechanisms or normal flow of fuel. Vapor lock is a serious condition and is the cause of many engine flame-outs in flight. *See also: Vapor Eliminator.*

VAPOR GUTTER (afterburner)
Another name for Flame Holder.

VAPORIZING COMBUSTOR
A Turbine Engine Combustor which utilizes Vaporizing Tube fuel distributors.

VAPORIZING TUBE
Locate under Fuel Vaporizing Tube.

VARI-RAMP (aircraft inlet)
A moveable ramp in a Variable Geometry Inlet. This ramp moves to reduce the flow area at high aircraft speeds allowing the area downstream of the ramp to properly control the diffusion of air into the engine. Also called: 1) Boundary Diverter, 2) Movable Wedge. *See also: Convergent-Divergent Inlet Duct.*

VARIABLE ANGLE GUIDE VANES
Another name for Variable Inlet Guide Vanes.

VARIABLE ANGLE STATOR VANES (VSV)
A compressor Anti-Stall System also called Anti-Surge System. A generic term to include inlet guide vanes and compressor stator vanes which have a capability of changing their geometry (angle) to the oncoming airstream. When the vane is at right angles to the compressor disc, the vane is said to be at zero angle. This occurs at approximately cruise RPM. Variable vanes open as the power lever is advanced and close down approximately 20 to 30 degrees as the power lever is retarded to idle position. At less than full power, the pumping characteristics of individual compressor stages change, the front stages being more efficient than the rear stages. Variable vanes control the angle of attack of airflow into the front stages of a compressor to give the engine a rapid and stall-free acceleration and also to optimize fuel consumption at part power operating conditions. *See also: Variable Inlet Guide Vanes.*

VARIABLE ANGLE STATOR VANE SCHEDULE CHECK
A maintenance check-run to determine whether the Variable Angle Stator Vane system is opening and closing at the correct Percent RPM (compressor speed).

VARIABLE AREA EXHAUST NOZZLE
An engine component. An exhaust nozzle that changes its geometry (flow area) as a function of power lever position. In this manner, a predetermined relationship between Exhaust Nozzle flow area and engine mass flow is maintained. The variable area exhaust nozzle is a component part of an afterburner. The nozzle, however, will function with or without the afterburner in operation. *See also: 1) Eyelid Exhaust Nozzle, 2) Flap Exhaust Nozzle, 3) Iris Exhaust Nozzle.*

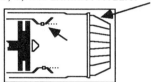

VARIABLE BLEED VALVE (VBV)
An engine accessory. A type of compressor Anti-Stall Bleed Valve System that progressively closes as the power lever is advanced, becoming fully closed at approximately cruise RPM. This valve functions to establish flow matching between the high and low pressure compressors. The VBV is more complex than older systems in which the valve had only two positions, fully open or fully closed. *See also: Compressor Bleed Air.*

VARIABLE BYPASS RATIO (gas turbine)
Another name for Variable Cycle Turbofan. Also called Variable Bypass Injection.

VARIABLE BYPASS VALVE
Another name for Variable Bleed Valve.

VARIABLE COMPRESSOR VANES
Locate under Variable Angle Vanes.

VARIABLE CYCLE ENGINE
A family of engine designs, mainly for future higher flight speeds beyond the capability of traditional Turbine Engines. Designs being currently proposed are as follows: 1) A Variable Cycle Turbofan, which contains a variable-angle fan to vary the fan bypass ratio allowing a reduction of the fan bypass in the Mach-2.5 region, 2) A Low bypass Turbofan which converts to a direct Turbojet at supersonic speeds of over Mach-3, 3) A Turbojet which converts to Ram-Jet or Rocket-Jet operation for speeds from Mach-5 to Mach-25. *See also: 1) Hypersonic Engine, 2) Inverse Variable Cycle Engine, 3) Turbo-Ramjet.*

VARIABLE CYCLE TURBOFAN ENGINE
A type of Turbofan Engine undergoing development.

An engine designed with the capability of varying the fan bypass ratio to match operational needs and to conserve fuel. A component called a Variable-Area Bypass Injector schedules the fan bypass condition.

VARIABLE DISPLACEMENT PUMP
A pump that can vary its displacement (volume of output) per revolution. A pump used mainly as a fuel pump or a hydraulic pump. The pumping output automatically increases when actuating units in the system are operating and decreases in volume when the actuating units are at rest. Also called a 1) Variable Delivery Pump, 2) Variable Volume Pump.

VARIABLE EXHAUST NOZZLE
Another name for Afterburner.

VARIABLE GEOMETRY ACTUATOR SYSTEM
A hydraulic servo system (generally fuel operated) that provides the necessary motive force to open and close compressor Variable Angle Stator Vanes. The actuator system operates at the command of the engine power lever via the fuel control.

VARIABLE GEOMETRY COMBUSTOR
A recent development in design which utilizes a combination liner fitted with adjustable swirl vanes. The vanes open to their maximum airflow area to obtain a high combustion temperature rise and close down at low temperature operation. By varying the primary and secondary airflow, as much as 50 percent of total airflow can be used for combustion at high engine power settings. *See also: 1) Combustion Zone, 2) Swirl Vanes.*

VARIABLE GEOMETRY INLET
An inlet used on supersonic aircraft which contains a computer assisted means of adjusting its airflow area. *See also: Convergent-Divergent Inlet Duct.*

VARIABLE GUIDE VANES (VGV)
Another name for Variable Inlet Guide Vanes.

VARIABLE INLET GUIDE VANES (VIGV)
A compressor inlet guide vane system with a capability of varying vane angles into the oncoming airstream. Some engines have only Variable Inlet Guide Vanes and some also have variable angle vanes in the compressor. *See also: 1) IGV Schedule Check, 2) Variable Angle Stator Vanes.*

VARIABLE METERING ORIFICE (fuel control)
Refers to the main metering valve where fuel is scheduled by varying the flow area, while maintaining a differential pressure between unmetered and metered fuel.

VARIABLE PITCH FAN (turbofan)
A ducted fan section of a turbofan engine with a capability of changing the angle of its blading to the oncoming airstream. Very few engines of this type have been developed. *See also: Turbofan Engine.*

VARIABLE PITCH SYSTEM
A blade pitch changing mechanism on Turboprop and Turbofan engines capable of automatically maintaining a preset engine speed (percent RPM) under varying load conditions. For example, whenever the engine speed tends to increase, the Variable Pitch System, through its governing system, will increase blade angle and keep the engine on-speed. Conversely, whenever the engine speed tends to decrease, the blades will change to a decreased angle to keep the engine on-speed.

VARIABLE POLARITY PLASMA ARC
A robotic type machine welding process especially suited for welding larger engine parts made of aluminum.

VARIABLE REDLINE LIMIT (EGT)
A cockpit engine condition gauge which provides a continuous display of operating limits based on altitude, airspeed, outside air temperature and RPM. *See also: Single Redline Limit.*

VARIABLE VANES
Locate under Variable Angle Vanes.

VARNISHING
Locate under Bearing Distress Terms and Inspection Terms in Appendix.

VBV
Abbreviation for Variable Bleed Valve.

VECTOR QUANTITY
A physics term. A force that has magnitude and direction and is shown as an arrow of scaled length on a diagram. For example, vectors that are used to conveniently describe compressor interstage airflow and compressor blade angle of attack.

VECTORED THRUST ENGINE
An engine that can produce thrust for both vertical lift and forward propulsion. Also called Liftfan Engine.

VECTORING EXHAUST NOZZLE (or duct)
Locate under Exhaust Nozzle (vectored thrust).

VEE-BAND COUPLING
A circular clamping device made of stainless steel, used to secure two mating flanges. Commonly seen on air tubes, accessory mountings, and tailpipes as a means of quick removal and replacement. Also called V-Band Coupling or Marman Clamp™.

VELOCITY
A physics term. Mathematically expressed as:

$$V = d \div t$$

Where V is velocity, d is distance, t is time. Air Velocity appears in the thrust formula in U.S. Units of feet per second (meters per second in Metric Units).

VELOCITY (calculated from velocity pressure)
When Velocity pressure is known, Velocity is calculated as follows:

$$V = 1,096.7 \sqrt{hv \div d}$$

Where (V) is Velocity in ft./min., (hv) is Velocity Pressure in inches of water, (d) is density in lb./cu.ft.

VELOCITY PRESSURE
Another name for Ram Pressure.

VENT PRESSURIZING VALVE
Another name for Breather Pressurizing and Vent Valve.

VENT SUBSYSTEM (lubrication system)
Locate under Oil Vent Subsystem.

VENTILATION (nacelle)
Locate under Nacelle Ventilation and Cooling.

VENTRAL NOZZLE
A recent development in design of vertical thrust exhaust nozzles.

VERNATHERM CONTROL VALVE™
A thermostatic valve that is heat sensitive and operated via a bi-metallic spring. This valve is found in fuel heaters and fuel-oil coolers. Also called Thermostatic Control Valve.

VERTICAL TAKE-OFF AND LANDING (VTOL)
An aircraft with vertical take-off, hover, and vertical landing capabilities. For example, rotary-wing aircraft, tilt-rotor aircraft, and tilt-wing aircraft.

VERY HIGH BYPASS TURBOFAN (VHB)
A type of Turbofan Engine under development in which the bypass ratio is expected to be in a range of 8 : 1 to 10 : 1. The VHB will more likely be of shrouded, gearless fixed pitch design. *See also: Turbofan Engine.*

VESPEL INSERT
A replaceable quill shaft liner constructed of Vespel material. The insert is designed to absorb wear and minimize quill shaft wear. It is used in some Accessory Drive Gearbox drive pads.

VIBRATION
Locate under 1) Forced Vibration, 2) Free Vibration, 3) Frequency Tuning, 4) Harmonics, 5) Natural Frequency, 6) Resonant Frequency, 7) Synchronous Vibration.

VIBRATION ACCELEROMETER
Another name for Vibration Pickup.

VIBRATION ISOLATOR
A type of shock-mount for small engine accessories in which rubber insulated mounting points prevent engine vibration from being transmitted into the accessory.

VIBRATION METER
A Test Cell or Aircraft installed unit used to measure engine vibration. The meter receives a vibration signal from Vibration Pickups mounted at selected locations on the engine. The meter reads out in MILS of inches or inches per second. A typical limit might be 3 to 4 mils (thousandths of inches) at the fan case or turbine case. *See also: Airborne Vibration Monitoring ystem System.*

VIBRATION PICKUP
A small engine-mounted, electrical generating unit which transmits an electrical signal proportional to the vibration present at a point on an engine. The signal is directed to a Vibration Meter in the cockpit or on a test cell console to indicate the state of balance of parts in real time. Also called 1) Accelerometer, 2) Vibration Transducer. *See also: Vibration Meter.*

VIBRATOR IGNITION SYSTEM
An older type ignition system similar to a conventional reciprocating engine ignition system, using only vibrating points and transformer coils with no storage capacitors.

VIFFING
Short for Vectoring-In-Forward-Flight. Refers to an aircraft with the capability of raising its nose without an altitude change. Viffing is accomplished with either: 1) a Lift-Fan engine, 2) a Turbofan Engine configured with a Vectored Exhaust Nozzle. *See also: Exhaust Nozzle (vectored thrust).*

VIGV SCHEDULE CHECK
Locate under IGV Schedule Check.

VISCO ELASTIC COMPOUND
A material used in construction of engine cowl panels because of its noise and vibration damping qualities.

VISCOSIMETER (oil)
An instrument used to determine the viscosity of lubricants. *See also: Saybolt Seconds Universal Viscosimeter.*

VISCOSITY (oil)
The cohesive force between molecules of an oil (the fluid resistance to flow). Petroleum oils are rated according to viscosity in SAE Numbers, such as SAE-20, SAE-50, etc. Synthetic oil viscosity ratings are given in Centistokes. *See also: 1) Centistoke, 2) Kinematic Viscosity.*

VISCOSITY INDEX NUMBER (VI)
A numeric system indicating the resistance to breakdown in viscosity when heat is added to an oil. Zero to 100 Viscosity Index Numbers are used primarily for petroleum oils and 100 to 200 VI's are used for synthetic oils. The numbers are plotted on a Nomograph using Saybolt Seconds or Centistokes. A typical turbine engine oil will plot out to approximately a 135 VI rating.

VISIBLE MOISTURE
Refers to the requirement to turn on the inlet Anti-Ice air system when flying in the presence of visible moisture. Visible moisture is present when flying in rain and snow and also though certain cloud formations. In general terms, clouds are considered visible moisture when the visibility is less than one mile.

VMO
Abbreviation for Variable Metering Orifice.

VOLVO FLYMOTOR (Sweden)
Manufacturer of Gas Turbine Engines. Locate in section entitled "Gas Turbine Engines for Aircraft".

VOLUTE COMBUSTOR
Another name for Scroll Combustor.

VOLUTED SHAPE
A coiled, spiral, or scroll like shape used in many turbine engine flow chambers for air and fuel mixing.

VON OHAIN, HANS
Dr. Von Ohain was the designer and developer of Germany's first Turbojet Engine while working for the Heinkel Company in the mid 1930's. His He(S-3b) centrifugal flow Turbojet Engine was installed in the HE-178 prototype fighter, the first aircraft to fly powered by a Gas Turbine Engine. The He-178 made a

total of only three flights. Dr. Von Ohain later became the Chief Scientist for the U.S. Air Force at its Wright-Patterson Engine Research Center. See also: Franz, Anselm.

VORTEX AIR SWIRLER (combustor)
Another name for Swirl Vanes (combustor).

VORTEX BLADING (turbine)
A technique of creating the correct twist (stagger-angle) of turbine blades. When the correct angles are present, the blading will perform uniform work from base to tip and be at optimum efficiency.

VORTEX CONTROLLER DEVICE (VCD)
A small engine accessory. A device which takes bleed air from the engines and flows it over aircraft surfaces to prevent airflow separation and subsequent reduction in lift at low aircraft speeds.

VORTEX DISSIPATOR (destroyer)
Locate under Vortex Spoiler.

VORTEX GENERATOR (engine)
A small component part of a flight inlet or tailpipe. Small drag reducing airfoils attached to the inner walls of engine inlets and exhausts that enhance airflow by reducing turbulence. These generators are small airfoils similar to the type used on outer aircraft surfaces. The generator sheds a vortex which penetrates through the slow moving boundary layer and down to the surface behind the generator. Drag is then reduced as the airflow is energized. *See also: 1) Boundary Layer Air, 2) Nacelle Chine.*

VORTEX GENERATOR (oil tank)
A component part of an oil tank. A spiral passageway inside an oil tank scavenge oil inlet. As the oil swirls through, a break-up of bubbles occurs before the oil floods into the main body of fluid in the tank. Also called a Swirler.

VORTEX SPOILER
A small engine accessory. A small air nozzle which directs a bleed air stream at the ground under the flight inlet to break up vortices. Such a vortex can lift small ground debris into the engine inlet, causing erosion and damage to fans and compressors. Also called 1) Vortex Dissipator, 2) Vortex Destroyer, 3) Vortex Separator.

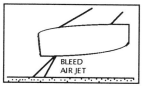

VORTEX TIP (compressor blade)
A very thin squealer-tip design that allows air to leave its trailing edge smoothly and pass to the next stage with minimum turbulence. The vortex tip counteracts tip clearance induced turbulence which tends to spill air to the upper, low pressure side of the blade, resulting in loss of blade efficiency.

VRL
Abbreviation for Variable Redline Limit.

V/STOL AIRCRAFT
Abbreviation for Vertical/Short Take-Off and Landing. Generally refers to fixed-wing aircraft, typically configured with Lift-Fan type Turbofan Engines. For example, the McDonnell AV-8B Harrier II Aircraft.

VSV
Abbreviation for Variable Stator Vane System.

VTOL AIRCRAFT
Abbreviation for Vertical Take-Off and Landing. Generally to mean Rotary Wing, Tilt-Rotor, and Tilt-Wing aircraft.

W

Wa
Symbol for Weight of Airflow. Another name for Mass Airflow (Ms) as used in the Thrust Formula.

WAFER OIL FILTER
Another name for Stacked Wafer Oil Filter.

WCF BLADE
Short for Wide Chord Fan Blade.

WARM COMPOSITES
Name given to certain composite materials with heat-strength properties. Carbon-epoxy cloth is one such material. It is used in sections of an engine nacelle where heat rejection from the engine causes residual temperatures to be relatively high.

WASH FILTER (fuel)
An auxiliary engine fuel filter. One in which micronic sized particles are filtered from fuel going to servo fuel systems, and then allows the remaining particles to be washed out through the main fuel system where tolerances are not as critical. Also called Wash Flow Filter.

WASH MANIFOLD (engine)
An engine nacelle installed spray ring to which a pressurized ground wash tank can be attached. Compressor washing is accomplished as a corrosion prevention measure for the compressor and its internal case surfaces and also for restoring lost thrust due to dirt accumulation on compressor airfoils. *See also: Wash Methods.*

WASH METHODS (compressor)
Locate under 1) Desalination Wash, 2) Performance Recovery Wash. *See also: Field Cleaning (compressor).*

WASPALOY™
A Pratt & Whitney developed nickel-base, high temperature strength alloy in the family of turbine engine super-alloys. *See also: Super Alloy.*

WATER (in fuel)
Water occurs in aviation fuels in two forms, described as dissolved water and free water.

1) Dissolved Water in Fuel - Water in solution which cannot be removed by filtration. Dissolved Water is not a problem for Gas Turbine Engine operation because it passes through the fuel system along with the fuel. A problem can arise, however, if temperature change allows Dissolved Water to become Free Water.

2) Free Water in Fuel - Water in excess of that which will dissolve in fuel. There are two classifications of Free Water in Fuel:

a) Entrained Water - Water in tiny droplets that may or may not be visible to the naked eye, but will make jet fuel appear cloudy or hazy (this does not occur in aviation gasoline).

b) Water Slugs in Fuel - Water that settles out in bulk fuel storage, amounts ranging from ounces in small fuel storage tanks to possibly gallons in large fuel storage tanks.

WATER IN FUEL TESTING
Testing for presence of water in jet fuels is generally accomplished by one of three methods:

1) Clean & Bright Test - A procedure using a clean and dry clear glass container of approximately ten ounces, filled with drained fuel. When fuel is acceptably dry (no water) it will appear bright, with a certain fluorescence and no cloudy or hazy appearance. When fuel is clean it has a clear color and no visible contaminants.

2) Detection Paste Test - A procedure in which detection paste is applied to a dip stick which is then lowered to the bottom of a fuel tank. The paste changes color in the presence of water. The highest point of color change along the paste line indicates the water depth in the bottom of the tank.

3) White Bucket Test - A procedure using a white porcelain, enameled, or stainless steel container of approximately ten quart size, containing 4 to 5 inches of drained fuel. The fuel is swirled with a mixing paddle, then when the paddle is removed, several drops of household red food dye is added to the vortex center. The dye will mix with water to indicate water presence in the fuel. If no water is present, the dye will settle to the bottom of the bucket.

WATER BRAKE UNIT
A unit of support equipment using a fluid resistance principle. A device similar in function to a Dynamometer to measure shaft horsepower (torque) of operating Torque Producing Engines.

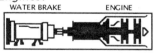

WATER INJECTION (engine)
An optional Gas Turbine Engine system in which water is sprayed into the compressor, the combustor, or both as a means of augmenting engine take-off thrust. The injection fluids add to the air mass passing though the engine and also absorb heat of compression. Additional mass and increased compressor efficiency boost the value of dry thrust output by approximately 10 to 15 percent. Water Injection can be thought of as a "hot day" thrust recovery system, in that as high ambient temperatures reduce thrust output, water injection tends to bring thrust back up to "cold day" values.

WATER INJECTION SYSTEM (compressor injection)
A fine spray of water introduced into the engine inlet. Water adds to mass of flow through the engine and also absorbs heat as it passes through the compressor, thereby increasing compressor efficiency.

WATER INJECTION SYSTEM (combustor injection)
A fine spray of water introduced into the combustor inlet. Not widely used, but said to be an efficient injection process because it increases the mass flow available to the turbine relative to the compressor. This water is generally introduced through a separate orifice in the main fuel nozzles.

WATER INJECTION SYSTEM ARM SWITCH
A safety switch. A cockpit switch that must be activated in order for the power lever micro-switch to initiate water injection to the engine.

WATER INJECTION FLUID
Either pure demineralized water or a mixture of water and alcohol. Large aircraft utilize pure water which is completely used up during take-off and climb, then the water system is blown dry with engine bleed air. Smaller aircraft that make frequent takeoffs and landings utilize a mixture of ethanol or methanol and pure water to prevent system freeze up at altitude.

WATER LINE (WL)
A reference location on an engine. The distance from the horizontal center-line, up or down, as viewed from the side of the engine. For example, when the center-line is designated as WL-100, WL-99.0 is one inch below the center-line and WL-101.0 is one inch above the center-line. Also called Water Level. *See also:* 1) Butt Line, 2) Engine Station.

WATER SHIELD (combustor)
A shield around a combustion liner to provide the combustor with a tolerance to flame-out when a high atmospheric moisture content is present in compressor discharge air.

WAVE DRAG
Supersonic drag on a surface due to compressibility of air after shock wave formation occurs. *See also:* 1) Induced Drag, 2) Parasite Drag, 3) Shock Drag.

WAVERIDER VEHICLE
A proposed high speed aircraft in the Mach 6 to 12 speed range. A design which allows the aircraft to fly with a bow shock wave attached to its leading edge, giving the appearance of riding on it own shock wave. It may possibly be powered by a type of combined Turbine and Scramjet at flight levels up to 100,000 feet.

WAVE SPRING
A washer type space which presents a wavy appearance. It is used to position shaft bushings and to provide an axial preload to gears in fuel and oil pumps. Also called Spring Washer.

WAVE WASHER
Another name for Spring Washer.

WAXING (jet fuel)
Separation of paraffin from kerosene type fuels at low fuel temperatures. A possible source of blockage in fuel systems. *See also: Fuel Types-Jet.*

WCF
Abbreviation for Wide Chord Fan.

WEAR
Locate under Bearing Distress Terms and Inspection Terms in Appendix.

WEAR RESISTANT COATINGS (metal)
Coatings which provide smooth surfaces which reduce air turbulence, and which resist abrasion, erosion and sulfidation. *See also: 1) Gator Guard, 2) Sermetal-W Coating.*

WEAR METALS (in oil)
The numerous types of metallic contaminants released into the engine oil through metal to metal contact within the oil wetted parts of an engine. *See also: Spectrometric Oil Analysis Program.*

WEB (turbine disc)
The portion of a disc between the blade root area and the hub.

WEDGE (inlet)
Another name for Vari-Ramp.

WEIBULL RISK ANALYSIS
Locate under Risk Analysis.

WEIGHT
A physics term. The force with which the earth's gravity pulls on a mass. The force is expressed in pounds (lbs.) in U.S. Units and Newtons (N) in the Metric Units.

WEIGHT OF FUEL (Wf)
Refers to the fact that in a Gas Turbine Engine, fuel is scheduled by weight via a Fuel Control to the engine combustor. Scheduling by weight is more accurate than by volume because the fuel's BTU value is constant per unit weight. Fuel weight of flow is in accordance with the formula:

$$Wf = KA\sqrt{\Delta P}$$

Where: Wf = weight of fuel, K = a constant value, A = area of the metering orifice, ΔP = differential pressure across the main fuel metering orifice. *See also: Main Metering Valve (fuel control).*

WELDING
Locate under: 1) Electron Beam, 2) Heli-Arc, 3) Inertia, 4) Laser, 5) Metal Inert Gas (MIG), 6) Tungsten Inert Gas (TIG).

WELDMENT
Refers to an assembly that is fabricated by welding together its various parts.

WEST GERMAN MANUFACTURED ENGINES
Locate under Motoren-Und Turbinen-Union and Kloeckner Humboldt-Deutz in section entitled "Gas Turbine Engines for Aircraft".

WESTINGHOUSE CORPORATION
One of the first manufacturers of Gas Turbine Engines for aircraft, who now produce only non-aircraft Gas Turbine Engines.

WET MOTORING
Locate under Motor Over (wet).

WET SPLINE (accessory drive gearbox)
Drive spline locations for engine accessories that are lubricated by the engine oil system. The traditional lubrication method is to use a graphite type grease. The newer, wet spline lubrication method is said to give more resistance to spline wear and seizing.

WET START
An engine start in which wet fuel is seen issuing from the tailpipe.

WET SUMP (lubrication system)
A lubrication system in a Gas Turbine Engine in which the Accessory Drive Gearbox also serves as the oil reservoir. Also called Integral Oil System. *See also: 1) Dry Sump (lubrication system), 2) Lubrication System.*

WET TAKE-OFF
Locate under Take-Off (wet).

WET THRUST
Thrust which results when Water Injection is used to augment power in Gas Turbine Engines. May also be used to indicate an afterburner assisted take-off. *See also: Take-Off (wet).*

Wf
Symbol and abbreviation for Weight of Fuel.

WHEEL CASE
British for the Fan Case of a Turbofan Engine.

WHIRL (compressor or turbine)
An aerodynamics term for angular deflection in airflow as follows: 1) The tangential component of velocity off the tip of a rotating blade, 2) The change in air velocity and direction from entrance to exit of a compressor or turbine stage. Note: Whirl is designed to be greatest in the turbine area for a greater energy transfer.

WHITTLE, SIR FRANK
British design engineer responsible for early development of Gas Turbine Engines for aircraft. Recognized as the "father" of the Gas Turbine Engine for aircraft propulsion. The holder of the first Gas Turbine Engine patent, issued in 1930. By 1936, he had developed the first Turbojet Engine to successfully operate on a Test Bench. *See also: Gloster E-28.*

WHITTLE, W-1 ENGINE
Locate under Gloster E-28.

WIDE-BODIED JET
The name given to the family of wide-bodied, Gas Turbine Engine powered aircraft such as the Boeing B-747, McDonnell Douglas DC-10, Lockheed L-1011, and Airbus A300. Wide-Bodied Jets are thought of as being over 15.5 feet in fuselage width. Also called Jumbo-Jets.

WIDE CHORD BLADES (fan and compressor)
Refers to a wide blade design as measured along its chord, from its leading to its trailing edge. Wide-Chord is a recent development in low aspect ratio designs to provide stiffness and eliminate the need for blade span-shrouds. Wider blades pump more air than narrow blades, and without the shrouds, flow area and mass airflow are increased for a given fan diameter. Wide-Chord blades are generally of hollow titanium construction and have approximately the same weight as

the older, solid forged titanium blades with span-shrouds. Noise levels are also stated to be lower in Wide-Cord Blades. This design also has improved resistance to erosion and foreign object damage. *See also: 1) Compressor Blade Shapes, 2) Controlled-Diffusion Airfoils, 3) Diffusion Bonding Process, 4) Low Aspect Ratio Rotor Blading, 5) Span-Shrouds.*

WIDE CHORD FAN (WCF)
Locate under Wide Cord Blades.

WIGGINS COUPLING™
A type of fluid line or air line coupling.

WIGGLE STRIP
A type of chafing pad to reduce wear-through of fluid lines and electrical leads, but which allows some movement.

WILLIAMS INTERNATIONAL COMPANY
Manufacturer of Gas Turbine Engines. Locate in section entitled "Gas Turbine Engines for Aircraft".

WINDAGE
An aerodynamics term referring to friction or braking effect on moving airfoils caused by air turbulence. Windage is also the effect that a rotating engine part has on air in proximity with the rotating unit.

WINDAGE SEAL
Air seal which controls air movement from one section to another within the pressurized inner chambers of the engine.

WINDBACK CHOPPER
Locate under Chopper (oil).

WINDBACK SEAL
A component part of a labyrinth air-oil seal. The Windback has a reverse thread-like surface designed to guide oil back to the oil sump. Often used in conjunction with a Windback Chopper.

WINDMILL BYPASS SHUT-OFF VALVE (fuel)
A valve which allows fuel to recirculate from the main fuel pump back to the pump inlet during an in-flight windmilling condition. This valve prevents excessive fuel pressure build-up in the system and the possibility of unmetered fuel being forced out through the fuel nozzles in the combustor.

WINDMILLING (engine)
The fan and/or compressor rotation of an inoperative engine caused by the aerodynamic forces of the airstream moving through the flight inlet and into the engine. Windmilling is used to bring an engine to proper RPM during an in-flight start procedure. Very few engines are configured with braking systems to prevent this type compressor rotation, because drag is considered negligible and not an aircraft control problem. *See also: Anti-Windmilling Brake.*

WINDMILLING (propeller)
The airstream induced rotation of a propeller on an inoperative engine during flight. Windmilling can be controlled by feathering the propeller.

WINDMILLING BRAKE (rotorcraft)
Locate under Rotor-Brake.

WINDMILLING ENGINE START
Another name for in-flight Air Start.

WING POD
An engine nacelle attached to a pylon in an under wing configuration.

WINGLET TIP (turbine blade)
A recent development in blade design with a semi-closed tip area to retard radial airflow. The winglet is smaller than the traditional tip shroud, covering only the area at the tip along the chord-line, between the leading and trailing edge.

WITNESS DRAIN (fuel or oil)
Drains located at lower points of an engine nacelle. Drains positioned such that they are visible for inspection with the engine running. These drains allow the trained observer to determine whether fluid seepage is normal or abnormal. *See also: 1) Gang Drain, 2) Seepage Checks.*

WITNESS SHIELD
A device used during destructive testing of engines. The shield is generally located outside of the plane of rotation of the fan or turbine wheel. After engine separation, the impact points on the shield provide visible evidence of amounts and direction of debris, information later used in analysis of the failure.

WOODRUF KEY
Another name for Machine Key.

WORK FORMULA
A physics formula. An important concept in describing the function of the turbine engine. The work formula is as follows:

$$W = F \times d$$

Where (W) is work, (F) is force and (d) is distance. Work is accomplished when a force is exerted over a distance. In U.S. Units Work is expressed in Inch-Pounds or Foot-Pounds; for example, 25 ft.lbs. of work. When time is involved, Work becomes Power; for example 25 ft.lb./min. Note: In Metric Units, Work is expressed as meter kilograms and Power is expressed in meter kilograms per second.

WORKING FLUID (engine)
Refers to the Mass Airflow (Ms) passing through a turbine engine.

WORKING PRESSURE (fuel system)
Refers to the fuel pressure head at the fuel nozzles which produces the correct spray pattern.

WROUGHT ALLOYS
Refers to alloys produced by a metal forging process. One of the early methods of producing Gas Turbine Engine parts. *See also: Forging.*

XYZ

X-RAY INSPECTION
A Non-Destructive Inspection. X-Rays are used to locate discontinuities in a structure, either surface or subsurface, and are especially effective when dealing with irregularly shaped parts. An inspection technique similar to medical type X-Ray procedures. *See also: Non-Destructive Inspection.*

X-RAY FLUORESCENT INSPECTION (oil)
An analysis of wear-metals in engine lubricating oil to determine the level of contamination. This technique serves the same purpose, but is quite unlike the traditional Spectrometric Oil Analysis Procedure. See also: Spectrometric Oil Analysis Program.

YEAGER, CHUCK
U.S. Air Force test pilot. The first man to break the sound barrier in a jet aircraft.

YIELD POINT (materials)
A point at approximately two-thirds of the Ultimate Strength, where the material starts to weaken. Also see listings under Material Strength (terms).

YIELD STRENGTH (metals)
The point at which a load starts to deform the metal or the metal reaches its elastic limit. See also: Material Strength.

Z-CUT BLADE
Refers to shrouded, inter-locking tips most often seen on turbine blades. Also called Z-Form. See also: Shroud Lock (turbine blade).

ZERO BALANCE MARK
A type of Matchmark. A "0" mark stamped or etched on two adjacent rotating parts. Zero Balance Marks are used as a reference for alignment purposes to ensure the unit retains its balance when rotating. *See also: Peen Mark.*

ZINC CHROMATE (MIL-L-P8585)
A yellow-green anti-corrosion treatment applied to metal surfaces. Generally applied from an aerosol container.

ZIRCONIA
A ceramic heat-barrier coating applied to turbine engine parts. *See also: Thermal Barrier Coating.*

ZTTRIA
A stabilizing material added to Zirconia. *See also: Thermal Barrier Coating.*

ZYGLO INSPECTION
A commonly used Non-Destructive Inspection. A procedure involving application of a green dye to a metal surface and the illuminating of the surface with a black light. Cracks appear as a visible green hairline under this type of lighting. *See also: Non-Destructive Inspection.*

APPENDIX SECTION

APPENDIX-1

EARLY SIGNIFICANT TURBINE ENGINES (1937 TO EARLY 1960'S)

NOTE: 1. THE YEAR COLUMN REFERS TO THE <u>ENGINE</u> YEAR OF FINAL DEVELOPMENT AND NOT TO THE AIRCRAFT YEAR-IN-SERVICE.
2. WHERE APPROPRIATE, THE ORIGINAL MANUFACTURER'S NAMES ARE USED IN THIS SECTION.

KEY: TJ-TURBOJET, TF-TURBOFAN, TS-TURBOSHAFT, TP-TURBOPROP

YEAR	DESIGNATION	TYPE	MANUFACTURER	COUNTRY	AIRCRAFT APPLICATIONS
1937	HeS	TJ	HEINKEL	GER.	BENCH MODEL
1937	WU	TJ	POWER JETS LTD.	U.K.	BENCH MODEL
1939	HeS 3	TP	HEINKEL	GER.	HEINKEL HE-178
1940	CS-1 (RM-1)	TP	JENDRASSIK	HUNGARY	DEVELOPMENT ENGINE
1940	W1-X	TJ	POWER JETS	U.K.	DEVELOPMENT ENGINE
1941	HeS 8	TJ	HEINKEL	GER.	HEINKEL HE-280
1941	WHITTLE W-1	TJ	POWER JETS	U.K.	GLOSTER E-28
1942	HeS 011	TJ	HEINKEL-HIRTH	GER.	JUNKERS JU-88
1942	GE I-A	TJ	GENERAL ELECTRIC	USA	DEVELOPMENT ENGINE
1943	BMW-003	TJ	BMW	GER.	MESSERSCHMITT ME-262
1943	BMW-018	TP	BMW	GER.	JUNKERS JU-287
1943	F-2/4	TJ	METROPOLITAN-VICKERS	U.K.	GLOSTER METEOR F9/40
1943	GOBLIN	TJ	deHAVILLAND	U.K.	GLOSTER METEOR, deHAVILLAND DH.108, deHAVILLAND "VAMPIRE", FIAT G.80, LOCKHEED XP-80, SAAB-21
1943	I-16 (J-31)	TJ	GENERAL ELECTRIC	USA	BELL XP-59A
1943	1-HS	TJ	DASSAULT	FRANCE	DASSAULT "MYSTERE"
1943	I-14	TJ	GENERAL ELECTRIC	USA	DEVELOPMENT ENGINE
1943	JUMO-004	TJ	JUNKERS	GER.	MESSERSCHMITT ME-262 "SWALLOW"
1943	VRD-2	TJ	LYULKA	USSR	DEVELOPMENT ENGINE
1943	W-2	TJ	POWER JETS	U.K.	GLOSTER METEOR F9/40
1943	YANKEE-9	TJ	WESTINGHOUSE	USA	DRONE AIRCRAFT
1944	109-012	TJ	JUNKERS	GER.	DEVELOPMENT ENGINE
1944	109-022	TP	JUNKERS	GER.	JUNKERS JU-287
1944	ASP /ASX	TP	ARMSTRONG-SIDDELEY	U.K.	LANCASTER TESTBED
1944	HeS-109	TJ	HEINKEL-HIRTH	GER.	ARADO AR-234
1944	J-30 (YAN-19)	TJ	WESTINGHOUSE	USA	McDONNELL FD-1
1944	TG-180	TJ	GENERAL ELECTRIC	USA	MARTIN XB-48, REPUBLIC XP-84
1944	W-2B/WELLAND	TJ	POWER JETS	U.K.	GLOSTER "METEOR II"
1945	W-2B/DERWENT-I	TJ	ROLLS-ROYCE	U.K.	FOKKER MACH-TRAINER, GLOSTER "METEOR III"
1945	J-33	TJ	ALLISON (GM)	USA	LOCKHEED P-80, GRUMMAN F9F, N.AMERICAN T-2
1945	J-33A (I-40)	TJ	GENERAL ELECTRIC	USA	LOCKHEED P-80, GRUMMAN F9F, N.AMERICAN T-2
1945	NENE	TJ	ROLLS-ROYCE	U.K.	LOCKHEED XP-80, HAWKER "SEA FURY" & "SEA HUNTER", VICKERS "SUPERMARINE"
1945	T-31	TP	GENERAL ELECTRIC	USA	VULTEE XP-81
1945	TRENT	TP	ROLLS-ROYCE	U.K.	TRENT-METEOR
1945	YANKEE-24	TJ	WESTINGHOUSE	USA	McDONNELL F2D, DOUGLAS F6U
1945	XT37	TP	NORTHROP	USA	DEVELOPMENT ENGINE
1946	CLYDE	TP	ROLLS-ROYCE	U.K.	ROYAL NAVY "WYVERN"
1946	B-502	TP	BOEING	USA	DRONE AIRCRAFT
1946	J-35	TJ	ALLISON (GM)	USA	BOEING XB-47, GRUMMAN F9F, NO.AMER. B-45, NORTHROP YB-49A, F-89; REPUBLIC F-84G
1946	RD-10	TJ	YAKOVLEV	USSR	YAKOVLEV YAK-15/17/23
1946	RD-20 (003)	TJ	YAKOVLEV/BMW	USSR	MIKOYAN MIG-9, YAKOVLEV YAK-15
1946	T-38	TP	ALLISON (GM)	USA	CONVAIR (G.D.) CV-240
1946	T-50 (502)	TS	BOEING	USA	FAIREY-WESTLAND QH-50D "GYRODYNE"
1946	THESEUS	TP	BRISTOL-SIDDELEY	U.K.	LINCOLN TESTBED & "HERMES"
1947	J-33 Ugrade	TJ	ALLISON	USA	LOCKHEED T-1A "SEASTAR"
1947	J-34	TJ	WESTINGHOUSE	USA	McDONNELL F2H, F-10, FH-1, P-2, T-2; CHANCE-VOUGHT F6U, F7U; DOUGLAS F3D, CONVAIR (GENERAL DYNAMICS) "SEA DART"
1947	J-47	TJ	GENERAL ELECTRIC	USA	BOEING B-47, CANADIAR CL-13, CONVAIR B-36, NORTH AMERICAN B-45, F-86, REPUBLIC XF-91
1947	MAMBA	TP	ARMSTRONG-SIDDELEY	U.K.	LANCASTER TESTBED
1947	RD-45 (NENE)	TJ	TUMANSKY/R-R	USSR	MIKOYAN MIG-17, LA-15
1947	TR-1	TJ	LYULKA	USSR	SUKHOI SU-7/ SU-10
1947	VK-1	TJ	KLIMOV	USSR	ILYUSHIN IL-28, MIKOYAN MIG-15/17
1948	ADDER	TJ	ARMSTRONG-SIDDELEY	U.K.	SAAB PICA, SAAB 210

YEAR	DESIGNATION	TYPE	MANUFACTURER	COUNTRY	AIRCRAFT APPLICATIONS
1948	JANUS	TP	BRISTOL AERO ENGINES	U.K.	DEVELOPMENT ENGINE
1948	TB.1000	TJ	SNECMA	FRANCE	DEVELOPMENT ENGINE
1949	DART	TP	ROLLS-ROYCE	U.K.	VICKERS "VISCOUNT", CONVAIR-240, FOKKER F-27, HANLEY-PAGE "HERALD", AVRO-748, NAMC YS-11, GULFSTREAM G-159
1949	DOUBLE-MAMBA	TP	ARMSTRONG-SIDDELEY	U.K.	FAIREY-WESTLAND "GANNET"
1949	GHOST	TJ	deHAVILLAND	U.K.	deHAVILLAND "COMET-1", "VAMPIRE", "VENOM"; SUD-AVIATION "AQUILON"
1948	J-42 (NENE)	TJ	P&W/R-R	USA	GRUMMAN F9F-6 "PANTHER"
1949	J-48	TJ	PRATT & WHITNEY	USA	GRUMMAN F9F-8 "COUGAR", LOCKHEED F-94
1949	J-71	TJ	ALLISON (GM)	USA	DOUGLAS B-66, RB-66, F3B, McDONNELL F3H
1949	J-73	TJ	GENERAL ELECTRIC	USA	NORTH AMERICAN F-86H, SAAB-29
1949	T-40	TP	ALLISON (GM)	USA	CONVAIR (G.D.) P5Y, DOUGLAS A2D "SKYSHARK" LOCKHEED XFV-1 "POGO"
1949	TR-3	TJ	TURBOMECA	FRANCE	DASSAULT "FOUGA-SYLPHE",
1950	AVON	TJ	ROLLS-ROYCE	U.K.	deHAVILLAND "COMET-2/4" & "SEA VIXEN", FAIREY FD-2, SUD AVIATION "CARAVELLE", VICKERS "SUPERMARINE-SCIMITAR" & "SWIFT"
1950	J-57	TJ	PRATT & WHITNEY	USA	BOEING B-52, KC-135; CHANCE-VOUGHT F-8, CONVAIR (G.D.) F-102, DOUGLAS A3D & F4D, LOCKHEED U-2, McDONNELL F-101
1950	NENE Upgrade	TJ	ROLLS-ROYCE	USSR	CANADAIR T-33, DASSAULT "OURAGAN", FIAT G-82, MIKOYAN MIG-15, SUD-EST "MISTRAL"
1950	TR-3	TJ	LYULKA	USSR	TUPOLEV TU-110
1951	ORENDA	TJ	deHAVILLAND	CANADA	AVRO CF-100, CANADAIR CF-86, CF-104 & CL-13B
1951	J-44	TJ	FAIRCHILD	USA	AUXILIARY ENGINE FAIRCHILD C-119 "PACKET"
1951	T-57	TP	PRATT & WHITNEY	USA	DEVELOPMENT ENGINE
1951	VERNON-350	TJ	HISPANO SUIZA	FRANCE	DASSAULT "MYSTERE-4"
1951	VIPER	TJ	BRISTOL-SIDDELEY	U.K.	BAC H.145 "JET PROVOST" & "STRIKEMASTER", FIAT G.91Y, HAWKER-SIDDELEY HS-125, HINDUSTAN HJT-16, MACCHI MB-326, OTHERS
1952	ASPIN	TF	TURBOMECA	FRANCE	DASSAULT "GEMEAUX"
1952	ATAR	TJ	SNECMA	FRANCE	DASSAULT "ETENDARD", "MYSTERE", "OURAGAN"; NORD "GRIFFON"
1952	J-46	TJ	FAIRCHILD	USA	DEVELOPMENT ENGINE
1952	J-65 (SAPPHIRE)	TJ	WRIGHT (R-R LICENSE)	USA	GRUMMAN F-11F, McDONNELL A-4, MARTIN B-57, REPUBLIC F-84 & RF-84, NORTH AMERICAN FJ-4
1952	NK-4	TP	KUZNETSOV	USSR	DEVELOPMENT ENGINE
1952	NK-12	TP	KUZNETSOV	USSR	TUPOLEV TU-4, TU-95
1952	PROTEUS	TP	BRISTOL-SIDDELEY (R-R)	U.K.	BRISTOL "BRITANNIA"
1952	PYTHON	TP	ARMSTRONG-SIDDELEY	U.K.	ROYAL NAVY "WYVERN"
1952	RD-3M	TJ	MILULKIN	USSR	TUPOLEV TU-16
1952	SAPPHIRE	TJ	ARMSTRONG-SIDDELEY	U.K.	BR. ELECTRIC "CANBERRA", GLOSTER "JAVELIN" HANLEY-PAGE B.1 "VICTOR"
1953	AM-3	TJ	MIKULIN	USSR	MYASISHCHEV M-4 "BISON"
1953	VERDON-350	TJ	HISPANO-SUIZA	FRANCE	DASSAULT "MYSTERE"
1953	VK-1 Upgrade	TJ	KLIMOV	USSR	MIKOYAN MIG-19, YAKOVLEV YAK-25
1953	VK-3	TJ	KLIMOV	USSR	DEVELOPMENT ENGINE
1953	RM2	TJ	VOLVO	SWEDEN	SAAB-29
1953	VK-5	TJ	KLIMOV	USSR	MIKOYAN MIG-19, YAKOVLEV YAK-25
1954	217A	TJ	CONTINENTAL-TELEDYNE	USA	BELL UH-1
1954	AL-7	TJ	LYULKA	USSR	SUKHOI SU-7
1954	ELAND	TS	NAPIER	U.K.	CANADAIR CL-66, FAIREY-WESTLAND "ROTODYNE"
1954	IROQUOIS	TJ	deHAVILLAND	CANADA	AVRO (HAWKER) CF-100
1954	J-69	TJ	CONTINENTAL-TELEDYNE	USA	CESSNA T-37
1954	J-79	TJ	GENERAL ELECTRIC	USA	CONVAIR (GEN. DYNAMICS) B-58 , LOCKHEED F-104 McDONNELL F-4, NORTH AMERICAN A3J & A-5
1954	NK-12 Upgrade	TP	KUZNETSOV	USSR	TUPOLEV TU-4
1954	OLYMPUS	TJ	BRISTOL-SIDDELEY	U.K.	AVRO (HAWKER) B.1 & B.2 "VULCAN BOMBERS"
1954	RD-3	TJ	MIKULIN	USSR	MYASISHCHEV M-4 "BISON", MIG-19
1954	T-65	TP	CONTINENTAL-TELEDYNE	USA	DEVELOPMENT ENGINE
1954	TV-2	TP	KUZNETSOV	USSR	TUPOLEV TU-91
1955	ARTOUSTE	TS	TURBOMECA	FRANCE	SUD AVIATION "ALOUETTE", MERCKLE SM-67
1955	DART Upgrade	TP	ROLLS-ROYCE	U.K.	ARMSTRONG "ARGOSY", BREGUET "ALIZE"
1955	D-15	TJ	SOLOVIEV	USSR	MYASISHCHEV M-50 "BOUNDER"
1955	FIAT 4002	TS	FIAT	U.K.	FIAT 7002 HELICOPTER
1955	GAZELLE	TS	NAPIER AERO (R-R)	U.K.	SIKORSKY S-58, WESTLAND "BELVEDERE & WESSEX"
1955	MARBORE	TJ	TURBOMECA	FRANCE	AEROSPATIALE-POTEZ CM-170 "FOUGA-MAGISTER", HISPANO HA-200 & HA-300, MORANE "PARIS"
1955	NK-12M	TP	KUZNETSOV	USSR	TUPOLEV TU-20, TU-114, TU-124
1955	PALAS	TS	TURBOMECA	FRANCE	HELICOPTERS
1955	RD-9	TJ	TUMANSKY	USSR	MIKOYAN MIG-21, SU-21
1955	RM5/6	TJ	VOLVO (AVON LICENSE)	SWEDEN	SAAB-32 "LANSEN", SAAB-35 "DRAKEN"
1956	AI-20	TP	IVCHENKO	USSR	ANTONOV AN-10 & AN-12, ILYUSHIN IL-18
1956	J3-1	TJ	NIPPON	JAPAN	FUJI T-1A

YEAR	DESIGNATION	TYPE	MANUFACTURER	COUNTRY	AIRCRAFT APPLICATIONS
1956	AM-3/5/9	TJ	MIKULIN	USSR	M-4, MIG-19, MIG-21, TUPOLEV TU-16, TU-104
1956	J-75	TJ	PRATT & WHITNEY	USA	REPUBLIC F-105
1956	NK-4	TJ	KUZNETSOV	USSR	ILYUSHIN IL-12, IL-18, YAK-42 "BACKFIN"
1956	PIMENE	TS	TURBOMECA	FRANCE	HELICOPTERS
1956	R-11	TJ	TUMANSKY	USSR	MIKOYAN MIG-15
1956	R-13	TJ	TUMANSKY	USSR	SUKHOI SU-15
1956	RD-45	TJ	TUMANSKY	USSR	SUKHOI SU-15/17
1956	T-53	TS	LYCOMING (TEXTRON)	USA	BELL UH-1, BELL 204, CANADAIR CL-84, KAMAN HH-43
1956	T-53	TP	LYCOMING (TEXTRON)	USA	GRUMMAN OV-1 "MOHAWK"
1956	T-56 (501)	TP	ALLISON (GM)	USA	CONVAIR CV-440, GRUMMAN E-2, LOCKHEED C-130, L-188 "ELECTRA", P-3 "ORION"
1956	TYNE	TP	ROLLS-ROYCE	U.K.	VICKERS "VANGUARD", CANADAIR CL-44, SHORTS "BELFAST", FAIREY-WESTLAND "ROTODYNE"
1956	VD-7	TJ	KOLIESOV	USSR	MYASISHCHEV M-50, M-52
1957	AVON Upgrade	TJ	ROLLS-ROYCE	U.K.	BAC "LIGHTNING", deHAVILLAND "COMET-4", HAWKER "HUNTER" SERIES, SAAB-35 "DRAKEN"
1957	D-25	TS	SOLOVIEV	USSR	MIL MI-6
1957	GYRON	TJ	BRISTOL-SIDDELEY	U.K.	BLACKBURN "BUCCANEER", BRISTOL T.188
1957	J93-GE-3	TJ	GENERAL ELECTRIC	USA	NORTH AMERICAN XB-70 "VALKYRIE"
1957	JT3C (J-57)	TJ	PRATT & WHITNEY	USA	BOEING 367-80, B-707, B-72
1957	JT4 (J-75)	TJ	PRATT & WHITNEY	USA	REPUBLIC F-105, CONVAIR (GEN. DYNAMICS) F-106
1957	MEDWAY	TF	ROLLS-ROYCE	U.K.	deHAVILLAND DH-121, HAWKER-SIDDELEY HS-681
1957	R-11	TJ	TUMANSKY	USSR	YAKOVLEV YAK-28
1957	T-58	TS	GENERAL ELECTRIC	USA	SIKORSKY H-52, SH-3, S-61, S-62; KAMAN UH-2A
1957	TURMO	TS/TP	TURBOMECA	FRANCE	SUD AVIATION "FRELON" & "PUMA", BREGUET 941
1958	ATAR Upgrade	TJ	SNECMA	FRANCE	DASSAULT "SUPER MYSTERE", "MIRAGE 30" & "ETENDARD"; SUD-AVIATION "VAUTOUR"
1958	CJ-805 (J79)	TJ	GENERAL ELECTRIC	USA	CONVAIR (GENERAL DYNAMICS) CV-880 "SKYLARK"
1958	CONWAY	TF	ROLLS-ROYCE	U.K.	BOEING 707-420, DOUGLAS DC-8-40, HANLEY-PAGE B.2 "VICTOR", SAAB-37 "VIGGEN", VICKERS VC-10, DOUGLAS C-133
1958	T-34	TP	PRATT & WHITNEY	USA	
1958	T-63 (250)	TS	ALLISON (GM)	USA	BELL OH-6
1959	AL-21	TJ	LYULKA	USSR	TUPOLEV TU-128, TU-138
1959	CT-58	TS	GENERAL ELECTRIC	USA	BOEING V-107, SIKORSKY S-61, S-62
1959	GNOME	TS	ROLLS-ROYCE	U.K.	AGUSTA-101, BELL-204, WESTLAND "SEA KING", "WESSEX" & "WHIRLWIND"
1959	J-52	TJ	PRATT & WHITNEY	USA	GRUMMAN EA-6A, McDONNELL A-4
1959	J-58	TJ	PRATT & WHITNEY	USA	LOCKHEED SR-71, YF-12A
1959	JT3D (TF33)	TF	PRATT & WHITNEY	USA	BOEING B-707, KC-135, DOUGLAS DC-8
1959	J-60 (JT12)	TJ	PRATT & WHITNEY	USA	CANADAIR CL-41, LOCKHEED C-140 "JETSTAR", McDONNELL-220, ROCKWELL T-2 & T-39
1959	J-97	TJ	GENERAL ELECTRIC	USA	DRONE AIRCRAFT
1959	NIMBUS	TS	BRISTOL-SIDDELEY (R-R)	U.K.	WESTLAND "WASP" & "SCOUT"
1959	ORPHEUS	TJ	BRISTOL AERO (R-R)	U.K.	FIAT G.91, FUJI T-1B, HAWKER-SIDDELEY "GNAT", HINDUSTAN HF-24
1959	TB-2BM	TS	SOLOVIEV	USSR	MIL MI-6
1959	TPE-331 (T76)	TP	GARRETT	USA	REPUBLIC "LARK" (ALOUETTE), MANY OTHERS
1959	WALTER M701	TP	WALTER	USSR	LET L-29
1960	AI-24	TP	IVCHENKO	USSR	YAKOVLEV YAK-40, ANTONOV AN-24
1960	D-20P	TJ	SOLOVIEV	USSR	TUPOLEV TU-124
1960	GTD-3	TP	KUZNETSOV		BERIEV BE-30
1960	JT8D	TF	PRATT & WHITNEY	USA	DOUGLAS DC-9, BOEING B-727, B-737
1960	JT9A	TF	PRATT & WHITNEY	USA	DEVELOPMENT ENGINE
1960	PT-6	TP	P & W CANADA	CANADA	DOUGLAS DC-3 TESTBED AIRCRAFT
1960	NK-8	TP	KUZNETSOV	USSR	ILYUSHIN IL-68
1960	SO-5	TJ	MIELEC	POLAND	MIELEC ISKRA TS-11
1960	T-64	TP	GENERAL ELECTRIC	USA	deHAVIL DHC-5 "CARIBOU", LTV/RYAN XC-142
1960	T-64	TS	GENERAL ELECTRIC	USA	SIKORSKY CH-53, OTHER HELICOPTERS
EARLY 1960'S	ASTAZOU	TS	TURBOMECA	FRANCE	AGUSTA-115, SUD "ALOUETTE", SHORTS "SKYVAN"
	AUBISQUE	TF	TURBOMECA	FRANCE	HISPANO-230 SAAB-105 & SK-60
	BASTAN	TP	TURBOMECA	FRANCE	DASSAULT "SPIRALE", HOLSTRE "SUPER-BROUSSARD"
	GE-4	TJ	GENERAL ELECTRIC	USA	BOEING 2707 SST
	CJ-805-23	TJ	GENERAL ELECTRIC	USA	CONVAIR CV-990, SUD AVIATION "SUPER-CARAVELLE"
	D-30	TJ	SOLOVIEV	USSR	TU-134
	HELWAN E-300	TJ	HINDUSTAN	INDIA	HINDUSTAN HF-24
	JTFD-12	TS	PRATT & WHITNEY	USA	SIKORSKY CH-54
	OLYMPUS-320	TJ	BRISTOL-SIDDELEY	U.K.	BAC TRS-2
	PEGASUS	TF	BRISTOL-SIDDELEY (R-R)	U.K.	HAWKER-SIDDELEY AV-8 "HARRIER"
	RB.162/175	TF	ROLLS-ROYCE	U.K.	DASSAULT "MIRAGE", FIAT G.222, DORNIER-31
	RB.172/178	TF	ROLLS-ROYCE	U.K.	DASSAULT "MYSTERE", VICKERS "SUPER VC-10"
	SPEY	TF	ROLLS-ROYCE	U.K.	BAC "BUCCANEER" & BAC 1-11; FOKKER F.28
	T-55	TS	LYCOMING (TEXTRON)	USA	BOEING CH-47
	T-55	TP	LYCOMING (TEXTRON)	USA	NORTH AMERICAN YAT-28
	TF-30	TF	PRATT & WHITNEY	USA	CHANCE-VOUGHT (LTV) A-7A
	TF-106	TF	SNECMA	FRANCE	DASSAULT "MIRAGE"

APPENDIX-2

ENGINE INSPECTION and DISTRESS TERMINOLOGY

INSPECTION TERMS (Gas Turbine Engine) - Words that clarify, describe or identify distressed areas of the engine, whether from normal wear or early failure.

Note: Bearings, because of their critical role in the engine, require even more precise inspection categories. Terminology for bearing inspection is listed separately under Bearing Inspection and Distress Terms in Appendix-3.

Abrasion (abraded): Roughened surface, varying from light to severe. Result of fine foreign material present between moving parts.

Arc (arced): visible effects (burned spots, fused metal); the result of an electrical discharge between two connections.

Banding (banded): Locate in Bearing Inspection and Distress Terms, Appendix-3.

Battered: Damage by repeated impact.

Bend (bent): Distortion in a part, meaning it differs from a local in original conformation. Result of application of excess heat or force.

Bind (binding): Restricted movement. Result of excessively tight fits or jammed mechanisms.

Blister (Blistered): Raised portion of a surface, usually where the surface has separated from the base. Generally found on surface treated parts, such as plated or painted surfaces. Result of poor original bond, excessive heat, or pressure.

Bow (bowed): Curved or gradual deviation from original line or plane. Result of heat or lateral forces.

Break (broken): Separation of a part. Result of severe force, pressure, or overload.

Brinell (brinelled)**:** Locate in Bearing Inspection and Distress Terms, Appendix-3.

Brittle: Resilience of material not to required specifications.

Broken: Parts separated by force into two or more pieces.

Buckle (buckling): Large scale deformation of part contours. Result of pressure or impact with a foreign object, unusual structural pressures, excessive localized heating, or any combination of these causes.

Bulge (bulged): Localized outward or inward swelling. Result of excessive localized heating or differential pressure.

Burn (burned): Discoloration of a part or a burn through hole. Result of localized excessive heat.

Burnishing: The smoothing of a metal surface by mechanical action, but without loss of material. Burnishing from normal service is not detrimental if coverage approximates the carrying load and if there is no evidence of burns. Result of normal wear or heat.

Burr (burred): Rough edge or sharp projection. Result of excessive wear or poor machining.

Carboned: Accumulation of carbon deposits.

Chafe (chafed): Frictional wear damage. Result of two parts rubbing together with limited motion.

Chip (chipped): Breaking away of small metallic particles. Result of heavy impact.

Collapsed (crushed): Inward deformation of the original contour of a part. Result of high pressure differentials.

Corrosion (corroded): Surface chemical action resulting in surface discoloration, a layer of oxide, or in advanced stages, pitting or removal of metal surface. Result of improper corrosion prevention or excessive moisture.

Crack (cracked): Partial separation of material. Result of residual cyclic or vibratory stresses, severe stress from overloading or shock, or possible extension of a minute material defect or scratch from improper grinding or improper heat treating.

Crazing: Craze lines or patches. Hairline cracks in a surface coating. A condition often found in plastic and epoxy resin materials. Result of normal wear over time, excessive heat, or deformation of parent metal.

Crossed Thread: Material damage to threaded fasteners as a result of improper assembly.

Curled Tips: A condition wherein the tips of compressor and turbine blades are curled over due to rubbing action.

Dent (dented) : A small smoothly rounded depression. Result of a sharp blow from a blunt object or excessive pressure.

Deposits: A build up of material on a part either from foreign material or materials within the engine itself.

Disintegrated: Separated or decomposed into fragments resulting in loss of original form.

Distortion (distorted): A change in original shape. Result of application of excessive heat.

Eccentric: Parts within an intended center that are significantly displaced.

Electrolytic Action: Surface breakdown. Result of galvanic action between two dissimilar metals.

Erosion (eroded): Wearing away of metal. Result of high velocity airflow, hot gases, airborne contaminants, or fluids over a surface.

Extruded: Plastic deformation caused by excessive pressure between two parts.

Fatigue Failure. Complete rupture or breakdown of a part. Result of normal wear over time, improper manufacture, or low material strength.

Feathered Edge: Thinning of material at an edge of a surface.

Flaking: Loose particles of metal or evidence of surface coating removal. Result of defective plating, imperfect bonding, severe overloading, or corrosion.

Flattened: Permanent deformation beyond tolerance limits. Result of compression forces.

Fracture: Same as Break.

Fret (fretted): Discoloration of contact surfaces, possibly rust deposits between surfaces, and removal of original surface material. Occurs between contacting surfaces but not between rolling surfaces. For example, a bearing outer race and its housing or other parts that have a static fit. Rust acts as a lapping compound to loosen the fit. Result of slight movement of mating parts that may or may not have moisture present.

Fuse (fused): Joining together of two materials not intended to be so. Result of heat or friction loads.

Gall (galling): Severe chafing, recognized by presence of metal from one part remaining attached to another. Generally occurs at poorly lubricated surfaces that are in sliding contact. Result of localized lubrication breakdown causing friction.

Gouge (gouged): Removal of surface material, typified by rough and deep depressions or furrowing. Result of contact with protruding object or misalignment.

Grain Boundary Oxidation: A stretch mark pattern following boundaries of a metal's grain structure. Result of loss of surface material.

Heat Oxidized: A film of surface discoloration material such as rust. Result of normal high temperature operation or in extreme cases excessive heat.

Hot Spot: A grey, brown, or black spot, in advanced stages often accompanied by small irregular cracking. Result of localized overheating.

Inclusion: Foreign matter enclosed in a metal. Result of contaminated metal mixture during initial manufacture.

Indentation (indented): Presence of cavities with smooth bottoms and sides. Some types of indents closely resemble nicks or pits but do not have the same ragged edges. Result of loose foreign particles between moving parts. Also located in Bearing Inspection and Distress Terms, Appendix-3.

Intergranular Oxidation: Separation in the grain structure. Results from high heat and where oxygen interacts with the metal to cause oxidation.

Metalize (metalized): Coating of a part with hot metal debris. Result of a failed part and molten metal being carried along the gas path.

Nick (nicked): A sharp bottomed or vee-shaped depression that may have rough edges. Result of impingement of an object on a surface such as from careless handling of tools or parts.

Out of Round: Diameter of part not constant.

Overheated: Discoloration or loss of hardness. Result of high temperature produced by excessive friction from overspeeding, overloading, lack of backlash, or faulty lubrication.

Peeled: Breaking away of surface finish, such as coating or plating.

Peen (peened): Blunt depression, flattening, or displacement of metal. Result of repeated blows on a surface from loose objects or foreign objects.

Pickup: Transfer (build-up) of one material onto another. Result of insufficient lubrication causing partial seizure of rotating parts.

Pit (pitted): Small cavities, usually with perceptible roughened edges on a surface. Result of chemicals causing oxidation or electrolytic action, or mechanical pressure of foreign materials on a surface.

Rub (rubbed): To move with pressure or friction against another part.

Rupture (ruptured): Extensive breaking apart of material, usually caused by high stresses, differential pressure, locally applied force, or a combination.

Score (scored): Deep scratches following the path of the part travel. Result of localized lubrication breakdown between sliding surfaces.

Scratch (scratched): Very shallow furrow or irregularity usually with more length than width. Result of movement of a sharp object across a surface.

Scuff (scuffed): Surface condition evidenced by transfer of material. Result of insufficient clearance or lubrication between moving parts.

Seizure (seized): Fusion or binding of two adjacent surfaces preventing continued movement. Result of improper lubrication or wear.

Sheared: A part divided by cutting action or by sliding of parts parallel to their plane of contact.

Shingle (shingling: An unsatisfactory condition found on fan blades when the span shrouds overlap due to excessive clearances caused by wear of their mating surfaces. A serious condition that can lead to fan vibration.

Spall (spalled): Sharply roughened area characterized by chipping or peeling of a surface material. It is more progressive than flaking. Also located in Bearing Inspection and Distress Terms, Appendix-3.

Stain (stained): Locate in Bearing Inspection and Distress Terms, Appendix-3.

Stress Failure: Failure of a metal. Result from: 1) Compression - Action of two opposed forces that tend to squeeze apart. 2) Shear - Action of two parallel forces acting in opposite directions. 3) Shock - Instantaneous application of stress. 4) Tension Action - two directly opposed forces that tend to stretch apart.

Stress Rupture: Failure of a metal airfoil in the hot section, characterized by small chordwise cracks along its leading and trailing edges. Result of excessive overheat.

Stripped Thread: Thread of a bolt, nut, screw. stud or other threaded fastener that is damaged by tearing away part of the thread structure. Result of improper installation (cross-threading), improper thread size one part to another, or seizure of a fastener to its mating part.

Tear: Parting of parent metal. Result of excessive tension load, created by an external force.

Unbalance: Uneven rotation of a part, a condition that usually results in vibration. Result of unequal distribution of load about a rotating axis.

Unbonded (composite material): Separation of two adhesively bonded surfaces. Result of adhesive failure at a surface interface or cohesive failure wherein the adhesive fails.

Varnished (varnishing): Discolored film on a surface that ranges from straw color to dark brown. Result of lubrication oxidizing (mixing with oxygen) and normal on some parts. Not normal on bearings. Also located in Bearing Inspection and Distress Terms, Appendix-3.

Wear: Rubbing away of a surface by mechanical action that occurs on all contact surfaces. May be indicated by a wavy path, slight discoloration, dull or shiny tracks, or a very finely pitted dull surface. Result of normal operation over time with abrasive substances, such as rust, acting as a lapping compound. Also results from movement between parts, caused by loose mountings, or off square mountings.

APPENDIX 3

BEARING INSPECTION AND DISTRESS TERMINOLOGY

Because bearing condition in the Gas Turbine Engine is critical to engine integrity, a special listing of terms is used to describe bearing wear and failure patterns.

Abrasion: A roughened area caused by the presence of fine foreign material between moving parts.

Banding: Parallel bands of discoloration, caused by oil varnish or oxide film formation. Generally this results from high temperature bearing operation.

Brinelling (false): Shallow depressions or indentations in the loaded side of a bearing race at each ball or roller location, occurring only at the rolling contact surfaces. The indentations are usually polished or satin finished in appearance. In roller bearings the indentation will frequently be slightly flatter than the roller radius. False brinelling is a specialized form of fretting and the result of continuous nonrotational shaft oscillation and does not occur when the bearing is in operation. Vibration during engine transportation may cause this condition.

Brinelling (true): A shallow flat spot or indentation found on the surface of ball or roller bearing races at only one location. True Brinelling can be caused by shock (impact) loads imposed when the bearing is not rotating or by skidding of the rolling elements when operating loads are quickly removed.

Brittleness: Loss of resiliency generally due to severe application of heat or cold.

Burning: An injury to the surface caused by excessive heat by lube film failure. This is evidenced by discoloration, or in severe cases, by loss of material.

Burnishing: A smoothing of the surface by mechanical action but without loss of material. Generally found on plain bearing surfaces. Surface discoloration is sometimes present around the outer edges of a burnished area. Normal burnishing occurs from operational service and is not detrimental if coverage approximates the carrying load and there is no evidence of burning. Abnormal burnishing is the result of high heat.

Cage Failure: Cracks, breaks, or severe wear on the cage working surface. It is the result of breakdown in lubrication or material fatigue.

Chafing: A rubbing action between two parts that have limited relative motion.

Chipping: Breaking out of small pieces of material.

Corrosion: Pitted or roughened surface due to chemical action.

Depression: Locate under Indentation.

Fretting: Discoloration that occurs on surfaces pressed together under high loading. Fretting does not occur on rolling surfaces, but at the inner race and shaft or the outer race and housing. Fretting will reduce the fit between surfaces. It is caused by slight movement between two contacting surfaces.

Frosting: Many minute indentations within a localized area on bearing races. Usually this is a condition that cannot be measured. Frosting can be thought of as an early stage of corrosion.

Galling: Transfer of metal from one part to another; caused by chafing.

Gouging: Displacement of material from a surface by cutting or tearing, caused by contact with a protruding object or by misalignment.

Grooving: Elongated depressions found on the working surfaces of balls and rollers. Grooving is caused by lack of lubrication or bearing skidding.

Inclusion: Foreign material enclosed in the metal causing a discontinuity. Surface inclusions appear as dark spots or lines.

Indentation: Depressions or cavities with smooth bottoms and sides. Occurs on rolling contact surfaces. The cavities are caused by metal displacement, not metal removal. The metal is thrown up as a rim around the cavity. This condition is caused by loose particles rolling between rotating parts.

Nick: Sharp indentation caused by striking one part against another.

Peening: Deformation of a surface by impact.

Pitting: Small, irregular cavities in a surface caused by corrosion or chipping. Pits from chipping can be distinguished from corrosive pitting. Corrosive pitting is usually accompanied by deposits formed by the agent and the base metal.

Roller End Wear: Indicated by irregular, sharp edges at roller ends. This condition is the result of misalignment of a bearing due to excessive strain imposed on it by other engine parts.

Scoring: Deep scratches or elongated gouges caused by contact with sharp edges.

Scratching: Narrow or shallow cuts caused by movement of a sharp object across a surface.

Skidding: Balls or rollers that do not roll normally, causing scratching or grooving and loss of material in the form of powdered metal.

Spalling: Sharply roughened edges caused by slight surface cracks or progressive chipping or peening on a surface. Spalling can be caused by over-loading, but is generally the result of normal wear.

Staining: Discolored areas caused by lubrication breakdown under high temperature.

Varnishing: A film of discolored oil residue, ranging from straw colored to dark brown, and the result of normal operation or, in severe cases, from high heat transfer to the oil.

Wear: Rubbing away of a surface by mechanical action, with wear occurring on all contact surfaces. May be indicated by a wavy path, slight discoloration, and/or dull or shiny tracks. Wear results from movement caused by looseness of parts, with abrasive substances trapped between surfaces acting as a lapping compound. Wear can also occur from an off square condition between inner and outer races or pressure on the bearing cage.

APPENDIX-4

CONVERSION TABLES COMMON TO THE TURBINE ENGINE

Multiply	By	To Obtain	Multiply	By	To Obtain
Amperes	1.04×10^{-5}	Faradays/Sec.	Grams/Liter	0.122	Ounces/Gallon (Troy)
Amperes/Sq. Ft.	0.00108	Amperes/Sq. Cm.	Grams/Liter	0.134	Ounces/Gallon (Avdp)
Ampere-hours	3,600	Coulombs	Grams/Liter	1,000	Parts Per Million
Amperes/Sq. Cm.	929	Amperes/Sq. Ft.	Grams/Liter	2.44	Pennyweights/Gallon
Angstrom Units	3.94×10^{-9}	Inches	Inches	2.54	Centimeters
Angstrom Units	1×10^{-10}	Meters	Inches	1/12	Feet
Angstrom Units	1×10^{-4}	Microns	Inches	1,000	Mils
Centimeters	0.394	Inches	Kilograms	15,432.4	Grains
Centimeters	393.7	Mils	Kilograms	1,000	Grams
Centimeters	0.0328	Feet	Kilograms	35.27	Ounces (Avdp)
Circular Mils	5.07×10^{-6}	Sq. Centimeters	Kilograms	32.15	Ounces (Troy)
Circular Mils	7.85×10^{-7}	Sq. Inches	Kilograms	643.01	Pennyweights
Cubic Centimeters	3.53×10^{-5}	Cubic Feet	Kilograms	2.205	Pounds (Avdp)
Cubic Centimeters	0.061	Cubic Inches	Kilograms	2.679	Pounds (Troy)
Cubic Centimeters	2.64×10^{-4}	Gallons	Liters	1,000.027	Cubic Centimeters
Cubic Centimeters	9.9997×1^{-4}	Liters	Liters	0.035	Cubic Feet
Cubic Centimeters	0.0338	Ounces (Fluid)	Liters	61.025	Cubic Inches
Cubic Centimeters	0.0021	Pints	Liters	0.264	Gallons
Cubic Centimeters	0.0011	Quarts (Liquid)	Liters	33.81	Ounces (Fluid)
Cubic Feet	28.317	Cubic Centimeters	Liters	2.11	Pints
Cubic Feet	1.728	Cubic Inches	Liters	1.057	Quarts (Liquid)
Cubic Feet	7.48	Gallons	Meters	100	Centimeters
Cubic Feet	28.32	Liters	Meters	3.281	Feet
Cubic Feet	29.92	Quarts	Meters	39.37	Inches
Cubic Feet of Water			Meters	0.001	Kilometers
60°F (16°C)	62.37	Pounds	Meters	1.094	Yards
Cubic Inches	16.39	Cubic Centimeters	Microhms	1×10^{-12}	Megohms
Cubic Inches	0.0043	Ounces (Fluid)	Microhms	1×10^{-6}	Ohms
Cubic Inches	0.0173	Quarts (Liquid)	Microns	3.9×10^{-5}	Inches
Faradays	9.65×10^{-4}	Coulombs	Microns	0.001	Millimeters
Faradays/Second	96,500	Amperes	Mile	1.15	Knots
Feet	30.48	Centimeters	Miles per hour (mph)	1.467	Feet/Second
Feet	12	Inches	Milligrams	0.0154	Grains
Feet	0.3048	Meters	Milligrams	0.001	Grams
Feet/Second	0.6817	Miles Per Hour	Milligrams	1×10^{-6}	Kilograms
Gallons	4	Quarts (Liquid)	Milligrams	3.5×10^{-5}	Ounces (Avdp)
Gallons	3,785.4	Cubic Centimeters	Milligrams	3.215×10^{-5}	Ounces (Troy)
Gallons	0.1337	Cubic Feet	Milligrams	6.43×10^{-4}	Pennyweights
Gallons	231	Cubic Inches	Milligrams	2.21×10^{-6}	Pounds (Avdp)
Gallons	3,785	Liters	Milligrams	2.68×10^{-6}	Pounds (Troy)
Gallons	128	Ounces (Fluid)	Milligrams/Liter	1.0	Parts Per Million
Gallons (Imperial)	1.201	Gallons	Milliliters	1.000027	Cubic Centimeters
Gallons	0.8327	Gallons (Imperial)	Milliliters	0.061	Cubic Inches
Gallons	8	Pints	Milliliters	0.001	Liters
Gallons	8.31	Pounds (Avdp of Water	Milliliters	0.034	Ounces (Fluid)
		at 62°F (179°C))	Millimeters	0.1	Centimeters
Grains	0.0648	Grams	Millimeters	0.039	Inches
Grains	0.0023	Ounces (Avdp)	Millimeters	0.001	Meters
Grains	0.0021	Ounces (Troy)	Millimeters	1,000	Microns
Grains	0.0417	Pennyweights (Troy)	Mils	0.0025	Centimeters
Grains	1/7,000	Pounds (Avdp)	Mils	0.001	Inches
Grains	1/5,760	Pounds (Troy)	Mils	25.4	Microns
Grams	15.43	Grains	Ounces (Avdp)	437.5	Grains
Grams	1,000	Milligrams	Ounces (Avdp)	28.35	Grams
Grams	0.0353	Ounces (Avdp)	Ounces (Avdp)	0.911	Ounces (Troy)
Grams	0.0321	Ounces (Troy)	Ounces (Avdp)	18.23	Pennyweights
Grams	0.643	Pennyweights	Ounces (Avdp)	1/16	Pounds (Avdp)
Grams	0.0022	Pounds (Avdp)	Ounces (Avdp)	0.076	Pounds (Troy)
Grams	0.0027	Pounds (Troy)	Ounces/Gallon (Avdp)	7.5	Grams/Liter

Multiply	By	To Obtain	Multiply	By	To Obtain
Ounces (Troy)	480	Grains	Pounds (Avdp)	14.58	Ounces (Troy)
Ounces (Troy)	31.1	Grams	Pounds (Avdp)	291.67	Pennyweights
Ounces (Troy)	31,103.5	Milligrams	Pounds (Avdp)	1.215	Pounds (Troy)
Ounces (Troy)	1.097	Ounces (Avdp)	Pounds (Troy)	5,760	Grains
Ounces (Troy)	20	Pennyweights	Pounds (Troy)	373.24	Grams
Ounces (Troy)	0.069	Pounds (Avdp)	Pounds (Troy)	0.373	Kilograms
Ounces (Troy)	1/12	Pounds (Troy)	Pounds (Troy)	13.17	Ounces (Avdp)
Ounces/Gallon (Troy)	8.2	Grams/Liter	Pounds (Troy)	12	Ounces (Troy)
Ounces (Fluid)	29.57	Cubic Centimeters	Pounds (Troy)	240	Pennyweights
Ounces (Fluid)	1.80	Cubic Inches	Pounds (Troy)	0.823	Pounds (Avdp)
Ounces (Fluid)	1/128	Gallons	Pounds of Water 62°F		
Ounces (Fluid)	0.0296	Liters	(17°C)	0.016	Cubic Feet
Ounces (Fluid)	29.57	Milliliters	Pounds of Water 62°F		
Ounces (Fluid)	1/16	Pints	(17°C)	27.68	Cubic Inches
Ounces (Fluid)	0.031	Quarts	Pounds of Water 62°F		
Ounces/Gallon (Fluid)	7.7	CC/Liter	(17°)	0.1198	Gallons
Pennyweights	24	Grains	Quarts (Liquid)	946.4	Cubic Centimeters
Pennyweights	1.56	Grams	Quarts (Liquid)	57.75	Cubic Inches
Pennyweights	1,555	Milligrams	Quarts (Liquid)	0.033	Cubic Feet
Pennyweights	0.0549	Ounces (Avdp)	Quarts (Liquid)	0.25	Gallons
Pennyweights	0.05	Ounces (Troy)	Quarts (Liquid)	0.946	Liters
Pennyweights	0.0034	Pounds (Avdp)	Quarts (Liquid)	32	Ounces (Fluid)
Pennyweights	0.0042	Pounds (Troy)	Quarts (Liquid)	2	Pints
Pennyweights/Gallon	0.41	Grams/Liter	Square Centimeters	127.32	Circular Millimeters
Pints	473.2	Cubic Centimeters	Square Centimeters	197,350	Circular Mils
Pints	0.017	Cubic Feet	Square Centimeters	0.001	Square Feet
Pints	28.88	Cubic Inches	Square Centimeters	0.155	Square Inches
Pints	0.125	Gallons	Square Centimeters	100	Square Millimeters
Pints	0.473	Liters	Square Feet	929.03	Square Centimeters
Pints	16	Ounces (Fluid)	Square Feet	144	Square Inches
Pints	0.5	Quarts	Square Inches	1.2732	Circular Mils
Pounds (Avdp)	7,000	Grains	Square Inches	6.45	Square Centimeters
Pounds (Avdp)	453.6	Grams	Square Inches	1/144	Square Feet
Pounds (Avdp)	0.454	Kilograms	Square Inches	1×10^{-6}	Square Mils
Pounds (Avdp)	16	Ounces (Avdp)	Square Inches	1/1,296	Square Yards

APPENDIX - 5

GAS TURBINE ENGINE SYMBOLS

A	ACCELERATION	Nh	SPEED-HP COMPRESSOR
Aj	AREA OF JET NOZZLE	NL	SPEED-LP COMPRESSOR
am	AMBIENT	P	PRESSURE
Cp	SPECIFIC HEAT-CONSTANT PRESSURE	Pam	PRESSURE-AMBIENT
Cs	LOCAL SPEED OF SOUND	Pb	PRESSURE-BURNER
Cv	SPECIFIC HEAT-CONSTANT VOLUME	Pe	POTENTIAL ENERGY
E	ENTROPY	Peff	PROPULSIVE EFFICIENCY
ESHP	EQUIVALENT SHAFT HORSEPOWER	Pj	PRESSURE-JET NOZZLE
F	FORCE (THRUST)	PLA	POWER LEVER ANGLE
FF	FUEL FLOW	Po	PRESSURE-STANDARD DAY
Fg	GROSS THRUST	Ps	PRESSURE-STATIC
Fn	NET THRUST	Pt	PRESSURE-TOTAL
g	ACCELERATION OF GRAVITY	Pt2	PRESSURE-ENGINE INLET
H	ENTHALPY	Pt7	PRESSURE-TURBINE DISCHARGE
HP	HORSEPOWER	SFC	SPECIFIC FUEL CONSUMPTION
HPC	HIGH PRESSURE COMPRESSOR	T	TEMPERATURE
IPC	INTERMEDIATE PRESSURE COMPRESSOR	THP	THRUST HORSEPOWER
IPT	INTERMEDIATE PRESSURE TURBINE	TLA	THROTTLE LEVER ANGLE
LPC	LOW PRESSURE COMPRESSOR	To	TEMPERATURE-STANDARD DAY
LPT	LOW PRESSURE TURBINE	Ts	TEMPERATURE-STATIC
Ke	KINETIC ENERGY	TSFC	THRUST SPECIFIC FUEL CONSUMPTION
m	MASS AIRFLOW	Tt	TEMPERATURE-TOTAL
M	MACH NUMBER	Tt2	TEMPERATURE-ENGINE INLET
N	ROTATIONAL SPEED	Tt7	TEMPERATURE-TURBINE EXHAUST
Nc	SPEED-COMPRESSOR	V	VELOCITY
N1	SPEED-FAN/LP COMPRESSOR	V1	VELOCITY-INITIAL (also Vi)
N2	SPEED-HP COMPRESSOR	V2	VELOCITY-FINAL (also Vf)
N2	SPEED-IP COMPRESSOR	Wf	WEIGHT OF FUEL (FUELFLOW)
N3	SPEED-HP COMPRESSOR	ΔP	Delta-P (Differential Pressure)
Nf	SPEED-FREE TURBINE	η	EFFICIENCY (Eta)
Ng	SPEED-GAS GENERATOR	μ	Micron (Mu)

AIRCRAFT MISSION SYMBOLS
(U.S. MILITARY)

- A - ATTACK
- B - BOMBER
- C - CARGO/TRANSPORT
- D - DIRECTOR
- E - ELECTRONIC
- F - FIGHTER
- H - SEARCH/RESCUE
- K - TANKER
- L - COLD WEATHER
- O - OBSERVATION
- P - PATROL
- Q - DRONE
- R - RECONNAISSANCE
- S - ANTI-SUBMARINE
- T - TRAINER
- U - UTILITY
- V - STAFF
- W - WEATHER
- X - RESEARCH
- Y - PROTOTYPE

U.S. MILITARY AIRCRAFT DESIGNATION SYMBOLS

DEPARTMENT OF DEFENSE DOCUMENT DOD 4120.15 PRESENTS THE STANDARD AIRCRAFT DESIGNATION SYSTEM APPLICABLE TO ALL MILITARY SERVICES. THIS SYSTEM CONSISTS OF A SERIES OF LETTERS AND NUMBERS THAT DESCRIBE THE BASIC MISSION/TYPE, MODIFIED MISSION, DESIGN NUMBER AND SERIES LETTER. THIS SYSTEM HAS BEEN IN EFFECT SINCE 1976. MOST AIRCRAFT ALSO HAVE A COMMON NAME, FOR EXAMPLE THE 'FIGHTING-FALCON'.

EXAMPLE
USAF/McDONNELL-DOUGLAS TF-16N "FIGHTING-FALCON"

(THE TF-16N IS A TRAINER AIRCRAFT DERIVATIVE OF A BASIC FIGHTER AIRCRAFT)

U.S. MILITARY ENGINE DESIGNATION SYMBOLS

MILITARY STANDARD MIL-STD-879 DEFINES THE U.S. MILITARY ENGINE DESIGNATION SYSTEM AS CONSISTING OF THREE PARTS, AS FOLLOWS: 1) TYPE INDICATOR, 2) MANUFACTURER'S SYMBOL, 3) MODEL INDICATOR.
1) TYPE INDICATOR - CONSISTS OF A TYPE-LETTER TOGETHER WITH A TYPE-NUMBER. TYPE LETTER "J" INDICATING TURBOJET, "T" INDICATING TURBOSHAFT OR TURBOPROP, "F" INDICATING TURBOFAN.
   TYPE NUMBER - IS ASSIGNED BY THE MILITARY SERVICE WITH 100 SERIES USED BY THE U.S. AIR FORCE, 400 SERIES USED BY THE U.S. NAVY, 700 & 800 SERIES USED BY THE U.S. ARMY.
2) MANUFACTURER'S SYMBOL - IS IDENTIFIED BY A TWO-LETTER SET ASSIGNED BY THE DEPARTMENT OF DEFENSE.
3) MODEL INDICATOR - CONSISTS OF A MODEL NUMBER OFTEN FOLLOWED BY A SUFFIX LETTER. MODEL NUMBERS ALSO START WITH 100 & 200 FOR THE U.S. AIR FORCE, 400 FOR THE U.S. NAVY, AND 700 & 800 FOR THE U.S. ARMY.

NOTE-1 - IT IS POSSIBLE TO SEE A TYPE NUMBER USED IN CONJUNCTION WITH ANOTHER SERVICE MODEL INDICATOR. EXAMPLE: F110-GE-400 BASIC USAF ENGINE ALSO PRODUCED FOR THE USN.
NOTE-2 - AN ENGINE PRESENTLY DESIGNATED WITH A TYPE INDICATOR AND MANUFACTURER'S SYMBOL WILL KEEP THIS DESIGNATION FOR THE REMAINDER OF ITS SERVICE LIFE.

EXAMPLE
PRATT AND WHITNEY F-100 TURBOFAN ENGINE

(THE EXAMPLE ENGINE IS A TURBOFAN TYPE, PRODUCED FOR THE USAF AND USED BY THE USAF)

U.S. MILITARY ASSIGNED MANUFACTURER'S SYMBOLS

AD ALLISON TURBINE ENGINES	**GE** GENERAL ELECTRIC	**RR** ROLLS ROYCE
CA TELEDYNE-CONTINENTAL	**LD** TEXTRON-LYCOMING	**WA** CURTIS WRIGHT
CF CFM INTERNATIONAL	**LHT** LIGHT HELICOPTER TURB. CO.	**WR** WILLIAMS RESEARCH
CP PRATT & WHITNEY CANADA	**PR** PRATT & ROLLS	**WV** PRATT & WHITNEY W.VA.
GA GARRETT	**PW** PRATT & WHITNEY	

INDEX
ENGINES, AIRCRAFT,
MANUFACTURERS

Author's note: In this Index, an attempt has been made to include every Gas Turbine Engine and every Gas Turbine powered Aircraft from the early beginnings of this industry in the 1930's to the present time.

KEY: (A) - BEFORE PAGE NUMBER INDICATES AN AIRCRAFT DESIGNATION.

(E) - BEFORE PAGE NUMBER INDICATES AN ENGINE DESIGNATION.

A

A-1 (AM-X)	(A) 255
A-1 (Ching-Kuo)	(A) 256
A-3	(A) 256
A3D (Skywarrior, early aircraft)	(A) 438
A3J (Vigilante, early aircraft)	(A) 438
A-4 Series	(A) 270
A-4 (Skyhawk, early aircraft)	(A) 438
A-5 (China)	(A) 256
A-5 (Vigilante)	(A) 257
A-5 (Vigilante, early aircraft)	(A) 438
A-6 Series	(A) 270
A-7 Series	(A) 270
A-7A (early aircraft)	(A) 439
A-10A	(A) 270
A-18	(A) 270
A-33	(A) 270
A-37	(A) 270
A-40	(A) 274
A-50	(A) 275
A.90	(A) 277
A-109	(A) 263
A-129 (Mangusta)	(A) 263
A 300 (Series)	(A) 256
A 310	(A) 256
A 320	(A) 256
A 321	(A) 257
A 330 (Series)	(A) 257
A 340	(A) 257
AB 205 (Iroquois)	(A) 263
AB 206 (Jet Ranger)	(A) 263
AB 212	(A) 263
AB 412	(A) 263
AC-130	(A) 270
ACADEME	(A) 273
ADDER (early turbojet)	(E) 438
ADDRESS-Aircraft Manufacturers	253
ADOUR	(E) 273
ADOUR 811/815	(E) 69, 152
ADOUR 871	(E) 153
ADP	(E) 227
AERITALIA (Alenia) SPA.	259, 270
AERMACCHI SPA.	259
AERO BUREAU (USSR)	274
AERO-COMMANDER	(A) 266
AERO L-29	(A) 256, 276
AERO L-39	(A) 256
AERO INDUSTRIES (Taiwan)	256
AERO MODS INC.	265
AERO SCOUT	(A) 269
AEROSPACE CENTER (Taiwan)	147
AERO SPACELINE CO.	256
AEROSPATIALE CORP.	256, 262, 268
AF-1	(A) 270
AGRICULTURAL AIRCRAFT	264
AG-CAT	(A) 264
AGUSTA SPA.	259, 263
AGUSTA (early aircraft)	(A) 438, 439
AH-1 (Lynx)	(A) 262
AH-1 Series	(A) 268
AH-6	(A) 269
AH-47	(A) 269
AH-56 (Cheyenne)	(A) 269
AH-64 (Apache)	(A) 269
AIRBORNE COMMAND POST	(A) 271
AI-20 (early turbojet)	(E) 438
AI-20M	(E) 129
AI-24 (early turboprop)	(E) 439
AI-24VT	(E) 130
AI-25A	(E) 131
AIRBUS INDUSTRIES	256
AIRCRAFT LISTINGS	255
AIRCRAFT Manufacturers	253
AIRCRAFT Designation Symbols	448
AIRCRAFT Mission Symbols	448
AIRLINER (B-99)	(A) 265
AIRTEC CORP.	265
AIRTECH CORP.	261
AIR TRACTOR INC.	264
AJ37 (Viggen)	(A) 261
AJEET	(A) 258
AKROBAT (L-29)	(A) 256
AL-7 (early turbojet)	(E) 438
AL-7F-1-100	(E) 121
AL-21 (early turbojet)	(E) 439
AL-21F-3	(E) 122
AL-31F	(E) 140
AL-34	(E) 122
AL5512	(E) 207
ALBATROS (L-39Z)	(A) 256, 276
ALBATROSS (BE-42)	(A) 274
ALF502L	(E) 208

ALFA ROMERO AVIO SPA.

AR.318	(E) 73
J85-GE-13A	(E) 74
T58-GE-10	(E) 75

ALH	(A) 263
ALLIED SIGNAL CORPORATION	9
ALLISON (early engines)	437,8,9

ALLISON TURBINE DIVISION (GM)

ALLISON 250-B17	(E) 159
ALLISON 250-C20R	(E) 160
ALLISON 250-C34	(E) 161
ALLISON 250-MTU-C20B	(E) 45
ALLISON 501 (early turboprop)	(E) 439
ALLISON 501-D22	(E) 62
ALLISON 501-M78	(E) 162
ALLISON 501-M80	(E) 162
GMA-2100	(E) 163
GMA-3007	(E) 164
T56-A-720	(E) 165
T63-A-700	(E) 166
T406-AD-400	(E) 167
T703-A-700	(E) 168
TF41-A-400	(E) 169
ALIZE	(A) 257
ALIZE (early aircraft)	(A) 438
ALOUETTE	(A) 262, 264
ALPHA JET	(A) 257
AM-3/5/9 (early turbojet)	(E) 438
AM-3	(E) 123
AMPHIBIAN (CL-215)	(A) 255
AM-X	(A) 255, 259
AN-10 (Cat, early aircraft)	(A) 438
AN-12	(A) 275
AN-12 (Cat-B, early aircraft)	(A) 438
AN-22	(A) 275, 277
AN-24	(A) 277
AN-24 (early aircraft)	(A) 439
AN-26	(A) 275, 277
AN-28	(A) 260, 277
AN-30	(A) 275, 277
AN-32	(A) 275, 277
AN-38	(A) 277
AN-72	(A) 275, 277
AN-74	(A) 275, 277
AN-120	(A) 277
AN-124	(A) 275, 277
AN-128	(A) 275
AN-180	(A) 277
AN-218	(A) 277
AN-225	(A) 275, 277
ANDOVER (BAe 748)	(A) 258
ANTONOV BUREAU	274
AP 68 (Sparticus)	(A) 259
AP 68 (Viator)	(A) 259
APACHE (AH-64)	(A) 269
APU's	11
AR-237 (early aircraft)	(A) 437
AR.318	(E) 73
ARA-3600	(A) 255
ARADO (early aircraft)	(A) 437
ARAVA (IAI)	(A) 258
ARGOSY (AW-660)	(A) 257
ARGOSY (early aircraft)	(A) 438
ARMSTRONG (early aircraft)	437,8,9
ARMSTRONG-SIDDELEY (early engines)	437,8
AROCET INC.	273
ARRIEL	(E) 35
ARRIUS (TM-319)	(E) 42
ARTOUSTE 3B	(A) 263
ARTOUSTE (early turboshaft)	(E) 439
AS 332 (Super Puma)	(A) 262
AS 350 (Ecureuil)	(A) 262
AS 355F (TwinStar)	(A) 262
AS 355N (Ecureuil-II)	(A) 262
AS 365 (Dauphin)	(A) 262
AS 365N (Dauphin-II)	(A) 262
AS 366 (Dauphin-II)	(A) 262
AS 532 (Cougar)	(A) 262
AS 550 (Fennecs)	(A) 262
AS 565 (Panther)	(A) 262
AS series also see "SA"	
ASH-3	(A) 263
ASP/ASX (early turboprop)	(E) 437
ASPIN (early turbojet)	(E) 438
ASTAFAN IV	(E) 37
ASTAZOU (early turboshaft)	(E) 439
ASTAZOU	(E) 38,70
ASTAZOU 16G	(E) 39
ASTRA (1125)	(A) 259, 266
ASTRA JET CORP.	259, 266
ASTROJET (BAC 1-11)	(A) 257, 260
ASUKA	(A) 260
AS-X (early turboprop)	(E) 437
AT-3 (Shengue)	(A) 256
AT-9	(A) 273
AT-26 (Xavante)	(A) 255
A/T-33	(A) 270
AT 400	(A) 264
AT 402	(A) 264
AT 502	(A) 264
AT 503	(A) 264
ATAR (early turbojet)	(E) 438
ATAR 9K50	(E) 32
ATF (Locate under F-22)	
ATF3-6	(E) 171
ATLANTIC (Atlantique)	(A) 257

ATLAS AIRCRAFT CORPORATION 260

VIPER 540	(E) 104
ATP	(A) 257
ATR 42	(A) 256, 259
ATR 72	(A) 256, 259
ATTA 3000	(A) 255
ATTACK Aircraft-USA	270
AU-24	(A) 273
AUBISQUE 1A	(E) 40
AUH-76	(A) 269
AUXILIARY Power Units	11
AV-8A	(A) 270
AV-8A (Harrier, early aircraft)	(A) 439
AVANTI	(A) 259, 266

AVIOCAR (C-212)	(A) 258, 260, 271
AVIOCAR (USAF)	(A) 271
AVIOJET (C-101)	(A) 260
AVIONS Marcel Dassault	257
AVON	(E) 256, 258
AVON (early turbojet)	(E) 438, 439
AVRO-748 (early aircraft)	(A) 438
AVTEC INC.	265
AVTEC-400	(A) 265
AW-660 (Argosy)	(A) 257
AYRES CORP.	264
AZOR (C-207)	(A) 260

B

B-1B	(A) 270
B-2A	(A) 270
B-6	(A) 256
B-7	(A) 256
B-12	(A) 274
B-17 (early aircraft)	(A) 437
B-45 (Tornado, early aircraft)	(A) 437
B-47 (Stratojet, early aircraft)	(A) 437
B-52	(A) 270
B-52 (early aircraft)	(A) 438
B-57 (early aircraft)	(A) 438
B-58 (Hustler, early aircraft)	(A) 438
B-66 (early aircraft)	(A) 438
B-99 (Airliner)	(A) 265
BO.108	(A) 262
B-367-80 (early aircraft)	(A) 439
B-502 (early engine)	(E) 437
B-707 (early aircraft)	(A) 439
B-707 Series	(A) 267
B-720 (early aircraft)	(A) 439
B-727 Series	(A) 267
B-737 Series	(A) 267
B-747 Series	(A) 267
B-757 Series	(A) 267
B-767 Series	(A) 267, 268
B-777	(A) 268
B-2707 (early aircraft)	(A) 439
BAC (early aircraft)	438, 439
BAC 1-11 (Astrojet Series)	(A) 257, 260
BAC-145 (Jet Provost)	(A) 257
BAC-167 (StrikeMaster)	(A) 257
BACKFIRE (TU-26)	(A) 274
BADGER (TU-16)	(A) 274, 275
BADGER (TU-16, early aircraft)	(A) 438
BAe 125 Series	(A) 257
BAe 125 (USAF)	(A) 271
BAe (HS-125, early aircraft)	(A) 438
BAe 146 Series	(A) 258
BAe 748 (Andover)	(A) 258
BAe 748 (early aircraft)	(A) 438
BAe 800	(A) 258

BAe 801	(A) 258
BAe 1000	(A) 258
BAe. CORP.	257
BAe J-31 (Jetstream)	(A) 258
BAe RJ-70	(A) 258
BAe RJ-85	(A) 258
BANDEIRANTE (EMB-110)	(A) 255
BANSHEE (F2H, early aircraft)	(A) 437
BR-700	(E) 59
BASLER AIRCRAFT	265
BASTAN (IA-50)	(E) 255, 256, 262
BASTAN (early engine)	(E) 439
BD-5J	(A) 266
BE-12	(A) 274
BE-30 (early aircraft)	(A) 439
BE-42 (Mermaid)	(A) 274
BE-100,200,300,350	(A) 265
BE-200 (CIS)	(A) 277
BE-1900,2000	(A) 265
BEAGLE (IL-28)	(A) 274
BEAR (TU-95)	(A) 274, 275
BEDE CORP.	266
BEECH CORP.	265, 266, 267, 270, 273
BEECH B-99	(A) 265
BEECH B-1900	(A) 265, 267
BEECHJET 400 (Diamond)	(A) 266, 270, 273
BEECH STARSHIP	(A) 265
BELFAST (Shorts)	(A) 258
BELFAST (early aircraft)	(A) 439
BELL (early aircraft)	437,8,9
BELL 47	(A) 269
BELL 200 Series (USA)	(A) 268, 269
BELL 205 (Taiwan/Turkey)	(A) 262,264
BELL 212 (Canada)	(A) 261
BELL 400 Series (USA)	(A) 269
BELL 901 (V-22)	(A) 269
BELL JAPAN	263
BELL TEXTRON Helicopter	268
BELL TURKEY	264
BELL XP-59 (early aircraft)	(A) 437
BELPHEGOR (M-15)	(A) 260
BELVEDERE (early aircraft)	(A) 438
BERIEV BUREAU	274
BET-SHEMESH ENGINES LTD.	
MARBORE-6.	(E) 71
BIG CROW (NKC-135)	(A) 271, 272
BIG LIFTER (Bell 214)	(A) 268
BISON (M-4)	(A) 274, 275
BISON (M-4, early aircraft)	(A) 438
BK.117	(A) 262, 263
BLACKBIRD (SR-71)	(A) 273
BLACKHAWK (UH-60)	(A) 263, 269
BLACKJACK (TU-160)	(A) 274
BLACK MAMBA (TR-3A)	(A) 273
BLACK TIGER (MD-500)	(A) 269
BLINDER (TU-22)	(A) 274,5,6
BMW (early engines)	437

BMW, ROLLS-ROYCE GMBH
T-53L-13	(E) 142
T117	(E) 143
BMW-003 (early turbojet)	(E) 437
BMW-018 (early turboprop)	(E) 437
BN-2T (Islander)	(A) 258
BN-2T (Defender)	(A) 258
BO.105	(A) 262
BO.108	(A) 262
BOEING AIRCRAFT	267, 270, 271
BOEING AIRCRAFT (USAF)	270, 271
BOEING (early aircraft)	437,8,9
BOEING (early engines)	437
BOEING 107/145/234/414	(A) 269
BOEING 412 (Japan)	(A) 263
BOEING CANADA	255, 271
BOEING HELICOPTER	269
BOEING VERTOL CORP.	261
BOEING-VERTOL (early aircraft)	(A) 439
BOMBARDIER CORP.	266
BOMBER AIRCRAFT-USA	270
BONANZA (Beech)	(A) 265
BOUNDER (M-50)	(A) 274
BOUNDER (TU-16, early aircraft)	(A) 438
BR-1050	(A) 257
BR-2004 (Alize)	(A) 265
BRASILIA (EMB-120)	(A) 255
BREDA NARDI (INDIA)	263
BREGUET (early aircraft)	(A) 438,9
BRENDA 369 (NH 500D)	(A) 263
BREWER (Yak-38)	(A) 275, 276
BRISTOL (early aircraft)	438,9
BRISTOL-AERO (early engines)	437,8,9
BRISTOL-SIDDELLEY (early engines)	437,8,9
BRITANNIA	(A) 258
BRITANNIA (early aircraft)	(A) 438
BRITISH AEROSPACE (BAe)	257, 270, 271, 273
BROMON AIRCRAFT	265
BRONCO (OV-10)	(A) 272
BUCCANEER	(A) 258
BUCCANEER (early aircraft)	(A) 439
BUCKEYE (T-2B)	(A) 259, 273
BUFFALO (DHC-5B)	(A) 255
BUSINESS TURBOPROP Aircraft-USA	265
BUSINESS JET Aircraft-USA	266
BUSINESS JET Aircraft - INT'L	255

C

C-1 (Kawasaki)	(A) 259
C-1 (Asuka)	(A) 260
C-2 (Kfir)	(A) 259
C-2A (Grumman)	(A) 271
C-4C	(A) 271
C-5 (Galaxy)	(A) 271
C-7 (Kfir)	(A) 259
C-8A	(A) 271
C-9	(A) 271
C-12	(A) 270
C-17	(A) 271
C-20 Series	(A) 271
C-21A	(A) 271
C-22B	(A) 270
C-22J	(A) 259
C-22J (Vantura)	(A) 259
C-23	(A) 271
C-26A	(A) 271
C-27	(A) 270
C-29	(A) 271
C-33B	(A) 271
C-101 (Aviojet)	(A) 260
C-130	(A) 271
C-130 (early aircraft)	(A) 439
C-131	(A) 271
C-133 (early aircraft)	(A) 439
C-135 Series	(A) 270
C-140	(A) 271
C-141	(A) 271
C-212 (Aviocar)	(A) 260, 271
C-160 (Transall)	(A) 257
C 207 (Azor)	(A) 260
C-550	(A) 271
CA-27 (early aircraft)	(A) 438
CAC 100	(A) 259
CAMBER (IL-86)	(A) 277
CAMEL (TU-104)	(A) 274, 438
CANADA BELL CORP.	261
CANADA-Engine Companies	1
CANADA IMP COMPANY	255
CANADAIR LTD.	255
CANADAIR (early aircraft)	438
CANADAIR-RJ	(A) 255
CANBERRA	(A) 258, 273
CANBERRA (early aircraft)	(A) 438
CANDID (IL-76)	(A) 275, 277
CANGURO	(A) 259
CAPRONI SPA.	259
CARAJA (NE-821)	(A) 255
CARAVAN (U-27)	(A) 257, 273
CARAVAN-I	(A) 265, 267
CARAVAN-II	(A) 257, 265
CARAVELLE (early aircraft)	(A) 438
CARAVELLE	(A) 256
CARELESS (TU-154)	(A) 278
CARGO Commercial Aircraft-USA	267
CARGO LINER (VOPAR)	(A) 266
CARGO Military Aircraft-USA	270
CARGOMASTER (205B)	(A) 265
CARIBOU (early aircraft)	(A) 438
CARIBOU	(A) 255, 258, 271
CASA (SPAIN)	260, 271
CASH (AN-28)	(A) 277
CAT (AN-120)	(A) 277
CAT (early aircraft)	(A) 438

CATIC (PRC) ... (Locate under CAREC)	
CAYUSE (MD-500)	(A) 269
CBA-123 (Vector)	(A) 255
CC-115 (DHC-5A)	(A) 255
CD-2 (Seastar)	(A) 257

CENTRAL NATIONAL AERONAUTICE

SPEY 512-14DW	(E) 101
VIPER 632	(E) 102
TURMO IVC	(E) 103
CESSNA .. 264, 265, 266, 267, 270, 272, 273	
CESSNA 205 (CargoMaster)	(A) 265
CESSNA 206/207	(A) 266
CESSNA 208 (Caravan-I)	(A) 265
CESSNA 208B (Grand Caravan)	(A) 265
CESSNA 402	(A) 159
CESSNA 406 (Caravan-II)	(A) 265
CESSNA 425 (Conquest-II)	(A) 265
CESSNA 441 (Conquest-I)	(A) 265
CESSNA (Citation Series)	(A) 266
CESSNA N-22 (Nomad)	(A) 265
CF-5	(A) 255
CF6-45A2	(E) 178
CF6-50	(E) 155
CF6-50	(E) 179
CF6-80	(E) 180
CF34-1A	(E) 186
CF-100 (early aircraft)	(A) 16, 438
CF-104	(A) 255
CF700-2D-2	(E) 187
CFE-738	(E) 170
CFM-88	(E) 34

CFM INTERNATIONAL

CFM56-5C2	(E) 144
CH-3	(A) 269
CH-46	(A) 269
CH-47 (Chinook)	(A 263, 269
CH-47 (early aircraft)	(A) 439
CH-53	(A) 269
CH-53 (early aircraft)	(A) 439
CH-54	(A) 269
CH-113 (Labrador)	(A) 261
CH-118 (SH-3B)	(A) 261
CH-135 (BELL-212)	(A) 261
CHALLENGER 600	(A) 255
CHALLENGER 601	(A) 255
CHANCE-VOUGHT (early aircraft)	437,8,9
CHARGER (TU-144)	(A) 275, 278
CHEETAH (SA-315)	(A) 260, 263
CHENGDU (J-7)	(A) 256
CHENGDU (WZ-6)	(E) 94
CHETAK (SA-316)	(A) 263
CHEYENNE Series	(A) 266
CHEYENNE (AH-56)	(A) 269
CHINA (TAIWAN)	41

CHINA NATIONAL ENGINE CORP. (CAREC) ... 5

CHENGDU WP-7A	(E) 94
HARBIN WJ-5	(E) 94
HARBIN WZ-8	(E) 94
LIYANG WP-7B	(E) 94
LIYANG WP-13A	(E) 94
SHENYANG WP-6A	(E) 94
SHENYANG WS-6	(E) 94
XIAN WP-8	(E) 94
XIAN SPEY 202	(E) 94
ZHUZHOW WJ-6	(E) 94
CHINA-National Aircraft Factory	261
CHING-KUO	(A) 256
CHINOOK (CH-47)	(A) 263, 269
CHS-2 (Rooivalk)	(A) 264
CHUNG-HSING (T-HC-1)	(A) 256
CINAIR CORP.	260
CIS Aircraft (Locate under USSR)	
CIS Engine Manufacturers (Locate under USSR)	
CITATION Series	(A) 266, 271
CITATION (USAF)	(A) 271, 273
CJ610-8	(E) 188
CJ-805 (early turbojet)	(E) 439
CJ805-3	(E) 181
CJ-805-23 (early turbofan)	(E) 439
CJ805-23	(E) 182
CL-13 (Sabre, early aircraft)	(A) 437
CL-41A (Tutor)	(A) 255
CL-41 (early aircraft)	(A) 439
CL-44 (Yukon)	(A) 255
CL-44 (early aircraft)	(A) 439
CL-84 (early tilt-wing aircraft)	(A) 439
CL-215T (Amphibian)	(A) 255
CLANK (AN-50)	(A) 275, 277
CLASSIC (IL-62)	(A) 275, 277
CLEAT (TU-114/126)	(A) 275, 278
CLEAT (TU-144, early aircraft)	(A) 438
CLINE (AN-52)	(A) 275, 277
CLOBBER (YAK-42)	(A) 278
CLYDE (early turboprop)	(E) 437
CM-170 (Magister)	(A) 256
CM-170 (early aircraft)	(A) 438
CN-235	(A) 258, 260
COALER (AN-72/74)	(A) 275, 277
COBRA (BELL-209)	(A) 268
COCK (AN-22)	(A) 275, 277
CODING (YAK-40)	(A) 278
COKE (AN-24)	(A) 277
COMBAT SCOUT (BELL 406)	(A) 269
COMET (early aircraft)	(A) 438, 439
COMANCHE (RAH-66)	(A) 269
COMANCHERO 500	(A) 266
COMANCHERO 750	(A) 266
COMMANDER 680	(A) 266
COMMANDER 690	(A) 266
COMMANDER 840	(A) 265, 266
COMMANDER 980	(A) 265, 266
COMMANDER 1000	(A) 265, 266
COMMANDER (II21)	(A) 258
COMMANDO (Helicopter)	(A) 262
COMMERCIAL Passenger Aircraft-USA	268
COMMERCIAL Passenger Aircraft-INT'L	255
COMMODORE (1123)	(A) 259

COMMONWEALTH OF INDEPENDENT STATES (SEE USSR)	
COMMUTER (340)	(A) 261
COMMUTER Aircraft Company	259
CONCORDE	(A) 256, 258
CONDOR (AN-124)	(A) 275, 277
CONQUEST 1/2	(A) 265
CONSTRUCCIONES AERO SPA.	260
CONTINENTAL-217 (early turbojet)	(E) 439
CONVAIR CORP.	267, 268, 271
CONVAIR-240,340,440 (early aircraft)	(A) 438
CONVAIR 340	(A) 271
CONWAY (early turbofan)	(E) 439
CONWAY 43	(E) 53
COOKPOT (TU-124)	(A) 278
COOT (IL-18)	(A) 275, 277
COOT (IL-18, early aircraft)	(A) 437, 438
CORONADO (CV-990, early aircraft)	(A) 439
CORSAIR-II	(A) 270, 271, 273
CORVETTE (SN-601)	(A) 256
COUGAR (AS-532)	(A) 262
COUGAR (F9, early aircraft)	(A) 437
COX AIRCRAFT	265
CRUSADER (F-8)	(A) 270
CRUSADER (F-8, early aircraft)	(A) 438
CRUSTY (USSR)	(A) 278
CS-1 (early turboprop)	(E) 437
CT4-CR	(A) 260
CT7-6	(E) 189
CT7-9	(E) 190
CT-39	(A) 271
CT58 (early turboshaft)	(E) 439
CT58-140	(E) 191
CT58-IHI-140	(E) 83
CT63-M-5A	(E) 92
CT64-800-4	(E) 192
CT-114 (Tutor)	(A) 255
CT-142 (DHC-8)	(A) 255
CT700-700	(E) 84
CTS-800	(E) 202
CU-21	(A) 271
CUB (AN-12)	(A) 275
CURL (AN-26)	(A) 275, 277
CUTLASS (F7U, early aircraft)	(A) 437
CV-7	(A) 255
CV-240 (early aircraft)	(A) 437
CV-540/580	(A) 267, 268
CV-600/640	(A) 267, 268
CV-880 (early aircraft)	(A) 439
CV-990	(A) 268
CV-990 (early aircraft)	(A) 439
CZEC Aviation Industries	256
CZECHOSLOVAKIA-Engine Companies	1

D

D-15	(E) 140
D-15 (early turbojet)	(E) 438
D-18T	(E) 132
D-20 (early turbojet)	(E) 439
D-20	(E) 124
D-25V	(E) 125
D-25 (early turboshaft)	(E) 439
D-27	(E) 140, 277
D-30	(E) 126
D-30F	(E) 140
D-30-KP	(E) 127
D-36	(E) 140
D-136	(E) 133
D-227	(E) 140
D-236	(E) 134
D-436T	(E) 134
DART (early turboprop)	(E) 438
DART MK.540	(E) 54
DASH-7	(A) 255
DASH-8	(A) 255, 271
DASSAULT (early aircraft)	437,8,9
DASSAULT-BREGUET	257
DAUPHIN (AS-365/66)	(A) 262
DC-3-65TP	(A) 266
DC-3 (Turbo)	(A) 265
DC-8 (early aircraft)	(A) 439
DC-8 Series	(A) 267, 268
DC-9 Series	(A) 267, 268
DC-10 Series	(A) 267, 268
DEE HOWARD INC.	257
DEFENDER (BN-2T)	(A) 258
DEFENDER (MD-530)	(A) 269
DeHAVILLAND AIRCRAFT	419,421
DeHAVILLAND/BOEING CANADA	255
DeHAVILLAND (early aircraft)	437,8,9
DELFIN (L-29)	(A) 256, 275, 276
DELTA DAGGER (F-102)	(A) 272
DELTA DAGGER (early aircraft)	(A) 438
DELTA DART (F-106)	(A) 272
DELTA DART (early aircraft)	(A) 438
DEMON (F3H, early aircraft)	(A) 438
DERWENT-1 (early engine)	(E) 437
DESTROYER (RB-66, early aircraft)	(A) 439
DH.108 (Swallow, early aircraft)	(A) 437
DHC (early aircraft)	(A) 439
DHC-2 (Turbo Beaver)	(A) 255
DHC-4 (Caribou)	(A) 258
DHC-5A (Caribou)	(A) 255
DHC-5D (Buffalo)	(A) 255
DHC-5E (Transporter)	(A) 255
DHC-6 (Twin Otter)	(A) 255
DHC-7 (DASH-7)	(A) 255
DHC-8 (DASH-8)	(A) 255, 271
DHC-8 (Triton)	(A) 255
DHC-8-300	(A) 255
DIAMOND I/II	(A) 259, 266
DO 128 (Sky Servant)	(A) 257
DO 228	(A) 257
DO 328	(A) 257
DOLPHIN (HH-65)	(A) 268
DORNIER GMBH.	257, 262

DOUBLE MAMBA (early turboprop)	(E) 438
DOUGLAS (early aircraft)	438,9
DRAGONFLY (A-37)	(A) 270,272
DRAKEN	(A) 260
DRAKEN (early aircraft)	(A) 438
DREAM (AN-255)	(A) 275
DV-2	(E) 29

E

E-2C	(A) 271
E-2C (Hawkeye, early aircraft)	(A) 439
E-3 Series	(A) 271
E-4	(A) 271
E-6A	(A) 271
E-8A	(A) 271
E-9A	(A) 271
E-28 (early aircraft)	(A) 342, 437
EA-6A/B	(A) 272
EA-6A (early aircraft)	(A) 439
EA-7D	(A) 271
EAGLE (F-15)	(A) 259, 269, 272
EAGLE (S-76)	(A) 269
EARLY TURBINE ENGINES	437
EARLY WARNING AIRCRAFT-USA	271
EC-1 (Japan)	(A) 259
EC-18A	(A) 271
EC-24	(A) 272
EC-130	(A) 271
EC-135 (B.108)	(A) 262
EC-135A	(A) 271
ECUREUIL (AS-350)	(A) 262
EF-111A	(A) 272
EFA (EuroJet)	(A) 257
EGYPT AVIATION	262
EH INDUSTRIES	262, 263
EH 101 (Heliliner)	(A) 262
EH 101 (Merlin)	(A) 262
EJ-200	(E) 145
EKRANOPLANE	(A) 277
ELAND (early turboshaft)	(A) 438
ELECTRA	(A) 267, 268
ELECTRA (L-188, early aircraft)	(A) 439
ELECTRONIC WARFARE AIRCRAFT-USA	271
EMB-110 (Bandeirante)	(A) 255
EMB-111 (Bandeirante)	(A) 255
EMB-120 (Brasilia)	(A) 255
EMB-121 (Xingu)	(A) 255
EMB-145	(A) 255
EMB-312 (Tucano)	(A) 255
EMB-326	(A) 255
EMBRAER CORP. (Brazil)	255
ENSTROM HELICOPTER	269
ENSTROM 480	(A) 269
EQUATOR GMBH.	257
ETENDARD	(A) 257
ETENDARD (early aircraft)	(A) 439

EUROCOPTER GMBH.	262
EUROFIGHTER	(A) 257
EUROJET TURBO GMBH	
EJ 200	(E) 145
EXEC-LINER (Beech)	(A) 265
EXHAUST DUCT	325
EXPEDITER (SA-227)	(A) 267
EXPLORER (MD-900)	(A) 269
EXTENDER (KC-10)	(A) 271

F

F-1 (Japan)	(A) 259
F-2	(E) 50
F-2/4 (early turbojet)	(E) 437
F2D (early aircraft)	(A) 437
F2H (Banshee, early aircraft)	(A) 437
F3D (Skyknight, early aircraft)	(A) 437
F3H (Demon, early aircraft)	(A) 438
F3-IHI-30	(E) 86
F4D (Skyray, early aircraft)	(A) 438
F-4 Series	(A) 272
F-4EJ (Phantom-II)	(A) 259
F-4 (Phantom, early aircraft)	(A) 438
F-5 (China)	(A) 256
F-5 Series	(A) 270, 272
F-6 (China)	(A) 256
F6U (early aircraft)	(A) 437
F7U (Cutlass, early aircraft)	(A) 437
F-8	(A) 270
F-8 (China)	(A) 256
F-8 (early aircraft)	(A) 438
F9/40 (early aircraft)	(A) 437
F9F (early aircraft)	(A) 437, 438
F-10 (early aircraft)	(A) 437
F-11F (Tiger, early aircraft)	(A) 438
F-14 Series	(A) 272
F-15J (Eagle)	(A) 259
F-15 Series	(A) 272
F-16 Series	(A) 272
F-18 Series	(A) 270, 272
F-20	(A) 272
F-21A	(A) 272
F-22A	(A) 272
F-27	260, 271
F-27 (Friendship, early aircraft)	(A) 438
F-27 (USAF)	(A) 271
F-28 (Fellowship)	(A) 260
F-50/100	(A) 260
F-84 (ThunderJet, early aircraft)	(A) 438
F-84F (ThunderStreak, early aircraft)	(A) 438
F-84K (Thunderflash, early aircraft)	(A) 438
F-86H (Sabre, early aircraft)	(A) 437
F-89 Scorpion (early aircraft)	(A) 437
F-94 (Starfire, early aircraft)	(A) 438
F-100 (Super-Sabre, early aircraft)	(A) 438

COMMONWEALTH OF INDEPENDENT STATES (SEE USSR)	
COMMUTER (340)	(A) 261
COMMUTER Aircraft Company	259
CONCORDE	(A) 256, 258
CONDOR (AN-124)	(A) 275, 277
CONQUEST 1/2	(A) 265
CONSTRUCCIONES AERO SPA.	260
CONTINENTAL-217 (early turbojet)	(E) 439
CONVAIR CORP.	267, 268, 271
CONVAIR-240,340,440 (early aircraft)	(A) 438
CONVAIR 340	(A) 271
CONWAY (early turbofan)	(E) 439
CONWAY 43	(E) 53
COOKPOT (TU-124)	(A) 278
COOT (IL-18)	(A) 275, 277
COOT (IL-18, early aircraft)	(A) 437, 438
CORONADO (CV-990, early aircraft)	(A) 439
CORSAIR-II	(A) 270, 271, 273
CORVETTE (SN-601)	(A) 256
COUGAR (AS-532)	(A) 262
COUGAR (F9, early aircraft)	(A) 437
COX AIRCRAFT	265
CRUSADER (F-8)	(A) 270
CRUSADER (F-8, early aircraft)	(A) 438
CRUSTY (USSR)	(A) 278
CS-1 (early turboprop)	(E) 437
CT4-CR	(A) 260
CT7-6	(E) 189
CT7-9	(E) 190
CT-39	(A) 271
CT58 (early turboshaft)	(E) 439
CT58-140	(E) 191
CT58-IHI-140	(E) 83
CT63-M-5A	(E) 92
CT64-800-4	(E) 192
CT-114 (Tutor)	(A) 255
CT-142 (DHC-8)	(A) 255
CT700-700	(E) 84
CTS-800	(E) 202
CU-21	(A) 271
CUB (AN-12)	(A) 275
CURL (AN-26)	(A) 275, 277
CUTLASS (F7U, early aircraft)	(A) 437
CV-7	(A) 255
CV-240 (early aircraft)	(A) 437
CV-540/580	(A) 267, 268
CV-600/640	(A) 267, 268
CV-880 (early aircraft)	(A) 439
CV-990	(A) 268
CV-990 (early aircraft)	(A) 439
CZEC Aviation Industries	256
CZECHOSLOVAKIA-Engine Companies	1

D

D-15	(E) 140
D-15 (early turbojet)	(E) 438
D-18T	(E) 132
D-20 (early turbojet)	(E) 439
D-20	(E) 124
D-25V	(E) 125
D-25 (early turboshaft)	(E) 439
D-27	(E) 140, 277
D-30	(E) 126
D-30F	(E) 140
D-30-KP	(E) 127
D-36	(E) 140
D-136	(E) 133
D-227	(E) 140
D-236	(E) 134
D-436T	(E) 134
DART (early turboprop)	(E) 438
DART MK.540	(E) 54
DASH-7	(A) 255
DASH-8	(A) 255, 271
DASSAULT (early aircraft)	437,8,9
DASSAULT-BREGUET	257
DAUPHIN (AS-365/66)	(A) 262
DC-3-65TP	(A) 266
DC-3 (Turbo)	(A) 265
DC-8 (early aircraft)	(A) 439
DC-8 Series	(A) 267, 268
DC-9 Series	(A) 267, 268
DC-10 Series	(A) 267, 268
DEE HOWARD INC.	257
DEFENDER (BN-2T)	(A) 258
DEFENDER (MD-530)	(A) 269
DeHAVILLAND AIRCRAFT	419,421
DeHAVILLAND/BOEING CANADA	255
DeHAVILLAND (early aircraft)	437,8,9
DELFIN (L-29)	(A) 256, 275, 276
DELTA DAGGER (F-102)	(A) 272
DELTA DAGGER (early aircraft)	(A) 438
DELTA DART (F-106)	(A) 272
DELTA DART (early aircraft)	(A) 438
DEMON (F3H, early aircraft)	(A) 438
DERWENT-1 (early engine)	(E) 437
DESTROYER (RB-66, early aircraft)	(A) 439
DH.108 (Swallow, early aircraft)	(A) 437
DHC (early aircraft)	(A) 439
DHC-2 (Turbo Beaver)	(A) 255
DHC-4 (Caribou)	(A) 258
DHC-5A (Caribou)	(A) 255
DHC-5D (Buffalo)	(A) 255
DHC-5E (Transporter)	(A) 255
DHC-6 (Twin Otter)	(A) 255
DHC-7 (DASH-7)	(A) 255
DHC-8 (DASH-8)	(A) 255, 271
DHC-8 (Triton)	(A) 255
DHC-8-300	(A) 255
DIAMOND I/II	(A) 259, 266
DO 128 (Sky Servant)	(A) 257
DO 228	(A) 257
DO 328	(A) 257
DOLPHIN (HH-65)	(A) 268
DORNIER GMBH.	257, 262

DOUBLE MAMBA (early turboprop)	(E) 438
DOUGLAS (early aircraft)	438,9
DRAGONFLY (A-37)	(A) 270,272
DRAKEN	(A) 260
DRAKEN (early aircraft)	(A) 438
DREAM (AN-255)	(A) 275
DV-2	(E) 29

E

E-2C	(A) 271
E-2C (Hawkeye, early aircraft)	(A) 439
E-3 Series	(A) 271
E-4	(A) 271
E-6A	(A) 271
E-8A	(A) 271
E-9A	(A) 271
E-28 (early aircraft)	(A) 342, 437
EA-6A/B	(A) 272
EA-6A (early aircraft)	(A) 439
EA-7D	(A) 271
EAGLE (F-15)	(A) 259, 269, 272
EAGLE (S-76)	(A) 269
EARLY TURBINE ENGINES	437
EARLY WARNING AIRCRAFT-USA	271
EC-1 (Japan)	(A) 259
EC-18A	(A) 271
EC-24	(A) 272
EC-130	(A) 271
EC-135 (B.108)	(A) 262
EC-135A	(A) 271
ECUREUIL (AS-350)	(A) 262
EF-111A	(A) 272
EFA (EuroJet)	(A) 257
EGYPT AVIATION	262
EH INDUSTRIES	262, 263
EH 101 (Heliliner)	(A) 262
EH 101 (Merlin)	(A) 262
EJ-200	(E) 145
EKRANOPLANE	(A) 277
ELAND (early turboshaft)	(A) 438
ELECTRA	(A) 267, 268
ELECTRA (L-188, early aircraft)	(A) 439
ELECTRONIC WARFARE AIRCRAFT-USA	271
EMB-110 (Bandeirante)	(A) 255
EMB-111 (Bandeirante)	(A) 255
EMB-120 (Brasilia)	(A) 255
EMB-121 (Xingu)	(A) 255
EMB-145	(A) 255
EMB-312 (Tucano)	(A) 255
EMB-326	(A) 255
EMBRAER CORP. (Brazil)	255
ENSTROM HELICOPTER	269
ENSTROM 480	(A) 269
EQUATOR GMBH.	257
ETENDARD	(A) 257
ETENDARD (early aircraft)	(A) 439

EUROCOPTER GMBH.	262
EUROFIGHTER	(A) 257
EUROJET TURBO GMBH	
EJ 200	(E) 145
EXEC-LINER (Beech)	(A) 265
EXHAUST DUCT	325
EXPEDITER (SA-227)	(A) 267
EXPLORER (MD-900)	(A) 269
EXTENDER (KC-10)	(A) 271

F

F-1 (Japan)	(A) 259
F-2	(E) 50
F-2/4 (early turbojet)	(E) 437
F2D (early aircraft)	(A) 437
F2H (Banshee, early aircraft)	(A) 437
F3D (Skyknight, early aircraft)	(A) 437
F3H (Demon, early aircraft)	(A) 438
F3-IHI-30	(E) 86
F4D (Skyray, early aircraft)	(A) 438
F-4 Series	(A) 272
F-4EJ (Phantom-II)	(A) 259
F-4 (Phantom, early aircraft)	(A) 438
F-5 (China)	(A) 256
F-5 Series	(A) 270, 272
F-6 (China)	(A) 256
F6U (early aircraft)	(A) 437
F7U (Cutlass, early aircraft)	(A) 437
F-8	(A) 270
F-8 (China)	(A) 256
F-8 (early aircraft)	(A) 438
F9/40 (early aircraft)	(A) 437
F9F (early aircraft)	(A) 437, 438
F-10 (early aircraft)	(A) 437
F-11F (Tiger, early aircraft)	(A) 438
F-14 Series	(A) 272
F-15J (Eagle)	(A) 259
F-15 Series	(A) 272
F-16 Series	(A) 272
F-18 Series	(A) 270, 272
F-20	(A) 272
F-21A	(A) 272
F-22A	(A) 272
F-27	260, 271
F-27 (Friendship, early aircraft)	(A) 438
F-27 (USAF)	(A) 271
F-28 (Fellowship)	(A) 260
F-50/100	(A) 260
F-84 (ThunderJet, early aircraft)	(A) 438
F-84F (ThunderStreak, early aircraft)	(A) 438
F-84K (Thunderflash, early aircraft)	(A) 438
F-86H (Sabre, early aircraft)	(A) 437
F-89 Scorpion (early aircraft)	(A) 437
F-94 (Starfire, early aircraft)	(A) 438
F-100 (Super-Sabre, early aircraft)	(A) 438

F100-IHI-100	(E) 85
F100-PW-100/200 series	(E) 230
F-101	(A) 272
F-101 (early aircraft)	(A) 438
F101-GE-102	(E) 193
F104-GA-100 (ATF3)	(E) 171
F-102	(A) 272
F-102 (early aircraft)	(A) 438
F102-LD-100	(E) 208
F103-GE-100	(E) 179
F-104	(A) 272
F-104 (early aircraft)	(A) 438
F-104S (Italy)	(A) 259
F-105	(A) 272
F-105 (early aircraft)	(A) 439
F105-PW-100	(E) 224
F-106	(A) 272
F-106 (early aircraft)	(A) 439
F108-CF-100	(E) 144
F109-GA-100	(E) 172
F110-GE-100	(E) 194
F110-GE-129	(E) 194
F110-GE-400	(E) 107, 194
F-111	(A) 272
F113-RR-100 (Spey)	(E) 59, 271
F-117 (Stealth Fighter)	(A) 272
F117-PW-100	(E) 231
F118-GE-100	(E) 195
F119-PW-100	(E) 272
F124-GA-101	(E) 173
F402-RR-406	(E) 51
F404-GE-400	(E) 196
F414-GE-400	(E) 196
F405-RR-100/400	(E) 152, 273
F406 (Caravan)	(A) 257
F-1300 (Jet Squalus)	(A) 255
FACEPLATE (MIG-21, early aircraft)	(A) 438
FAGOT (MIG-15)	(A) 276
FAGOT (MIG-15, early aircraft)	(A) 438
FAIRCHILD AIRCRAFT	265,7,8; 271, 273
FAIRCHILD (early engines)	438
FAIRCHILD-REPUBLIC	270
FAIREY (Gannett, early aircraft)	(A) 438
FAIREY (FD-2, early aircraft)	(A) 438
FALCON DASSAULT Series	(A) 257
FALCON (Northrop)	(A) 272
FALCON (Vopar)	(A) 267
FALCON (W-3 Sokol)	(A) 264
FAMA CORPORATION (Argentina)	255
FANJET (SJ-30)	(A) 267
FANTRAINER	(A) 257
FARMER (MIG-19, early aircraft)	(A) 438
FB-111	(A) 270
FB-204	(A) 263
FENCER (SU-24)	(A) 274,5
FENNES	(A) 262
FD-1 (early aircraft)	(A) 437
FD-2 (early aircraft)	(A) 438
FH-227	(A) 265
FIAT AVIAZIONE SPA.	259
J79-GE-19	(E) 76
SPEY MK.807	(E) 77
T64-P4D	(E) 78
FIAT G.80 (early aircraft)	(A) 437
FIAT G.91 (early aircraft)	(A) 439
FIAT 4002 (early turboshaft)	(E) 438
FIAT 7002 (early aircraft)	(A) 438
FIDDLER (TU-28P)	(A) 274,5
FIELDMASTER (NDN-6)	(A) 258
FIGHTER AIRCRAFT-USA	272
FIGHTER AIRCRAFT-Int'l	255
FIGHTER AIRCRAFT-USSR	274
FIGHTING FALCON	(A) 272, 273
FIREBAR (YAK-28P)	(A) 275
FIRECRACKER (NDN-1)	(A) 258
FISHBED (MIG-21)	(A) 274, 275
FISHPOT (SU-9/11)	(A) 274
FISHPOT (SU-7, early aircraft)	(A) 438
FITTER (SU-17/22)	(A) 274, 275, 276
FJ-4 (Fury, early aircraft)	(A) 438
FJ44	(E) 158
FJR-710-600S	(E) 93
FLAGON (SU-15/21)	(A) 274, 276
FLANKER (SU-27)	(A) 275
FLASHLIGHT (YAK-25, early aircraft)	(A) 438
FLETCHER 1060	(A) 265
FLOGGER (MIG-23)	(A) 274
FLYING BOAT (PS-1)	(A) 260
FLYING WING (early aircraft)	(A) 437
FOKKER CORP.	260, 264, 271
FOKKER Series	(A) 260
FORGER (YAK-38)	(A) 275, 276
FOUGA-90	(A) 37
FOUGA (early aircraft)	(A) 438
FOXBAT (MIG-25)	(A) 274, 275
FOXHOUND (MIG-31)	(A) 274
FRANCE-Engine Companies	1
FRAKES INC.	264, 265
FREEDOM FIGHTER	(A) 260, 272, 273
FREESTYLE (Yak-141)	(A) 275
FREGATE (N262)	(A) 256
FRELON (early aircraft)	(A) 438
FRELON (SA-321)	(A) 262
FRESCO (MIG-17, early aircraft)	(A) 438
FRIENDSHIP (F-28)	(A) 271
FROGFOOT (SU-25/28)	(A) 274
FT-7	(A) 256
FT-400 (FanTrainer-400)	(A) 257
FT-600 (FanTrainer-600)	(A) 257
FURY (FJ-4, early aircraft)	(A) 438
FUJI INDUSTRIES (Japan)	259, 263
FUJI T-1 (early aircraft)	(A) 438, 439
FULCRUM (MIG-29)	(A) 274

G

G-1	(A) 265
G-II, III, IV	(A) 266, 271
G-2A (Galeb)	(A) 261
G-4 (Super Galeb)	(A) 261
G-21G (Turbo Goose)	(A) 255
G.80 (early aircraft)	(A) 437
G.91 (early aircraft)	(A) 439
G.91T/Y	(A) 259
G-164 (Ag-Cat)	(A) 264
G.222	(A) 259, 270
G.222 (early aircraft)	(A) 439
GAF (Australia)	255
GALAXY (C-5)	(A) 271
GALEB (G-2A)	(A) 261

GARRETT TURBINE ENGINE COMPANY

APU'S	244
ATF3-6A	(E) 171
F109-GA-101	(E) 172
F124-GA-101	(E) 173
T76-G10	(E) 174
TFE-731-5	(E) 175
TPE-331-15	(E) 176
TPF-351-20	(E) 177

GARRETT AND ALLISON COMPANY

CFE 738	(E) 170
GAS TURBINE Engine Symbols	448
GAZELLE (early turboshaft)	(E) 438
GAZELLE (SA-342)	(A) 262, 263
GAZELLE MK.165	(E) 55
GAZELLE (SA 342)	(A) 262
GE IA/I-14/I-16 (early turbojet)	(E) 437
GE-4 (early turbojet)	(E) 439
GE-36 UDF™)	(E) 183
GE-90	(E) 180
GEM-60-3	(E) 65
GEM/RR.1004	(E) 66, 79
GEMEAUX (early aircraft)	(A) 438
GEMINI (Bell-206L)	(A) 268
GENERAL DYNAMICS CORP.	272, 273
GENERAL ELECTRIC (early engines)	437,8,9

GENERAL ELECTRIC ENGINE CO. (OHIO)

CF6-45A2	(E) 178
CF6-50E2F	(E) 179
CF6-80C2	(E) 180
CJ805-3	(E) 181
CJ805-23	(E) 182
GE-36 (UDF™)	(E) 183
GE-90	(E) 180
J79-GE-119	(E) 184
TF39-GE-1C	(E) 185

GENERAL ELECTRIC ENGINE CO. (MASS)

CF34-1A	(E) 186
CF700-2D-2	(E) 187
CJ610-8	(E) 188
CT7-6	(E) 189
CT7-9	(E) 190
CT58-140	(E) 191
CT64-800-4	(E) 192
F101-GE-102	(E) 193
F110-GE-400	(E) 194
F118-GE-100	(E) 195
F404-GE-400	(E) 196
J85-GE-17A	(E) 197
J85-GE-21	(E) 198
T58-GE-16	(E) 199
T64-GE-419	(E) 200
T700-GE-701A	(E) 201
GERMANY-ENGINE COMPANIES	2
GHOST (early turboprop)	(E) 438
GLOSTER E-28 (early aircraft)	(A) 342, 437
GLOSTER METEOR (early aircraft)	(A) 437

GLUSHENKOV DESIGN BUREAU 6

GTD-3BM	(E) 108
GTD-3F	(E) 108
TVD-10B	(E) 109
TVD-20	(E) 140
TVD-1500	(E) 110
GMA-2100	(E) 163
GMA-3007	(E) 164
GMA-3014	(E) 164
GNAT	(A) 258
GNAT (early aircraft)	(A) 439
GNOME H.1400-1	(E) 67
GNOME (early turboshaft)	(E) 439
GOBLIN (early turbojet)	(E) 437
GOSHAWK	(A) 273
GOVERNMENT ACFT. FACTORIES (AUSTRALIA)	255
GREAT BRITAIN-ENGINE COMPANIES	2
GREYHOUND (C-2)	(A) 271
GRIFFON (early aircraft)	(A) 438
GRIPEN (JAS-39)	(A) 261
GRUMMAN CO.	264, 265, 270, 271, 272
GRUMMAN (early aircraft)	438,9
GTD-3 (early turboprop)	(E) 439
GTD-3BM	(E) 111
GTD-3F	(E) 108
GTD-350	(E) 95
GTP & GTPC (Locate under APU'S)	(E) 11
GTX-35	(A) 256
GUARANI (IA-50)	(A) 255
GUARDIAN (Falcon)	(A) 257, 272
GUARDRAIL (RC-12)	(A) 272
GULFSTREAM CORP.	265, 266
GULFSTREAM (USAF)	(A) 271, 273
GULFSTREAM G-II, III, IV	(A) 266
GULFSTREAM G-159 (early aircraft.)	(A) 438
GUNSHIP (AC-130)	(A) 270
GUPPY	(A) 256
GYRON (early turbojet)	(E) 439

H

H-2 (K860)	(A) 269
H-5	(A) 256
H-6	(A) 256
H-7	(A) 256
H-46	(A) 269
H.145 (Provost, early aircraft)	(A) 438
H-600 (Stallion)	(A) 265
H-1100	(A) 269
HA 220 (Super Saeta)	(A) 260
HA-200 (early aircraft)	(A) 438
HA-300 (early aircraft)	(A) 438
HALO (MI-26)	(A) 276
HAMILTON STANDARD APU'S	247
HANLEY-PAGE CORP.	258, 421
HANLEY-PAGE (early aircraft)	437,8,9
HANSAJET (HFB-320)	(A) 257
HARBIN (PRC)	(A) 256
HARBIN WJ-5	(E) 94
HARBIN WZ-8	(E) 94
HARKE (MI-10)	(A) 276
HARRIER	(A) 258, 270, 273
HARRIER (early aircraft)	(A) 439
HAS MK.2	(A) 263
HAVOC (MI-26)	(A) 276
HAWK 100/200	(A) 258
HAWK IAR.99	(A) 260
HAWKEYE (E-2)	(A) 271
HAWKEYE (E-2, early aircraft)	(A) 439
HAWKER SIDDELEY (early aircraft)	438
HAWKER SIDDELEY	
J79-11A	(E) 13
J85-CAN-15	(E) 14
J-85-CAN-40	(E) 15
ORENDA	(E) 16
HAZE (MI-8/14)	(A) 276
HBT-320 (HansaJet)	(A) 257
HC-2 (Helicopter)	(A) 262
HCC MK.4 (Helicopter)	(A) 262
HE-178 (early aircraft)	(A) 437
HE-280 (early aircraft)	(A) 437
HEINKEL HE-178	(A) 344, 437
HELILINER (EH-101)	(A) 262
HELIO AIRCRAFT	265
HELIX (KA-27)	(A) 276
HERALD	(A) 258
HERALD (early aircraft)	(A) 438
HERCULES	(A) 267, 271, 273
HERCULES (C-130, early aircraft)	(A) 439
HERCULES 730	(E) 260
HERMES (early aircraft)	(A) 437
HeS (early turbojet)	(E) 437
HF-24 (Mahut)	(A) 258
HF-24 (early aircraft)	(A) 439
HH-1	(A) 268
HH-3F (Pelican)	(A) 263
HH-43 (Husky)	(A) 269
HH-43 (early aircraft)	(A) 439
HH-53	(A) 269
HH-60	(A) 269
HH-65	(A) 268
HILLER (UH-12)	(A) 269
HIND (MI-24/25/35)	(A) 276
HINDUSTAN AERONAUTICS LTD.	258
ADOUR MK.811	(E) 69
ASTAZOU IIIB	(E) 70
HINDUSTAN AIRCRAFT	258, 263
HINDUSTAN (early aircraft)	438,9
HIP-E (MI-8)	(A) 275, 276
HIP-H (MI-17)	(A) 276
HISPANO (early aircraft)	438,9
HISPANO SUZIA (early engines)	438
HJT-16 (Kiran)	(A) 258
HJT-16 (early aircraft)	(A) 438
HK-500	(A) 263
HOKUM (KA-41)	(A) 276
HOMER (MI-12)	(A) 276
HOODLUM (KA-26)	(A) 276
HOOK (MI-6)	(A) 276
HOPLITE (MI-2)	(A) 276
HORMONE (KA-25)	(A) 276
HORNET (F/A-18)	(A) 272, 273
HORNET (RH-100)	(A) 269
HORNET (TwinJet)	(A) 269
HS.681 (early aircraft)	(A) 439
HS.748	(A) 258
HSS-2 (S-61)	(A) 263
HT-34	(A) 258
HU-25A	(A) 272
HUEY COBRA (Bell-209)	(A) 268
HUNTER	(A) 258
HURON (C-12)	(A) 270
HUSTLER (B-58, early aircraft)	(A) 438
HUSKY Series	(A) 269

I

I-A (early turbojet)	(E) 437
I-14 (early turbojet)	(E) 437
I-16 (early turbojet)	(E) 437
I-22 (Iryda)	(A) 260
I-40 (early turbojet)	(E) 437
I-93 (early turbojet)	(E) 439
IA-50 (Guarani)	(A) 255
IA-58 (Pucara)	(A) 255
IA-63 (Pampa)	(A) 255
IA-66 (Pucara-C)	(A) 255
IAI 101B (Lavi)	(A) 258
IAI 201 (Arava)	(A) 258
IAI 202 (Arava)	(A) 258
IAI 1121 (Commander)	(A) 258
IAI 1123 (Commodore)	(A) 259
IAI 1124 (SeaScan)	(A) 259
IAI 1125 (Astra)	(A) 259
IAR.93A (Yurom)	(A) 260, 261

IAR.99 (Soim)	(A) 260
IAR.825 (Triumf)	(A) 260
ICA CORP.	260
I-HS (early turbojet)	(E) 437
ILYUSHIN BUREAU (USSR)	274
IL-18	(A) 277
IL-18 (Coot, early aircraft)	(A) 437, 438
IL-20	(A) 275
IL-28	(A) 274
IL-28 (early aircraft)	(A) 437
IL-38	(A) 274, 275
IL-62	(A) 275, 277
IL-68 (early aircraft)	(A) 439
IL-76	(A) 275, 277
IL-78	(A) 275
IL-86	(A) 277
IL-96	(A) 277
IL-108	(A) 277
IL-114	(A) 277
IMP CORP.	265
IMPALA (MB 326M)	(A) 260
INDIA-Engine Companies	3

INDUSTRIE AERONAUTICHE MECCANICHE

GEM/RR.1004	(E) 79
T53-L-13	(E) 80
T55-L-712	(E) 81
VIPER 632-43	(E) 82
INTERCEPTOR CORP.	265
INTERCEPTOR 400	(A) 265

INTERNATIONAL AERO ENGINES LTD.

V2500	(E) 146
INTERNATIONAL Fixed Wing Aircraft	255
INTERNATIONAL Rotary Wing Aircraft	261

INTERNATIONAL TURBINE ENGINE CO.

TFE 1042	(E) 147
INTRUDER (A-6)	(A) 270-272
INTRUDER TANKER (KA-6E)	(A) 271
IPTN (INDONESIA)	258, 263
IRIDIUM	(A) 260
IROQUOIS (Bell 205)	(A) 262, 263, 268
IROQUOIS (early turbojet)	(E) 438
IROQUOIS (early aircraft)	(A) 439
IRYDA (Iridium)	(A) 260
ISH-60J (S-70)	(A) 263
ISHIDA CORP.	259

ISHIKAWAJIMA-HARIMA INDUSTRIES

CT58-IHI-140	(E) 83
CT700-700	(E) 84
F100-IHI-100	(E) 85
F3-IHI-30	(E) 86
T56-IHI-14	(E) 87
T64-IHI-10J	(E) 88
TF40-IHI-801	(E) 89
ISKRA (TS-11)	(A) 260
ISKRA (early aircraft)	(A) 439
ISLANDER (BN-2T)	(A) 258
ISOTOV (Locate under Soyuz)	

ISRAEL AIRCRAFT INDUSTRIES-ENGINES

J79-IAI-J1E	(E) 72
ISRAEL AIRCRAFT IND. (IAI)	258, 272
ISRAEL-ENGINE COMPANIES	3
ITALY-ENGINE COMPANIES	4
ITP (Industria de Turbo Propulsores-Spain)	145
IVCHENKO DESIGN BUREAU	7
IVCHENKO ENGINES (Locate under Progress)	

J

J-1 (Jastreb)	(A) 261
J-5	(A) 256
J-6	(A) 256
J-7	(A) 256
J-8	(A) 256
J-12	(A) 256
J-30 (early turbojet)	(E) 437
J-31 (Jetstream)	(E) 438
J-44 (early turbojet)	(E) 438
J-46 (early turbojet)	(E) 438
J-47 (early turbojet)	(E) 437
J-48 (early turbojet)	(E) 438
J52 (early turbojet)	(E) 439
J52-P-408	(E) 232
J57 (early turbojet)	(E) 438
J57-P-59W	(E) 233
J57-P-420	(E) 234
J58	(E) 235
J58 (early turbojet)	(E) 439
J60 (early turbojet)	(E) 439
J60-P-6	(E) 218
J65 (early turbojet)	(E) 438
J69 (early turbojet)	(E) 438
J69-T-25	(E) 204
J71 (early turbojet)	(E) 438
J73 (early turbojet)	(E) 438
J75 (early turbojet)	(E) 439
J75-P-19W	(E) 236
J79 (early turbojet)	(E) 438
J79-11A	(E) 13
J79-GE-19	(E) 76
J79-GE-119	(E) 184
J79-IAI-J1E	(E) 72
J79-IHI-17	(E) 259
J79-MTU-17A	(E) 46
J85 Series	(E) 197, 198
J85-CAN-15	(E) 14
J85-CAN-40	(E) 15
J85-GE-13A	(E) 74
J97 (early turbojet)	(E) 439
J402-CA-702	(E) 205
JAFFE CORP.	269, 273
JAFFE-BELL 222	(A) 269
JAGUAR (BAe/Dassault)	(A) 257, 258
JAPAN Engine Companies	4

JAMES AVIATION	265
JANUS (early turboprop)	(E) 438
JAS 39 (Gripen)	(A) 261
JASTREB (J-1)	(A) 261
JAVELIN (early aircraft)	(A) 438
JAYHAWK (HH-60J)	(A) 269
JAYHAWK (TA-1)	(A) 270, 273
JET CRUZER	(A) 265
JET PROVOST (BAC-145)	(A) 257
JET PROVOST (early aircraft)	(A) 438
JET RANGER (Bell 206)	(A) 263, 268
JET SQUALUS (F-1300)	(A) 255
JETSTAR (Lockheed)	(A) 266, 271
JETSTAR (USAF)	(A) 271
JETSTAR (early aircraft)	(A) 439
JETSTREAM Series	(A) 258
JET TRADER (DC-8 Series)	(A) 267
JH-7	(A) 256
JOLLY GREEN (SH-3)	(A) 269
JT3C (early turbojet)	(E) 439
JT3C-7	(E) 219
JT3D (early turbofan)	(E) 439
JT3D-7	(E) 220
JT4 (early turbojet)	(E) 439
JT4A-11	(E) 221
JT8D (early turbofan)	(E) 439
JT8D-17AR	(E) 222
JT8D-219	(E) 223
JT9A (early turbofan)	(E) 439
JT9D-7	(E) 224
JT12A-8	(E) 225
JT12 (early turbojet)	(E) 439
JT15D-5	(E) 17
JTFD12	(E) 226
JTFD12 (early turboshaft)	(E) 439
JU-88 (early aircraft)	(A) 437
JU-287 (early aircraft)	(A) 437
JUMO (early turbojet)	(E) 437
JUNKERS-109 (early turbojet)	(E) 437
JUNKERS-109 (early turboprop)	(E) 437

K

K-8	(A) 256
K-15	(E) 96
K-800 Series (Kaman)	(A) 269
KA-6E	(A) 271
KA-25	(A) 276
KA-26	(A) 276
KA-27	(A) 276
KA-28/29	(A) 276
KA-32	(A) 276
KA-41	(A) 276
KA-50	(A) 276
KA-62	(A) 276
KA-126	(A) 276
KA-226	(A) 276
KAKORUM (China)	(A) 256
KAMAN Helicopter	269
KAMAN (early aircraft)	(A) 439
KAMOV BUREAU (USSR)	274
KAWASAKI AKASHI INDUSTRIES	
T53-K-703	(E) 90
T55-K-712	(E) 91
KAWASAKI HEAVY INDUSTRIES	263
KAWASAKI 414	(A) 263
KC-10	(A) 271
KC-135 (early aircraft)	(A) 438, 439
KC-135 Series	(A) 270
KE-130	(A) 271
KFIR	(A) 259
KFIR (F-21)	(A) 272
KING AIR	(A) 265, 270
KING AIR (USAF)	(A) 270,1,2,3
KIOWA (OH-58)	(A) 268
KIRAN (HJT-16)	(A) 258
KLIMOV DESIGN BUREAU	6
KLIMOV (early engines)	437,8,9
GTD-350	(E) 111
GTD-35OP	(E) 95,111
RD-33	(E) 112
TV2-117A	(E) 113
TV3-117V	(E) 114
TV7-117V	(E) 115
VK-1	(E) 140
KOLIESOV DESIGN BUREAU	6
KOLIESOV (early engines)	438
VD-7	(E) 140
VD-57	(E) 140
KOPTCHVENCO (Locate under Sozyz)	
KLIMOV TV-0-100	(E) 116
KS-3	(A) 271
KUNIA (MI-2)	(A) 264
KUZNETSOV-TRUD DESIGN BUREAU	6
KUZNETSOV (early engines)	438,9
NK-8-4	(E) 117
NK-12	(E) 118
NK-86	(E) 140
NK-93	(E) 140
NK-321	(E) 140
NK-144	(E) 119
NK-144 UPGRADE	(E) 119
KV-107 (UH-46)	(A) 263

L

L-29 (Delfin)	(A) 276
L-29 (Maya)	(A) 256
L-29A (Akrobat)	(A) 256
L-29 (early aircraft)	(A) 439
L-39Z (Abatros)	(A) 256, 276
L-59	(A) 256

461

L-159	(A) 256
L-80TP	(A) 256
L-90TP (Redigo)	(A) 256
L100 (Hercules)	(A) 267, 268
L188 (Electra)	(A) 267, 268
L-188 (Electra, early aircraft)	(A) 439
L-410 (Turbolet)	(A) 256
L-610	(A) 256
L-1011 Series	(A) 268
LABRADOR (CH-115)	(A) 261
LAMA (SA-315)	(A) 262, 263
LANCER (B-1B)	(A) 270
LARK (early aircraft)	(A) 439
LARZAC	(E) 156
LARZAC 04-C20	(E) 156
LASER 300	(A) 265
LAVI (IAI)	(A) 258
LEAR CORP.	265, 266
LEARFAN 2100	(A) 265
LEARJET Series	(A) 266, 271
LEARJET (USAF)	(A) 271
LET/OMNIPOL CORP. (CZEC)	256
LF-507	(E) 209
LIGHT FIGHTER (HAL)	(A) 258
LIGHT HELICOPTER TURBINE COMPANY	
CTS-800	(E) 202
T-800	(E) 203
LIGHTNING (early aircraft)	(A) 439
LINCOLN (early aircraft)	(A) 437
LITTLE BIRD (MD-500)	(A) 269
LIYANG WP-7B	(E) 94
LIYANG WP-13A	(E) 94
LOCKHEED ACFT.	266, 267, 270, 271, 272, 273
LOCKHEED HELICOPTER	269
LOCKHEED MODEL 86	(A) 269
LOCKHEED (early aircraft)	437,8,9
LONGHORN (Lear-28/55)	(A) 266
LONG RANGER (Bell 206B/8L)	(A) 263, 268
LOTAREV ENGINES (Locate under Progress)	
LR-1 (MU-2)	(A) 259
LTP101-700	(E) 210
LTS101-750	(E) 211
LTV CORP.	270, 273
LUI-MENG (A-3)	(A) 256
LYNX (AH-1)	(A) 262
LYNX MK.9	(A) 262
LYULKA DESIGN BUREAU	6
LYULKA (early engines)	437,8,9
AL-7F-1-100	(E) 121
AL-21F-3	(E) 122
AL-31F	(E) 140

M

M-4	(A) 274, 275
M-4 (Bison, early aircraft)	(A) 438
M7-420	(A) 265
M15 (Belphegor)	(A) 260
M-17	(A) 275
M18 (Turbo Dromander)	(A) 260
M-50/M-52 (early aircraft)	(A) 438, 439
M-50	(A) 274
M53-P2	(E) 33
M-55	(A) 275
M88-2	(E) 34
M 601	(E) 26
M 602	(E) 27
M 701 (early turboprop)	(E) 438
M 701	(E) 28
M701C500	(E) 140
MACH-TRAINER (early aircraft)	(A) 437
MADCAP (AN-74)	(A) 275
MAESTRO (Yak-28)	(A) 276
MAGISTER (CM-700)	(A) 256, 438
MAGISTER (early aircraft)	(A) 438
MAHUT (HF-24)	(A) 258
MAIL (BE-12)	(A) 274
MAINSTAY (IL-76)	(A) 275
MAKILA	(E) 41
MALIBU (TP-400)	(A) 266
MALLARD	(A) 265
MAMBA (early turboprop)	(E) 437
MANGUSTA (A-129)	(A) 263
MARBORE (early turbojet)	(E) 438
MARBORE-4	(E) 256
MARBORE-6	(E) 71
MARQUISE (MU-2N)	(A) 259
MARTIN AIRCRAFT	273
MARTIN (early aircraft)	437,8,9
MAUL CORP.	265
MAY (IL-38)	(A) 274, 275
MAYA	(A) 256
MB 326 (early aircraft)	(A) 438
MB 326K	(A) 259
MB 326M (Impala)	(A) 260
MB 329A	(A) 259
MB 339K (Veltro-II)	(A) 259
MBB GMBH.	257, 262
McDONNELL AIRCRAFT	273
McDONNELL (early aircraft)	437,8,9
McDONNELL/D	267, 268, 270, 271, 272, 273
McDONNELL DOUGLAS HELICOPTER	269
McKINNON VIKING CORP.	255
MD-11 Series	(A) 267, 268
MD-12	(A) 268
MD-80 Series	(A) 268
MD-90 Series	(A) 268
MD-500 Series	(A) 269
MD-900 (MDX)	(A) 269
ME-262 (early aircraft)	(A) 366, 437
MEDWAY (early turbofan)	(E) 439
MERCKLE (early aircraft)	(A) 438
MERCURE	(A) 257
MERLIN (EH 101)	(A) 262
MERLIN II,III,IV	(A) 265
MERMAID (BE-42)	(A) 274
MESSERSCHMITT (early ME-262)	366, 437

MESSERSCHMITT-BOELKOW-BLOHM (MBB)	262
METEOR (early aircraft)	(A) 437
METRO Series	(A) 265
METRO (USAF)	(A) 271
METRO-VICKERS (early engines)	437
MH-47	(A) 269
MH-53	(A) 269
MI-2	(A) 264
MI-2	(A) 276
MI-6	(A) 276, 277
MI-8	(A) 276, 277
MI-10	(A) 276
MI-12	(A) 276
MI-14	(A) 276, 277
MI-17	(A) 276
MI-24	(A) 276
MI-26	(A) 276, 277
MI-28	(A) 276
MI-38	(A) 276
MICROJET 200	(A) 257
MICROJET CORP.	256

MICROTURBO CORPORATION
TRI 60-1	(E) 30
TRS 18-1	(E) 31
MIDAS (IL-78)	(A) 275
MIDGET (MIG-15)	(A) 276

MIKOYAN DESIGN BUREAU
MIG (early aircraft)	(A) 437, 438, 439
MIG-15U (Midget)	(A) 276
MIG-15 (Fagot, early aircraft)	(A) 437/8
MIG-17 (Fresco, early aircraft)	(A) 437
MIG-19 (Farmer, early aircraft)	(A) 438
MIG-21 (Faceplate, early aircraft)	(A) 438
MIG-21	(A) 274, 275
MIG-21U	(A) 276
MIG-23	(A) 274
MIG-25	(A) 274, 275
MIG-27	(A) 275
MIG-29	(A) 274
MIG-31	(A) 274

MIKULIN (Locate under Soyuz)
MIKULIN (early engines)	438
RD-3M	(E) 123

MIL DESIGN BUREAU (USSR)
MIL (early aircraft)	(A) 439
MIL MI-2	(A) 276
MIL MI-6	(A) 276, 277
MIL MI-8	(A) 276
MIL MI-10	(A) 276
MIL MI-12	(A) 276
MIL MI-14	(A) 276, 277
MIL MI-17	(A) 276
MIL MI-24	(A) 276
MIL MI-26	(A) 276, 277
MIL MI-28	(A) 276
MIL MI-38	(A) 276
MILITARY AIRCRAFT-USA	270
MIRAGE Series	(A) 257
MIRAGE (early aircraft)	(A) 439
MISSION-MASTER (N-22C)	(A) 255

MITSUBISHI NAGASAKI INDUSTRIES ... 259
CT63-M-5A	(E) 92
MITSUBISHI Heavy Industries	263
MOGOL (MIG-21)	(A) 276
MOHAWK (N 262)	(A) 256
MOHAWK-298	(A) 264
MOHAWK (OV-1B)	(A) 272
MOHAWK (OV-1, early aircraft)	(A) 439
MOONEY CORP.	265
MOSCOW SCIENTIFIC CORP.	6
MOSS (TU-126)	(A) 275

MOTOREN TURBINEN-UNION (MTU)
250-MTU-C20B	(E) 45
J79-MTU-17A	(E) 46
MTR-390	(E) 148
T64-MTU-7	(E) 47
TYNE-MTU-MK.21/22	(E) 48

MOTORLET KONCERN
WALTER M601	(E) 26
WALTER M602	(E) 27
WALTER M701C	(E) 28
MOUJIK (SU-7)	(A) 274
MRIYA (AN-225)	(A) 275, 277
MTR 390	(E) 148

MTU, ROLLS-ROYCE, TURBOMECA LTD.
MTR 390	(E) 148
MU-2	(A) 259
MU-300	(A) 259
MULTI-National Companies	7
MX7	(A) 265
MYA (early aircraft)	(A) 438

MYASISHCHEV BUREAU (USSR) ... 274
MYSTERE (early aircraft)	(A) 437, 438, 439
MYSTIC (M-17)	(A) 275

N

N 22B (Nomad)	(A) 255
N 22C (Nomad)	(A) 255, 265
N 24A (Nomad)	(A) 255
N-250 (IPTN)	(A) 258
N 262 (Fregate)	(A) 256
N 262 (Mohawk)	(A) 256
N 262 NORD	(A) 262
NAF (China)	256
NAL-KAWASAKI CORP.	260
NAMMER	(A) 258
NANCHANG Aircraft Factory	256
NAPIER (early engines)	438

NATIONAL AEROSPACE LABORATORY
FJR-710-600S	(E) 93
NBO-105	(A) 263
NC-212 (Aviocar)	(A) 258
NDN-1 (Fire Cracker)	(A) 258

NDN-6 (Field Master)	(A) 258
NE-821 (Caraja)	(A) 255
NEACP	(A) 271
NENE (early engine)	(E) 437, 438
NEPTUNE (P-2)	(A) 259
NESHER	(A) 259
NEW-CAL AVIATION	258, 264
NH-90	(A) 262,3,4
NH 500D	(A) 263
NIGHTFOX (MD-530)	(A) 269
NIGHT HAWK (HH-60)	(A) 269
NIGHTINGALE (C-9)	(A) 271
NIHON CORP.	259
NIMBUS	(E) 68
NIMBUS (early turboshaft)	(E) 439
NIMROD (BAC 801)	(A) 258
NK-4 (early turboprop)	(E) 439
NK-8 (early turboprop)	(E) 438
NK-8-4	(E) 117
NK-12 (early turboprop)	(E) 438
NK-12M (early turboprop)	(E) 438
NK-12	(E) 118
NK-86/87	(E) 140
NK-144	(E) 119
NK-144 UPGRADE	(E) 120
NK-Series (engines in development)	(E) 140
NKC-135	(A) 271
NOEL PENNY TURBINES LTD.	
NTP-401B	(E) 49
NOMAD (N-22)	(A) 255, 265
NORD N 262	(A) 262
NORMAN AEROSPACE	258, 262
NORTH AMERICAN AIRCRAFT	273
NORTH AMERICAN (early aircraft)	437,8,9
NORTHROP AIRCRAFT	270, 272, 273
NORTHROP (early aircraft)	437,8,9
NOS	(A) 272
NOTAR (MD series "no tail rotor"	(A) 269
NPT-401	(E) 49

O

OA-37A	(A) 272
OBSERVATION AIRCRAFT-USA	272
OCEANHAWK (SF-60)	(A) 269
OH-5A	(A) 269
OH-6 (500D)	(A) 263, 269
OH-6 (early aircraft)	(A) 438
OH-58 Series	(A) 268, 269
OLYMPUS (early turbojet)	(E) 438, 439
OLYMPUS-593	(E) 150
OMAC AVIATION	265
OMEGA (TB-31)	(A) 256
OMNI AVIATION	265
OMNIPOL/LET CORP. (CZEC)	256
OMSK DESIGN BUREAU	6
ORAO 1/2	(A) 261
ORAO AIR FORCE DEPOT	
VIPER 632	(E) 141
ORENDA	(E) 16
ORENDA (early turbojet)	(E) 438
ORION (P-3)	(A) 259, 272, 273
ORION (P-3, early aircraft)	(A) 439
ORPHEUS (early turbojet)	(E) 439
ORPHEUS 805	(E) 50
ORYX (PUMA)	(A) 264
OSPREY	(A) 269
OURAGAN (early aircraft)	(A) 438
OV-1	(A) 272
OV-1 (early aircraft)	(A) 439
OV-10	(A) 272

P

P-2J (Neptune)	(A) 259
P-2 (Neptune, early aircraft)	(A) 437
P-3C (Orion)	(A) 259, 272
P-3 (Orion, early aircraft)	(A) 439
P5Y (Tradewind, (early aircraft)	(A) 438
P-7A	(A) 272
P-16 (early aircraft)	(A) 438
P-80 (Shooting-Star, early aircraft)	(A) 437
P-95 (EMB-111)	(A) 255
P.135 (BAe-125)	(A) 258
P-163 (Grumman, S-2)	(A) 255
P.166	(A) 259
P.180 (Avanti)	(A) 259
P-550 (Turbo-Equator)	(A) 257
PACARA (IA-58)	(A) 255
PACIFIC AEROSPACE	260
PACKET (C-119, early aircraft)	(A) 438
PALAS (early turboshaft)	(E) 438
PAH-1	(A) 262
PAH-2	(A) 262
PALAS (early turboshaft)	(E) 438
PAMPA (IA-63)	(A) 255
PANAVIA CORP.	257
PANTHER (F9F)	(A) 262
PANTHER (F9F, early aircraft)	(A) 438
PANTHER (Helicopter)	(A) 262
PARTENAVIA CORP.	259
PATROL AIRCRAFT-USA	272
PAVE HAWK (HH-60)	(A) 269
PAVE LOW (HH-53)	(A) 269
PC-6 (Porter)	(A) 261
PC-7 (Turbo-Trainer)	(A) 261
PC-9	(A) 261
PC-12	(A) 261
PC-X	(A) 261
PD.808	(A) 259
PEGASUS (F-402)	(E) 51
PEGASUS (early lift-fan)	(E) 439
PELICAN (HH-3F)	(A) 263
PENZEL CORP.	260
PEOPLES REPUBLIC OF CHINA	5, 256, 261

PERM SCIENTIFIC MFR. ASSOCIATION ... 7	PS-5 .... (A) 256
SOLOVIEV D-15 .... (E) 140	PS-90A .... (E) 128
SOLOVIEV D-20 .... (E) 124	PT6 (early turboprop) .... (E) 439
SOLOVIEV D-25V .... (E) 125	PT6A-27 .... (E) 18
SOLOVIEV D-30 .... (E) 126	PT6A-65AG .... (E) 19
SOLOVIEV D-30-KP .... (E) 127	PT6B-36A .... (E) 20
PS-90A .... (E) 128	PT6T-6 .... (E) 21
PHANTOM-1 (FH-1, early aircraft) .... (A) 437	PUCARA (IA-58) .... (A) 255
PHANTOM-II (F-4) .... (A) 259, 272, 273	PUCARA-C (IA-66) .... (A) 255
PHANTOM-II (early aircraft) .... (A) 438	PUMA (AS-332) .... (A) 262,3,4
PIAGGIO SPA. 266	PW Designation System 11
PIRATE (See F6U - early aircraft)	PW 100 Series .... (E) 22, 23
PILATUS AIRCRAFT LTD. 261	PW 118 .... (E) 22
PILATUS BRITTEN-NORMAN 258	PW 119 .... (E) 22
PIMENE (early turboshaft) .... (E) 439	PW 126 .... (E) 23
PIPER CORP. 266	PW 205 .... (E) 24
POGO XFY-1 (early aircraft) .... (A) 437	PW 305 .... (E) 25
POLAND-ENGINE COMPANIES 5	PW 1115 .... (E) 237
POLSKI ZAKLADY LOTNICZE (PZL)	PW-1119 .... (E) 272
KLIMOV GTD-35OP .... (E) 95	PW 1120 .... (E) 238
K-15 .... (E) 96	PW 1212 .... (E) 239
PZL-10W .... (E) 97	PW 1216 .... (E) 240
SO-1/2 .... (E) 98	PW 2037 .... (E) 227
SO-3W .... (E) 99	PW 3005 .... (E) 241
TWD-10B .... (E) 100	PW 4152 .... (E) 228
PORTER .... (A) 261	PW 4460 .... (E) 229
PORTER (USAF) .... (A) 272, 273	PYTHON (early turboprop) .... (E) 438
POTEZ 841 .... (A) 266	PZL-10W .... (E) 97
POTEZ (early aircraft) .... (A) 437	PZL 106 (Turbo Kruk) .... (A) 260
POWER JET LTD (early engines) .... (E) 437	PZL 130 (Turbo Orlik) .... (A) 260
PRATT & WHITNEY USA (See United Technologies)	PZL M15 (Belphegor) .... (A) 260
PRATT & WHITNEY (early engines) 437,8,9	PZL M18 (Turbo Dromander) .... (A) 260
PRATT & WHITNEY CANADA	
APU'S 248	**Q**
JT15D-5 .... (E) 17	
PT6A-27 .... (E) 18	Q-5 .... (A) 256
PT6A-65AG .... (E) 19	
PT6B-36A .... (E) 20	
PT6T-6 .... (E) 21	**R**
PW 118 .... (E) 22	
PW 126 .... (E) 23	R-11-300 .... (E) 135
PW 205 .... (E) 24	R-11 (early turbojet) .... (E) 439
PW 305 .... (E) 25	R-13F2-300 .... (E) 136
PRECISION AIRMOTIVE INC. 266	R-13 (early turbojet) .... (E) 439
PROGRESS DESIGN BUREAU 7	R-15BD .... (E) 140
IVCHENKO AI-20M .... (E) 129	R-25 .... (E) 140
IVCHENKO AI-24VT .... (E) 130	R-27-22A .... (E) 140
IVCHENKO AI-25A .... (E) 131	R27V-300 .... (E) 140
LOTAREV D-18T .... (E) 132	R-29B .... (E) 137
LOTAREV D-136 .... (E) 133	R-31 .... (E) 138
LOTAREV D-436T .... (E) 134	R-31 UPGRADE .... (E) 139
DV-2 .... (E) 29	R-35F .... (E) 140
PROMAVIA SA. 255	R-45 .... (E) 140
PROPJET XP99 .... (A) 266	R-79 .... (E) 140
PROTEUS (early turboprop) .... (E) 438	R-195 .... (E) 140
PROTEUS .... (E) 258	RA-5 .... (A) 273
PROWLER (EA-6) .... (A) 272	RAFALE .... (A) 257
PS-1 (Flying Boat) .... (A) 260	

RANGER (Bell-47)	(A) 269
RAVEN (FB-111)	(A) 270, 271, 272
RB-57	(A) 273
RB-66 (early aircraft)	(A) 439
RB-Series (early turbofans)	(E) 439
RB-160	(E) 258
RB-168	(E) 59
RB-169	(E) 149
RB.183	(E) 56
RB.199	(E) 157
RB.211	(E) 57, 58
RC-12	(A) 272
RD-3 (early turbojet)	(E) 438
RD-3M	(E) 123
RD-9 (early turbojet)	(E) 438
RD-10 (early turbojet)	(E) 437
RD-20 (early turbojet)	(E) 437
RD-33	(E) 112
RD-36	(E) 140
RD-38	(E) 278
RD-41	(E) 140
RD-45 (early turbojet)	(E) 437, 439
RD-45F	(E) 140
RD-60	(E) 274
RECONNAISSANCE AIRCRAFT-USA	273
RED FALCON (CHS-2)	(A) 264
REDIGO (L-90TP)	(A) 256
REIMS AVIATION	257
REPUBLIC AIRCRAFT	272
REPUBLIC (early aircraft)	437,8,9
RF-4	(A) 273
RF-5	(A) 273
RH-1100 (OH-5)	(A) 269
RHEIN FLUGBAGBAU GMBH.	257
RILEY AIRCRAFT	264
RILEY 421 (Turbine Rocket/Eagle)	(A) 264
RINALDO PIAGGIO SPA.	259
RJ-70 (BAe)	(A) 258
RJ-85 (BAe)	(A) 258
RM2/5/6 (early turbojet)	(E) 438
RM6B	(E) 105
RM6B/C	(E) 260
RM8B	(E) 105
RM12	(E) 106
ROCKWELL INT'L	418,419,420,421
ROCKWELL (early aircraft)	439
ROGERSON- HILLER AIRCRAFT	269
ROGERSON 1099	(A) 269
ROLLS-ROYCE (early engines)	437,8,9

ROLLS-ROYCE (BRISTOL)
OLYMPUS-593	(E) 150
ORPHEUS 805	(E) 50
PEGASUS (F402)	(E) 51
VIPER 601	(E) 52

ROLLS-ROYCE (DERBY)
ADOUR 811/815	(E) 152
ADOUR 871	(E) 153
CONWAY 43	(E) 53
DART MK.540	(E) 54
GAZELLE MK.165	(E) 55
RB.183	(E) 56
RB.199	(E) 157
RB.211	(E) 57, 58
SPEY	(E) 59, 149
TAY	(E) 60
TRENT	(E) 61, 62, 63
TYNE	(E) 64, 151

ROLLS-ROYCE (LEAVENSDEN)
GEM-60-3	(E) 65
GEM/RR.1004	(E) 66
GNOME H.1400-1	(E) 67
NIMBUS	(E) 68
RTM-322	(E) 154

ROLLS-ROYCE, ALLISON TURBINE CO.
SPEY RR.168-66	(E) 149

ROLLS-ROYCE, SNECMA LTD.
OLYMPUS 593	(E) 150

ROLLS-ROYCE, SNECMA, MTU LTD.
TYNE-20	(E) 151

ROLLS-ROYCE, TURBOMECA LTD.
ADOUR MK.811/815	(E) 152
ADOUR MK.871	(E) 153
RTM 322	(E) 154

ROOIVALK (CHS-2)	(A) 264
ROTODYNE (early aircraft)	(A) 438
ROTARY WING AIRCRAFT-USA	268
RTF-180	(E) 223
RTM 321	(E) 154
RTM 322	(E) 154
RU-21	(A) 272
RUMANIA AIRCRAFT	264
RUMANIA-ENGINE COMPANIES	5
RUSLAN (AN-124)	(A) 275, 277
RUSSIAN AIRCRAFT	274
RYBINSK MOTORS (CIS)	6

S

S-2 (Turbo-Tracker)	(A) 255, 265
S2 R (T-11)	(A) 264
S2 R (T-15)	(A) 264
S2 R (T-34)	(A) 264
S2 R (T-65)	(A) 264
S-3A	(A) 272
S-58 (Wessex)	(A) 263
S-61 (Japan)	(A) 263
S-61 (Sea King)	(A) 263
S-61R	(A) 269
S-62	(A) 263, 269
S-64	(A) 269
S-70	(A) 263, 269
S-76	(A) 269
S-80	(A) 269
S.211	(A) 259, 270
S-312 (Tucano)	(A) 258
SA-32T	(A) 273

SA-204C	(A) 266
SA 227 (Expediter)	(A) 267, 268
SA 300 (Helicopters also locate under "AS")	
SA 315 (Cheetah)	(A) 263
SA 315 (Lama)	(A) 262, 263
SA 316 (Chetak)	(A) 263
SA 318 (Alouette-II)	(A) 262
SA 319 (Alouette-III)	(A) 262, 264
SA 321 (Super Frelon)	(A) 262
SA 330 (Puma)	(A) 262
SA 332 (Super Puma)	(A) 262
SA 341 (Gazelle)	(A) 262
SA 342 (Gazelle)	(A) 262, 264
SA 500 (Helicopters - Locate under "AS")	
SAAB-SCANIA AB.	260
SAAB 21 (early aircraft)	(A) 437
SAAB-105	(A) 260
SAAB 210 (early aircraft)	(A) 438
SAAB 340 (Commuter)	(A) 261, 265
SAAB 2000	(A) 261
SAAB SK-60 (early aircraft)	(A) 439
SAAB-105 (early aircraft)	(A) 439
SAAB-105	(A) 260
SAAB-29 (early aircraft)	(A) 438
SAAB-32 (Lansen, early aircraft)	(A) 438
SAAB-35 (Draken, early aircraft)	(A) 438
SAAB-37 (Viggen, early aircraft)	(A) 439
SABRE (F-86H, early aircraft)	(A) 437
SABRELINER CORP.	266, 270, 273
SABRELINER Series	(A) 266
SABRELINER (T-39, early aircraft)	(A) 439
SABRELINER (USAF)	(A) 271, 273
SAC (China)	256
SAETA (HA-200/300, early aircraft)	(A) 438
SALMON (XFV-1, early aircraft)	(A) 438
SAMARA SCIENTIFIC (CIS)	6
SAPPHIRE (early turbojet)	(E) 438
SATURN DESIGN BUREAU (See Lyulka)	6
SAUNDERS AVIATION	255
SCHAFER AVIATION	266
SCHWEIZER AIRCRAFT	264, 269
SCHWEIZER MODEL 330	(A) 269
SCIMITAR (early aircraft)	(A) 438
SCOUT	(A) 263
SD 330 (Sherpa)	(A) 258
SD 360 (Advanced)	(A) 258
SEA COBRA (Bell-209)	(A) 268
SEA DRAGON (MH-53)	(A) 269
SEAHAWK (SH-60)	(A) 269
SEA KING (S-61)	(A) 263, 269
SEA KING (early aircraft)	(A) 439
SEA KNIGHT (H-46)	(A) 269
SEA SCAN (1124)	(A) 259
SEA RANGER (Bell 205)	(A) 268
SEASPRITE (H-2)	(A) 269
SEA STALLION (CH-53)	(A) 269
SEASTAR (T-1A, early aircraft)	(A) 437
SEA STAR (CD-2)	(A) 257
SEA VIXEN (early aircraft)	(A) 438
SEARCH-MASTER (N-22B)	(A) 255
SENTRY (E-3A)	(A) 271
SEPECAT CORP.	261
SF-5A (Freedom Fighter)	(A) 260
SF.260	(A) 259
SF.260TP	(A) 259
SF.600 (Canguro)	(A) 259
SFA CORP.	256
SH-2 (Super Seasprite)	(A) 269
SH-3B (Sea King)	(A) 261, 263, 269
SH-5	(A) 256
SH-60	(A) 269
SH-65	(A) 269
SHANGHI FAREAST AERO (SFA)	256
SHENGUE (AT-3)	(A) 256
SHENYANG AIRCRAFT	256
SHENYANG WP-6A	(E) 94
SHENYANG WS-6	(E) 94
SHERPA (Advanced)	(A) 258
SHERPA (SD-330)	(A) 258
SHERPA USAF)	(A) 271
SHIN MEIWA CORP.	260
SHOOTING STAR (T-33)	(A) 270, 273
SHORTS (early aircraft)	438
SHORTS 330	(A) 271
SHORTS BROTHERS	258, 271
SHORTS SKYLINER	(A) 258
SHORTS SKYVAN	(A) 258
SHORTS TUCANO	(A) 258
SHUISHANG CORP.	256
SIAI MARCHETTI SPA.	259
SIKORSKY AIRCRAFT CORP.	261, 269
SIKORSKY (early aircraft)	(A) 438,9
SILVER (S-61)	(A) 269
SILVER STAR (T-33, early aircraft)	(A) 438
SIOUX (Bell-47)	(A) 269
SJ 30 (FanJet)	(A) 267
SK35C (Draken)	(A) 260
SK37	(A) 261
SK60	(A) 260
SKYCRANE	(A) 269, 276
SKYHAWK (A-4)	(A) 270, 273
SKYHAWK (F2H, early aircraft)	(A) 438
SKYKNIGHT (F3D, early aircraft)	(A) 437
SKY KNIGHT (F-10, early aircraft)	(A) 437
SKYLARK (CV-880, early aircraft)	(A) 438
SKY LINER (Shorts)	(A) 258
SKYRAY (F4D, early aircraft)	(A) 438
SKY SERVANT (DO 128)	(A) 257
SKY TRADER (1400/1700)	(A) 266
SKYTRAIN-II (C-9B)	(A) 271
SKY VAN (Shorts)	(A) 258
SKYWARRIOR (A3D, early aircraft)	(A) 438
SM-109	(A) 259
SMITH AERO CORP.	266
SN-601 (Corvette)	(A) 256

SNECMA (Society Nationale d'Etude et de Construction de Moteurs d'Aviation)
- SNECMA (early engines) 437,8,9
  - ATAR 9K50 ...... (E) 32
  - CF6-50 ...... (E) 155
  - LARZAC ...... (E) 156
  - M53-P2 ...... (E) 33
  - M88-2 ...... (E) 34
  - OLYMPUS-593 ...... (E) 150
  - TYNE-20 ...... (E) 151

SNECMA, GENERAL ELECTRIC, USA
- CF6-50 ...... (E) 155

SNOW AVIATION 266
SO-1/2 ...... (E) 98
SO-3W ...... (E) 99
SOKO AIRCRAFT CORP. 261
SOKOL ...... (A) 264
SOIM (IAR-99) ...... (A) 260
SOLITARE (MU-2P) ...... (A) 259
SOLOVIEV (early engines) 438,9
SOLOVIEV 7
SOLOVIEV D-15 ...... (E) 140
SOLOVIEV D-20 ...... (E) 124
SOLOVIEV D-25V ...... (E) 12
SOLOVIEV D-30-KP ...... (E) 126
SOLOVIEV D-30-KU ...... (E) 127
SOLOY CONVERSIONS 266, 269
SOUTH AFRICA-ENGINE COMPANIES 5
SOYUZ DESIGN BUREAU 7
SPARK (TS-11) ...... (A) 260
SPARTICUS ...... (A) 259
SPECTRE-GUNSHIP ...... (A) 270
SPEY Series (early turbofans) ...... (E) 439
SPEY 202 (PRC) ...... (E) 94
SPEY MK.512 (G.B.) ...... (E) 59
SPEY 512-14DW (RUM.) ...... (E) 101
SPEY MK.807 (ITALY) ...... (E) 77
SPEY RB.168 (USA) ...... (E) 149
SPIRAL (early aircraft) ...... (A) 439
SPIRIT (S-76) ...... (A) 269
SR-30 ...... (E) 217
SR-71 ...... (A) 273
SR-71 (early aircraft) ...... (A) 439
ST-27 (Saunders) ...... (A) 255
ST-1400 ...... (A) 266
ST-1700 ...... (A) 266
ST. PETERSBURG SCIENTIFIC ASSOCIATION 6
STALLION ...... (A) 265, 273
STARFIGHTER (F-104) ...... (A) 272
STARFIGHTER (early aircraft) ...... (A) 438
STARFIRE (F-94, early aircraft) ...... (A) 438
STARLIFTER (C-141) ...... (A) 271
STARSHIP (Beech) ...... (A) 265
STEALTH BOMBER (B-2A) ...... (A) 270
STEALTH FIGHTER (F-117) ...... (A) 272
STRATO-FORTRESS (B-52) ...... (A) 270
STRATO-LIFTER (C-135) ...... (A) 270, 271
STRATO-LINER (VC-137) ...... (A) 270

STRATO-TANKER (KC-135) ...... (A) 270
STRIKEMASTER (BAC-167) ...... (A) 257
STRIKEMASTER (early aircraft) ...... (A) 438
SUD AVIATION (early aircraft) ...... (A) 438, 439
SUKHOI BUREAU 274
SU Series (early aircraft) ...... (A) 437, 438
SU-7 ...... (A) 274, 276
SU-9/11 ...... (A) 274
SU-15/21 ...... (A) 274, 276
SU-17 ...... (A) 274,5,6
SU-19/22 ...... (A) 274
SU-24 ...... (A) 274, 275
SU-25/28 ...... (A) 274
SU-27 ...... (A) 275
SU-35 ...... (A) 275
SU-51 ...... (A) 277
SUNDSTRAND APU'S 248
SUPER COBRA (Bell-206) ...... (A) 258
SUPER COMMUTER (ATR 42) ...... (A) 256, 259
SUPER BROUSSARD (early aircraft) ...... (A) 439
SUPER FRELON (SA-321) ...... (A) 262
SUPER GALEB (G-4) ...... (A) 261
SUPER GUPPY ...... (A) 257
SUPER HERCULES (L-100) ...... (A) 267,8,271
SUPERJET (B-747) ...... (A) 267
SUPER JOLLY (HH-53) ...... (A) 269
SUPER KING (BE-200) ...... (A) 265, 271
SUPER-LYNX (MK-99) ...... (A) 262
SUPERMARINE (early aircraft) ...... (A) 437, 438
SUPER PUMA (AS-332) ...... (A) 258
SUPER SAETA (HA-220) ...... (A) 260
SUPER SEAPLANE (Transport) ...... (A) 277
SUPER SEASPRITE (K-888) ...... (A) 269
SUPER-SABRE (F-100) ...... (A) 438
SUPER SHERPA (SD-330) ...... (A) 258
SUPER SKYHAWK ...... (A) 270
SUPER STALLION (CH-53) ...... (A) 269
SUPER TOMCAT (F-14AD) ...... (A) 272
SUPER TRANSPORTER (Bell 214) ...... (A) 268
SWALLOW (DH.10, (early aircraft) ...... (A) 437
SWALLOW (ME-262, early aircraft) ...... (A) 437
SWEARINGEN AIRCRAFT 266
SWEDEN-Engine Companies 5
SWIFT (early aircraft) ...... (A) 437

T

T-1 ...... (A) 50
T-1 (early aircraft) ...... (A) 438, 439
T-1 (Sabre) ...... (A) 259
T-1A ...... (A) 270, 273
T-1A (Seastar, early aircraft) ...... (A) 437
T-2 (Buckeye) ...... (A) 259, 273
T-2 (early aircraft) ...... (A) 437, 439
T-4 ...... (A) 259
T-5 (Japan) ...... (A) 259
T-5 (early aircraft) ...... (A) 438

T-11 (S2R)	(A) 264
T-15 (S2R)	(A) 264
T.19 (CN 235)	(A) 258, 260
T-31 (early turboprop)	(E) 437
T-33A (Shooting Star)	(A) 270, 273
T-33 (Silver Star, early aircraft)	(A) 438
T-34 (early turboprop)	(E) 439
T-34 (S2R)	(A) 264, 273
T-37	(A) 273
T-37 (early aircraft)	(A) 438
T-38B	(A) 273
T-38 (early turboprop)	(E) 437
T-39 (CT-39)	(A) 271, 273
T-39 (early aircraft)	(A) 438
T-40 (early turboprop)	(E) 438
T-44A	(A) 273
T-43	(A) 273
T-45A	(A) 273
T-46A	(A) 273
T-47A	(A) 273
T-48	(E) 437
T-50 (early turboshaft)	(E) 437
T53 (early turboshaft)	(E) 438
T53-K-703 (Japan)	(E) 90
T53-L-13 (Jap./Ger.)	(E) 80, 142
T53-L-701A (USA)	(E) 212
T53-L-703 (USA)	(E) 213
T5508D (USA)	(E) 214
T55-K-712 (Japan)	(E) 91
T55-L-712 (Japan/USA)	(E) 81, 215
T55-L-714 (USA)	(E) 216
T-56 (early turboprop)	(E) 439
T56-IHI-14	(E) 87
T56-A-720	(E) 165
T-57 (early turboprop)	(E) 438
T-58	(E) 199
T-58 (early turboshaft)	(E) 439
T-58-GE-10	(E) 75
T63 (early turboshaft)	(E) 439
T63-A-700	(E) 166
T64	(E) 200
T64 (early turboprop)	(E) 439
T64-IHI-10J	(E) 88
T64-MTU-7	(E) 47
T64-P4D	(E) 78
T-65 (S2R)	(A) 264
T-71 (early turboprop)	(E) 437
T-74	(E) 18
T-76	(E) 174
T-101	(A) 277
T 117	(E) 143
T.188 (early aircraft)	(A) 439
T-400	(E) 21
T406-AD-400	(E) 167
T700	(E) 201
T703-A-700	(E) 168
T800	(E) 203
TA-4	(A) 273
TA-7D	(A) 273
TACANO (E-6A)	(A) 271
TALON (T-38B)	(A) 273
TARHE-SKYCRANE	(A) 269
TAV-8	(A) 273
TAY Series	(E) 60
TB-2 (early turboshaft)	(E) 439
TB.31 (Omega)	(A) 256
TB.1000 (early turbojet)	(E) 438
TBM-700	(A) 256, 257
TBM INT'L	256, 266
TC-2 (Kfir)	(A) 259
TC-4	(A) 273
TC-7 (Kfir)	(A) 259
TCHAIKA	(A) 274
TELEDYNE CAE CORPORATION	
J69-T-25	(E) 204
J402-CA-702	(E) 205
TELEDYNE CONTINENTAL MOTORS	
TP-500	(E) 206
TEXTRON LYCOMING CORPORATION	
TEXTRON LYCOMING (early engines)	438,9
AL5512	(E) 207
ALF502L	(E) 208
LF-507	(E) 209
LTP101-700	(E) 210
LTS101-750	(E) 211
T53-L-701A	(E) 212
T53-L-703	(E) 213
T5508D	(E) 214
T55-L-712	(E) 215
T55-L-714	(E) 216
TF-5A/B	(A) 273
TF-16N	(A) 273
TF-18	(A) 273
TF30 (early turbofan)	(E) 439
TF30-P-100	(E) 242
TF33 (early turbofan)	(E) 439
TF33-PW-102	(E) 243
TF39-GE-1C	(E) 185
TF-34	(E) 186
TF40-IHI-801	(E) 89
TF41-A-400	(E) 169
TF-106 (early turbofan)	(E) 439
TFE-731-5	(E) 175
TFE 1042	(E) 147
TG-180 (early turboprop)	(E) 437
THC-1 (Chung-Hsing)	(A) 256
TH-28	(A) 269
TH-57	(A) 268
TH-206 (Training Helicopter - see Bell-206)	
TH-330	(A) 269
THESEUS (early turboprop)	(E) 437
THUNDERBOLT-II (A-10)	(A) 270
THUNDERCHIEF (F-105)	(A) 272
THUNDERJET (F-84, early aircraft)	(A) 438
THUNDERFLASH (RF-8, early)	(A) 438
THUNDERSTREAK (F-84F, early)	(A) 437,8

TIGER (F-11F, early aircraft)	(A) 438
TIGER (Helicopter)	(A) 262
TIGER-II (F-5)	(A) 270, 272
TIGEREYE (RF-5)	(A) 273
TIGERSHARK (F-20)	(A) 272
TJ-1 (Jastreb)	(A) 261
TM-319 (ARRIUS)	(E) 42
TM-333	(E) 43
TOMCAT (F-14)	(A) 272
TOPAZ	(E) 264
TORNADO	(A) 257, 258, 259
TP-3 (Orion)	(A) 273
TP-400 (Turbine-Malibu)	(A) 266
TP-500	(E) 20
TPE-331 (early turboprop)	(E) 438
TPE-331-14	(E) 176
TPF-351-20	(E) 177
TR-1A	(A) 273
TR-1 (early turbojet)	(E) 437
TR-3A (Black Mamba)	(A) 273
TR-3 (early turbojet, France)	(E) 438
TR-3 (early turbojet, USSR)	(E) 438
TRADEWIND (P5Y, early aircraft)	(A) 438
TRS-2 (early aircraft)	(E) 439
TRAINER AIRCRAFT-USA	273
TRANSALL (C-160)	(A) 256, 257
TRANSPORTER (DHC-5E)	(A) 255
TRENT (turbofan)	(E) 61, 62, 63
TRENT (early turboprop)	(E) 437
TRI 60-1	(E) 30
TRIDENT	(A) 258
TRISTAR (L-1011)	(A) 268
TRITON (DHC-8)	(A) 255
TRIUMF (IAR-825)	(A) 260
TRS 18-1	(E) 31
TRUD DESIGN BUREAU (LOCATE UNDER KUZNETSOV)	
TS-11 (Iskra)	(A) 260
TS-11 (ISKRA, early aircraft)	(A) 439
TSCP'S (Locate under APU'S)	11
TSU-CHIANG	(A) 256
TU Series (early aircraft)	(A) 438
TU-16	(A) 274, 275
TU-22	(A) 274, 275
TU-22U	(A) 276
TU-26	(A) 274
TU-28P	(A) 274
TU-95	(A) 274, 275
TU-104	(A) 274, 278
TU-104 (early aircraft)	(A) 438
TU-105	(A) 274
TU-114	(A) 278
TU-114 (early aircraft)	(A) 438
TU-124	(A) 278
TU-124 (early aircraft)	(A) 438
TU-126	(A) 275, 278
TU-128	(A) 274
TU-128 (early aircraft)	(A) 437
TU-134	(A) 278
TU-138 (early aircraft)	(A) 437
TU-142	(A) 274, 275
TU-144	(A) 275, 278
TU-154	(A) 278
TU-160	(A) 274
TU-204	(A) 278
TU-334	(A) 275, 278
TUCANO (EMB-312)	(A) 255
TUCANO (Shorts)	(A) 258
TUTOR (CL-41)	(A) 255
TUTOR (CL-4, early aircraft)	(A) 439
TUMANSKY (Locate under Soyuz)	
TUMANSKY (early engines)	437,8,9
D-30F	(E) 140
M701C500	(E) 140
R-11-300	(E) 135
R-13F2-300	(E) 136
R-15BD	(E) 140
R-25	(E) 140
R-27-22A	(E) 140
R-29B	(E) 137
R-31	(E) 138
R-31 Upgrade	(E) 139
R-79	(E) 140
RD-36	(E) 140
RD-35F	(E) 140
RD-45F	(E) 140
TUPOLEV BUREAU	274
TUPOLEV (early aircraft)	438,9
TURBINE ROTOR SYSTEMS	385
TURBINE EAGLE (Riley 421)	(A) 264
TURBINE MALIBU (TP-400)	(A) 266
TURBINE ROCKET (Riley 421)	(A) 264
TURBINE TECHNOLOGIES LTD.	
SR-30	(E) 217
TURBO-18 (Vopar)	(A) 266
TURBO-BEAVER (DHC-2)	(A) 255
TURBO-CAT (Frakes)	(A) 264
TURBO DC-3	(A) 265
TURBO-DROMANDER	(A) 260
TURBO-EQUATOR (P-550)	(A) 257
TURBO-FIRECAT (Conair)	(A) 255
TURBO-GOOSE (G-21G)	(A) 255
TURBOJET ENGINE	425
TURBO-KRUK (PZL-106)	(A) 260
TURBOLET	(A) 256
TURBO-MENTOR (T-34)	(A) 273
TURBO-ORLIK (PZL-130)	(A) 260
TURBO-OTTER (Cox)	(A) 265
TURBOPROP ENGINE	425
TURBOSHAFT ENGINE	426
TURBO-TITAN (Omni)	(A) 265
TURBO-TRACKER (S-2T)	(A) 255, 265
TURBO-TRAINER (PC-7)	(A) 261
TURBOMECA CORPORATION	
TURBOMECA (early engines)	438,9
ARRIEL 1B	(E) 35
ARTOUSE 3B	(E) 36

ASTAFAN IV	(E) 37
ASTAZOU 3A	(E) 38
ASTAZOU 16C	(E) 39
AUBISQUE 1A	(E) 40
LARZAC	(E) 156
MAKILA	(E) 41
TM 319	(E) 42
TM 333-1A	(E) 43
TURMO 4C	(E) 44, 103

TURBOMECA, SNECMA, MTU LTD.
LARZAC 04	(E) 156

TURBO-UNION (FIAT,MTU,R-R) LTD.
RB.199-34R	(E) 157
TURKEY Engine Companies	6
TURMO	(E) 44, 103
TURMO (early turboshaft)	(E) 439

TUSAS ENGINE INDUSTRIES
F100-GE-100	(E) 107
TUTOR (CL-41)	(A) 255
TV-0-100	(E) 116
TV2 (early turboprop)	(E) 438
TV2-117A	(E) 113
TV3-117V	(E) 114
TV7-117 Series	(E) 115
TVD-10B	(E) 109
TVD-20	(E) 140
TVD-1500	(E) 110, 277
TW-68	(A) 259
TWD-10	(E) 100
TWEET (T-37)	(A) 273
TWIN HUEY (Bell 212)	(A) 261, 268
TWINJET (Rogerson)	(A) 269
TWIN OTTER (DHC-6)	(A) 255, 271
TWIN STAR (AS-355F)	(A) 262
TYNE (early turboprop)	(E) 439
TYNE	(E) 48, 64, 151
TYNE-20	(E) 151
TYNE-MTU-MK.21/22	(E) 48

U

U-2R	(A) 273
U-21 Series	(A) 273
U-23	(A) 272
U-27	(A) 273
UC-12	(A) 273
UDF™	(E) 183
UH-1	(A) 263, 268
UH-1 (early aircraft)	(A) 439
UH-1S (Germany)	(A) 262
UH-12	(A) 269
UH-46	(A) 263, 269
UH-60	(A) 263
UH-12	(A) 269

UNITED TECHNOLOGIES INC. (CONN.)
J60-P-6	(E) 218
JT3C-7	(E) 219
JT3D-7	(E) 220
JT4A-11	(E) 221
JT8D-17AR	(E) 222
JT8D-219	(E) 223
JT9D-7	(E) 224
JT12A-8	(E) 225
JTFD12	(E) 226
PW 2037	(E) 227
PW 4152	(E) 228
PW 4460	(E) 229

UNITED TECHNOLOGIES INC. (FLORIDA)
F100-PW-220	(E) 230
F117-PW-100	(E) 231
J52-P-408	(E) 232
J57-P-59W	(E) 233
J57-P-420	(E) 234
J58	(E) 235
J75-P-19W	(E) 236
PW 1115	(E) 237
PW 1120	(E) 238
PW 1212	(E) 239
PW 1216	(E) 240
PW 3005	(E) 241
TP30-P-100	(E) 242
TF33-PW-102	(E) 243
UPGUN COBRA (Bell 209)	(A) 439
US-1	(A) 260
USA CIVIL AIRCRAFT	264
USA MILITARY AIRCAFT	270
USSR AIRCRAFT	274
USSR-ENGINE COMPANIES	108
UTE	(A) 272, 273
UTILITY AIRCRAFT-USA	273
UV-18A	(A) 271, 273
UV-20	(A) 273

V

V-22 (BELL 900)	(A) 269
V-107 (early aircraft)	(A) 439
V2500 Series	(E) 146
VALKYRIE (XB-70, early aircraft)	(A) 439
VALIANT (early aircraft)	(A) 438
VALMET CORP. (Finland)	256
VAMPIRE (early aircraft)	(A) 438
VANGUARD	(A) 258
VANGUARD (early aircraft)	(A) 439
VAUTOUR (early aircraft)	(A) 439
VANTRUA (C-22J)	(A) 259
VC-6B	(A) 270
VC-9A	(A) 271
VC-10	(A) 258
VC-10 (early aircraft)	(A) 437
VC-10 Super (early aircraft)	(A) 439
VC-25	(A) 270
VC-137	(A) 270

VD-7	(E) 140
VD-7 (early turbojet)	(E) 439
VD-57	(E) 140
VECTOR (CBA-123)	(A) 255
VELTRO-II (MB-339k)	(A) 259
VENOM (early aircraft)	(A) 438
VERDON (early turbojet)	(E) 438
VFW-614 (Fokker)	(A) 260
VH-60	(A) 269
VIATOR (P-60)	(A) 259
VICKERS (early aircraft)	437,8,9
VICTOR (Bomber)	(A) 258
VICTOR (early aircraft)	(A) 439
VIGGEN (Thunderbolt)	(A) 261
VIGGEN (early aircraft)	(A) 439
VIGILANTE (A-5)	(A) 273
VIGILANTE (A-5, early aircraft)	(A) 438
VIKING (S-3/KS-3)	(A) 271, 272
VIKING TANKER (KS-3)	(A) 271
VIPER Series	(E) 52
VIPER (early turbojet)	(E) 438
VIPER 540	(E) 104
VIPER 632	82, 102, 141
VISCOUNT	(A) 258
VISCOUNT (early aircraft)	(A) 438
VK-1 (early turbojet)	(E) 140, 437
VK-3 (early turbojet)	(E) 438
VK-5 (early turbojet)	(E) 438
VOLVO FLYGMOTOR AB.	
RM8B	(E) 105
RM12	(E) 106
VOLVO (early engines)	438,9
VOODOO (F-101)	(A) 272
VON OHAIN, HANS	431
VOPAR INC.	266
VOUGHT (early aircraft)	437,8,9
VRD-2 (early turbojet)	(E) 437
VULCAN	(A) 438
VULTEE (XP-81, early aircraft)	(A) 437

W

W1X (early turbojet)	(E) 437
W-2 (early turbojet)	(E) 437
W-3 (Sokol)	(A) 264
W-2B (early turbojet)	(E) 437
W30-100	(A) 263
W30-140 (Westland-30)	(A) 263
W30-160	(A) 263
W30-200	(A) 263
W30-300	(A) 263
WALTER M601	(E) 26
WALTER M602	(E) 27
WALTER M701C	(E) 28
WALTER M701 (early turboprop)	(E) 439
WASP	(A) 263

WASP (early aircraft)	(A) 439
WC-130	(A) 273
WEATHERLY AIRCRAFT	264
WEATHERLY 620TP	264
WESSEX	(A) 55, 263
WESSEX (early aircraft)	(A) 438
WESTINGHOUSE (early engines)	437
WESTLAND HELICOPTER LTD.	262, 263
WESTLAND (early aircraft)	(A) 437,8,9
WESTWIND	(A) 259
WHITTLE, SIR FRANK	434
WHITTLE W-1/W-2 (early turbojet)	(E) 437
WILLIAMS INTERNATIONAL CORP.	
APU'S	249
FJ44	(E) 158
WJ-5/6/7	(E) 94
WP-6A	(E) 94
WP-7	(E) 94
WP-8	(E) 94
WP-13	(E) 94
WS-6	(E) 94
WS-9	(E) 94
WS-70 (UH-60)	(A) 263
WSK-PZL MIELEC	260
WSK-PZL-SWIDNIK	264
WU (early turbojet)	(E) 437
WYVERN (early aircraft)	(A) 437,438
WZ-6	(E) 94
WZ-8	(E) 94

X

X-29	(A) 272
X-31A	(A) 272
XAC (China)	256
XAVANTE (AT-26)	(A) 255
XB-48 (early aircraft)	(A) 437
XB-70 (early aircraft)	(A) 439
XC-2	(A) 256
XC-142 (early aircraft)	(A) 439
XF-5 (early aircraft)	(A) 437
XFV-1 (Salmon, early aircraft)	(A) 438
XFY-1 (POGO, early aircraft)	(A) 437
XIAN AIRCRAFT CORP. (XAC)	256
XIAN SPEY 202	(E) 94
XIAN WP-8	(E) 94
XINGU (EMB-121)	(A) 255
XP-47 (early aircraft)	(A) 437
XP-5Y (early aircraft)	(A) 437
XP-59 (early aircraft)	(A) 437
XP-80 (early aircraft)	(A) 437
XP-81 (early aircraft)	(A) 437
XP-84 (early aircraft)	(A) 437
XT-37 (early turboprop)	(E) 437
XT-45 (early turboprop)	(E) 437
XT-57 (early turboprop)	(E) 437

Y

Y-7	(A) 256
Y-7-200	(A) 256
Y-8	(A) 256
Y-12	(A) 256
Y-14	(A) 256
YAKOVLEV BUREAU	274
YAKOVLEV (early engines)	437
YAK (early aircraft)	(A) 437, 438, 439
YAK-28	(A) 275, 276
YAK-38	(A) 275, 276
YAK-40	(A) 278
YAK-41	(A) 275
YAK-141	(A) 275
YAK-42 (Clobber)	(A) 278
YAK-42 (Backfin, early aircraft)	(A) 438
YANKEE (early turbojet)	(E) 437
YAT-28 (early aircraft)	(A) 439
YB-49A (Flying-Wing, early aircraft)	(A) 437
YF-12A (early aircraft)	(A) 439
YS-11A	(A) 259
YS-11 (early aircraft)	(A) 438
YUGOSLAVIA AIRCRAFT	264
YUGOSLAVIA ENGINE COMPANIES	141
YUKON (CL-44)	(A) 255
YUROM	(A) 260, 261

Z

Z-8	(A) 261
Z-9	(A) 261

ZAVODY NAVYROBU LOZISK KONCERN

PROGRESS DV-2	(E) 29
ZLIN-Z 37	(A) 256
ZHUZHOW WJ-6	(E) 94